m.s.u.
1964

Gettysburg College
Musselman Library
Gettysburg, PA

Gift of

Peter Fong
Department of Biology

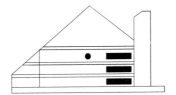

THE INVERTEBRATA

THE INVERTEBRATA
A MANUAL FOR THE USE OF STUDENTS

BY THE LATE
L. A. BORRADAILE & F. A. POTTS

WITH CHAPTERS BY
L. E. S. EASTHAM
Professor of Zoology in the University of Sheffield

AND

J. T. SAUNDERS
Fellow of Christ's College, Cambridge

FOURTH EDITION REVISED BY
G. A. KERKUT
Lecturer at the University of Southampton
Sometime Fellow of Pembroke College, Cambridge

CAMBRIDGE
AT THE UNIVERSITY PRESS
1961

PUBLISHED BY
THE SYNDICS OF THE CAMBRIDGE UNIVERSITY PRESS
Bentley House, 200 Euston Road, London, N.W. 1
American Branch: 32 East 57th Street, New York 22, N.Y.
West African Office: P.O. Box 33, Ibadan, Nigeria

©
CAMBRIDGE UNIVERSITY PRESS
1958

©
NEW MATERIAL IN FOURTH EDITION
CAMBRIDGE UNIVERSITY PRESS
1961

First Edition	1932
Second Edition	1935
Reprinted	1941
,,	1946
,,	1948
,,	1951
,,	1956
Third Edition, Revised, Reset	1958
Reprinted	1959
Fourth Edition	1961

*Printed in Great Britain
by Spottiswoode, Ballantyne and Co. Ltd.,
London and Colchester*

PREFACE TO THE FIRST EDITION

This book is intended for the use of students who have completed a year's study of the principles of zoology and of the anatomy and physiology of a series of invertebrate types such as is provided by any of several elementary text-books in use in this country. The types commonly included in these books—various Protozoa, *Hydra*, *Ascaris*, and the Liver Fluke, Earthworm, Leech, Crayfish, Cockroach, Pond Mussel, and Starfish—are not here described in detail. We have endeavoured to provide the student with a classification of the Invertebrata which proceeds as far as is usual in an honours course, with a concise statement of the characteristic features of each of the groups mentioned, and with a more detailed statement and discussion of matters of importance or interest concerning them. The choice of examples has been difficult, and we have not always been able to include all those we should have wished, but a fairly full account of certain representative genera has been given.

The writing of the book has been shared among us as follows: Chapters I–IV, X, XI (except Onychophora), XII, XVIII and XIX have been written by L. A. Borradaile, Chapters V (except Ctenophora), VII–IX, XIII, XV–XVII, and the Onychophora in Chapter XI by F. A. Potts, Chapter VI and the Ctenophora in Chapter V by J. T. Saunders, and Chapter XIV jointly by F. A. Potts and L. E. S. Eastham, but each of us has read and criticized the work of the others.

We desire to express our grateful thanks for valuable advice and criticism to Dr S. J. Hickson, Professor D. Keilin and Dr S. M. Manton; for much care bestowed upon the illustrations to Messrs A. P. Hayle, J. F. Henderson, and C. F. Pond; and for valuable assistance in the preparation of the index and in other matters to Mr B. Newman.

For permission to reproduce illustrations acknowledgment is due to Professor G. H. F. Nuttall; Messrs Geo. Allen & Unwin, Ltd. (*Textbook of Zoology*, Sedgwick); Messrs A. & C. Black, Ltd. (*Treatise on Zoology*, Lankester); the Council of the Cambridge Philosophical Society (*Biological Reviews*); Cambridge University Press (*The Determination of Sex*, Doncaster, *Plant Biology*, Godwin, *Ciliary Movement*, Gray, *Zoology*, Shipley and Mac-Bride, *Primitive Animals*, Smith, *Palæontology*, Wood); Herr Gustav Fischer, Jena (*Ergebnisse u. Fortschritte der Zoologie, Lehrbuch der Protozoenkunde*); Herren Walter de Gruyter & Co. (*Handbuch der Zoologie*); the Council of the Linnean Society of London (*Zoological Journal*); Messrs Macmillan & Co., Ltd. (*Cambridge Natural History*, Harmer and Shipley, *Human Protozoology*, Hegner and Taliaferro, *Textbook of Comparative Anatomy*, Lang, *Textbook of Zoology*, Parker and Haswell); Messrs Methuen & Co., Ltd. (*Textbook of Entomology*, Imms); Oxford University Press (*The Animal and its Environment* and *Manual of Zoology*, Borradaile). Acknowledgment to the authors of the works from which these illustrations are taken is made in the legends.

THE AUTHORS

CAMBRIDGE
February, 1932

NOTE TO THE SECOND EDITION

The book is now eighty pages longer. The additional matter is chiefly in Chapters IV and XIV. In other chapters a number of smaller additions and corrections have been made to the text and to figures.

Each chapter has been revised by its writer, but the revision has been submitted to the other authors of the book.

Our grateful thanks for various assistance are due to Drs A. M. Bidder, O. M. B. Bulman, S. M. Manton, and C. F. A. Pantin, and to Messrs L. E. R. Picken, J. D. Robertson, and P. Ullyott. Dr A. M. Bidder very kindly communicated to us certain facts, as yet unpublished, which are stated in the second half of p. 595.

THE AUTHORS

CAMBRIDGE
February, 1935

PREFACE TO THE THIRD EDITION

This edition differs from the preceding one in several respects. The labels to the diagrams have wherever possible been incorporated into the figure. Many of the diagrams have been altered, usually to make them more simple. Over a hundred new drawings have been made. The text has been revised and several alterations made, such as the removal of the Graptolita from the Coelenterata to the Protochordata, the expansion of the sections of Nemertea, Nematoda, Araneida, Onychophora. The Polyzoa have been split into Endoprocta and Ectoprocta, one being placed in the minor Acoelomata, the other in the minor Coelomata. Professor Eastham has revised the chapter on the Insects. A chapter has been added on zoological literature. It is not intended to be a comprehensive bibliography but instead should help the student to become acquainted with the literature. I should like to thank Drs J. O. Corliss, R. C. Fisher, M. S. Laverack, A. D. Lees, J. E. Morton, K. A. Munday, and J. D. Robertson for their help and criticism.

For permission to reproduce certain figures I am grateful to the following: *Biological Bulletin* for figs. 195, 196, 471; Blakiston Publishing Co. for fig. 64 (*The Biology of Paramecium*, Wichterman); the Royal Society for figs. 401, 408, 409, 419; Cambridge University Press for figs. 513, 514 (*Biological Reviews*); Comstock Publishing Co. for figs. 368, 369, 371, 373, 374, 375 (*The Spider Book*, Comstock and Gertsch); Masson et Cie for figs. 2, 63, 170, 363, 377, 378, 382, 383, 500, 502 (*Traite de Zoologie*, Grassé); McGraw-Hill Book Co. for figs. 114, 117, 133, 134, 136, 147, 148, 150, 151, 232 (*The Invertebrates*, Hyman), 396, 421, 431, 503 (*Principles of Invertebrate Paleontology*, Shrock and Twenhofel); Methuen & Co. for fig. 405A (*Nature*); Oxford University Press for figs. 25 (*Orientation of Animals*, Fraenkel and Gunn), 78, 174, 180, 187, 188, 189, 400 (*Quarterly Journal of Microscopical Science*), 202 (*The Oligochaeta*, Stephenson); Princeton University Press for fig. 54 (*Morphogenesis*, Bonner); the Ray Society for fig. 380 (*Comity of Spiders*, Bristowe); John Wiley & Sons for fig. 11 (*Problems of Morphogenesis in Ciliates*, Lwoff), 164, 165 (*Freshwater Biology*, Ward and Whipple).

G. A. K.

SOUTHAMPTON
1957

NOTE TO FOURTH EDITION

I have taken the opportunity of this new edition to make several minor corrections in the text; to replace several of the figures, and to bring the last chapter on zoological literature up-to-date. I should like to thank the many workers and teachers who have offered me their kind comments, and in particular Professor J. Lever for his constructive help.

G. A. K.

CONTENTS

Chapter I. INTRODUCTION   *page* 1

Chapter II. THE SUBKINGDOM PROTOZOA   10
  Phylum Protozoa
  Class Mastigophora (Flagellata)   52
    Subclass Phytomastigina   54
      Order Chrysomonadina   57
      Order Cryptomonadina   58
      Order Euglenoidina   58
      Order Chloromonadina   60
      Order Dinoflagellata   60
      Order Volvocina   62
    Subclass Zoomastigina   65
      Order Rhizomastigina   65
      Order Holomastigina   65
      Order Protomonadina   65
      Order Polymastigina   70
        Suborder Polymastigina *s. str.*
        Suborder Diplomonadina
        Suborder Hypermastigina
      Order Opalina   73
  Class Sarcodina (Rhizopoda)   74
    Order Amoebina   74
    Order Foraminifera   75
      Suborder Monothalamia   78
      Suborder Polythalamia   80
    Order Radiolaria   84
      Suborder Peripylea   84
      Suborder Actipylea   84
      Suborder Tripylea   84
      Suborder Monopylea   84
    Order Heliozoa   87
    Order Mycetozoa   89
      Suborder Acrasina   89
      Suborder Plasmodiophorina   89
      Suborder Eumycetozoina   89
  Class Sporozoa   90
    Subclass Telosporidia   91
      Order Coccidiomorpha   92
        Suborder Coccidia   92
        Suborder Haemosporidia   94
      Order Gregarinidea   96
        Suborder Schizogregarinaria   96

CONTENTS

Suborder Eugregarinaria	page 97
Suborder Piroplasmidea	100
Subclass Neosporidia	101
Order Cnidosporidia	101
Suborder Myxosporidia	102
Suborder Microsporidia	102
Suborder Actinomyxidia	103
Order Acnidosporidia	103
Suborder Haplosporidia	103
Suborder Sarcosporidia	103
Class Ciliophora	103
Subclass Ciliata	104
Order Holotricha	108
Suborder Astomata	108
Suborder Gymnostomata	108
Suborder Hymenostomata (Vestibulata)	108
Order Spirotricha	110
Suborder Polytricha	110
Suborder Oligotricha	110
Suborder Hypotricha	110
Order Peritricha	112
Order Chonotricha	113
Subclass Suctoria	113

Chapter III. THE SUBKINGDOM PARAZOA (PORIFERA)   116
Phylum Porifera
Class Calcarea	122
Class Hexactinellida	123
Class Demospongiae	124
Order Monaxonida	124
Order Keratosa	125
Order Myxospongiae	125
Order Tetractinellida	125

Chapter IV. THE SUBKINGDOM METAZOA   126

Chapter V. THE PHYLUM COELENTERATA   144
Phylum Coelenterata
Subphylum Cnidaria	152
Class Hydrozoa	153
Order Calyptoblastea	160
Order Gymnoblastea	161
Order Hydrida	162
Order Trachylina	162
Suborder Trachomedusae	162
Suborder Narcomedusae	162
Order Hydrocorallina	162
Order Siphonophora	163

Class Scyphozoa (Scyphomedusae)	page 169
Subclass Stauromedusae	170
Subclass Discomedusae	170
Order Cubomedusae	170
Order Coronatae	170
Order Semaeostomeae	170
Order Rhizostomeae	170
Class Actinozoa (Anthozoa)	176
Order Alcyonaria (Octoradiata)	176
Order Zoantharia (Hexaradiata)	181
Subphylum Ctenophora	187
Class Tentaculata	
Class Nuda	

Chapter VI. THE ACOELOMATA: PLATYHELMINTHES   191

Phylum Platyhelminthes	
Class Turbellaria	207
Order Acoela	207
Order Rhabdocoela	207
Order Alloiocoela	207
Order Tricladida	208
Suborder Maricola	
Suborder Paludicola	
Suborder Terricola	
Order Polycladida	208
Suborder Acotylea	
Suborder Cotylea	
Order Temnocephalea	210
Class Trematoda	211
Order Heterocotylea (Monogenea)	211
Order Malacocotylea (Digenea)	215
Class Cestoda	219
Order Cestodaria	221
Order Eucestoda	221
Suborder Tetraphyllidea	224
Suborder Diphyllidea	224
Suborder Tetrarhynchidea	224
Suborder Pseudophyllidea	225
Suborder Cyclophyllidea	225

Chapter VII. THE MINOR ACOELOMATE PHYLA   228

Phylum Nemertea	228
Class Anopla	228
Order Palaeonemertini	228
Order Heteronemertini	228
Class Enopla	228
Order Hoplonemertini	228
Order Bdellonemertini	228

Phylum Nematoda	page 235
Class Aphasmidia	236
Order Trichurata	236
Order Dioctophymata	236
Class Phasmidia	236
Order Rhabditata	236
Order Ascaridata	236
Order Strongylata	236
Order Spirurata	236
Order Camallanata	236
Phylum Nematomorpha	252
Phylum Acanthocephala	253
Phylum Rotifera	254
Phylum Gastrotricha	260
Phylum Kinorhynchia	261
Phylum Priapulida	262
Phylum Endoprocta	263
Chapter VIII. THE PHYLUM ANNELIDA	266
Phylum Annelida	
Class Polychaeta	269
Class Oligochaeta	296
Order Terricolae	298
Order Limicolae	302
Class Hirudinea	305
Family Acanthobdellidae	311
Family Rhynchobdellidae	311
Family Gnathobdellidae	311
Class Archiannelida	311
Class Echiuroidea	314
Class Sipunculoidea	316
Chapter IX. THE PHYLUM ARTHROPODA	317
Chapter X. THE CLASSES ONYCHOPHORA AND TRILOBITA	329
Class Onychophora	329
Family Peripatidae	
Family Peripatopsidae	
Class Trilobita	338
Chapter XI. THE CLASS CRUSTACEA	340
Class Crustacea	
Subclass Branchiopoda	368
Order Anostraca	370
Order Lipostraca	373
Order Notostraca	373
Order Diplostraca	375
Suborder Conchostraca	375

Suborder Cladocera	page 375
Tribe Ctenopoda	375
Tribe Anomopoda	375
Tribe Onychopoda	380
Tribe Haplopoda	380
Subclass Ostracoda	382
Subclass Copepoda	383
Subclass Branchiura	388
Subclass Cirripedia	389
Order Thoracica	389
Order Acrothoracica	393
Order Apoda	393
Order Rhizocephala	393
Order Ascothoracica	396
Subclass Malacostraca	396
Order Leptostraca	399
Order Hoplocarida (Stomatopoda)	400
Order Syncarida	400
Order Peracarida	401
Suborder Mysidacea	402
Suborder Cumacea	402
Suborder Tanaidacea	402
Suborder Isopoda	403
Suborder Amphipoda	405
Order Eucarida	408
Suborder Euphausiacea	408
Suborder Decapoda	408
Tribe Penaeidea	
Tribe Caridea	
Tribe Palinura	
Tribe Astacura	
Tribe Anomura	
Tribe Brachyura	
Chapter XII. THE CLASS MYRIAPODA	420
Class Myriapoda	
Subclass Chilopoda	420
Subclass Diplopoda	423
Chapter XIII. THE CLASS INSECTA (HEXAPODA)	427
Class Insecta	
Subclass Apterygota (Ametabola)	475
Super-order Entotropha	475
Order Collembola	475
Order Protura	476
Order Diplura	476
Super-order Ectotropha	476
Order Thysanura	476

CONTENTS

Subclass Pterygota (Metabola) — page 479
Section I Palaeoptera (Exopterygota) (Hemimetabola) — 479
  Super-order Ephemeropteroidea — 479
    Order Ephemeroptera — 479
  Super-order Odonatopteroidea — 481
    Order Odonata — 481
Section II Polyneuroptera (Exopterygota) (Heterometabola) — 483
  Super-order Blattopteroidea — 483
    Order Dictyoptera — 483
    Order Isoptera — 483
    Order Zoraptera — 485
  Super-order Orthopteroidea — 485
    Order Plecoptera — 485
    Order Notoptera — 486
    Order Cheleutoptera — 486
    Order Orthoptera — 486
    Order Embioptera — 488
  Super-order Dermapteroidea — 488
    Order Dermaptera — 488
Section III Oligoneuroptera (Endopterygota) (Holometabola) — 489
  Super-order Coleopteroidea — 489
    Order Coleoptera — 489
  Super-order Neuropteroidea — 492
    Order Megaloptera — 492
    Order Raphidioptera — 493
    Order Planipennia — 493
  Super-order Mecopteroidea — 494
    Order Mecoptera — 494
    Order Trichoptera — 494
    Order Lepidoptera — 495
    Order Diptera — 500
  Super-order Siphonapteroidea — 511
    Order Siphonaptera — 511
  Super-order Hymenopteroidea — 513
    Order Hymenoptera — 513
    Order Strepsiptera — 523
Section IV Paraneoptera (Exopterygota) (Heterometabola) — 525
  Super-order Psocopteroidea — 525
    Order Psocoptera — 525
    Order Mallophaga — 525
    Order Anoplura — 526
  Super-order Thysanopteroidea — 527
    Order Thysanoptera — 527
  Super-order Rhynchota — 527
    Order Homoptera — 531
    Order Heteroptera — 534

CONTENTS

Chapter XIV. THE CLASS ARACHNIDA — *page* 535

Class Arachnida
- *Order* Scorpionidea — 541
- *Order* Pseudoscorpionidea — 544
- *Order* Eurypterida — 544
- *Order* Xiphosura — 546
- *Order* Araneida — 549
- *Order* Palpigrada — 567
- *Order* Solifuga — 567
- *Order* Acarina — 567
  - *Suborder* Onchopalpida — 568
  - *Suborder* Mesostigmata — 568
  - *Suborder* Ixodides — 568
  - *Suborder* Trombidiformes — 568
  - *Suborder* Sarcoptiformes — 568
- *Order* Phalangida — 573
- *Order* Pantopoda (Pycnogonida) — 574
- *Order* Tardigrada — 576
- *Order* Pentastomida — 577

Chapter XV. THE PHYLUM MOLLUSCA — 578

Phylum Mollusca
- *Class* Amphineura (Loricata) — 589
  - *Order* Polyplacophora — 589
  - *Order* Monoplacophora — 593
  - *Order* Aplacophora — 594
- *Class* Gasteropoda — 596
  - *Order* Prosobranchiata — 610
    - *Suborder* Diotocardia (Aspidobranchiata) — 611
    - *Suborder* Monotocardia (Pectinibranchiata) — 612
  - *Order* Opisthobranchiata — 614
    - *Suborder* Tectibranchiata — 615
    - *Suborder* Nudibranchiata — 615
  - *Order* Pulmonata — 618
    - *Suborder* Basommatophora — 618
    - *Suborder* Stylommatophora — 618
- *Class* Scaphopoda — 621
- *Class* Lamellibranchiata (Pelecypoda) — 622
  - *Order* Protobranchiata — 630
  - *Order* Filibranchiata — 631
  - *Order* Eulamellibranchiata — 633
  - *Order* Septibranchiata
- *Class* Cephalopoda (Siphonopoda) — 636
  - *Order* Dibranchiata — 636
    - *Suborder* Decapoda — 636
      - *Tribe* Belemnoidea — 636
      - *Tribe* Sepioidea — 636

CONTENTS

Tribe Oegopsida	page 636
Tribe Myopsida	636
Suborder Octopoda	637
Order Tetrabranchiata	647
Suborder Nautiloidea	647
Suborder Ammonoidea	647

Chapter XVI. THE MINOR COELOMATE PHYLA — 652

Phylum Ectoprocta	652
Class Phylactolaemata	652
Class Gymnolaemata	652
Order Cyclostomata	652
Order Cheilostomata	652
Order Ctenostomata	652
Phylum Brachiopoda	660
Class Ecardines	660
Class Testicardines	660
Phylum Chaetognatha	666
Phylum Phoronidea	668

Chapter XVII. THE PHYLUM ECHINODERMATA — 669

Phylum Echinodermata	
Subphylum Eleutherozoa	681
Class Asteroidea	681
Class Ophiuroidea	685
Class Auluroidae	
Class Somasteroidea	
Class Echinoidea	687
Order Endocyclica	693
Order Clypeastroida	693
Order Spatangoida	693
Class Holothuroidea	694
Order Aspidochirotae	697
Order Pelagothurida	697
Order Elasipoda	697
Order Dendrochirotae	697
Order Molpadida	697
Order Synaptida (Paractinopoda)	
Subphylum Pelmatozoa	699
Class Crinoidea	699
Class Cystoidea	704
Subclass Hydrophoridea	704
Subclass Blastoidea	704
Class Eocrinoidea	704
Class Paracrinoidea	704
Class Edrioasteroidea	704
Class Carpoidea	704

Class Machaeridea	*page* 705
Class Cyamoidea	705
Class Cycloidea	705
Chapter XVIII. THE PROTOCHORDATA	706
Phylum Chordata	
Subphylum Hemichorda	708
Class Enteropneusta	709
Class Pterobranchiata	714
Class Graptolita	716
Order Dendroidea	718
Order Tuboidea	718
Order Camaroidea	718
Order Stolonoidea	718
Order Graptoloidea	718
Subphylum Urochorda (Tunicata)	720
Class Larvacea (Appendicularia)	732
Class Ascidiacea	733
Class Thaliacea	735
Order Pyrosomatidae (Luciae)	737
Order Salpidae (Hemimyaria)	738
Order Doliolidae (Cyclomyaria)	739
Subphylum Cephalochorda	
Subphylum Vertebrata	
Chapter XIX. LITERATURE	740
Index	790

CHAPTER I

INTRODUCTION

(1) SUBKINGDOMS AND PHYLA

The Invertebrata have long since ceased to constitute one of the primary divisions in the scientific classification of the Animal Kingdom. Their name is now no more than a convenience for designating a group of phyla with which it is often necessary to deal as a whole. The primary lines of real cleavage in the Animal Kingdom divide it, not into Vertebrata and Invertebrata, but into three unequal sections, the Protozoa, Parazoa and Metazoa, which are ranked in the following chapters as subkingdoms.

Between the Protozoa, which are without cellular differentiation and contain a large group of photosynthetic members, and the Metazoa, in which such differentiation is always strongly marked and photosynthesis is absent, there is a gulf which is in fact far deeper than that which sunders the Protozoa from the lower plants. The view, indeed, has been put forward that these two components of the Animal Kingdom are not, as is usually held, directly related to one another, but arose, with the Plants, as entirely distinct branches of an ancestral stock of living beings. The Parazoa or sponges—unique among many-celled organisms in possessing collared flagellate cells—are probably derived from the Protozoa by an origin distinct from that by which the latter group gave rise (if they did so indeed) to the Metazoa.

DIPLOBLASTICA AND TRIPLOBLASTICA. Within the Metazoa, the most significant difference is that which exists between the Coelenterata or Diploblastica and the triploblastic phyla which constitute the rest of the subkingdom. There is a view that the most primitive of the Metazoa are the Platyhelminthes and that the coelenterates were derived from these by the reduction of the mesoderm. The more orthodox view is that the ancestral metazoan gave rise to both the ancestral coelenterate and the ancestral platyhelminth. The Coelenterata, which typically start life as a simple, two-layered, ciliate larva, the *planula*, either retain throughout life the two-layered condition, or depart from it only by the immigration, late in development, of cells from the two primary layers (ectoderm and endoderm, p. 126) into the space (blastocoele) between those layers. The triploblastic animals always possess a true third layer (mesoderm) which is early developed and forms important organs. They are the great majority of animals, and compose a number of phyla.

SUPERPHYLA. The brigading of these phyla is a difficult task—one, indeed, which is at present impossible to effect completely. Two main stocks, however, stand out fairly clearly. The annelid superphylum which contains the annelids, molluscs, and arthropods, and the echinoderm superphylum which contains the echinoderms, protochordates, and chordates. The differences between these two superphyla are as follows (Fig. 1):

1. *Cleavage*. The annelid superphylum has eggs that develop by means of

spiral cleavage. When the blastula divides from the four-cell stage to the eight-cell stage the second quartet lie on top of and between the cells of the first quartet. In the echinoderm superphylum the second quartet of cells lie immediately above and in line with the cells of the first quartet. This is called radial cleavage.

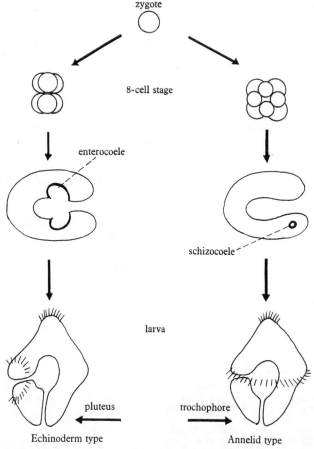

Fig. 1. The two superphyla. The different phyla are classified into two superphyla, the echinoderm superphylum and the annelid superphylum. These groups differ in their form of cleavage, origin of the coelom, type of larva and other instances discussed in the text.

2. *Origin of the coelom.* In the annelid superphylum the coelom arises by a split in the mesoderm. This is called schizocoelic development. In the echinoderm superphylum the coelom develops as a pouch or series of pouches from the gut. This is called enterocoelic development.

3. *Larvae.* The annelids have a trochosphere larva with its main ciliated band around the circumference of the body. In the echinoderm pluteus larva the main ciliated band runs around the mouth. There is a close resemblance

between the annelid and the mollusc larva on the one hand and the echinoderm and the balanoglossid larva on the other.

4. *Development of the nervous system.* The annelid develops its internal central nervous system by delamination from the ectoderm. The echinoderms separate the nervous system from the ectoderm by invagination.

5. *Potency.* The annelid eggs are mosaic eggs. If part of the developing blastula is removed then the adult develops without certain parts that would have been formed by the removed cells. The echinoderm blastula, if it has some cells removed at an early stage, can compensate for the loss and develop into a complete adult. This type of egg is called equipotential.

These differences between the two superphyla are based on generalizations and many exceptions are known. Further details will be found in the appropriate sections concerning the embryology of the different groups.

Table 1. *Mayr's estimation of the number of known species in the different groups*

Protozoa	30,000	Onychophora	65
Porifera	4,500	Linguatula	70
Coelenterata	9,000	Chelicerata	35,000
Ctenophora	90	Crustacea	25,000
Platyhelminthes	6,000	Other arthropods (excl. insects)	13,000
Acanthocephala	300	Insecta	850,000
Rotifera	1,500	Mollusca	80,000
Gastrotricha	175	Ectoprocta	3,300
Kinorhynchia	100	Brachiopoda	250
Nematomorpha	100	Echinodermata	4,000
Nematoda	10,000	Phoronidea	4
Priapulida	5	Chaetognatha	30
Entoprocta	60	Hemichorda	80
Nemertina	750	Tunicata	1,600
Annelida	7,000	Fishes	20,000
Echiuroidea	60	Amphibia and Reptiles	6,000
Sipunculoidea	250	Birds	8,590
Tardigrada	180	Mammals	3,200

Invertebrates have over 97% of the total number of species.

MINOR PHYLA. The remaining phyla, smaller and less important, are hard to relate either to the foregoing groups or to one another. By the type of cleavage of their ova and the possession of flame cells (p. 200), the Platyhelminthes and Nemertea seem to be akin to the annelid stock. Their lack of coelom is a difficulty in this respect. The structure of the adults of the Rotifera and of the larva of the Ectoprocta, which has the character of a trochosphere, might link these groups to the same stock. Some other small phyla (Brachiopoda, Chaetognatha) have possibly distant relationship to the echinoderm-chordate grouping. Others, notably the Nematoda, are more difficult to place.

CHORDATA. In the great assemblage of triploblastic phyla, the backboned animals, or Vertebrata properly so-called, stand as a branch of one phylum, the Chordata. Yet their considerable numbers, the size, high organization, and intelligent activity of their members, and the fact that Man is one of them, give them an importance so great that they have always been the subject of a distinct department of zoological study, and were at one time regarded as a primary branch of the Animal Kingdom. That standing they have lost; but it is still necessary for many purposes to treat them apart.

NOMENCLATURE. The term 'Invertebrata' is retained to cover all the non-chordate phyla and the chordates other than the Vertebrata. In that sense it is used in this book. Only the Cephalochorda (*Amphioxus*), which, though they are not vertebrates, have much in common with those animals, are left aside as best studied with them.

The limits of the several phyla are, with one or two exceptions, agreed among zoologists. As much cannot be said for the lower grades of the classification. Different views upon phylogeny, and considerations of convenience, lead to many divergences as to the extent and rank of the various divisions in the systems preferred by different authorities; and even when there is agreement as to the limits of a group different names may be applied to it. In no two works will quite the same arrangement be found. This fact should be borne in mind by the student in using the tables of classification.

(2) PRINCIPLES

In surveying the diverse organisms which constitute the Invertebrata, the student should bear in mind the following principles.

ADJUSTMENT. The most fundamental of the characteristics of living organisms is the way in which, in the face of an environment which presents as many dangers as opportunities, they hold their own by making adjustments within themselves. This statement applies equally to the struggle for existence of the individual and to the slow racial adjustments which we know as evolution.

ENVIRONMENT. The term environment is a collective name for all the external things which affect any living being. The student may occasionally be puzzled by the phrase 'internal environment'. This bizarre contradiction in terms is sometimes applied to what we shall presently call the 'internal medium'. Four principal factors constitute the environment—the ground or 'substratum' (if any) upon which the organism stands, the 'medium' (water or air) which bathes it, the heat and light which it receives from the sun's rays or can lose to its surroundings, and the other organisms in its neighbourhood. Of these factors the *substratum* has in most cases relatively little importance, and we may dismiss it now.

MEDIUM. The medium, on the other hand, is of enormous importance. Meeting all parts of the surface of objects that it contains, it exerts everywhere a pressure upon them, supports them, may transport them, affects the movements they execute, and controls all exchange, whether of matter or of energy, between them and the world about them; and from it animals obtain their supplies of free oxygen, often of water, and sometimes of food. If it be liquid, according as the concentration of substances dissolved in it be greater or less than that within the organism, water and solutes will tend to pass to or from the body of any animal which is not covered by an impermeable cuticle. This exchange is of the utmost importance, both as a danger by upsetting equilibria within the body and as an advantage by facilitating the excretion of substances which are harmful in the organism. It is controlled by the surface layer of protoplasm, which either is, or owing to surface tension behaves as, a delicate membrane that has the power of actively regulating, to some extent, the passage of substances through itself, and by the activity of the organs of excretion. If the medium be gaseous, according to the amount of water vapour it contains, water will tend to evaporate to it from the surface of the body. This is important owing to the necessity for the intake of water by the mouth to compensate for it, and also because the latent heat of the evaporation lowers the temperature of the body. Whether the medium be liquid or gaseous, it offers, according to the free gases it contains, varying possibilities of interchange of oxygen and carbon dioxide with organisms. This has naturally extremely important effects upon respiration.

HEAT AND LIGHT. The loss or gain of heat tends, of course, to affect the temperature within organisms, and with this the chemical processes of the latter vary, being, as is usual in such processes, slowed as the protoplasm becomes colder and quickened when it is warmed, and being brought finally to a stop when its minute organization is destroyed either by the coagulation of certain of its proteins by heat or by the freezing of its water. Every organism is tuned to work within a range of temperature peculiar to it. 'Warm-blooded' animals keep their temperature within proper limits by active chemical and physical means; 'cold-blooded' animals (to which all invertebrata belong) are in this respect at the mercy of their surroundings except in so far as they can circumvent them by their habits. *Light*, while it is essential for photosynthetic organisms, has chemical effects of importance in many others, and all animals which possess light-sensitive organs appreciate it as a source of stimuli from the external world.

ANIMAL RELATIONSHIPS. Relations between an animal and other organisms in its surroundings are almost always based in the long run on nutrition. Either such organisms serve the animal for food, or they attack it to make it their food, or they are competitors for a common food-supply, or in rarer cases they assist it, or obtain its assistance, in the quest for food or in defence against enemies which would use it or them for food. Only between members of opposite sexes of the same species are there relations of another kind, namely those which are concerned with reproduction. The coming of organisms into relation with one another usually involves the receipt of stimuli and more or less complicated behaviour, with the use of organs of locomotion and prehension.

ENVIRONMENTAL EFFECTS. The action of the environment upon the organism will be seen to be threefold:

(1) it affects it mechanically, as by transporting it from place to place, by the impact of adjacent objects, or by the attacks of enemies;

(2) it affects the working of the living machine by the compulsory introduction or abstraction of materials (water, salts, etc.) or of energy;

(3) it directly stimulates it to activity, which may be an inevitable response,

such as the movement of certain organisms towards light (phototaxis), or be dependent upon conditions existing at the moment in the organism; or it may inhibit such activity.

Besides such action the environment may affect the organism negatively, by failure in respect of food, oxygen, or some other necessity which the organism is dependent upon obtaining from its surroundings. Where such failures occur from time to time the organisms have usually means of enduring them (reserve stores, resting stages, etc.).

REACTIONS TO ENVIRONMENT. In proportion as the organism is unable to resist these influences of the environment it is liable upon occasion to be harmed by them. The process of evolution has been the development of organisms in such a way as to set them free from such influences in respect of their proper environments. Its results may be classed under three heads.

(1) Some, such as the acquirement of a cuticle or of a habit of burrowing or of hibernation, merely fend off or avoid the action of the environment: these involve the least increase in the complexity of the organism.

(2) Others, such as the formation of organs for the excretion of the excess of water, provide for remedial action: in these, as a rule, more complicated machinery is formed.

(3) Others, such as the development of a nervous system or of organs of locomotion or weapons of offence, bring it about that the action which results from the receipt of stimuli is turned to the best advantage by the organism: these cause a considerable, often a very great, complication of the living machine.

Thus a general outcome of evolution is the forming of more complex, that is of 'higher', organisms. But a relatively simple organism may, in its proper environment, enjoy as much autonomy as in other circumstances is possessed by one that is more highly organized. This is notably true of many parasites.

Some of the results of evolution, as for instance the formation of a nervous system or of a cuticle, are such as to increase the independence of the organism in *all* circumstances. Others, however, such as the substitution of pulmonary for branchial respiration, or of absorption for ingestion of food, are of value only in particular environments or modes of life, and even unfit the organism for other ways of living. Thus two distinct phenomena underlie the diversity of the Animal Kingdom—an increase in the autonomy of the individual, and the specialization of animals for particular modes of life.

DEATH AND REPRODUCTION. Every species, however good a fight it maintains, is threatened with extinction owing to the continual loss of individuals, always by the action of its environment and usually also by that 'natural death' which appears to await all organized protoplasm that is not periodically reorganized. It is possible that in some of the least highly organized Metazoa natural death either is no more inevitable than in Protozoa or is long delayed. In *reproduction*, however, the individual provides by fission for the maintenance of its race. In the lower organisms the protoplasm of the body retains a certain plasticity, and in these there is very often an *asexual* process of reproduction in which the new construction that is necessary to organize at least one of the products of fission, and often goes on in both, is carried out with cells (or, in protozoa, with organized protoplasm) which existed as such in the parent. In more highly organized animals the only protoplasm which retains the required plasticity is that of the germ cells, and consequently such animals have only the *sexual* reproduction which these cells perform. The germ cells (gametes), before they reconstitute the adult body, normally undergo the process known as *conjugation* or *syngamy*, which is not an essential part of the reproductive process but a provision for heritable variation whereby the race becomes adaptable to its surroundings. Conjugation can only be performed by uninucleate individuals and therefore, while in Protozoa it sometimes takes place between adults (hologamy, p. 36), in Metazoa it always requires the production of uninucleate young (the ova and spermatozoa). The lower Metazoa reproduce both by means of these gametes and also asexually. In the higher animals, as we have seen, reproduction is solely by gametes, though conjugation may be suspended for one or more generations by the development of unfertilized ova (parthenogenesis).

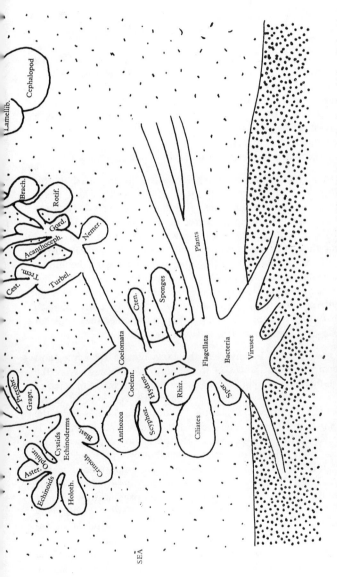

Fig. 2. The phylogenetic tree. This diagram shows the relationships of the main groups of animals. (After Cuénot.)

CHAPTER II

THE SUBKINGDOM PROTOZOA

DEFINITION. The Protozoa are sundered from the rest of the Animal Kingdom by a perfectly sharp distinction. The distinction consists in this: that in the body of a protozoon, whether there be one nucleus, or a few, or many, no nucleus ever has charge solely of a specialized part of the cytoplasm; whereas in other animals there are always many nuclei, each in charge of a portion of cytoplasm which is specialized for a particular function, such as contraction, or conduction, or secretion.

Stated thus, the definition of the Protozoa is quite unambiguous.

CELL CONCEPT. Unfortunately, ambiguity is usually imparted to this subject by the introduction of a concept, that of the 'cell', which has a different extension for different authorities. If that concept, primarily of use in other connexions, is to be introduced here, we must frame our definition in one of two ways, according to the meaning which we attach to the word 'cell'. If we define the cell as the unit of the animal body then the Protozoa are for the most part unicellular. If on the other hand we regard the animal body as being divisible into units called cells, then the Protozoa are not divisible into any such units and so can be called acellular. It is more convenient perhaps to use the term 'energid'. It will then be convenient to employ this term to any nucleus with its cytoplasm, whether they together constitute the body of a protozoon or a cell of a metazoon.

PROTOZOAN FEATURES. In any case the facts remain the same, and they provide one of the main sources of the interest which the study of the Protozoa offers, namely the carrying-out of the processes of life, and often of a complex life, by an organization which, though it may visibly be of corresponding complexity, is purely cytoplasmic. Considered in this light the structure of, for instance, the more complicated ciliates and flagellates is exceedingly instructive. In three other respects the Protozoa are peculiarly interesting. In their bodies dead 'formed' material, however plentiful it be as a covering or scaffold for the body, never assumes the importance which it has as ground substance or skeleton in the Metazoa, where the size of the body is such that the protoplasm cannot maintain its organization without support against forces that tend to deform it. Consequently, in observing the physiology and behaviour of a protozoon, we are seeing in the actual protoplasm of an intact organism processes which in an intact metazoon we observe as the activities of a complex in which protoplasm is masked and conditioned by other components of the body: in short, in the Protozoa we observe the normal activities of protoplasm more directly than in the Metazoa. Again, a life cycle comprising more than one generation, which is comparatively rare among metazoa, is universal among protozoa, and its varieties are extraordinarily interesting. Finally, while every metazoon is thoroughly an animal, the Protozoa present an unbroken series from wholly plant-like organisms, through various intermediates, to members whose nutrition and behaviour are those of animals—or rather, as we shall see, there are several such series.

SIZE. The Protozoa are all of small size. Most of them are minute, ranging from a few thousandths of a millimetre to a little over one millimetre in length: a few reach dimensions of several, or even of many centimetres, but these for the most part consist of a relatively thin layer of protoplasm (certain Mycetozoa). With the small size of Protozoa is probably to be connected, not only, as we have seen, the relative unimportance to them of dead skeletons, but also their characteristic type of organization. In larger organisms, the regions differentiated for special purposes must usually be correspondingly larger, and therefore require the services of nuclei of their own, the absence of which is the hall-mark of a protozoon. The actual size varies very much in each group. It is, on the average, least in the Mastigophora. The order of magnitude of certain representative species may be gathered from the approximate magnifications stated for figures below.

SHAPE. The bodies of the Protozoa vary greatly in shape. Whereas in each of the metazoan phyla there is a fundamental type of body form to which the members of the phylum conform in essentials, however aberrant from it they be, the Protozoa have no such type. When the surface of the protoplasm is virtually fluid and is not retained by a shell, it takes, while it is at rest, a spherical form. When there is a firm surface layer, the individual tends, if it be a flagellate, to have an egg or spindle shape, if it be a ciliate to be bilateral with a spiral twist at one end, in the Suctoria to be cup-shaped. Concerning the body form of the Sporozoa, which are parasitic, no generalization can be made. Bodies of any of these shapes may be anchored, and have then usually a stalk, which may be of dead material as, for instance, in *Acineta* and *Codosiga* (Figs. 3, 38B), or a part of the living protoplasm. In the latter case it has generally a cuticular covering, as in *Vorticella* (Fig. 3) but may be naked (various flagellates).

COLONIAL FORMS. Stalked forms, and occasionally others, may be colonial; that is to say, a number of *zooids*, each having a nucleus and the shape and complete organization of an individual of related solitary species, are united by protoplasmic connections to form a single living being. The zooids of a colony are usually all alike, but differentiation may exist between them, in that certain of them are specialized for the production of new colonies, which is not performed by the other zooids (various Volvocina, Figs. 4, 34, etc.). Colonies arise by the division of a single primary zooid, whose fission is not carried to completion, so that its products do not entirely separate. Their origin is therefore usually said to be a form of asexual reproduction. It may, however, also be looked upon from another point of view, as the repetition, within a continuous mass of protoplasm, of the nucleus and the other organs coincidentally. In this aspect, the colony is seen to have features in common with other multinucleate conditions of Protozoa, such as (1) that of *Hexamita* (Fig. 3), etc., in which a unitary body has two similar

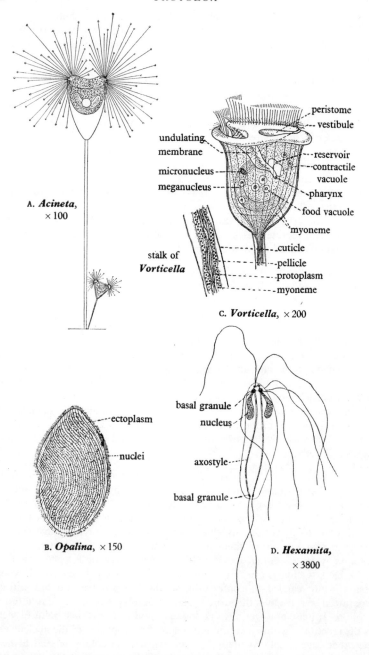

Fig. 3. Morphological form in the Protozoa. (A, after Saville-Kent; B, after Bronn; D, after Dobell.)

sets of organs, one on each side of the body, or several sets, with a nucleus assigned to each, (2) that of *Polykrikos* (Fig. 31), etc., in which there are several nuclei, and several sets of the other organs of the body, but the repetition (merism) of the nuclei and that of the other organs do not correspond, and (3) that of *Opalina* (Fig. 3), *Actinosphaerium* (Fig. 27), etc., in which there are numerous nuclei, but only one set of the other organs of the body. Multinucleate masses of protoplasm are known as *syncytia*. Syncytia which, like those cited above, arise by the division of an original nucleus in the mass of protoplasm are known as *symplasts*. An entirely different kind of syncytium arises by the union of uninucleate individuals, whose nuclei remain distinct in the resulting body. Such syncytia are known as *plasmodia*. They are found in the Mycetozoa (Fig. 53) and occasionally elsewhere.

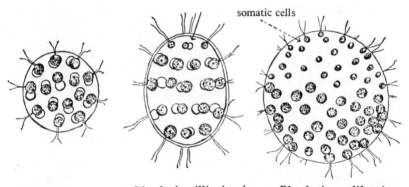

A. *Eudorina*   B. *Pleodorina illinoiensis*   C. *Pleodorina californica*

Fig. 4. Colonial Volvocina: *Eudorina*, a spherical motile colony of thirty-two similar zooids all capable of division; *Pleodorina illinoiensis*, a spherical motile colony of thirty-two zooids of which four at one end constitute a 'soma' which dies when the other twenty-eight zooids divide; *Pleodorina californica*, in which the 'somatic cells' constitute about half of the colony. (After West and Fritsch.)

Pseudo-colonies, consisting of distinct individuals united only by stalks, tubes, etc., of dead material, are formed by various mastigophora (Fig. 29) and vorticellids.

PROTOPLASM. After fixation protoplasm may show many diverse structures, though it is difficult to tell the extent to which these reflect the structure of living protoplasm and to what extent they are artifacts. Protoplasm contains long chain molecules that can at times become orientated with respect to one another though it is a mistake to imagine that protoplasm is entirely fibrillar. Equally it is a mistake to assume that because small vacuoles or alveoli are present in fixed protoplasm, living protoplasm is entirely made up of alveoli. The submicroscopic patterns in protoplasm are capable of great variation, this being correlated with changes that take place in the environment.

Some of the larger alveoli in the Protozoa are called vacuoles and these may be used for storage, as a site for chemical processes such as digestion, for drainage, for hydrostatic function, etc. The largest of the vacuoles have a definite wall of their own and it is probable that in all these vacuoles there is a concentration of mitochondria around the vacuole and that this is mainly responsible for the chemical work performed by the vacuole.

SURFACE. The surface of the protoplasm is protected in various ways.

(1) Sometimes, as in some amoebae, it is apparently quite fluid. Then, however, there exists upon it an extremely thin membrane, known as the *plasmalemma*, which has the power of regulating the exchange of materials between the organism and the watery medium in which it lives. Without this power the protoplasm would soon be poisoned or dissolved.

(2) In other cases, the surface layer is semi-solidified (gelated) as a visible, firm, but living *pellicle*. This is often 'sculptured' in a pattern, as in *Paramecium* (Fig. 64).

(3) Intermediate conditions connect the pellicle with the *cuticle*, a close-fitting dead membrane which may be nitrogenous, as in *Monocystis*, or of carbohydrate, as in many plant-like flagellates. In typical dino-flagellates (Fig. 31) it is composed of stout plates of cellulose.

(4) Again, there may be a *shell* from which protoplasm can issue through an opening. Such a shell may be nitrogenous, as in *Arcella* (Fig. 46), etc., of a nitrogenous basis with foreign bodies built into it, as in *Difflugia* and *Rhabdammina* (Fig. 5), of siliceous plates as in *Euglypha* (Fig. 46), calcareous, as in most foraminifera (Fig. 5), or of cellulose, as in the spores and sclerotium of the Mycetozoa. It is said that mineral shells always contain a groundwork of organic material. They are often composed of several chambers, and may be perforated by numerous small pores. *Houses* are loose-fitting, wide-mouthed shells (Fig. 29). *Cysts* are temporary shells without opening.

(5) Finally, there may be an external lattice, which is pseudochitinous in *Clathrulina* (Fig. 5), and siliceous in the Silicoflagellata (Fig. 29), or a case of calcareous pieces (Coccolithophoridae, Fig. 29). The siliceous lattice of many radiolarians is part of an internal skeleton.

ECTOPLASM: ENDOPLASM. The term ectoplasm is applied to any conspicuously differentiated outer layer of the protoplasm, and denotes very different conditions in different organisms—in *Amoeba*, a stratum which, save at its surface, is only unlike that below it in not containing granules; in various planktonic protozoa (Figs. 26, 27) a highly vacuolated layer whose low specific gravity confers buoyancy; in the Ciliophora and many flagellates and sporozoa a stout pellicle with an underlying layer, the *cortex*, which is said to be stiffer than the internal protoplasm (*endoplasm*) and may exhibit differentiations of various kinds.

NEMATOCYSTS: TRICHOCYSTS. Occasionally the protoplasm contains structures (trichocysts of ciliates and mastigophora, Fig. 6, so-called 'nematocysts' in certain dinoflagellates, Fig. 6, pole capsules of Neosporidia, Fig. 62), from which threads can be shot out upon the surface of the body. The function of these threads is often doubtful, but it has been shown that the trichocysts of *Paramecium* are fixing organs, others which lie around the mouth of their possessor (*Cyathomonas*, Fig. 30, etc.) seize prey, and the pole capsules serve to anchor spores to the lining of the host's gut. The threads of 'nematocysts'

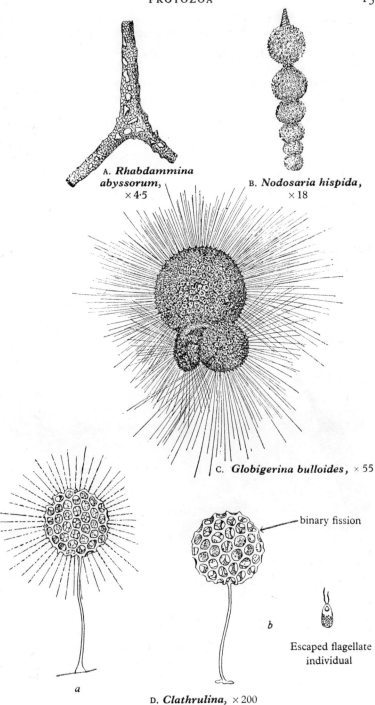

Fig. 5. Shells of Protoza. A, B and C are foraminiferan shells; D is the heliozoan *Clathrulina*: *a*, normal individual, *b*, binary fission within the lattice.

and pole capsules are coiled up in vesicles before they are shot out; those of trichocysts are formed by the stiffening of an extruded secretion.

MOTILE ORGANS. The motile organs of the Protozoa are of several kinds, each of which is mainly found in one of the classes of the phylum. *Pseudopodia* are temporary protrusions of protoplasm. They are of various types—blunt

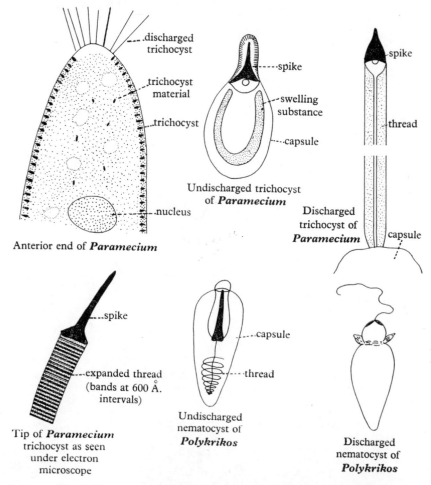

Fig. 6. Protozoan nematocysts and trichocysts. (After various authors.)

lobopodia (Figs. 43, 46), fine *filopodia* (Fig. 46), branching and anastomosing *rhizopodia* (Fig. 48), and *axopodia* (Fig. 51c), with an internal supporting filament. They are used in various ways and for various purposes. Their mode of formation is not fully understood, but it is clear that, at least in many cases, they do not arise, as has been alleged, by alterations in the surface tension of the protoplasm, and it is probable that the movement (*amoeboid movement*) in the course of which they are formed is not fundamentally different from the movements of muscles, or cilia, or flagella. Granules may often be seen to stream up and down the axopodia and rhizopodia.

Flagella are lashes, long and usually few in number, which by a rowing (Fig. 7A) or by an undulating motion (Fig. 7B) draw or propel the body or attract particles to it. In the rowing stroke the flagellum is held rigid and slightly concave in the direction of the stroke; in recovering its position it bends as it is drawn back, so that less resistance is offered to the medium. When, as is usually the case, the flagellum beats obliquely, or the undulations pass around as well as along it, the body rotates as it advances, or if it be fixed a whirlpool is set up.

Down each flagellum or cilium run many internal threads which fuse on entering the body, or at some distance within it, with the basal granule. The structure of the flagellum or cilium is fairly constant throughout the animal kingdom. The simplest ones such as are present in some bacteria and a few Protozoa are single filaments. Most, however, are made up of an outer ring of nine filaments and an inner ring of two filaments. These fuse together before they enter the body (Fig. 7c). The basal granule is in most cases connected to the nucleus by a thread or threads known as *rhizoplasts* (Fig. 36). Sometimes it lies against the nucleus. Rhizoplasts may connect it to other structures, notably in many parasitic flagellates to a body of unknown function called the *parabasal body*. The parabasal body of *Trypanosoma* (Fig. 36) is a body of this class, which possibly includes structures of more than one kind.

Sometimes, as in *Trypanosoma*, a flagellum runs for some distance parallel with the surface of the body and is connected to it by a film of protoplasm known as an *undulating membrane*, which must be distinguished from the structures of the same name which are formed by the fusion of cilia. When there are two flagella, it often happens that one is trailed behind the body and the other directed forwards (Figs. 36, 42, 51B). Flagella are often used for anchoring, and sometimes appear to have a sensory function.

Cilia are smaller and more numerous lashes which by a rowing action repeated by one after another of them in 'metachronal rhythm' (Fig. 8A) cause movements of the animal or of the water near it. Like flagella they have each internal filaments, a basal granule, and a rhizoplast, which, however, does not connect with the nucleus. Often cilia are united into compound organs, which may be conical *cirri*, paddle-like *membranellae* (Fig. 8B), or *undulating membranes* (Fig. 65A). Many Protozoa which possess a definite body form are able temporarily to alter it by contractions of the protoplasm stretching the pellicle (*metaboly*), and in various cases this contractility is localized in fibrils, known as *myonemes*, situated in the ectoplasm.

INFRACILIATURE. Within recent years our knowledge of the structure and life cycles of the ciliates and mastigophorans has been considerably increased, mainly due to the work of French and American protozoologists. They have made a careful study of the structure of the cilia and flagella together with their associated organelles, and have shown that the pattern is not a random one but that there is usually a precise and regular relationship

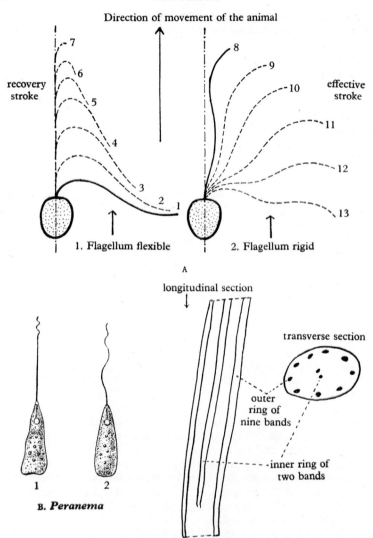

Fig. 7. Flagellar movement and structure. A, simplest movement of flagellum of *Monas* during rapid forward movement. Note that flexure in the recovery stroke begins at the base and spreads to the tip. In the effective stroke note the rigidity of the flagellum. (After Krijgsman.) B, the movement of the flagellum in *Peranema*. In 1 the animal is gliding slowly forward with undulations restricted to the tip. In 2 the animal is moving rapidly and the undulations are along the whole length of the flagellum. C, diagram of flagellar and ciliary structure based on electron microscopy. There is an outer circle of nine bands and an inner circle of two bands. These usually have a spiral twist along the length of the flagellum.

between the cilia, flagella and the underlying organelles. This pattern can be called the infraciliature though the term is usually restricted to the more superficial organelles related to the cilia. It is important to distinguish this system from the 'silver line' system, the latter being the more variable systems of lines that take up silver stains and in many cases include the sculpturing on

A. Metachronal rhythm

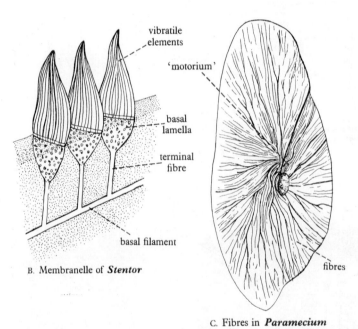

B. Membranelle of *Stentor*

C. Fibres in *Paramecium*

Fig. 8. Ciliary structure. (A, after Verworn; B, after Doflein; C, after Rees.)

the protozoan pellicle. In many protozoans throughout their life history there are regular specific changes in the pattern of the infraciliature, indicating a different stage in the life history.

MASTIGOPHORAN STRUCTURE. In the polymastigine *Calonympha* found in the gut of termites, the flagella are associated with a basal granule or centrosome just inside the pellicle (Fig. 9). Lateral to the centrosome is a body, the parabasal body, which is identified by its staining reactions with silver and osmium tetroxide. A filament runs from the basal granule into the central region of the body forming the axial filament. If a nucleus is associated with these three units (basal granule, parabasal body and axial filament) then

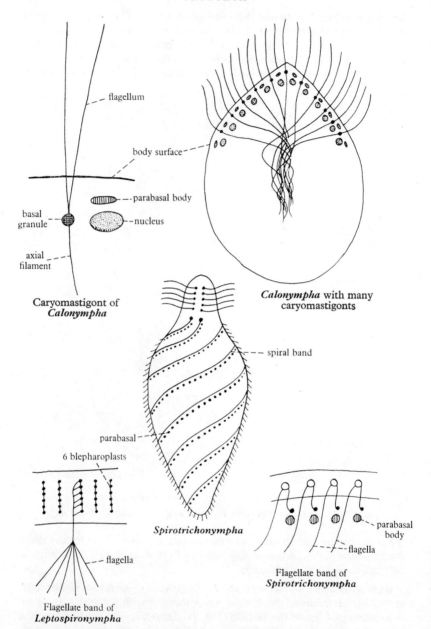

Fig. 9. Flagellate structure; see the text for further details. (After various authors.)

the unit is called a caryomastigont. If there is no nucleus then the unit is called an acaryomastigont. In *Calonympha* there are many caryomastigonts as shown in Fig. 9. The acaryomastigont structure is seen in *Spirotrychonympha* and *Leptospironympha*. There is only one nucleus in these animals and it is situated in the central region of the body. The flagella are arranged in a

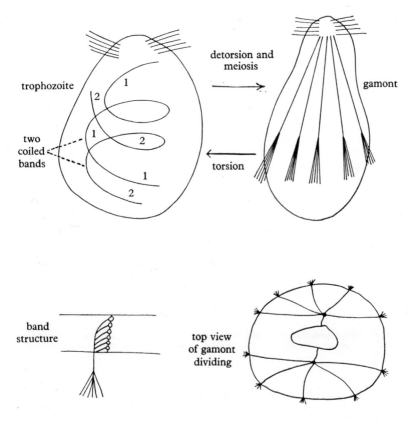

Leptospironympha

Fig. 10. Flagellate structure. *Leptospironympha* shows untwisting in the gamete stage the spiral bands of cilia uncoiling prior to division of the adult to form the gamonts.

series which spiral round the animal. Each band consists of a series of basal granules arranged in a definite pattern (Figs. 9, 10), each giving off a flagellum. Near the basal granule are a series of parabasal bodies. The adult form shows two such ciliated bands, but when the animal divides to give rise to the gametes, the bands unwind and it and the gametes have simple straight bands of flagella. Two such gametes conjugate and the resulting zygote undergoes twisting or torsion of the bands leading to the normal adult form. (It is interesting to note that meiosis and conjugation and encystment in *Calonympha* are influenced by the moulting hormones of the host termite, and the protozoan changes into its resting stage at the same time as the host starts to moult, the result being that the ectodermal lining of the gut is cast off containing the protozoans in the protected encysted stage.)

CILIATE STRUCTURE. In the ciliates each cilium is associated with a basal granule. These basal granules are called kinetosomes (Fig. 11). If the observer looks towards the anterior of the animal he will find that to the right of the kinetosomes runs a fine strand, the kinetodesma. Since the kinetodesma is always to the right of the kinetosome it provides a means of orientating the stained preparation. The kinetosomes and the kinetodesma form a unit called the kinety, and many kinety make up the infraciliature. The infraciliature

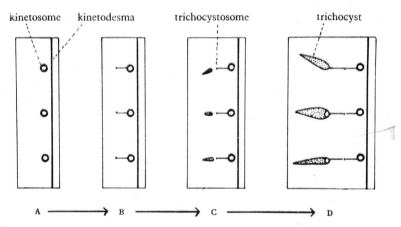

Fig. 11. Development of the trichocyst. It has been shown that in certain Protozoa the trichocysts are formed from the small bodies (kinetosomes) which bud and give rise to trichocysts. (Kinetosome + kinetodesma = kinety.)

is sometimes visible in the living animals but can be more easily studied in Bouin- or Champy-fixed material stained with haematoxylin or silver. The ciliate infraciliature differs from that of the flagellates in that it is not connected in any way to the parabasal body, nor are the kinety interconnected. Instead each kinety remains as a distinct entity. In both the ciliates and flagellates the kinetosomes (parabasal body or centrosomes) arise by binary fission and are never formed *de novo*.

Another major difference between the ciliates' and the flagellates' infraciliature lies in the manner of cell division. When the flagellates divide they do so along a plane parallel to the lines of the infraciliature. This is called symmetrical division, and the kinety are not broken in half, but instead are divided into two groups. In the ciliates, however, the plane of cleavage cuts right across all the kinety, this type of division being called percentien division (Fig. 12). In this way one can accurately place certain problematical animals such as *Opalina* or *Multicilia* definitely into one group or the other.

Considerable variation can occur in the infraciliature of the ciliates.

(1) The kinetosomes can become aligned so that they form parallel rows as well as longitudinal rows.

(2) This tendency can become exaggerated till the kinetosomes become packed into very close rows.

(3) The animal can become twisted so that the kinety take on a spiral or twisted pattern (Fig. 13).

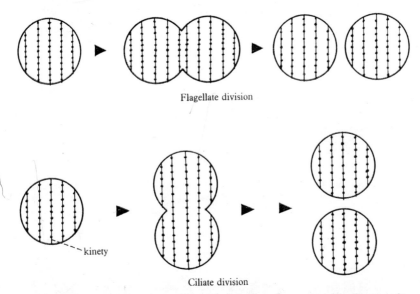

Fig. 12. Division in the ciliates and flagellates. Both groups possess cilia. They can be differentiated from each other by their manner of dividing: the ciliates divide across the line of the kinety whilst the flagellates divide parallel to the kinety.

(4) A few of the kinety become specialized to form organelles such as the buccal or peristomial cilia. In these cases the specialization is such that only those particular kinety can form the buccal ciliature (Fig. 14).

(5) The kinety can have some of their cilia united to form cirri which act as walking legs, or they can form little triangular membranellae.

MORPHOGENESIS. Experiments have shown that the infraciliature plays an important role in morphogenesis of the protozoans. For example, in *Tetrahymena* the mouth cirri always develop from one particular row of kinety and during cell division these kinety can be seen to form the new buccal cirri (Fig. 14). Similarly in *Stentor* the mouth is formed by kinetosomes on what is called the first kinety. This kinety in some way prevents the adjacent kinety from developing mouth structure (stomatogenesis), since if the first kinety is removed the adjacent kinety form a new mouth. In *Paramecium* as in *Tetrahymena*, however, stomatogenesis is restricted to one set of kinety alone and if these are removed a new mouth cannot be formed.

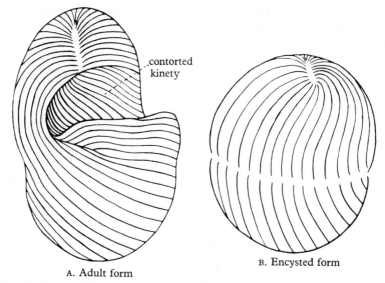

Fig. 13. Torsion of the kinetodesma. Certain Protozoa such as *Tillina* become twisted in the adult stage. This is seen when their kinety are compared to that of the encysted untwisted state and the adult state. (After Corliss.)

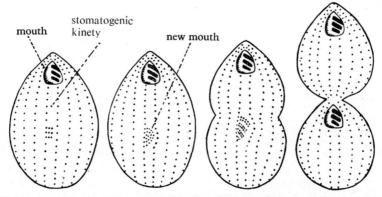

Fig. 14. Development of the mouth in *Tetrahymena*. The mouth has specially developed ciliary apparatus; this is derived from one particular kinety, and in division it is seen that this kinety divides to form the new oral apparatus (stomatogenesis). (After Corliss.)

The kinetosomes in *Foettingeria* divide to form small granules which do not develop cilia but instead develop into trichocysts (Fig. 11). It is not clear whether all trichocysts are derived from kinetosomes, some of the more complex ones such as are present in *Polykrikos* have been shown to have a more central origin.

NERVOUS SYSTEM. Systems of fibres which ramify from a central mass known as the 'motorium' and have been thought to be of the nature of a nervous system have been described in various ciliates; in some of these cutting the fibres is said to destroy the co-ordination between different sets of ciliary organs. It is possible also that the rhizoplast system of flagellates may have a conducting function. *Sense organs* are possessed by various protozoa in the form of specialized flagella and cilia in which the tactile sense is highly developed, and by many of the plant-like flagellates as pigment spots (eyespots), which may be provided with a lens. A chemical sense seems to be indicated by the fact that food is often recognized at a distance, and also probably in some of the cases of discrimination in ingestion (p. 26).

INTERNAL SKELETON. Internal skeletal structures are found in many members of the phylum. They may be part of the living protoplasm, as the axial fibres of axopodia and the *axostyles* which lie in the midst of the body of various mastigophora (*Trichomonas*, Fig. 39) and probably the central capsules of the Radiolaria, or of dead inorganic matter, as the skeletons of the Radiolaria (Fig. 50).

NUTRITION. The Protozoa present every type of nutrition exhibited by organisms, except that of the 'prototrophic' bacteria, which perform chemosynthesis by the use of energy obtained from reactions between inorganic substances.

The nutrition of an organism is said to be *holophytic* when it is effected, as in typical plants, by the building up of complex organic substances from simple inorganic ones by use of the energy of certain of the sun's rays (photosynthesis). The radiant energy is obtained by means of the green, yellow, or brown structures known as 'chromatophores' or 'chromoplasts' (e.g. the chloroplasts of green plants). In this mode of nutrition the simple materials of the food are absorbed through the surface of the body. In *holozoic* nutrition complex organic substances are swallowed through temporary or permanent openings as in the majority of animals. In *saprophytic* (or *saprozoic*) nutrition, practised by certain organisms, including among others various parasites, which are in contact with solutions of organic matter, relatively complex carbon compounds are taken, but these are absorbed through the body surface. The modes of nutrition classed under this head vary greatly in the complexity of the substances they require.

Holophytic nutrition, however, is found, among Protozoa, only in certain of the Mastigophora. Of the *holozoic* members of the phylum, some feed by amoeboid action. Usually this can be done at any point of the surface, as in the familiar case of *Amoeba*, but most of those flagellates which perform amoeboid ingestion do so in a particular region only. Other Protozoa swallow through a permanent *mouth*.

MOUTH. The true mouth is the spot at which the food passes below the ectoplasm. It may be (1) a bare patch of endoplasm, (2) the opening of an excavation (*oesophagus*) in the endoplasm, (3) the bottom of a depression (*vestibule*) in the ectoplasm, (4) the junction of a vestibule and an oesophagus. Any passage, whether oesophagus, or vestibule, or compounded of both, through which food enters is called a *gullet*, though not all cavities to which this name is applied are actually used in feeding. The opening of a gullet is the *cytostome*, which when there is a vestibule is not the true mouth. Gullets are found in many of the Mastigophora and most of the Ciliata. In ciliates either of the kinds may be present (p. 106). In the Mastigophora the gullet is at least sometimes ectoplasmic, but its morphology needs further investigation. A gullet may be supported by skeletal rods (Figs. 30, 66A), and is then often dilatable: a vestibule may have ciliary apparatus, trichocysts, etc., for taking food (Figs. 30, 65A). The Suctoria (Fig. 70) draw the protoplasm of their prey into their bodies through tentacles. The details of ingestion into the protoplasm differ considerably in different organisms. In some amoeboid forms the cytoplasm comes into contact with the food at once, either by flowing over it or by its adhering to the surface and being drawn in; others enclose the particles to be swallowed without touching them, either by arching over them, as *Amoeba proteus* does, or by excavating a vacuole for their reception. In some at least of the organisms whose food is driven into a gullet, a vacuole forms for it, apparently by the pressure of the water forced in, and on reaching a certain size nips off. Often, but by no means always, discrimination is exercised between particles which appear equally capable of being swallowed. It is doubtful whether this discrimination is concerned solely with such properties as the size and shape of the particles or also with their chemical qualities.

DIGESTION. Solid food is digested in food vacuoles, which usually contain visible fluid and in which the reaction is often first acid and then alkaline. Live food dies during the acid phase, and protein is digested during the alkaline phase. Protozoa can digest fat, and can usually dissolve starch, and sometimes cellulose. The latter faculty becomes of great importance when they are symbionts in the alimentary canal of animals whose food consists of plant tissues (pp. 70, 110). In a few cases (*Balantidium*, some *Amoebae*) contractions of the protoplasm divide large morsels into fragments. Often, but not, for instance, in foraminifera, the food vacuoles circulate in the cytoplasm; sometimes they do this along a regular track. Their circulation is often due to streaming of the endoplasm, but sometimes (ciliates, etc.) it is brought about by peristalsis of the cytoplasm. Defaecation of the indigestible remains of food takes place at any part of the surface when there is no pellicle, but in pelliculate forms at a fixed spot. Sometimes there is a permanent rectal passage lined by ectoplasm (Fig. 68B). *Saprophytic* forms range from some which can subsist on mixtures of substances as simple as aminoacids and acetates (or even, as *Polytoma* can, upon ammonium acetate alone), to parasites whose food probably differs chemically but little from that of holozoic forms.

RESERVE MATERIALS, used at times when nutriment is not being taken or when some process, such as rapid multiplication, is making heavy demands upon the resources of the organism, are stored by most protozoa, and are often conspicuous, as granules, vacuoles, crystals, etc., in the cytoplasm. The carbohydrates starch, paramylum (in the Euglenoidina), and leucosin (in the Chrysomonadina) are formed by holophytic organisms and by some colourless forms related to these (as by *Polytoma*, Fig. 18, *Peranema*, etc.). Glycogen is stored by parasitic and other anaerobic forms, in which it is perhaps split with evolution of energy, as in various anaerobic metazoa. Protein reserves are common in holozoic species. Nucleic acid ('volutin') is widespread, probably as a reserve for the nucleus. Oil reserves also occur in practically all groups. In *phosphorescent* forms (dinoflagellates, radiolarians) the oxidation of fats is the source of the emission of light.

OSMOREGULATION AND EXCRETION. The nitrogenous excreta of the Protozoa appear to be most often ammonia compounds, less often urea, and

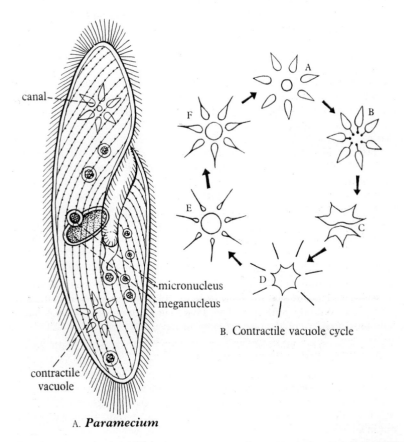

A. *Paramecium*

B. Contractile vacuole cycle

Fig. 15. A, *Paramecium caudatum* from the ventral side, ×375. (After Doflein.) B, the cycle action of the contractile vacuole and its canals in *Paramecium*. (From Lloyd after Pütter.)

occasionally urates. Excretion doubtless frequently takes place from the general surface of the body. Sometimes there are recognizable in the cytoplasm granules or crystals of urates or phosphates, which may be expelled with the faeces but appear in other cases to be redissolved. Their material is then perhaps passed into the *contractile vacuoles*. The latter are spaces filled with water which periodically undergo collapse with expulsion of their contents to the exterior. In the simplest cases, as in the familiar laboratory types *Amoeba, Chlamydomonas* and *Actinosphaerium* (Figs. 18, 27), the contractile vacuoles are solitary, spherical cavities, one or more in number according to the organism; over these in pelliculate genera there is a soft patch in the pellicle through which discharge takes place. Sometimes, as in *Euglena* and *Paramecium*, they are accompanied by accessory vacuoles by whose contents they are reconstituted (Figs. 15A, 15B, 30) and which in some ciliates (Fig. 66c) extend as long canals through the cytoplasm. Another complication sometimes exists in the presence of a 'reservoir' through which the vacuole communicates with the exterior, either directly, as in *Peranema* (Fig. 7), or by way of the gullet, as in *Euglena* and *Vorticella* (Figs. 30, 3). At least some contractile vacuoles appear to have a lining membrane, and it is probable that they are not entirely abolished at systole. The fact that these organs are commoner in freshwater protozoa than in marine or parasitic species suggests that their primary function may be the discharge of water, which must enter the body when the surrounding medium has a lower osmotic pressure than the protoplasm.

Further evidence of this is provided by the following tests: (1) if the external medium has its osmotic pressure raised, the rate of pulsation of the contractile vacuole falls off; (2) if the animals are placed in a dilute solution of KCN the contractile vacuole becomes paralysed and cannot work. The animal then swells. This ceases if the animal is now placed in a fresh solution not containing KCN; the contractile vacuole starts to beat once more and the animal returns to normal size. There is, however, some possibility that the contractile vacuoles may assist in excretion though there is no evidence that the excretory material is specifically concentrated or secreted into the contractile vacuoles.

RESPIRATION no doubt takes place upon the whole surface of the body. It has been supposed that the contractile vacuoles subserve this function, but, while they no doubt remove carbon dioxide in solution, it is difficult to see how their activity could cause the entry of oxygen.

ENCYSTMENT. Many Protozoa either regularly or occasionally pass a period of their lives in a cyst. The cysts may be coats of jelly or stronger coverings, usually organic, but sometimes, as in the Chrysomonadina, chiefly composed of inorganic material. The function of the cyst is nearly always to shield the organism, either from unfavourable circumstances or from stimuli which would interfere with some process, such as reproduction or digestion, but in a few cases it facilitates syngamy by keeping gametes together. Encystment is less common among species which live in the relatively equable conditions of the sea, than in fresh-water and parasitic forms. Cysts which do not subserve reproduction may be *resistance cysts*, against drought, alterations of concentration, or the appearance of poisonous substances in the surrounding medium, either in the habitat in which encystment takes place or in those encountered in the course of distribution. They may on the other hand be *resting cysts*, which enable the organism to proceed undisturbed with digestion or photosynthesis or by quiescence to conserve its energy during starvation. Cysts which subserve reproduction may be *gamocysts*, in which union of gametes takes place (gregarines, Figs. 57–59), *oocysts*, containing a zygote, or *sporocysts* containing several small individuals produced by fission. The oocyst frequently becomes a sporocyst by fission of the zygote. Reproductive cysts are often also resistance cysts.

NUCLEI. The nuclei of the Protozoa (Fig. 16) vary greatly in structure. They usually contain masses of some size composed of various materials. But precise chemical data concerning their structure are not available. Two

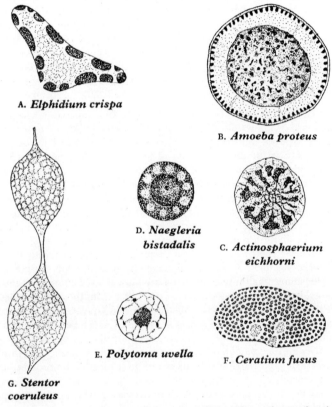

Fig. 16. Nuclei of Protozoa; all highly magnified. (After various authors.)

descriptive terms have been applied to the nuclear material: plastin, to material taking up acid dyes, and chromatin, to material adsorbing basic dyes. The latter substances include the two types of nucleic acids, the polymerizing deoxyribose nucleic acid, and the non-polymerizing ribose nucleic acid. The nucleic acids are not restricted to the nuclei alone but are also found throughout the cytoplasm in small basically staining granules, the mitochondria and microsomes.

A single central mass is an *endosome*: it may be a temporary aggregation, as, for instance, in *Actinosphaerium*, but more often is permanent except, sometimes, at division. Such a permanent endosome is usually a nucleolus or an amphinucleolus, but is said sometimes to consist solely of chromatin or of achromatic matter. A permanent endosome consisting of plastin or chromatin, or both, is known as a *caryosome*. Two principal types of protozoan nuclei—the *dense* and the *vesicular*—may be distinguished; there are, however, intermediates between them, and they do not characterize each a distinct branch of the phylum, but the dense appears to have arisen more than once from the vesicular. In nuclei of the dense type the achromatic part has a relatively firm consistency, and often exhibits, at least in fixed specimens, a fine meshwork. The plastin is in masses scattered through the nucleus, or occasionally in a single excentric mass. The shape is often not spherical (Figs. 67–70). The meganuclei of ciliophora and dinoflagellate nuclei belong to this type, which otherwise is rare. In vesicular nuclei the achromatic part is more fluid and its meshwork, if any, is coarse. The plastin may be in several masses under the nuclear membrane, but usually is in a caryosome

NUCLEAR DIVISION. The modes of division of protozoan nuclei are also very various. Many were formerly classed as *amitoses* but are now regarded as unusual types of mitosis. True amitoses are rare, and perhaps occur only in the meganuclei of the Ciliophora. The *mitoses* are sometimes (Fig. 46) practically identical with those of the Metazoa, but are usually more or less aberrant. The 'division centre' by which mitosis is initiated may be a centrosome consisting of centrosphere and centriole, or may be either of the latter two entities alone. The centrosphere often forms a plate or cap at each pole of the nucleus. Most often the nuclear membrane remains intact throughout the process. The division centre may be intranuclear or extranuclear; when it is an extranuclear centriole, it is often identical or associated with the basal granule of a flagellum. Cases in which the chromosomes are distinct and on the whole behave like those of metazoa are known as *eumitoses*. Another set of mitoses, known as *paramitoses* (Fig. 17), differ from those of the Metazoa in that the chromosomes do not shorten in the metaphase, and are not symmetrically arranged on the equator of the spindle (if such be visible); and their longitudinal halves, when they separate, hang together to the last at one end so that they appear, though deceptively, to divide transversely. In a third set, known as *cryptomitoses* (Fig. 17), there are no distinct chromosomes but the chromatin merely concentrates as a mass upon the equator of a spindle, whose fibres may not be visible, and divides into two halves which travel to opposite poles. Intermediate cases connect cryptomitoses with eumitoses. Paramitoses occur in coccidians, dinoflagellates, and the spore formation of radiolarians, cryptomitoses for the most part in parasitic and coprozoic forms, as in *Haplosporidium* and *Naegleria*.

POLYENERGID NUCLEUS. In certain cases mitoses repeated several times without dissolution of the nuclear membrane give rise to polyenergid nuclei which possess numerous sets of chromosomes, the sets being finally liberated to form each a daughter nucleus. The polyenergid condition is probably always a provision for spore formation, and may (as in the coccidian *Aggregata*) occur only as a transient phase before sporulation, but in other cases (radiolarians) it persists for a long time, the nucleus dividing meanwhile as a

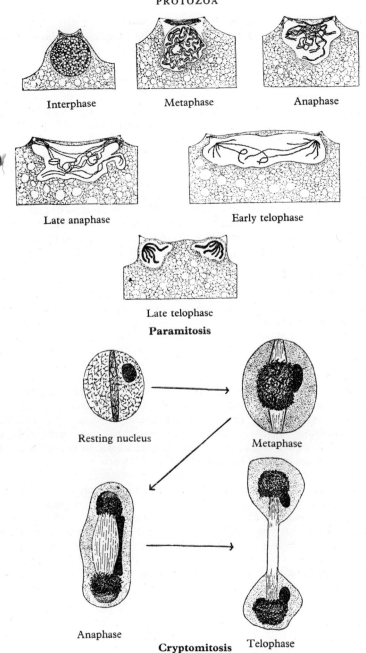

Fig. 17. Paramitosis in *Aggregata eberthi*. (After Belar.) Cryptomitosis in *Haplosporidium limnodrili*. (After Granata.)

whole by 'giant mitoses' in which all the chromosomes take part. A remarkable process in *Amoeba proteus* is possibly to be interpreted as the multiplication of units in a polyenergid nucleus by cryptomitoses.

TROPHOCHROMATIN AND IDIOCHROMATIN. In the Ciliphora (other than the Chonotricha) the nucleoplasm, which in other Protozoa is contained in one nucleus or in several which are all alike, is divided into two portions: a large amitotic *meganucleus* which breaks up periodically in 'endomixis' (see p. 39) and also at conjugation, and one or more small *micronuclei*, by division of one of which the pronuclei of conjugation are provided and the meganucleus replaced when the latter disintegrates. Individuals without micronuclei have been observed and kept through several asexual generations. Thus the meganucleus is capable of conducting by itself the normal vegetative existence of the individual, though the absence of this nucleus at syngamy shows that it does not establish the characters of the race. That function must be performed by the micronucleus, but, since the latter does not exist without the meganucleus, save for a brief period during conjugation, it presumably does not regulate the life of the individual. The chromatin of the meganucleus is known as *trophochromatin*, that of the micronucleus as *idiochromatin*.

A similar distinction between trophochromatin and idiochromatin is discernible in various other Protozoa. In the Chonotricha there are two sets of chromosomes, an outer and an inner, which divide successively at mitosis. The members of the outer set (*megachromosomes*), larger and less regular than those of the inner, are held to represent the meganucleus of other Ciliophora. The members of the inner set (*microchromosomes*) represent the micronucleus. In various cases of gamete and spore formation by members of other classes, especially of the Sporozoa, there is a destruction, or a casting out from the body, of a portion of nuclear substance which is probably trophochromatin. It has been suggested also that the obscurity of cryptomitoses is due to a veil of trophochromatin dividing amitotically around the idiochromosomes. It may be that all Protozoa contain chromatin in both these conditions; and it is perhaps in this respect, as well as in restriction of function, that the cells of Metazoa differ from Protozoa.

From the fact that, in many cases at least, the trophochromatin is periodically destroyed and replaced, and from further facts which we shall cite in discussing the significance of conjugation, it would appear that trophochromatin, or some part of the protoplasm associated with it, in the course of its regulative activity eventually becomes effete and is replaced from the idiochromatin, which is not liable to that fate. Perhaps the possession by Protozoa of the facility for this replacement, and the lack of such facility in the body cells of Metazoa, is the explanation of the fact that protozoa are not subject to the 'natural death' which eventually overtakes the body of a metazoon.

MEIOSIS. The loss of trophochromatin during the formation of gametes is not to be confused with the *reduction division* (meiosis) of maturation. Reduction divisions, however, have been seen in members of all classes of the Protozoa, and it may be suspected that a process of this kind occurs in all species in which there is syngamy. Such divisions are sometimes (*Actinophrys* etc.) strikingly similar to those of the Metazoa, but in other cases (*Paramecium*, etc.) are peculiar. The reduction division usually closely precedes syngamy, as in the Metazoa, but in the Telosporidia and Volvocina it takes place in the first division of the zygote, so that for the whole of the rest of its life history the organism is haploid (Fig. 23).

CHROMIDIA. In many protozoa there are present in the cytoplasm, scattered or massed into a group, numerous granules which, like chromatin, take basic stains and are known collectively as the chromidium (Fig. 39). They appear to arise from the nucleus, and have been said, but probably incorrectly, in some cases to give rise to nuclei by condensation. They may appear upon occasion or be present through the greater part of the life cycle. Their function is uncertain and probably not always the same.

FISSION. The fission of the Protozoa takes place in several ways. Whether in asexual reproduction or in the formation of gametes, it may be: (1) equal *binary fission*, the familiar mode of division of *Amoeba*, *Paramecium*, and a vast number of other cases; (2) *budding*, in which one or more small products separate from a parent body, as in *Arcella* (Fig. 46), the Suctoria (Fig. 70), etc.; (3) *repeated fission*, in which equal divisions give rise to four or more young which do not separate till the process is completed, as in *Chlamydomonas* (Fig. 18), the microgamete formation of *Volvox* (Fig. 35), etc.; or (4) *multiple fission*, in which the nucleus divides several times without division of the cytoplasm, which finally falls into as many parts as there are nuclei, usually leaving behind some residual protoplasm, which may contain nuclear matter. Multiple fission is seen in the spore formation of numerous protozoa, as *Amoeba*, sporozoa (Fig. 55), etc. The fission of multinucleate protozoa, such as *Actinosphaerium*, *Opalina* (Fig. 41), etc., to form multinucleate offspring by division of the cytoplasm without relation to that of the nuclei, is known as *plasmotomy*. It is usually binary, but occasionally takes place by budding or its multiple. The plane of simple binary or of repeated fission is often transverse to the principal axis—if there be one—of the body, but in most flagellates it is longitudinal. Repeated longitudinal fission in which the daughter individuals remain in position is called *radial*; such fission is common in the green flagellates of the order Volvocina (e.g. some species of *Chlamydomonas*, Fig. 18). Sometimes an individual in longitudinal fission shifts in its cuticle during the process, till the plane of division becomes transverse. Fission of this kind is said to be *pseudotransverse*: it is seen, for instance, in some *Chlamydomonas* (Fig. 18). In *Polytoma* (Fig. 18) the only vestige of longitudinal fission consists in a slight obliquity of the first division of the nucleus.

Each type of fission takes place in some cases in a cyst and in others without encystment.

The fate of flagella at fission varies. Sometimes, as in *Chlamydomonas* and *Polytoma*, they are lost, early or late in the process. In other cases they are retained. When this happens in an organism with a single flagellum, that organ has been said sometimes to be split longitudinally, but usually, if not always, a second flagellum grows out from the basal granule, which divides. When several flagella are present and persist, they are distributed between the

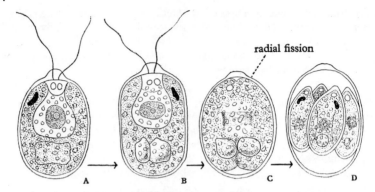

Chlamydomonas angulosa, ×1000

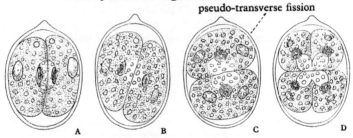

Chlamydomonas longistigma, ×1000

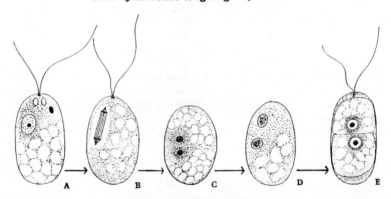

***Polytoma uvella*,** ×1300

Fig. 18. Forms of cleavage in *Clamydomonas* sp. and *Polytoma*.
(After Dill and Dangeard.)

offspring, each of which grows new flagella to complete its equipment. Probably, a new flagellum always grows from a basal granule. Chromatophores divide, and if numerous may do so independently of the fission of the body. If the animal divides very rapidly the dividing chromatophores may fail to be shared amongst the progeny. Those without are unable to develop

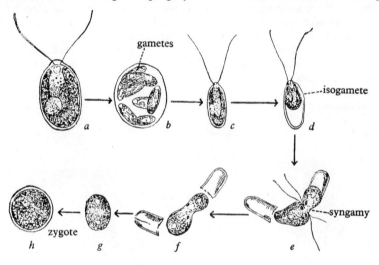

A. *Chlamydomonas media*

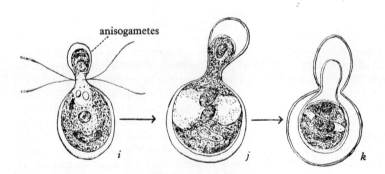

B. *Chlamydomonas brauni*

Fig. 19. Anisogamous syngamy of *Chlamydomonas*. (After Klebs.)

new chromatophores and so remain colourless. In this way *Euglena* can be transmuted to *Astasia* and *Chlamydomonas* to *Polytoma*, the names of the achloritic forms being more grades than generic terms.

Contractile vacuoles and other organs rarely (*Euglena*) divide, but are usually shared as the flagella. Complex organs, however, are often destroyed (dedifferentiated) and remade by the individual that receives them.

PROTOZOA

CONJUGATION. Conjugation or syngamy, the union of two nuclei, accompanied by the fusion of such cytoplasm as each may possess, is, so far as our knowledge goes at present, by no means universal in the Protozoa. Especially among the Mastigophora, but also in other groups as in the Amoebina, there are many cases in which it appears not to occur. Probably, however, in the majority of species it either is known or may reasonably be inferred to take place. The energids by which it is performed, known here, as in all organisms, as *gametes*, may be either *merogametes*, formed by special acts of fission and smaller than the ordinary energids of the species, or *hologametes*, not formed by special fissions, and as large as, or larger than, the ordinary energids. Syngamy between like gametes is known as *isogamy*, that between unlike gametes as *anisogamy*. The simplest cases of the process are those instances of isogamy in which two full-sized ordinary individuals unite. Such unions are known as *hologamy* and are rare, though they occur in *Copromonas* (Fig. 30) and a few other species. The fact that nearly all protozoa

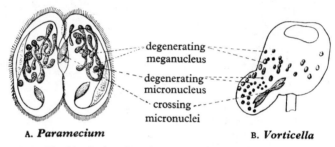

A. *Paramecium* B. *Vorticella*

Fig. 20. Conjugation of ciliates; both at the moment of the exchange of the male pronuclei.

in which it is certainly known to occur are coprozoic (p. 47) suggests that it is an adaptation to special conditions—perhaps to brief duration of the active stage—and is not, as might be assumed, the primitive form of syngamy. In all other cases the gametes are special individuals, and one at least is a merogamete. They may be isogamous or anisogamous, and in the latter case one—the *female gamete*—is less active than the other, which is the *male gamete*. Nearly always the female gamete (*macrogamete*) is larger than the male (*microgamete*), and often it is a hologamete. In the latter case the process is known as *oogamy*. As examples of isogamy of merogametes we may cite the syngamy of *Elphidium* (Fig. 49) and of some *Chlamydomonas* (e.g. *C. steini*).

Anisogamy occurs independently in many genera, and has more than once become oogamy. An interesting series of grades in this respect is provided by the Volvocina. *Chlamydomonas euchlora* exhibits the transition from isogamy to anisogamy. By undergoing different numbers (2–6) of divisions, its individuals form merogametes of several different sizes, but these pair indifferently, some unions being isogamous, some anisogamous. *C. brauni* and other species (Fig. 19) form merogametes of two sizes and are definitely anisogamous. *Volvox* (Fig. 34) and related forms have an anisogamy in which the female gamete is a hologamete (oogamy). A similar series is shown by the Sporozoa. The syngamy of some species of *Monocystis*, for instance, is isogamy of merogametes, that of others is anisogamy of merogametes with various degrees of unlikeness between the gametes, and that of the malaria parasite (Fig. 56) and its relations is anisogamy between a hologamete and a merogamete.

Reproductive Terms

Gamete	marrying cell
Merogamete	gamete formed by division of the adult, usually smaller than the adult
Hologamete	gamete as large as the adult
Isogamete	gametes of equal size
Anisogamete	gametes of unequal size
Oogamete	gamete large and immobile
Microgamete	small gamete
Macrogamete	large gamete
Autogamy	nuclei divide into two and then fuse again
Syngamy / Conjugation	fusion of two cells and their nuclei
Caryogamy	fusion of nuclei
Plasmogamy	fusion of cytoplasm

SEX. Syngamy, whether isogamous or anisogamous, nearly always is *exogamous*, that is, takes place between the offspring of different parents.

Since the male and female gametes are usually formed by distinct parents, sex may be said to exist among protozoa, but it is rarely that (as in the sporozoon *Cyclospora*, etc.) the sexes may be distinguished by other features. In many of the Telosporidia (e.g. *Monocystis*, Fig. 59) *sexual congress* may be held to occur, in that individuals, male and female in cases of anisogamy, apply themselves together at the time of gamete formation, and their gametes unite each with one from the other parent. Hermaphroditism appears in the Ciliophora (except in some of the Chonotricha). Here congress (Fig. 20) takes place between two individuals (*conjugants*) in each of which the meganucleus (see above) disintegrates, and the micronuclei divide to form a number of nuclei—perhaps a reminiscence of the formation of numerous merogametes. All but one of these nuclei disappear, and the survivor divides to form a male pronucleus, which passes over into the partner, and a female pronucleus which, in possession of the cytoplasm of the parent, awaits the arrival of the male pronucleus of the partner. Fusion now takes place between the male and female pronuclei in each of the pair of conjugants, the latter separate, and by the division of their zygote nuclei mega- and micronuclei arise. Two hermaphrodites have formed each a male and a female gamete and cross-fertilization has taken place. Occasionally (*Collinia, Dendrocometes*) the conjugants also exchange halves of their meganuclei. The latter, however, always disintegrate.

In the Vorticellidae (Fig. 20) the individuals which enter into congress differ, one being of the ordinary size and fixed, the other small and free-swimming. The smaller arises from an ordinary individual, as a bud or by repeated fission. After reciprocal fertilization of the type just described, the smaller partner perishes, its endoplasm being sucked into the larger. This curious simulation of sexual dimorphism by hermaphrodites occurs in a less marked form in other ciliates.

Amongst the Sarcodina *Actinophrys* (p. 88) may be said to be hermaphrodite, and so perhaps are many of the Radiolaria. But it is not certain that the 'gametes' of this group are not parasitic dinoflagellates (see p. 84).

CONJUGATION. During conjugation of a ciliate such as *Paramecium*, the following stages occur (Fig. 21).

(1) Two paramecia come together and adhere at the oral region. The membrane between them breaks down and the cytoplasm can intermingle.

(2) The meganucleus of each paramecium breaks down and dissolves into the cytoplasm.

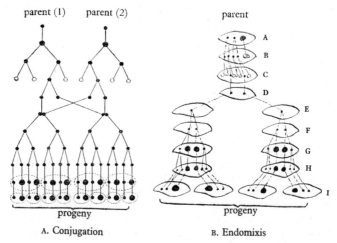

Fig. 21. Diagram of the behaviour of the nuclei during conjugation of *Paramecium caudatum* (after Borradaile), and endomixis in *P. aurelia*. (From Robertson, after Jennings.)

(3) The micronuclei of each paramecium divides many times, one of these being a reduction division. There are usually three divisions but the number varies according to the species studied. All but one of these nuclei dissolve into the cytoplasm.

(4) This single nucleus divides into two pronuclei, one of which migrates across to the foreign paramecium.

(5) Each paramecium then contains two nuclei, one of which is derived from the other animal. The nuclei fuse.

(6) This nucleus divides many times. The animals separate and commence to divide. They distribute their nuclei so that: (*a*) One becomes the new micronucleus. (*b*) Others fuse and form the meganucleus. (*c*) The remainder are dissolved into the cytoplasm.

It will be seen that during conjugation the following changes have occurred. The cytoplasm becomes enriched by the break-down of the micronuclear and meganuclear material. The cytoplasm of the different paramecia become intermingled. A new meganucleus is formed. Some species such as *Dendrocometes* and *Collinia* even interchange part of the meganucleus between the conjugants, though this later breaks down in the cytoplasm.

ABERRANT REPRODUCTIVE BEHAVIOUR IN PARAMECIUM. *Paramecium* can show certain variations in its nuclear behaviour during fission and conjugation. Some of these aberrations have been classified as follows:

Hemixis is an aberrant behaviour of the meganucleus independent of cell division or syngamy. It has been found to occur in *P. aurelia*, *P. multimicronucleatum* and *P. caudatum*. The meganucleus may either divide and so provide the animal for a short time with two meganuclei, or else it may protrude masses of nuclear material into the cell.

Endomixis occurs in *P. aurelia*. It resembles conjugation but it is carried out by only one paramecium and there is no meiosis or fusion of pronuclei. The resemblance lies in that a new meganucleus is formed from micronuclear material. This may be a specialized case of autogamy, though its occurrence is open to some doubt.

Autogamy occurs in *P. aurelia*. The animal undergoes all the processes of conjugation, i.e. meiosis and fusion of pronuclei, but only one individual takes part in the process and it provides both the pronuclei. The advantage of this, besides the formation of a new meganucleus, is that it establishes different combinations of allelomorphic genes.

Cytogamy occurs under experimental conditions. It is possible to compress two paramecia together in a compressorium so that the two animals can be induced to show autogamy. It differs from syngamy in that there is no breakdown of the membrane between the two animals and there is no nuclear material interchanged between the two. It differs from autogamy in that the two animals are in contact with each other.

Regeneration experiments on the meganucleus indicate that about one-fortieth of the meganucleus is still able to form the whole complete meganucleus. It appears that the meganucleus is a polyploid containing at least forty sets of nuclear material. The number of chromosomes in the micronucleus varies from eighty to several hundred depending on the species. Polyploidy is common and there appears to be no correlation between the chromosome number and the mating types.

PARTHENOGENESIS. Parthenogenesis is known to occur in members of at least three of the four classes of the phylum. The clearest case is presented by *Actinophrys*, when gametes which have failed in attempt at cross-fertilization develop parthenogenetically (p. 87): it is interesting that one of these gametes is a functional male. Individuals of *Polytoma* which are potential gametes will grow and divide, and the same is true of the gametes of some species of *Chlamydomonas* and *Haematococcus* when syngamy has been missed. The endomixis of ciliates (Fig. 21 B) is a phenomenon of this kind.

RESULTS OF SYNGAMY. Since it is comparatively easy to observe the conditions which precede and the results which follow syngamy in the Protozoa, many experiments and observations have been made upon those creatures, with a view to discovering the significance which the process has for organisms in general. Most of these researches have been carried out upon ciliata. They have led to two theories: (1) that syngamy effects a periodical rejuvenescence of the organism; (2) that it produces new types of individual and therefore gives the species more chances of survival in the struggle for existence. The facts are as follows:

(1) Cultures of protozoa in which conjugation is prevented are liable after a time to fall into an unhealthy condition known as *depression*, in which the nucleus (in ciliophora the meganucleus) is overgrown, the body stunted, division retarded, and the various organs and functions increasingly degenerate, until finally digestion ceases and the organisms die. From this condition conjugation will recover a culture which is not too far gone. It was held that depression was the senility of the organism—ultimately of the same nature as that which in the Metazoa destroys the parent body, while the gametes, after syngamy, continue the existence of the species—and the conclusion was drawn that in both cases the effect of the union of nuclei was rejuvenation. Now, however, it is known that depression is a disease, which by more natural methods of culture can be avoided without conjugation. It is true that in cultures of ciliates there has been observed a periodical waxing and waning of vitality of which the low points in some cases coincide with conjugation; but in other cases there occurs at these points not conjugation but a process known as *endomixis*, which closely resembles the procedure in conjugation, but takes place in solitary individuals and does not involve syngamy. In this process (Fig. 21) the meganucleus is destroyed and replaced by one of the products of the division of the surviving micronucleus, as in conjugation. It would appear from these facts that the invigorating effects of conjugation are due not to the true syngamy (union of nuclei) but to the accompanying replacement of the meganucleus, which probably has become effete (see p. 32). If, as has been suggested (p. 32), those protozoa which have no meganucleus have in their nuclei trophochromatin which is destroyed at syngamy, this conclusion may be extended to them also.

(2) Variety in a protozoan species is of three kinds: (*a*) that which results from the production of different *combinations* of genes at syngamy, and is permanent, forming races (*pure lines*) like those which exist in higher organisms in the absence of cross-fertilization; such pure lines have, for instance, been found in respect of body-length in cultures of *Paramecium*, each line in the culture breeding true so long as asexual reproduction continues; (*b*) that which results from the spontaneous appearance of *mutations*; this also is permanent; it has been studied in *Ceratium* and other genera; (*c*) that which results from modification of the individual by the direct action of the environment; this, like mutation, produces differences between individuals of a pure line, but it is not permanent, though it may be inherited for several generations before it disappears. It would seem that, apart from the occasional appearance of mutations, the permanent varieties in a species are produced only by syngamy.

Here may be mentioned the union of individuals by fusion of their cytoplasm, the nuclei remaining distinct, which is practised by the Mycetozoa (Fig. 53) and in some other cases. This process, which is not syngamy, is known as *plasmogamy*, and its product as a *plasmodium*.

SENESCENCE AND SEX. A general interpretation of mitosis, meiosis, and conjugation in the Protozoa is as follows. The chemicals present in the protoplasm can be arbitrarily classified in two main groups: A-chemicals that are able to synthesize themselves from raw materials, and B-chemicals that are unable to synthesize themselves but instead take part in other reactions and are themselves synthesized by A-chemicals. In the hypothetical ancestral cell these A- and B-chemicals were scattered throughout the whole of the cell and there were so many of each kind that when the cell divided into two, there was a representative of each type of A-chemical in each of the daughter cells so that metabolism continued as it had done in the parent cell. However, if the cell divided very rapidly some of the A-chemicals might become grouped so that after division one daughter cell got all the A-chemical of a certain type whilst the other cell got none. This might lead to serious consequences and even death of the daughter cell lacking the A-chemical. A method of overcoming this problem is to group all the more important A-chemicals into the centre of the cell and prior to cell division to make them go through a complex procedure that ensures each daughter cell will receive its full compliment of A-chemicals. These A-chemicals are also often arranged in linear chains. This has certain advantages during the synthesis of A- and B-chemicals since the chains can act in a manner similar to a factory assembly line and enable different raw materials to flow in a definite sequence over the different synthetic stages. This then explains the development of the A-chemicals into chains (chromosomes) and their collection into a central position (the nucleus) but has not yet considered why cell division takes place. The answer is not known but several theories exist. One is that as the cell grows its volume increases as the cube of the radius whilst its area only increases as the square of the radius. The result is that above a certain size the interior of the cell receives less and less oxygen diffusing in from the outside as the surface area of the cell is no longer adequate to allow sufficient diffusion. One answer is for the cell to change its shape and become flat. Another answer is for the cell to divide and thus reduce the cell volume. A second explanation of cell division is that during the life of the cell many of the A- and B-chemicals in the protoplasm become used up or poisoned. This leads to senescence of the protoplasm, the protoplasmic reactions slow up and are no longer as efficient as they were in the young cell. It is therefore necessary that these chemicals should be replaced. One method of doing this is to eat another such cell on the chance that it has the chemicals that are in short supply. An alternative to such cannibalism is to undergo cell division, for during this process the A-chemicals in the nucleus proceed to manufacture large quantities of new A- and B-chemicals which diffuse out from the nucleus into the protoplasm and rejuvenate it. Evidence for this is found in *Amoeba* where the cytoplasmic concentration of SH-groups falls off as the animal ages but increases immediately after cell division. This then explains how it is one old amoeba can turn into two young amoebae. This degree of rejuvenation appears to be sufficient for amoeba, though at times an extra system comes into use. Groups of amoebae come together and their cytoplasm fuses. The chemicals in the cytoplasm then get mixed up and it is likely that the loss of B-chemicals in the cytoplasm of one amoeba will differ from the shortages in another amoeba, so by a system of swapping each amoeba finally emerges with a complete set of B-chemicals (plasmogamy). Such cytoplasmic interchange leading to rejuvenation is also seen during the conjugation of ciliates such as *Paramecium*. Here the system is more complex since there is also a complete interchange of all the A-chemicals (transference of the pronuclei) leading to increased variation, as well as the development of a new centre for the distribution of B-chemicals and minor A-ones (meganucleus). Ciliates are able to live after removal of the micronucleus but are unable to do so when the meganucleus is removed. This and other evidence indicates that the meganucleus plays an important role in metabolism and anabolic processes. In fact the meganucleus would seem to provide the main source of A- and B-chemicals for the cyto-

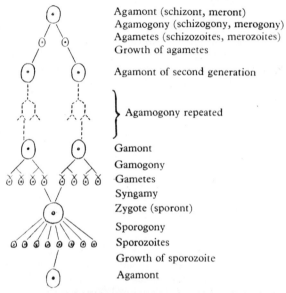

Agamont (schizont, meront)
Agamogony (schizogony, merogony)
Agametes (schizozoites, merozoites)
Growth of agametes

Agamont of second generation

} Agamogony repeated

Gamont
Gamogony
Gametes
Syngamy
Zygote (sporont)

Sporogony
Sporozoites
Growth of sporozoite
Agamont

Fig. 22. A table of the life history of a protozoon.

plasm. At times when the cytoplasm lacks these chemicals the meganucleus may go through a series of divisions, many of the meganuclei so produced autolysing and dissolving into the cytoplasm. Such a process is seen in *endomixis*. On other occasions the meganuclei have insufficient material to rejuvenate the cytoplasm and nuclear material is brought in from a foreign source by conjugation. This new micronucleus then makes a new meganucleus which can continue its function. Thus rejuvenation can be established in the ciliate by cell division, by cytoplasmic interchange, by the meganucleus breaking down, and by the importation of new micronuclear material.

LIFE CYCLE. The life of a protozoon passes in the course of generations through a cycle in which individuals of different kinds succeed one another. The life cycles of various protozoa differ greatly, being related to the vicissitudes of the environment of the species and to the need for distribution as well as to the recurrence at intervals of conjugation, but it is possible to formulate a type of which all of them may be regarded as variants. After a period of 'vegetative' existence and increase by asexual reproduction, during which the individuals are known as *agamonts*, there appears a generation known as *gamonts* because they produce *gametes*; the latter unite in pairs, and the *zygote* or *sporont* gives rise to a generation of *sporozoites* which, becoming agamonts, repeat the asexual part of the cycle. Fig. 22 shows this typical life history.

In comparing this table with the actual course of the cycle in any species, the student should remember:

(1) that in each part of the cycle fission may take place in any of the modes described above, and that the agamogony of a species may proceed in more than one of these ways (as, e.g., that of *Amoeba proteus* by binary or multiple fission);

(2) that in the cycle of most protozoa there is a point at which adjustment must be made to unfavourable conditions, either recurring in the local habitat or met with in the course of the distribution of the species (e.g. freshwater forms and parasites); and that at this time (*a*) the ordinary agamogony is suspended, (*b*) the syngamy, if any, usually takes place, (*c*) there is often a phase of protective encystment, (*d*) there is often very rapid multiplication by multiple or repeated fission, which may be the sporogony, the gamogony (eugregarines), or an agamogony;

(3) that any part of the cycle may be omitted; in such cases it is most often the sporogony which is dropped, but many species appear to omit gamogony, and in a few (e.g. *Monocystis* and the other eugregarines) there is no agamogony;

(4) that a reduction division may occur at either of two points in the cycle—shortly before syngamy (most cases), or directly after the formation of the zygote (the Telosporidia and Volvocina)—and that correspondingly either the diploid or the haploid phase may extend over the greater part of the life history.

SPORE. The term spore is applied to various phases of the life history in a way which is liable to cause confusion. (1) Strictly speaking, perhaps, it should be applied only to the products of repeated or multiple fission of a zygote (sporont). (2) Most often, however, it is used to denote the products of any repeated or multiple fission. (3) In a few cases (e.g. the 'ciliospores' of the Suctoria) it is applied also to products of budding. A cyst in which several spores are enclosed is a *sporocyst*. Individual spores may be enclosed in *spore cases*, when they are *chlamydospores* (as those of the Mycetozoa, Fig. 53), or naked, when they are *gymnospores*. The latter may be amoeboid (*amoebulae* or *pseudopodiospores*, e.g. *Entamoeba*, Figs. 45, 49), flagellate (*flagellalulae* or *flagellispores*, e.g. Fig. 18, *Chlamydomonas*), or ciliate (*ciliospores*, e.g. the Suctoria, Fig. 70). Spores may be gametes (e.g. the Mycetozoa, *Chlamydomonas*), or serve for the distribution of the species, when, if they are motile, they are known as 'swarm spores'. The *sporoblasts* of many telosporidia (e.g. *Plasmodium*, Fig. 56) are spore-like bodies which are not set free, but give rise under cover to another generation of spores. The so-called spores of such sporozoa as *Monocystis* (Fig. 59) are really minute sporocysts, enclosing several spores ('falciform young').

INDIVIDUAL LIFE HISTORY. The life history of the individual protozoon usually exhibits little change save increase in size. Sporozoites and other spores, however, may differ considerably from the adults into which they grow. This difference reaches its height in the ciliospores of the Suctoria.

BEHAVIOUR. The behaviour of a living being is that part of its life which consists in action upon the outer world. Like the rest of life it comprises activity of various kinds—mechanical, chemical, etc.—which in some cases,

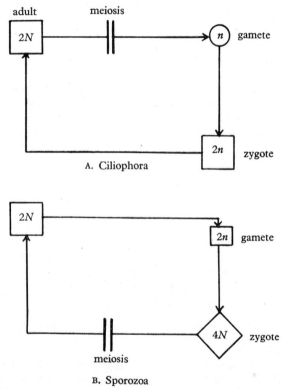

Fig. 23. Two types of nuclear life cycle found in the Protozoa. The difference depends on the position of meiosis in the life cycle, either before gametogenesis or after formation of the zygote.

as in the direction of locomotion to or from the light or the shooting out of trichocysts, is immediately due to external circumstances (*stimuli*), while in others, as in the beating of cilia which continues even when the organism is encysted, it is not. Both these sorts of activity are so ordered that in normal circumstances they conduce to the welfare of the organism. The reactions of the Protozoa to stimuli are at least superficially analogous to the reflexes of the Metazoa. Study of them has chiefly been directed to those which result in locomotion.

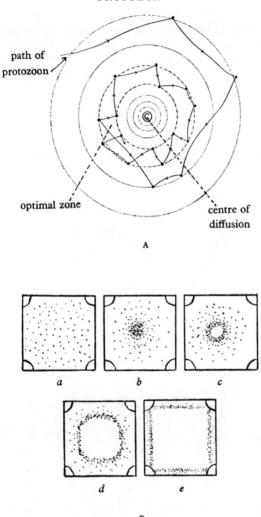

Fig. 24. Behaviour of Protozoa: A, diagram of the path by which a protozoon is directed into the zone of optimum concentration of a substance diffusing from a particle. (After Kuhn.) B, kinesis of *Bodo sulcatus*; *a–e* positions taken up by members of a culture placed under a coverslip. The position in which the individuals gather in each case is that of the optimum concentration of oxygen, which alters as the supply of the element is lessened by the action of the flagellates. (From Fraenkel, after Fox.)

Such reactions have been classified into two main groups: *kineses* in which the stimulus leads to an increase in the random movement of the animal, *taxes* in which the reaction is directed to the stimulus. An example of a kinesis is seen in the avoiding reaction of *Paramecium*; the animal turns through a finite angle each time it is stimulated till finally the animal is free from the stimulus. The response is random, the initial turns may just as well bring the animal into the stimulus as well as take it away from the stimulus. *Bodo* shows kinetic responses to oxygen tensions, the stimulus to random movement being a chemical one resulting in the animal moving till it once more is in a favourable oxygen tension (Fig. 24). Taxes depend on the possession of a sense organ: in *Euglena* the base of the flagellum is sensitive to changes in light concentration, and when a crawling *Euglena* is stimulated by light from one side, the animal makes rapid adjustments in its movements so

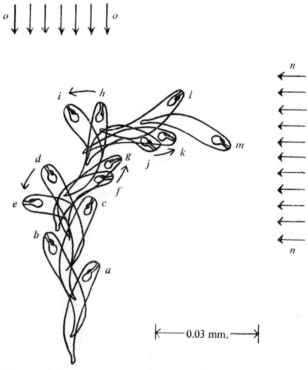

Fig. 25. Diagram of *Euglena* crawling. When the position of the light is changed at *c* from *o* to *n* so the animal changes its direction of crawling. (From Fraenkel and Gunn, after Mast.)

that it is once more directed towards the light. The response is related to the position of the stimulus and is not a simple increase in the random movement of the animal (Fig. 25).

The number of ways in which a protozoon can respond to stimuli is at most small, but the response to a stimulus by an individual in many cases depends not only upon the nature of the stimulus but also upon the condition of the individual at the moment (hunger, fatigue, etc.).

ECOLOGY. The *relation of protozoa to their environment* is governed primarily by the fact that, owing to their small size, any cuticle which is thick enough to protect their protoplasm from loss of water or poisoning by substances in the medium has the effect of immobilizing the organism. Hence in the active phase they are only found in water or in damp places on land, and are peculiarly susceptible to variations in the composition of the medium. Purely holophytic protozoa are also dependent upon the presence of sunlight. Save for these restrictions, members of the phylum are found in every environment in which any other species of organism can exist.

(1) *In the sea* they are plentiful alike in the plankton and in the benthos, and occur at all depths. Their planktonic members are liable to possess the same peculiarities which appear in members of other phyla in the same conditions—spininess (Figs. 5c, 26), phosphorescence, buoyancy, etc. In attaining a low specific gravity they often show an expedient of their own, namely the presence in their protoplasm of vacuoles of water of lower saline content than the medium in which they are suspended (radiolarians, *Globigerina*, heliozoa; Figs. 26, 27, 50, 51 c).

(2) *In fresh waters* their species have the same cosmopolitan distribution as other small fresh-water organisms. Most of them, however, are severely restricted, in all the localities in which they are found, by the necessity for conditions which only occur in some one type of environment, and often even there only during certain seasons or (as in the case of the dung fauna) for yet shorter periods. In this matter protozoa are particularly subject to the pH of the medium, its dissolved organic contents, and its saline contents. Thus *Polytoma* flourishes in an acid medium, *Spirostomum* requires a slightly alkaline one, and *Acanthocystis* pronounced alkalinity. *Euglena viridis* and *Polytoma* live in highly nitrogenous infusions, *Actinosphaerium* and *Paramecium caudatum* in less highly organic infusions, *Volvox* and *Amoeba proteus* in much purer waters, *Haematococcus* in rain water. As a rule the marine and freshwater faunas are restricted by conditions of salinity, but *Elphidium* ranges from the sea into brackish waters. For many holophytic protozoa the amount of sunlight is important. Others, as *Euglena gracilis*, bleach in the absence of light, but can still flourish if the presence of organic matter in solution makes saprophytic nutrition possible. Holozoic species must of course have their proper food; in infusions they appear as this becomes plentiful, first, after the bacteria, those whose diet is purely bacterial, such as *Monas* and *Colpoda*, then those, such as *Stylonychia*, that feed upon the first comers, and so on; though some bacterial feeders, as *Paramecium*, are rather late to appear. Temperature has also an influence upon protozoan faunas. The powers possessed by freshwater protozoa of distribution across inhospitable regions and of surviving unfavourable conditions at home are no doubt due to the facility with which they form resistance cysts (p. 28). In various cases all the unsuitable conditions of the environment indicated above have been found to induce encystment, and encysted protozoa have been discovered in dust from the most remote desert regions.

(3) The Protozoa which live in dung (*coprozoic* species) and in decaying bodies, and those of very foul waters, are branches of the aquatic fauna: they include many flagellates, *limax* amoebae (p. 75), and ciliates, and the conditions in which they are in the active state may exist only for a very short period. These faunas merge on the one hand into that of intestinal parasites,

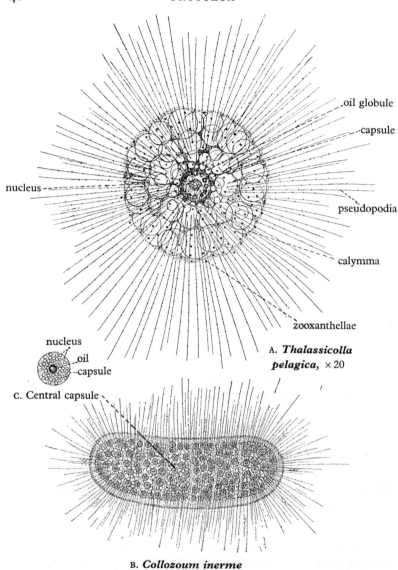

Fig. 26. Radiolaria without skeleton. (A, after Haeckel; C, after Doflein.)

and on the other into that of damp earth. In the latter there is a large population, some of whose members (*Euglena, Arcella, Paramecium*, etc.) are of common occurrence elsewhere. It has important effects upon the fertility of the soil, by devouring valuable bacteria. Perhaps the only truly subaerial members of the phylum are certain Mycetozoa.

(4) *Parasitic* members are included in nearly all the principal divisions of the phylum, but not in the Radiolaria or Volvocina. The Sporozoa are exclusively parasitic. The relations of parasitic protozoa to their hosts are of all degrees of intimacy: they may be merely epizoic (as *Spirochona*, p. 113), ectoparasitic (as *Oodinium*, p. 62), inhabitants of internal cavities (as *Opalina*, p. 73), tissue parasites (as *Myxobolus*, p. 102), or intracellular (as *Plasmodium*, p. 95). They show, according to their degree of parasitism, the same peculiarities as other parasites—reduction of organs of locomotion, simplicity of form,

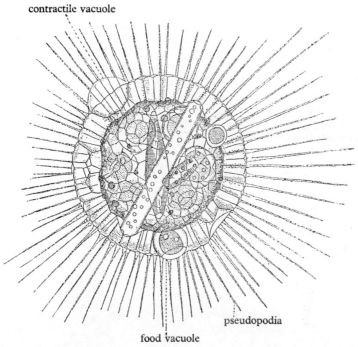

Fig. 27. *Actinosphaerium eichorni*, ×180. The endoplasm is crowded with food vacuoles containing diatoms, and the nuclei are represented in the figure by the dark areas. (From Leidy.)

means of fixation, the liberation of numerous young (in the Sporozoa), etc. Some, as *Entamoeba histolytica*, are harmful by destroying for their own nutriment the tissues of the host: more by secreting poisonous substances, as the malaria parasites do. Many are specific to a particular host or hosts. Not infrequently there are two successive hosts belonging to different phyla: both of these may be invertebrates, as with *Aggregata*, which passes from the crab to the octopus, but more often one is a vertebrate and the other an invertebrate. In such cases it is often possible to decide which was the original host, and this proves sometimes to be the vertebrate and sometimes the invertebrate. It is interesting that the two most dangerous protozoan parasites of man, the sleeping-sickness and malaria parasites, differ in this way (pp. 68, 95).

(5) *Symbiosis* of various kinds is practised by both holophytic and holozoic Protozoa. Instances of this are described below, on pp. 55, 70, 110, 187. The term 'symbiosis' has been used in various senses. It is here applied to all

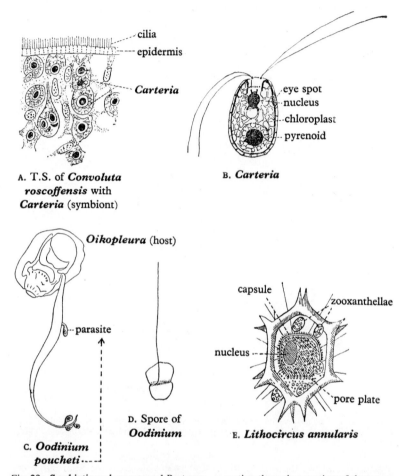

Fig. 28. Symbiotic and commensal Protozoa. A, section through a portion of the superficial tissues of *Convoluta roscoffensis* showing symbionts belonging to a species of *Carteria* (Chlamydomonadidae, Volvocina). (From Keeble.) B, a free individual of the genus *Carteria* which is symbiotic in the resting stage with *Convoluta roscoffensis*. (From Keeble.) C, *Oodinium poucheti*, parasitic on *Oikopleura*. D, Spore of *Oodinium*. E, *Lithocircus annularis* with enclosed zooxanthellae.

cases of partnership between two organisms of which one lives within the body of the other and both derive benefit from the association. It is sometimes restricted to cases, such as those described on p. 57, in which the infesting partner is photosynthetic.

PHYLOGENY. The division of the phylum into the four classes, Sarcodina, Mastigophora, Ciliophora, and Sporozoa, characterized by the presence or absence in the predominant phase of the life history of the several types of motile organs, will be familiar to the student. Two attempts have been made to brigade these classes into subphyla. One contrasts the Sarcodina under the name of *Gymnomyxa* with the other classes, or *Corticata*, on the ground that the latter possess a firm ectoplasm. The other contrasts the *Ciliophora* with the rest of the classes (*Plasmodroma*), which lack cilia and a meganucleus. Neither of these systems is satisfactory, for each is probably grounded, not upon a fundamental cleavage of the phylum, but upon the specialization of one branch of it.

The *ancestral group* of the Protozoa is probably the Mastigophora. This is fairly evident as concerns the Sporozoa—a class highly adapted to parasitism, and often possessing a flagellated phase—and the Ciliophora, also a greatly specialized group, which possesses in the cilia organs easy to derive from flagella. The Sarcodina, on the other hand, were formerly held to be ancestral to all protozoa, on account of the supposedly primitive condition of their protoplasm. But neither the structure nor the behaviour of amoeboid organisms is really simple; their holozoic nutrition is a less easy process and is much less likely to be primitive than photosynthesis, which is common in the Mastigophora; the sporadic occurrence of amoeboid forms in various groups of the Mastigophora probably indicates that the latter have more than once given rise to organisms resembling the Sarcodina; and, finally, the Sarcodina very commonly have flagellate young, but the Mastigophora do not have amoeboid young. The Mastigophora, indeed, are probably not only the basal group of the Protozoa but also not far removed from the ancestors of all organisms, for they alone present (and often can alternate) the modes of nutrition both of plants and of animals; and their characteristic organ, the flagellum, occurs in the zoospores of plants, in bacteria, and in the spermatozoa of Metazoa.

The *connexion between the Protozoa and the Metazoa* in the family tree of the Animal Kingdom is an interesting but a very obscure problem. Concerning it three theories are held.

(1) The first, supported by the morphological resemblance of the uninucleate protozoon to a cell in the body of a metazoon, and of *Volvox* to the blastosphere stage in the development of such a body, holds that the metazoon is a colony of Protozoa, each differentiated as a whole for some function in the body which they compose.

(2) The second, based on the fact that the protozoon, which performs equally all the processes of life, is thus physiologically equivalent not to one cell but to the whole body of a metazoon, holds that the Metazoa arose from multinucleate protozoa by the nuclei taking in charge each a local, differentiated portion of the cytoplasm.

(3) The third, based on the fact that, save for their mode of nutrition, the Metazoa have—in their cellular structure, nuclear division, maturation of gametes, etc.—more in common with multicellular plants than with the Protozoa, holds that the earliest organism we can as yet envisage was multinuclear and photosynthetic, and gave rise independently to the Metazoa and, by

reduction of the body, to flagellates, and so to the Protozoa, which on this view are not truly members of the Animal Kingdom.

Table 2. *Classes of the Protozoa*

Character	Mastigophora	Rhizopoda	Sporozoa	Ciliophora
Flagella / Cilia	Yes	Sometimes	Sometimes	Yes
Amoeboid	Yes	Yes	Sometimes	No
Parasitic	Often	Sometimes	Yes	Sometimes
Intracellular parasite	Rarely	No	Usually	No
Meganucleus	No	No	No	Yes
Large number of spores after syngamy	No	No	Yes	No

CLASS 1 MASTIGOPHORA (FLAGELLATA)

DIAGNOSIS. Protozoa which in the principal phase possess one or more flagella; may be amoeboid, but are usually pelliculate or cuticulate; are often parasitic but rarely intracellular; have no meganucleus; and do not form very large numbers of spores after syngamy.

The Mastigophora fall into a number of fairly well-defined orders. It is convenient to group these by their nutrition into two subclasses—the *Phytomastigina*, containing orders most of whose members are holophytic (see p. 25), and the *Zoomastigina*, which have no holophytic members—but all the orders of the Phytomastigina contain some colourless members, whose nutrition is purely saprophytic, and all except the Volvocina include colourless holozoic forms. Owing to this fact it is impossible to frame a definition which will enable every member of each subclass to be recognized as such without comparison with other species. Certain characteristics, however, distinguish most members of the Zoomastigina from most of the colourless Phytomastigina. These characteristics are stated below, in the section which deals with the Zoomastigina.

REPRODUCTION. The reproduction of the Mastigophora is in most cases by equal longitudinal fission. The way in which in many of the solitary Volvocina this becomes transverse has been described above (p. 33). In the Dinoflagellata fission is oblique or transverse. The fission may be simply binary or repeated. The number of fissions often varies in the same species, and is usually greater in the formation of gametes than in asexual reproduction. Binary fission in forms which have not a stout cuticle usually occurs in the free-swimming stage, but may take place in a cyst or jelly case, as, for instance, occasionally in *Euglena viridis*. In forms with a stout cuticle, as in the Volvocina, the protoplasm shrinks from the cuticle, which serves as a cyst. Repeated fission usually occurs in a cyst. The fate of the flagella at fission has been dealt with on p. 33. The mitoses (see p. 30) in this group range from beautiful eumitoses to the extremest cryptomitoses, the latter generally in parasitic forms. Paramitosis occurs in the Dinoflagellata.

In many genera *syngamy* is not known to occur. Among those in which it does, all degrees of difference between gametes are found, and in particular among the Volvocina there are interesting cases intermediate between holo-

CLASSIFICATION

Table 3. *Classification of the Mastigophora*

Mastigophora
- Phytomastigina — have chromatophores
  1. Dinoflagellata
     Have two flagella, one around the body, the other longitudinally
     Ceratium
  2. Cryptomonadina
     Have a gullet
     Food reserve: starch
     Chilomonas
  3. Euglenoidina
     Food reserve: paramylum
     Euglena
  4. Chloromonadina
     Food reserve: oil
     Vacuolaria
  5. Volvocina
     No gullet
     Green
     Volvox
  6. Chrysomonadina
     Brown
     Dinobryon
- Zoomastigina — no chromatophores
  1. Rhizomastigina
     1–2 flagella, all body surface can be amoeboid
     Mastigamoeba
  2. Holomastigina
     Many flagella, all body surface can be amoeboid
     Multicilia
  3. Protomonadina
     1–2 flagella, only part of the body amoeboid
     Bodo
  4. Polymastigina
     2–many flagella, not amoeboid
     Have extra nuclear division centre
     Trichonympha
  5. Opalina
     Many flagella, never amoeboid
     No extra nuclear division centre
     Opalina

gamy and merogamy, and between isogamy and anisogamy. Thus in *Polytoma* the age at which the products of fission unite varies in a species, so that some are merogametes while others, delaying, become hologametes; in *Pandorina* (p. 62) isogamy and anisogamy are facultative; and various species of *Chlamydomonas* (see p. 35) make up a series in which there is a transition from complete isogamy to a pronounced anisogamy which rises to oogamy in *Volvox* and other colonial forms.

The zygote is very commonly encysted.

SUBCLASS 1. PHYTOMASTIGINA

DIAGNOSIS. Mastigophora which possess chromatophores, and species without chromatophores which closely resemble such forms.

CLASSIFICATION. Of the *orders of the Phytomastigina*, that which contains the most highly organized members is the large and protean group *Dinoflagellata*, characterized by the possession of two flagella, one longitudinally directed and the other transverse, usually in a groove around the body but in a few cases twisted about the base of the longitudinal flagellum. Three of the remaining orders differ from the rest in the possession, in the anterior part of the body, of a pit ('gullet') or groove, from which the flagella usually arise. One of these, the *Cryptomonadina*, has simple contractile vacuoles and its carbohydrate reserves are of starch: it is held by some authorities to be related to the ancestors of the dinoflagellates. The second, the *Euglenoidina*, has a more complex contractile vacuole system, and its reserves are of paramylum. The third is the little group *Chloromonadina*, which differs from the Euglenoidina in having oil reserves only and in the delicacy of its pellicle. The orders without groove or gullet are the *Volvocina*, the most plant-like of the Mastigophora, with green chromatophores (except in a few colourless genera) and starch reserves; and the *Chrysomonadina*, by some regarded as the most primitive members of the class, which have yellow or brown chromatophores and no starch reserves and are often capable of becoming amoeboid.

Each of these groups exhibits most or all of the varieties of nutrition and motility. Each of them possesses (1) coloured flagellate, solitary forms which constitute most of its membership, (2) coloured species, whose individuals pass most of their time in a non-flagellate condition, as a palmella, which is sometimes of branched, plant-like form, (3) colourless saprophytic forms, and (4), except in the Volvocina, colourless, holozoic forms. More than one order has purely amoeboid members, non-flagellate throughout the greater part or all of their existence. The support which this versatility gives to the view that the Mastigophora, and in particular the Phytomonadina, are near the base of the genealogical tree of organisms has already been mentioned.

NUTRITION. There can be no doubt, for reasons which have been given above, that this subclass contains the most primitive members of the phylum. Its nutrition is extraordinarily interesting from that point of view. Some of its species, notably among the Volvocina, are purely holophytic. Others are normally also saprophytic, and some of these, like *Euglena*, can upon occasion practise this mode of nutrition alone. Yet others, like *Polytoma*, have become colourless, and are purely saprophytic. Others again are both holophytic and, by amoeboid ingestion, holozoic. These lead insensibly to similar forms, members of the Zoomastigina (*Monas*, etc.), which, being without chromatophores, have not the faculty of photosynthesis, but are purely animal in their nutrition. Some of the coloured forms which possess a pit that is called a gullet are said to take food with it, and thus to combine holophytic and holozoic nutrition. In any case certain of their relatives which have lost the chromatophores (*Cyathomonas, Peranema*, etc.) take solid food through a similar gullet. Most of the holozoic forms are probably also saprophytic. Certain species (*Ochromonas*, etc.) are known to make use of all three modes of nutrition. Thus all ways of obtaining nutriment meet in this group.

CHROMATOPHORE. The species which practise photosynthesis do so, like plants, by means of chromatophores, of which they may possess one, two, or many. The chromatophores are plate- or cup-shaped masses of protoplasm of a green, yellow, or brownish colour, owing to the presence in various proportions of the pigments chlorophyll, xanthophyll, carotin, etc. The chlorophyll absorbs the rays of sunlight whose energy is used in photosynthesis. The green chromatophores are known as *chloroplasts*, the yellow as *xanthoplasts*. Often there are to be seen in or on the chloroplasts the protein bodies known as *pyrenoids*, which act as centres of starch formation. A red pigment, *haematochrome*, is frequently present, diffused through the cytoplasm. In bright light it spreads over the surface and is believed to shield the chloroplasts from excess of certain rays. A small red spot of carotin, sometimes darkened by another pigment, is generally present in photosynthetic species, and probably acts as a rudimentary eye, making the organism sensitive to light, which is of such importance in its nutrition.

RESTING PHASE. The holophytic forms are usually capable of passing into a resting phase, in which the flagella are withdrawn, the body rounded off, a cyst or jelly case secreted, and the organism closely resembles a plant cell. Division may take place in that condition, establishing a pseudocolonial stage known as the *palmella* and from this there may be built up a branched body (Fig. 29) which simulates those of the lower algae. Plant-like forms of this kind occur in every order of the group. It is indeed impossible to define the Phytomastigina from the Algae, and the members of this subclass are regarded both by botanists and by zoologists as coming within the scope of their sciences.

ACHLORITIC FORMS. Many of the coloured species are liable to produce colourless individuals. This happens in two ways: the chromatophores may become bleached owing to the animal living in darkness; or the rate of division of the chromatophores may lag behind that of the body, so that eventually there are produced offspring in which there are no chromatophores ('apoplastid' individuals). These facts show how the colourless species may have arisen.

SYMBIOSIS. Members of various orders of the Phytomastigina (cryptomonads, a chrysomonad, a chlamydomonad, and perhaps dinoflagellates) are known to live in the resting stage as symbionts in holozoic organisms

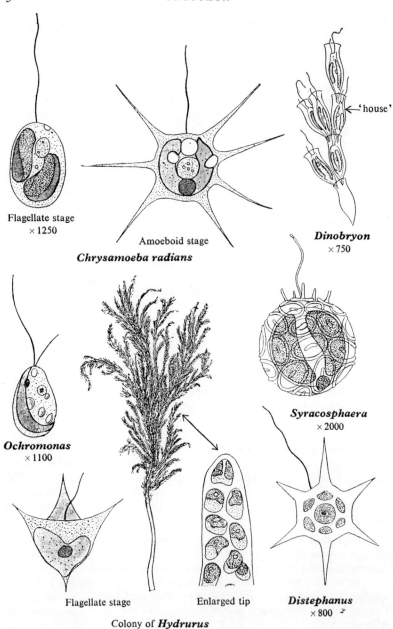

Fig. Chrysomonadina.

(other protozoa, sponges, coelenterates, worms, etc.). Nearly all are yellow or brown (*Zooxanthellae*); most green symbionts (*Zoochlorellae*) are algae belonging to the Protococcaceae. An exception to this is the chlamydomonad of the genus *Carteria* which lives as a zoochlorella in the tissues of the turbellarian worm *Convoluta roscoffensis* (Figs. 28 A, 28 B). The photosynthetic partner in these symbioses benefits by a supply of carbon dioxide and the nitrogenous excreta of its host; the latter has waste matters removed, is supplied with oxygen, and sometimes draws on the supply of carbohydrates manufactured by the guest, though it is rarely, as *Convoluta*, unable to dispense with this nutriment, and often, as the reef corals (p. 187), makes no use of it. If kept in the dark it is apt to devour the guest. A photosynthetic organism is specific to a particular host species. In some cases the two partners are capable of living apart; in others, they are mutually dependent. The plant organism usually enters the host by being ingested but not digested. It may be passed on from one generation to the next in asexual reproduction or even, as with the green *Hydra*, in the ovum, but is often lost in the gametes of its host, so that the zygote must be reinfected. Protozoan hosts in symbiosis are usually members of the Radiolaria (Fig. 28E) or Foraminifera, but various ciliates, *Noctiluca*, etc., also harbour holophytic symbionts. Zooxanthellae are commonest in marine hosts, zoochlorellae in fresh water.

AMOEBOID FORMS. The amoeboid faculty possessed by some members of the group may be limited to ingestion, but is often exhibited also in locomotion. Certain forms with such locomotion lose their flagella for shorter or longer periods: some may have done so altogether. When species with amoeboid movement become colourless they are only to be separated from the Sarcodina by certain features (of their nuclei, cysts, swarm spores, etc.) which prove them to be related to various Mastigophora.

1. *Order* **Chrysomonadina**

Yellow, brown, or colourless Phytomastigina; without starch reserves, but usually with leucosin and oil; without gullet or transverse groove; often amoeboid.

The genera briefly mentioned under this and the following orders illustrate the range of variety within the group.

Chrysamoeba (Fig. 29). One flagellum; two yellow chromatophores; no skeleton. Egg-shaped when swimming, but on the substratum becomes amoeboid and may lose flagellum. Ingests food by pseudopodia. In fresh waters.

Ochromonas (Fig. 29). As *Chrysamoeba*, but with two unequal flagella; and usually one chromatophore.

Dinobryon (Fig. 29). Two unequal flagella; two yellow chromatophores. Secretes a flask-shaped house, which in some species adheres to those of other individuals to form a pseudocolony. In fresh waters.

Hydrurus (Fig. 29). One flagellum; one chromatophore. Passes most of its life in the resting stage, which by division forms a plant-like growth (see p. 55). In fresh waters.

Rhizochrysis. Flagella normally lacking; one chromatophore; body naked and permanently amoeboid.

Leucochrysis. As *Rhizochrysis*, but colourless.

Silicoflagellata (or *Silicoflagellidae*). One flagellum; numerous yellow chromatophores; a lattice-work case of hollow, siliceous bars. Marine, planktonic, e.g. *Distephanus* (Fig. 29).

Coccolithophoridae. One or two equal flagella; two chromatophores (sometimes green); a case composed of calcareous plates (*coccoliths*) or rods (*rhabdoliths*) enclosing the body. Marine, planktonic, e.g. *Syracosphaera* (Fig. 29).

2. Order Cryptomonadina

Green, yellow, brown, or colourless Phytomastigina; with starch (and occasionally also oil) reserves; with gullet or with longitudinal groove, without transverse groove; very rarely amoeboid.

Many of the yellow members of this group live in the resting stage as symbionts in other organisms.

Owing to certain features of their nucleus and its mode of division these symbionts have been held to be related to the Dinoflagellata. Their other features, however, are those of the Cryptomonadina.

Cryptomonas (Fig. 30). Two flagella; two chromatophores, usually green; a gullet. Marine and in fresh waters.

Chrysidella (Fig. 30). Two flagella; two yellow chromatophores; a groove anteriorly. Symbiotic in Foraminifera, radiolarians, etc.

Cyathomonas (Fig. 30). Two flagella; chromatophores absent. Holozoic, seizing food by trichocysts in the gullet. In fresh waters.

Chilomonas. Two flagella; chromatophores absent; gullet very deep and narrow. Saprophytic. In foul fresh waters.

Phaeococcus. Normally in the palmella phase. Marine and in fresh waters.

3. Order Euglenoidina

Phytomastigina which have numerous green chromatophores or are colourless; with reserves of paramylum and sometimes also oil; with gullet; with contractile vacuole opening by a 'reservoir', usually into the gullet; without transverse groove; with stout pellicle, usually with metaboly ('euglenoid movement').

Euglena (Fig. 30). A typical member of the group, with chromatophores; one flagellum, arising from the bottom of the gullet, double at base, and connected by two rhizoplasts to a basal granule behind the nucleus; pyrenoids present only in a few species; paramylum reserves; and contractile vacuole fed by accessory vacuoles. The nutrition is interesting. Most species, at least, can live and multiply, with purely holophytic nutrition. All, however, flourish better if traces of amino acids be present. If the medium be rich in organic substances, the use which is made of these varies with the species. Most, including *E. viridis*, can take in organic combination nitrogen, but not carbon; a minority, including *E. gracilis*, can also obtain carbon in that way. In the dark, if suitable compounds, especially peptones, be present, the latter set of species bleach and live as saprophytes. It has not been established that *Euglena* uses its gullet to take solid food. Fresh waters, and infusions.

Peranema (Fig. 30). Without chromatophores; gullet supported by rods

PHYTOMASTIGINA

and can open or close. Saprophytic and holozoic. Paramylum reserves formed. In infusions.

Copromonas (=*Scytomonas*, Fig. 30). Without chromatophores; body pear-shaped; no metaboly; gullet long and narrow. Nutrition holozoic,

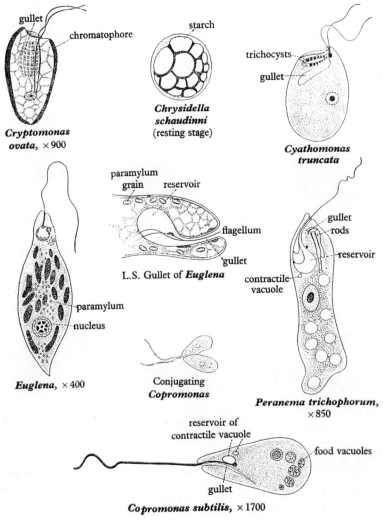

Fig. 30. Cryptomonadina and Euglenoidina.

chiefly by bacteria. Coprozoic in dung of frogs. After some days of binary fission syngamy takes place between ordinary individuals (hologamy), the nuclei first throwing out two 'polar bodies'. Some zygotes encyst; others continue to divide. Finally all encyst. The cysts are washed away and swallowed by a frog or toad with its food. They pass uninjured through the gut and hatch in the moist faeces, where alone the active stage exists.

Colacium. Normally in the palmella phase, forming branched, plant-like growths.

4. Order **Chloromonadina**

Phytomastigina which have numerous green chromatophores or are colourless; with reserves of oil; gullet; and complex contractile vacuole; without transverse groove; possessing a delicate pellicle, or amoeboid.

Vacuolaria. Typical, bright green members of the group, which pass much of the life history in the palmella stage. In fresh waters.

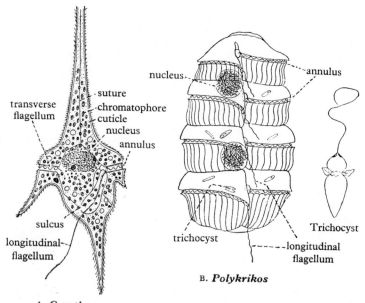

A. *Ceratium*

B. *Polykrikos*

Fig. 31. Dinoflagellata: A, *Ceratium macroceras*, × c. 300. B, *Polykrikos schwarzi*, × 250, with much enlarged trichocyst by its side.

5. Order **Dinoflagellata**

Phytomastigina which have numerous yellow, brown, or green chromatophores or are colourless; with reserves of starch or oil or both; with complex vacuole system; with two flagella, one directed backwards and usually in a longitudinal groove (*sulcus*) and the other transverse, usually in a more or less spiral groove (*annulus*); usually with an armour of cellulose plates, but sometimes amoeboid.

The complex *vacuoles* of dinoflagellates are not, as was held, contractile, but contain water driven into them through their external pores by the action of the flagella. Their function is unknown. Possibly they are hydrostatic, or alimentary, or both.

The plane of *fission* is oblique, but resembles the longitudinal fission of other Mastigophora in passing between the two flagella. Fission may be within or without a cyst; in either case it may be simply binary or repeated; within a cyst it is sometimes multiple. The products of repeated binary fission of pelagic forms sometimes hang together for a considerable time as a chain. The occurrence of *syngamy* is suspected but has not yet been proved beyond doubt.

The typical members of this order are free-living and highly organized, but it includes forms which are greatly degenerate and only recognizable as

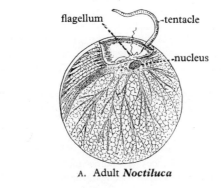

A. Adult *Noctiluca*

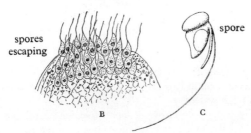

Fig. 32. *Noctiluca*, ×65: A, ordinary individual. B, spore formation. C, a spore. (After various authors, with modifications.)

belonging to it while they are spores. The members may be holophytic, saprophytic, or holozoic, feeding in the latter case by pseudopodia either from a spot on the sulcus or at any point. They are usually pelagic, sometimes parasitic, and for the most part marine.

Ceratium (Fig. 31). Typical, armoured, holophytic species; with three long spines. In fresh-water forms the chromatophores are green; in marine species they are yellow or brown.

Dinophysinae. Pelagic genera, often of bizarre form, with the annulus at one end of the body, and the shell in two lateral plates.

Polykrikos (Fig. 31). Soft-bodied species; colourless and holozoic; with the flagella and other external features repeated several times along the axis of the body, and the nucleus also repeated, but not in correspondence with the other features (see p. 13). The protoplasm contains peculiar nematocyst-like organs. Holozoic.

Oodinium (Fig. 28 c). Thin-cuticled; pear-shaped; colourless; living as an ectoparasite on marine pelagic animals, and possessing the typical dinoflagellate organization only in the spore stage.

Dinamoebidium. Colourless and holozoic; completely amoeba-like in the ordinary phase, but forming dinoflagellate swarm spores in a fusiform cyst.

Noctiluca (Fig. 32). (Formerly placed in an independent order—*Cystoflagellata*.) Large, peach-shaped forms; colourless and holozoic; with highly vacuolated protoplasm; a stout pellicle; and, in the groove of the peach, an elongate mouth, a small flagellum, a structure known as the tooth which is said to represent the transverse flagellum, and a strong tentacle, homologous with a similar structure in certain more normal dinoflagellates. *Noctiluca* is phosphorescent. Like other dinoflagellates it reproduces by binary fission and by spore formation after multiple fission. The spores are more dinoflagellate-like than the adult. Marine, pelagic.

Dinothrix. Normally in the palmella phase, forming thread-like growths. Marine.

6. Order **Volvocina**

Phytomastigina which have usually a flask-shaped, green chromatophore, with one or more pyrenoids, but are sometimes colourless, though never holozoic; form starch reserves, even when colourless; have no gullet or transverse groove; possess usually a cellulose cuticle and often haematochrome; and regularly undergo syngamy.

Of all the Mastigophora, the members of this order most closely resemble the typical plants.

Chlamydomonas (Figs. 18, 19). Typical solitary members of the order, with two flagella; an eye-spot; a close-fitting cellulose cuticle; and one pyrenoid. The various species exhibit isogamy, anisogamy, and intermediate conditions (see p. 35). In fresh waters.

Polytoma (Fig. 18). A colourless *Chlamydomonas*; retaining the eye-spot (usually) and the habit of starch formation; but with the cuticle composed of some substance which does not give the cellulose reaction. Nutrition saprophytic by means of simple substances (fatty acids, amino acids, etc.). Syngamy is facultatively hologamy or merogamy, isogamous or anisogamous, according to the age of the gametes. In infusions of decaying animal substances. Its relation to *Chlamydomonas* is discussed on p. 35.

Carteria (Fig. 28 B). Differs from *Chlamydomonas* in having four flagella. It is probably a species of this genus that is symbiotic in the turbellarian *Convoluta roscoffensis*.

Haematococcus (=*Sphaerella*). Differs from *Chlamydomonas* in that there is a wide space, traversed by protoplasmic threads, between body and cuticle; several pyrenoids. Much haematochrome is often present. Isogamous. Common in collections of rainwater.

Pandorina (Fig. 33). Spherical, free-swimming colonies of 16 or 32 green pear-shaped zooids, each with the organization of the solitary members of the order, closely pressed together with the narrow end inwards and the flagella outwards. An additional cellulose envelope containing mucilage encloses the whole colony. The colonies are reproduced in two ways: (1) asexually, by the repeated fission of each zooid to form a group of 16 like the parent colony, the dissolution of the colonial and zooid envelopes, and the setting free of

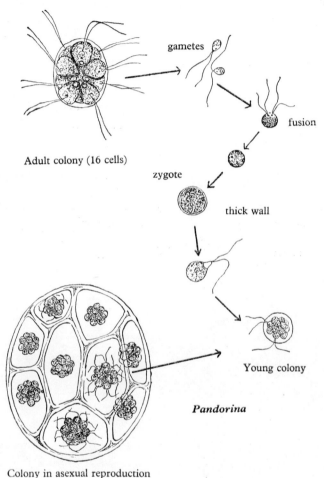

Fig. 33. *Pandorina*. (From Godwin.)

16 young colonies; (2) sexually, by the division of each zooid and the setting free of its products as gametes which, except in size, resemble ordinary zooids. Since the number of fissions in the formation of gametes differs in different colonies, the gametes differ in size. They unite indifferently, so that some of the unions are isogamous, though most are anisogamous. The zygote, after a period of encystment, becomes a free flagellate and divides to form a colony. In fresh waters.

Eudorina (Fig. 4A). Colonies which differ from those of *Pandorina* in that: (*a*) the zooids are spaced on the inside of the common envelope, though connected by strands of protoplasm; (*b*) the sexual reproduction is strongly

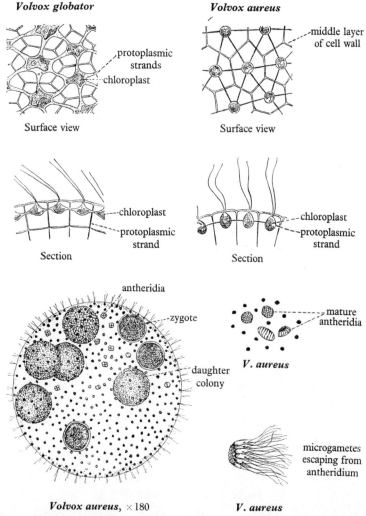

Fig. 34. The structure of *Volvox*, showing the different arrangement of the cells in the two species *Volvox globator* and *Volvox aureus*. (After Janet and Klein.)

anisogamous, since in some colonies the zooids do not divide but, becoming somewhat larger, act as macrogametes, while in others each zooid divides into a bundle of 16–64 slender individuals (microgametes), which are set free and fertilize the individuals of a macrogamete (female) colony.

Pleodorina (Fig. 4B). Rather larger colonies which differ from those of *Eudorina* in that some of the zooids do not perform reproduction. These zooids, which are smaller than the rest, are termed 'somatic'.

Volvox (Figs. 34, 35). Large, subspherical colonies resembling in general features those of *Pleodorina* but with smaller and more numerous zooids, of which a much smaller proportion is reproductive. Those zooids which perform asexual reproduction are known as *parthenogonidia*: the plates of young zooids which arise by their radial fission, curving into spheres to form the new colonies, bulge into the hollow of the parent colony, where they remain for a time before they are set free. The clusters (*antheridia*) of microgametes arise in the same way. In some species the microgametes are considerably modified, being pale, very slender, and bearing their flagella in the middle of their length. Male, female, and asexual reproductive zooids may be found in any combination in a colony. Details of the structure of the colonies are shown in Figs. 34, 35.

SUBCLASS 2. ZOOMASTIGINA

Mastigophora which do not possess chromatophores and are not otherwise practically identical with coloured forms.

By one or more of the following peculiarities of the Zoomastigina most members of the group are distinguished from most colourless members of the Phytomastigina.

(1) The Zoomastigina never have starch or other amyloid reserves.

(2) They often have more than two flagella. This is very rare in the Phytomastigina.

(3) With a single exception (*Helkesimastix*), it has not yet been established that syngamy occurs in any of them.

(4) Many of their parasitic members possess parabasal bodies.

1. *Order* **Rhizomastigina**

Zoomastigina with one or two flagella, and the whole surface of the body permanently amoeboid.

Mastigamoeba (Fig. 36). One flagellum; numerous, finger-like pseudopodia. In fresh waters.

2. *Order* **Holomastigina**

Zoomastigina with numerous flagella, and the whole surface of the body capable of amoeboid action.

Multicilia. Spherical, with 40 or 50 flagella scattered evenly over the whole surface, at any point on which food can be ingested by amoeboid action. A marine species with one nucleus; fresh-water species multinucleate.

3. *Order* **Protomonadina**

Zoomastigina with one or two flagella; amoeboid movement, if present, not active over the whole surface of the body; and no extra-nuclear division centre.

Monas (Fig. 36). Two unequal flagella. Ingestion at base of flagella. Except for absence of chromatophores much resembles *Ochromonas* among the

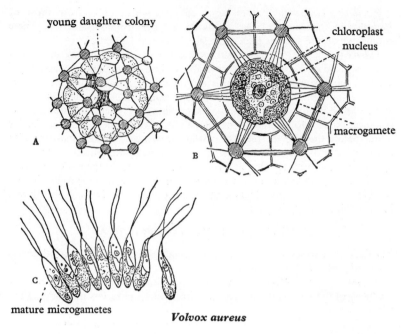

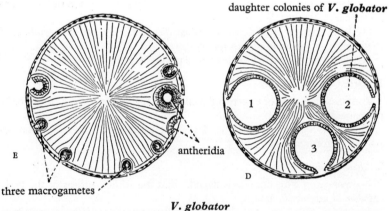

Fig. 35. *Volvox*. A, a young daughter-colony of small size as seen through the layer of zooids of the parent colony. The opening left in the young colony at its formation is shaded. B, a single macrogamete of *V. aureus* amongst the ordinary somatic zooids. Abundant protoplasmic strands connect it with the surrounding zooids and it contains a large nucleus and chloroplasts. C, mature microgametes of *V. aureus* just liberated from an antheridium and now beginning to separate. D, *V. globator*; diagrammatic section through an old colony showing three large daughter-colonies projecting into the interior of the parent colony which is full of thin mucilage with a radiating structure. E, *V. globator*; similar section to D showing three antheridia in differing stages of maturity and three large macrogametes. (After Janet and Klein.)

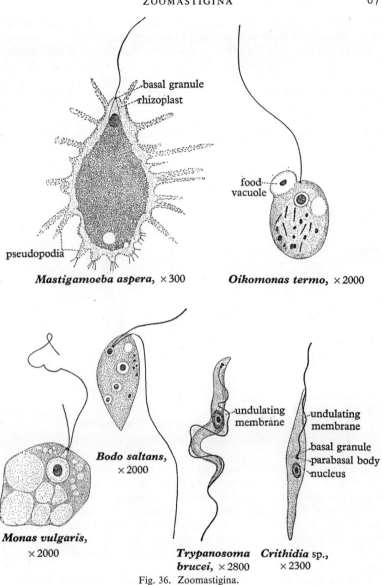

Fig. 36. Zoomastigina.

Phytomastigina and is probably related to that genus. In fresh waters and infusions.

Bodo (Fig. 36). Two rather unequal flagella, of which one trails freely behind and is used for temporary anchoring. Ingestion at a spot near the base of the flagella. In infusions and coprozoic.

Oikomonas (Fig. 36). One flagellum. Ingestion of food as in *Monas*. This genus bears the same relation to certain uniflagellate Chrysomonadina that *Monas* bears to *Ochromonas*. In fresh waters and soil.

Trypanosomidae (Fig. 36). Parasites, with one flagellum; a slender, usually pointed shape; a strong pellicle without ingestion spot; a parabasal body; and no contractile vacuole. This family, which contains many dangerous parasites of man and domestic animals, appears to have originally infested invertebrates and to have obtained access to vertebrates owing to the latter being subject to attack by the original hosts. The original mode of infection was by faeces. The species of each genus assume, in certain circumstances, the forms characteristic of other genera. The following are the principal genera.

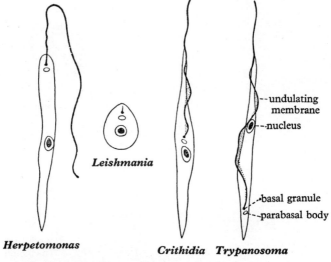

Fig. 37. A diagrammatic comparison of various Trypanosomidae.

Herpetomonas (= *Leptomonas*) (Fig. 37). Basal granule and parabasal body at one end, near the origin of the flagellum. Parasitic in the gut, principally of insects, but also of other invertebrates and of reptiles.

Leishmania. Oval bodies containing a nucleus, parabasal body, basal granule and rhizoplast, but with no flagellum, infesting the tissues of vertebrates, and transferred by flies of the genus *Phlebotomus*, in whose gut they assume the form of *Herpetomonas*. In man they cause kala-azar and Oriental sore.

Crithidia (Fig. 37). Flagellum starts from a basal granule near the middle of the long, slender body, to which the flagellum is united by an undulating membrane; parabasal body placed between the basal granule and the nucleus. Parasitic in the gut of insects.

Trypanosoma (Fig. 37). As *Crithidia*, but the basal granule of the undulating membrane and the parabasal body are beyond the nucleus, towards the non-flagellate end. Many species, all parasitic in the blood and other fluids of vertebrates, and nearly all (not *T. equiperdum*) distributed by a second, invertebrate host, which is usually an insect for terrestrial species and a leech for aquatic species. In the invertebrate the trypanosome passes for a time into a condition in which it resembles *Crithidia*, and during which it is incapable of reinfecting the vertebrate. Reinfection is in some species (e.g. *T. lewisi* in the rat, transmitted by a flea) by the invertebrate or its faeces being swallowed by the vertebrate; this is probably the original mode of obtaining entry to the vertebrate host. Other species (e.g. *T. gambiense*, transmitted by a tsetse fly) are reintroduced to the vertebrate by the bite of the invertebrate. *T. equiperdum*, parasitic in horses, in which it is the cause of 'dourine', is transmitted by coitus and has dispensed with the invertebrate host.

Most, if not all, of the pathogenic species have a wild host with which they are in equilibrium and in which they are non-pathogenic. *T. lewisi*, non-pathogenic in the blood of the rat, has a period of intracellular multiple fission

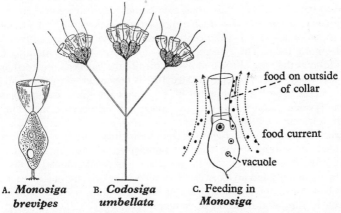

A. *Monosiga brevipes* B. *Codosiga umbellata* C. Feeding in *Monosiga*

Fig. 38. Choanoflagellata: A, ×1200; B, ×310. (Both after Saville-Kent.)

in the stomach of the flea and then passes into the rectum of the latter, where it changes from the crithidial to the trypanosome form and becomes capable of reinfecting the vertebrate, which it accomplishes in the manner mentioned above. *T. cruzi*, the cause of Chagas's disease in man in South America, is non-pathogenic in the armadillo. It is transmitted by the bug *Triatoma*, in which it probably has an intracellular stage, and becomes infective in the faeces. In the vertebrate host it passes most of its time, and reproduces as a *Leishmania* form, in the tissues. *T. gambiense* and *T. rhodesiense*, causes of sleeping sickness in man when they have passed into the cerebrospinal fluid, and *T. brucei*, the cause of African cattle sickness, are non-pathogenic in antelopes. Their crithidial stage is passed in the salivary glands of the tsetse (*Glossina*), reproduces by binary fission, and is not intracellular. They are transmitted to the vertebrate host by the bite of the fly.

Choanoflagellata (or *Choanoflagellidae*). Uniflagellate, generally fixed, forms; with a protoplasmic collar around the base of the flagellum. Ingestion by attraction of particles by the flagellum to the outside of the collar, adherence to this, and transference by streaming of protoplasm to the base of the collar, where they are received by a vacuole which is formed between the cuticle, if present, and the protoplasm (Fig. 38): defaecation within the collar. There is usually a stalk, generally not of living matter. This may branch, and thus unite numerous zooids. Examples are *Monosiga* (Fig. 38), solitary, with protoplasmic stalk; *Codosiga* (Fig. 38), branched, with cuticular stalk.

4. Order Polymastigina

Zoomastigina with two to many, generally with more than three, flagella, and an extra-nuclear division centre.

The genera here placed in one order are usually separated as Polymastigina, Hypermastigina, and Diplomonadina. They are the most highly organized members of the Mastigophora.

Trichomonas (Fig. 39). (One of the Polymastigina *sensu stricto*.) Body roughly egg-shaped; with four flagella, of which one is directed backward and united to the body by an undulating membrane; a cytostome near the broad anterior end; and an axostyle which projects from the posterior end. The united basal granules act as a division centre, possibly in virtue of a centriole concealed among them. The cytostome is used for ingestion. A staining body which follows the base of the undulating membrane has been regarded as the parabasal body, but a deeper-lying structure is now asserted to represent that organ. In the cytoplasm, a number of 'chromatic granules' are also present. Several species, parasitic in various cavities of vertebrates, including the mouth, intestine, and vagina of man.

Hexamita (= *Octomitus*, Fig. 3). (Diplomonadina.) Body elongate; without gullet; presenting strong bilateral symmetry; and bearing on each side four flagella, three anterior and one posterior, the basal granules of the foremost being united; and an axostyle. Two nuclei are present, one on each side of the body, near the anterior group of basal granules, with which they are connected. Intestinal parasites of vertebrates.

Giardia (= *Lamblia*, Fig. 39). (Diplomonadina.) Shaped like a half-pear, broad end forwards, with, on flat side, a concavity for adhesion. Organization as *Hexamita* but all flagella in middle or hinder region. Parasitic in intestine of man and other mammals.

Trichonympha (Fig. 40). (Hypermastigina.) Body narrower in front than behind; provided with very numerous flagella arranged in three distinct sets; without gullet. At the front end is a papilla. The ectoplasm, thin behind, is strong and complex in the fore part of the body, where it is composed of the following layers: (1) a pellicle, sculptured into longitudinal ridges, (2) a layer containing longitudinal rows of the basal granules of the flagella, (3) a layer containing a network of rhizoplasts ('oblique fibres'), (4) an alveolar layer, (5) a layer of transverse myonemes, (6) a layer of longitudinal myonemes. In the conical front region on which the first set of flagella stand, the rhizoplasts and basal granules are merged to form converging strands with which the flagella are connected. At division this conical apparatus acts as a division centre, dividing first and forming the spindle between its halves as they separate. Possibly it does so in virtue of a concealed centriole. *Trichonympha* is symbiotic with termites, in whose gut it lives (p. 485). The termite devours wood but is unable itself to digest it. The digestion is performed by the protozoon,

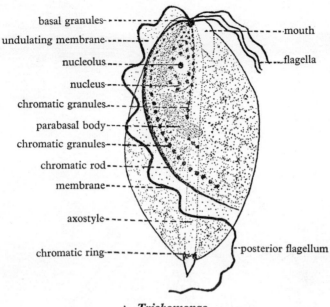

A. *Trichomonas*

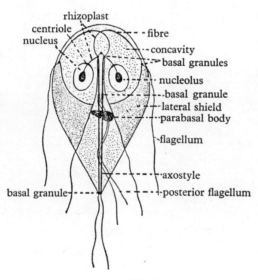

B. *Giardia*

Fig. 39. A, *Trichomonas muris*; semidiagrammatic. (From Hegener and Taliaferro, after Wenrich.) B, *Giardia intestinalis* from the intestine of man; semidiagrammatic.

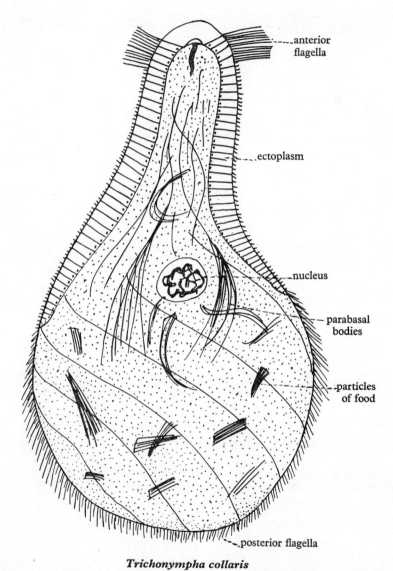

Trichonympha collaris

Fig. 40. A diagram of the structure of *Trichonympha collaris*. (After Cleveland.)

which obtains in return food and lodging. Wood particles are contained in the endoplasm of the hinder part of the body of *Trichonympha*, into which they are ingested by the cupping-in of this region under the action of the myonemes of the fore part.

5. Order Opalina

Many flagella, never amoeboid, no extra nuclear division centre. *Opalina* (Figs. 3, 41). Found in the rectum of frog or toad. It reproduces by binary division, the plane of cleavage being parallel to the lines of the kinetia. In spring the cell division outruns the nuclear division rate so that there are small individuals with few nuclei. These encyst and pass out of the host. They are then swallowed by a tadpole and hatch and give rise to uni-

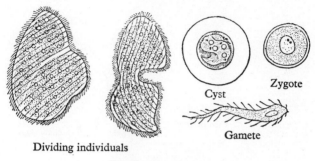

Dividing individuals

Cyst

Zygote

Gamete

Opalina ranarum

Fig. 41. *Opalina ranarum*. (From Borradaile.)

nucleate gametes of two sizes (anisogamous). After fusion of the gametes the zygote encysts for a while, issues forth and by nuclear division develops into an adult.

Other opalinids are *Protoopalina*; *Zelleriella*, *Cepedea*.

SYSTEMATIC POSITION. Previously *Opalina* had been placed in the ciliates. It differed from the other ciliates in two respects; first it had several nuclei which were all alike, there being no mega- and micronucleus. In this respect it resembled the Chonotricha but in the Chonotricha there is good evidence that the nucleus contains two sets of nuclear material corresponding to macro- and micronuclear material. Though such a system has been described for *Opalina* it has not been confirmed. A second difference from the ciliates lies in the relation between the plane of cleavage during cell division and the infraciliature (p. 22). In the flagellates the cleavage plane is parallel to the kinetia whilst in the ciliates the cleavage plane is across the kinetia. *Opalina* cleaves parallel to the kinetia and thus resembles the flagellates. A transverse cleavage has been described but has not been confirmed. The balance of evidence indicates that *Opalina* has slightly greater affinities with the flagellates than it has with the ciliates.

CLASS 2. SARCODINA (RHIZOPODA)

DIAGNOSIS. Protozoa which in the principal phase are amoeboid, without flagella; are usually not parasitic; have no meganucleus; and, though they may have a phase of sporulation, do not form large numbers of spores after syngamy.

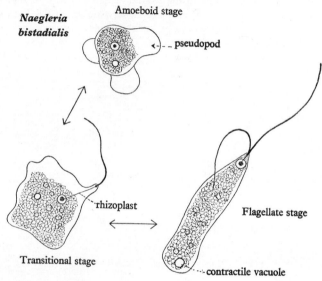

Fig. 42. *Naegleria bistadialis*, × 800. (Partly after Kuhn in Doflein.)

With the exception of the Amoebina and Foraminifera, which are undoubtedly closely related, the orders of this class have much less affinity with one another than have those of the Mastigophora. In all of them flagellate young and gametes are common.

CLASSIFICATION

Table 4. *Sarcodina*

Order	Shell	Skeleton	Capsule	Pseudopodia	Example
Amoebina	0	0	0	Lobose	*Amoeba*
Foraminifera	+	0	0	Lobose Reticulate	*Arcella*
Radiolaria	0	spicules	+	Reticulate Filopod	*Acanthometra*
Heliozoa	0	spicules	0	Axopoda	*Actinophrys*
Mycetozoa (colonial)	0	0	0	Lobose	*Badhamia*

1. Order Amoebina

Sarcodina which have no shell, skeleton, or central capsule; whose pseudopodia never form a reticulum and are usually lobose; and whose ectoplasm is never vacuolated.

Thus defined, the group excludes forms such as *Arcella* which differ from its members practically only in the possession of a shell. These forms, however, are also connected with the typical Foraminifera by intermediates (as *Lieberkühnia* and *Allogromia*). There is, indeed, a continuous series from naked amoebae to such foraminifera as *Elphidium*, and the drawing of a boundary line between the groups of which they are typical is a matter of convenience.

Naegleria (Fig. 42). Small amoebae which live in various foul infusions; possess a contractile vacuole; and in certain conditions pass into a biflagellate phase. *Naegleria* is placed here rather than among the Rhizomastigina because it is most often in the non-flagellate condition, its flagellate phase, though fully grown, is not known to perform reproduction, and the general features of the amoeboid phase are those of the amoebina of the *limax* group, most of whose members appear to have no flagellate phase. These organisms form one or two broad pseudopodia, are given to assuming a slug-like shape with one pseudopodium at the foremost end, and have a very simple nucleus with a large caryosome.

Vahlkampfia, also found in foul infusions, is a typical member of the *limax* group.

Amoeba (Fig. 43). Typical amoebae, with numerous pseudopodia; contractile vacuole; and no flagellate phase. Various species, of which the commonest three are shown in the figure. The true *A. proteus* is the largest of the common *Amoebae*, has a lens-shaped nucleus and longitudinal ridges on the ectoplasm, forms spores endogenously in the unencysted condition, and does not normally feed on diatoms, which form a great part of the food of *A. dubia*.

Entamoeba (Fig. 44 A, B). Parasitic amoebae; without contractile vacuole. Reproduction during most of the life history is by binary fission. Finally encystment takes place and in the cyst the nucleus divides several times. The cysts pass out of the host and infect a new individual, in which they are dissolved and set free their contents, which divide into uninucleate young. The cysts must remain in a fluid medium if they are to cause reinfection. Several species exist, occurring in various vertebrates and invertebrates. *E. coli* is a harmless commensal in the colon of man, feeding on bacteria, etc. *E. histolytica* (= *E. dysenteriae*), a parasite which often causes dysentery and occasionally abscesses of the liver and other organs, differs from *E. coli* in having a distinct ectoplasm, in the central position of the karyosome and in certain other features of the nucleus (Fig. 44) and in forming only four, instead of eight, nuclei in the cyst. This species breaks up by digestion cells of the intestinal epithelium and other tissues, absorbs the soluble products, and ingests portions of the destroyed cells and also red corpuscles.

Pelomyxa (Fig. 44c). Large, multinucleate species, living in, and feeding by ingesting, the mud of stagnant fresh waters rich in vegetable debris. The cytoplasm contains glycogen granules (see p. 26).

2. Order **Foraminifera**

Sarcodina which have either a shell or reticulate pseudopodia or, usually, both these features; and in pelagic species a vacuolated outer layer of protoplasm.

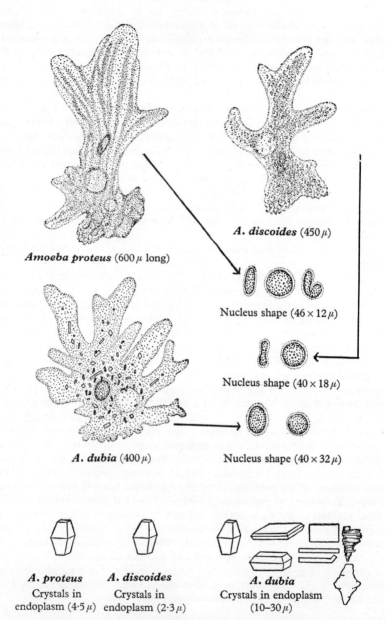

Fig. 43. *Amoeba*, showing features of three different species.
(From Hegener and Taliaferro, after Shaeffer.)

SARCODINA

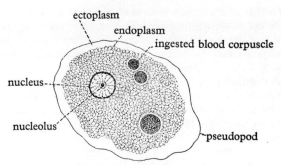

A. *Entamoeba histolytica*

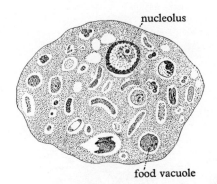

B. *Entamoeba coli*

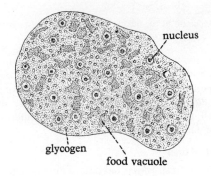

C. *Pelomyxa palustris*

Fig. 44. A, B, *Entamoeba*, × c. 2000. (After Dobell and O'Connor.) C, *Pelomyxa palustris* (Partly after Doflein.)

The *shell* may be of one or of several chambers, and is composed in different cases of different materials, nitrogenous, calcareous, siliceous, or of foreign particles.

The *pseudopodia* may be lobose, filose, reticulate without streaming of particles along them, or reticulate with streaming. The latter type alone is found in the Polythalamia.

The *reproduction* of the single-chambered forms (Monothalamia) is both by binary and by multiple fission. In binary fission, *Lieberkühnia* and *Trichosphaerium* divide the shell. In the rest, a portion of the protoplasm emerges from the old shell and secretes a new one (Fig. 46), the nucleus or nuclei divide, one of the products of each passing into the protruded protoplasm while the other remains in the old shell, and the two portions of protoplasm

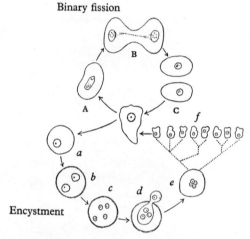

Fig. 45. Diagram of the life cycle of *Entamoeba histolytica*.

break apart. Multiple fission usually produces amoebulae, sometimes flagellulae. The latter are known or suspected to be gametes. In these forms there does not usually appear to be a regular alternation of sexual and asexual reproduction. In the Polythalamia binary fission does not occur, and in some of them, perhaps in all, there is a more or less regularly alternate production of asexual amoebulae and flagellate gametes.

Suborder 1. Monothalamia

Foraminifera, usually of fresh-water habitat; with non-calcareous, single-chambered shells; whose pseudopodia are rarely reticulate; and whose protoplasm does not extend as a layer over the shells.

Arcella (Fig. 46). Shell pseudochitinous, shaped like a tam-o'-shanter cap, finely sculptured; pseudopodia lobose; two or several nuclei and a chromidium present. Gas vacuoles in the protoplasm are said to contain oxygen and to have a hydrostatic function. Reproduction by binary fission,

SARCODINA

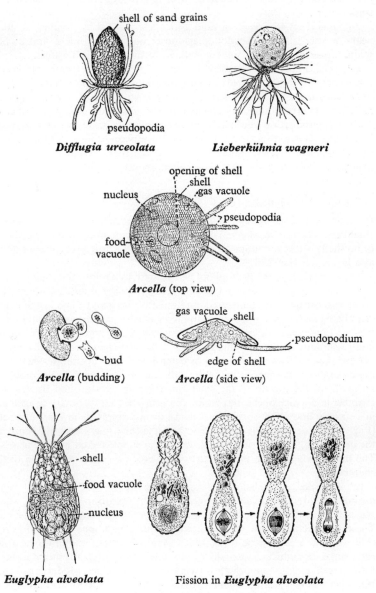

Fig. 46. Foraminifera. (After various authors.)

or by budding to form amoebulae with fine pseudopodia (*Nucleariae*). In fresh waters.

Difflugia (Fig. 46). Shell of sand grains, etc., united by organic secretion, pear- or vase-shaped; pseudopodia lobose; one or two nuclei and a chromidium present. Gas vacuoles sometimes formed. In fresh waters.

Euglypha (Fig. 46). Shell resembling that of *Difflugia* but formed of siliceous plates secreted by the animal; pseudopodia filose. In fresh waters.

Trichosphaerium. Flat, encrusting forms, with a jelly coat; finger-like pseudopodia protruding through separate openings in the coat; and numerous nuclei. Reproduction alternately by escape of amoebulae and of biflagellate isogametes; but both generations can perform plasmotomy. Marine.

Lieberkühnia (Fig. 46). Shell thin, flexible, egg-shaped, with mouth directed to one side; pseudopodia reticulate. Shell divided at binary fission. Marine and in fresh waters.

Suborder 2. Polythalamia

Foraminifera, nearly always of marine habitat; usually with a shell of several chambers, which is most often calcareous, but sometimes with one chamber or no shell; whose pseudopodia are reticulate; and whose protoplasm extends as a layer over the shell.

The external layer of protoplasm can be withdrawn into the shell.

The *shells* of this group are typically many-chambered and calcareous, but a fair number are one-chambered, and most of these and some of the many-chambered shells are composed of foreign particles (*arenaceous*). Either kind may be *imperforate* or *perforate* by numerous small pores, but most of the non-calcareous shells are imperforate. The one-chambered shells are of various shapes. They usually grow by extension at their openings. Shells with more than one chamber grow by the addition of chambers. The protoplasm bulges from the mouth of the shell and there secretes around itself a new chamber into which opens the previous mouth. The chambers may be arranged in a straight line, as in *Nodosaria* (Fig. 5), or in a spiral, as in *Elphidium*, etc. (Fig. 48), or occasionally irregularly; and the shell may be strengthened by the deposition, upon their original walls, of a supplemental layer (Fig. 48B). The nuclei, where there is more than one, bear no constant relation to the chambers.

In many species the shells are *dimorphic*, the two forms (Figs. 48 C, 49) being distinguished by the size and arrangement of the first-formed chamber, which is small in one (the *microspheric* form) and larger in the other (*megalospheric*). These forms correspond to the alternation of generations in the life cycle (Fig. 49), the microspheric form, which usually becomes multinucleate at an early stage, reproducing asexually by multiple fission, while the megalospheric form, which remains uninucleate till it is about to reproduce, produces gametes. It has been shown that in some Polythalamia the microspheric form is diploid and the megalospheric form haploid.

Most foraminifera are creeping organisms, but the Globigerinidae are planktonic and have, correspondingly, vacuolated ectoplasm and long slender spines on the shell. The shells of such forms, falling to the bottom, form an important constituent of many deep-sea oozes.

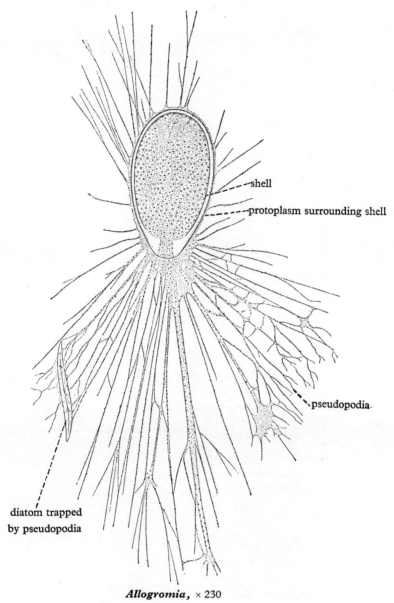

Allogromia, × 230

Fig. 47. *Allogromia oviformis*, × 230. The pseudopodia are less than one-third their relative natural length. (From M. S. Schultze.)

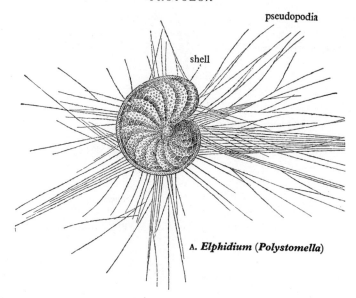

A. *Elphidium* (**Polystomella**)

B. Foraminiferan structure (in section)

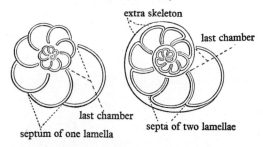

C. *Nummulites laevigatus*

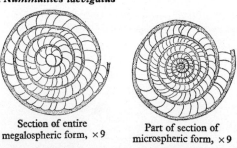

Section of entire megalospheric form, ×9

Part of section of microspheric form, ×9

Fig. 48. Polythalamian Foraminifera. B, One lamella and two lamella shells. C, Dimorphism of *Nummulites laevigatus*, Bracklesham Beds (Eocene), Selsea.

Allogromia (Fig. 47). Shell one-chambered, egg-shaped, pseudochitinous. Marine and in fresh waters.

Rhabdammina (Fig. 5). Shell one-chambered, straight or forked, tubular, composed of foreign particles. Marine.

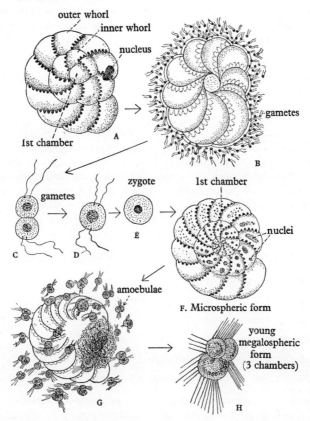

Fig. 49. Stages in the life cycle of *Elphidium* (*Polystomella*); semidiagrammatic. Note the alternation of the microspheric form and the megalospheric form.

Nodosaria (Fig. 5). Shell perforate, calcareous, consisting of several chambers arranged in a longitudinal row, the mouth of each chamber opening into the next younger and larger. Marine.

Elphidium (Fig. 48A). Shell perforate, calcareous, consisting of numerous chambers, arranged in a flat spiral, and complicated as follows in the details of their architecture: each whorl is *equitant*, i.e. overlaps the previous whorl at the sides and thus hides it; the mouth is replaced by a row of large pores; backward pockets (*retral processes*) stand along the hinder edge of each chamber; the supplemental layer contains a system of canals filled with protoplasm. Marine. The life cycle of this genus, which shows the alternation of generations described above, has been followed in detail (Fig. 49).

Nummulites (Fig. 48c). As *Elphidium* but with more chambers. Marine. Includes, besides recent forms, large fossil species in Eocene limestones.

Globigerina (Fig. 5). Shell perforate, calcareous, chambers fewer and less compact than in *Elphidium*, arranged in a rising (helicoid) spiral, and bearing long spines. External layer of protoplasm frothy, with large vacuoles by which the specific gravity is reduced. Marine, pelagic. Its shells are common in oceanic oozes and in chalk.

3. Order Radiolaria

Marine, planktonic Sarcodina, which have no shell but possess a central capsule and usually a skeleton of spicules; whose pseudopodia are fine and radial and usually without conspicuous axial filament; and the outer layer of whose protoplasm is highly vacuolated.

The *pseudopodia* branch, and to some extent join; they are said to contain an axial filament and they show streaming of granules. The *central capsule* is a pseudochitinous structure, of varying shape according to the species, which encloses the nucleus and some cytoplasm containing oil globules. It is perforated by pores, which by their arrangement characterize the suborders.

The Radiolaria are divided into four suborders on the basis of the arrangements of the pores in the central capsule.

Suborder 1. **Peripylea (Spumellaria)**

Pores evenly distributed over the capsule. *Thalasicolla, Spumellaria.*

Suborder 2. **Actipylea (Acantharia)**

Pores grouped into several groups. *Acanthometra.*

Suborder 3. **Tripylea (Phaeodaria)**

Pores in three groups. *Aulactinium, Challengeron.*

Suborder 4. **Monopylea (Nassellaria)**

Pores concentrated into one pore plate. *Lithocircus.*

The *spicules* are usually siliceous, but in one group (*Acantharia*) they are said to be of strontium sulphate. They are rarely absent, occasionally loose, but usually united into a lattice-work (Fig. 50 B, C, D, F), which is often very complicated, with projecting spines. The latter may be radial but do not meet at a central point except in the Acantharia. The *outer layer* of the body differs from that of the pelagic Foraminifera in that the vacuoles are contained in a layer of jelly (*calymma*) traversed by strands of protoplasm, which secrete it and the vacuoles, and in that it cannot be withdrawn.

There is no contractile vacuole.

The Radiolaria *reproduce* by binary fission and by spore formation. The spores found in them are sometimes alike (*isospores*) and sometimes of two kinds (*anisospores*). The latter are held to be gametes, and it is said that union between them has been observed. On account of their resemblance to the Dinoflagellata it has been suggested that they belong to parasitic members of that group. It is possible, on the other hand, that the Radiolaria have an alternation of generations like that of the Foraminifera.

Peculiarities of the *mitoses* in this group have been mentioned above (p. 30).

Symbiotic flagellates, known as 'yellow cells' (*Zooxanthellae*, see pp. 50, 57), are present in large numbers in the cytoplasm of many of the Radiolaria.

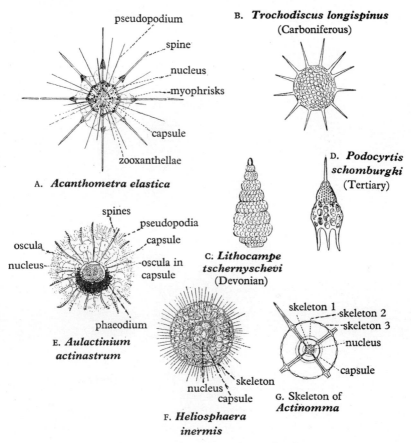

Fig. 50. Radiolaria. (After various authors.)

Thalassicolla (Fig. 26A). Skeleton absent or represented by some loose siliceous spicules; one nucleus; yellow cells in extracapsular protoplasm.

Collozoum (Fig. 26B). As *Thalassicolla*, but with central capsules united by their extracapsular protoplasm into a colony; and each capsule contains several nuclei.

Heliosphaera (Fig. 50F). As *Thalassicolla*, but the skeleton has the form of a lattice-work on the surface of the body.

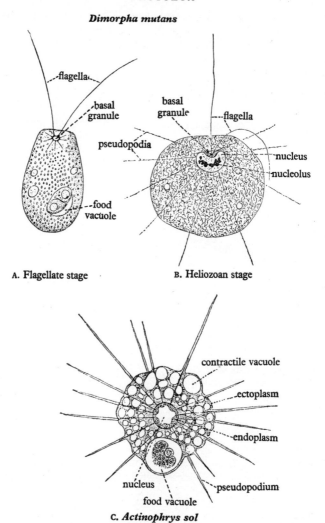

Fig. 51. A, B, *Dimorpha mutans*, showing the two different stages, the heliozoan stage taken from a stained specimen. (Partly after Blochman.) C, *Actinophrys sol*, ×800. (From Bronn.)

Actinomma (Fig. 50G). As *Heliosphaera*, but the skeleton consists of several lattice spheres, formed successively at the surface as the animal grows, with radial struts joining them. Ultimately the innermost sphere may lie in the nucleus.

Acanthometra (Fig. 50A). A skeleton of radial spicules of strontium sulphate meeting centrally in the central capsule; nuclei numerous; yellow cells intracapsular. Remarkable structures known as 'myophrisks', surrounding the spines of this genus at their junction with the calymma, are contractile and are used in the regulation of the diameter of the body.

Lithocircus (Fig. 28 E). A siliceous skeleton in the form of a ring, bearing spines. Yellow cells extracapsular.

Aulactinium (Fig. 50 E). A skeleton of hollow, radial, compound, siliceous spicules, not meeting in the centre; nuclei two; central capsule with three oscula, one of which is surrounded by a mass of coloured granules (the *phaeodium*, from which the suborder is named). Like the rest of the Phaeodaria this is a deep-sea form and does not possess yellow cells.

4. Order Heliozoa

Sarcodina, generally of floating habit and fresh-water habitat; without shell or central capsule; sometimes with siliceous skeleton; with spherical bodies; typical axopodia; and usually a highly vacuolated outer layer of protoplasm.

The *locomotion* of members of this group, in the ordinary phase, is effected as rolling, due to contraction of successive pseudopodia in contact with the ground so that the body is pulled over. The pseudopodia usually show streaming of granules. When they bend, which they do to enclose prey which has adhered to one of them, their axial filaments are temporarily absorbed at the bend. Protoplasm from the pseudopodia then surrounds the prey and streams with it inward to the ectoplasm, where a food vacuole is secreted around it.

Contractile vacuoles are present.

Asexual *reproduction* is usually by binary fission (or plasmotomy in multinucleate forms), sometimes by budding. Sexual processes have been thoroughly investigated only in *Actinophrys* and *Actinosphaerium*, where they take the form of autogamy (see below).

Dimorpha (Fig. 51 A, B). One of the Helioflagellata, a small group of organisms which is usually appended to the Heliozoa, bears somewhat the same relation to that order that *Naegleria* bears to the Amoebina. It has a biflagellate and a heliozoan phase, and can pass from one to the other. In the latter it retains the flagella, whose filaments share a common basal granule with those of the axopodia, and has no vacuolated layer or protecting case. In fresh waters.

Actinophrys (Figs. 51 C, 52). Unprotected; with one nucleus, against which the central filaments of the axopodia end; no skeleton. *Autogamy* (or more correctly *paedogamy*) takes place as follows: the pseudopodia are withdrawn and a jelly cyst formed. Binary fission now takes place, so that two individuals lie side by side in the cyst. Each divides mitotically twice, throwing out as a polar body one product of each division. The first of these two divisions is a reduction division. The two individuals now fuse, one behaving as a male by sending out a pseudopodium towards the other, and a strong inner cyst forms around the zygote. After a while the latter undergoes binary fission and the two products escape from the cyst. Occasionally two individuals enter a jelly cyst together and then either the two gametes of each undergo cross-conjugation with those of the other, or there is one cross-conjugation and the remaining gamete of each of the two original individuals performs parthenogenesis. In fresh and marine waters.

Actinosphaerium (Fig. 27). Unprotected; with many nuclei, against which

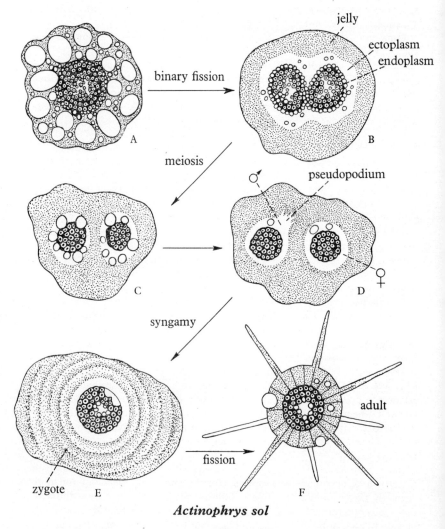

Actinophrys sol

Fig. 52. Successive stages in the autogamy of *Actinophrys sol*. (After Belar.)

the central filaments of the axopodia do not end. In preparation for autogamy the nuclei are reduced in number and the cytoplasm divides into as many corpuscles as there are nuclei. Each of these then undergoes a process similar to that which occurs in *Actinophrys*, forming a zygote which hatches as an independent individual. In fresh waters.

Clathrulina (Fig. 5). Animal enclosed in a stalked, pseudochitinous lattice sphere; one nucleus. At binary fission, one product becomes a biflagellula and swims away. In fresh waters.

5. Order **Mycetozoa**

Plasmodial Sarcodina, living usually in damp places on land; which have in the active phase no shell, skeleton, or central capsule, but in the quiescent phase a cyst of cellulose; possess numerous, blunt pseudopodia; and are usually distributed by air-borne, cellulose-coated spores.

There are three suborders, the differences between them being based on their morphology and life histories.

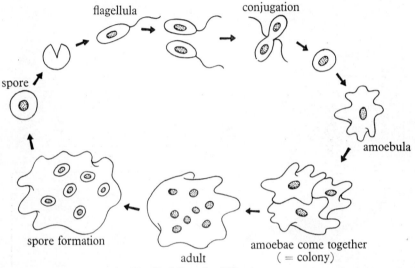

Fig. 53. Life cycle of Mycetozoan.

Suborder 1. **Acrasina**

The structural unit is the unicell but they can form plasmodia, though the cytoplasm between the unicells does not fuse. *Dictyostelium*.

Suborder 2. **Plasmodiophorina**

Parasitic forms which at maturity are plasmodia. They do not form spores. *Plasmodiophora*.

Suborder 3. **Eumycetozoina**

Free-living forms with a migratory plasmodium; produce spores. *Badhamia*.

The life history of a typical mycetozoon is as follows. The adult plasmodium is a sheet of protoplasm containing many thousands of nuclei and numerous contractile vacuoles. In it there are to be seen veins along which streaming takes place, alternately towards and from the periphery. It feeds in a holozoic manner, usually upon decaying vegetable matter, sometimes (*Badhamia*) on a living plant. In drought it breaks up into numerous multinucleate cellulose cysts which constitute the *sclerotium*. It prepares for reproduction by condensing at certain points, at each of which it forms a cellulose *sporangium*, often stalked. In the sporangium is a *capillitium* of cellulose threads and entangled in the capillitium are uninucleate, cellulose-coated spores, whose formation is preceded by a reduction division. When the sporangium is ripe it bursts and the spores are disseminated by wind, etc. In damp surroundings they often open and liberate each an amoebula which becomes a flagellula. The flagellulae perform syngamy and the zygote again becomes an amoebula. The amoebulae tend to fuse and form small plasmodia. By multiplication of their nuclei the adults arise (Figs. 53, 54).

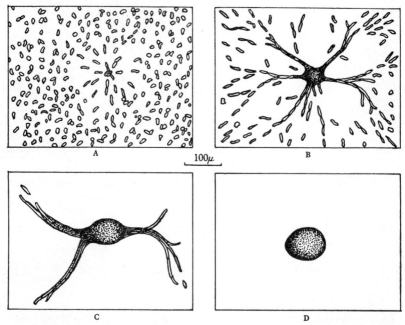

Fig. 54. Mycetozoa. Patterned movements of amoebulae of *Dictyostelium*. The central cells give off a coordinating chemical to which the amoebae show a positive chemotaxis. (From Bonner.)

CLASS 3. SPOROZOA

DIAGNOSIS. Protozoa which in the principal phase have no external organs of locomotion or are amoeboid; are parasitic, and nearly always at some stage intracellular; have no meganucleus; and form after syngamy large numbers of spores, which may be sporozoites or undivided zygotes.

The two subclasses, Telosporidia and Neosporidia, of this class have little in common, and their association in classification is a matter of convenience.

CLASSIFICATION

Subclass 1. **TELOSPORIDIA.** Sporozoans in which the adult has only one nucleus.

    Order 1. **Coccidiomorpha.** Adult intracellular; female gamete hologamous.

Suborder 1. **Coccidia.** Gut parasites. Zygote non-locomotory.
Eimeria
Suborder 2. **Haemosporidia.** Blood parasites. Zygote locomotory.
Plasmodium
Order 2. **Gregarinidea.** Adult extracellular, female and male gamete merogamous.
Suborder 1. **Schizogregarinaria.** Have a schizogonous stage during life cycle. *Schizocystis*
Suborder 2. **Eugregarinaria.** No schizogony during life cycle.
Monocystis
Suborder 3. **Piroplasmidea.** Hologamous, trophozoite has basal granule. *Piroplasma*

Subclass 2. **NEOSPORIDIA.** Sporozoans in which the adult has more than one nucleus.

Order 1. **Cnidosporidia.** Neosporidia whose spores possess pole capsules.
Suborder 1. **Myxosporidia.** Spore capsules bivalve usually with two, but sometimes four, polar capsules. *Myxobolus*
Suborder 2. **Microsporidia.** Small spores with only one polar capsule. *Nosema, Glugea*
Suborder 3. **Actinomyxidia.** Three polar capsules. *Sphaeractinomyxon*
Suborder 4. **Helicosporidia.** Spores have one coiled filament but no polar capsules. *Helicosporidium*
Order 2. **Acnidosporidia.** Neosporidia whose spores do not possess polar capsules.
Suborder 1. **Haplosporidia.** Have spore cases. *Haplosporidium*
Suborder 2. **Sarcosporidia.** No spore cases. *Sarcocystis*

LIFE HISTORY. Though upon analysis the type of life history characteristic of the Telosporidia is found to differ profoundly from those of the Neosporidia, all sporozoan life histories are complicated. Usually they comprise all the phases indicated in the scheme on p. 42, though in the Eugregarinaria (and perhaps in the Actinomyxidea) agamogony is omitted. Each phase, moreover, is liable to be elaborated. The term *sporoblast* is applied to certain stages in various life histories, but unfortunately the stages so named are not all comparable with one another. In the Telosporidia it denotes either the zygote or the products of the first of two successive multiple fissions whereby the sporozoites and other spore-like stages often arise. In the Neosporidia it denotes the syncytia (of different origins in different groups) from which by differentiation of cells complex spores are formed.

SUBCLASS 1. TELOSPORIDIA

Sporozoa in which the adult of the vegetative stage has only one nucleus; and comes to an end with spore formation; and the spore cases, if present, are simple structures, which nearly always contain several sporozoites.

The vegetative stage (*trophozoite*) has usually a definite shape, but in some haemosporidia is amoeboid. Its fission (agamogony), if such occur, is multiple, and is usually known as *schizogony*, the term *schizozoites* or *merozoites* being applied to the offspring. Its single nucleus only divides to form those of the young into which this stage breaks up, but owing to such division the body may be for a while multinucleate. The trophozoite of one of the two orders (the Coccidiomorpha) remains intracellular: in the other order (the Gregarinidea) it after a time outgrows its cell host. Save in one suborder (Eugregarinaria), it passes through the usual phase of agamogony before giving rise to gamonts, but in the Eugregarinaria agamogony is omitted, and the members of the single vegetative generation become gamonts, which provide for the increase of the species by the formation of many gametes in both sexes. The gamonts may be free or intracellular. Free individuals are often able to adhere by a sticky secretion, forming what is known as a *syzygy*. When gamonts so adhere (Figs. 57F, 58) they do so in pairs whose members are to be the parents of gametes that will unite reciprocally. Syngamy is isogamous in a few of the Gregarinidea, but is usually anisogamous, and in the Coccidiomorpha becomes an oogamy (p. 36). In some cases, perhaps in all, the first division of the zygote is a reduction division, so that nearly the whole of the cycle is haploid.

The little group Piroplasmidea, whose members in some respects resemble the Telosporidia, are best placed as a suborder to this subclass.

1. Order **Coccidiomorpha**

Telosporidia in which the adult trophozoite remains intracellular; and the female gamete is a hologamete.

Typically the members of this order are parasites of the gut, but more than once they have come to infest the blood. One such invasion gave rise to the suborder Haemosporidia. The rest of the group constitute the Coccidia.

Suborder 1. **Coccidia**

Coccidiomorpha, for the most part gut parasites; of which the zygote is non-locomotory; the sporozoites are nearly always encased; and the gamonts often form a syzygy.

Eimeria (Fig. 55) is parasitic in the intestinal epithelium of various vertebrates and invertebrates. *E. schubergi*, from the intestine of the centipede *Lithobius*, may be described as a type of the suborder. The spherical trophozoite (agamont) undergoes schizogony (agamogony) by multiple fission within the epithelial cell which it inhabits. The spindle-shaped schizozoites (agametes) being set free into the cavity of the organ, each infects another cell in which it grows like its parent. After some days of this there occur fissions in which the young on invading a host cell grow into adults unlike their parents and of two kinds—male and female gamonts. Each female gamont extrudes stainable matter from its nucleus and thus becomes a female hologamete. In the male gamont the nucleus divides several times, and the daughter nuclei are set free with portions of the cytoplasm as biflagellate male gametes, which are thus merogametes. The gametes leave the host cell and unite while free in the gut cavity. The zygote encysts and its nucleus undergoes what is probably a reduction division. Within its cyst (the oocyst) it divides by multiple fission into four sporoblasts each of which forms a cyst of its own (a secondary sporocyst) in which it divides into two sporozoites. Thus sporogony takes place in two stages. In each of these there is some residual protoplasm. Meanwhile the oocyst has passed out of the host in the faeces. Infection of a new host takes place by contamination of food by the encysted spores, which hatch in the intestine.

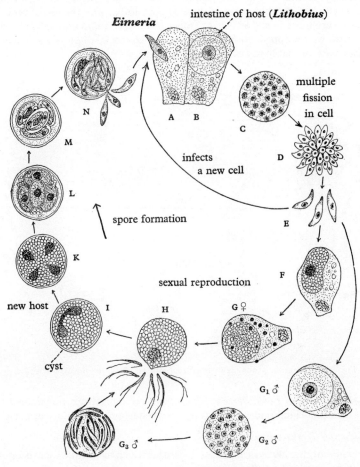

Fig. 55. A diagram of the life cycle of *Eimeria schubergi*.

Aggregata is remarkable among coccidians for having two hosts. Its agamogony takes place in crabs and involves a generation of sporoblasts, but is not repeated. A cuttlefish, devouring a crab, ingests the agametes, which in the new host proceed to become gamonts. After gamogony with flagellate male gametes, fertilization, and sporogony, the spores, containing four or more sporozoites, are passed with the faeces of the mollusc and swallowed by a crab.

94 PROTOZOA

Adelea is parasitic in the epithelium of the gut of *Lithobius*. Its life history resembles that of *Eimeria*, but the gamonts, which differ considerably in size, the male being the smaller, become free and form a syzygy in the gut, though without encystment. The male gametes are consequently not under the necessity of reaching the female by swimming, and are not flagellated.

Haemogregarina has become completely a blood parasite, and has a life history closely resembling that of the Haemosporidia, with the sexual process in an invertebrate host (see below). Since, however, it undergoes syzygy, the organism would appear to belong to the *Adelea* stock, whereas the Haemosporidia are probably related to *Eimeria*.

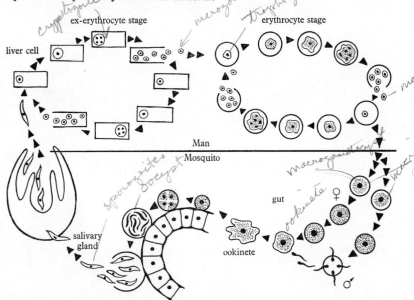

Fig. 56. A diagram of the life cycle of *Plasmodium vivax*, the malarial parasite.

Schellackia and *Lankesterella*, which have no syzygy, are transitional to the Haemosporidia, under which (on p. 95) their life histories are described.

Suborder 2. Haemosporidia

Coccidiomorpha, always true blood parasites; which have naked sporozoites; a locomotory zygote (*ookinete*); and no syzygy.

The members of this suborder are intracellular blood parasites of vertebrates and contain granules of pigment (melanin) derived from the haemoglobin of the host—a feature which is lacking in the blood parasites that belong to the Coccidia. They are transmitted from one vertebrate host to the next by a blood-sucking invertebrate. Their agamogony and the formation of their gamonts take place in blood cells of the vertebrate host, but their gametes are formed, and fertilization takes place, in the invertebrate. A series of intermediate cases shows how this condition may have arisen.

(1) *Schellackia* (suborder Coccidia), parasitic in the gut of a lizard, leaves the gut epithelium after schizogony and completes its cycle in the subepithelial tissues. In order to reach a new host it has therefore to rely on transference by a carrier instead of passing out with the faeces. To accomplish this, the sporozoites enter blood vessels, get into red corpuscles, and are sucked up by a mite. The blood-sucker, however, does not inject the parasite into the new vertebrate host, but is swallowed, so that the parasite infects the host through the gut epithelium, in which its schizogony is still performed.

(2) *Lankesterella* (suborder Coccidia), parasitic in frogs, passes its whole cycle in the epithelioid lining of blood vessels, the sporozoites being transferred, as in *Schellackia*, in red corpuscles, which are sucked up by a leech. Infection is still through the gut of the vertebrate, whose wall the sporozoites pierce on their way to the blood vessels.

(3) *Haemoproteus* (Haemosporidia), parasitic in birds, has its schizogony alone in the blood vessel walls, the sexual part of the cycle being remitted to the invertebrate host. The parasite enters the red corpuscles not as a sporozoite but earlier, as the young stage of the gamont, which grows up in the corpuscle. At the same time a change in the mode of infection has taken place, the blood-sucker injecting the sporozoites into the blood vessels of the vertebrate host. Thus the parasite has completely abandoned the gut wall and become a true blood parasite.

(4) *Plasmodium* (Haemosporidia), the cause of malaria and ague in man, is parasitic in the red blood-corpuscles of mammals and transmitted by the mosquito *Anopheles*. Its schizonts (trophozoites), as well as its gamonts, inhabit red corpuscles (erythrocyte stage).

The trophozoites of *Plasmodium* (Fig. 56) are amoeboid. In the young stage they are rounded and each contains a large vacuole which gives it the appearance of a ring. They undergo schizogony in the red corpuscles, which then break up, setting free the schizozoites (merozoites) and also products of the metabolism of the parasite which cause fever. After some generations, gamonts similar to those of *Eimeria* appear, but remain quiescent unless sucked up by a mosquito, in whose gut the female gamont becomes a spherical macrogamete, the male gamont throws off whip-like microgametes, and syngamy takes place. The zygote becomes elongate and active (an ookinete), and bores its way through the wall of the mosquito's stomach, on the outside of which it becomes encysted (oocyst). Here its nucleus divides and it breaks up into sporoblasts which in turn produce spindle-shaped sporozoites. The oocyst now bursts, setting the sporozoites free in the blood of the insect. They make their way into the salivary glands and are injected with the saliva into a mammal where they go first of all to the liver. There in the liver cells they undergo schizogony. This is the pre-erythrocyte or ex-erythrocyte stage. The schizozoites so produced are now capable of infecting the red blood corpuscles.

Three species of *Plasmodium* infest man—*P. vivax* which sets free a generation of schizozoites in forty-eight hours, *P. malariae* which does so in seventy-two hours, and *P. falciparum* whose schizogony occurs at more irregular intervals. Since the attacks of fever take place when the corpuscles break up and set free the toxins formed by the parasites, the fever caused by *P. vivax* returns every third day and is known as 'tertian ague', and that caused by

96   PROTOZOA

P. malariae ('quartan ague') recurs every fourth day, while *P. falciparum* causes irregular (quotidian) fevers which are more or less continuous. These latter are the 'pernicious malaria' of the tropics. The morphological differences between the species are small, but *P. vivax* is distinguished by the active movement of its pigment granules and the large number (15–24) of its schizozoites, *P. malariae* by the sluggishness and often quadrilateral form of its amoeboid stage, *P. falciparum* by the paucity of its pigment and by its curved, sausage-shaped gamonts.

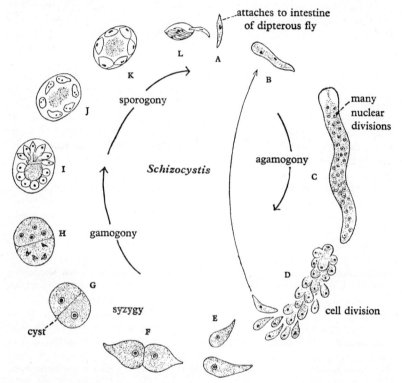

Fig. 57. A diagram of the life cycle of *Schizocystis*.

2. Order **Gregarinidea**

Telosporidia in which the adult trophozoite becomes extracellular; and the female (as well as the male) gametes are merogametes.

Intestinal and coelomic parasites of invertebrates, especially of arthropods and annelids.

Suborder 1. **Schizogregarinaria**

Gregarinidea which undergo schizogony.

Schizocystis (Fig. 57). Parasitic in the intestine of the larvae of dipterous flies. The young trophozoite attaches by one end to the gut epithelium of the host. Its nuclei multiply. When ripe it undergoes multiple fission. The products (schizozoites) either repeat asexual reproduction or become gamonts. These undergo syzygy, coencystment, and gamogony. The gametes unite, and the zygotes form small oocysts ('spore cases') within the gamocyst. In its case each zygote divides into a bundle of sporozoites. The spores are set free and swallowed by new members of the host species, in whose intestine the spore cases are digested and the process repeated.

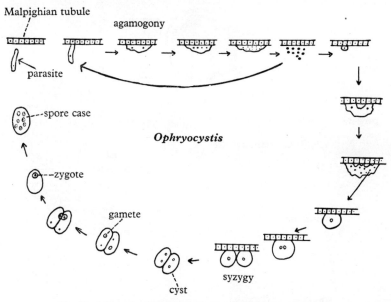

Fig. 58. Life cycle of *Ophryocystis mesnili* (Schizogregarinaria).

Ophryocystis (Fig. 58). Parasitic in the Malpighian tubules of beetles. The cushion-shaped trophozoites are attached to the host's cells by branched processes. After several generations of schizogony, they become free gamonts, enter into syzygies, encyst, and within the gamocyst undergo two divisions, whereby each forms one definitive gamete and a binucleate enveloping cell which perhaps represents abortive gametes. Syngamy then takes place, and the zygote divides to form within the enveloping cells a parcel of eight sporozoites in a case. Thus each syzygy produces only one pair of gametes and results in only a single spore.

Suborder 2. **Eugregarinaria**

Gregarinidea which have no schizogony.

The adult trophozoite has a stout cuticle and the ectoplasm contains myonemes, longitudinal or transverse, or both. Partitions of the ectoplasm without myonemes may (Fig. 60) divide the body into three segments—the *epimerite* or fixing organ, *protomerite*, and *deutomerite*, which latter contains the nucleus. When ripe the trophozoites become gamonts, joining in syzygies of two which together form a gamocyst and give rise to gametes (iso- or aniso-gametes according to species) by multiple fission in which residual protoplasm remains. Syngamy takes place within the cyst between the gametes of one parent and those of the other. The zygotes secrete small oocysts (*pseudonavicella*) of their own, and within these divide into several sporozoites ('falciform young'). Passing out of the host these are swallowed by another

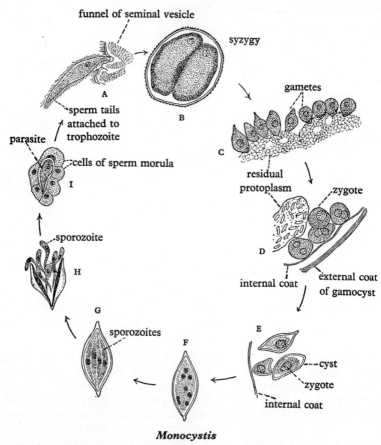

Fig. 59. A diagram of the life cycle of *Monocystis*.

of the same species, within which their cysts are digested and a new infection begins by the sporozoites invading cells of the host. These they eventually outgrow, and lie in a cavity of the host, either entirely free or attached by an epimerite.

In comparing this life cycle with that of *Eimeria*, given above, it should be noted that in the gregarines, whose female gametes are merogametes and numerous, the 'spores' (small sporocysts each containing several sporozoites) are each the whole product of a zygote (i.e. are oocysts), whereas in the coccidians, where the female gamete is a hologamete, the zygote forms, by means of a generation of sporoblasts, several such spores in its oocyst.

Monocystis (Figs. 59, 60). Without divisions of the body. Parasitic in seminal vesicles of earthworms. Several species, some isogamous, others anisogamous. The spores escape either down the vasa deferentia of the host

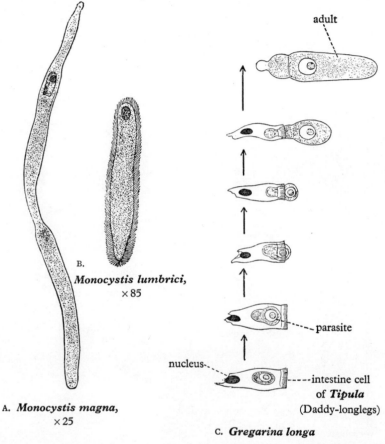

Fig. 60. A. *Monocystis magna*, ×25. (From Borradaile.) B, *Monocystis lumbrici*, ×85. (From Borradaile.) C, stages in the development of *Gregarina longa*. (After Léger.)

or by the latter being eaten by a bird, whose faeces contain them intact. Swallowed by another worm, their cases are digested and the sporozoites traverse the intestinal wall to reach the vesiculae seminales, where they enter sperm mother-cells, in which they pass their earlier stages.

Gregarina (Fig. 60). All three divisions of the body present. Parasitic in the alimentary canals of cockroaches and other insects. The gamocyst develops into a complicated structure with ducts for the discharge of the pseudonavicellae.

Suborder 3. Piroplasmidea

Protozoa, parasitic in red blood-corpuscles and transmitted by ticks; which have no external organs of locomotion; perform agamogony by binary fission; conjugate as hologametes; and after syngamy become motile zygotes which divide in a cyst into numerous, naked sporozoites (Fig. 61).

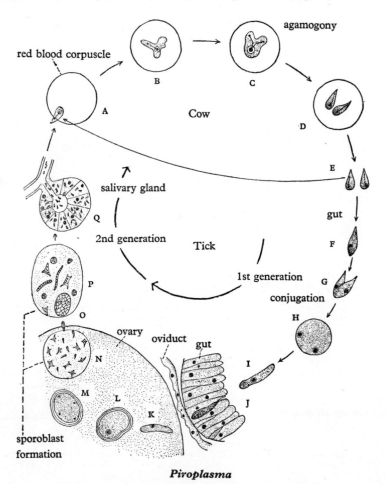

Fig. 61. A diagram of the life cycle of *Piroplasma* (*Babesia*). (After Dennis, with modifications.)

The members of this group are of doubtful affinity. In the general course of the life-cycle they resemble the Telosporidia, but in the possession by the trophozoite of part of a flagellar apparatus, and in that the gametes are both hologametes, they differ from the other members of that subclass. An interesting feature of their life history is that they are transmitted in the ovum from one generation of the invertebrate host to the next. They are at present only known from mammals and ticks.

Piroplasma (= *Babesia*). Infests various mammals (cattle, dogs, monkeys) and causes the red-water fever of cattle and other diseases. The trophozoites, in red corpuscles, are pear-shaped and unpigmented, and have a rhizoplast and basal granule as if for a flagellum. When taken into the alimentary canal of a tick they become gametes and form zygotes, which are ookinetes (p. 94), bore through the gut wall of the host, and reach its ovary. There they enter ova in which they are transmitted to the next generation of the tick. They encyst in the ova and divide into amoeboid sporoblasts (sporokinetes) which are distributed as the cells of the host divide and by their own active migration. Thus some reach the salivary glands. There they become multinucleate and break up into sporozoites, which are injected with the saliva into a new mammalian host.

SUBCLASS 2. NEOSPORIDIA

Sporozoa in which the adult of the vegetative stage is a syncytium; which usually forms spores continuously within itself; and the spore cases are usually complex structures, which, except in the Actinomyxidea, contain only one germ.

1. *Order* **Cnidosporidia**

Neosporidia whose spores possess pole capsules.

The formation of the spores in this group is a complex process of which the details and the relation to the typical life cycle of the Protozoa have not yet been completely elucidated. The following scheme provisionally co-ordinates the facts that have been established concerning it. In the syncytium (Fig. 62), which is the agamont and which often multiplies by plasmotomy, there arise, perhaps by the coming together of nuclei, bodies known as *pansporoblasts*, each composed of a couple of *envelope cells* with one or more cells known as *sporoblasts*. The nucleus of each sporoblast divides and the sporoblast gives rise to a complex, multicellular spore, composed of a *case* of two or three pieces, each with an underlying nucleus, one to five nematocyst-like *pole capsules*, each with a nucleus, and one or more *germs*. In most cases the germ is single and at first has two nuclei, which later fuse. Here we may regard the sporoblast as a gamont and the products of its division as homologues of gametes, of which some become the accessory cells of the spore and two (those which the germ at first possesses) the definitive gametes. In one group, however (the Actinomyxidea), there are several germs (often as a syncytium), and syngamy takes place not between nuclei in a germ but at an earlier stage, between pairs of cells in the pansporoblast, each zygote becoming a sporoblast. Here the sporoblast is a true sporont, and the products of its division are homologues of sporozoites, of which some become the accessory cells of the spore and the others (the germs) are the definitive sporozoites. It is a remarkable, but apparently an established, fact, that syngamy thus takes place at different stages in the formation of essentially similar spores.

Infection of new hosts is by the mouth, and the function of the pole capsules is, by discharging their threads, to anchor the spore to the gut wall. A schizogony may precede pansporoblast formation.

Suborder 1. Myxosporidia

Myxobolus (Fig. 62). Large syncytia in the tissues of various fresh-water fishes. Some species are harmless, others dangerous pests.

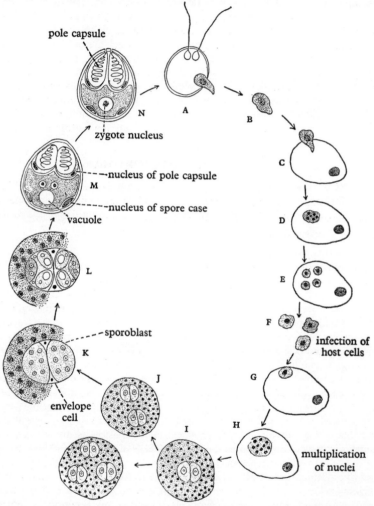

Fig. 62. A diagram of the life cycle of a typical member of the Myxosporidia. The schizogony shown here (D–F) probably does not often occur.

Suborder 2. Microsporidia

Nosema. The syncytium early breaks up, first into binucleate forms and finally into single sporoblasts. In the intestinal epithelium of insects. A serious pest of the silkworm, causing the disease known as pébrine, and of the bee.

NEOSPORIDIA

Suborder 3. **Actinomyxidea**

Sphaeractinomyxon. The whole body is reduced to a single pansporoblast, as in all members of the suborder. The spores are without the spines found in related genera. In annelids.

2. Order Acnidosporidia

Neosporidia whose spores do not possess polar capsules.

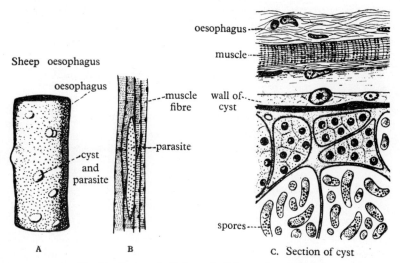

Fig. 63. *Sarcocystis lindemanni.* (After various authors.)

Suborder 1. **Haplosporidia**

Neosporidia whose spores possess cases with a lid, but have no polar capsules.

This order contains certain parasites which infest aquatic invertebrates. They are perhaps derived from the Cnidosporidia by loss of the pole capsules.

Haplosporidium, parasitic chiefly in annelids, is the typical genus.

Suborder 2. **Sarcosporidia**

Neosporidia whose spores do not possess cases or polar capsules.

These organisms are tubular syncytia with a radially striped ectoplasm, parasitic in the muscle fibres of mammals, and reproducing by simple, sickle-shaped spores.

Sarcocystis (Fig. 63). In various mammals, occasionally in man.

CLASS 4. CILIOPHORA

DIAGNOSIS. Protozoa which, at least as young, possess cilia; are never amoeboid; if parasitic are very rarely intracellular; nearly always possess a meganucleus; and do not, after syngamy, form large numbers of spores.

This class, though some of its parasitic members are of comparatively simple structure, contains the most highly organized Protozoa. Facts concerning sundry of the organs and processes in its members (the ciliary apparatus, p. 22; the contractile vacuole system, p. 27; the nucleus, p. 29; conjugation, p. 36; etc.) have been stated above. The life history, except for the remarkable process of conjugation undergone by most of the class, is relatively uncomplicated. In particular, though the nuclear peculiarities of the typical members of the group render inevitable certain special features in the metagamic divisions, there is no true sporogony.

CLASSIFICATION. Ciliophora. Protozoa which, at least when young, possess cilia; are never amoeboid.

Subclass 1. CILIATA. Ciliophora which as adults possess cilia and which do not possess suctorial tentacles.

Order 1. **Holotricha.** Ciliata without an oral wreath, have uniform ciliation.
Suborder 1. **Astomata.** No mouth. *Collinia*
Suborder 2. **Gymnostomata.** With mouth, simple gullet. *Prorodon*
Suborder 3. **Hymenostomata.** With mouth, complex gullet. *Paramecium*

Order 2. **Spirotricha.** Ciliata which possess a gullet with an undulating membrane.
Suborder 1. **Polytricha.** Uniform ciliation. *Stentor*
Suborder 2. **Oligotricha.** Reduced ciliation. *Entodinium*
Suborder 3. **Hypotricha.** Flattened body, have cirri. *Stylonychia*

Order 3. **Peritricha.** Ciliata usually permanently fixed by aboral surface, cilia reduced over the body. *Vorticella, Trichodina*

Order 4. **Chonotricha.** Ciliata permanently fixed to body of crustaceans. Only one type of nucleus. *Spirochona*

Subclass 2. SUCTORIA. Ciliophora of which all but a few primitive forms lose their cilia in the adult stage and possess one or more suctorial tentacles. *Acineta, Dendrocometes*

SUBCLASS 1. CILIATA

Ciliophora which as adults possess cilia; and which do not possess suctorial tentacles.

The *morphology* of this group is much affected by the disposition of the apparatus used in obtaining nutriment. The food may be absorbed through the surface: the shape of the body is then simple (Fig. 67 B). Nearly always, however, there is a mouth. In some of the lower genera this is anterior and terminal, or nearly so (Fig. 66 A), but usually it is removed to one side of the body (Fig. 67 E). This side is then said to be 'ventral', and that opposite to it is 'dorsal'.

The *mouth*, either is merely a soft patch of exposed endoplasm or possesses a *gullet* (p. 25). In a relatively few cases (including all those in which the mouth is terminal and a few of those in which it is ventral) the mouth is at the surface of the body: in such cases the gullet, if there be one, is an *oesophagus*, excavated in the endoplasm and capable of being opened and closed to seize

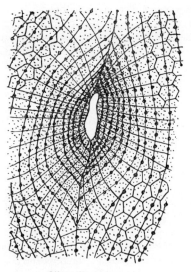

A. Silver line system

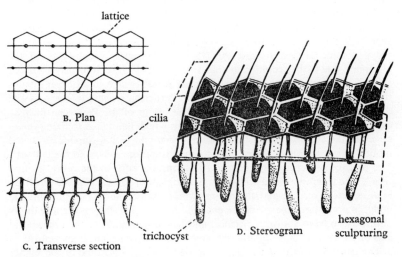

B. Plan
C. Transverse section
D. Stereogram

Fig. 64. Structure of *Paramecium*. A, silver staining around the oral region showing the kinetodesma. B, plan view of the pellicular lattice. C, diagrammatic transverse section of the pellicular structure. D, stereogram of the pellicle. (From Wichterman, after various authors.)

the prey, which is of some size. Most often, however, there is a *vestibule*. This, to which also the name 'gullet' is often applied, is a depression leading to the mouth, incapable of being closed, lined by inturned ectoplasm, and containing a ciliary apparatus, which usually includes one or more *undulating membranes*. By this apparatus the minute objects which constitute the food of all ciliates that have a vestibule are drawn in, being meanwhile, in some cases at least, entangled by a mucous secretion. At the bottom of the vestibule lies the true mouth; sometimes an oesophagus is present (*Stentor*) or is represented by a cleft in the endoplasm (*Paramecium*). The inner part of the vestibule may be free from cilia, and so simulate an oesophagus (*Paramecium*, *Vorticella*). The vestibule is usually approached by a *peristome*. This is a groove, of varying dimensions, which leads from the front end along the ventral side to the opening (cytostome) of the gullet. It is not straight, but runs in a longer or shorter spiral round the body, so that the anterior end of the latter is spirally deformed (Figs. 15 A, 65 A). The higher forms have along what is primarily the outer edge of the peristome a food-gathering row of cirri or membranellae, the *adoral wreath* (Fig. 65 A). Typically, the spiral is open as in *Paramecium*, but in some cases, as in *Stentor* (Fig. 66 c), it has contracted, so that it lies coiled as a crown at the anterior end. In such cases the animal is usually fixed temporarily or permanently by the opposite end.

The members of the suborder Hypotricha are depressed dorso-ventrally, and have a flat ventral side, along which the peristome runs and which is usually provided with a complex apparatus of cirri (Figs. 65A, B). The animal applies this side to the substratum, and certain of the cirri are used in its locomotion. The dorsal side is naked save for a few 'sensory' cilia. It is probably from such forms that the familiar bell-animalcules and their relations (Peritricha) are derived. In these, the shape of the body and the position of the peristome at first suggest that the morphological peculiarities of the group are due to an evolution similar to that by which such forms as *Stentor* came into being—but the fact that the peristome, which in all other ciliates that possess it curves clockwise, is in the Peritricha twisted in the opposite direction, makes this view impossible. The origin of the Peritricha may be explained as follows. In hypotrichous forms which had taken to fixing themselves to the substratum by that (ventral) side which they applied to it, the mouth, being no longer of use in its ventral situation, moved to the left side. The peristome accordingly came to run along the edge of the body, around which it became continued on the dorsal surface. In dorsal aspect its direction is of course reversed; and the adoral wreath has come to be internal. The body, in correspondence with the changed habit of life, has shortened, till its outline, seen from above, is circular, and has deepened. Thus the oral-aboral axis of the Peritricha is not anteroposterior as in *Stentor*, but dorsoventral.

The general *surface of the body* is in the lower and in some of the higher genera uniformly covered with cilia, but most of the more highly organized forms are naked save where there stand certain special pieces of ciliary apparatus. The ectoplasm (Fig. 64) has a definite and often complicated structure. There is always a tough pellicle, which is frequently sculptured. Under it is often an *alveolar layer* of minute, regular vacuoles. When there are myonemes, these lie on the inner walls of larger canal vacuoles of this layer.

Under it again is usually a layer, the *cortex*, whose firm consistency prevents the granules, vacuoles, etc., of the endoplasm from entering it, though it may possess small granules of its own. The basal granules of the cilia lie

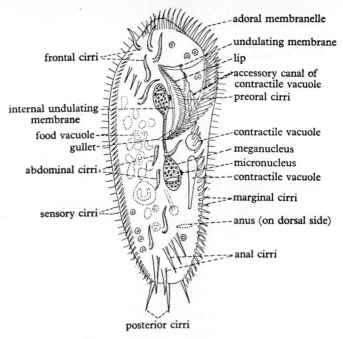

A. *Stylonychia mytilus* (ventral view), ×200

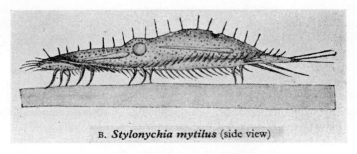

B. *Stylonychia mytilus* (side view)

Fig. 65. *Stylonychia mytilus* (Ciliata).

immediately below the alveolar layer; trichocysts are imbedded in the cortex. Either the cortex or both it and the alveolar layer may be absent. In *Paramecium* the cortex is covered by a thick pellicle which possibly contains a minute alveolar layer.

1. Order Holotricha

Ciliata which do not possess an adoral wreath; and which nearly all have uniform ciliation of the whole surface of the body.

This order is a collection of relatively simply organized ciliates, some of which are primitive while others are degenerate through parasitism.

Suborder 1. Astomata

Holotricha without mouth.
The members of this group are probably not primitive but degenerate through parasitism.

Collinia. Parasitic in the blood-spaces of the gills of *Gammarus* and other crustaceans.

Anoplophrya (Fig. 67B). Reproduction by repeated budding at one end of the elongate body, forming a chain. Parasitic in various annelids.

Suborder 2. Gymnostomata

Holotricha with a mouth, whose gullet, if any, is without ciliary apparatus (i.e. an oesophagus).

Ichthyophthirius. Subspherical, with a mouth at one pole and short gullet; numerous contractile vacuoles near the surface of the body; and a meganucleus, but no micronuclei visible in the adult. Parasitic in various fresh-water fishes, where it lies in blisters in the skin. When it is full-grown, it falls out of the host, encysts, and forms by repeated fission a number of small ciliospores, each of which has a mega- and a micronucleus, the latter having appeared during the process, perhaps from within the meganucleus. The spores infect new hosts. A sexual process of the nature of autogamy has been described, but is very doubtful.

Prorodon (Fig. 66A). Ovoidal, with mouth at one pole, a deep gullet which is supported by skeletal rods and is capable of opening and closing; one mega- and one micronucleus. In fresh waters.

Loxodes. Compressed, with mouth as a mere slit in the pellicle on the ventral edge of the body, overhung by the beak-like anterior end; numerous mega- and micronuclei; a row of vacuoles containing excreta along the dorsal border, and a contractile vacuole at the hinder end. In fresh waters.

Suborder 3. Hymenostomata (Vestibulata)

Holotricha with a mouth and a gullet (vestibule) which is permanently open and usually possesses an undulating membrane.

Colpoda (Fig. 67E). Kidney-shaped; with large vestibule on concave side; but no undulating membrane; and no peristome. Fission, binary or repeated, takes place in a cyst. Common in infusions, fresh-water and marine.

Colpidium. As *Colpoda*; but with undulating membrane. Common in infusions, fresh-water and marine.

Paramecium (Figs. 15, 64). Slipper- or pear-shaped according to species; with undulating membranes; and peristome. Common in infusions, fresh-water and marine.

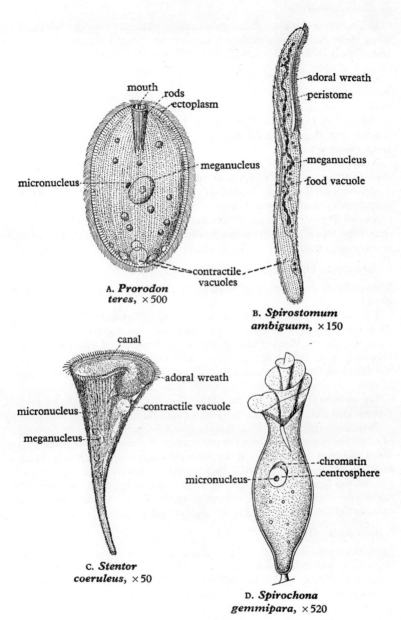

Fig. 66. Ciliates.

2. Order Spirotricha

Ciliata which possess a gullet, permanently open and provided with undulating membrane; an adoral wreath, curving clockwise; and most often on the rest of the body a uniform covering of cilia; and whose body is not depressed.

Suborder 1. Polytricha

Spirotricha which retain the uniform ciliation of the general surface of the body.

Balantidium (Fig. 68 A). Egg-shaped; the peristome a deep groove at the anterior end. Parasitic in the rectum of frogs, the intestine of man (where it is occasionally harmful), etc.

Nyctotherus (Fig. 68 B). Kidney-shaped; with permanent anus. Parasitic in the rectum of frogs, the intestine of man, etc.

Spirostomum (Fig. 66 B). Rod-shaped; with the peristome as a long groove; meganucleus beaded; several micronuclei. In fresh waters and marine.

Stentor (Fig. 66 C). Long and funnel-shaped; attached by the base, but often frees itself to swim; meganucleus beaded; several micronuclei. The animal is very highly contractile. In fresh waters.

Suborder 2. Oligotricha

Spirotricha of shortened form; with the body cilia reduced to a few rows or absent.

This suborder contains two tribes of very different habits, the pelagic Tintinnina, and the Entodiniomorpha, forms of bizarre shape parasitic in the alimentary canal of mammals, chiefly in the stomach of ruminants. Both suborders have an anterior peristome with very strong membranellae, and are naked on the rest of the body, save sometimes for a few cilia or patches of cirri.

Tintinnidium (Fig. 67 C). (Tintinnina.) Cup-shaped; anchored by an aboral process into a chitinoid case. In fresh waters and marine.

Entodinium (Fig. 67 A). (Entodiniomorpha.) With three posterior processes, of which the largest is said to serve as a rudder. In the rumen and reticulum of sheep and oxen. Like others of the tribe, these organisms are present in such numbers that they are believed to be symbionts which play a part in the nutrition of the host, rendering the vegetable food more easily assimilable by feeding on it and being in turn digested farther on in the alimentary canal. Infection of the host is probably by cysts on grass.

Suborder 3. Hypotricha

Spirotricha with depressed body; a gullet, permanently open and provided undulating membranes; an adoral wreath, curving clockwise; the dorsal cilia represented only by a few stiff hairs; and on the ventral side usually an elaborate system of cirri and other ciliary organs.

The animals can swim but spend much of their time crawling over solid objects by means of the cirri.

Stylonychia (Fig. 65). A typical example. Common in infusions.

Kerona. With a less highly developed ciliary system than *Stylonychia*. Ectoparasitic on *Hydra*.

CILIATA

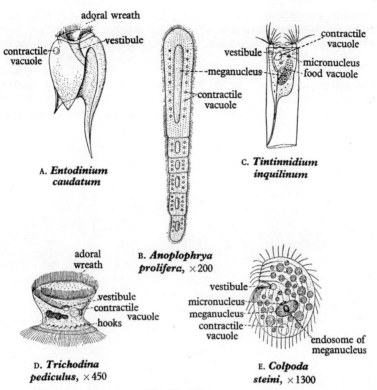

Fig. 67. Ciliates.

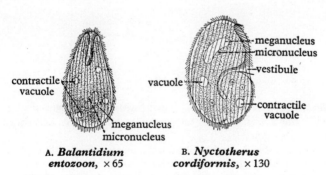

Fig. 68. Ciliates from the rectum of the frog. (From Borradaile.)

3. Order **Peritricha**

Ciliata, for the most part permanently fixed by the aboral surface; with a gullet, permanently open and provided with undulating membrane; an adoral wreath, curving counter-clockwise; and on the rest of the body no cilia, save those of an aboral ring in the free-swimming species and stages.

The conjugation of members of this group has been discussed on p. 37, their morphology on pp. 104, 106. The anus and contractile vacuole open into the deep vestibule, perhaps owing to an extension of the depression of the ectoplasm which forms the latter. The meganucleus is horseshoe-shaped.

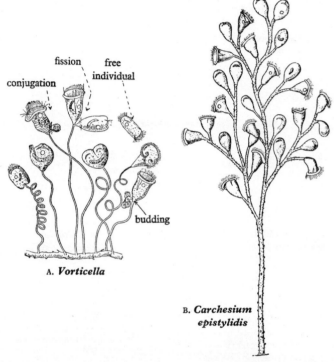

Fig. 69. Colonial ciliates. (A, after Borradaile; B, after Saville-Kent.)

Trichodina (Fig. 67 D). Dice-box shaped; with aboral ring of cilia for swimming, enclosing a ring of hooks for temporary attachment. Ectoparasitic on *Hydra* and other animals.

Vorticella (Fig. 69 A). Shaped like a solid, inverted bell, with, in place of the handle, a stalk which consists of a prolongation of the body, and is clad in a cuticle and contractile by means of a myoneme. Solitary. In fresh waters and marine.

Carchesium (Fig. 69 B). As *Vorticella*, but colonial. In fresh waters.

Epistylis. As *Carchesium*, but the stalk is purely cuticular and non-contractile. In fresh waters and marine.

4. Order **Chonotricha**

Ciliata, permanently sessile by the posterior end upon the bodies of crustacea; with the peristome represented by a spiral funnel at the anterior end, coiled clockwise, ciliated inside, and leading to the mouth; and the rest of the body naked.

A small but very interesting group which has two characteristics not found elsewhere in the class, namely (1) that their nuclei are of one kind only and at mitosis form two sets of chromosomes (see p. 32), (2) that they form numerous gametes, which unite in the same way as those of members of the other classes of the phylum. In the Chonotricha the reproduction, both sexual and asexual, is carried out by buds. The nucleus contains a large achromatic mass which acts as a division centre.

Spirochona (Fig. 66D). Shaped like a slender vase. On the gills of *Gammarus*, etc., in fresh and marine waters.

SUBCLASS 2. SUCTORIA

DIAGNOSIS. Ciliophora of which all but a few primitive forms lose their cilia in the adult; and which possess one or more suctorial tentacles.

A few members of the group are free; a few are endoparasitic; most are attached, and these have usually a cuticular *stalk*, which is often expanded at the end to form a shallow cup in which the animal sits or a deep one which encloses it.

The suctorial *tentacles* contain a tube, lined by ectoplasm, which opens at the end, where there is often a knob. In some species there are also solid, sticky tentacles, used to capture prey.

REPRODUCTION by simple binary fission does not occur. In a few cases fission is equal or almost so (*Podophrya*, *Sphaerophrya*, Fig. 70B) but here one of the products differs from the parent in losing its tentacles and acquiring cilia and thus resembles the buds of other species. This happens whether the parent be a stalked or a floating form. Most species multiply by typical budding. The buds may be external (Fig. 70A) or formed in brood pouches (Fig. 70F) from which they escape when they are ripe. External budding is the more primitive, internal the commoner process. In either, one bud or more than one may be formed at a time. The buds (Fig. 70D), whether external or internal, are usually ciliated and at first without tentacles; the cilia form a girdle round the body, with sometimes the vestige of an adoral wreath. Certain species form also unciliate and often tentaculate offspring by external budding. Some species will, in unfavourable circumstances, resolve practically the whole body into one internal bud which swims away, leaving the pellicle and stalk behind.

CONJUGATION is of the same nature as in the Ciliata. Two individuals become united by pseudopodia-like processes of protoplasm, their meganuclei break up, and their micronuclei form pronuclei which unite reciprocally. Often, however, the conjugants do not break apart, but one detaches itself from its stalk to unite permanently with the other. It is not known what happens to the two zygote nuclei in these cases.

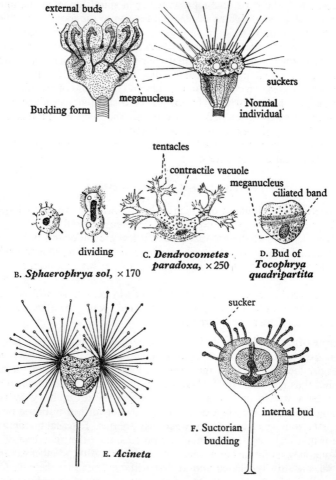

Fig. 70. Suctoria (A, after Hertwig; B, C and D, after Butschli; E, after Saville-Kent.)

The arrangement of the larval cilia in rings, the prevalence of a sessile habit, the frequent inequality of conjugants, and sometimes the absorption of one of these by its partner, suggest the derivation of this subclass from a form which resembled the Peritricha.

Sphaerophrya (Fig. 70B). Spherical species; which are at first free and provided with knobbed tentacles on all sides; afterwards become endoparasites in ciliates; and are then without tentacles. Fission equal or somewhat unequal; in the parasitic stage it is repeated before the young escape. Parasitic in *Paramecium*, etc.

Ephelota (Fig. 70A). Stalked; not seated in a cup; bearing tentacles distally. Reproduction by external, usually multiple, budding. Marine.

Acineta (Fig. 70E). Stalked; the stalk expanding to form a shallow cup. Reproduction by internal budding. In fresh waters and marine.

Dendrocometes (Fig. 70C). Body lens-shaped; without stalk; with branched arms which end in several pointed tentacles. Reproduction by formation of one internal bud. Sessile upon the gills of *Gammarus*.

CHAPTER III

THE SUBKINGDOM PARAZOA (PORIFERA)

THE PHYLUM PORIFERA

DEFINITION. Multicellular organisms; invariably sessile and aquatic; with a single cavity in the body, lined in part or almost wholly by collared flagellate cells; with numerous pores in the body wall through which water passes in, and one or more larger openings through which it passes out; and generally with a skeleton, calcareous, siliceous, or horny.
The members of this phylum are the sponges.

CLASSIFICATION

CLASS 1. CALCAREA

Sponges with skeletons consisting solely of calcareous spicules and with large choanocytes. *Grantia, Sycon.*

CLASS 2. HEXACTINELLIDA

Sponges with a purely siliceous skeleton composed of six-rayed spicules. A deep-sea group. *Euplectella.*

CLASS 3. DEMOSPONGIAE

Sponges whose skeleton if present does not contain six-rayed spicules of silica though the skeleton may be siliceous. *Spongilla, Halichondria.*

Order 1. **Monaxonida**
Order 2. **Keratosa**
Order 3. **Myxospongiae**
Order 4. **Tetractinellida**

STRUCTURE. The simplest sponge is a little creature, known as the *Olynthus* (Fig. 71), which is found only as a fleeting stage in the development of a few of those members of the group which possess calcareous skeletons; but the bodies of all sponges may be regarded as derived from it, even though it may not appear as a stage in their life history. It is a hollow vase, perforated by many *pores*, and having at the summit a single large opening, the *osculum*. Through the pores water constantly enters it, to pass out through the osculum. Herein it and its kind differ from all the Metazoa, using the principal opening not for intaking—as a mouth—but for casting out. The wall (Fig. 72) of the vase consists of two layers, (1) a *gastral layer*, composed of collared flagellate cells resembling the Choanoflagellata (p. 69) and known as *choanocytes*, standing side by side but not touching, which lines the internal cavity or *paragaster* except for a short distance within the rim; and (2) a *dermal layer*, which makes up the greater part of the thickness of the wall and is turned in a little way at the rim. This layer again consists of two parts, (*a*) a *covering layer* of flattened cells, known as *pinacocytes*, rather like those of a pavement epithelium, but with the power of changing their shape; and (*b*) the *skeletogenous layer*, between the covering layer and the gastral layer. The skeletogenous layer consists of scattered cells, with a jelly in which they are imbedded. The most numerous of these cells are engaged in secreting spicules of calcium carbonate by which the wall is supported. They wander from the covering layer into the jelly, and then each divides into two, and the resulting pair secrete in their protoplasm, which is continuous, a needle-like spicule which presently outgrows them.

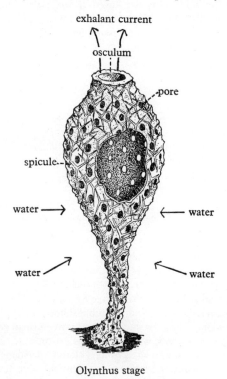

Olynthus stage

Fig. 71. The olynthus of a simple calcareous sponge with part of the wall cut away to expose the paragaster.

Most often the original spicule cells come together in threes before this process, so that the three spicules which they secrete become the rays of a three-rayed compound spicule. This lies in the wall with two rays towards the osculum and one away from it. Sometimes a fourth cell joins the others later, and forms a fourth ray which projects inwards towards the paragaster. Often there are simple spicules which project from the surface of the sponge. Other cells, known as *porocytes*, of a conical shape, extend through the jelly, having their base in the covering layer while their apex reaches the paragaster between the choanocytes. Each is pierced from base to apex by a tube, which is one of the pores. Besides these cells of the dermal layer, there are in the jelly wandering amoeboid cells which appear, in some cases at least, to belong neither to the gastral nor to the dermal layer, but to be descended independently from blastomeres of the embryo. Some of them become ova; others, it is believed, give rise to male gametes; the rest are occupied in transporting nutriment and excreta about the sponge. There are no nerve or sense cells in this or any other sponge.

The current which flows through the body is set up by the working of the flagella of the choanocytes. It carries with it various minute organisms which serve the sponge for food, being swallowed, in some way which is still in dispute, by the collar cells. These digest the food, rejecting the indigestible parts into the space within the collar; and passing on the digested food to amoebocytes, which visit them to obtain it.

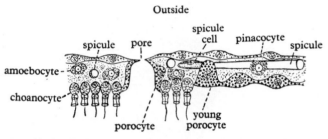

Fig. 72. Part of a longitudinal section of the wall of an olynthus including a portion of the rim of the osculum. (From Borradaile.)

THREE GRADES. No sponge remains at this simple stage throughout its life. At the least the body branches and thus complicates its shape, and then often new oscula appear at the ends of the branches (Fig. 77A). A higher grade is reached when, as in the calcareous sponge *Sycon* (Fig. 77B), the greater part of the vase is covered with blind, thimble-shaped outgrowths, regularly arranged, and touching in places, but leaving between them channels, known as *inhalant* (or *afferent*) *canals*, whose openings on the surface of the sponge are often narrowed and are known as *ostia*. The thimble-shaped chambers are known as *flagellated chambers*, and are lined by choanocytes, but these are now lacking from the paragaster, where they are replaced by pinacocytes. Water enters by the ostia, passes along the inhalant canals and through the pores, now known as *prosopyles*, into the excurrent canals, leaves these through the openings, known as *apopyles*, by which they communicate with the paragaster, and flows outwards through the osculum. A third grade is found in sponges such as the calcareous sponge *Leucandra* (Fig. 74A), where the wall of the paragaster is folded a second time, so that the flagellated chambers, instead of opening direct into the paragaster, communicate with it by *exhalant* (or *efferent*) *canals* lined with pinacocytes.

The three grades of sponge structure (Fig. 73), in which successively the choanocytes line the whole paragaster, are restricted to flagellated chambers, or are still further removed by the presence of exhalant canals, are known as the 'ascon', 'sycon', and 'leucon' grades. In many of the sponges whose canal systems are of the third grade, the flagellated chambers are no longer thimble-shaped, but small and round. As the canal system has grown more intricate, complication has taken place also in the skeletogenous layer. It has grown thicker, forming outside the flagellated chambers a layer known as the *cortex*, in which the inhalant canals ramify; and there appear in it branched connective tissue cells which can change their shape.

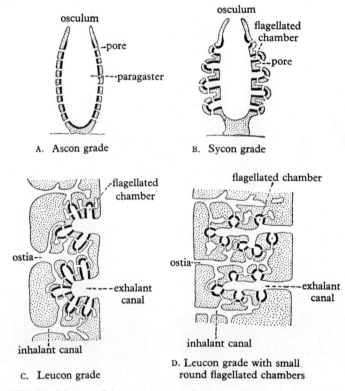

Fig. 73. Diagram of the canal systems of sponges. (Partly after Minchin.)

SKELETON. The sponges which we have so far considered have skeletons composed solely of calcareous spicules, and their choanocytes are relatively large. They constitute a comparatively small group, the class *Calcarea*. The majority of the phylum are without calcareous spicules and have relatively small choanocytes. They have usually siliceous spicules, of which there exist many different types (Fig. 76), characteristic of various groups of sponges, while minor differences distinguish those of the species, which are often only separable by this means. A horny substance, *spongin*, may occur as a cement uniting spicules, as fibres in which spicules are imbedded, or as a fibrous skeleton from which spicules are absent. The sponges in which the skeleton is in the latter condition constitute the horny sponges (Keratosa), of which the bath sponge (*Euspongia*, Fig. 74B) is an example. Foreign bodies (sand grains, etc.) are often imbedded in the spongin fibres. In a few cases (Myxospongiae) there is no skeleton. The choanocytes of non-calcareous sponges are always restricted to flagellated chambers. Almost without exception these are arranged as in calcareous sponges of the leucon type, and in most cases the

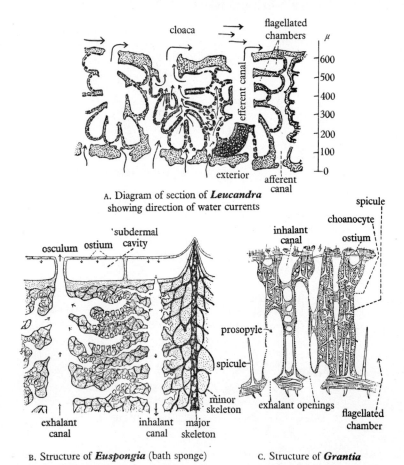

A. Diagram of section of **Leucandra** showing direction of water currents

B. Structure of **Euspongia** (bath sponge)

C. Structure of **Grantia**

Fig. 74. Sponge structure. (A, after Bidder; B, from Borradaile; C, from Dendy.)

system is made still more intricate by ramifications of the paragaster, the irregular appearance of numerous oscula, which put it into communication with the water at many points, and the appearance of 'subdermal cavities' and other complications in the outer part of the body.

The non-calcareous sponges fall into two very distinct classes—the *Hexactinellida*, in which there is always a siliceous skeleton of six-rayed spicules (Fig. 76F), the jelly is absent, and the flagellated chambers are thimble-shaped, as in the simpler sycons; and the *Demospongiae*, in which the skeleton, if present, does not contain six-rayed spicules of silica, jelly is present, and the flagellated chambers are almost invariably small and rounded (Fig. 75 C).

DEVELOPMENT. Sponges have free larvae, of several different kinds, but all covered, wholly or in part, with flagellate cells, by which they swim. The

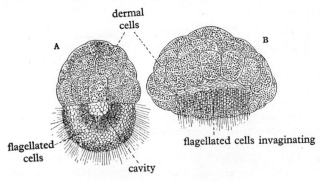

Amphiblastula of *Sycon*

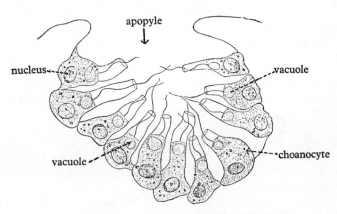

C. Flagellated chamber of *Spongilla*

Fig. 75. Development of the sponge: A, larva (amphiblastula) of *Sycon raphanus*. B, the same with flagellated cells invaginating. C, section of the flagellated chamber of *Spongilla lacustris*. (A, B after Schultze; C from Vosmaer.)

remarkable feature of the metamorphoses by which these larvae become the fixed adults is that the flagellated cells pass into the interior, develop collars, and become the choanocytes (Fig. 75).

Asexual reproduction is found throughout the group. It takes place by the outgrowth and separation of external buds, or by the formation of internal buds or *gemmules*, enclosed in stout coats. In some cases (Spongillidae) the gemmules are remarkable in that they originate as clumps of the amoeboid cells of the parent. They will stand freezing or drought, and carry the species through unfavourable conditions. The power of regeneration and repair is possessed by sponges in a high degree, and they can be propagated artificially by cuttings.

DISTRIBUTION. Sponges are found in all parts and at all depths of the sea. Only one family, the Spongillidae, occurs in fresh water, but its members are plentiful and widespread.

AFFINITIES. The affinities, and therefore the systematic position, of the phylum Porifera have been the subject of much dispute. In that their bodies consist of many 'cells', they might seem to be Metazoa. But they differ from all members of that group in several important respects. In no metazoon are choanocytes found. In none is the principal opening exhalant. In none is there during development an inversion whereby a flagellated outer covering becomes internal. Lastly, and perhaps most significantly, in a sponge the 'cells' are far less specialized and dependent upon one another than the cells of a metazoon. Many of them can assume various forms, becoming amoeboid, collared, etc. Many are isolated in the jelly, and when they touch they are often not continuous. No nervous system co-ordinates their activities. Even the choanocytes, though the sum of their efforts produces a current, do not keep time in their working. In short, the Porifera are practically colonies of Protozoa. Moreover, it would seem that they took origin from choanoflagellate Mastigophora. Now opinion is, as we have seen, not unanimous that the Metazoa arose as colonies of Protozoa, and in any case it is unlikely that they sprang from choanoflagellates. Thus the sponges, in spite of certain superficial resemblances to the Metazoa, have no real similarity to, and probably no genetic affinity with, that subkingdom. For this reason it is best that, in a classification of animals, they should be given, under the name of Parazoa, the same rank as the Protozoa and the Metazoa.

CLASS 1. CALCAREA

Sponges with skeletons consisting solely of calcareous spicules; and with large choanocytes.

Clathrina. A meshwork of ascon tubes. The nuclei of the choanocytes are at the bases of the cells. British.

Leucosolenia. A clump of erect ascon tubes, each of which may be branched, connected at their bases. The nuclei of the choanocytes are apical. British.

Sycon (Fig. 77B). A simple vase with a canal system of the second type, having the thimble-shaped outgrowths little adherent to one another. The nuclei of the choanocytes are apical. British.

Grantia. Differs from *Sycon* in that the outgrowths which contain the flagellated chambers adhere in many places and are covered by a cortex (Fig. 74c). British.

Leucandra. Canal system of the third type (Fig. 74A). Nuclei of choanocytes basal. British.

CALCAREA

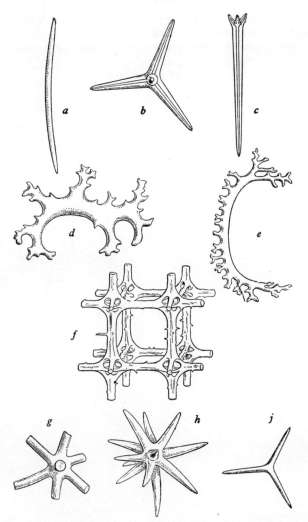

Fig. 76. Various types of sponge spicules: *a–e*, from Demospongiae; *f*, from a hexactinellid; *g–h* from extinct groups. (From Woods.)

CLASS 2. HEXACTINELLIDA

Sponges with a purely siliceous skeleton composed of six-rayed spicules; with small choanocytes and thimble-shaped flagellated chambers; and without jelly, the soft parts of the body being united solely by a meshwork of trabeculae furnished by branching cells of the dermal layer. A deep-sea group.

Euplectella, Venus' flower basket, and *Hyalonema*, the glass-rope sponge, have both been dredged in British waters. Both harbour various commensal crustaceans. On the rooting-tuft of long, fine spicules, which is the 'glass-rope' of *Hyalonema,* grows an epizoic anemone of the genus *Epizoanthus.*

CLASS 3. DEMOSPONGIAE

Sponges whose skeleton, if present, does not contain six-rayed spicules of silica, and may be purely siliceous, or composed of silica and spongin, or of spongin alone; whose flagellated chambers have small choanocytes and are usually small and rounded; and which possess jelly.

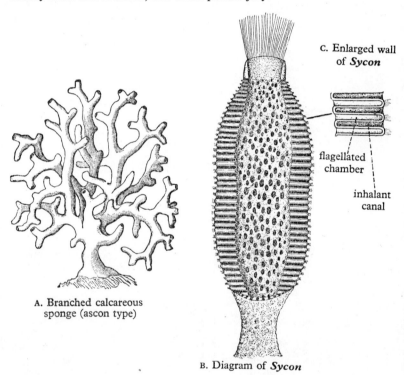

A. Branched calcareous sponge (ascon type)

B. Diagram of *Sycon*

C. Enlarged wall of *Sycon*

flagellated chamber

inhalant canal

Fig. 77. Sponge structure. (A, from Sedgwick, after Haeckel.)

1. *Order* **Monaxonida**
With monaxons.

Cliona. A cosmopolitan genus, which bores into the shells of molluscs and into calcareous rocks.

Halichondria, the crumb-of-bread sponge. A common British littoral form, usually of encrusting growth.

Spongilla. A member of the family of fresh-water sponges mentioned on p. 122. Cosmopolitan.

2. *Order* **Keratosa**
   With spongin skeleton.

   Euspongia, the bath sponge. Medit., W. Indies, etc.

   Hippospongia. A sponge of the same kind with a coarser texture due to the inclusion of much foreign matter in its skeleton.

3. *Order* **Myxospongiae**
   Without skeleton.

   Oscarella. British, has no skeleton.

4. *Order* **Tetractinellida**
   With tetraxon spicules (Fig. 76*b*).

   Plakina.

CHAPTER IV

THE SUBKINGDOM METAZOA

The fundamental difference in histology which distinguishes the Metazoa from the Protozoa has already been described in ch. II. Something must here be said concerning the main features of the organization of the Metazoa.

DIPLOBLASTIC STAGE: THE TISSUE GRADE. The simplest type of bodily architecture in this subkingdom is that with which the student is familiar in *Hydra*, where the body consists of a sac with one opening, and with the wall composed of two cellular layers and a layer of secreted jelly between them. The inner layer is the *endoderm*. It consists of cells specialized for the processes of digestion, and the cavity which it lines is for the reception of food. The outer layer is the *ectoderm*: by its cells relations with the environment are regulated. Some of these cells form a protective and retaining sheet; among them stand others which are sensitive; others—nerve-cells—lying below the sheet, are branched so as to serve for the transmission in various directions of the stimuli received by the sense cells: together they form a *nerve-net*. At the base of both ectoderm and endoderm there lie muscle fibres—which in *Hydra* are elongate contractile processes of the retaining cells but in other animals of this type are often whole cells that have left the surface. Lastly, from certain undifferentiated cells at the base of the ectoderm there are formed the generative cells.

PROTOZOA AND METAZOA. When we compare this organization with that of a protozoon we observe that the cellular structure of the metazoon, primarily, perhaps, necessitated by its size (p. 11), has the following result: by isolating the units specialized for the performance of particular functions it (1) removes most of them from the direct action of the outer world, (2) makes it possible that groups of them should constitute independent organs, and (3) enables the relations of such organs, both with the environment and with one another, to be regulated by intervening cells and internal media. Already in the simple case we have examined these facts are turned to advantage. Under the protection of the layer which remains in contact with the outer world there are established a special organ of digestion and a system for distributing stimuli which are received by distinct units on the surfaces. Other elements (muscular, genital) are beginning to separate. In the following pages we shall see this process of separation and differentiation carried much further. Its result is that the activities of the organism are less and less liable to interference from or suppression by the environment, either through the unregulated exchange of substances or by unregulated stimuli. We shall see, also, how the machinery which is fashioned in this way varies in correspondence with the environment.

TRIPLOBLASTICA—THE ORGAN GRADE. In the phylum to which *Hydra* belongs, the Coelenterata, the body is always of the type just described, whatever form the sac or its layers may assume, though the jelly may contain cells, sometimes plentiful, of various kinds—muscle fibres, skeleton forming cells, and amoeboid corpuscles—which have migrated into it from the ectoderm or endoderm. In all other metazoan phyla there is between ectoderm and endoderm a third layer, the *mesoderm*, which usually is more bulky than either of the other layers and forms the greater part of the body. The phyla which possess this layer are known as *Triploblastica*—three-layered animals—while the Coelenterata are *Diploblastica*. It is true that the mesoderm is partly foreshadowed by the cells which are present in the jelly of many coelenterates, but mesoderm is more plentiful than the cells in the jelly generally are, it contains important organs and usually definite systems of spaces (see p. 129) and its rudiment appears very early in the development of the individual.

EARLY EMBRYOLOGY. Every triploblastic animal, however, passes through a stage—the *gastrula*—in which it consists only of ectoderm and endoderm. Save in this essential feature, the gastrulae of different animals may be extraordinarily unlike, and, especially when the animal is developed from a very yolky egg, they are sometimes very difficult to recognize as such; but where the gastrula is well formed, as in the familiar development of *Amphioxus* or in that of a starfish (Fig. 464), its two-layered wall may always be found to contain a cavity, the *archenteron*, which possesses a single opening, the *blastopore*. The ectoderm and endoderm are separated by a space, which is often a mere crack, but may be much wider, and contains a fluid or a slight jelly. This space is known as the *blastocoele*, and when, as in the cases cited above, the gastrula arises by the dimpling-in (invagination) of the wall of a one-layered hollow vesicle or *blastula*, the blastocoele begins as the cavity of the blastula.

ORIGIN OF MESODERM. The mesoderm, whose appearance converts the gastrula into a triploblastic body, is not a single entity, but contains components which originate in two different ways, namely:

(1) Cells which migrate from ectoderm or endoderm, or from mesoderm of type 2, into the blastocoele; this kind of mesoderm (Fig. 464) is known as *mesenchyme*, and is comparable to the cells which invade the jelly of coelenterates.

(2) Cells which constitute the wall of the cavity known as the coelom. This kind of mesoderm is called *mesothelium*. In some cases, as in *Amphioxus*, the starfish, *Sagitta*, and the Brachiopoda (Figs. 494, 464, 460, 459), it arises as pouches of the archenteron which separate from the latter, their cavity becoming the coelom and their wall the mesothelium. In other cases it arises as solid outgrowths or layers shed off from the wall of the archenteron, and coelomic cavities afterwards appear in it. This happens, for instance, in the tadpole. In yet other cases a single *pole cell* or *teloblast*, as in annelids (Fig. 194) and molluscs, or a group of a few cells, as in arthropods, separate, on each side of the embryo, from the rudiment of the endoderm, and multiply so as to form a band of cells in which coelomic cavities appear. A coelom which arises as a pouch from the archenteron is known as an *enterocoele*; one which arises in a mass of mesothelium is a *schizocoele*.

In the lower triploblastic phyla (Platyhelminthes, Nemertea, Nematoda, etc., p. 195) there is no mesothelium. Chaetognatha have no mesenchyme. In most phyla, both kinds of mesoderm develop.

We must now consider the organs formed by each of the three layers.

(1) ENDODERMAL ORGANS

After giving rise to mesoderm, the archenteron becomes the rudiment of the alimentary canal. Except in Platyhelminthes, its blastopore is in various ways replaced by two openings, so that it has both mouth and anus.

The most primitive way is probably that of *Peripatus* (Fig. 229), in which the middle of the blastopore closes and the ends become mouth and anus.

The wall of the archenteron, the endoderm, forms the lining of the alimentary canal, except in those regions, known as *fore gut* or *stomodaeum* and *hind gut* or *proctodaeum*, which are formed by a tucking-in of the ectoderm at the mouth and anus. The endoderm also gives rise to the various diverticula of the mid gut, such as the liver and other digestive glands, the lungs of vertebrata, etc. A true stomach is an enlargement of the mid gut.

Digestion was perhaps originally entirely *intracellular* in the endoderm cells, and many of the lower animals still have intracellular digestion, though this is usually preceded by an *extracellular* process which by dissolving certain components of the food enables the remainder to be reduced to particles small enough to be taken up by the cells. In the annelids, arthropods (except certain ticks, p. 570), cuttlefishes, and Chordata, digestion is entirely extracellular. The enzymes secreted vary with the food: in carnivorous animals such as cephalopods and starfishes they are principally proteases, in feeders on vegetable tissues they are largely carbohydrases, in omnivores such as the crayfish and cockroach and holothurians they are adapted to deal with all classes of food-stuffs. Considering the importance of cellulose both as a potential foodstuff and as cell walls which enclose more valuable foods, it is remarkable that cellulases should be rare (pp. 443, 608, 635), though those few cases demonstrating cellulases have usually had associated bacteria, capable themselves of secreting cellulases.

Both intracellular ingestion and absorption are not always confined to the alimentary canal proper but may take place in digestive glands or 'livers', as for instance in those of the mussel, the snail, and the crayfish, but not in those of cuttlefishes or vertebrates. It is said that in various bivalve molluscs and in holothurians amoeboid corpuscles pass through the endoderm, take up particles in the gut, digest them, and, returning, distribute the products. The presence of a cuticle in the ectodermal portions of an alimentary canal does not always prevent absorption there (e.g. in the fore gut of some insects). Finally it should be noted that some animals perform a part of their digestion *externally to the body*, as the starfish by extruding its stomach (p. 682), and various insects, mites, earthworms, etc., by pouring out saliva; and that in other cases bacterial or protozoan symbionts (pp. 70, 110) play a part in the digestion of food—particularly of celluloses—in the gut.

The food of all animals contains amino acids, usually as protein, for the manufacture of the proteins needed in the repair and growth of protoplasm. Much amino acid, however, is *deaminated,* the carbonaceous residue being oxidized, together with the carbohydrate and fat which the food usually also contains, for the liberation of energy, and the ammonia excreted in various forms by various organs presently to be mentioned.

(2) MESODERMAL ORGANS

Since mesothelium gives rise to mesenchyme, it is often difficult to distinguish between the two and to decide what part each plays in the formation of organs; but broadly speaking it can be said that the connective and endoskeletal, the vascular, and some muscular tissues arise from mesenchyme, while in coelomata the peritoneum and the organs derived from it—gonads (ovaries and testes), mesodermal kidneys, etc.—and the principal muscles arise from mesothelium.

Within the massive layer of mesoderm, cavities are necessary for sundry purposes. Channels must be provided for the transport of various materials—the products of the digestion of food, the gases of respiration, water, the waste products of metabolism, which are usually eliminated with the excess of water, and the substances known as hormones which are secreted by certain organs as messengers to regulate the activity of others. The germ cells, which are sheltered in this layer, must be given access to the exterior. Often there must also be spaces to give play to movements of the viscera. Such facilities are provided by two systems of cavities, the *primary and secondary body cavities*, of which either or both may be present.

Primary Body Cavity

The primary body cavity, sometimes known as the *haemocoele*, is to be regarded, morphologically, as representing that part of the blastocoele which is not obliterated by the mesenchyme cells or by a solid matrix or fibres secreted by them. Its fluid contents, containing free mesenchyme cells ('corpuscles'), are the blood and lymph, and it has usually the form of a branching system of vessels ('vascular system') through which the fluid is caused to circulate by the contraction of muscular fibres in the wall of some portion of it which is known as a *heart*. In some cases, however, the haemocoele forms large 'perivisceral' sinuses around the internal organs. It never contains germ cells or communicates with the exterior.

Since the haemocoele fluid is in intimate relation with the tissue, its composition is a matter of very great importance to the animal. It bears to the tissues much the same relation that the external medium bears to the body as a whole and is on that account often spoken of as an *internal medium*. If it be changed the working of the organism is influenced. It is liable to be fouled by poisonous waste products of metabolism and these must be removed from it and excreted or so changed as to be harmless. It is liable to alteration by diffusion between it and the external medium, and in proportion as this can take place the animal will be at the mercy of its surroundings. To maintain it in a constant condition in respect of the substances which it might exchange with a particular external medium two agencies are at work—the guardianship, active or passive, of the protective sheet of ectoderm and of any cuticle or other covering which the latter may secrete, and the activity of the excretory organs, especially in the excretion of water. The effectiveness of these agencies varies. The independence of the body fluids from the external medium is least in some marine animals, such as echinoderms and certain molluscs: in these the fluids closely resemble sea water both in the ions present and in the total osmotic pressure. In a series of others, independence grows, and it is highest, in the sea, in teleostean fishes. In fresh-water animals the composition of the blood is kept entirely different from that of the external medium. In land animals there is of course no question of the exchange of solutes, and unless the loss of water were reduced to a minimum life would be impossible. It is an interesting fact that, though the resemblance of the body fluids in fresh-water and land animals to sea water is much less than that of marine animals, something of it still remains, no doubt because protoplasm came into being in sea water and still requires to be bathed by a fluid which somewhat resembles the latter. The principal differences are an increase in potassium and a decrease in magnesium and SO_4 ions and a lower total osmotic pressure.

RESPIRATION. The blood is the principal means of transport within the body. A very important part of its freight is oxygen. Its capacity for this gas, however, would be quite insufficient for it to maintain the metabolism of an active animal if the gas were carried in mere solution. This deficiency is met, when necessary, by the presence in the blood of *respiratory pigments*. These bodies are compounds of a protein with a nitrogenous pigment which contains a metal. They are related to one another, to chlorophyll, and to the colourless substance *cytochrome* which is very widely distributed in the protoplasm of animals and plants, where it plays a part in bringing about oxidations. They form very labile addition compounds with oxygen, which they can thus take up in the organs of respiration and carry to the tissues, where they yield it up by dissociating under the lower oxygen tension, undergoing at the same time a change in colour. The most important of them are *haemoglobin*, which contains iron and is red, *chlorocruorin*, also containing iron, which is green, and *haemocyanin*, containing copper, which is blue when oxygenated. Haemoglobin is present in Vertebrata, where it is carried in the 'red corpuscles', and sporadically in many invertebrates, as in the earthworm, where it is in solution in the plasma. Chlorocruorin is found in solution in the blood of various polychaete worms, haemocyanin in solution in the blood of the higher crustacea, the king-crab (*Limulus*), and various molluscs. Both haemoglobin and haemocyanin are slightly different compounds in different animals, and with these differences are associated differences in the pressure at which they take up or yield oxygen. Broadly speaking, the blood pigments of animals which live under conditions of low oxygen pressure take up the gas at a lower pressure than those which live under high oxygen pressure. On the other hand they do not maintain so high a pressure in the tissues. Independently of such differences, the haemocyanins are less efficient oxygen carriers than the haemoglobins. In tracheate arthropods, where air is brought direct to the tissues by a system of tubes, there are no blood pigments.

COLLOID OSMOTIC PRESSURE. The blood of the higher invertebrates contains in solution a considerable amount of protein, of which the respiratory pigment, if present, is only a part. This protein is comparable with the organic ground substance of a skeletal tissue. It is not a food for the tissues but by maintaining the osmotic pressure of the blood it is of importance in regulating the distribution of water between that fluid and the tissues, and, since proteins combine with both acids and alkalis, it helps to neutralize excess of either of these. In vertebrates some of this protein provides the material for clotting, by which loss of blood or injury is prevented; but invertebrates, when they form a clot, do so from material furnished by corpuscles.

Secondary body cavities

One of the secondary body cavities, the coelom, is from the first completely surrounded and separated from the blastocoele by the mesothelium, which is derived from the endoderm (Fig. 1). This cavity has various forms, but is rarely tubular and never possesses a heart. Usually it constitutes one or more large perivisceral spaces around the heart, alimentary canal, and other organs. It will be noted that the *perivisceral cavity* which surrounds the internal organs of most triploblastic animals, so that these organs are unaffected by the movements of the body wall and are able freely to perform movements of their own, may be either coelomic or haemocoelic, but is usually coelomic. In the Arthropoda, where the perivisceral function of the coelom is entirely usurped by the haemocoele (Fig. 219), the former space is reduced to small cavities in the gonads and excretory organs.

In animals which possess a coelom, the gonads are derived from its walls, and either the germ cells are shed into a coelomic perivisceral cavity or the gonad itself contains a cavity which is a separated portion of the coelom.

COELOMIC DUCTS. The coelom communicates with the exterior. The communication is usually made through organs belonging to one or other of the types known as 'nephridia' and 'coelomoducts', though it occasionally takes place through openings of other kinds, such as the dorsal pores of the earthworm and the abdominal pores of fishes.

Nephridia and coelomoducts are organs which meet the need for the passage to the exterior of products of organs derived from or imbedded in the mesoderm. Their characteristic features are as follows:

The *nephridial system* is primarily an organ which serves the mesenchyme, though it may come to lie in the coelom, and in certain annelids communicates with that space. It is for the most part intracellular, and consists of tubes, of ectodermal origin, usually branched and bearing at the end of each branch a *solenocyte* or *flame cell* (see p. 200). It may be continuous or divided into segmental units, the *nephridia*. Water, probably containing excreta, is shed by the protoplasm of the tubes, and passes out in the current set up by the action of the flame cells or by cilia.

Coelomoducts are mesodermal passages which open at one end to the exterior and at the other usually into the coelom, though the coelomic opening may lead only into a minute vesicle of the coelom, or even be lost altogether. They may (1) be solely excretory, the excreta being shed into them by gland cells in their walls, or borne into them by a current of fluid from the coelom through the coelomic opening of the organ, or derived from both these sources (see p. 138); (2) combine excretion with the function of conducting the germ cells to the exterior; (3) be simply gonoducts, which was perhaps their original function.

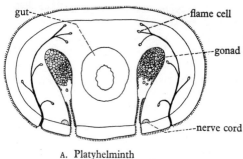

A. Platyhelminth

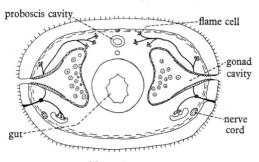

B. Nemertine

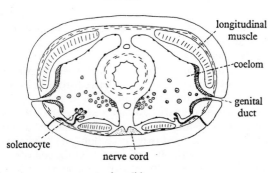

C. Annelid

Fig. 78. Origin of the coelom. One view is that the coelom is an enlarged gonad cavity; three possible stages in the development of the coelom from a gonad cavity are shown here. (After Goodrich.)

Many annelida possess compound excretory organs formed by the union in various ways of nephridia with coelomoducts or other mesodermal elements (see p. 285). In such cases the nephridia acquire a communication with the coelom, and excreta or germ cells may pass from it through them. In other groups, as in some Crustacea, a coelomoduct is supplemented or in great part replaced by an ectodermal component.

(3) ECTODERMAL ORGANS

The ectoderm gives rise to the epidermis (epithelium which covers the body), to certain glands, to the nephridia, to the principal external organs of sense, and to the nervous system (in nearly all cases; there are nerve cells under the endoderm of certain coelenterates, and a part of the nervous system of the Echinodermata is remarkable in being formed from the peritoneum and therefore mesodermal).

The *epidermis* with some underlying mesodermal connective tissue known as the *dermis* constitutes the *skin*. In invertebrates it is columnar or syncytial, in vertebrates it is stratified. In the lower invertebrates its cells are usually ciliated, which was probably the original condition. The cilia subserve locomotion, the taking of food, or respiration (p. 135). When unciliated its protective function is often increased by a *cuticle*. To it belong various glands, especially, in naked epithelia, mucous glands whose secretion is protective, in aquatic animals against parasites, in terrestrial against desiccation. Others form cuticular structures, cement, poisons, etc.

The *nervous system* was no doubt primitively situated immediately below epithelia, having arisen by specialization of epithelial cells for the transmission of impulses due to stimuli received upon the surface—for the most part, presumably, upon the ectoderm. In many cases (the Coelenterata, Echinodermata, Hemichorda, some annelids, etc.) it remains there, but usually it is in a deeper, more protected situation. All Triploblastica possess a *central nervous system*. This arose as a condensation of the primitive nerve-net of branched cells, portions of which may remain unchanged. The central nervous system was formed in different positions in different animals. In those which have a long axis it has the form of cords along that axis. The cords may be paired or unpaired, lateral, ventral, or dorsal. Anteriorly they pass into an enlargement, the 'brain' or cerebral ganglion, connected with the principal organs of distant sense. In Chordata the central nervous system is hollow, its removal from the surface being not, as usual, by separation from the epithelium, but by the folding-in of the strip of epithelium which it adjoins and which remains to line its cavity. A similar condition is seen in some echinoderms. From the central nervous system *nerves* proceed to various parts of the body.

The nerve-net is joined by processes from the bases of the sense cells. Probably at an early stage in the evolution of the Metazoa stimuli were transmitted only by such processes, running directly from the sense cells (*receptor* cells) to end against the muscle or other cells which are set in action through them (*effector* cells). This condition, however, is now rare, occurring only in the tentacles of some coelenterates: nearly always there are nerve-cells which have left the surface layer, whose processes continue those of the sense cells and so extend and complicate the system of communications. In the primitive condition of the nervous system, as seen, for instance, in *Hydra*, the nerve-cells have numerous similar branches, forming a network over which messages pass in all directions from any point of stimulation. That there is co-ordination in the action which results is due only to the fact that the messages do not evoke responses equally in all the effector cells.

UNITS OF THE NERVOUS SYSTEM. The condensation of nerves and a central nervous system out of this network is due to a change in the form and arrangement of the elements of which it is composed. The change, which is already foreshadowed in certain parts of the bodies of coelenterates (p. 150), consists in processes of the nerve cells elongating in particular directions and thus forming paths of conduction which in the higher triploblastica are isolated by the loss of the rest of the network. As this system is perfected its elements become *neurones*—cells with one main process, the *axon* or *nerve fibre*, along which the impulse passes from the cell body. The axon ends by breaking up into a tuft of branches, the *terminal arborization*, from which a stimulus is given either to another nerve cell or to an effector cell. Thus instead of spreading in all directions the impulse is conducted to a definite destination: the interference of the environment in the affairs of the organism is regulated. The cell body may be a sense cell in the epithelium, or it may be internal. In the latter case it possesses other processes—the *dendrons*—which by fine branches—the *dendrites*—receive stimuli from the axons of other neurones. Two neurones at least are concerned in the transmission of an impulse. In the simplest case the axon of a sense cell (or of a cell whose dendrites receive stimuli from a sense cell) transmits the impulse to a neurone whose axon conducts it to the effector cell. This process is known as a *reflex* and the arrangement of neurones as a *reflex arc*. The impulse is passed from the first neurone to the second in an exchange station—the central nervous system. The nerve fibres run to and from this station in bundles which are the nerves. This arrangement is not only, as we have seen, more precise but, since one efferent neurone can serve several afferent, more economical of fibres than the nerve-net. Usually, moreover, the system is complicated, and in the highest animals it is enormously complicated, by the branching of axons, which increases the number of efferent fibres an afferent fibre can affect, and by the intervention, in the central nervous system, of intermediate neurones between those which are directly afferent and efferent. By this the number of afferent fibres which an efferent fibre can serve is increased. For the efficient working of this system it is essential that impulses should pass over it in one direction only, and thus should not leak from one path to another and affect organs for which they were not destined. That is provided for in the following way. Where the terminal branches of an axon meet the dendrites of another neurone the two are not continuous but interlace without joining, making what is called a *synapse*. Their discontinuity makes an obstacle, over which impulses can pass in one direction only, from axon to dendrites. The mode of passage of impulses from the one to the other and again from efferent neurone to effector cell is not at present known. It is perhaps an electrical process, but it involves the production of chemicals that probably affect the sensitivity of the recipient cell. In that case the transmission of a nervous impulse includes a process which recalls that other mode of communication, mentioned above, in which the chemical messengers known as hormones are distributed through the vascular system.

INHIBITION. Stimuli received from the nervous system, like other stimuli, may inhibit as well as cause activity. This is very important, because when an action is to be performed activity which hinders it must be abolished. Thus, for instance, a contracting muscle may by a reflex inhibit contraction in an opposing muscle: the circular and longitudinal muscles of the earthworm are an example of such a system (p. 270). A similar end is obtained in a different way in the muscles which open and close the claws of crabs and lobsters, where each muscle fibre has two nerve fibres, one excitatory and the other inhibitory, and impulses from the central nervous system pass simultaneously to the excitatory fibres of one muscle and the inhibitory fibres of the other. Further, one neurone may inhibit another, and thus inhibition may be effected not only through but in the central nervous system.

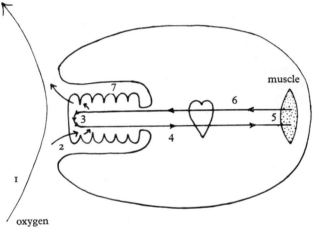

Fig. 79. Diagram of the process of gas transfer to and from the tissues. $1 = O_2$ carried to gill; $2 = O_2$ diffusing through gill; $3 = O_2$ combining with blood; $4 = O_2$ carried to tissue; $5 = O_2$ transferred to tissue; $6 = CO_2$ carried to gill; $7 = CO_2$ liberated. Note the 'principle of counterflow', i.e. the way in which the blood in the gills (3) flows in the opposite direction to the respiratory current (1).

Concerning the way in which the central nervous system, which in the lowest animals that possess it is merely a relay station where impulses from the principal sense organs are multiplied and distributed, later takes on inhibitory functions, and later still develops 'functional units' for co-ordination, and concerning the control of the latter by the brain, something is said on pp. 197, 270, 458.

RESPIRATORY ORGANS. (Fig. 79). Certain organs are formed in different animals from different layers. *Organs of respiration* may be covered or lined by ectoderm, as are the gills of crustaceans and annelids, the external gills of the tadpole, and the lungs of snails; or by endoderm, as are the gills of a fish or the lungs of vertebrates. The skin, when it is naked or covered only by thin cuticle, is always respiratory, and in many small animals is the only organ of respiration. Cilia may keep water in movement over it so as to renew the supply of oxygen, and when there is a vascular system a rich blood-plexus may increase the efficiency of the skin, as in earthworms. In larger animals there are usually localized organs of respiration. In these the surface is increased by folding or branching either outwards or inwards, the blood supply is richer than elsewhere, and there is some means of constantly renewing the medium. Organs of aquatic respiration are usually projections, known as gills: the water around them is renewed, either by muscular movements of the body, of limbs which bear the gills or of structures in their neighbourhood, or by ciliary action. Organs of aerial respiration may be those that were originally used for aquatic respiration: this is especially the case when they are enclosed in a chamber which protects them; sometimes, as in snails and land crabs, such a chamber becomes itself converted into a respiratory organ by the vascularizing of its lining. In other cases, as in terrestrial vertebrates, there are developed for this function cavities in the body into which air is drawn. In order to prevent damage to the epithelium by desiccation, a layer of moisture is always maintained over air-breathing surfaces, either by exudation or by special glands or by water-retaining hairs, etc. Consequently aerial respiration is in the long run aquatic respiration, with the difference that the supply of oxygen in the layer of water over the respiratory epithelium is maintained not by renewal of the layer but by diffusion from adjacent air. It is maintained that on that account aerial respiration is less efficient than aquatic, and this argument is supported by the fact that the respiratory area of air-breathing animals is greater than that of related forms which have aquatic respiration.

What has been said in the foregoing paragraph does not apply to the tracheate arthropods, whose respiration does not take place through an epithelium with the intermediation of the blood, but by the bringing of air directly to the tissues by a system of fine ectodermal tubes—the tracheal system (p. 448).

The relative effect upon respiration of the pressures of carbon dioxide and of oxygen differs according as the animal is living in water or in air. In clean waters the pressure of carbon dioxide varies little, because any excess of the gas is removed by the formation of carbonates and bicarbonates, but the pressure of oxygen is easily lowered as the amount in solution is used up. In air, on the other hand, free carbon dioxide will accumulate, but there is a large supply of oxygen. Consequently aquatic respiration will sooner be affected by changes in the pressure of oxygen in the medium, aerial respiration by changes in the pressure of carbon dioxide. In foul waters, however, free carbon dioxide may be present in such quantities as to be an important factor.

METABOLISM. In foul waters, and in the habitat of many internal parasites, free oxygen may be practically absent. Many animals which live in such circumstances obtain energy by an anaerobic process, the complex molecules of carbohydrates being decomposed to form simpler ones without the intervention of free oxygen. To this end the animals in question lay up in their tissues large quantities of the starch-like substance glycogen of which carbohydrate stores in animals are usually composed. Thirty per cent of the dry weight of an *Ascaris*, and nearly half that of a tape-worm, consist of this substance. The glycogen is converted into glucose (dextrose) and then decomposed, according to the following equations:

(i) $(C_6H_{10}O_5)n + nH_2O = nC_6H_{12}O_6$

(ii) $C_6H_{12}O_6 = 2C_3H_6O_3$
(Glucose) (Lactic acid)

This process of course yields considerably less energy than would be obtained by total oxidation. Apparently, at least in many animals, it cannot go on indefinitely unless the lactic acid be removed. This may happen either by the acid being discharged into the surrounding fluid and swept away by movements of the latter (which would occur, for instance, in the host's intestine), or by an access of oxygen with which some of the lactic acid is oxidized so as to give energy for building the rest back into glycogen. In the latter case the

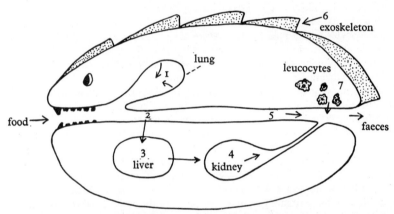

Fig. 80. Diagram of excretion. $1 = CO_2$ and H_2O; $2 =$ selective secretion (bile) and absorption; $3 =$ deamination; $4 =$ secretion and filtration; $5 =$ faeces + bacteria; $6 =$ storage; $7 =$ leucocytes.

process becomes ultimately aerobic. It is probable, indeed, that even in aerobic animals the process by which energy is liberated is at first anaerobic, but that this phase is quickly followed by one in which, by the use of oxygen, a part of the product is destroyed and the rest built back, so that the process as a whole appears aerobic. Thus anaerobic animals differ from those that are aerobic only in the length of time for which the anaerobic process goes on.

EXCRETION. Organs of excretion are even more various, in kind and in origin, than those of respiration (Fig. 80). If the removal of carbon dioxide from the body be disregarded, there are two processes to be considered here—the excretion of water and that of solids. In the lower aquatic animals, whose surface is in various degrees permeable to water, the removal of the latter from the body is, as we have seen, a matter of very great importance: it was probably the original function of the nephridial system and is an essential part of that of excretory coelomoducts. But the removal of solids—both of solutes which have entered from without and of the nitrogenous products of metabolism—is also essential, and in most animals advantage is taken of the outgoing water to remove the solids. It is to provide for this as well as to meet the loss due to evaporation, that terrestrial animals must take in water by the mouth.

EXCRETORY ORGANS. In coelenterates excretion probably takes place from the general surface of the ectoderm, and perhaps also from the endoderm. In triploblastic animals without a perivisceral cavity ectodermal ingrowths—the nephridia, mentioned above—permeate the mesenchyme and perform excretion. In the Nematoda there are lateral ducts in the ectoderm, which probably subserve excretion. In animals with a coelomic perivisceral cavity excreta are shed into the cavity (or carried into it by such cells as the 'yellow cells' of the earthworm); and removed to the exterior by nephridia (which may, as in the earthworm, open to the coelom), by the mesodermal coelomoducts, or in other ways, as in echinoderms, which shed excreta from the coelomic fluid through the respiratory organs (gills, respiratory trees). In echinoderms also solid excreta, perhaps not nitrogenous, are removed by amoebocytes which pass to the exterior through the gills. In the Vertebrata the excretory portions of the coelom are in the adult separated, and imbedded as the Malpighian corpuscles in the mass of coelomoducts which forms the kidney. During its passage along the nephridial tube or coelomoduct the fluid containing excreta receives additional substances secreted by the walls of the tube; and in the terrestrial vertebrates, in which it originates as an exudation filtered out under blood-pressure in the Malpighian capsules, water and some of its solid contents are regained by absorption from it. In the Arthropoda, where the perivisceral cavity is haemocoelic, the excretory organs are still often coelomoducts (segmental organs of *Peripatus*, antennal and maxillary glands of crustaceans, coxal glands of arachnids): attached to or imbedded in these are vestiges of the coelom (end-sac). Instead of, or in addition to, these organs, tubular diverticula of the ectodermal or endodermal parts of the alimentary canal often perform excretion in this phylum (Malpighian tubes, certain of the 'hepatic' caeca of crustaceans). In the insects the 'fatty body' contains a temporary or permanent deposit of excreta removed from the circulation. In ascidians excreta are similarly laid up as concretions by mesodermal cells. Various other organs which are known or supposed to have an excretory function will be mentioned in later chapters. The nitrogenous excreta vary in chemical constitution in different animals. Their variety appears to depend partly on the fact that the products of the decomposition of protein, ammonia compounds, are toxic and accordingly, unless they can be speedily discharged from the body, are converted into such substances as urea, guanin, and uric acid, which are relatively harmless. In aquatic animals, where plenty of water is available to carry off the excreta rapidly, the latter are principally ammonia compounds. In terrestrial animals it is necessary to expend energy in converting them into substances such as those mentioned above.

SKELETAL SYSTEMS. The only triploblastic animals which have a rigid skeleton of great importance are the Echinodermata and Vertebrata, in which it is internal and mesodermal, and the Arthropoda, in which it is a cuticle secreted by the ectoderm and is therefore primarily external, though ingrowths of it may form a kind of internal skeleton to which muscles are attached (Fig. 81).

The *muscular system* is in coelenterates derived from ectoderm or endoderm, in triploblastica almost entirely from mesoderm, though some muscles of crustacea arise from ectoderm. In Coelomata fibres from the mesenchyme form only minor muscular structures, such as the walls of blood vessels; the great masses of muscle are mesothelial. In the lower animals the fibres mostly lie parallel to the layers from which they arose, forming a sheet in the gut wall and another, known as the *dermomuscular tube*, under the skin. In these sheets

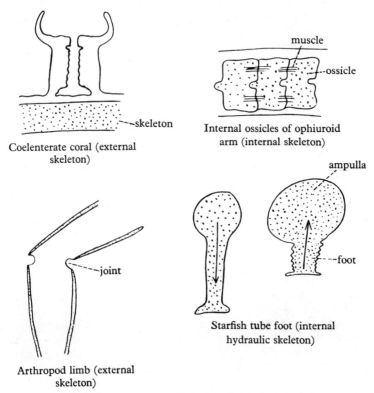

Fig. 81. Diagram of the different types of skeletal systems.

there is always a longitudinal and usually also a circular layer; sometimes diagonal fibres are added. The movements which such layers bring about are changes of size and shape of regions of the body or gut wall by the contraction of one set of fibres with relaxation of the other, their action being aided by the changes in turgor which are caused by the compression of the fluids they enclose. This fluid acts as a hydraulic skeleton. When there is a true skeleton muscular action is different. The dermomuscular tube is now broken up into muscles which pull upon pieces of the skeleton and so move parts of the body. When the skeleton is internal and the body wall remains flexible, more or less of the dermomuscular sheet remains. It is lost when there is a stiff cuticle. The muscles of limbs are provided by outgrowth from the dermomuscular layer.

MUSCLE. In the lower animals the muscular fibres are usually varieties of the unstriped kind. In other cases, chiefly in higher animals (vertebrates, *Amphioxus*, arthropods, part of the adductor muscle in the scallop, etc.), there appears a new type, the striped fibre, more swift and powerful in action but more dependent upon the nervous system; it has lost the power of automatic rhythmical contraction and of retaining without nervous stimuli a certain degree of contraction, known as 'tone'. Some striped fibres, however, retain one or other of these powers; thus the fibres of the heart of vertebrates contract automatically and those of the adductor of the claw of crabs and lobsters automatically maintain tone. Tone may also be maintained in striped fibres by the nervous system. In some cases (adductors of the scallop and of crustacean claws, spines of sea-urchins, etc.) a muscle contains two sets of fibres, one of which by rapid contraction brings an organ into a certain posture, in which it is held by tonic contraction of the other (the 'catch' fibres).

Most of the energy expended by an animal is liberated in its contractile tissues. It is obtained, normally from carbohydrates, by a process which, as we have seen (p. 137), is at first anaerobic and then aerobic.

GONADS. The rudiments of the gonads may be situated either in ectoderm, endoderm, or mesoderm. In Coelomata they always arise in mesothelium. However, since they are often recognizable as early in development as the layers, and the cells of which they are composed may migrate from one layer to another, and they do not form tissues, they are best regarded as an independent entity.

BODY PLAN. The body constituted by the elements described above has usually a bilateral *symmetry*, though this is rarely exhibited completely by all the systems. In the Coelenterata and Echinodermata, however, there is a radial symmetry. It is interesting to find that a sessile life, for which such symmetry seems particularly advantageous, is characteristic of the Coelenterata, and was probably adopted by the ancestors of all the Echinodermata. The terms *ventral* and *dorsal*, which belong by right respectively to those aspects of a bilateral animal which are normally turned to and from the ground or sub-stratum, are sometimes conveniently applied to a pair of structures by which two sides may be distinguished in the body of an animal whose symmetry is predominantly radial. They should, however, never be applied to the oral and aboral aspects of such an animal.

Meristic repetition of organs of the body is common in Metazoa. It may, as in parts of the body of annelids, affect practically all systems, so that there is a complete *segmentation* of the body into similar *somites*, or may be confined to certain organs. In the latter case it is important to distinguish between (1) the repetition of single organs in an unsegmented animal, as the ctenidia and shell plates are independently repeated in the mollusc *Chiton*, and (2) the condition, presented for instance by the Vertebrata and by much of the body of many arthropods, in which a formerly more complete segmentation now affects only some of the systems to which it at one time extended. The student should beware of thinking that the segmentation of all animals which present the phenomenon is derived from that of a common ancestor. The strobilation of the Cestoda in preparation for the detachment of reproductive units is a very different matter from the segmentation of the Annelida, and that again is far from being, as is sometimes assumed, certainly the same thing as the segmentation of the Vertebrata.

The anterior end of a bilateral animal is the site of the principal sense organs, of the 'brain', and usually also of the mouth, and is often obviously differentiated as a *head*. In a segmented animal this *cephalization* may extend to one or more of the anterior somites; and these usually become part of the head, losing their individuality in the way mentioned in the preceding paragraph, and only betraying their existence by the presence of certain of their organs (ganglia, appendages, etc.).

ONTOGENY, PHYLOGENY AND CAENOGENESIS. In the process of development by which the body peculiar to the species is reconstituted from the ovum, the early stages are of necessity much unlike the adult; but because the general features must arise before the more special ones, and because general features are shared by animals, the young resembles other young animals which have reached the same stage. Since the more special a feature is the fewer are the animals which share it, as the young approaches the adult form the circle of other animals whose young it resembles narrows. Since the evolution of its species consisted in the appearance of the same special features, its development (*ontogeny*) roughly recapitulates its evolution (*phylogeny*), but its features at any moment are not those of the adult of some ancestor but those of the corresponding young stage of that ancestor, and it is only because that stage was preparing the features of its own adult that there is recapitulation of the latter. Not all features of young animals, however, are anticipatory of those of their adults. Some of them—the embryonic membranes of the higher vertebrates and the ciliated bands of echinoderm larvae for instance—are adaptations to the needs of the young only and disappear in the adult. Such features are said to be *caenogenetic*. In respect of them development in no sense recapitulates adult phylogeny.

NEOTENY. Now it is held that in some cases a young animal, becoming sexually mature at an early stage (as in the well-known instance of the axolotl which may breed as a tadpole), has cut out permanently its later stages and started a new course of evolution from a young stage of an ancestor. This is known as *neoteny*, and in it caenogenetic features may be taken up into the new adult form: it may, for instance, account for some of the peculiarities of the Larvacea (p. 732), the Cladocera (p. 365), and *Leucifer* (p. 416).

EMBRYO AND LARVA. A young animal which is developing within an egg shell or in the womb of its mother is known as an *embryo*: one which is fending for itself is a *larva*. A stage which is larval in one animal has often become embryonic in another. Embryonic development is said to be 'direct'. Actually it is no more so than that which is larval. Caenogenetic features are found in both, those which are most conspicuous being in larvae organs of locomotion and feeding, in embryos the presence of yolk or means of obtaining from the mother the nutriment which the embryo cannot acquire from the outer world.

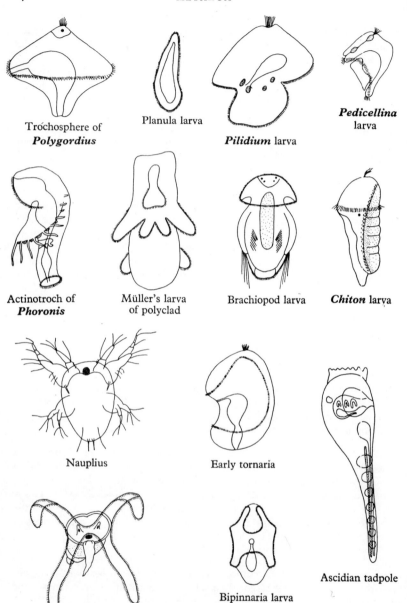

Fig. 82. Larval forms in the invertebrates.

EFFECTS OF YOLK. The yolk, which varies in amount in all phyla, is responsible for much of the difference between animals at corresponding stages, especially the youngest. The student will recall that from this cause the cleavage of the ovum which in *Amphioxus* is complete and equal, is in the frog complete but unequal and in the chick incomplete. The blastula has in *Amphioxus* a large cavity, in the frog a small one, in the crayfish (Fig. 223) is full of yolk, and in the chick is a disc upon the yolk. The mode in which the establishment of the two-layer stage (gastrulation) takes place is also partly affected by yolk, though evidently other factors are concerned. In the *Planula* (Fig. 87) it is by immigration, in *Amphioxus* and the crayfish by invagination, in the frog largely by overgrowth (epiboly), in the chick by delamination.

CLEAVAGE. All these modes of development are repeated sporadically in various groups of Metazoa: thus the early stages of the mollusc *Paludina* and the crustacean *Leucifer* are analogous to those of *Amphioxus*, those of the squid and the scorpion, members of the same phyla, to those of the chick. On the other hand, certain features of cleavage are constant through whole phyla and groups of phyla. The cleavage of coelenterates and echinoderms is radial, that of chordates is bilateral, that of polyclads, nemerteans, annelids, and molluscs is spiral (p. 291). Determinate cleavage, in which the part of the body to be formed by each blastomere is fixed from the first, as in the case described on p. 294, is common to the spirally cleaving groups but occurs in a quite different manner in the tunicates. Again, while the mesothelium of annelids, arthropods, and molluscs is laid down as a pair of ventral bands proliferated from behind, that of other coelomata arises from the wall of the definitive enteron (p. 127).

PLANKTONIC LARVAE. An important function of many larvae is the distribution of the species. This is often effected by their being planktonic. Among the larval types adapted to that existence is a series whose members have delicate tissues and a large blastocoele, whereby their buoyancy is increased, and strongly ciliated bands, often drawn out into processes, whereby swimming and feeding take place (Fig. 82). To this series belong Müller's larva, the *pilidium*, the trochospheres, the *actinotrocha*, the *dipleurulae*, and the *tornaria*. The student should beware of supposing that these types are phylogenetically related. With one or two exceptions, their resemblance is probably an instance of convergent adaptation.

CHAPTER V

THE PHYLUM COELENTERATA

DIAGNOSIS. Metazoa, either sedentary or free-swimming, with primarily radial structure; the body wall composed of two layers of cells, the *ectoderm* and *endoderm*, and between these a layer secreted by them which is originally a structureless lamella (*mesogloea*) but usually contains cells derived from the primary layers; within the body wall a single cavity, the *enteron*, corresponding to the archenteron of the gastrula, having a single opening for ingestion and egestion, and often complicated by the presence of partitions or by the formation of diverticula or canals; digestion partly intracellular; the nervous system a network of cells; commonly with the power of budding, by which either free individuals or colonial zooids may be formed; and whose sexual reproduction typically produces an ovoidal, uniformly ciliated larva, known as the *planula*, which has at first a solid core of endoderm.

CNIDARIA AND CTENOPHORA. Thus defined, this phylum contains the whole of the diploblastic animals, that is, those in which the space (blastocoele) between ectoderm and endoderm is either devoid of cells, or contains only cells derived late in development by immigration from ectoderm or endoderm. Of such animals there are two very distinct stocks—the *Cnidaria*, characterized by muscular movements, which possess nematocysts (p. 147), and are reducible either to the polyp or to the medusa type (p. 150); and the *Ctenophora*, which retain the ciliary locomotion of the planula, are without nematocysts, and are not to be assigned either to the polyp or to the medusa type.

The ctenophores are now regarded by some authorities as a distinct phylum from the coelenterates and there is much to be said in favour of this view. They have a different body shape and pattern and the more aberrant forms such as *Coeloplana* show similarities with the turbellarians, though this seems due to convergence. Nematocysts have been found in ctenophores but this does not indicate coelenterate affinities; instead the nematocysts seem to have come from the ctenophores' coelenterate food. Generally speaking the grade of organization of the ctenophores is similar to that of the coelenterates and, though in fact the two groups do not seem to be closely related, they can, as a matter of convenience, be left in the same phylum.

CLASSIFICATION

Subphylum 1. **CNIDARIA.** Coelenterates with nematocysts.

   Class 1. **HYDROZOA.** Cnidaria with both polyp and medusoid forms in their life cycle.

     Order 1. **Calyptoblastea.** (Leptomedusae.) Hydrozoa in which the coenosarc is covered with a horny perisarc produced over the nutritive polyp as a hydrotheca. Polyp and medusa of equal importance. *Obelia, Plumularia*

Order 2. **Gymnoblastea.** (Anthomedusae.) Hydrozoa in which the coenosarc is covered with a horny perisarc which stops short at the base of the polyps. Polyps and medusa of equal importance. *Pennaria, Hydractinia*

Order 3. **Hydrida.** Hydrozoa existing as solitary polyps without a medusoid stage. *Hydra*

Order 4. **Trachylina.** Hydrozoa in which the medusoid stage is large and the hydroid stage minute. *Geryonia, Limnocnida*

Order 5. **Hydrocorallina.** Hydrozoa existing as fixed colonies with an external calcareous skeleton. *Millepora*

Order 6. **Siphonophora.** Hydrozoa existing as free-swimming polymorphic colonies. *Physalia, Halistemma*

Class 2. **SCYPHOZOA.** Cnidaria with the medusoid stage as the main stage in the life cycle.

Subclass 1. **STAUROMEDUSAE.** Sessile polyp-like Scyphozoa. *Lucernaria*

Subclass 2. **DISCOMEDUSAE.** Scyphozoa whose main stage is as a free-swimming medusa; sessile only during the larval stage.

Order 1. **Cubomedusae.** Discomedusae of cuboid shape with four flat sides. *Carybdea*

Order 2. **Coronatae.** Discomedusae circumscribed by a groove on the upper surface. *Nausithoe*

Order 3. **Semaeostomeae.** Discomedusae with no septa or gastric pouches in the adult. *Aurelia*

Order 4. **Rhizostomeae.** Discomedusae without tentacles, manubrium branched to form sucking mouths. *Cassiopeia*

Class 3. **ACTINOZOA (ANTHOZOA).** Cnidaria with only the polyp stage represented in their life cycle.

Order 1. **Alcyonaria.** (Octoradiata.) Actinozoa with eight mesenteries and pinnate tentacles. *Alcyonium*

Order 2. **Zoantharia.** (Hexaradiata.) Actinozoa with six or multiples of six mesenteries and simple tentacles. *Metridium*

Subphylum 2. **CTENOPHORA.** Coelenterates without nematocysts.

Class 1. **TENTACULATA.** Ctenophores with tentacles.
Pleurobrachia

Class 2. **NUDA.** Ctenophores without tentacles. *Beroe*

TISSUE-GRADE. In the Coelenterata the Metazoa are at the beginning of their evolution and we have a primitive type with great potentialities, though these animals have also already acquired specialized features. The tissues consist of two single layers of cells, the ectoderm and endoderm, which constitute a thin body wall surrounding the central cavity (Fig. 83): the only increase in thickness and complexity of the body wall that is possible is by development of a gelatinous intermediate layer. Thus, while the typical polyps like *Hydra* have a very thin layer of this kind, it has become thicker, very much folded and penetrated by cells in the actinozoan polyps and exceedingly thick in the larger jellyfish, forming not only a kind of internal skeleton but even a reservoir of food. All the functions of the body are carried out by tissues and never by organs.

ECTODERM AND ENDODERM. The principal type of cell found in the tissues, both ectoderm and endoderm, of the primitive coelenterate is the *musculo-epithelial* cell which is columnar in shape and only differs from similar epithelial cells in the higher Metazoa in the fact that it is produced into one or two contractile fibres, which are imbedded in the mesogloea. Such a tissue

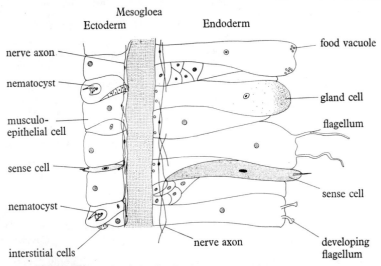

Fig. 83. Diagrammatic longitudinal section through the body of *Hydra*.

unit resembles a protozoon in the fact that different parts of the cytoplasm carry on different functions although they are not separated by any partition from each other nor provided with separate nuclei. An endodermal cell of *Hydra* has an inner border which can be produced into flagella, by means of which the fluid of the body cavity is kept in motion: or these may be retracted and the cell instead puts out pseudopodia to engulf particles of food. In the interior of the cell, beyond the border, the food is contained in vacuoles where it is digested, and finally the external border of the cell is produced into permanent cell organs, the muscle fibres or tails already mentioned, in which the cytoplasm can contract with much greater force and rapidity than in any other part of the cell. Among the endodermal cells, however, some are met with of a more specialized type: gland cells which pour into the cavity a digestive secretion (for the preparatory or extracellular digestion), and sense cells, found also in the ectoderm, which are thread-like, with a short projecting process. Both these kinds have no muscle tails. There are also interstitial cells, cells which preserve an embryonic character and may develop into germ cells, musculo-epithelial cells and cnidoblasts. They are important in regeneration.

NEMATOCYSTS (Fig. 84). These are distinctive organs of the coelenterates though it will be remembered that similar organs are present in other groups of animals. For example, they are present in protozoans, such as *Polykrikos* (p. 14), and they are also present in turbellarians, such as *Microstomum*, and

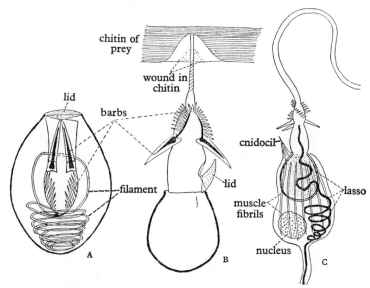

Fig. 84. Nematocysts of *Hydra*, large penetrating type: A, undischarged; B, discharged; rom *H. attenuata*. (After P. Schulze); C, discharged but retained within its thread cell. (After Will.)

in nudibranch molluscs, such as *Eolis*, but in the latter two cases it has definitely been shown that the nematocysts are derived from the food, since the animals feed on coelenterates and the type of nematocysts that they possess are identical with those present in their food. When the animals are starved they loose their nematocysts and can only regain them if they feed on coelenterates. Though they do not make their own nematocysts it is nevertheless surprising that they are able to take up the undischarged nematocysts from their food, incorporate them into their tissues and then orientate them peripherally so that the filament will shoot out and hit their prey.

A similar problem exists in the manufacture of the nematocysts in the coelenterates themselves. The nematocysts are formed from interstitial cells in the basal region of the column. These undifferentiated cells form the cnidoblast which later contains the nematocyst and its internally coiled filament. The cnidoblast is ejected into the enteron of the animal, where it is churned about with the rest of the food till it happens to come near the endoderm cells at the base of the tentacles. The cnidoblast is then seized by amoeboid processes of these cells and taken into the mesogloea, from where it is carried to the tentacles and orientated so that the cnidocil points outwards.

The cnidocil is a short process from the nematocyst, which bores through the cuticle of the blastocyst in which the nematocyst lies and comes into contact with the water. Stimulation of the cnidocil at a sufficient intensity causes a disturbance of the lid on the nematocyst. Water enters the fine filament and causes it to expand rapidly. This effects the ejection and inversion of the filament, which then hits the source of the disturbance. As the cnidocils can be 'detonated' by pressure waves in the water set up by a passing crustacean, it is clear that many nematocysts will be discharged at the crustacean and that some should hit it.

The nematocyst is an *independent effector organ*, a unit which possesses the receptor and effector organs necessary for a remarkable 'reflex' action, the whole being quite independent of the nervous system. There is no indication of any nervous control of the nematocysts, the coelenterate nematocysts being functional in the bodies of other animals, such as turbellarians and nudibranchs. From the high degree of differentiation and the independence of action of the nematocysts they might almost be considered as separate parasitic organisms (Cnidosporidia, p. 101) within the coelenterates, if their development had not been traced from interstitial cells. *Hydra* has four different types of nematocyst: a penetrant that pierces the prey and injects a poison; a volvant that wraps around any hairs that the prey might have; and two types of glutinants that stick on to smooth surfaces. It has been a problem for many years that animals swimming near *Hydra* can immediately discharge the nematocysts, yet animals such as *Kerona* or *Trichodina* can run all over *Hydra*, touch its cnidocils and yet not discharge the nematocysts. This problem was solved by experiments on the nematocysts of *Anemonia* where it was shown there were two factors responsible for nematocyst detonation: first the presence of certain lipoid-like substances in the water which lowered the threshold to tactile stimulation, second the mechanical contact of the animal against the cnidocil. Quick detonation of the nematocyst requires both stimuli, the chemical and the tactile. If only one of these is present then the threshold for detonation is very much higher, and it is possible that *Kerona* and *Trichodina* neither possess the lipoid substance nor have sufficient strength to detonate the nematocyst at its high mechanical threshold.

The threshold of the different types of nematocyst of *Hydra* varies according to the condition of the animal. The penetrant and the volvant have this combined chemical and tactile effect. The glutinants have a very high threshold and are not used for catching prey but are used either as a means of attaching the tentacles to the ground when the animal stands on its head, or they may be a means of protection against browsing animals. In either case the glutinants appear to have their threshold to mechanical shock actually raised by the presence of food in the water.

Seventeen different types of nematocyst have been described in the coelenterates and they have been used as a taxonomic character in classification. For example, the Gymnoblastea can be separated from the Calyptoblastea by many characters, one of which is the presence of a theca around the hydranth in the Calyptoblastea. But *Halecium* and *Cuspidella* have no hydrotheca yet are placed in the Calyptoblastea, whilst *Bimeria* and *Atractylis* are placed in the Calyptoblastea even though they have a small theca. Their phylogenetic position is supported by their nematocysts. The Gymnoblastea have many different types of nematocysts which indicates that the group is polyphyletic.

NERVOUS SYSTEM. The nervous system of the coelenterates is one of their most diagnostic systems, composed of small nerve cells which are only demonstrated by difficult staining techniques that often fail to differentiate

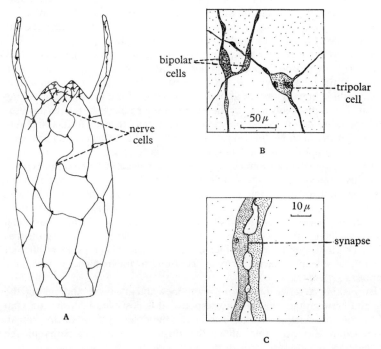

Fig. 85. Nervous system. A, diagram of *Hydra* to show the nerve net. B, nerve cells in *Metridium*. C, synapse in nerve cells of *Metridium*. (B, C, from Batham and Pantin.)

between nerve cells and connective tissue. Over the surface of the mesogloea is a network of nerve cells, there being a separate net each side of the mesogloea (Fig. 85A). The cells give off axons, some cells having three or more axons, though many have only two. The cells and their axons are discrete, i.e. the axons of one cell do not fuse with the axons of another cell. Instead there is a junction between them, the synapse. There are sensory cells both in the ectoderm and endoderm, a sense cell being shown in Fig. 83. It has a rod-like process projecting from the surface and the other end terminates in fine branches which synapse with other nerve cells. Such nerve cells are sensitive to touch and probably also to light and chemical stimulation.

The nervous system has been studied in detail in *Metridium*. Here, as in most coelenterates, there is one nerve net in the ectoderm and another in the endoderm. There are no nerves running through the mesogloea, the nets making contact with each other at the oral cone and at the region of the mesenteries. The ectodermal net is best developed in the region of the tentacles and most poorly developed in the middle region of the column.

The mesenteries show special development of long bipolar nerves, some of which have single axons extending for 7–8 mm. These form the *through conduction system* of the anemone and are responsible for conducting the rapid impulses that lead to contraction of the sphincter and column muscles after strong stimulation. These bipolar cells are comparable to the giant nerve fibres seen in earthworms, tubicolous polychaetes and squids, where they all facilitate rapid conduction leading to an escape reaction.

The nerve net is the most primitive type of nervous system. The cells which compose it are morphologically simpler than the nerves in the central nervous system of the higher Metazoa, but more particularly they conduct in a non-polar fashion equally well in almost all directions from the source of stimulation. This is especially true for the most primitive polyps, but in the medusae and the more differentiated polyps the nerve cells tend to form tracts that conduct better in some directions than in others; the *through conduction paths* being an example of such a concentration of nerves.

The responses of anemones vary considerably according to the type of stimulus applied. If a battery of electrical shocks are rapidly sent into the disc, the whole of the animal retracts. If the shocks are sent in more slowly, the tentacles twitch and the response gradually spreads to tentacles farther away from the electrodes. If the shocks are given more slowly still, the animal may open its mouth and start to swallow the electrodes, i.e. treating them as food. This indicates the way in which the anemone might analyse their different stimuli, i.e. chemical, tactile or pain according to the frequency at which the impulses arrive.

By making a series of cuts in anemones and medusae and timing the interval between the stimulus and response, it is possible to construct a map of the rate at which the stimuli pass along the different parts of the animal. Conduction is most rapid along the through conduction paths of the mesenteries and slowest through the body wall.

POLYP AND MEDUSA. Much of the interest of the coelenterates lies in the conflict between the two modes of life: an easy sedentary existence and a wandering or rather freely drifting life which demands a larger measure of activity and a greater elaboration of structure and physiological development. The two types of individual which correspond to these modes of life are the polyp and the medusa. There are large divisions of the coelenterates in which only one type is present, while in the others they may even be united in the same species and the same colony of that species. A survey of the phylum is very largely concerned with the variations of these types and the combination of them in the life histories of the different coelenterates.

The polyp (Fig. 86A) is an attached cylindrical organism with a thin body wall consisting of two single layers of ectoderm and endoderm separated by a narrow structureless lamella. At the free end an *oral cone* occurs and at its apex the mouth opening into the enteron. The oral cone (in the Hydrozoa) is surrounded by a number of tentacles, which are usually very extensible and armed with batteries of nematocysts, by which the living animals, on which the coelenterate feeds, are caught. Tentacles contain a prolongation of the endoderm which may form a tubular diverticulum of the enteron or a solid core.

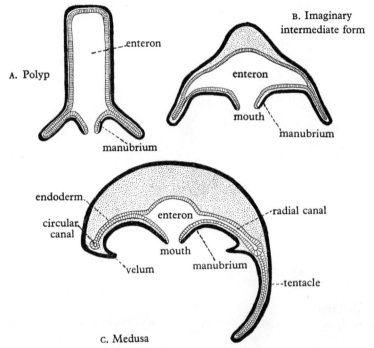

Fig. 86. Diagram to show the relation between the polyp and medusa. The velum, present in many medusae, is absent in *Obelia*.

The medusa (Fig. 86c) is a free-living organism differing from the polyp in the great widening of the body, especially along the oral surface, and the restriction of the enteron by the increase in thickness of the structureless lamella on the aboral side of the endoderm, so that while a central *gastric cavity* remains, the two endodermal surfaces have come together peripherally to form a solid two-layered *endoderm lamella* except along certain lines, where the *canal system* is developed radiating from the gastric cavity. The oral cone becomes the *manubrium*; the rim which bears the original tentacles of the polyp is now separated widely from the mouth by differential growth and drawn downwards in the formation of the bell. Very often a secondary set of oral tentacles is developed on the manubrium. The radial symmetry of the polyp is more strongly emphasized in the medusa by the radial development of the canal system. The muscular system of the bell is greatly developed by the substitution of a type of cell in which the muscular processes form a long striated fibre while the epithelial part is greatly reduced; such a cell is capable of rapid rhythmical contraction. The nervous system may be partially concentrated to form a nerve ring and well-defined sense organs occur in connexion with this.

REPRODUCTION. In this phylum, the lowest of the Metazoa, the gametes are of the type which is found throughout that great animal division; the maturation divisions make their typical appearance here. Eggs and spermatozoa respectively are nearly always borne by different individuals or colonies. After fertilization the egg segments by equal divisions until, first, a single layer of cells (ectoderm) arranged to enclose a central cavity constitutes the *blastula*. Then, by the migration of cells into this cavity, it becomes filled up with tissue

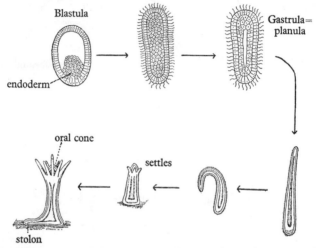

Fig. 87. Development of a hydrozoan. All the figures are not at the same magnification. (After Merejkowsky and Allman.)

(endoderm) while the ectoderm becomes ciliated. Such a larva with a solid core of endoderm is a *planula* (Fig. 87). It is capable of wide distribution by currents and may live for a considerable period before settling down. A split appears in the endoderm, the first appearance of the enteron, and the larva sinks to the bottom and attaches itself by one end. At the other end a mouth and tentacles appear and the creature becomes a polyp. There are a few exceptions to this in the phylum in which the egg develops directly into a medusa.

SUBPHYLUM CNIDARIA

DIAGNOSIS. Coelenterata referable to two types, the fixed *polyp* and the free *medusa*; locomotion usually by muscular action; possessing nematocysts.

CLASSIFICATION. They are divided into the following classes:

Class 1. **HYDROZOA.** Cnidaria, nearly always colonial; typically with free or sessile medusoid phase, arising as buds from the polyp-colony: no

CNIDARIA

vertical partitions in the enteron; medusae with a velum and nerve ring; tentacles of polyp usually solid; ectodermal gonads; and an external skeleton.

Class 2. **SCYPHOZOA.** Cnidaria in which the polyp stage is inconspicuous and may be absent altogether: the polyp, where present, gives rise to medusae by transverse fission (strobilation); with vertical partitions (mesenteries) in the enteron of some forms and larval enteron of others; velum and nerve ring generally absent; endodermal gonads; skeleton absent.

Class 3. **ACTINOZOA.** Solitary or colonial cnidarian polyps without medusoid phase; vertical partitions (mesenteries) in the enteron; endodermal gonads; with or without a skeleton.

CLASS 1. HYDROZOA

The most typical life histories of the 'hydroids' are those in which the phenomenon of 'alternation of generations' is presented. That is, there is a regular alternation of phases, hydroid colonies giving rise to free-swimming medusae and the fertilized eggs laid by the medusae each giving rise to a new colony of polyps. In the first two orders of the Hydrozoa, the Calyptoblastea and the Gymnoblastea, alternation of generations is well shown in the typical genera. As will be shown, there is a progressive suppression of the medusoid 'generation' in other members of these orders. In the other orders there is, however, complete suppression of the polyp phase in the Trachomedusae and Narcomedusae, and in the Siphonophora remarkable colonies are found which appear to have originated by budding from a medusa.

CLASSIFICATION. The following orders are contained in the class:

Order 1. **Calyptoblastea (Leptomedusae).** Hydrozoa in which the coenosarc is covered by a horny perisarc, produced over the nutritive polyps as hydrothecae and over the reproductive individuals as gonothecae; the medusae flattened, with gonads on the radial canals, and usually statocysts.

Order 2. **Gymnoblastea (Anthomedusae).** Hydrozoa in which the coenosarc is covered by a horny perisarc which stops short at the base of the polyps and reproductive individuals; the medusae bud-shaped, the depth of the bell greater than the width, with gonads on the manubrium and eyes, but not statocysts.

Order 3. **Hydrida.** Hydrozoa existing as solitary polyps without medusoid stage; tentacles hollow; without perisarc, the polyps being capable of locomotion; gastrula forms a resting stage encased in an egg shell.

Order 4. **Trachylina.** Hydrozoa in which the medusoid is large and the hydroid phase minute. The latter either forms medusa buds or, being represented by the planula larva, metamorphoses into a medusa. Statocysts with endodermal concretions: generative organs lying on the radial canals or on the floor of the gastric cavity.

Order 5. **Hydrocorallina.** Hydrozoa existing as fixed colonies with an external calcareous skeleton into which the usually dimorphic polyps can be retracted.

154 COELENTERATA

Order 6. **Siphonophora.** Hydrozoa existing as free-swimming, polymorphic colonies, without perisarc, derived by budding from an original medusiform individual.

GENERAL ACCOUNT
Orders Calyptoblastea, Gymnoblastea, Hydrida

We will take as examples of these orders *Obelia*, belonging to the Calyptoblastea, and *Bougainvillea*, to the Gymnoblastea, both of which produce free-swimming medusae, and then describe *Tubularia* with its sessile gonophores. The series ends with *Hydra* (Hydrida).

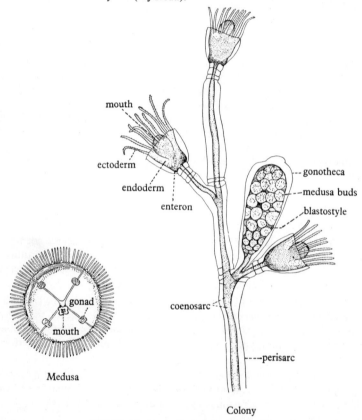

Fig. 88. Medusa and colony of *Obelia* sp.

OBELIA. In a colony of *Obelia* (Fig. 88) root-like hollow tubes (the *hydrorhiza*) run over the surface of attachment, such as a seaweed, and from these spring free stems, which branch in a cymose fashion giving off the polyp heads (hydranths) on alternate sides. At the growing ends of the main branches are produced buds which develop into hydranths, and towards the base of the branches in the axils of the hydranths, polyps modified for reproduction, the *blastostyles*, occur. The whole system of tubes which connect up the individual polyps is the *coenosarc*, and it must be understood that the enteron or cavity of the colony is continuous and common to all its members. The rhythmical contraction of the hydranths causes currents which distribute the food obtained by some individuals to those parts of the colony where feeding is not taking place. As in all Calyptoblastea the coenosarc is completely invested by the cuticular secretion, the *perisarc*, composed of chitin and produced to form cups round the hydranths (*hydrothecae*) and the blastostyles (*gonothecae*). The hydranth of *Obelia* is an expansion of the coenosarc, ending in a prominent oral cone, surrounded by a single ring of rather numerous tentacles, which have a solid core of endoderm cells. The blastostyle has neither mouth nor tentacles; the body wall proliferates to form distinct individuals, the medusae. Those nearest the mouth of the gonotheca mature first, and they are liberated as they mature. The medusa of *Obelia* (Fig. 88), the type of the Leptomedusae, is like a shallow saucer, the middle of the concave (subumbrellar) surface being produced into a short manubrium. The rim of the medusa bell is furnished with a large number of short tentacles. Like all medusae belonging to the Hydromedusae, it has four radial canals, running from the gastric cavity to the circular canal. On the course of the radial canals, and at the end of a short branch, patches of the subumbrellar ectoderm are modified to form the gonads. The germ mother-cells originate in the ectoderm of the manubrium, pass through the endoderm and along the radial canals to the gonads and then migrate into the ectoderm again. Only male or female germ cells are produced by each medusa. At regular intervals in the circumference are eight sense organs, the *statocysts*. They are tiny closed vesicles, lined with ectoderm and filled with fluid in which minute calcareous grains occur. The epithelial lining not only secretes these but is also sensory: the impact of the grains on the cells produces a stimulus which is transmitted through the nerves to the muscles, and if the position of the medusa should be abnormal the muscles contract in such a way as to right the bell of the animal.

Another characteristic of the hydrozoan medusa is the *velum* (which is practically absent in *Obelia*), a narrow internal shelf running inside the border of the subumbrellar cavity. This is largely composed of ectodermal circular muscles, separated by a horizontal partition of structureless lamella. At its base is a double nerve ring: the inner half of this is concerned with the subumbrellar musculature (and, in the Trachylina only, the outer with the sense organs).

The ripe ova are shed into the water by the rupture of the gonad, and fertilization takes place here. Segmentation leads to the formation, first, of a hollow blastula, and from this, by the immigration of cells at one pole, the elongated planula larva (Fig. 87) with a solid core of endoderm is formed. It is ciliated and swims freely for a time, eventually settling down by its broader end, while the other end develops a mouth and tentacles surrounding it. The endoderm delaminates to form the enteron. From the base of this first-formed polyp there is an outgrowth along the surface of attachment which is the beginning of the hydrorhiza. From this the rest of the colony is developed.

In *Bougainvillea* (Fig. 89) the polyps belong to the gymnoblastic type to be described for *Tubularia*. The creeping hydrorhiza gives off branches, one of which is seen in the figure, and from these numerous individuals are budded. Most of these are polyps (hydranths), distinguished from those of the Calyptoblastea by the fact that the perisarc stops short at the base of the polyp and does not form a hydrotheca. The medusoid individuals take their origin

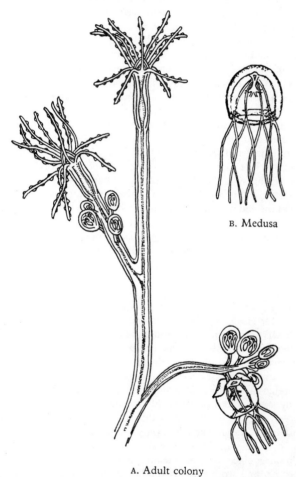

A. Adult colony

Fig. 89. *Bougainvillea fructuosa*. A, the fixed hydroid form with medusae in various stages of development. B, the free-swimming sexual medusa has broken away from the colony. (From Allman.)

directly from the coenosarc each as a simple bud, within which is developed a single medusa which eventually divests itself of a thin covering, breaks from its stalk and swims away. Several may spring from the same stem, but this may also bear normal polyps. There is here no blastostyle, or polyp modified for budding off medusae, and this condition, in which polyps and medusae belong to the same grade of differentiation from the coenosarc, is possibly to be regarded as primitive, that of *Obelia* as secondary.

In *Eudendrium* an intermediate stage occurs. Medusae are budded off from the stalk of a normal polyp, and as soon as this budding commences the polyp loses its tentacles, diminishes in length and may be said to become a blastostyle.

Tubularia (Fig. 90A) occurs as a colony of large polyps with long stalks springing from a hydrorhiza of insignificant extent. At the base of the polyp the stalk forms a swelling; there the perisarc stops. There is an oral cone surrounded by a ring of tentacles and also a ring of larger (aboral) tentacles at the broadest part of the polyp. Both kinds of tentacles are solid, with an axis of vacuolated endoderm cells placed end to end, which have a skeletal value. In Fig. 90B, C part of the phenomenon of digestion is illustrated. A crustacean has been swallowed and lies in the stomach. After preliminary digestion a fluid mass of half-digested material is formed and, by alternate contraction of A the stomach and B the spadix (manubrium) of the gonophore together with the basal swelling of the polyp, the food is forced into contact with all the absorptive epithelium of the polyp and gonophore and also pipetted along the cavity of the stalk.

SESSILE MEDUSAE. The reproductive individuals of *Tubularia* originate from hollow branched structures springing from the polyp itself between the oral and aboral tentacles. Each polyp has several of these branches, and from each branch a number of reproductive individuals arise. The branch is usually termed a *blastostyle*, although it is only part of an individual and not a modified polyp as in *Obelia*. Each of the buds it produces, however, has the structure of a medusa but remains attached to the parent polyp as long as it lives. Like the free-swimming medusa of *Bougainvillea* it conforms to the anthomedusan type, the depth of the medusa bell exceeding the width and the gonads being situated on the manubrium (spadix). This sessile medusa is called a *gonophore*. As seen in Fig. 91, the radial and circular canals are formed as in *Obelia*, and four very short tentacles occur opposite the radial canals on the margin of the bell; but the entrance to the subumbrellar cavity is very much constricted compared to *Obelia* or a free-swimming anthomedusa. Another modification is that the eggs, which are large and yolky when liberated from the gonad, are fertilized in the subumbrellar cavity and develop there through the planula stage into an advanced larva called the *actinula* (Fig. 91B) which is really a polyp of *Tubularia* with a short stem. At this stage it makes its way out of the shelter of the gonophore and fixes by its aboral end. As a rule, only one of these large eggs can be produced at one time and a ripe gonophore generally contains two larvae of different ages, one a planula and the other an actinula, which may be seen protruding from the aperture of the bell.

ANTHOMEDUSAE. In such gonophores the neuromuscular structures of the bell are hardly developed at all, the mouth never opens and there are no evident sense organs. In the medusae called *Lizzia* and *Margellium*, common plankton forms whose polyp stages are not known, we see the normal anthomedusan type. In both of these there are a number of short tentacles, arranged in groups round the margin of the bell, and four double tentacles at the end of

158   COELENTERATA

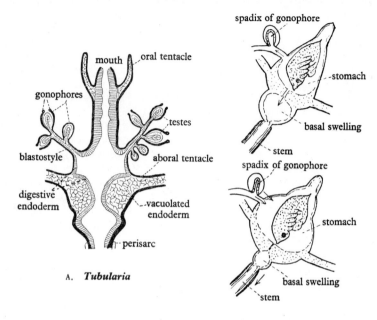

B, C. *Tubularia* feeding

Fig. 90. A, median section through a polyp of *Tubularia*. (After Kükenthal.) B, *Tubularia* with food, the food being pumped through the enteron by contractions of the basal swelling. Arrows denote direction of water movement. (After Beutler.)

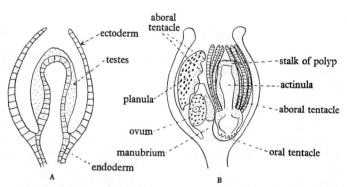

Fig. 91. Longitudinal section through gonophores of *Tubularia*. A, young male B, female with actinula larvae. Details of tissues are not shown in B.

the manubrium. *Lizzia* possesses eight 'eyes' (Fig. 92 A) which are little patches of ectoderm, in which some of the cells develop pigment while others elongate and end in rods. The latter are concluded to be the light-perceiving cells. There is also an outer enlargement of the cuticle which serves to concentrate light on the organ and may be called a lens. Though there is no direct evidence that these organs have a relation to light, they have in a simple form all the structural elements of the eye of higher animals. *Margellium* (Fig. 93) has no eyes but apparently suffers no disability from their absence: probably the light-perceiving cells are scattered over the general surface of the ectoderm. 'Eyes' are however a general character of the Anthomedusa as 'ears' (as statocysts may be broadly termed) are of the Leptomedusa.

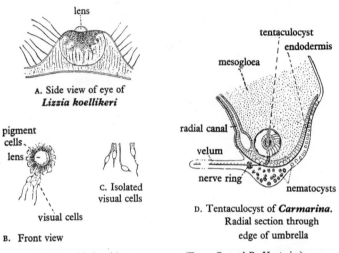

Fig. 92. Medusoid sense organs. (From O. and R. Hertwig.)

LOSS OF MEDUSOID PHASE. Among the hydroids with sessile medusoids or gonophores there are many forms in which the medusoid structure is lost, and a bud-like structure is found in which a transverse section shows simply successive layers of ectoderm with generative cells, structureless lamella and endoderm round an enteron which does not open by a mouth. In forms like this the migrations of germ cells, mentioned as occurring in *Obelia*, are very noticeable. Thus in *Eudendrium* (Fig. 94 D) the germ cells are often to be distinguished making their way along the coenosarc towards developing gonophores. If this degeneration of medusae is followed to its conclusion, a stage is arrived at in which there are no special reproductive buds at all, but the generative cells occur in the body of the hydroid. This is the condition in *Hydra*, where the multiplication of the interstitial cells at different positions produces testes or ovaries.

HYDRA. Each ovary contains a single egg of a size unusual in the Hydrozoa, which grows by the ingestion of its sister oocytes and the conversion of their protoplasm into yolk spherules as in *Tubularia*. This phenomenon appears to be a consequence of the habitat of the genus. As in so many other fresh-water animals, a free-swimming stage is omitted from the early history and the period of larval development is passed in the shelter of the egg shell; when the gastrula stage has been arrived at and the yolk is mostly absorbed, development is suspended during a resting stage of three or four weeks. After this the young *Hydra* pokes its oral end out of the shell and, after creeping about for a short time, frees itself and develops a mouth and tentacles. Other characters which differentiate *Hydra* from the majority of hydroids are the solitary habit, which it shares with some Gymnoblastea, the hollow tentacles and the complete absence of a stiff perisarc, this enabling the animal to execute its characteristic looping movements. It is often pointed out that the presence of a distinct migratory phase, the medusa, would entail a serious disadvantage on *Hydra*;

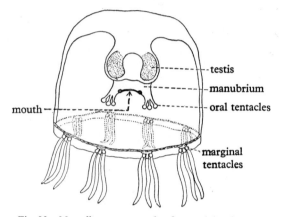

Fig. 93. *Margellium*, an example of an anthomedusan.

it is suggested that the medusae might be swept out to sea and lost. *Hydra* usually lives in ponds and is therefore hardly subject to this danger, but at the same time the embryo in its horny egg shell is admirably fitted for dispersal, for example in mud on the feet of migratory birds. This modification of reproductive habits in *Hydra* is paralleled in the fresh-water sponges with their gemmules, the fresh-water polyzoa with their statoblasts and the cladoceran crustacea with their ephippial eggs. It must, however, be mentioned that a remarkable group of fresh-water medusae occur which belong to the Trachylina, and a stage occurs in their life history which has sometimes been compared with *Hydra* and named a separate genus (*Microhydra*) of hydroid polyps. This is, however, an interesting case of convergence.

1. Order **Calyptoblastea**

Plumularia (Fig. 94A) with a creeping hydrorhiza, giving off plume-like branches, each of which bears a series of hydrothecae on one side only; hydrothecae small, so that the polyps cannot be completely retracted within them; beside the nutritive polyps a second smaller kind (nematophore), without mouth, but with long amoeboid processes which engulf decaying

HYDROZOA

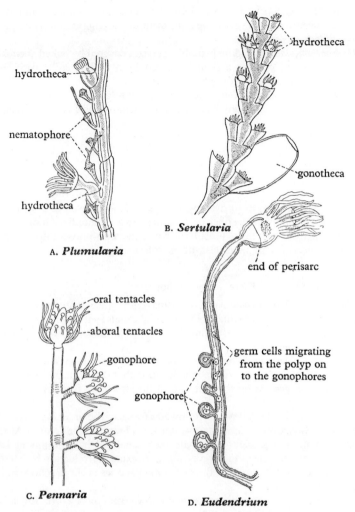

Fig. 94. Examples of Hydrozoa: A, B, Calyptoblastea, C and D, Gymnoblastea.
(A, after Allman; B, C, original; D, after Weismann.)

polyps, epizoic organisms such as diatoms and protozoa and larvae of other epizoic forms.

Sertularia (Fig. 94B) with a creeping hydrorhiza, more or less branching stems which bear opposite hydrothecae; hydrothecae large, so that the polyps can completely retract within them.

2. *Order* Gymnoblastea

Cordylophora, living in fresh or brackish water (Norfolk Broads), polyps with scattered filiform tentacles.

Pennaria (Fig. 94c) with two kinds of tentacles, oral capitate and aboral filiform; nematocysts of very large size; medusae degenerate but become free when gonads are mature.

Hydractinia, with spreading plate-like perisarc covered by naked coenosarc, very often found coating a shell inhabited by a hermit crab; with spiral dactylozooids and sessile gonophores.

Podocoryne, as *Hydractinia*, but with free medusae.

The polyp forms of many medusae, both Antho- and Leptomedusae, are unknown.

3. Order **Hydrida**

Hydra.

4. Order **Trachylina**

This group consists of forms in which the medusoid develops directly from the egg and the polyp has either been reduced to a minute fixed individual or is represented only by the planula larva which metamorphoses into a medusa. The possession of sense tentacles with endodermal concretions is an important character.

CLASSIFICATION There are two suborders:

Suborder 1. **Trachomedusae.** Trachylina with sense tentacles in pits or vesicles and with gonads situated in the radial canals; with marginal tentacles on the edge of the umbrella. Examples: *Geryonia, Limnocodium, Carmarina* (Fig. 92D), *Limnocnida*.

Suborder 2. **Narcomedusae.** Trachylina with sense tentacles not enclosed and marginal tentacles inserted some distance aborally from the edge of the umbrella; with gonads on the oral wall of the stomach. Example: *Cunina*.

The inclusion of the following fresh-water forms in the order is provisional:

Limnocnida is a remarkable fresh-water form found in the Central African lakes. Up till the present only male medusae have been found in Lake Tanganyika and female in Victoria Nyanza. Asexual reproduction by budding takes place from the margin of the bell. Other species occur in Rhodesia and the Indian rivers.

Craspedacusta (*Limnocodium*) was first known from the Victoria Regia tank in the Royal Botanic Gardens at Kew, but has now been discovered in various North American rivers and has even colonized ponds and canals in England. It has a polyp-like stage, *Microhydra*, which has a certain likeness to *Hydra*.

5. *Order* **Hydrocorallina**

The forms included in this group are mostly associated with reef corals in tropical seas. The main part of the colony consists of a much-branched hydrorhiza with frequent anastomoses. Instead of secreting a horny perisarc as the Calyptoblastea and the Gymnoblastea do, the ectoderm lays down an exoskeleton consisting of calcareous grains, which becomes bulky and solid. It may be either massive or encrusting or branching. From pits in the surface of the colony arise the polyps. These are of two types (Fig. 95). First there are the individuals of normal structure with a mouth surrounded by tentacles (*gastrozooids*): these nourish the colony. Then there are the *dactylozooids* which are much longer and more slender. They have no mouth but they possess scattered capitate tentacles and may form a ring round a gastrozooid, in which case it is readily observed that their function is to catch prey and hand it to the central gastrozooid for digestion. Besides the polyps there are the medusae, which, as in *Bougainvillea*, are budded directly off from the coenosarc: they are lodged in pits of the skeleton called *ampullae*, but their liberation has been observed in *Millepora*. It is supposed, however, that their free-living existence is very brief.

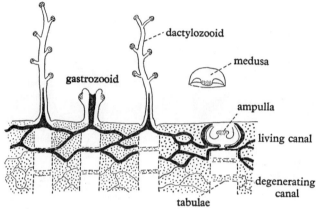

Fig. 95. Diagrammatic section through *Millepora* showing a gastrozooid, a dactylozooid, and an ampulla enclosing a medusa. (Slightly altered from Hickson.)

6. *Order* **Siphonophora**

The Siphonophora are colonial animals which exhibit the maximum development of polymorphism found in the Coelenterata or indeed in any group of the Animal Kingdom. They are pelagic and each colony originates from a planula which metamorphoses to form a single medusiform individual, which later drops off from the colony, from the exumbrellar side of which springs a coenosarcal tube budding off all the other members of the colony (Fig. 96B). It usually happens that those which are developed first are needed to buoy up and propel the young colony. Consequently the first individual is either medusiform or else forms an apical float or *pneumatophore*, the epithelium of which secretes gas. There may also be formed from the ectoderm of the first-formed individual an *oleocyst* containing a drop of oil. The succeeding medusiform individuals resemble the bell of an anthomedusa, with velum, musculature and canal system but lacking the manubrium, and they are called *nectocalyces*: while the most primitive siphonophores have only a single one there may be a series of them.

LINEAR CORMIDIUM. Following these the coenosarc in one type of colony (Fig. 96A) grows to a great length and buds off at intervals along its length similar assemblages of individuals. Such an assemblage is known as a *cormidium*, and may consist of (1) a shield-shaped *hydrophyllium* which covers the rest of the cormidium, (2) a *gastrozooid* resembling the manubrium of a medusa, with a mouth, and a tentacle usually branched, (3) a mouthless individual, the *dactylozooid*, with a tentacle usually of great length and provided with strong longitudinal muscles, and (4) a *gonozooid* (or individual bearing gonophores) which may or may not have a mouth. The gonophores often resemble those found in some of the Gymnoblastea like *Tubularia*. Such forms as those described above are the genera *Halistemma, Diphyes* and *Muggiaea*.

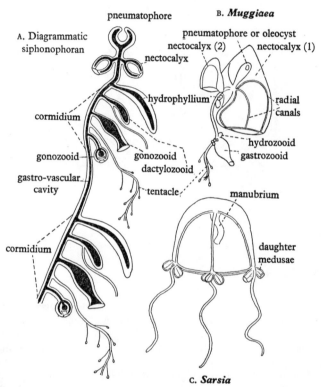

Fig. 96. Development of a siphonophore colony and an anthomedusa (*Sarsia*) for comparison. (A, altered from Hertwig; B, after Chun; C, after Allman.)

NON-LINEAR CORMIDIUM. In other cases the coenosarc is not a linear stolon but a massive body from which are budded off innumerable cormidia, in which gastrozooids, dactylozooids and gonozooids are all crowded together to form a compact colony. In *Physalia* (Fig. 97 B), the 'Portuguese man-of-war', there is an enormous cap-shaped pneumatophore which floats above the surface of the water. There are no nectocalyces, but the colony is borne hither and thither by the wind and countless numbers are cast up on the lee shores. The dactylozooids of *Physalia* hang suspended from the colony and form a drift net; when they are touched by a fish the nematocysts discharge and the fish is captured. The tentacles contract and the prey is drawn up until the gastrozooids can reach it. The lips of these are spread out over the surface of the fish until it is enclosed in a sort of bag in which it undergoes the first stage

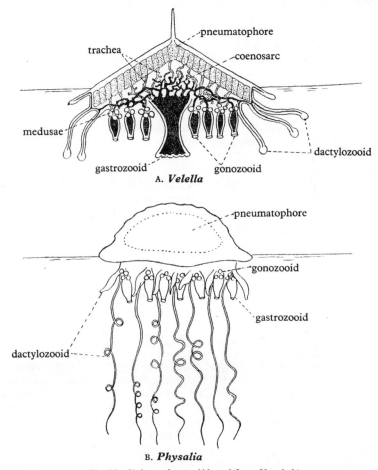

Fig. 97. Siphonophora. (Altered from Haeckel.)

of its digestion. *Physalia* can catch and devour a full-grown mackerel, and the poison of its nematocysts is so virulent as to endanger human life. In *Velella* (Fig. 97A) the disc-shaped colony has a superficial resemblance to a single medusa. The pneumatophore consists of a chitinous disc containing a number of chambers and raised into a vertical ridge which forms a sail. On the undersurface there is a single large gastrozooid in the centre, a larger number of gonozooids surrounding it and a fringe of dactylozooids at the margin. The gonozooids produce buds which actually escape as free medusae. The coenosarc consists of a mass of tissue which is traversed by endodermal tubes placing in communication the cavities of the gastrozooid and the gonozooids, and ectodermal tubes (tracheae) which are prolongations of the gas cavity of the pneumatophore. This tropical form is often brought in large numbers to the shores of Devon and Cornwall by the Gulf Stream.

The medusae and nectocalyces of the Siphonophora are very similar to the Anthomedusae. Medusae like *Sarsia* (Fig. 96c) may bud off other medusae either from the bell or the manubrium, but the Siphonophora are probably not to be regarded simply as a colony of medusae connected by coenosarc. A further change has gone on in which organs have been displaced from their original position. The manubrium has come to lie outside the primary medusa bell, forming a gastrozooid (Fig. 96B) at the beginning of the main coenosarcal axis. No manubria corresponding with the medusa bells of the nectocalyces are present. In the cormidia the hydrophyllium, which may be a modified bell, the gastrozooid, and the tentacles may be quite separate from one another while the complete medusoid form is shown only by the fixed gonophores (Fig. 96A, B).

In more specialized siphonophores owing to the shortening of the main axis the displacement of parts is more extreme and the component parts of the cormidia no longer recur in the typical groups, all kinds of organs being crowded together. Lastly, with the great development of the gas-secreting pneumatophore, the medusa bell is suppressed.

While the above description gives an impression of the order regarded as colonial animals the siphonophores must be primarily considered as coelenterates exhibiting growth variability to such an extent that the identification of the component structures as organs or individuals is difficult and of purely academic interest.

BODY FORM AND REGENERATION IN THE HYDROZOA

In the previous section the different morphological parts of the hydrozoans have been described as if they were fixed static systems. This is not correct, since even when the embryonic development of the colony is finished there is still a continuous change going on within the colony and new parts are kept from developing by the presence of the already developed parts. This is made more clear by the study of regeneration in the coelenterates.

The Hydrozoa have a marked ability to regenerate lost parts and have formed the subject for a great deal of experimental work into the nature of regeneration and the way in which an animal maintains its specific form. If the head of *Hydra* is removed a new one will soon develop in the old position (Fig. 98 A). It is also possible to separate the ectoderm from the endoderm and make small groups of pure cells, but these will not form polyps unless both ectoderm and endoderm are represented in the same cell mass. Thus there is a definite difference between the two tissues and both are essential for regeneration. If *Hydra* is turned inside out by running a needle in through the base and out of the mouth, the ectoderm will then lie inwards whilst the endoderm lies outwards. After a little while the cell layers will be found back in their correct position (the ectoderm outside and the endoderm inside) even though the animal has not turned itself outside in. Instead the cells have migrated past each other, the ectoderm cells moving outwards and the endoderm cells moving inwards. It would seem that not only are the ectoderm and endoderm

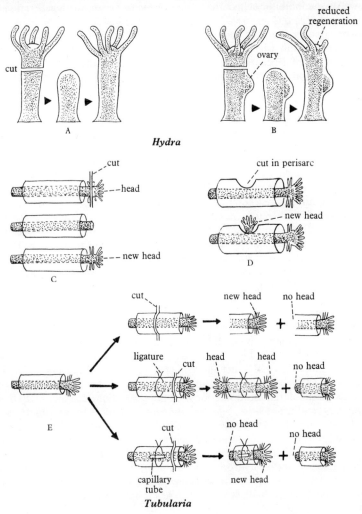

Fig. 98. Diagram to show experiments on regeneration in Hydrozoa. For details see the text.

cells different and each necessary for regeneration but they also have specific positions in the animal which they are able in some way to maintain.

In addition to these ectoderm and endoderm cells there are important undifferentiated cells called interstitial cells that can be selectively stained with toluidine blue. When the distribution of these cells was studied it was found that they were in their highest concentration at the oral end of *Hydra* but when the animal started to develop a lateral bud or gonad there was an increase in the number of interstitial cells at that particular region. In some way these cells appear to be correlated with regeneration and growth as is shown by the inhibitory effect that a bud or ovary has on the regeneration of tentacles in *Hydra* (Fig. 98 B). The presence of the bud attracts the interstitial cells from one side of the head, and no tentacles develop on that side. It is difficult to explain why these cells move about in the animal though they appear to be influenced by the presence of injured tissue. This injury effect and its chemical relationship can be seen in more detail in the regeneration patterns of *Tubularia*.

Tubularia is able to regenerate a new head if the old one is cut off and it was generally thought that this new growth was due to the stimulation of the adult tissue by the injury (Fig. 98 C). Similarly if a region of the stem of *Tubularia* is damaged a new head can develop at the site of the injury. However, when the experiment was repeated but this time with the damage limited to the perisarc around the stem (the living tissue being left untouched) a new head was still seen to regenerate at the affected site (Fig. 98 D). Another indication that the stimulus to regeneration was not due to injury is seen when the stem of *Tubularia* is transected. A new head develops on the headless piece but no head develops at the other cut surface (Fig. 98 E). If one ties a ligature around the stem that still has a head attached to it, a new head will be found to regenerate at this cut surface too (Fig. 98 E). This is not due to mechanical abrasion of the stem by the ligature since if the experiment is repeated with a piece of capillary tubing inserted inside the ligature, then the new head does not develop. The general conclusion is that the head gives off some chemical that prevents adjacent tissues from developing into a head. This chemical is prevented from diffusing away by the perisarc but when the animal is cut or the perisarc is injured the chemical diffuses away and allows the tissue to develop into a head. In Fig. 98 E a head could only develop at one end since the other piece was still under the influence of its own head but when this influence was removed by the ligature then a new head could develop.

A second tubularian head will normally develop a finite distance away from an already established head, but if an oil droplet is injected into the enteron just below the established head a new head can develop nearer than usual, apparently because the old head's influence is transmitted through the enteron and is diminished by the oil droplet. This is of particular interest since it demonstrates that the hydrozoan structure is not a static affair; instead, everything is in a condition of dynamic flux, and since regenerative ability in these animals is so high, strict control has to be kept over the organization of the whole animal.

It is possible that some differences in the property of the perisarc may account for the differences in pattern of such genera as *Sertularia* and *Sertularella* or *Plumularia*. In some as yet unknown way, these chemicals affect the distribution of the interstitial cells. Probably a similar system is at work in the siphonophoran colonies such as *Halistemma* or *Physalia*, where the different individuals develop at a finite interval from previous individuals.

This pattern control is not confined to the coelenterates. A similar system is found in the colonial ciliate *Zoothamnion*, and it indicates that in such animals the observed adult form is not just a static end product but is instead a collection of parts with a pattern impressed upon them. Should the restraint of this pattern be removed then the units immediately grow into a new pattern that gradually reaches a new equilibrium and thus another stable pattern.

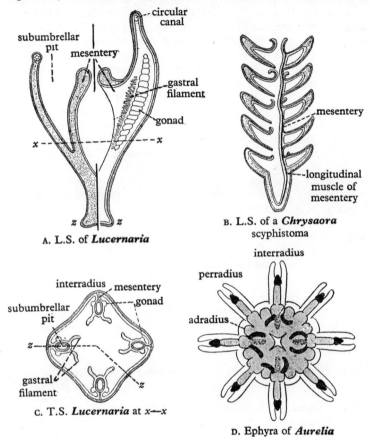

Fig. 99. Scyphozoan structure. (A, C, altered from Bourne; B, after Herič; D, original.)

CLASS 2. SCYPHOZOA

This class contains the common jellyfishes of temperate and colder seas, some of which are of extraordinary size, like *Cyanea arctica*, the diameter of whose disc is a couple of yards.

CLASSIFICATION. There are two subclasses:

SUBCLASS 1. STAUROMEDUSAE

Sessile polyp-like Scyphozoa. *Lucernaria*.

SUBCLASS 2. DISCOMEDUSAE

Free-swimming medusae. *Aurelia*.

SUBCLASS 1. STAUROMEDUSAE

The simplest type of Scyphozoa is found in the division known as the *Stauromedusae*, two members of which, *Haliclystus* and *Lucernaria* (Fig. 99), are not uncommon on the British coasts, adhering to the blades of *Zostera* or *Laminaria*. It has a narrow stem arising from its exumbrellar surface, by which it attaches itself temporarily to seaweed. The edge of the bell is divided into eight lobes, on each of which are several short tentacles and the adhesive organs which are called *marginal anchors* (absent in *Lucernaria*). There is no velum and tentaculocysts are absent. The manubrium is well developed and the mouth opens into a spacious gastric cavity which is divided by four partitions, the *interradial mesenteries*, into four broad chambers which are said to be *perradial*. The mesenteries are vertical walls projecting from the body wall and composed of endoderm with an internal layer of mesogloea. They have a free edge centrally, while on each side a vertical series of *gastric filaments* projects into the enteron, and a parallel series of gonads stand nearer the body wall. The perradial chambers do not quite extend to the edge of the bell: a circular canal is cut off from the rest of the enteron. Also in the interradial position and penetrating the whole length of the mesentery is an ectodermal invagination, the *subumbral pit*.

The Stauromedusae only exist as individuals of this structural type, superficially more like a polyp than a medusa, but usually supposed to be a medusa, and the egg develops into an individual exactly resembling the parent.

Other examples of Stauromedusae are *Craterolophus*, *Lipkea*, *Depastrum*.

SUBCLASS 2. DISCOMEDUSAE

DIAGNOSIS. Free-living medusae, only sessile during the larval stage.

CLASSIFICATION

Order 1. **Cubomedusae.** Cuboid medusae with four flat sides.
Carybdea, Tripedalia

Order 2. **Coronatae.** Medusae circumscribed by a groove on the upper surface. *Nausithoe*

Order 3. **Semaeostomeae.** Medusae with no septa or gastric pouches in the adult. *Aurelia, Cyanea, Pelagia*

Order 4. **Rhizostomeae.** Medusae without tentacles, manubrium branched and forms sucking mouths. *Rhizostoma, Cassiopeia*

STRUCTURE. The vast majority of the Scyphozoa belong to the subdivision *Discomedusae*, which includes our type *Aurelia aurita* (Fig. 100), the commonest British jellyfish, but one whose distribution is world-wide.

It has a similar external appearance to that of *Obelia*, save for the difference in size, the margin of the bell being surrounded by very numerous short tentacles. The manubrium is well developed and the corners of the mouth are drawn out into four long frilled lips along the inside of which are ciliated grooves leading into the gullet. The gullet is very short and opens into the endodermal stomach. This is produced into four interradial pouches in the lining of which the genital organs develop as pink horseshoe-shaped bodies. Parallel to the internal border of the gonads there is a line of *gastric filaments* which project freely into the lumen of the pouch. The endodermal cells of which they are composed contain batteries of thread cells which kill any living prey taken into the stomach. The gastric pouches of *Aurelia* occupy the

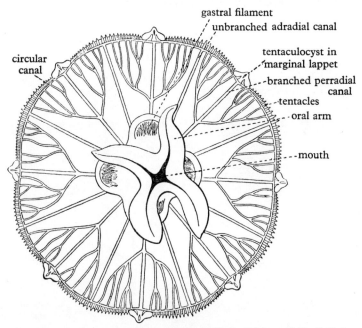

Fig. 100. *Aurelia aurita*, somewhat reduced. (From Shipley and MacBride.)

position of the mesenteries of *Lucernaria*, and the *subgenital pits* occurring underneath the gonads and lined by ectoderm correspond to the subumbral pits of the simpler form. The broad perradial pouches in *Lucernaria* have disappeared owing to the great growths of the mesogloea and the restriction of the gastric cavity to a central position. There is, however, an extensive canal system running from the gastric cavity to the circular canal which is all that represents the former extension of the gastric cavity. It consists of eight branched and eight unbranched canals: four of the branched canals are interradial and four perradial: the eight alternating unbranched canals are called *adradial*.

In this elaborate 'vascular' system there is a circulation of fluid produced by the cilia of the lining epithelium working in definite directions (Fig. 101). The water drawn in by the mouth passes first into the gastric cavity and then the gastric pouches; thence by the adradial canals to the circular canal. It returns thence by the branched interradial and perradial canals to exhalant grooves on the oral arms. The whole circulation takes about twenty minutes, and it serves to maintain a constant supply of food to all parts of the body. Food undergoes its preparatory digestion in the stomach: the half-digested fragments are swept by the cilia on the round described above and may be ingested by any of the endodermal cells of the canal system and become available for local needs. The gastrovascular system thus at once fulfils the functions of the digestive and circulatory systems of higher animals.

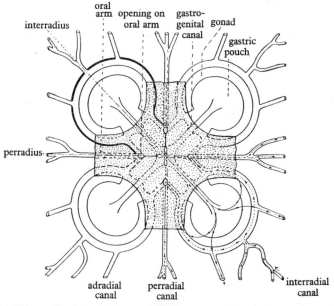

Fig. 101. Diagram to show the course of ciliary circulation (see arrows) in the genital pits and other organs of an adult *Aurelia*. (After Widmark.)

NERVOUS SYSTEM. The neuromuscular system is further developed than in even the medusoid individuals of the Hydrozoa. The muscles are ectodermal, and each cell is almost entirely converted into contractile protoplasm with a cross-striated pattern forming an elongated fibre; physiologically they are capable of rapid rhythmic contraction and not of slow tonic contraction like the muscle of a sea anemone (p. 185). The fibres are arranged as a circular musculature over the peripheral part of the subumbrella. The nerve net is also confined to the ectoderm and is concentrated in the neighbourhood of the tentaculocysts. There is no true velum, but a *pseudovelum* consisting of an internal flange which is not occupied by muscles and a nerve ring as in the Hydrozoa.

The tentaculocysts are the characteristic sense organs of the Scyphozoa (but are present also in some Trachylina in the Hydrozoa). They are minute tentacles which project at the end of the interradial and perradial canals, which are continued into them. The edge of the bell projects over them as a hood. In each apical endoderm cell of the tentacle there is a crystal which according to some authors is calcium sulphate. On one side of the tentacle is a pigment spot which may be an ocellus, and near it are two pits lined with sensory epithelium and said to be olfactory. In the neighbourhood of these tentacles, then, all the senses appear to be localized. The tentaculocyst (Fig. 102) is made up of two parts, a club-shaped projection heavy at its distal end, and a pad of sensory epithelium immediately beneath it. If the medusa is tilted from the normal horizontal position the club of the highest tentaculocyst will press more firmly against its sensory pad, and the club of the lowest tentaculocyst less firmly. Whatever tentaculocyst is highest produces greatest stimulation: this alone controls the rate of beating of the bell, which has been shown to be 50% greater than normal when the animal is tilted through 90°. Further, the state of excitation of the highest tentaculocyst does not allow

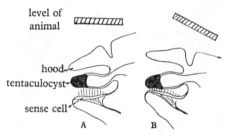

Fig. 102. Diagram of a tentaculocyst of *Aurelia*: A, in horizontal position. B, with medusa tilted, the tentaculocyst being pressed down on the sensory epithelium.

complete relaxation of the musculature of the section of the bell nearest to it between successive beats. This means that less water is driven downwards at each beat from the uppermost half of the bell than from the lower half, with the result that the bell automatically rights itself. The Scyphozoa are excellent subjects for experiment, and if cut into ribbons will still live and their muscles function. If the tentaculocysts are cut out one by one the rhythmic movements of the bell continue until the last is removed when they suddenly cease. After that, drastic stimulation, tactile or chemical, is necessary to make the muscles contract.

REPRODUCTION. The gonads are situated, as has been already stated, in the floor of the stomach, and the ripe gametes are liberated into the genital pouch. The eggs are fertilized as soon as they become free by spermatozoa from another individual which are drawn into the mouth along with the food. They pass through the canals to the opening on the oral arms (Fig. 101) and undergo the first stage of their development enclosed in pouches at the side of the oral grooves. Little opaque patches along the side of the lips are to be seen with a lens, and when dissected out they prove to be masses of planula larvae. The planula is eventually set free, but soon attaches itself to stone or weed and develops into a small polyp, without perisarc, the *hydratuba*, which eventually grows sixteen long and slender tentacles. Internally this stage has the same structure as *Lucernaria* with four interradial mesenteries, which are invaded by vertical ectodermal pits, and form perradial pouches between. At the base of the hydratuba a horizontal stolon grows out, and off this fresh hydratubae may be budded (Fig. 103 A). They may separate from the parent as in *Hydra*. During the winter the whole hydratuba is segmented by transverse horizontal furrows. This process is termed strobilation (Fig. 103 B, C). In each of the disc-like segments so produced, marginal growth at once begins, eight notched lobes being formed, four of which are interradial and four perradial. In each notch there is a short tentacle and this becomes a tentaculocyst. Each lobe is provided also with two short lateral tentacles, but these disappear. A prolongation of the gastric cavity into each lobe indicates the beginning of

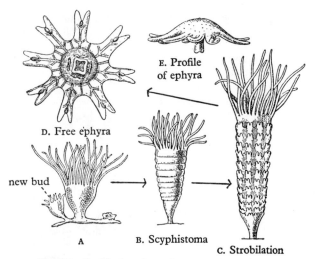

Fig. 103. Strobilation of *Aurelia aurita*. (From Sars.)

the branched perradial and interradial canals, and at a little later stage the adradial canals also appear (Fig. 103 D). The gastric filaments are also seen as four pairs in the interradial mesenteries.

The *scyphistoma* is the name given to the segmented body and each of the segments is an *Ephyra* larva (Figs. 99 D, 103 D). They lie upon each other like a pile of saucers, connected, however, by strands of tissue in which run the muscles of the interradial mesenteries continuous throughout the pile of individuals. These muscles contract violently at intervals until the communicating strands snap and one by one the ephyrae swim away. The ephyra develops into the adult by the filling up of the adradial notches in the margins as well as by the growth of the bell as a whole. The mesogloea increases enormously in thickness, causing the two layers of the endoderm to come together as a solid lamella except where the canals occur. The mesenteries lose their attachment and cease to exist as partitions with the collapse of the enteron, but their position is marked by the gastric filaments. The basal part of the scyphistoma remains and grows new tentacles, and after a resting period as a hydratuba may strobilate again.

The life history of the sessile form may thus be summarized. The hydratuba feeds and buds in the summer, continues to feed and stores food in the autumn but ceases to bud, strobilates in the winter, grows new tentacles in the spring and feeds and buds again. In this the Scyphozoa show features in common with the life history of the hydroid colonies and the fresh-water *Hydra*.

The mode of development described above is typical in the Discomedusae. There are, however, certain exceptions. In the genus *Pelagia* the medusa develops directly from the egg into an ephyra larva, and in *Cassiopeia* the hydratuba only produces a single ephyra at a time, a condition which is obviously primitive compared with *Aurelia*; 'polydisc' strobilation being a secondary adaptation for the more effective spread of the species.

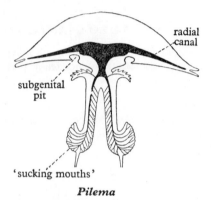

Pilema

Fig. 104. Diagrammatic section through *Pilema*.

The Rhizostomeae are a division of the Discomedusae in which the four lips around the mouth are vastly developed and folded, and the central mouth itself is narrowed and in a number of forms entirely closed. It is replaced by thousands of small 'sucking mouths' which lie along the course of the closed-in grooves of the lips. These lips now constitute organs of external digestion. Small copepods and even fish are enclosed by the lips, digested and the fluid absorbed through the 'sucking mouths' which are too small to admit solid particles of any size. The young medusa of *Rhizostoma* still has a central mouth, but in the adult of that and other forms, e.g. *Pilema* here figured (Fig. 104), it is entirely closed. *Cassiopeia* is a semi-sedentary form, which lies with its subumbrellar surface upwards on the mud of mangrove swamps. The bell pulsates gently and brings in a constant stream of plankton organisms which are seized by the lips.

LIFE CYCLES. The Scyphozoa show considerable variation in their life cycles. Some lay eggs that form actinula larvae which immediately turn into ephyrae. Others have eggs that turn into planula larvae which settle on the ground and bud off ephyrae one at a time (monodisc strobilation). Others bud off many ephyrae at a time (polydisc strobilation). Most of these phases exist within one genus, *Aurelia*; very large yolky eggs turn into actinula larvae and thence directly into ephyrae. Smaller eggs turn into planulae first of all and then into scyphistomae which bud or strobilate. The onset of strobilation is caused by a period of intensive feeding being followed by a drop in temperature. *Aurelia* normally strobilates in the winter months and the amount of food and the prevalent temperature determine the type of strobilation. If there is a lot of food and the temperature is low then the animal shows polydisc strobilation. If there is less food and the temperature is higher then the animal shows monodisc strobilation. Under certain conditions some genera of medusae have been found to have scyphistoma larva that never strobilate and they may even contain mature gonads. This onset of neoteny caused by a high uniform temperature indicates a possible way in which the Stauromedusae might have arisen.

CLASS 3. ACTINOZOA (ANTHOZOA)

DIAGNOSIS. Solitary or colonial coelenterates with polyp individuals only: coelenteron divided by mesenteries: stomodaeum present: genital cells derived from endoderm.

CLASSIFICATION

Order 1. **Alcyonaria (Octoradiata).** Actinozoa with eight mesenteries and pinnate tentacles. *Alcyonium, Corallium.*

Order 2. **Zoantharia (Hexaradiata).** Actinozoa with six or multiples of six mesenteries and simple tentacles. *Metridium, Tealia.*

1. *Order* Alcyonaria

DIAGNOSIS. Actinozoa with eight mesenteries and eight pinnate tentacles; stomodaeum with a single siphonoglyph (ciliated groove); skeleton internal, consisting of spicules in the mesogloea, occasionally supplemented by an external skeleton; longitudinal muscles on the ventral faces of the mesenteries.

ALCYONIUM. As a type of the order we will describe *Alcyonium digitatum*, 'Dead men's fingers', a colonial form which occurs below low-tide mark, attached to stones, in various sizes and shapes, but usually in broad-lobed masses. A small portion or lobe of a colony is shown in Fig. 105, and it is seen that the polyps project in life from the general surface of the colony. The ectoderm, mesogloea and endoderm of the polyps are of course continuous with the same layers in the coenosarc of the colony, but while the ectoderm is only a thin skin composed of a single layer of cells spread over the surface of the whole colony, the mesogloea is expanded to form a bulky mass of jelly which is traversed by the endodermal tubes of the polyps. These run parallel with each other without joining for considerable distances, but they are connected by other endodermal tubes which are much more slender, so that, like a hydroid colony, the alcyonarian colony has a common coelenteric system.

The polyps are delicate and withdraw on the slightest stimulus, the oral disc with its crown of tentacles being pulled inside the enteron by the contraction of longitudinal muscles running in the mesenteries and attached to the oral disc. By a continuation of this contraction the whole column of the polyp is introverted ('turned outside in', as with the finger of a glove). This is the condition in which preserved colonies of *Alcyonium* are nearly always found, and tangential sections through the superficial layers of the colony are rather difficult to interpret in consequence.

There is no oral cone in the actinozoan polyp, but the *mouth* is an elongated slit and is situated in the middle of a circular flattened area, the *oral disc*, which is surrounded by the tentacles. It does not open directly into the enteron

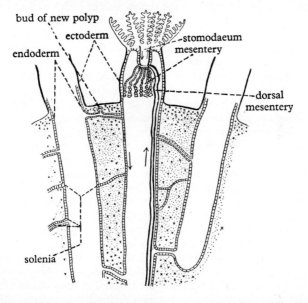

Alcyonium

Fig. 105. Diagram of section through colony of *Alcyonium*, showing extended polyps with pinnate tentacles and coenosarc. The direction of the water current is shown by arrows. (Original.)

but into a tube lined with ectoderm, the *stomodaeum* or *gullet*, which communicates with that cavity. The whole of the stomodaeum is ciliated, but at one end of it there is a groove which is lined with specially strong cilia which draw water in at the mouth. This is the *siphonoglyph*, and it is said to occupy a *ventral* position, but the student must be warned that there is no homology between surfaces so termed in the coelenterates and in the higher Metazoa.

Internally the enteron is divided up by eight vertical folds of the body wall, the *mesenteries*, which project so far into the cavity of the enteron that their upper parts join with the stomodaeum. Below the level of this organ they end in an enlarged free edge, the *mesenteric filament*. The foundation of the mesentery is the mesogloea, which is not much thicker here than in the body wall but is folded in the muscular region of the mesentery. On both sides it is covered with endodermal epithelium. While in the hydroid polyp there is little differentiation into regions, in the actinozoan polyp the endodermal cells specialized for various functions are arranged in strips of tissue occupying definite positions on the mesenteries. This may be seen in the sections of a polyp in Figs. 109 and 110. It must in the first place be explained that the presence of the siphonoglyph and the elongation of the stomodaeum are an indication that on the original *radial* symmetry of the polyp a bilateral symmetry has been imposed, and on each side of the axis of the stomodaeum the mesenteries correspond exactly in arrangement. Now the muscular endodermal cells are concentrated on the ventral side of each mesentery and into a narrow part of it to form a longitudinal retractor muscle. In the section below the siphonoglyph the mesenteric filament is seen, and this consists of different elements in the different mesenteries. One pair of mesenteries, which are 'dorsal' in position, are distinguished from the rest in having a filament which is flattened in cross-section, and is covered by very large ciliated cells (Fig. 109 E). They work in concert with the cells of the siphonoglyph to produce a current of water which is drawn in at the mouth and flows right along the ventral side of the tubes through the system, bearing with it oxygen and food for the tissues which are contained in the depths of the colony. The cilia of the dorsal mesenteries are responsible for the return current which makes its way out of the polyp by the dorsal side of the stomodaeum. These two mesenteries are much longer than the rest, as may be seen in Fig. 105, and their persistence throughout the endodermal tubes is necessary for the maintenance of the exhalant current. In contrast with this the remaining six mesenteries have rounded filaments covered with an epithelium consisting largely of gland cells. Also the germ cells arise near the free border (Fig. 109 E). Small organisms caught by the tentacles and introduced into the enteron are embraced by these mesenteric filaments and held fast while the fluid from the glands brings about a disintegration and partial digestion of the tissues. Solid fragments of food resulting from this are ingested by individual endodermal cells and the digestion completed. Not only do the dorsal mesenteric filaments differ from the others in function but they are ectodermal while all the rest are endodermal.

The mesogloea of *Alcyonium* is invaded by cells from the ectoderm which form in their cytoplasm aggregations of calcium carbonate with a characteristic shape which are called spicules. As the spicules develop the secretory cells migrate into the deeper parts of the colony. They are present in such numbers as to give a certain quality of solidity to the colony, and on its death the spicules it contains remain behind as a not inconsiderable mass. The part which alcyonarians consequently play in the formation of coral reefs, though secondary, is not unimportant. The mesogloea, as has been mentioned above, is traversed by hollow strands of endoderm (*solenia*) which communicate between the polyp tubes and also by solid endodermal strands which may play some part in the secretion of the jelly of the mesogloea. From the solenia, where they approach the surface, small buds are formed which develop into new polyps.

The gonads are developed at the breeding season, from groups of endodermal cells near the filaments, but they only occur on the six ventral mesenteries. The eggs are comparatively large and pass very slowly up the enteron and out of the stomodaeum, being fertilized outside the polyp and developing into a planula larva. After a free-swimming period this fixes and becomes a single individual which by budding gives rise to a colony.

STOLONS. Variation in the Alcyonaria occurs mostly in the method of formation of the colonies and the skeleton. The simplest form is found in *Cornularia* and *Clavularia*. From the original polyp a creeping stolon with a single endodermal tube is given off, and this gives rise at intervals to polyp buds, which may in turn produce fresh stolons. The coenosarc of the colony thus forms a network like a hydroid colony. In *Alcyonium*, as already described, the elongated polyps are crowded together in bundles and fused along nearly the whole of their length, the ectoderm and mesogloea of adjacent polyps being continuous, and the endodermal tubes in frequent communication. The mesogloea thickens enormously.

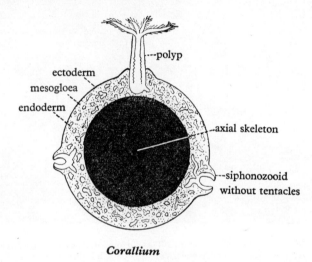

Corallium

Fig. 106. Section transverse to the axis of *Corallium*. (After Hickson.)

CORALLIUM. In the red coral *Corallium rubrum* (Fig. 106) there is an upright branched colony with a rigid axis composed of spicules compacted together which is the precious coral of commerce. This is clothed by the delicate tissue of the coenosarc from which the short polyps arise and which contains a network of endodermal tubes, some of which run along the parallel grooves which are sometimes to be seen on the surface of a piece of precious coral. The mesogloea contains spicule-forming cells derived from the ectoderm, and these travel inwards and add their secretion to the central skeleton. This form occurs at considerable depths in the Mediterranean and the seas of Japan. Dimorphism, as described below for *Pennatula*, also occurs here.

GORGONACEA. The gorgonians also have upright branching colonies. The supporting axis has, however, an origin, different to the last, being horny and not calcareous and secreted by the ectoderm on what is really the outer surface of the animal, As secretion is confined to an invagination of the basal epithelium which burrows into the whole length of the colony, it appears to be an internal skeleton. The gorgonians are a remarkable feature in shallow tropical seas, forming groves and thickets which challenge comparison with the plant forms of the land (Fig. 107).

Fig. 107. Gorgonians (two species on the left) and hydrocorallines (on the right) growing on a coral reef in Florida. (From an underwater photograph by Professor W. H. Longley.)

PENNATULACEA. In *Pennatula* and its relations a single axial polyp grows to a relatively enormous length, sometimes as much as three or four metres, and contains a long horny axis which is possibly endodermal. The secondary polyps are budded off from endodermal tubes which ramify in the much thickened mesogloea of the body wall of the primary polyp, and belong to two types of individuals, the normal *autozooids* which feed the colony and the *siphonozooids*, with reduced mesenteries and enlarged siphonoglyph, whose only function is to maintain the circulation of water in the canals of the colony. The autozooids in *Pennatula* are arranged in rows side by side to form equal and regular lateral branches on each side of the axis giving the colony its feather-like form, and the siphonozooids are mainly found on the back of the axis. A colony has a limited but remarkable power of movement and can burrow into sand or mud by its basal stalk.

TUBIPORA: HELIOPORA. In two genera, *Tubipora* (the organ-pipe coral) and *Heliopora* (the blue coral), which are widely distributed on coral reefs, a continuous calcareous skeleton is developed resembling that of reef corals. The polyps of *Tubipora* are elongated and parallel and connected by stony platforms which are traversed by the endodermal tubes. But while in *Tubipora* there is an internal skeleton developed as in *Corallium*, by the fusion of spicules in the mesogloea, in *Heliopora* the skeleton is secreted by a layer of ectodermal cells and not composed of spicules. In *Heliopora* (Fig. 108) there are on the surface of the colony larger pits (thecae) occupied by the polyps and smaller pits which lodge tubular processes of the network of solenia: the same skeletal characters also occur in the fossil *Heliolites* which closely resembles it and was a dominant type in Palaeozoic coral reefs. *Tubipora* too has a Palaeozoic representative in *Syringopora*.

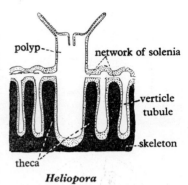

Fig. 108. Diagrammatic section through the edge of a colony of *Heliopora*. (After Kükenthal.)

2. Order Zoantharia

DIAGNOSIS. Actinozoa with mesenteries varying greatly in number, typically arranged in pairs, the longitudinal muscles of which face each other, except in the case of two opposite pairs, the *directives*, in which the muscles are on opposite sides; tentacles usually simple, six or some multiple of six in number; mesenteric filaments trefoil-shaped in section; stomodaeum with two ciliated grooves; typically a calcareous exoskeleton, but this may be entirely absent.

ANEMONES AND CORALS. The coelenterate animals which are included in this group fall into two apparently different categories, the sea anemones, which are usually single individuals and never possess any kind of skeleton, and the madreporarian corals, which are usually colonial animals and always have an ectodermal exoskeleton. The polyps, however, may all be referred to the same type of structure, and the presence or absence of a skeleton or of the colonial habit are matters of secondary importance compared with this.

POLYP. In its main structural lines the zoantharian polyp resembles the alcyonarian type. The stomodaeum is elongated in the same plane but possesses two siphonoglyphs instead of one. There are tentacles which are hollow, unbranched, and often very numerous. The mesenteries are like those of *Alcyonium*, but their arrangement and the structure of the mesenteric filament is very different. Numbers and grouping of mesenteries vary greatly within the limits of the Zoantharia itself. The simplest form, and that most like *Alcyonium* (Fig. 109A), is found in the small burrowing sea-anemone, *Edwardsia* (Fig. 109c). Here there are eight mesenteries with bilateral symmetry, as in *Alcyonium*. In six of these the longitudinal muscles are on the same side, facing ventrally, while the remaining pair have the muscles facing outwards and dorsally, so that the arrangement is different from that in the Alcyonaria.

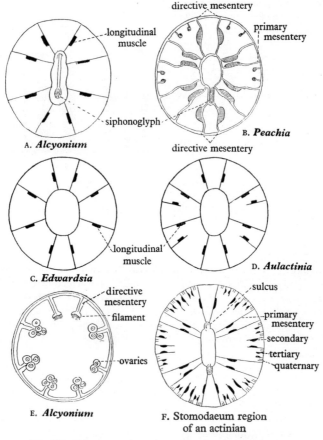

Fig. 109. Diagrammatic sections of corals and sea-anemones. The direction of the top of the page is dorsal.

MESENTERIES. In the typical sea-anemone, such as *Actinia*, and in coral polyps, the mesenteries are arranged in cycles (or generations). There are six couples of primary mesenteries in the first cycle, and these are the largest and alone reach as far as the stomodaeum. In four of these pairs the muscles face each other; in the other two pairs, the directives, they face away from each other. The secondary mesenteries, which are much smaller, are situated in the spaces between two adjacent pairs (exocoeles), never between two members of a pair (entocoeles). Finally, there may be tertiary and even quaternary mesenteries, always in exocoelic spaces of the generation preceding, making third and fourth cycles. This 'hexactinian' type, in which the mesenteries are present in multiples of twelve, is derived from that in *Edwardsia*, as may be seen in the development of some of the Zoantharia, for example another small burrowing anemone, *Halcampa*. In this there is first of all an *Edwardsia* stage (Fig. 109 C) with eight mesenteries. From this the hexactinian type is derived quite simply by the subsequent growth of four additional mesenteries with muscles on their dorsal faces. These belong to the first cycle and join up with the stomodaeum, and they arise in such positions as to complete, with pre-existing mesenteries, four pairs with muscles facing each other. These four mesenteries in *Halcampa* never develop a mesenteric filament, but the com-

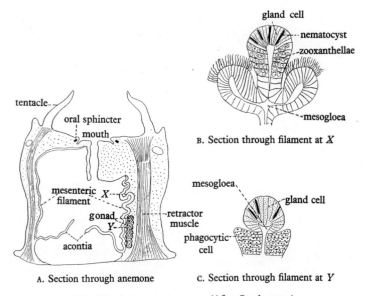

Fig. 110. Anemone structure. (After Stephenson.)

plete adult arrangement, as seen for instance in *Actinia equina*, the commonest of our British anemones, is given in Fig. 109 F. In such a form as *Peachia*, often used in laboratories on account of its simplicity, there are slight deviations from the type. There is no second siphonoglyph (sulcus) and the second cycle of mesenteries is incomplete, none of them having a mesenteric filament, while the pairs in two exocoeles are completely absent (Fig. 109 B).

The mesenteric filament of the Zoantharia (Fig. 110 B, C) is trefoil-shaped in section, and while the functions of digestion and water-circulation are in the Alcyonaria performed by different filaments, here they are performed by different parts of the same filament. Thus, near the stomodaeum, the central part of the filament of a sea-anemone or coral is crowded with digestive gland cells and also with nematocysts, while the wings are covered with strongly ciliated epithelium which maintains a current. In the lower part of the mesentery the filament is exclusively digestive in function: the cells of the wings are phagocytic, as is shown by feeding with carmine. From the central part of the filament free threads called *acontia* are produced in some anemones, which are loaded with nematocysts and may be shot out of the mouth or out of special pores in the body wall when the polyp is stimulated.

SKELETON. In the corals the skeleton is secreted by the ectoderm, but only by that part of it which forms the *basal disc*. A flat plate of calcium carbonate is laid down first of all by the whole of the disc, but almost at once the epithelium is thrown into radial folds and into a circular fold which encloses them, and in these are formed vertical walls which rise from the plate; the circular wall is called the *theca* and the radial wall *septa* (Fig. 111 A). The latter are

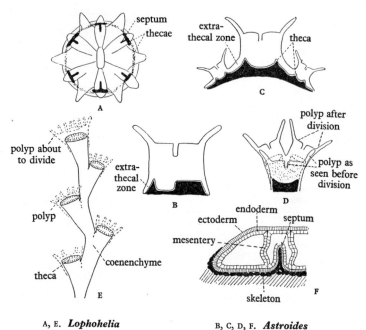

A, E. **Lophohelia**   B, C, D, F. **Astroides**

Fig. 111. Skeleton formation in the Zoantharia: A, oral view of a young coral polyp with the beginning of the skeleton seen through the transparent tissues. B, vertical section through a later stage. C, development of a colony showing budding from the extra-thecal zone. D, division of polyp. E, *Lophohelia*. F, *Astroides*; tangential section of young colony fixed to cork. (After von Koch.)

formed in spaces between the mesenteries. The continued secretion by such a form as the English solitary coral *Caryophyllia* produces a cup of limestone, of which the tapering basal portion is solid but which has a shallow apical depression, which is traversed by the radiating vertical septa and contains in the centre a more or less regular vertical rod, the *columella*. The depression always tends to become filled up by the secretory activity of the general surface of the basal disc, but the building up of the theca and septa keeps pace with this. It is difficult at first to realize that this is an exoskeleton and that in a massive structure like a brain coral the actual living tissue is a mere film on the surface of a great hemispherical mass of calcium carbonate which it has secreted. It is not surprising to learn that such colonies with a diameter of a yard or more have a life span of a hundred years or so.

SECRETION. With regard to the actual mechanism of lime secretion the view most generally held is that illustrated by Fig. 111 F, which shows a coral larva which has fixed upon a piece of cork. The skeleton as shown in section is, when first laid down, a series of spheroidal masses of calcium carbonate, which thus appear to be a secretion of the ectoderm cells, issuing from the cells as a solution and immediately crystallizing out as irregular masses. Another suggestion is that ammonium carbonate excreted by the coral meets the calcium salts of the sea water and calcium carbonate is precipitated round the ectoderm; and still another, that calcium carbonate is stored up in the ectoderm cells and when the cells are full they drop out of the epithelium and are added to the skeleton.

CORAL COLONIES exist in the most diverse shapes and forms (Fig. 112), from the slender tree-like colonies of many *Madrepora* to the massive rounded forms like *Porites*. Each colony is formed from a single planula which settles down and forms a polyp. From this first individual the hundreds of thousands of polyps in a large colony are formed by division or gemmation. An example of division is given in Fig. 111 D. In such a case when the polyp has reached a certain size the oral disc becomes elongated in the direction of the long axis of the mouth, tentacles and mesenteries increase in number, and finally a transverse constriction divides first the mouth, then the disc and lastly the whole polyp. The division of the polyp is followed by that of the theca. In the Meandrine corals (brain corals) the polyp elongates enormously and the mouth divides but not the theca, and so we get the curious thecae running more or less parallel to each other which recall the convolutions of the human brain. In *Lophohelia* (Fig. 111 E) division is equal, but while one of the polyps resulting from it continues to grow the other marks time; the axis of growth changes sides at each division and the result is a colony showing cymose branching.

In Fig. 111 B it is shown that part of the coral polyp overlaps the theca. It is this *extrathecal zone* which gives rise to young polyps when a colony is formed by gemmation (Fig. 111 C). The bud and the parent remain connected by their extrathecal portions, and this constitutes the coenosarc of the colony. The gaps between the thecae of the colony are filled up by calcareous material secreted by the coenosarc and called *coenenchyme*.

BEHAVIOUR. The polyps of the Zoantharia attain a higher physiological grade than those found elsewhere in the coelenterates. The sea anemones, like *Hydra*, in the absence of any external skeleton, are capable of locomotion, especially in the case of burrowing forms. The muscles of the body are arranged in such a way as to bring about many different kinds of movements. Thus, while the longitudinal muscles of the mesenteries cause a longitudinal retraction of the polyp, the transverse muscles of the mesenteries in the neighbourhood of the stomodaeum open the mouth when they contract, and the longitudinal muscles of the tentacles when these are touched by particles of food contract so that the tentacle bends towards the mouth and helps to push the food inside it. The muscular system is for the most part under the control of the nerve net. If a sea-anemone is violently stimulated, e.g. touched by a glass rod in any part, the stimulus is transmitted to every muscle and the whole animal shrinks to a shapeless lump. The process of feeding is extremely com-

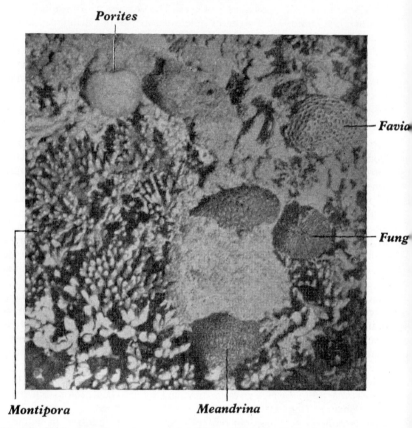

Fig. 112. Photograph of a pool on a coral reef (Great Barrier Reef) showing various types of zooantharian corals. (Photograph by Dr S. M. Manton.)

plex and involves the action of the muscles, the cilia and the glands. In a sea-anemone like *Metridium*, which lives on the minute animals of the plankton, when these approach the oral disc they are stunned by the nematocysts, snared by the mucus of the glands of the tentacles, transported by cilia to the tips of the tentacles, and pushed by the tentacles towards the mouth, which gapes to receive them. Most remarkable of all, the cilia of the lips, which normally maintain the outwardly flowing respiratory current, reverse their beat to sweep the food into the enteron. While there is this remarkable co-ordination of activities in feeding the nerve net preserves the individuality in action of the parts so that the severed tentacle of a sea-anemone is able to execute movements just as if it were still in place, on the appropriate stimulation. In another common anemone, *Tealia*, there are no cilia on the tentacles and oral disc, and feeding takes place entirely by the muscular movement of the tentacles.

Sea-anemones and corals are often nocturnal, remaining contracted by day, expanding and feeding at night. In such corals as *Lobophyllium* the tentacles are capable of enormous extension. In the forms which feed by day like *Fungia* the tentacles are shorter and the food is collected more by the action of cilia on the tentacles and oral disc and less by the seizing of organisms by the arms and withdrawal to the mouth. A remarkable biological feature is the frequent presence of commensal cells (compare *Hydra viridis*) in the tissues (Fig. 110B). This is especially the case in reef corals, in which the most recent investigations show that the cells are of no nutritive value while the oxygen they liberate in the tissues has no relation to the needs of the coral. On the other hand the fact that they remove excreta from the coral tissues is of importance.

SUBPHYLUM CTENOPHORA

DIAGNOSIS. Free and solitary Coelenterata; whose active locomotion takes place by ciliary action; which are not reducible either to the polyp or to the medusoid type; and are without nematocysts, but possess 'lasso cells'.

CLASSIFICATION. The Ctenophora are divided into two classes: (1) *Tentaculata*, possessing tentacles, to which the majority of forms belong; (2) *Nuda*, without tentacles, to which belongs only the genus *Beroe* (Fig. 114B).

The Ctenophora, apart from certain aberrant forms, are globular, pelagic, transparent animals living in the surface waters of the sea. They may be classed with the Coelenterata, but they differ from other members of that phylum in several important respects, notably in the entire absence of nematocysts.

STRUCTURE. Two British forms are easily procurable, *Pleurobrachia pileus* and *Hormiphora plumosa*. *Pleurobrachia pileus* is about the size of a small hazel nut, while *Hormiphora plumosa* (Fig. 113) is rather smaller. They are transparent and ovoid. At one pole is the mouth; the only other openings into the alimentary canal are two small pores near the sense organ. At the other pole is the sense organ marked as a small spot lying in a slight depression. The surface of the body is beset by eight meridional rows of comb plates formed of strong cilia borne upon modified ectodermal cells. The general surface of the body is not ciliated.

On opposite sides of the body are two tentacles set in pouches. The tentacles have muscular bases and are capable of being protruded from the pouches or withdrawn again. They are usually about half as long again as the body when fully extended. The tentacles are armed with cells of a special type called 'lasso cells' or colloblasts, which take the place of nematocysts. Each colloblast consists of a sticky head having at its base a spiral thread wound round a stiff central filament. The tentacles are used for catching the prey which is entangled by the sticky heads of the colloblasts.

The mouth leads through a stomodaeum lined with ectoderm into a space, the infundibulum, lined with endoderm. From the infundibulum four canals radiate outwards; each of these divides into two and then runs under the comb plates as the subcostal canals. Two more canals lead out from the infundibulum and run directly without branching to the base of the tentacles. There are also two paragastric canals running alongside the stomodaeum.

At the opposite pole to the mouth, the aboral pole, is the elaborate sense organ formed of small round calcareous bodies united into a morula. This morula is supported on four pillars of fused cilia and is covered by a roof also

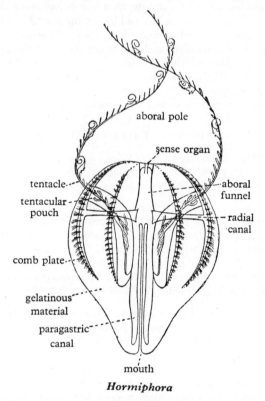

Hormiphora

Fig. 113. *Hormiphora plumosa*, side view. (After Chun.)

formed of fused cilia. Ciliated furrows lead out from the sense organ to the comb plates and are believed to assist in carrying stimuli to the comb plates from the sense organ.

The comb plates are the locomotor organs. When at rest the tips of the cilia are directed towards the oral pole. In movement a rapid beat of the cilia is directed aborally and the cilia then return slowly to rest. The ctenophore therefore moves slowly through the water with the oral end in front. Each plate of the comb beats in succession, the first plate to beat being the one at the aboral end and the remainder following in succession. This type of beating, which is common in ciliary movement, is termed 'metachronal' (see p. 17). It gives the appearance of waves travelling down the comb from the aboral to the oral pole. Ordinarily all the eight rows of comb plates beat in unison, but interference with the aboral sense organ destroys this unison.

The main substance of the ctenophore, which fills the space between the ectoderm and the endoderm, is a gelatinous material in which are found strands of muscle. Immediately beneath the ectoderm lies a subcuticular layer of muscle and nerve fibres which, in appearance, closely resembles the arrangement found in the Turbellaria. It is important to note that the whole musculature of the Ctenophora is derived from the mesenchyme. There are no musculo-epithelial cells.

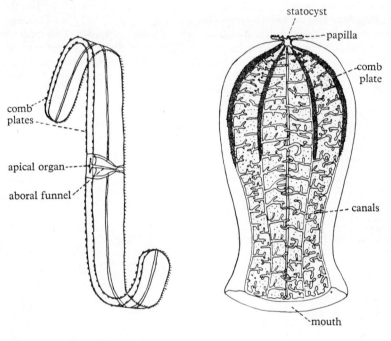

A. *Cestus veneris* B. *Beroe*
Fig. 114. A, *Cestus veneris*; B, *Beroe*.

REPRODUCTION. Ctenophores are hermaphrodite; the male and female gonads occur close to each other in the subcostal canals. Self-fertilization probably occurs. It is a remarkable fact that, if the first two segments of the dividing egg of a ctenophore be separated, a half-larva will develop from each segment. In the egg, therefore, the organ-forming substances must be localized. If these half-larvae be kept until generative organs develop, the missing half is then regenerated. In contrast to this behaviour in the Ctenophora, the separated blastomeres of the cnidarian egg as far as the sixteen-celled stage will develop each into a complete animal.

Most of the Tentaculata have the ovoid shape, similar to that seen in *Pleurobrachia*, but some are flattened in a peculiar manner. *Cestus veneris*, Venus' Girdle, is flattened laterally and the body is drawn out into a narrow band, two inches wide and nearly a yard long. It is found in the surface waters of the Mediterranean (Fig. 114 A).

The Platyctenea, a group of Tentaculata to which belong the forms *Coeloplana* and *Ctenoplana*, are flattened dorsoventrally. The flattening is produced by the expansion outwards of the stomodaeum so that the whole of the ventral surface corresponds to the stomodaeum of the normal types. *Ctenoplana* lives in the surface waters of the sea and retains traces of the swimming plates, but *Coeloplana* crawls over the rocks and seaweed, and resembles a turbellarian. It has lost the swimming plates and developed pigment, but it still retains the sense organ and the two tentacles. The gut system is irregularly branched and the muscular system is highly developed for crawling purposes. One member of the group, *Gastrodes*, is a parasite in the body of *Salpa*. Its chief interest, however, is in the larva, which is a *planula*, found nowhere else among the Ctenophora, and thus provides the strongest piece of evidence for the close relationship of the Ctenophora with the Coelenterata.

CHAPTER VI

THE ACOELOMATA

Under this title are grouped the phyla Platyhelminthes, Nemertea, Rotifera, Nematoda, Gastrotricha, Acanthocephala and Nematomorpha (the last three of which are very small groups). The animals contained in these are unsegmented forms with mesenchyme (p. 127) and the space between the gut and the body wall (when it exists) is a primary body cavity filled with fluid (e.g. Rotifera). The turgor of the body cavity fluid when present has a determining role in the preservation of the form of the body (e.g. Nematoda and Rotifera). Generally speaking this space with its contained fluid plays the part of a circulatory system, but in the Nemertea the body cavity is reduced to a series of canals which constitute the first vascular system in the animal kingdom. This primary body cavity has no definite epithelial boundaries and so can be easily distinguished from a true coelom. It tends to be invaded by mesenchyme cells; in the Platyhelminthes these completely fill it, forming a characteristic tissue (parenchyma), and in the Nematoda the cavity appears to be completely occupied by a very few enormous vacuolated cells whose vacuoles simulate a body cavity.

The excretory organ is of *nephridial* type (or it may be derived from this as in Nematoda). It is a canal, closed at the internal end, intracellular or intercellular, with some hydromotor arrangement which maintains a flow of fluid to the exterior. In the simplest cases there is a continuous ciliation of the inner wall of the canal (some Turbellaria). Usually, however, the ciliation has disappeared over most of the canal but is strengthened and differentiated in others; the characteristic units of the system, the flame cells, being now found. Flame cells may be situated in the course of the canal in some forms but usually constitute the *terminal organ* (Fig. 120B). This system, though usually spoken of as 'excretory', is primarily concerned with the regulation of fluid content and is sometimes absent in marine forms (e.g. Turbellaria Acoela, p. 207). A nerve net is usually present and from this are differentiated an anterior 'brain' and some longitudinal nerves. The reproductive system is that in which differences between and within the groups principally occur: these differences are to be regarded as adaptations to the varying conditions of life.

PHYLUM PLATYHELMINTHES

DIAGNOSIS. Free-living, and parasitic, bilaterally symmetrical, triploblastic Metazoa; usually flattened dorsoventrally; without anus, coelom or haemocoele; with a flame-cell system; and with complicated, usually hermaphrodite, organs of reproduction.

VERMES. The name Platyhelminthes is given to a division of that heterogeneous collection of animals which in Linnaeus's time were called Vermes. The Vermes included everything that looked like a worm, but appearances have since been found to be deceptive and the collection has been broken up into separate phyla, one of which is the Platyhelminthes or flatworms. Of all the worm-like animals the flat-worms are undoubtedly the most primitive, for they alone show relationships to the Coelenterata. Some authors have suggested that the Turbellaria are the most primitive of the Metazoa, and that the Coelenterates are derived from the Platyhelminthes.

CLASSIFICATION

Class 1. **TURBELLARIA.** Free-living platyhelminthes, with a gut, a cellular ciliated outer covering to the body, usually having rhabdites, not forming proglottides. Suckers are rarely present. The systematics are based primarily on the arrangement and structure of the gut.

Order 1. **Acoela.** The gut is not hollow but is a syncytium formed by the union of endodermal cells. There is no muscular pharynx. *Convoluta, Otocelis*

Order 2. **Rhabdocoela.** The gut is straight with the mouth at the anterior end. *Microstomum, Rhynchoscolex, Dalyellia*

Order 3. **Alloiocoela.** The gut has small diverticula arising from it. *Plagiostomum, Hofstenia, Otoplana*

Order 4. **Tricladida.** Gut with three branches, one directed forwards, two directed backwards.
Suborder 1. **Maricola.** Marine forms. *Procerodes, Bdelloura*
Suborder 2. **Paludicola.** Fresh-water forms.
Phagocata, Polycelis, Planaria
Suborder 3. **Terricola.** Terrestrial forms. *Bipalium, Cotyloplana*

Order 5. **Polycladida.** Gut has many branches radiating out from central mouth.
Suborder 1. **Acotylea.** No sucker. *Euplana, Leptoplana*
Suborder 2. **Cotylea.** Have a sucker. *Thysanozoon, Yungia*

Order 6. **Temnocephalea.** Ectocommensals on fresh-water crustaceans, reduced ciliation, develop prolongations on the anterior end, have suckers. *Temnocephala, Actinodactylella*

Class 2. **TREMATODA.** Parasitic platyhelminthes with a gut, a thick cuticle, and suckers that may be thickened by a series of chitinous ridges.

Order 1. **Heterocotylea** *or* **Monogenea.** Oral suckers usually absent or poorly developed, posterior suckers usually well developed and complex. No alternation of hosts. *Polystomum, Octobothrium*

Order 2. **Malacocotylea** *or* **Digenea.** Anterior sucker well developed, alternation of hosts. *Distomum, Schistosoma*

Class 3. **CESTODA.** Endoparasitic platyhelminthes, no gut, adult has lost ciliated ectoderm and replaced it by a thick cuticle; proglottides usually formed.

Order 1. **Cestodaria.** Tapeworms with undivided bodies, do not form proglottides. *Amphilina*

Order 2. **Eucestoda.** Tapeworms with body divided into proglottides.
Taenia, Diphyllobothrium, Moniezia

GENERAL ACCOUNT

Of these the Turbellaria are with few exceptions free-living, while the Trematoda and Cestoda are all, without exception, parasites. It is in the Turbellaria that we see most clearly the typical organization of a platyhelminth, for in the Trematoda and Cestoda the parasitic habit has induced a considerable departure from the structure of the free-living ancestor.

In shape the Platyhelminthes are flattened, they are not segmented and do not possess a coelom. The ectoderm is ciliated in the Turbellaria, but the ciliation is lost in the two parasitic groups and there are further modifications.

The gut, which is present only in the Turbellaria and Trematoda, has but one opening which serves both as mouth and anus, and in this respect reminds us of the Coelenterata. Between the ectoderm and the endoderm which constitutes the lining of the gut there exist a large number of star-shaped cells with large intercellular spaces forming a mass of *parenchymatous tissue*.

The nervous system consists essentially of a network as in the Coelenterata, with the important difference that there is an aggregation of nerve cells at the anterior end which, in the free-living forms almost always takes the form of a pair of *cerebral ganglia*, and that certain of the strands of the network stretching backwards from these cerebral ganglia are often more distinct than others and merit the name of nerve cords (Figs. 117, 118). There is, therefore, the beginning of a definite central nervous system. There are no ganglia other than the cerebral, but in the general nervous network nerve cells and nerve fibres are mixed together.

ECTODERM. The outer covering of a platyhelminth differs according to the group to which it belongs. In the Turbellaria the outer covering is formed of ectodermal cells. These are usually large and flat, sometimes with peculiar branched nuclei as in *Mesostomum*, or smaller and with round nuclei as in the majority of forms. Externally the cells are ciliated, the cilia being arranged in tracts over the surface of the body. Inside the cells are seen a number of crystalline, rod-shaped bodies, known as *rhabdites*. Although much has been written about rhabdites their function remains obscure. They are a secretion, more or less firm, which dissolves and becomes liquid in contact with water. They are formed in special cells, lying either between the ectoderm cells or just beneath them in the parenchyma, and distributed thence to the ectoderm cells. Rhabdites are usually absent from the ectoderm cells in the neighbourhood of sense organs. It will be noticed that when Turbellaria are placed for preservation in an irritant fluid such as acetic acid the body becomes covered with an opaque white layer. Whether this opaque layer is produced from the rhabdites or from the slime glands which occur in certain regions of the body is not certain.

BASEMENT MEMBRANE. Immediately below the ectoderm lies the basement membrane. This is a thin transparent structureless layer, which probably assists in preserving the general shape of the body and serves as an attachment for the muscles which lie immediately beneath it.

The basement membrane is continuous over the body except where it is penetrated by the openings of gland cells. It is absent beneath the ectoderm,

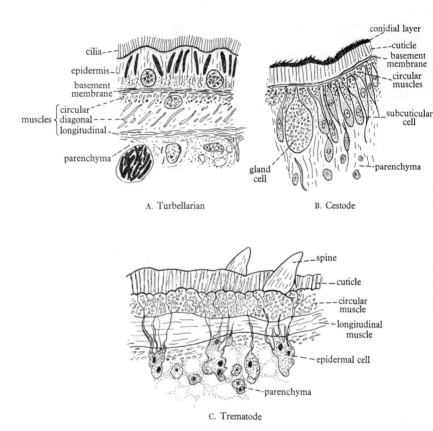

Fig. 115. Diagram of the structure of the ectodermal and mesodermal layers of a turbellarian, trematode and cestode.

overlying the sensory areas. In certain parts of the ectoderm, notably in the pharynx of the Tricladida, the nuclei of the ectoderm cells sink through the basal membrane and its underlying muscle layer and come to lie in the parenchyma attached to the cells by long strands of protoplasm. In the Trematoda and the Cestoda, the ectoderm cells have all sunk into the parenchyma, and the body is covered by a thick cuticle secreted by the ectoderm cells (Fig. 115).

PARENCHYMA. The parenchyma (also called the *mesenchyme*), which fills the interior of the body, is of very different structure in different Platyhelminthes. It is generally formed of cells with long irregular processes and much intercellular space. Within these cells are small granules and particles, which stain readily. Their appearance and number vary according to the state of health of the animal, whether it is starved or fed, and they are probably, therefore, products of secretory activity formed after the assimilation of food and destined eventually to be converted into rhabdites or the slime which flows from the slime glands. The parenchyma is no mere padding tissue. It probably serves for the transport of food materials, and certain cells in it provide for the repair of lost parts of the body. These free cells of the parenchyma retain their embryonic condition and do not become vacuolated or branched. They are smaller than the branched cells of the parenchyma and scattered among them in normal circumstances, but when an injury occurs they migrate to the cut surface, where they collect in large numbers and proceed to regenerate the tissues lost by injury.

MUSCLES. Passing through the parenchyma and running dorsoventrally are strands of muscle which are attached at either end to the dorsal and the ventral muscle layers. The muscles themselves consist of fibres formed of a homogeneous transparent material that shows no trace of any structure. These fibres are produced by a special cell, the *myoblast*, which is often to be seen lying alongside the fibre it has produced.

DIGESTIVE SYSTEM. The digestive system of the platyhelminth differs entirely from that of the higher animals in that it is a sac with one opening only, which serves both for the entry of the food and the exit of the faeces, and not a tube with a mouth and anus serving separately for the entry and exit of food. In the simplest forms, in many of the Rhabdocoela, the sac is a straight wide tube with no diverticula (Fig. 116), while in others the gut is branched. In the Tricladida the gut has three main branches. A muscular structure lined by an inturning of the ectoderm surrounding the mouth forms the *pharynx*. The pharynx itself may lie in a pit of the ventral body wall, called the *pharynx pouch*, from which it can be protruded or withdrawn. The epithelial lining of the gut cavity consists of large cells without cilia, the cell walls of which are often difficult to distinguish. A muscular wall to the gut is present, but is so exiguous as to avoid identification in many forms, and it appears therefore as if nothing separates the cells of the gut from the parenchyma. It is possible for food substances to pass not only from the lumen of the gut into the cells lining it, but also from the parenchyma. Thus when Turbellaria are starved they can consume certain organs lying in the parenchyma (ovaries, testes, etc.) by passing these into the gut cells or into the lumen of the gut for digestion.

The Turbellaria are carnivorous and will eat small living crustacea or worms which are caught by the protrusion of the pharynx. A sticky secretion, derived from the slime glands and perhaps the rhabdites, is immediately poured over the prey, which is thus wrapped up in slime. If the object is small enough it is ingested whole into the gut. Here digestion proceeds. Fat is digested in the lumen of the gut, but the digestion of other substances takes place in vacuoles in the cells of the gut wall. Animals which have recently died are also eaten by Turbellaria, and an effective trap can be made by placing a freshly killed worm or a *Gammarus* or two in a jampot and lowering it to the bottom of the stream or pond. The Turbellaria are able to 'scent out' the food, and all those within a wide area collect in the pot for the feast. When the animal is too large to be ingested whole, the pharynx is attached to the prey and worked backwards and forwards with a pumping motion, while at the same time a disinte-

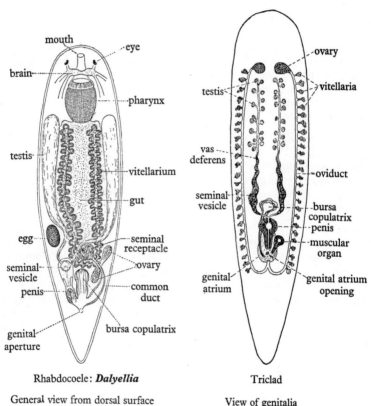

Fig. 116. Turbellarians: A, *Dalyellia viridis*. (From Bresslau.) B, Genitalia of a planarian. (After Steinmann.)

grating digestive fluid is poured out from the walls of the pharynx. Particles of food are thus pumped up into the gut cavity and digested in the same way as the living prey. In the Trematoda, also, the cells lining the gut have a certain limited power of amoeboid movement at their exposed edges, and intracellular digestion is apparently the usual method.

STARVATION AND REJUVENATION. The Turbellaria are able to go without food for long periods, but during starvation they grow smaller and smaller. Stoppenbrink starved *Planaria alpina*, keeping them entirely without food, while as a control he kept a similar collection supplied with food. His results are given in the table below. The measurements are in millimetres.

	Fed				Starved			
	Largest		Smallest		Largest		Smallest	
Date	Length	Breadth	Length	Breadth	Length	Breadth	Length	Breadth
16.iii.03	13	2	10	1	13	2	10	1
15.vi.03	17	2·5	12	1½	10	1½	6	⅔
15.ix.03	17	2·5	13	2	7	1	4	½
15.xii.03	17	2·5	14	2	3½	½	2⅓	⅓

This reduction in size is accompanied by the absorption and digestion of the internal organs, which disappear in a regular order, the animal using these as food in the manner already described. The first things to go are the eggs which are ready for laying, then follow the yolk glands and the remainder of the generative apparatus. Finally the ovaries and the testes disappear, so that the animal is reduced to sexual immaturity. Next the parenchyma, the gut and the muscles of the body wall are reduced and consumed. The nervous system alone holds out and is not reduced so that starved planarians differ in shape from the normal forms in having a disproportionately large head end, the bulk of which is the unreduced cerebral ganglion. On feeding these starved forms will regenerate all the lost organs and return to the normal size, like Alice when she ate the right half of the mushroom.

NERVOUS SYSTEM. The nervous system consists essentially of a network as in the coelenterates with the important difference that there is an aggregation of nerve cells at the anterior end which, in the free-living forms, almost always takes the form of a pair of cerebral ganglia, and that certain of the strands stretching backwards from these cerebral ganglia are often more distinct than others and merit the name of nerve cords. There is, therefore, the beginning of a definite central nervous system. There are no ganglia other than the cerebral but in the general nervous network nerve cells and nerve fibres are mixed together (Figs 117, 118 A).

By operating on the animals in different ways it is possible to show what functions the different parts of the nervous system have. If the cerebral ganglion of a Polyclad is removed, the body of the animal remains permanently quiescent after the operation. This state of quiescence is not, however, due to a loss of co-ordination in the motor system. Stimulation of the anterior end can evoke all the normal forms of locomotion, and this shows that the nerve net and not the cerebral ganglion is responsible for the correlation of the different parts of the musculature. The primitive central nervous system which here takes the form of a cerebral ganglion is best regarded as a development in connexion with the special sense organs, from which it receives stimuli. The cerebral ganglion functions as a relay system in which the stimuli received from the special sense organs are reinforced, often extended in time, and then passed on to the nerve net. When this sensory relay has been destroyed by removing the cerebral ganglia, the nerve net is no longer excited to bring the muscular system into action, although this may still be done by artificial stimuli.

198 ACOELOMATA

SENSE ORGANS. Sense organs occur in adults only in the free-living Turbellaria, where they may take the form of eyes, otocysts, tentacles and ciliated pits in the ectoderm. They may also occur in the free stages in the life history of the Trematoda and Cestoda. The *eyes* occur on the dorsal surface

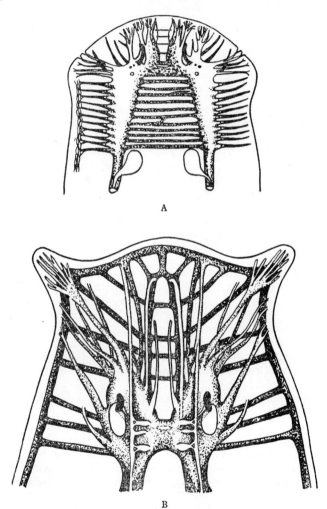

Fig. 117. Brain and anterior nervous system of triclad turbellarians.
A. *Dugesia*. B. *Planaria*. (After Hyman.)

where they are visible as dark spots. The retina is formed of cup-shaped cells, which are heavily pigmented. The interior of the cup is filled with special nerve cells, varying in number from two to thirty, the fibrillae of which touch the retina, and the fibres at the other end are joined together to form an optic nerve leading to the brain. There is no lens, but the ectoderm over the eye is not pigmented and so permits light to pass through it (Fig. 118 C). It should be noted that in this simple eye, as in the extremely complicated organ found in the vertebrates, the light has to pass through the sensory cells of the nervous system before it reaches the retina, for they are in front of, not behind, the retina. This type of eye is easily seen and studied in the common fresh-water planarians. In *Planaria lugubris*, the eye has only two sight cells, while in *Planaria lactea* there are thirty.

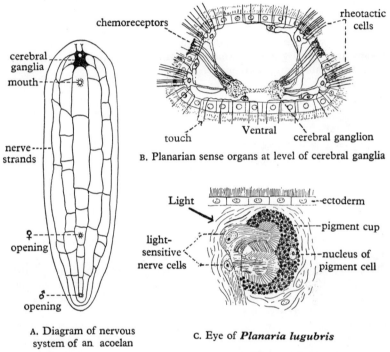

A. Diagram of nervous system of an acoelan

B. Planarian sense organs at level of cerebral ganglia

C. Eye of *Planaria lugubris*

Fig. 118. The nervous system and sense organs of turbellarians. (A, after Steinmann; B, after von Gelei; C, from Hesse-Doflein.)

Special sensory cells which act as receptors for the appreciation of changes in the composition of the surrounding medium (chemo-sensory receptors) or to changes in the flow of water past the surface of the body (rheotactic receptors) are situated just below the ectoderm. Their endings project through the ectoderm and form the actual receptor organ. The touch receptors are spread uniformly over the surface of the body in the Rhabdocoelida, but tend to be more numerous near the mouth. The endings of the touch receptors project among the cilia and are of the same length as these. The rheotactic receptors are confined to certain areas; their endings project among the cilia and are slightly longer than these. Special chemo-sensory receptors with short nerve endings that project only just above the surface of the ectoderm occur in definite areas or grooves on the head. Here the cilia and rhabdites are absent. These areas are known as auricular organs. These sensory organs may also be sunk into pits which, as they are provided with long cilia for driving the water into them, are known as *ciliated pits* (see Fig. 119).

The *tentacles* are projections of the body wall near the anterior end. They are found in the Turbellaria only, but are not present in all these. When present they are quite distinct and have very long cilia which, by their motion, set up currents which pass the water over special sensory areas and so lead us to suppose that their use is for water-testing, or searching for food. Occasionally these tentacles may be sunk into pits.

A statocyst occurs in primitive forms of the Turbellaria. It is situated above the brain and suggests a connexion with the Coelenterata where such sense organs are common, but as we know nothing of its nervous supply it is difficult to make a proper comparison.

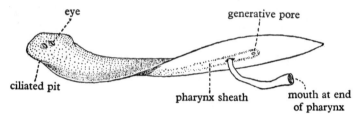

Fig. 119. *Planaria*, × *c*. 4. (From Shipley and MacBride.)

LOCOMOTION. Movement in the Platyhelminthes is effected in two ways. The animal may creep over a surface by the motion of the ectodermal cilia, the surface being freely lubricated when necessary, as is the case in land forms by the discharge of slime from the ectodermal slime glands. More rapid movement is effected by the general musculature of the body which causes a series of undulations to pass backwards along the flat body and urges it forward (Fig. 119). The *musculature* of a platyhelminth consists of a covering of muscle lying just below the ectoderm and composed of two layers, an outer circular and an inner longitudinal layer, except in the Cestoda and in the pharynx of the Turbellaria where the outer muscles are the longitudinal and the inner the circular.

EXCRETORY SYSTEM. An excretory system exists in nearly all Platyhelminthes. In the Acoela, however, it is absent. The excretory system usually consists of main canals, running down either side of the body (Fig. 120 A). The position of the openings of these main canals to the exterior varies. The main canals are fed by smaller branches which are ciliated, while the main canals are not. These smaller branches again branch many times and finally end in an organ known as a *flame cell* (Fig. 120 B). The large canals are often quite easily visible in living specimens, but the flame cell is exceedingly small and can only be seen in transparent forms as in the cercaria larvae of the Trematoda. The flame cell itself consists of a cell with branched processes extending amongst the parenchyma cells. Attached to the cell are a number of cilia which move together in the lumen of the canal with a flickering movement. It is from this flickering motion that the cell derives its name. It is generally believed that excretion of substances into the lumen of the tube is performed by the cells forming the wall of the tube itself. The flame cells represent concentrations of the originally complete ciliary lining of the canal and their function is to maintain a hydrostatic pressure which will cause the excreted substances to move down the lumen of the tube to the exterior (see also p. 191).

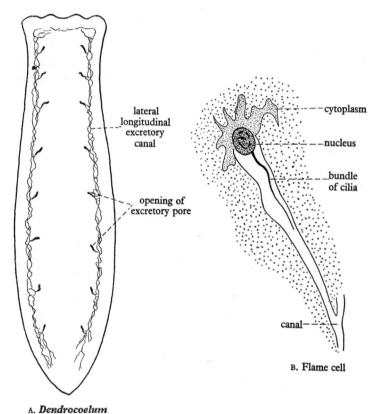

A. *Dendrocoelum*

Fig. 120. Excretion: A, the excretory system in *Dendrocoelum lacteum*. (After Wilhelmi.) B, terminal organ of an excretory canal, the flame cell. (After Wilhelmi.)

The cilia in the flame cells work against the colloid osmotic pressure of the body fluids and let water, salts and sugars into the flame cells. The required materials are later readsorbed in the proximal tubules and it is possible though not proven that specific excretory materials may be secreted into the tubules. In this way the animals are definitely able to osmoregulate and possibly excrete via their nephridia. It is conceivable that the lack of excretory organs in some of the Acoela is correlated with the presence of a symbiotic protistan.

REPRODUCTIVE ORGANS. It is in the generative organs that the Platyhelminthes show the greatest complexity of organization (Fig. 138). With rare exceptions the Platyhelminthes are hermaphrodite. The *generative pore* is variably placed but it is usually to be found in the middle line of the ventral surface not nearer to the anterior or posterior end than one-quarter or one-fifth the length of the body. This pore leads into a space known as the *genital atrium*. Into the genital atrium open the separate ducts leading from the male and female portions of the generative system, together with other accessory organs. The homologies of the various accessory portions of the generative organs in the three different groups are difficult to ascertain. Names are often used which were applied to organs before their homologies were ascertained, and this increases the confusion.

In studying the generative systems in actual specimens elaborate reconstruction from sections is often necessary, as the heavy pigmentation obscures them when the animal is viewed by transmitted light. In transparent specimens careful staining will bring to light most of the parts, but it often requires considerable skill and practice to identify these parts.

The organization of the platyhelminth generative system may be reduced to a general plan as follows. The *testes* are round bodies, often very numerous, having a lining of cells which give rise to the spermatozoa. From the testes lead out ducts, the *vasa efferentia*, which, uniting, form the *vas deferens*. There are usually two vasa deferentia collecting the sperm from the testes on either side of the body. The ends of the vasa deferentia are often distended and act as *vesiculae seminales*. The vasa deferentia unite and lead into a pear-shaped bag with very muscular walls. This is the *penis*. At rest it opens into the genital atrium, but during copulation it is extruded through the genital pore to the exterior and pushed into the genital pore of another individual. The penis is usually seen very easily, being one of the most conspicuous parts of the genital apparatus.

The female portion of the generative system consists of the *ovary*, which produces the ova, and the *vitellarium*, which supplies the ova with yolk and a shell. The shell substance is liquid and hardens later. This division into ovarium and vitellarium (or 'yolk gland' as it is sometimes called) occurs throughout the Platyhelminthes, but it is probably an elaboration of the more usual arrangement of forming the yolk in the ovary, an arrangement which occurs in the primitive Acoela and in the Polycladida. The ovaries discharge their ova into an *oviduct* which is enlarged near the point of this discharge and thus forms a *receptaculum seminis*. Here fertilization occurs. The oviduct next receives the opening of the *vitelline ducts*. After the opening of the vitelline ducts the duct continues as the *ductus communis*, and leads into the genital atrium. At the junction of the oviducts and vitelline ducts there is a thickening of the walls of the duct and certain glands, the 'shell' glands, pour a secretion on to the egg which probably assists in hardening the shell. This thickening is indistinct in the Turbellaria but is very marked in the Trematoda, and the structure there receives the name of *ootype*, because it is the place where the egg is shaped before being passed into the uterus for storage. In the Trematoda the ductus communis is long and coiled and serves for the storage of eggs. It is called the 'uterus', but it is not of course homologous with the 'uterus' of the Rhabdocoela which will be described shortly, nor with the 'uterus' of the Cestoda which is again probably a different organ.

The genital atrium receives not only the openings of the male and female organs but also certain accessory organs. In the Rhabdocoela, of which *Mesostoma* is an example, there open out from the genital atrium on either side the paired *uteri* (Fig. 138) in which the eggs are stored before laying. In *Dalyellia* (Fig. 116) the fertilized eggs pass into the parenchyma. There is another opening which leads into a short muscular receptacle, the *bursa copulatrix*. The bursa copulatrix receives the penis of another individual during copulation. Sperm is deposited here but remains only for a short time before being expelled by muscular contractions and received into the oviduct where

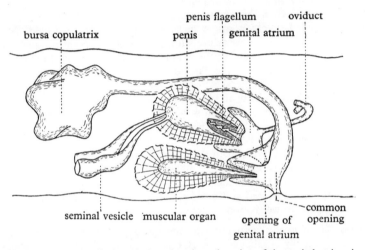

Fig. 121. Longitudinal vertical section through region of the genital atrium in *Dendrocoelum lacteum*. (After Ullyott and Beauchamp.)

it is collected near the ovary in the true receptaculum seminis. In the Tricladida the uterus and the bursa copulatrix are replaced by organs, the homologies of which are doubtful. These are the unpaired *stalked gland organ* and the unpaired *muscular gland organ*. The stalked gland organ is often called the 'uterus' but it has not been observed to contain eggs. It is regularly present, whereas the muscular gland organ is often absent. It has recently been shown that the stalked organ serves as a bursa copulatrix and receives temporarily the penis and the sperm of another individual.

COPULATION. During copulation the ventral surfaces of two animals are applied together so that the genital openings lie opposite to each other. The penes are extruded through the genital opening of one copulant into the genital opening of the other. There is a mutual exchange of sperm. Since the ova are ripe at the same time as the sperm, and as, in many forms, there is only one common genital opening to the exterior, special precautions are necessary to prevent self-fertilization. To ensure that cross-fertilization shall take place a great elaboration of the structures surrounding the genital atrium has occurred, resulting in that complication of the genitalia which is so characteristic of the Platyhelminthes.

In fresh-water Tricladida copulation occurs fairly freely among animals kept in glass jars, where they are easily observed. When the penis is retracted its lumen is closed so that sperm cannot escape into the genital atrium, whence it might find its way up the oviduct (Fig. 121). When the penis is thrust out through the genital opening during copulation it is dilated on extrusion, so that the lumen is opened. This dilation also causes the penis to fill completely the genital atrium and opening, so that the opening of the oviduct into the genital atrium is blocked and no sperm can enter or ova escape. At copulation the penis of one animal is squeezed past the penis of the other into the genital atrium. It cannot enter the oviduct, since this is blocked and so it is received into the stalked gland organ, where the sperm is temporarily deposited. After copulation is finished, the penes are withdrawn and the sperm is transferred from the stalked gland organ to the oviduct. The arrangement of the organs round the genital atrium in the Tricladida varies considerably. In *Bdellocephala*, for example, the penis is reduced and, when extruded, does not fill the genital atrium sufficiently to block the opening of the oviduct. In this case a flap of skin has developed which is drawn over the opening of the oviduct when the penis is extruded.

After the sperm is transferred to the oviduct, it moves up to the receptaculum seminis at the top, near to the point of discharge of the ova. The ova are fertilized in the oviduct and then move down towards the genital atrium, receiving on the way the products of the vitellaria. On arrival in the genital atrium a cocoon is shaped and made ready to be deposited. When laid it is usually attached to weeds, sometimes by a stalk.

EGGS. The parasitic Trematoda and Cestoda are unaffected by the seasons and are perpetually producing eggs. But in the Turbellaria the season of egg-laying varies. In some, for example *Dendrocoelum lacteum*, the generative system is in full working order all the year round, in others, for example *Planaria alpina*, the eggs are only produced during the winter months. *Mesostoma* produces two kinds of eggs which are called 'summer' and 'winter' eggs. The 'winter' eggs have a thick shell and are well supplied with yolk; they remain in the uterus and escape only with the death of the parent. The 'winter' egg can remain dormant for a long period. The 'summer' egg is very thin-shelled and has very little yolk. The development is very rapid and the young embryos are seen moving in the uterus of the parent seventy-two hours after the appearance of the eggs. They escape by the genital pore and their formation does not involve the death of the parent. The term 'winter' and 'summer' egg is not entirely apposite, for 'winter' eggs are often found in midsummer. The 'winter' egg is a method of carrying the species over unfavourable conditions which may develop in winter or in summer. The 'summer' egg is a means for rapid multiplication when conditions are favourable.

EMBRYOLOGY. With the exception of the Acoela and the Polycladida the eggs are provided with special cells that look after the nutrition of the developing embryos, but in the Acoela and Polycladida the yolk is enclosed in the egg, the egg being endolecithal. The egg divides into four blastomeres which then divide to form a total of eight cells. This division is unequal in that there are four large cells and four small ones, the macromeres and micromeres. The micromeres do not lie immediately above the macromeres, instead they lie above and at an angle of 45° to the macromeres. This type of cleavage is called spiral cleavage in contrast to the radial cleavage seen in echinoderms and the bilateral cleavage found in ctenophores.

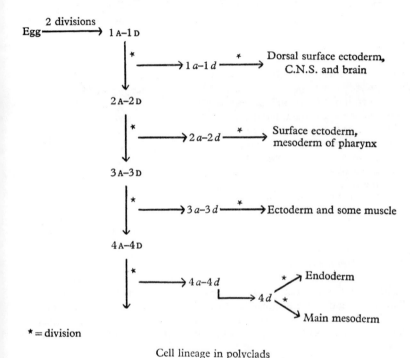

* = division

Cell lineage in polyclads

Fig. 122. Cell lineage in the polyclads, indicating the fate of the cells following cleavage (see also Fig. 195).

The development of the turbellarian embryo is very much like that of the annelids and reference should be made to p. 291 for details of the development and the terms used to describe the development. The cells of the first quartet give rise to the main body ectoderm as do also $3a$–$3d$; $4d$ forms the body of the mesoderm and also the endoderm, the situation being similar to that found in the annelid trochoblast cell $4d$. In addition, during the early stages of development it is possible to pick out a small rosette and quartet much like that seen in the polychaetes. The larva develops a temporary frontal tuft homologous with the apical tuft of the trochophore; this disappears at a later stage of development.

Gastrulation is epibolic, forming small gastrulae with no enteron. Near the blastopore a small invagination arises that gives rise to the pharynx and intestinal cells, the latter moving to form a small cavity, the gut. The embryo then develops to become a small planarian. This is the direct development.

In some Polyclads, i.e. *Hoploplana*, and in all the Cotylea, instead of direct development into an adult the embryo develops into a pelagic larva, 'Müller's larva'. This larva has eight ciliated lappets by which it swims for a few days. These lappets are then absorbed and the larva stops being planktonic and [settles down to a life of crawling on the sea bottom. In *Stylochus* the embryo only develops four lappets and the larval form is then called 'Götte's larva'. It is interesting to note that *Planocera reticulata* goes through a Müller's larva stage whilst still retained within the egg case.

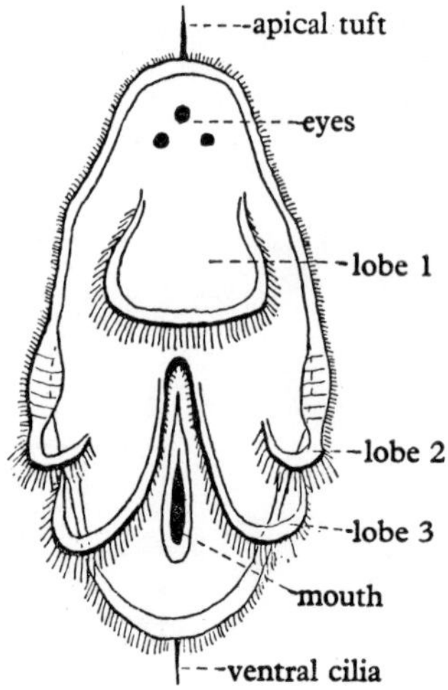

As mentioned on p. 143, projecting processes forming arms and bands of cilia are common and belong to many different phyla. Their presence is probably an adaptive feature and it is unwise to base phylogenetic speculations on them. Müller's larva is planktonic and acts as a distributive phase in the life of the animal (Fig. 123).

Fig. 123. Müller's larva of a polyclad, *Cycloporus papillosus* Lang. (Altered from Kükenthal.)

ASEXUAL REPRODUCTION AND REGENERATION. Asexual reproduction occurs commonly in the Turbellaria. In *Microstomum lineare* the hinder end buds off new individuals which remain attached for some time so that chains of three or four individuals in different stages of development are often seen. Planarians undergo autotomy, cutting themselves in two by a ragged line which traverses the middle of the body. Lost parts are easily regenerated in the Tricladida and the group is a favourite one for experimental work on regeneration.

The interstitial cells play an important role in regeneration of planarians. If one takes a white planarian and cuts off its head a new head will soon be regenerated. If one repeats the experiment but this time irradiates the body with X-rays then a new head will not grow. If one takes such an irradiated decapitated animal and implants a small piece of tissue from a black non-irradiated animal, the white animal starts to regenerate a new head. The tissue in the head is largely formed from the black implanted tissue as can be seen from its pigmentation. The cells that are most responsible are the interstitial cells, and it will be recollected that these cells are important in regeneration of the coelenterates too.

Having thus provided the reader with a general account of the organization of a platyhelminth it will now be possible for us to follow the systematic arrangement of the phylum, to define the divisions and to point out features of interest in various forms and life histories.

CLASS 1. TURBELLARIA

The Turbellaria may be defined as Platyhelminthes which are nearly all free living and not parasitic, which retain the enteron; which have a cellular, ciliated outer covering to the body; which usually have rhabdites; and which do not form proglottides. Suckers are very rarely present.

The systematic arrangement of the Turbellaria is based primarily on the structure of the gut. There are six orders: (1) Acoela, (2) Rhabdocoela, (3) Alloiocoela, (4) Tricladida, (5) Polycladida, (6) Temnocephalea.

1. Order Acoela

In these the gut is not hollow but consists of a syncytium formed by the union of endodermal cells. There is no muscular pharynx. Primitive features are the nerve net and the fact that the germarium and vitellarium are not separated. *Convoluta roscoffensis* is the best-known member of this division. It lives between the tidemarks on sandy shores. Imbedded in the parenchyma are algal cells which live in a symbiosis (p. 50) with the turbellarian. The photo-synthetic products of these algal cells provide a source of nourishment for the animal. *Convoluta henseni*, another member of this order, is a rare platyhelminth that has adopted a planktonic habitat.

2. Order Rhabdocoela

In these forms (Fig. 116) the gut is straight and the mouth is near the anterior end. The gut may or may not have lateral pouches. In the more primitive members of this order, of which *Microstomum lineare* is a common example, found in fresh water, the germarium and the vitellarium are not separated. Another well-known member of this group is *Dalyellia viridis*, common in fresh-water ponds in Britain and remarkable for the elaborate chitinous structure of the penis. *Mesostoma ehrenbergi* and *M. quadrangulare*, the latter X-shaped in cross-section, both occur in fresh-water ponds. They are large and transparent and form the best objects for studying the structure of the group. Another example is *Rhynchoscolex*.

The rhabdocoel *Microstomum* is of interest in that it has nematocysts in its ectoderm. These are derived from the undischarged nematocysts from coelenterates on which the rhabdocoel feeds. The types of nematocyst closely correspond to the types of nematocyst present in the coelenterates on which the *Microstomum* has recently fed. Similarly, starved animals loose their nematocysts. The undischarged nematocysts in the food are taken up by special cells in the gut and then carried to the epidermis and orientated so that they can be shot out at predators.

The Rhabdocoela occur in both fresh and salt water; marine forms are, however, very small, one marine form, *Fecampia* being an internal parasite in the lobster.

3. Order Alloiocoela

Turbellaria with small diverticula arising from the straight gut. It is often difficult to separate these animals off from the Rhabdocoela on the one hand and the Tricladida on the other. They also show close affinities with the

Acoela. Examples of alloiocoels are *Plagiostomum* with its large bulbous pharynx, *Vorticeros*, a small worm with two long tentacles or horns, *Otoplana* with its large otocyst above the brain, and *Hofstenia* and *Prorhynchus*.

4. Order **Tricladida**

In this group the gut is divided into three main divisions with numerous lateral diverticula from each division. The mouth has shifted backwards to the middle of the body. There are three well-recognized divisions of this order, separated according to habitat: the *Paludicola* or fresh-water forms, the *Maricola* or marine forms, and the *Terricola* or land forms. The Paludicola are all fairly large forms in contrast with the Maricola which are small, no more than 2–4 mm. long. To the Paludicola belong the three commonest fresh-water Turbellaria in Britain: *Dendrocoelum lacteum*, a white form, *Planaria lugubris*, a black form, and *Polycelis nigra*, a rather smaller black form easily recognized by the ring of eyes round the anterior edge of the body. Perhaps the best-known member of the Maricola is *Procerodes lobata* (= *Gunda segmentata*) in which the side diverticula of the gut are regularly arranged, with testes and excretory openings between them, giving the appearance of a segmented animal. The Terricola often reach a very large size—as long as 50 cm. They are often brightly coloured with stripes down the dorsal surface. *Bipalium kewense* is a cosmopolitan tropical form that often turns up in greenhouses. It is often a foot long and is easily recognized by the axe-shaped head. *Rhynchodemus terrestris*, a small form 6–8 mm. long, is a British representative of this division. It is found in damp situations under the bark of decaying trees and fallen timber.

Other examples of the Terricola are *Geoplana* and *Geodesmus*. It should be noted that the genus *Planaria* has recently been restricted so that the familiar planarian is placed in the genus *Dugesia*, whilst *Planaria alpina* is renamed *Crenobia alpina*.

5. Order **Polycladida**

These are entirely marine. The gut has many diverticulae leading out from a not very conspicuous main stem. The mouth has shifted to the posterior end. Some polyclads achieve considerable size, 6 in. in length or more. They are generally very much flattened. There are two suborders. The Acotylea do not possess suckers whilst the Cotylea do. Amongst the Acotylea are *Stylochus*—often found feeding on oysters—*Leptoplana* and *Hoploplana*. The Cotylea have the sucker placed just behind the genital pore. In *Thysanozoon* there are many small papillae on the dorsal surface of the body. *Yungia* has similar papillae which also contain diverticula of the gut, some of which open to the exterior. *Cycloporus* has a series of such pores opening all around the perimeter of the body.

The polyclads resemble the Acoela, the resemblance being mostly in the reproductive and embryological systems. In the Acoela and polyclads the entry of the sperm into the ovum takes place after the extrusion of the polar bodies whilst in the other turbellarians this follows the entry of the sperm. Similarly they are the only ones to have yolk inside the eggs, the other turbellarians having special yolk cells to provide the ovum with food. Development

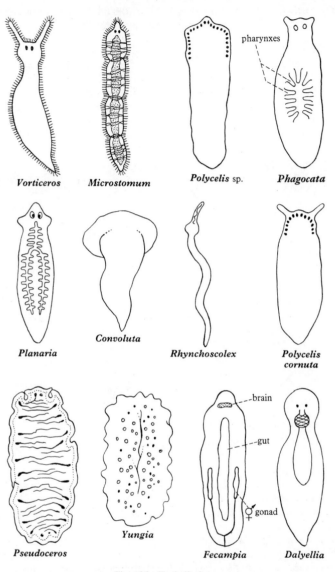

Fig. 124. Turbellarians.

of the fertilized egg is similar, there being only two macromeres in the Acoela but four in the polyclads.

The embryos often develop into a motile larva, 'Müller's larva' (Fig. 123). This is characterized by its series of projecting ciliated bands, the arms. At metamorphosis when the free-swimming larva looses the projecting arms and ciliated bands it adopts the crawling habit of the adult, and at the same time it loses its rotundity and becomes flattened and elongated.

6. *Order* **Temnocephalea**

The Turbellaria are in some ways linked to the Trematodes by a small group of animals which constitute the order Temnocephalea. These animals have a discontinuous distribution, being found mainly in tropical and subtropical waters. They are attached to fresh-water animals, usually crustaceans. They do not feed on their hosts' tissues but merely use them as a site from which they can catch rotifers, *Cyclops* and other small water-animals. They can live

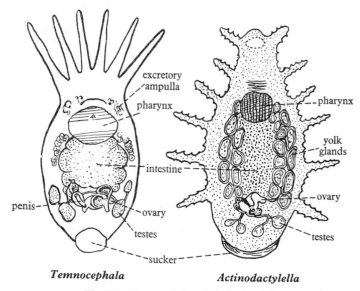

Fig. 125. *Temnocephala* and *Actinodactylella*.

for several weeks when isolated from their hosts. The animals usually have a series of tentacles at their head end, the number ranging from twelve in the case of *Actinodactylella*, five or six in *Temnocephala*, and two in *Scutariella* (Fig. 125).

The epidermis is a nucleated syncytium which secretes outside it a thick cuticle. Rhabdites occur in the region of the tentacles. The mouth is anterior, the gut has the same shape as that of the rhabdocoels. There is a large sucker at the posterior end with the common male and female opening in front of it. The ovary and vitellarium are separate. The nervous system is a primitive network. The Temnocephalea are placed in the Turbellaria on account of their cilia, rhabdites, basal membrane and the absence of any chitinous thickening to the suckers, and the absence of Laurer's canal. (Suckers are not rare amongst the Turbellaria, being extensively found in the Cotylea, a suborder of the polyclads.) Some authors place the Temnocephalea as a suborder of the

Rhabdocoela, with close affinities to the Dalyellidae. Other authors have considered them more as trematodes.

Examples. *Temnocephala, Actinodactylella, Scutariella, Monodiscus.*

CLASS 2. TREMATODA

DIAGNOSIS. The Trematoda may be defined as Platyhelminthes which are parasitic; which retain the enteron; which in the adult have outside the ectoderm a thick cuticle; which have suckers; usually, but not always, a sucker on the ventral surface in addition to one surrounding the mouth; the ventral sucker is subdivided in some forms and may also be stiffened with a ring-like scleroproteinous skeleton.

The Trematoda are all parasitic but they resemble in general shape the Turbellaria. They have retained the mouth, which is anteriorly placed, and the gut, which, however, is bifid—a shape not found in the Turbellaria. As in the Turbellaria, the gut may have lateral diverticula which branch freely. The Trematoda have, however, lost the external ciliation of the Turbellaria (Fig. 115 c). The ectoderm is represented by cells sunk into the parenchyma in much the same way as nuclei of the ectodermal cells in the pharynx of the Tricladida. But the outer portion of the cell is lost in the Trematoda and its place is taken by a thick *cuticle*, which is often armed with spines. Suckers are always present for attachment to the host and are of large size. The presence of these suckers and their shape makes it possible to divide the Trematoda proper into two orders: (1) Heterocotylea, (2) Malacocotylea.

CLASSIFICATION

1. *Order* **Heterocotylea.** Trematodes with only one host; suckers sometimes stiffened with scleroprotein supports. *Octobothrium, Polystomum.*

2. *Order* **Malacocotylea.** Trematodes with two or more hosts in their life cycles; suckers simple. *Fasciola, Schistosoma, Wedlia.*

1. *Order* **Heterocotylea (Monogenea.)**

DIFFERENCES BETWEEN HETEROCOTYLEA AND MALACOCOTYLEA. In the Heterocotylea there is a large posterior sucker stiffened with proteinous supports. It is often subdivided, as in *Octobothrium* or *Polystomum* (Fig. 126). In the Malacocotylea the sucker is not always posterior, it often moves forward on the ventral surface so that, as in *Fasciola*, it comes to lie one-third of the body-length from the anterior end. It is never provided with proteinous supports. All the Heterocotylea are ectoparasites with the single exception of *Polystomum* which occurs in the bladder of the common frog, of which from 3 to 10% are infected by it. They are confined to one host only. The Malacocotylea are all internal parasites and pass from one host to another at certain stages in their life history. In the Heterocotylea the excretory pores are paired and lie near the anterior end of the body, whereas in the Malacocotylea the excretory system discharges to the exterior through a single median pore placed at the posterior end of the body. In the Heterocotylea there are separate openings for the male and female portions of the generative system, while in the Malacocotylea there is but one common opening. In the

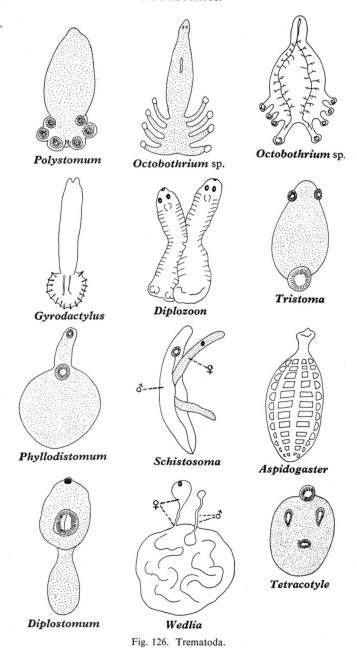

Fig. 126. Trematoda.

Heterocotylea there is a pair of ducts leading from the ootype to the exterior independently from the male and female ducts, usually called the *vaginae*. The vaginae are inconspicuous as a rule, but in *Polystomum* their openings are very clearly marked by two prominences on either side of the body about one-fifth of the body-length from the anterior end (Fig. 127). Corresponding ducts do not occur in the Malacocotylea. The nervous system of the Heterocotylea is more primitive than that of the Malacocotylea, but in both groups it is

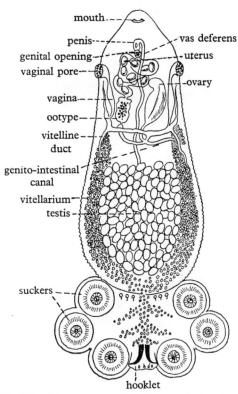

Fig. 127. *Polystomum integerrimum*, ventral view showing the reproductive system. (After Zeller.)

stereotyped and does not vary as it does in the Turbellaria. In both groups it consists of a cerebral ganglion with six cords leading posteriorly. In the Heterocotylea there are irregular commissures between the cords, while in the Malacocotylea the commissures are few in number and regular.

LIFE HISTORY OF THE HETEROCOTYLEA. The habitat of this order is on the gills of fishes where they often live isolated. Self-fertilization must therefore be practised, but copulation has been observed in *Polystomum* and also in *Diplozoon*, where it is permanent. The members of this order probably cause considerable inconvenience to their hosts, but the numbers infesting one host is seldom very considerable and they have no economic importance as parasites. The eggs when laid are sometimes attached to the body of the host.

Polystomum is exceptional in laying the eggs in the bladder whence they pass out to the exterior into water (Figs. 127, 128). The egg hatches as a larva with eyespots and a large ventral posterior sucker. It swims by means of cilia which are arranged in bands round the body. These larvae make their way to

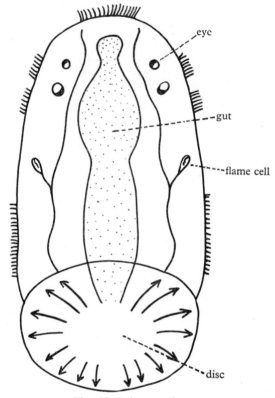

Fig. 128. *Polystomum* larva.

some particular spot on the host after being free-swimming for a time. As soon as they attach themselves the ciliary covering is cast off and the generative organs develop. The larva of *Polystomum* seeks out a tadpole, dying within twenty-four hours if one is not found. If a tadpole is reached, the parasite fastens itself on to the gills, where its ciliary covering is cast and it then creeps into the bladder to wait for three years before becoming sexually mature. The larvae may, however, attach themselves to the external gills, where a copious supply of nourishment induces such rapid growth that the animal becomes sexually mature in five weeks and produces eggs. But it dies when the tadpole metamorphoses, and thus it never reaches the bladder (Fig. 129).

In *Diplozoon*, which lives attached to the gills of the minnow, the larvae attach themselves to the gills of the host, but they do not develop generative organs until they meet another larva. If such a meeting occurs the larvae fuse across the middle. After fusion the generative organs develop and the animals grow in such a manner that the vas deferens of one form is permanently connected to the genital atrium of the other. They thus remain throughout their lives in permanent copulation (Fig. 127).

Another form which displays a variation of the usual type of history is *Gyrodactylus* which occurs in the gills of fresh-water fish. In *Gyrodactylus* the ovary and the vitellarium are not separated, as is the general rule in the Trematoda, but constitute one organ. A single egg ripens at a time and, after fertiliza-

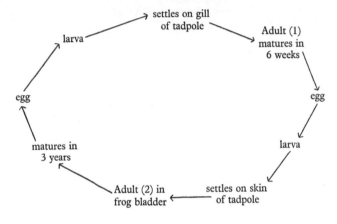

Polystomum

Fig. 129. Suggested life cycle for *Polystomum* showing the alternation of the hosts and ectoparasitic and endoparasitic stages. (After Gallien.)

tion, develops into an embryo in the uterus. Before the first embryo leaves the mother a second younger one appears inside it so that we thus have a condition of three generations one inside the other, and the conditions are such that the youngest embryo must develop without fertilization. This feature of the development of one larva with another without the agency of fertilization is common in the life histories of the Malacocotylea but *Gyrodactylus* is the one member of the Heterocotylea in which it occurs.

2. *Order* **Malacocotylea (Digenea.)**

The life history of *Fasciola* (Fig. 130) may be taken as the type of life history commonly found in the group. For details of this life history the reader is referred to elementary text-books.

In the Malacocotylea the adult is always, with rare exceptions, parasitic in some vertebrate host, the sporocyst and redia stages are always parasitic in a mollusc. Two hosts are always, and three may be, necessary for complete development. Divergence from the type of life history recorded for *Fasciola* may

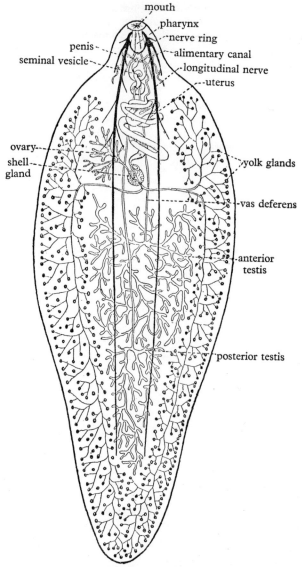

Fig. 130. Diagram of the reproductive and nervous system of *Fasciola hepatica*, × c. 8. (From Leuckart.)

come about by (1) a generation, the redia stage, being omitted, (2) the sporocyst forming by budding a second generation of sporocysts within which the cercariae arise, (3) the cercaria requiring to encyst in a host and to await this host being eaten by the final host before reaching sexual maturity as in the case of *Gasterostomum fimbriatum*, where the sporocyst develops in the liver of

Anodonta, the cercaria encysts in the roof of the mouth of the roach and only reaches sexual maturity when the roach is swallowed by a perch.

In *Distomum macrostomum*, which is parasitic in the gut of thrushes, there is no free-living stage in the life history. The eggs, passed out with the faeces of the bird, are eaten by a snail, inside which the sporocyst develops. The sporocyst finds its way into one of the tentacles. It there develops pigment, being brightly coloured in bands of green and red, while its presence stops the snail from withdrawing this tentacle. Presumably this brightly coloured object attracts the bird which devours the snail and infects itself by setting free the cercariae from the sporocyst.

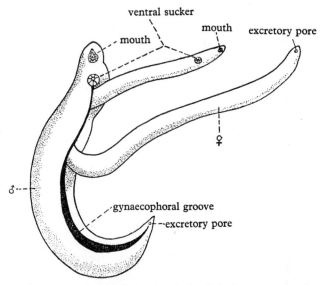

Fig. 131. *Schistosoma*; the male is clasping the female in the gynaecophoral groove.

Schistosoma (= *Bilharzia*) is a parasite of man, living as an adult in the abdominal veins (Fig. 131). It is long and thin and well adapted for this habitat. It is one of the rare examples of dioecious trematodes. The male, however, does not lose touch with the female once he has found her, but carries her permanently in a fold of the ventral body wall. The eggs are laid in the blood vessels and, being provided with a sharp spike, they lacerate the walls of the capillaries and pass into the bladder. Immediately the urine is diluted the miracidia hatch, but they wait for dilution before hatching. The second host is a water-snail. The cercariae swim freely in the water, and in districts in China and Egypt where the disease is common they swarm. Bathing, washing or drinking the infected water allows the cercaria to enter the final host. The cercariae penetrate the skin with great rapidity and, entering the blood system, make their way to the abdominal veins where they become mature. The disease can be prevented by strict sanitary measures in regard to water, and it can be cured by the administration of compounds of antimony to infected patients. That the disease is a very old one in Egypt is shown by the discovery of *Schistosoma* eggs in the kidneys of mummies of the twentieth dynasty (1250–1000 B.C.).

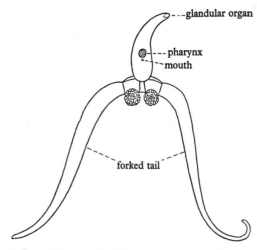

A. Bucephalus cercaria of *Gasterostomum fimbriatum*

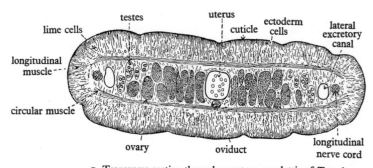

B. Transverse section through a mature proglottis of *Taenia*

Fig. 132. A, Bucephalus larva (cercaria of *Gasterostomum fimbriatum*). (After Benham.) B, transverse section through a mature proglottis of *Taenia*, × c. 12. (From Shipley and MacBride.)

The hatching of miracidia from the egg of *Schistosoma* is dependent on the dilution of the urine by fresh water and this serves to emphasize the fact that the stages in the life history of all parasites are ultimately connected with environmental conditions. The egg of *Fasciola hepatica* does not hatch unless the pH of the water in which it is deposited is below 7·5, the optimum point apparently being about pH 6·5. If the eggs are kept in water more alkaline than pH 7·5 the embryo remains within the shell and eventually dies.

The identification of a cercaria with an adult is a task which requires great patience, and many cercaria are known which have not been as yet connected with an adult. Almost any mollusc, if dissected carefully under a hand lens, will provide specimens of rediae and cercariae, although infected specimens may be more common in some localities than in others. The tail of a cercaria is often an elaborate structure. Some have rings and chitinous stiffenings, while the well-known Bucephalus larva of *Gasterostomum* is a cercaria with a forked tail (Fig. 132 A).

Other examples of the Malacocotylea are *Wedlia*, which is parasitic on fish and has the male carried in the female, *Diplostomum*, parasitic in birds, and *Phyllodistomum*, parasitic in amphibia (Fig. 126).

CLASS 3. CESTODA

The Cestoda may be defined as endoparasitic Platyhelminthes in which the enteron is absent and the ciliated ectoderm has, in the adult, been replaced by a thick cuticle. In the parenchyma lime cells occur (see Fig. 132 B). Proglottides are usually formed.

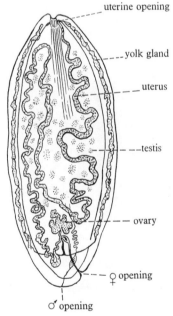

Amphilina

Fig. 133. *Amphilina*, a cestodarian. (After Hein, from Hyman.)

The Cestoda as a group have felt the influence of the parasitic habit more than the Trematoda. They have dispensed altogether with a gut, there is no mouth, and they absorb their food through the skin. As they live always in the alimentary canal of vertebrates they are conveniently situated for this

220   ACOELOMATA

purpose and the amount of food available to them probably counterbalances the difficulties attendant on dispensing with the usual method of digesting and assimilating food. The ectoderm cells have sunk into the parenchyma after secreting a cuticle as in the Trematoda, but this cuticle is thicker and divided into layers. Immediately beneath the cuticle are the longitudinal

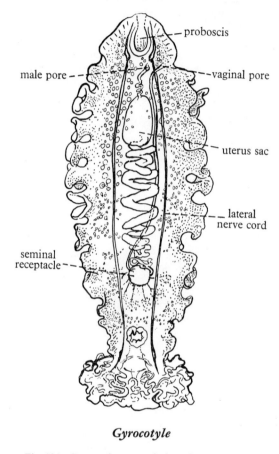

Gyrocotyle

Fig. 134. *Gyrocotyle*, a cestodarian. (From Hyman.)

muscles. The circular muscles are incomplete at the edges. In transverse sections the circular muscles appear to divide the parenchyma into two regions, an outer cortical zone, where occur the cut ends of the longitudinal muscle together with calcareous bodies, and an inner or medullary zone, where the generative system lies (Fig. 132 B).

The Cestoda may be divided into two orders: (1) Cestodaria, (2) Eucestoda.

1. Order Cestodaria

These are small forms which live in the gut of fishes, usually Elasmobranchs. They resemble a trematode in shape and in the fact that they do not form proglottides, but they have no gut. They have at one end a 'frilled' organ which serves for attachment, and a small sucker at the other end. Examples of this order are *Amphilina* and *Gyrocotyle*. It is difficult from the structure to say which end is the anterior and which the posterior, for the nervous system

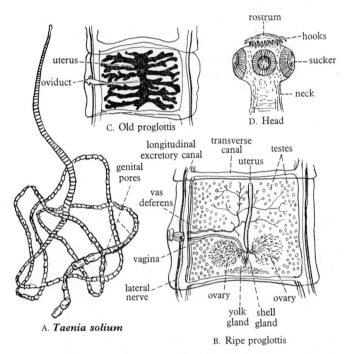

Fig. 135. *Taenia solium*: A, diagram of anterior portion of the animal. B, diagram of ripe proglottis, ×10. (A, after Shipley and MacBride; B, from Cholodkowsky.)

consists of two cords running down either side of the body with a single similar commissure at either end. But when the animal moves it has the 'frilled' organ in front so that is spoken of as the anterior end (Figs. 133, 134).

2. Order Eucestoda

These are distinguished from the Cestodaria by the fact that they all have the power of budding and so reproducing asexually, resembling in this respect the turbellarian *Microstomum lineare*.

The adult worm has a *scolex* which is provided with organs of fixation such as hooks, suckers or folds (Fig. 136). The scolex is usually buried in the intestinal mucosa of the host. Behind the scolex comes the *neck*, the most slender portion of the body, which may or may not be sharply marked off from the scolex. It is in the neck that asexual reproduction occurs, fresh segments being continually cut off and, as they grow larger, pushed by the formation of new segments away from the scolex. The segment so formed is called a *proglottis*. The proglottis is not truly comparable with the new individuals produced in *Microstomum lineare*. Through each proglottis run the excretory canals and the nervous strands which are common to all (Fig. 132 B). The proglottis when first cut off from the neck region is devoid of generative organs, but these develop as it becomes more mature. When the generative organs are mature, fertilization of the ova occurs, the ovaries and the testes disappear, and the uterus alone remains to store the eggs. When the proglottis reaches this stage it is 'ripe' and breaks off to pass out with the faeces (Fig. 135 B). Despite its connection with the scolex, each proglottis must be regarded as an individual for it contains a full set of generative organs both male and female.

SCOLEX. The structure of the scolex is of importance, for it forms the basis of the classification of the Eucestoda. In the tape-worms occurring in the gut of fishes the scolex may have two or four suckers and the neck may be sharply separated from the region where budding occurs. In these tape-worms the scolex is often armoured with proteinous projections and hooks, and the number of the proglottides is usually small. The tape-worms occurring in the mammals (*Cyclophyllidea*) are, with one exception, characterized by a head which bears four suckers at the sides, and, on a projection at the top, called the rostellum, is a crown of hooks.

PRIMITIVE CESTODES. As a general rule the more primitive cestodes are found in the lower vertebrates, while the advanced types are found in the mammals. The evolutionary stage of the parasite is therefore closely related to that of its host. A notable exception to this rule is *Diphyllobothrium latum*, the Broad Tape-worm of man, which belongs to a group of tape-worms occurring more commonly in the guts of fishes. The scolex of *Diphyllobothrium* has two suckers on either side of the head. These suckers are of the nature of flabby folds sharply distinct from the well-defined cuplike suckers of the Cyclophyllidea.

REPRODUCTIVE SYSTEM. The generative organs are of the same type as is found generally throughout the Platyhelminthes. There is a single opening for both male and female organs. From the ootype there leads out a duct which is called the *uterus* and is used for the storage of eggs, but it is doubtful whether it is homologous with the uterus of the Trematoda.

LIFE HISTORY. The life history of a cestode is a complicated combination of sexual and asexual reproduction. One, two or three hosts may be necessary. The egg passes to the exterior with the faeces. It contains inside it an embryo armed with six hooks called an 'onchosphere'. The egg case takes different shapes; in *Diphyllobothrium latum*, which is a more primitive type of cestode, the covering of the embryo is ciliated. In the Cyclophyllid tape-worms, which constitute the most advanced group of the Cestoda, the ciliary covering is lost.

In *Dipylidium caninum*, the adult of which occurs in the alimentary canal of the cat or dog, it is replaced by an albuminous coat with a proteinous lining inside, while in most of the other forms only the proteinous covering persists. The egg hatches as an onchosphere after being swallowed by the first host. The onchosphere then penetrates the wall of the alimentary canal using its hooks for this purpose and lodges somewhere in the peritoneal cavity of the host. Here it develops suckers and a scolex.

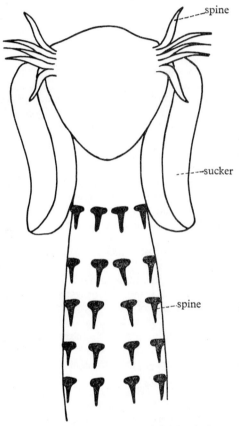

Fig. 136. Anterior end of scolex of *Echinobothrum*.

In primitive forms, such as *Diphyllobothrium*, the larval cestode rests inside the first host, a *Cyclops*, at a stage of its development known as the plerocercoid stage. This stage is ovate in shape and the generative organs are undeveloped and there are no signs of proglottides. The *Cyclops* is then eaten by a fresh-water fish, after which the larva, or plerocercus, bores through the wall of the alimentary canal and rests in the body cavity where it grows still further, reaching the metacestode stage. Proglottides can be distinguished in the metacestode stage but the generative organs are not fully mature. Growth now ceases but the metacestode stage is often inconveniently large for the body cavity, causing it to bulge. Sticklebacks thus infected with the metacestode of *Schistocephalus gasterostei* are commonly found. The adult in this case reaches maturity when eaten by a bird. Man acquires *Diphyllobothrium latum*, a nearly related form, by eating pike infected with the metacestode.

In the Cyclophyllidea the resting stage in the first host is the 'bladder-worm' (or cysticercus). The onchosphere on reaching its resting place becomes hollowed out into a ball filled with fluid. A depression then forms in the wall of the sphere and becomes an inverted scolex. In *Taenia serrata*, the common tape-worm of the dog, the bladder stage in the rabbit (to which the name *Cysticercus pisiformis* was given before the connexion with the adult was discovered) has but one head inverted into the cyst. In the bladder-worm stage of *Taenia coenurus*, which is found in the brain of the sheep and causes the disease known as 'gid' or 'staggers', many heads are formed and invaginated into the cyst so that multiple infection may occur when a sheep is devoured and torn to pieces by dogs or wolves. In *Taenia echinococcus*, the adult of which lives in the alimentary canal of the dog and is remarkable for having but three proglottides, the cysticercus stage is found in domestic animals and also in man in countries where men live in close association with dogs. The cyst stage is very large and the bladder may contain a gallon or more of fluid. Such a cyst, known as a 'hydatid', rapidly proves to be fatal. It is particularly dangerous and difficult to eradicate because the walls of the cyst have the power of budding off asexually daughter cysts. A still further development of asexual budding in the cysticercus stage occurs in *Staphylocystis*, where the onchosphere imbeds itself in the liver and then develops a stalk or stolon which buds off cysts which are detached and fall into the body cavity of the host.

Where the cysticercus is swallowed by the final host the head is everted from the bladder, the bladder is digested and proglottides forthwith make their appearance from the neck region of the scolex. So far as is known the production of proglottides continues for the duration of the life of the host.

The subdivision of the Eucestoda depends on the shape of the scolex. There are five divisions, the last of which contains the forms commonly found as adults in the alimentary canal of the Mammalia and is the only group of economic importance.

1. Tetraphyllidea. The four suckers are usually stalked outgrowths of the scolex. Parasitic in fish, amphibia and reptiles. Onchosphere enters a copepod and develops into a larva known as a plerocercoid, in which condition it remains until the copepod is eaten, when it develops into the adult. Size moderate, usually 20–30 cm. long, but occasionally as small as 1 cm. or as large as 1 m. An example is *Anthobothrium* (Fig. 137).

2. Diphyllidea. There are two suckers only and the scolex has a long neck armed with spines. There is only one family and one genus, *Echinobothrium*, which is found in the spiral intestine of Selachians. The larva, which is of cysticercoid form, is found in the prawn *Hippolyte*.

3. Tetrarhynchidea. These have four suckers each provided with a long spiniferous rectractile process. The adult is parasitic in the alimentary canal of Elasmobranchs and especially Ganoids. The larva, which may be of either the procercoid or cysticercoid type, occurs in marine invertebrates of many kinds, fish and occasionally reptiles.

4. **Pseudophyllidea.** The scolex has two suckers which may be absent in some forms, there is no clearly marked neck and hooks are usually absent. Occasionally as in *Trianephorus*, a common parasite of fresh-water fish, the external divisions between the proglottides are indistinct and these are only indicated by the regularly placed openings of the uterine birth pores. The majority of these are parasitic as adults in fresh-water fishes, but *Diphyllo-*

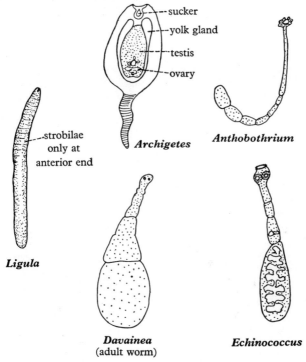

Fig. 137. Cestodes. (From Hyman.)

bothrium latus occurs in man and *Bothriotaenia* in birds. *Archigetes* is parasitic as an adult in body of *Tubifex,* an oligochaete worm living in fresh water. The larva is a plerocercoid which in some forms, *Caryophyllaceus* and *Archigetes*, develops gonads paedogenetically so that there is no adult with proglottides. These paedogenetic forms closely resemble the Cestodaria in appearance (Fig. 137).

5. **Cyclophyllidea.** The scolex bears four cup-shaped suckers and has a rostellum with a crown of hooks.

The Cyclophyllidea comprise the majority of the common tape-worms. Those infesting the gut of mammals all have a scolex closely resembling that of *Taenia*, with four well-defined suckers and a circlet of hooks. Those found in the gut of fish have a more elaborate scolex. The number of proglottides varies considerably, the smallest number is found in the genus *Echinococcus*, while many forms have hundreds of proglottides and are several yards in length. The proglottides never drop off before they are mature, as they may do in the other groups, and develop generative organs later, consequently the separated proglottides always contain fully developed onchospheres. Two interesting forms may be mentioned. *Dipylidium caninum* is a tape-worm

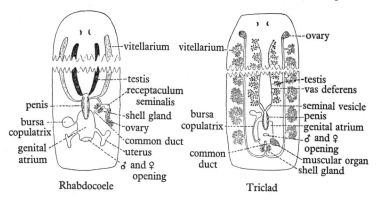

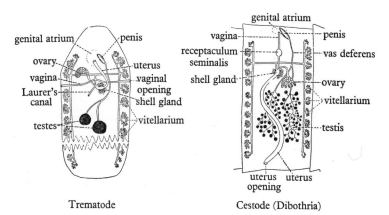

Fig. 138. Diagram of the arrangement of the genital organs and ducts in the Platyhelminthes.

infesting the alimentary canal of dogs and cats. Each proglottis has a double set of generative organs with two separate generative openings, a feature which gives the animal its name, but which may occur in other forms. The first host is the flea, and puppies and kittens are early infected by catching and eating these insects. *Hymenolepis nana* is one of the smallest tape-worms. The adult has ten to twenty proglottides and only measures half an inch in length. It occurs in children in certain places, particularly Lisbon and New York, where it is said to be increasing. It is remarkable among tape-worms for being the only one known to go through all its life history in one host. The embryos bore into the intestinal wall where they pass through the cysticercus stage and emerge again into the alimentary canal when adult.

GENITALIA HOMOLOGIES. The homologies of the various ducts of the genitalia of the Platyhelminthes (Fig. 138) present great difficulties. While one or two, the oviduct and the vas deferens for example, are quite clearly homologous throughout, the homologies of others, particularly the accessory organs such as uterus, bursa copulatrix, vagina, are very doubtful. The 'uterus' of the Trematoda is clearly the ductus communis of the Turbellaria greatly elongated and used for egg storage, while the vagina of the Cestoda is the same, but the relation of the 'vagina' of the Heterocotylea or the 'uterus' of the Cestoda remains at present obscure.

If the vagina of the Cestoda is homologous with the uterus of the Trematoda, the uterus of the Cestoda, which is a single duct, may correspond to the vagina of the Trematoda, which is, however, a paired structure. The homologies of the ducts in the Trematoda are further complicated by the presence of Laurer's canal, a duct leading out of the ductus communis and opening to the exterior in the Malacocotylea but into the gut in the Heterocotylea. The bursa copulatrix and the muscular pear-shaped organ, which open into the genital atrium in the Turbellaria, are accessory reproductive organs which are probably not represented in the parasitic forms.

CHAPTER VII

THE MINOR ACOELOMATE PHYLA

THERE are nine phyla that go to make up the minor Acoelomata. Some authorities link various of these phyla together to form single phyla, the groups mentioned here as phyla then having class status. Since the interrelationships are not clear the more conservative practice is followed here of making them all phyla of equal status: Nemertea, Nematoda, Nematomorpha, Acanthocephala, Rotifera, Gastrotricha, Kinorhynchia, Priapulida, Endoprocta.

PHYLUM NEMERTEA

DIAGNOSIS. Elongated flattened unsegmented worms with a ciliated ectoderm and an eversible proboscis lying in a sheath on the dorsal side of the alimentary canal, with which it is not connected; no perivisceral body cavity, the spaces between the organs being filled with parenchyma; alimentary canal with mouth and anus; excretory system with flame cells; a blood vascular system; gonads simple, repeated; sexes separate; sometimes a larval form (*Pilidium*).

CLASSIFICATION. The classification of the nemerteans is based upon the position of the longitudinal nerves relative to the muscles and also on the structure of the proboscis (Fig. 139).

Class 1. **ANOPLA**. Mouth posterior to the brain, the central nervous system being just below the epidermis. Proboscis unarmed.

Order 1. **Palaeonemertini**. Body wall muscles of two or three layers, the innermost being circular. *Carinoma, Cephalothrix, Carinella*

Order 2. **Heteronemertini**. Body wall muscles three layered, the innermost being longitudinal. *Lineus, Cerebratulus, Baseodiscus*

Class 2. **ENOPLA**. Mouth anterior to the brain, central nervous system inside the body wall muscles, proboscis often armed.

Order 1. **Hoplonemertini**. Proboscis armed.
Emplectonema, Amphiporus, Geonemertes, Pelagonemertes

Order 2. **Bdellonemertini**. Proboscis unarmed. *Malacobdella*

The Nemertea in their general organization resemble the Platyhelminthes very strongly. In certain positive features they have advanced, e.g. in the development of a proboscis independent of the gut, in the presence of a vascular system, and a second opening, the anus, into the alimentary canal, but in the simplicity of the gonads and absence of hermaphroditism the Nemertea are less specialized than the Platyhelminthes. There can be no doubt, however, that the two phyla are very closely connected, although the presence of an anus and a vascular system is an enormous advance.

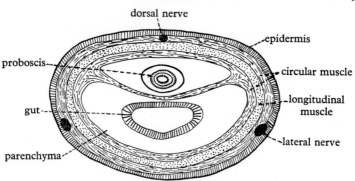

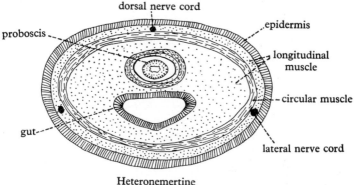

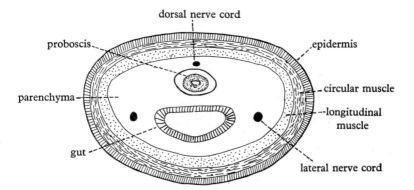

Fig. 139. Diagrammatic transverse sections of nemertines. The different groups of nemertines are differentiated on the number of rings of longitudinal muscles and the position of the nerve cords. (From Benham.)

230 MINOR ACOELOMATA

PROBOSCIS. The proboscis (Fig. 140) is the most characteristic organ of the nemerteans. It lies in a cavity (*rhynchocoel*), completely shut off from the exterior, which has muscular walls (the *proboscis sheath*), and is attached to the posterior end of the sheath by a retractor muscle which is really the solid

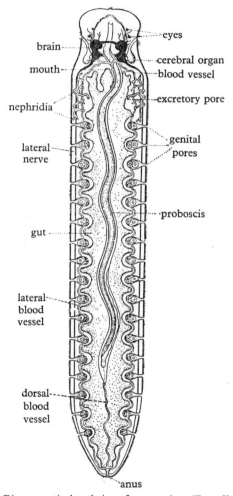

Fig. 140. Diagrammatic dorsal view of a nemertine. (From Kükenthal.)

end of the proboscis. The proboscis may be compared with the finger of a glove with a string tied to the inside of the tip; when the proboscis is at rest the string, i.e. the retractor muscle, keeps it turned inside out within the sheath; when the muscles of the proboscis sheath contract and press upon the fluid in the rhynchocoel the proboscis is everted, but never completely, because the retractor muscle keeps it from going beyond a certain point. At this point, in the Hoplonemertini is a diaphragm cutting off the apical part of the proboscis cavity, and mounted on this is a spike or *stylet* with reserve stylets in pouches at the side (Fig. 142 c). This part of the cavity probably contains a poisonous fluid which is ejected through a canal in the diaphragm into wounds caused by the stylets. The proboscis in this class of nemerteans is thus a formidable weapon. In other nemerteans, though the stylet is not developed, the proboscis is prehensile and can be first coiled round its prey and then retracted to bring it within reach of the mouth. Some forms use the proboscis to aid in burrowing. The part of the proboscis in front of the brain is called the *rhynchodaeum*.

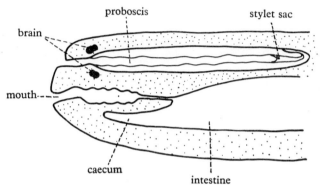

Fig. 141. Diagram of a longitudinal vertical section of a nemertine to show the relationship of the various cavities. (After Benham.)

Often if the proboscis does not possess stylets it will have small rhabdites imbedded in the epidermis wall which may play some defensive role. In *Drepanophorus* the accessory stylets are placed one behind the other giving the whole unit the appearance of a radula. In *Gorgonorhynchus* the proboscis shows complex dichotomous branching. The proboscis is used in attacking prey and possibly also helps in locomotion.

BODY CAVITY. The region between the gut and the body wall musculature is normally filled with parenchyma but some forms such as *Drepanophorus* and *Cerebratulus* have large cavernous vacuities in their gonads, the interiors of which are sometimes lined with epithelium. These cavities occupy the region between the body muscles and the gut, in some cases even encircling the gut. The cavities thus have many of the attributes of the coelomic cavities seen in the annelids. This is particularly the case in *Cerebratulus* where the hindermost cavities are often sterile.

NERVOUS SYSTEM. The ectoderm is completely ciliated: there are gland cells amongst the ciliated epithelium; within this are layers of, first, circular, and then longitudinal, muscles. There is a nerve net which in the most primitive nemerteans lies at the base of the ectoderm cells, in others between the circular and longitudinal muscles, and in the most advanced forms within both layers of muscle. While the nervous system is thus extremely primitive there are concentrations of the nerve net to form lateral nerve cords and a pair of *cerebral ganglia* above the mouth, each cerebral ganglion being divided into a dorsal and ventral lobe and connected by commissures above and below the proboscis sheath. (Fig. 140). The dorsal lobe is subdivided into an anterior and posterior part: the posterior part is in close relation with an ectodermal pit, the cerebral organ, which is situated in some forms in a lateral slit. As yet, however, the control of the movements of the organism is not dependent on the cerebral ganglia. There are occasionally eyes of simple structure.

Inside the muscle layers the body is filled with parenchyma like that of the Platyhelminthes (Fig. 139), but in it are one, two or three longitudinal vessels,

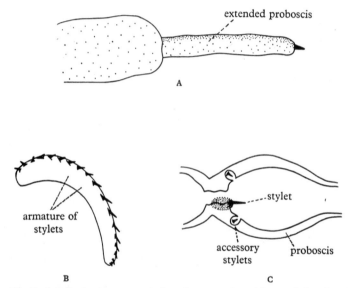

Fig. 142. Proboscis structure. A, anterior of a nemertine with extended proboscis. B, many stylets on the proboscis of *Drepanophorus*. C, base of the inverted proboscis of a metanemertine showing accessory stylets. (C, after Bresslau.)

connected together by transverse vessels with contractile walls, which constitute the vascular system. The blood is generally colourless, but has corpuscles which sometimes contain haemoglobin. The circulation is assisted by the movements of the body. It can hardly be supposed that the blood system, situated so deeply in the body, can be respiratory in function.

The alimentary canal is a straight tube, the mouth and anus being nearly or quite terminal (Fig. 140). The excretory system is formed by a pair of canals situated laterally, each of which communicates with the exterior by one or several pores and gives off many branches, ending internally in flame cells like those of the Platyhelminthes. In some cases the end organs come into contact with the blood vessels. The generative organs are a series of paired sacs alternating with the pouches of the mid gut and these each develop at the time of maturity a short duct to the exterior.

DEVELOPMENT. Most nemerteans develop directly, but in some a pelagic larva with a remarkable form of metamorphosis is found. This larva is known as the pilidium (Fig. 144 A). A conical gastrula with a flattened base is first formed by invagination and it passes into the pilidium by the following

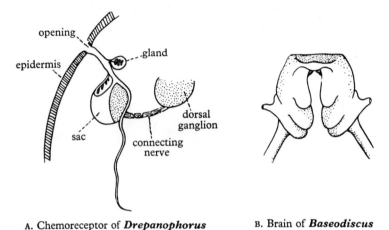

A. Chemoreceptor of *Drepanophorus*   B. Brain of *Baseodiscus*

Fig. 143. Nervous system and sense organs.

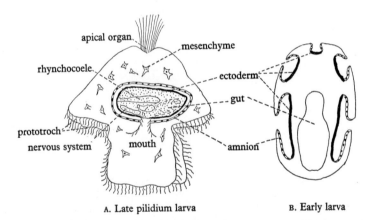

A. Late pilidium larva   B. Early larva

Fig. 144. Pilidium larva: A, side view of late form enclosing young nemertine. (After Korscheldt and Heider.) B, frontal view of earlier stage showing imaginal discs. The anterior unpaired invagination is continued to form the proboscis. (After Burger.)

changes. A band of cilia round the base constitutes the *prototroch* and forms the locomotory organ of the larva; it is drawn out into two lateral lappets. An apical sense organ is formed by a thickening of the ectoderm. Two cells migrate into the blastocoele and break up into a tissue called *mesenchyme*, which is partly converted into larval musculature and partly remains undifferentiated until needed as raw material for the adult organs. The gut is connected with the exterior by an ectodermal oesophagus, ending in a large mouth on the flattened base between the lappets. Thus a creature appears which has many resemblances to the trochosphere larva to be described later.

Inside this larva the young nemertean is produced (Fig. 144 B). Five ectodermal plates (imaginal discs) sink below the surface and each forms the floor of a sac. Eventually these sacs join round the gut and a continuous cavity is formed separating the adult inside from the larval skin (sometimes known as the *amnion*) which is thus its protecting husk while it develops. The imaginal discs join together and form the secondary or adult ectoderm. The pilidium continues to swim about with the little nemertean inside it, even when the

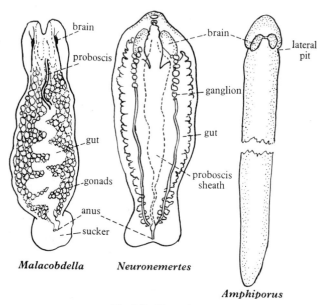

Fig. 145. Nemertines.

organs of the latter are developed and cilia cover its surface so that the adult moves freely as if a parasite of the larva. At length it bursts through the tissues of the amnion and the latter sinks like a discarded mantle.

Malacobdella (Fig. 145). A small nemertean found commensal in the mantle cavity of lamellibranchs. It attaches to the host by means of a thick posterior sucker. There is a proboscis without stylets, a simple digestive system and a well-developed reproductive system. The sense organs and brain are poorly developed.

Cerebratulus. A common marine nemertean with a flattened body well adapted for swimming. It has marked cephalic pits. Some species live in mud burrows.

Lineus. This nemertean is often found growing to very great length (20 m.).

Neuronemertes. A pelagic nemertean usually found at depths greater than a thousand metres below the surface. It has a broad flattened body, no eyes nor nephridia. It is peculiar in that the dorsal nerve develops a series of ganglia along its length (Fig. 145).

Geonemertes. A land nemertean living on the under-surface of damp logs and stones, thus subjected to considerable osmotic stress after rain. Has a well-developed nephridial system.

PHYLUM NEMATODA

DIAGNOSIS. Unsegmented worms, with an elongated body pointed at both ends; ectoderm represented by a thin sheet of non-cellular hypodermis, concentrated to form two *lateral lines* and to a less degree *dorsal* and *ventral midlines*, secreting an elastic cuticle, made of protein, not chitin, usually moulted four times in the life of the individual; cilia absent from both external and internal surfaces; a single layer of muscle cells underneath the hypodermis, divided into four quadrants, each muscle cell being elongated in the same direction as the body and composed of a peripheral portion of contractile protoplasm and a larger internal core of unmodified protoplasm which sends a process to a nerve; the space between the body wall and the gut sometimes filled by a small number of highly vacuolated cells, the vacuoles joining together and simulating a perivisceral cavity; excretory system consisting of two intracellular tubes running in the lateral lines; nervous system made up of a number of nerve cells rather diffusely arranged but forming a circumpharyngeal ring and a number of longitudinal cords of which the mid dorsal and mid ventral are the most important; sense organs of the simplest type; sexes usually separate, gonads tubular, continuous with ducts, the female organs usually paired, uniting to open to the exterior by a ventral vulva, the male organ single, opening into the hind gut, thus forming a cloaca, in a diverticulum on which lie the copulatory spicules; spermatozoa rounded and amoeboid, fertilization internal; alimentary canal straight and composed of two ectodermal parts, the suctorial fore gut and the hind gut and an endodermal mid gut without glands or muscles; segmentation of egg complete and bilateral in type, development direct, larvae only differing slightly from adult.

AFFINITIES. The nematodes appear to occupy an isolated position, but many of their characters, though more specialized, resemble those of the Platyhelminthes and Rotifera. They are certainly closely related to the Acanthocephala, Gastrotricha, and the Nematomorpha. One of their peculiar features is certainly secondary, namely the absence of cilia. There are in some nematodes cilium-like processes to the internal border of the endoderm cells; in one case active movement has been reported. The excretory canals, when the absence of flame cells is taken into account, are seen to resemble those of the Platyhelminthes. Nearly all the other characters may be called primitive. The simplicity of organization, the absence of segmentation at all stages and a vascular system, the diffuse nature of the nervous system and the structure of the muscle cells are all signs of a lowly origin. But it is still maintained by some that these features are not primitive but degenerate and that the origin of the phylum is to be sought in the arthropods, probably in the parasitic forms of that group (the degenerate arachnids called linguatulids). If this view is taken it must be supposed that the parasitic nematodes are the most primitive members of the phylum and that some of their descendants became less and less parasitic, until entirely free-living forms came into existence. This would be an extraordinary reversal of evolution for assuming which, at present, there are no grounds.

The view taken in this book is that the free-living nematodes are ancestral to the parasitic forms and that there is no real connection between the arthropods and the nematodes. Not only do the nematodes present no indications of segments or appendages at any point of the life history but also the cuticle is of an entirely different chemical composition in the two phyla, and the loss of cilia most likely a phylogenetically recent phenomenon in the nematodes as in the parasitic platyhelminthes.

CLASSIFICATION

Class 1. **APHASMIDIA**. No phasmids (caudal sensory organs), excretory system rudimentary or poorly developed. Males have one spicule.

Order 1. **Trichurata**. Have a long fine oesophagus. *Trichuris, Capillaria*

Order 2. **Dioctophymata**. Oesophagus cylindrical. *Dioctophyma*

Class 2. **PHASMIDIA**. Have phasmids, excretory system present and well developed. Males usually have two spicules.

Order 1. **Rhabditata**. Small, mainly free-living worms, oesophagus with one or two bulbs, simple mouth or may have six papillae. *Rhabditis, Strongyloides*

Order 2. **Ascaridata**. Oesophagus bulbed or cylindrical, long vagina, mouth with 3–6 papillae; males usually have two spicules but no true bursa, tail curls ventrally. *Ascaris, Enterobius*

Order 3. **Strongylata**. Simple mouth with no papillae. Males with two spicules and a true bursa. Oesophagus club-shaped or cylindrical. *Ancylostoma, Strongylus*

Order 4. **Spirurata**. Oesophagus cylindrical, often partly muscular and partly glandular; males have two spicules with well-developed alae and papillae; vagina elongate and tubular; require an intermediate host.

Wuchereria, Onchocerca, Gnathostoma

Order 5. **Camallanata**. Mouth simple or with lateral jaws, posterior part of oesophagus has from one to three large nuclei; requires an intermediate host. *Dracunculus*

ANATOMY. The anatomy of the nematodes is best known from the study of *Ascaris* which is one of the largest members of the group and the only one adapted for dissection in class. Full accounts of this form are given elsewhere, but the following points must be emphasized. In *Ascaris* (Fig. 146) there appears to be a wide space between the muscle layer and the endoderm cells, with no epithelial boundary walls, but on closer examination it is seen to be occupied by a very small number of greatly vacuolated cells, and what appears to be a continuous cavity is really the confluent vacuoles of adjacent cells, and so the term 'intracellular' may be applied to it. This arrangement has not been verified in many other nematodes but connective tissue cells can usually be demonstrated in the space. They may be phagocytic; the enormous branched cells of *Ascaris* (Fig. 146), lying on the lateral lines, take up in their tiny corpuscle-like divisions such substances as carmine and indigo which are injected into the body.

A striking feature of the histology of *Ascaris* is the presence of greatly enlarged cells. Not only do the body cavity cells show this, but in the excretory

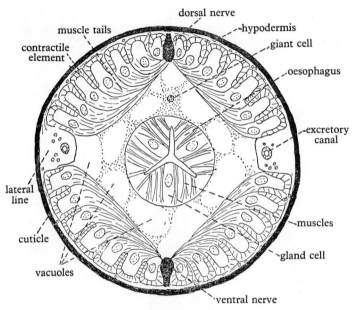

Fig. 146. Diagrammatic transverse section through *Ascaris* in the region of the oesophagus to show the single large cell occupying the space between the body wall and the gut. The number of muscle cells in each quadrant is very much greater than shown here. (Original.)

system the greater part of the canal is contained in the body of one cell which divides into two limbs each running the whole length of the body on opposite sides.

As a simple type of nematode the genus *Rhabditis* (Fig. 149) will be described, as it is seen alive as a transparent object under the microscope. Most species are free-living. They are obtained by allowing small pieces of meat to decay in moist earth. The larvae, which exist in an 'encysted' condition in the soil, are attracted by the products of decay, and in a few days become sexually mature. Great numbers of adults and young can then be scraped off the surface of the meat in the liquefied matter formed by bacterial decomposition.

The cuticle in *Rhabditis* is smooth and bears no protuberances. In other nematodes there is often considerable sculpturing and lobing of the cuticle.

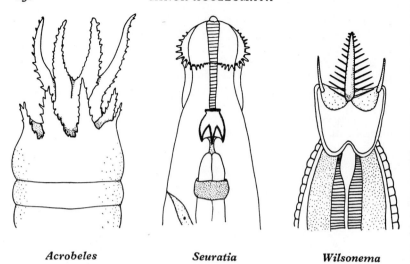

Acrobeles   *Seuratia*   *Wilsonema*

Fig. 147. Nematode head structure. (From Hyman.)

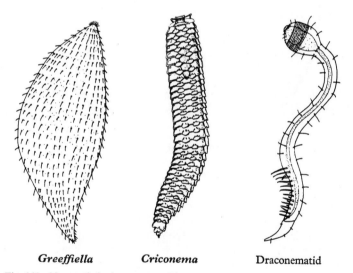

Greeffiella   *Criconema*   Draconematid

Fig. 148. Nematode body structure. These examples are chosen to show forms other than the usual *Ascaris* type. (From Hyman.)

Even in *Rhabditis* the mouth is surrounded with six oral lobes or papillae. In *Enoplus* each of these lobes possesses a sensory bristle, there being, in all, three rings of sensory bristles around the mouth. Sometimes, as in *Wilsonema*, the oral lobes have marked forked protuberances. *Greeffiella* has a series of cuticular spines covering the whole of the body. Spines are present in *Desmoscolex*, and in the draconematids they are used in locomotion (Fig. 148).

The cuticle is thin, tenacious but elastic. It enables the animal to keep an almost constant round cross-section and length; it will be remembered that there are no circular muscles in the nematodes, thus the cuticular strength acts as the antagonist to the longitudinal muscles. These longitudinal muscles run the whole length of the body and are not segmented. If the animals

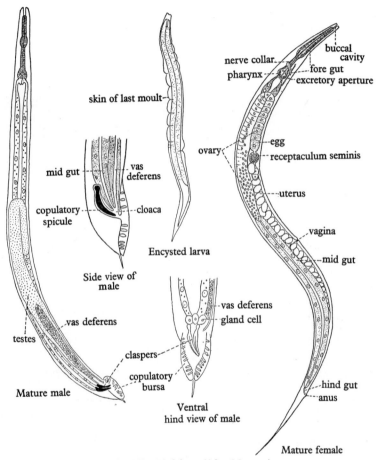

Fig. 149. *Rhabditis*. (After Maupas.)

are allowed to swim in clean water they exhibit alternate contractions of the muscles on each side of the body. However, if they are placed in a muddy, viscous medium or a solution containing many small particles, they show multiple ripples passing along the length of the body. The mechanism of muscular control is not understood. There are usually only a few muscles, the number being constant for any given species (sixty-five in *Enterobius*).

A cross-section through *Rhabditis* shows a similar structure to *Ascaris*, though the muscle cells are much less numerous (only two to each quadrant): each cell contains a number of contractile fibrils arranged in a different way to those in the *Ascaris* cell. The body cavity has not been investigated; that of *Ascaris* has therefore been described above.

GUT. The alimentary canal consists first of all of an ectodermal *fore gut* lined by cuticle in which the following parts can be distinguished: (1) a *mouth*, surrounded by *papillae*, opening into a narrow *buccal cavity* with parallel sides. In some free-living nematodes which are carnivorous (e.g. *Mononchus*) the buccal cavity is very wide and rotifers and other animals are taken into it;

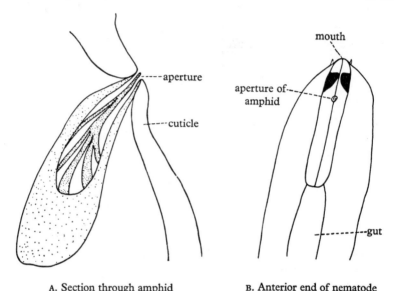

A. Section through amphid   B. Anterior end of nematode

Fig. 150. Amphids (sense organs). A, diagrammatic transverse section of an amphid to show sense cell. B, anterior end of nematode showing aperture of amphid.

(2) an *oesophagus*, with muscular walls and a small number of unicellular glands, forming two swellings, the *oesophageal bulbs*. The posterior of these (the so-called *pharynx*) exhibits rhythmical pumping movements, caused by the contraction of the radial muscles which enlarge the cavity of the bulb and open the valve formed by the thickened cuticle. In this way the surrounding fluid is drawn into the oesophagus: no solid particles much larger than bacteria can be admitted through the narrow lumen. When the muscles relax and the cavity disappears the fluid is driven on into the *mid gut*. This is composed of a single layer of cells, which internally are naked but externally have a fine cuticle. These are entirely absorptive in function, gland cells being absent. There are no muscles, but the gut contents are circulated by the locomotory movements of the animal. The hind gut which follows is lined with cuticle and opens at the ventrally situated anus. Near the anus is a sphincter muscle, but there are also dilator muscles running from the hind gut to the body wall,

and during the periodic contraction of these the gut contents are evacuated. The alimentary canal of the nematodes as thus seen in action represents a type simplified because the animal usually lives on food which has been split up into easily assimilable substances—in this case by bacterial action, in the case of *Ascaris* by the ferments of the living host—and this is passed with great rapidity through the alimentary canal by the pumping action of the oesophagus.

SENSE ORGANS (Fig. 150). The nematodes have poorly developed sense organs. Papillae of a sensory nature are commonly present on the surface of the cuticle. In addition there are spines and scales present in some of the free-living marine forms. There are a pair of specialized sense organs, the

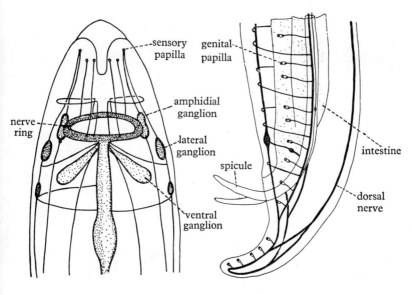

A. Anterior end   B. Posterior end, lateral view

Fig. 151. Central nervous system of *Ascaris*. (From Hyman, after Goldschmidt.)

amphids, present on the head of most nematodes; though their function is not clear they are probably chemoreceptors. The amphids are reduced in the parasitic forms. Posteriorly there are a pair of unicellular glands called phasmids. These are best developed in the parasitic worms. Fresh-water and marine nematodes often have a pair of eyes.

NERVOUS SYSTEM (Fig. 151). This consists of a circum-enteric ring encircling the pharynx, a pair of lateral ganglia, a dorsal and a ventral ganglion. From these ganglia emerge a dorsal, ventral and several lateral nerves which run the length of the body. Cross connexions between these nerves occur along the length of the body. Anteriorly from the nerve ring run six nerves to the oral papillae. There is apparently a fixed number of cells taking part in the nervous system, in *Ascaris* the number is 162.

EXCRETORY SYSTEM. There are two types of excretory systems, the glandular type and the H-type. The glandular type is found in the free-living nematodes where there is a single large cell on the ventral part of the junction of the pharynx and mid gut. The duct from this cell runs forwards and opens into the mid line of the cuticle. In other forms such as *Rhabdias* there are two such kidney-shaped cells. These cells may have their posterior part extended to run the length of the body, the two cells forming an H (Fig. 152). In *Ascaris* such an H-system is present but there is only one cell body.

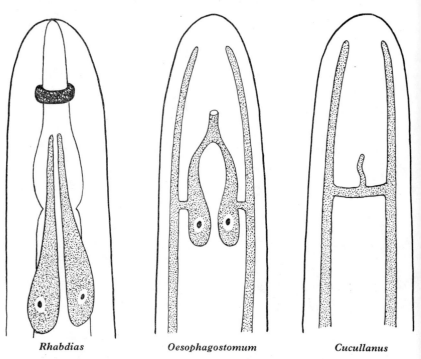

Fig. 152. Excretory organs of nematodes. (After Hyman.)

In addition there are easily seen in living *Rhabditis* the ventral *aperture of the excretory canal*, not far behind the mouth, and, when the animal is compressed under the coverslip, the coiled line of the excretory canal. The only part of the nervous system which can be so seen is the ring round the oesophagus.

GENITALIA. The genital organs are of the type seen in *Ascaris* but simpler. In the female there are two tubular gonads bent once on themselves, discharging by a single genital aperture, situated about half-way between the head and the tail. The *ovary* is a short syncytial tube, the nuclei becoming larger and larger and the centre of more definite and larger aggregations of cytoplasm and yolk nearer the uterus. Finally, there is a single ovum discharged at a time into the *oviduct*; as soon as this happens another ripens in its place. To reach the uterus the egg has first to pass through a portion of the oviduct (*receptaculum seminis*) filled with the amoeboid spermatozoa of the male. Fertilization takes place, a shell is formed and at the same time maturation proceeds. The two uteri join to form the median *vagina*. In this the fertilized egg develops and the young larva is formed and may hatch within the vagina. The stages of segmentation are seen nowhere with such ease or clearness as in a small transparent nematode of this kind.

The male, on the other hand, has only a single gonad. The apical testis is syncytial like the ovary. Nearing the vas deferens a zone may be seen of free spermatocytes and in the vas deferens itself can be seen large numbers of rounded spermatozoa. The genital duct opens into the gut to form a *cloaca*. This contains a dorsal pocket in which is secreted a chitinous apparatus consisting of two converging rods, the *copulatory spicules*, with a grooved connecting piece to hold the points together. The pocket has a special muscle which protrudes the spicules from the anus (cloacal aperture). To each side of this aperture is a lateral cuticular flange, supported by ribs, which meets its fellow at the root of the drawn-out tail. This acts as a sucker (*copulatory bursa*), by which the male retains its position on the body of the female until the spicules are thrust through the female aperture and keep the female and male apertures both apposed and open. Then by the contraction of the muscles of the cloaca the spermatozoa are expelled and passed into the vagina of the female. Here they become amoeboid and travel up the uteri so that they can meet the ova as the latter are discharged.

Besides the normal condition in which males and females are produced in equal numbers, many species of *Rhabditis* occur in which there is a remarkable disparity in numbers of the sexes. For a thousand females there may be only ten or twenty males, and they are lethargic in their sexual activities. The females, on the other hand, have developed a curious kind of hermaphroditism. When the gonad first becomes ripe a number of spermatozoa are produced. Afterwards the gonad produces nothing but eggs which are fertilized by the individual's own spermatozoa, and after these are exhausted nothing but sterile eggs are laid. Experiment has proved that in these animals self-fertilization may occur for an immense number of generations without any deterioration of the species.

EMBRYOLOGY. The eggs are fertilized internally and are usually covered with a chitinous cuticle. The precise stage at which the egg leaves the parent varies amongst the nematodes. In *Ascaris* the unsegmented egg is deposited, in *Ancylostoma* the eggs are already at a state of early cleavage on emergence, in *Enterobius* the eggs are at a late embryonic stage whilst those of *Wuchereria* are fully developed little embryos.

Cleavage is a much modified spiral cleavage with each stage being highly determinate, the embryo being unable to readjust or make good any cellular losses that may occur under experimental conditions. The first two cleavages lead to a T-shaped configuration which later rearranges to a rhomboid shape. These four cells are termed, A, B, P_2 and S_2. A and B give rise to the ectoderm, whilst S_2 gives rise to endoderm, mesoderm and the stomodaeum. Further cleavage gives rise to a blastula; gastrulation is by epiboly, the ectodermal cells growing over the endodermal cells (Fig. 153). The perivisceral region of the adult is related to the embryonic blastocoel which remains filled with large vacuolated cells.

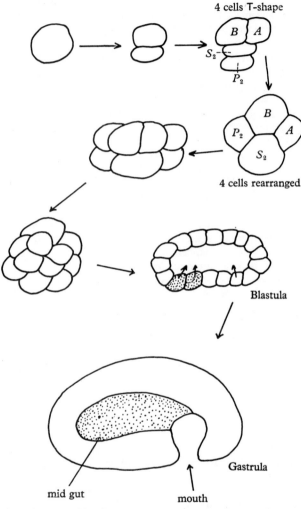

Fig. 153. Cleavage of fertilized nematode egg. (From Hyman, after Boveri.)

Two important points should be noted. The first is that the germinal cells soon become differentiated from the somatic cells on account of the special behaviour of their chromosomes. In the somatic cells the chromosomal material is continually being broken and lost, the germ cells being the only ones to retain their full chromosomal complement. It was this observation that helped Weissman to his celebrated germ plasm theory. Secondly, and related to the first point, in the embryo all cell division soon ceases except for that seen in the reproductive cells. Growth consists of vacuolation and extension of the cells already present. Thus like the Rotifers there are a fixed number of very large cells carrying out all the major functions of life. In *Rhabditis* there are sixty-eight muscle cells, some 200 nerve cells, 120 epidermal cells and 172 cells in the digestive tract.

LIFE CYCLES. In *Rhabditis*, as in the majority of nematodes, there are four moults. After the second moult the animal may remain within the loosely fitting skin as a so-called 'encysted' larva which possesses, however, the power

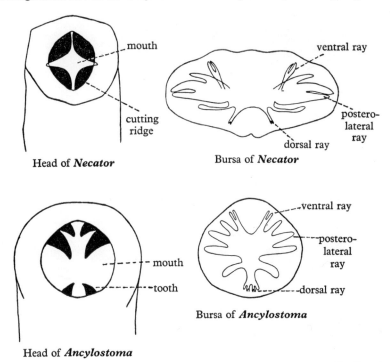

Fig. 154. Jaws and bursa of the hookworms, *Necator* and *Ancylostoma*. (After Loos.)

of movement. The protection of the cast skin and possibly other factors enables this stage in the life history to resist desiccation and to remain in a state of dormant metabolism until some odour of decaying substances attracts the larvae and the opportunity of rapid reproduction is given for a brief period.

This third larval period is characteristically the period of wandering in many nematodes, and this is seen in a remarkable manner in the classical life history of *Ancylostoma* (Figs. 154, 155). These animals live attached in the adult stage to the mucous membrane of the human small intestine, sometimes in such numbers as to present an aspect comparable to the pile of a carpet.

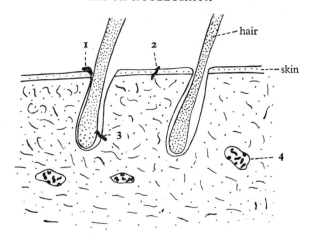

A. Larvae entering through the skin

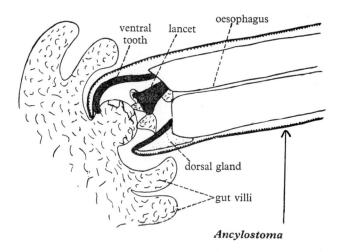

B. *Ancylostoma* attached to the gut

Fig. 155. A, *Ancylostoma* larva penetrating the epidermis of a mammal; 1–4 are larvae. B, *Ancylostoma* attached to the epithelium of the small intestine. (After Loos.)

They feed on the intestinal tissues and only accidentally rupture the blood vessels, causing anaemia in the host. The females are fertilized *in situ* and eggs are laid, which begin to segment before they pass out into the faeces. The rest of the life history may be shown as follows:

(1) First larval form (*rhabditoid*) with a buccal cavity like *Rhabditis*. This lives in the soil for three days before the first moult, which produces the

(2) Second larval form which moults after two days, the skin remaining as a cyst round this *strongyloid* larva.

(3) In this stage the animal becomes negatively geotropic and thigmotropic, ascending through the soil and being specially attracted to the moist skin of human beings. This they penetrate by way of the hair follicles, though occasionally the larva enters the gut by the mouth. In the former event, the minute larva is able to make its way through the skin to lymph spaces and to blood vessels, eventually being swept into the circulation by the vena cavae to the right auricle, thence to the right ventricle and then to the lung. In the pulmonary capillaries this career is ended and the larvae make their way into

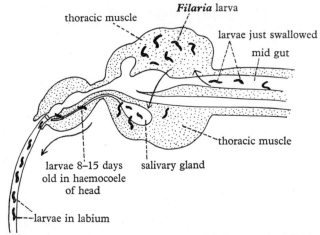

Fig. 156. Diagram of *Wuchereria* (*Filaria*) larvae wandering in the body of the mosquito *Stegomyia*. (After Bahr.)

the alveolar cavities of the lung. They then travel by the bronchi and the trachea to the oesophagus and so to the intestine. Here the animal is freed from the second skin, producing the larva *without buccal capsule*. The third moult produces the last larval stage towards the fifth to seventh day and this is termed the larva *with provisional buccal capsule* (4).

Finally, about the fifteenth day the fourth moult produces the worm with the *definitive buccal capsule* (5), and in three to four weeks from hatching the parasite has become sexually mature and is attached to the epithelium of the intestine.

This most important human parasite shows in its earliest stages the structure and the free-living habit of the primitive form *Rhabditis*, and it is noteworthy that there are many species of the latter genus which have already become parasites.

It may, however, be supposed that a less specialized life history is that of the species of *Enterobius* in which the egg is swallowed by the host and the remaining

Table 5. *Life histories of nematode parasites*

	Adult	Eggs	Early larvae	Larval wanderings	Pathology
Haemonchus and other parasites of ruminants	In small intestine of animal	Pass to exterior with faeces	Hatch in faeces and moult twice	Crawl up grass blades and remain there until swallowed by host: then develop *directly* in gut	Cause anaemia
Ancylostoma and other hookworms	In human small intestine attached to epithelium	Pass to exterior with faeces	Hatch in soil and moult twice	Crawl up to surface, and infect host by passing through skin—venous system—heart—pulmonary artery—lung capillaries—cavity of lung—trachea—gut where they become adult	Cause anaemia in adult stage by destroying intestinal epithelium: inflammation of skin and lung in larval wanderings
Enterobius (Oxyuris) vermicularis	Free in large intestine of man	Female lays eggs in the neighbourhood of the anus. Eggs are swallowed by the host and develop directly in intestine			Irritation and inflammation of anus, occasional penetration into mucosa of small intestine and appendix where they cause inflammation
Ascaris lumbricoides	Free in small intestine of man and pig	Pass to exterior with faeces and are eaten by new host	Hatch in intestine	Pass through intestinal epithelium into venous system—rest as in *Ancylostoma*: never by direct infection	Cause inflammation of lung in larval wanderings (in mass infections)
Strongyloides stercoralis	In wall of small intestine (females only)	Pass to exterior and hatch (development *parthenogenetic*) (1) at temperature 15–20° C., moult once; (2) at temperature 20–25° C., larva develops directly in faeces into *male* and *female* which produce eggs, hatching to larvae which moult once		Infects host through skin as in *Ancylostoma* Infects host through skin as in *Ancylostoma*	

Trichinella spiralis	In small intestine of man: females viviparous, penetrate into lymphatic spaces of intestinal wall and larviposit there	Larvae pass into blood system	Distributed by arterial capillaries to musculature—here they enter muscle cells and encyst. First host eaten by second: encysted larvae liberated by digestion and quickly mature in intestine	Cause great disturbance in traversing intestinal epithelium in mass infection, degeneration of muscle fibres
Wuchereria bancrofti	In lymphatic system of man	Larvae live in the blood (particularly showing themselves in the peripheral vessels during the night, but disappearing at dawn, *Microfilaria nocturna*)	Distributed to fresh hosts by mosquitoes which feed at night. In the stomach of the mosquito larvae lose their protective sheath; migrate through the gut wall into the thoracic muscles. After further development they pass into the proboscis of the mosquito and are reintroduced into the human host, where they rapidly mature	Adults are said to be the cause of elephantiasis
Loa loa	In subcutaneous tissues of man, which it traverses very actively	Larvae live in the blood (appearing in it at early morning and disappearing in the evening, *Microfilaria diurna*)	Distributed to fresh hosts by the blood-sucking fly, *Chrysops*. Details similar to *W. bancrofti* in mosquito	Cause 'Calabar swellings'
Dracunculus medinensis	In subcutaneous tissues of man	When ready to lay the female produces lesions or abscesses usually in the legs of the host. She protrudes her vagina when the skin of the host is in contact with water, liberating thousands of embryos into the water	Larvae penetrate skin of *Cyclops* and remain in body cavity. The *Cyclops* is swallowed by man and dies, but the larva is liberated and passes from the gut to the subcutaneous tissue where it becomes mature	Cause dracontiasis

stages of development take place in the gut. It is said that several successive generations of the parasite may occur within the same host. On the other hand, the wandering habit of nematodes is a fundamental character and even forms in the first stage of parasitism (facultative) may penetrate host tissues.

The life histories of the principal nematode parasites of man and domestic animals are summarized on pp. 248–9 (Fig. 157). They are arranged in a definite order passing from the simplest type in *Haemonchus* to the most specialized life histories in *Wuchereria* (*Filaria*), *Loa* and *Dracunculus*.

Two other classes of nematode parasites merit particular attention. They are, respectively, parasites of plants and insects.

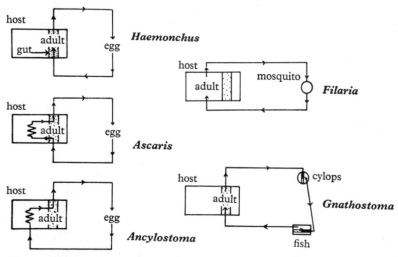

Fig. 157. Diagram of life cycles of nematodes to show increase of complexity. Note that the larval stages have been ignored.

PLANT PARASITES. Nematodes are particularly fitted for a parasitic life in plants by reason of their form and activity and their capacity (at the end of the second larval stage) for resisting desiccation and other unfavourable conditions. They are small enough, as larvae, to obtain entrance through the stomata of leaves, and sometimes possess dart-like projections of the buccal lining which enable them to penetrate the cell walls of plants. They feed on cell sap and by their interference with the life of the host plant cause the formation of galls, wilting and withering of the leaves, and stunting of the plant.

Tylenchus tritici passes through a single generation in the course of the year, and infects wheat. The animal becomes adult when the grain is ripening and a pair, inhabiting a single flower, produce several hundred larvae. Instead of the grain a brown gall is produced, and in this the larvae (after moulting twice) may survive for at least twenty years. If the grain falls to the ground the larvae may remain there over the winter or may escape into the soil. When the corn begins to grow in the spring they enter the tissues of the plant and make their way up the stem to the flower, where they speedily mature. The great interest of this life history lies in the easy adaptation of the parasitic life history to the annual cycle of the wheat plant and the extreme capacity for survival in a dormant and desiccated condition until the right plant host becomes available.

Tylenchus devastatrix, on the other hand, may pass through several generations in the year and attacks indiscriminately clover, narcissi bulbs and onions, and many other useful plants. *Heterodera* (Fig. 158 C) is a parasite of the roots of tomatoes, cucumbers and beets, and is remarkable because the female attaches herself in larval life to a rootlet from which she sucks a continuous flow of sap. She is fertilized by wandering males and grows enormously, becoming lemon-shaped. Inside the body thousands of larvae are produced, which escape into the soil and live there until the opportunity arises for infection of fresh roots.

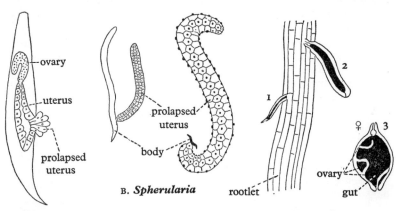

A. *Atractonema* B. *Spherularia* C. *Heterodera*

Fig. 158. Insect and plant nematodes: 1, 2, larvae attached to a rootlet with their heads imbedded in its tissues; 3, adult female (smaller scale) removed from plant. (A, B, after Leukart; C, after Strubell.)

INSECT PARASITES. Four of these may be mentioned, though other life histories are also of great interest.

In *Mermis* a curious reversal of the typical nematode life cycle occurs. The sexual forms are all free-living either in the soil or fresh water. On summer days after showers the sexual forms of *Mermis nigrescens* exhibit a curious tropism, leaving their haunts two or three feet in the ground and crawling up the stems of plants, but disappearing when the sun grows warm. The eggs are laid in the ground and when the larvae hatch they pierce the skin of insect larvae and wander into the body cavity where they nourish themselves by absorbing fluid food through the cuticle. The mid gut has become a solid body, having no connexion with the mouth and anus, and in it fat is stored up which serves as raw material for the production of eggs. When the animals become sexually mature they escape into the soil.

In *Tylenchus dispar* (a form which is thus placed in the same genus as the well-known plant parasites) the adult female and innumerable larvae are found in the body cavity of the bark-beetle, *Ips*, during the winter. *Allantonema* has similar relations to another bark-beetle, *Hylobius*. The female is enormously developed; the uterus and other female organs occupy the whole of the body, the gut having entirely disappeared. In the spring the larvae (having undergone two moults) bore through the walls of the end gut and undergo further development in the 'frass' (faeces of the beetle). The male develops precociously and fertilizes the female which, when it becomes mature, is still of normal proportions. After fertilization the females (only) infect the beetle larvae which by this time have appeared. Entrance is obtained by means of a 'dart' exactly like the similar organ in the plant parasites. In the body cavity the female *Allantonema* grows rapidly, and when metamorphosis occurs and the mature bark-beetle seeks another tree to form a new colony, it is full of larvae.

Spherularia (Fig. 158 B) is a parasite of the humble-bee. In the summer the moss and soil near the bee's nest is inhabited by the sexually mature worms, and after fertilization has taken place the female wanders into the body cavity of the insect, as in the preceding life histories. Though the number of cells in the somatic tissues of the bee is said not to increase in number, there is an enormous growth in size of the vagina which becomes prolapsed and forms eventually an organ many times the size of the rest of the body, which remains attached for some time but eventually disappears. The parasitized humble bees, after passing the winter in their nests, tend to emerge early. In the spring very often inactive bees may be caught which prove, on dissection, to contain one or more of these enormous sausage-shaped bodies, each of them full of eggs and larvae, which escape through the gut wall and become free-living.

Atractonema (Fig. 158 A), a parasite of the Cecidomyidae (p. 509), has a similar life history.

PHYLUM NEMATOMORPHA

DIAGNOSIS. As in Nematoda but lateral line and 'excretory' canal absent, nervous system consisting of a dorsal 'brain' and a single ventral cord, genital ducts in both sexes opening into the hind gut to form a cloaca, development very characteristic—gastrulation by invagination and a larva with peculiar boring organ which infects insects.

In addition it should be mentioned that the alimentary canal is always more or less degenerate and the body cavity may either be occupied by parenchymatous tissue or by reduction of this becomes more or less empty of cells.

An example of this group is *Gordius robustus* with the following life history. The adults are found in brooks from October till May when they copulate. The sperm is not directly introduced into the cloaca, but placed in masses on the body near it. The eggs are laid in the water and the larvae soon hatch. By using the boring organ which they possess they find their way into the body cavity of crickets which live near the water. There they remain and grow until the autumn when they leave the host and enter the water again as mature animals. Other forms have similar life histories. *Parachordodes* first infects chironomid larvae and then these are eaten by the second host, the beetle *Calathus* in which they grow to maturity (Fig. 159).

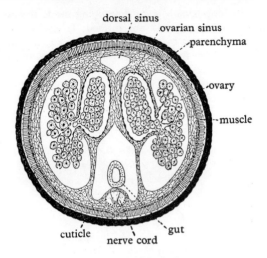

A. T.S. of *Parachordodes*

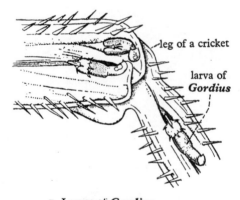

B. Larvae of *Gordius*

Fig. 159. Nematomorpha: A, transverse section through *Parachordodes*. B, larvae of *Gordius* in the leg of an insect.

PHYLUM ACANTHOCEPHALA

As in Nematoda, but possessing an eversible proboscis provided with hooks for attachment and glandular organs (the *lemnisci*), cuticle delicate, hypodermis containing a peculiar lacunar system, a layer of circular muscles as well as longitudinal, nervous system consisting of a brain and two lateral cords, excretory organs, which when they occur are nephridia with modified flame cells; body cavity without parenchyma but traversed by a tubular organ, the *ligament*, containing gonads, eggs developing inside the body until the provisional hooks of the pharynx are formed; larvae developing further when laid in water and eaten by a crustacean, becoming mature when eaten by a vertebrate after which the animals attach themselves to the wall of the intestine by their proboscis.

An example of these is *Neoechinorhynchus*, the adult of which lives in fish and the larvae in *Sialis* larvae (Fig. 160).

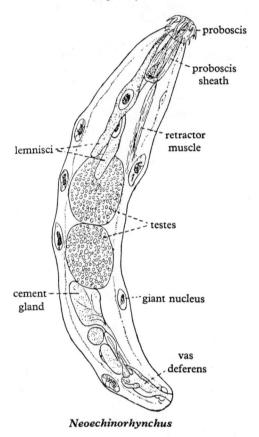

Neoechinorhynchus

Fig. 160. *Neoechinorhynchus* (Acanthocephala).

PHYLUM ROTIFERA

DIAGNOSIS. Minute animals, unsegmented and non-coelomate, typically with a ciliated trochal disc for locomotion and food collection, a complete alimentary canal with anterior mouth and posterior anus, and a muscular pharynx with jaws of a special type; excretory system with flame cells joining the hind gut to form a cloaca; no blood system or respiratory organ; very simple nervous system; sexes separate, two kinds of eggs, one developing immediately without fertilization and the other, which is fertilized, thick-shelled and developing only after a resting period.

AFFINITIES. This group contains a large number of forms of great interest to the microscopist which are easily obtained from many kinds of fresh water. They are, generally speaking, the smallest of all metazoa. They vary little in structure and present a remarkable similarity to the trochosphere larva. It must be admitted that the Rotifera are on a lower stage of organization than the annelids and molluscs which possess this larva and may even be related to a common ancestor of these phyla. On the other hand, the Rotifera come near to the Platyhelminthes, the Gastrotricha and Nematoda.

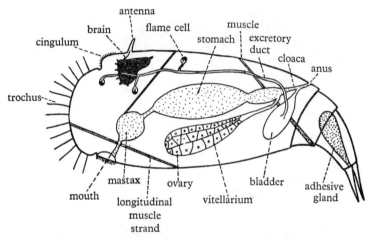

Fig. 161. Diagram of a rotifer.

ANATOMY. An elastic external cuticle covers most of the body. Under this is a syncytial ectoderm; a continuous layer of muscles forming a body wall is absent (as in the Arthropoda), but isolated bands of muscle, chiefly longitudinal, traverse the body (or perivisceral) cavity (Figs. 161, 162).

"What is the true nature of the body cavity?" is a question which has never been properly answered. It is a wide space between ectoderm and endoderm, traversed by muscles, and is neither a coelom nor a haemocoele in the narrower sense, but probably only a derivation of the segmentation cavity of the gastrula (the blastocoele), as in the trochosphere larva. But they do possess a body cavity and not a solid parenchyma, and so differ from the Platyhelminthes. Their excretory system is, however, very similar to that of the latter phylum, and in the union of the excretory duct with the gut the rotifers resemble certain specialized trematodes.

Like the Nematoda they consist of a small number of cells and all the tissues, except the cells of the velum, may lose their cell boundaries and become syncytial. Not only is there a superficial resemblance to heterotrichous ciliates in the Protozoa but the tendency to the acellular condition carries this a step further.

Hydatina senta may be taken as a type of the group (Fig. 163). The female is pear-shaped, the posterior end being the stalk. The anterior end is flattened and forms the *trochal disc*. This is, in many rotifers, bordered by a double ciliated ring, the *velum*, the outer part of which (the *cingulum*) is the original velum and is composed of strong cilia. The inner is called the *trochus*. Between the two rings, which are thus preoral and postoral respectively, is a ciliated groove in which is situated the mouth. The velum in life gives the impression of revolving wheels, the reason for the scientific name of the group. In *Hydatina* the cingulum forms a complete ring and the trochus is reduced to a double transverse row of cilia; in the groove between them is situated a number of

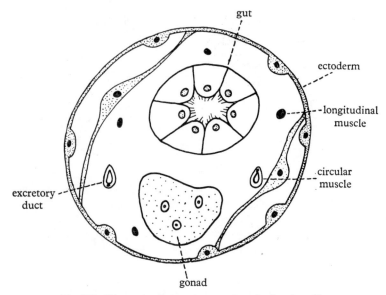

Fig. 162. Diagrammatic transverse section of a female rotifer.

papillae on which are stiff cilia. (In *Copeus* and other creeping forms there are no trochus and cingulum but cilia cover the trochal region and part of the ventral surface. This is said to be a primitive arrangement.) The posterior end is called the *foot* and it terminates in a pincer-shaped appendage, on which open glands with a sticky secretion. By means of this apparatus the rotifer can anchor itself in the intervals of its free-swimming life. The *dorsal* surface of the rotifer is marked out by the position of the *cloacal aperture* just in front of the foot; on this surface immediately behind the velum is a sense organ, the *dorsal antenna*, and below it the *brain*. There are also two *lateral antennae*; all three are prominences bearing stiff sense hairs. Elsewhere the body is covered by a thin, smooth, transparent cuticle secreted by the ectoderm.

The food, which consists of micro-organisms of various kinds, is swept by means of the ciliary currents of the disc into the mouth and then through the oesophagus into the muscular pharynx or *mastax* which is provided with chitinous jaws, the *trophi*, which are in constant movement and, in *Hydatina*, masticate the food as it passes through. This first part of the alimentary canal is ectodermal and constitutes the *stomodaeum*. Then follows the endodermal *stomach*, lined with ciliated epithelium, in which digestion takes place. Digestion is usually extracellular, but in *Ascopus* and other rotifers it is intracellular.

Two *gastric glands* open anteriorly into the stomach. A narrow *intestine* leads into the *cloaca*, into which the *excretory system* also opens. The latter consists of two lateral ducts, coiled at intervals, consisting of perforated cells placed end to end into which flame cells (vibratile tags) open frequently but irregularly. Anteriorly the ducts communicate by a transverse vessel just

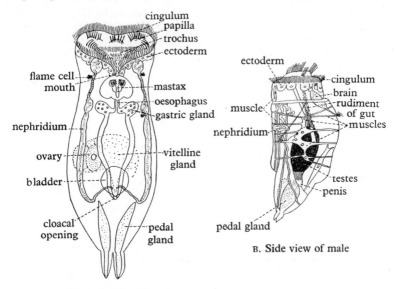

Fig. 163. *Hydatina senta*. (A, original; B, after Wesenberg Lund.)

behind the disc and posteriorly they open into a pulsating vesicle which expels its contents into the cloaca. It has been calculated that in some species this bladder expels a bulk of fluid equal to that of the animal about every ten minutes.

The single ovary is a bulky organ: it is divided into a small *germarium* (the ovary proper) and a much larger *vitellarium* or yolk gland which occupies much of the space between the stomach and the body wall. The ovary is continued into a duct which opens into the cloaca.

The female is still the only individual known in many kinds of rotifers. It was not until 1848 that a male rotifer of any kind was described. In only a few species is the male equal in size and organization to the female. In all the rest there is a more or less pronounced sexual dimorphism. In *Hydatina* (Fig. 163) the male has no alimentary canal, but the ciliated disc, musculature and excretory system are well developed. Usually the male is not only smaller but its ciliated disc and the alimentary canal are very much reduced and the excretory system may be absent. The chief organ is the large *testis*, usually filled with ripe spermatozoa, which opens by a median dorsal *penis* in many cases. Where the penis is absent the tapering hinder end may be inserted in the cloaca of the female. Finally, it may be mentioned that in one large family, the Philodinidae, which includes the genus *Rotifer*, no male has ever been found.

REPRODUCTION. Two kinds of reproduction occur in the rotifers as in the cladoceran crustacea (p. 375), but in this case there are two kinds of females, one of which always reproduces parthenogenetically, the eggs developing to form females (female producers), while the other may reproduce bisexually. In this second type (male producers) there are eggs, often smaller than the female eggs, which develop quickly by parthenogenesis into males. At various seasons

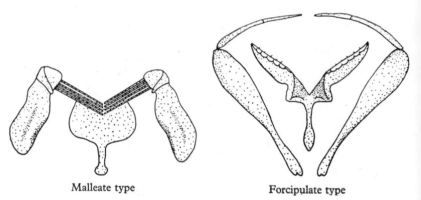

Malleate type   Forcipulate type

Fig. 164. Jaws of rotifers. (After Ward and Whipple.)

after the appearance of these male eggs there are produced by the same individual also other eggs, distinguished by a thicker shell, and these have been fertilized by the spermatozoa of the just-hatched males injected through the skin. These 'resting' eggs are fertilized 'male eggs' and they only develop after a dormant period into females.

The reproduction of a rotifer runs through a cycle in which at first only parthenogenesis occurs but which is terminated by sexual reproduction. In rotifers which are typical members of fresh-water plankton, the cycles run to a time-table. There are 'dicyclical' rotifers like *Asplanchna*, which have two sexual periods, one in spring and the other in autumn, while other forms like *Pedalion* are 'mono-cyclical' and have only a sexual period in the autumn, passing the winter as resting eggs. In rotifers like *Hydatina*, which inhabit puddles and ponds, the sexual periods are very frequent and begin soon after the resting eggs have hatched. The resting egg is a stage in which the species can survive when the puddle dries up. Sexual reproduction can be brought on in cultures by alteration of the external conditions (Fig. 166).

Besides the environmental types which have been mentioned above as free-

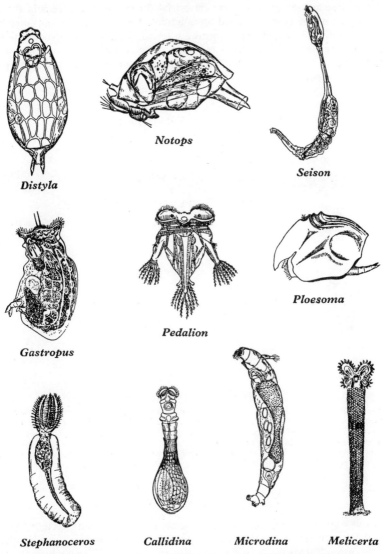

Fig. 165. Range of body form in the rotifers. (From Ward and Whipple.)

swimming and inhabiting larger and smaller bodies of water, the following rotifers may also be mentioned:

Stephanoceros (Fig. 165) and *Floscularia* are sedentary forms which secrete a protecting gelatinous tube into which they can withdraw rapidly. *Melicerta* is another sedentary form which produces a tube formed out of mud particles or its own faeces.

Callidina and other genera are terrestrial forms which can remain for a great part of the year in a dried-up condition but come to life immediately when moistened by rain. Such forms are found, for instance, in roof gutters and amongst moss. The group to which these forms belong is called the 'bdelloid' or leech-like rotifers, because they not only swim, but progress by a looping method like that of *Hydra* or a leech.

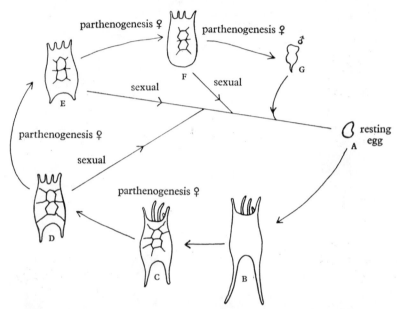

Fig. 166. Life cycle of *Anuraea aculeata*. From the resting egg (A) issues a long-spined generation (B). In this and the succeeding generation (C) which also has fairly long spines, reproduction is by parthenogenesis and only females are produced. In the parthenogenetic series of generations at some period or other, small forms with short spines (D) begin to appear; some of them may continue the parthenogenetic production of females, but others produce males (G) or, if fertilized, the resting egg (A). (From Berg, after Kratschmar.)

PHYLUM GASTROTRICHA

DIAGNOSIS. Minute, worm-like, unsegmented animals, with certain tracts of the skin ciliated, the cuticle often forming bristles and scales; a non-cellular hypodermis, forming adhesive papillae, longitudinal muscle cells which do not form a continuous sheath; straight alimentary canal consisting of a muscular pharynx like that of the nematodes and a mid gut without diverticula; a pair of nephridia in fresh-water representatives; a nervous system consisting of a cerebral ganglion and two lateral cords; hermaphrodite individuals in one division of the phylum (Macrodasyoidea) and parthenogenetic females in the other (Chaetonotoidea); the single female aperture opening near the anus, and the male aperture when present variable in position. Development direct and cleavage total.

These small animals (Fig. 167) are usually elongated and creep or swim by means of their cilia or move in a leech-like manner using their musculature. They feed on minute animals and plants which are sucked in by the pharynx.

The Gastrotricha have features in common with the Rotifera, such as the external ciliation, the bifid foot and the excretory system with flame cells, but in the character of the gut they recall the Nematoda. Examples are: *Macrodasys, Cephalodasys, Chaetonotus, Neodasys.*

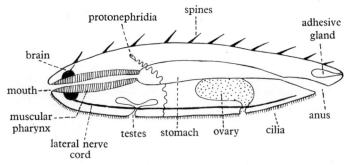

Gastrotricha

Fig. 167. Generalized diagram of a gastrotrich.

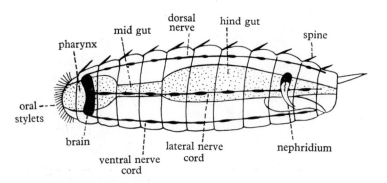

Kinorhyncha

Fig. 168. Generalized diagram of a kinorhynchian.

PHYLUM KINORHYNCHIA

DIAGNOSIS. Minute, worm-like, marine, unsegmented animals with regularly metameric bodies of fourteen metameres, a retractile head; devoid of cilia on the outside of the body; cuticle covered with bristles; a syncytial hypodermis; longitudinal muscles in cords running the length of the body; straight alimentary canal containing a muscular pharynx like that of the nematodes; one pair of protonephridia each with a single flame bulb; a nervous system consisting of a circumenteric ring and four cords running the length of the body; usually dioecious (Fig. 168).

The embryology is unknown, the larvae lack the normal divisions of the body into head and abdomen, and also lack a pharynx and anus but after a series of moults these develop. The animals are found living in slime and mud, moving in a worm-like manner. Examples are *Echinoderes*, *Pycnophyes*, *Centroderes*.

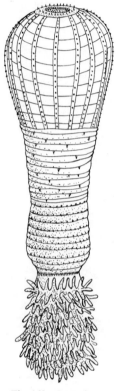

Fig. 169. *Priapulus*.

Fig. 170. *Halicryptus* larva. (From Dawydoff, after Hammarsten.)

The affinities of the group are not clear. They show certain similarities to the Rotifera and Gastrotricha in their protonephridia, cuticle and adhesive tubes. They also show certain affinities to the Nematoda in their digestive system, circumenteric nervous system, and longitudinal muscle bands.

PHYLUM PRIAPULIDA

DIAGNOSIS. Unsegmented worm-shaped animals with a large haemocoelic body cavity, a straight gut with anterior mouth and posterior anus; the posterior end of the body bears a series of caudal appendages; the central nervous system is not separated from the epidermis; urino-genital system simple with solenocytes.

GENERAL ACCOUNT. These are marine worms living in the mud. There are two common examples, *Priapulus* and *Halicryptus* (Figs. 169, 170). The outer surface of the body is covered with a series of rings that give the appearance of segmentation. The surface is also covered with small papillae and spines. There is a short introvert lined with spiny teeth, a straight gut ending in a terminal anus. There is a spacious body cavity of uncertain embryological origin. There is no vascular system. There are a pair of urino-genital ducts each lying on either side of the gut. The sexes are separate and the gonad contains

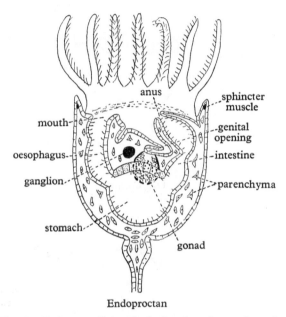

Fig. 171. Endoprocta (Polyzoa). Section through an endoproctan. (Altered from Ehlers.)

a series of solenocytes. The nervous system consists of a circumpharyngeal commissure and a ventral nerve cord, both being imbedded in the epidermis. There are no well-developed ganglionic swellings. Development leads to a larva similar to the adult in basic structure.

The affinities of this group are not yet clear. They are variously linked with the nematodes-rotifer-gastrotrich group on the one hand and with the echiuroids on the other.

PHYLUM ENDOPROCTA

DIAGNOSIS. Endoprocta (Figs. 171, 172). Simple and archaic animals in which the anus is situated inside the lophophore; without coelom, the space between gut and body wall being filled with parenchymatous tissue; with non-retractile tentacles which can be covered by a circular flap of the body wall, provided with a sphincter muscle; with a pair of protonephridia ending in flame cells, and gonads with a duct of their own; with a trochosphere larva. *Pedicellina, Loxosoma, Urnatella.*

GENERAL ACCOUNT. The Endoprocta are sometimes grouped with the Ectoprocta to form the Bryozoa or Polyzoa. There are, however, several differences between the two groups. The Endoprocta have the anus opening

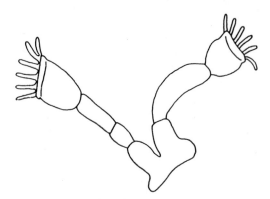

Fig. 172. *Urnatella.*

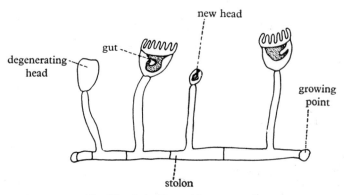

Fig. 173. *Pedicellina.* (After Sedgwick.)

within the circle of the tentacles and lack a coelomic body cavity; the Ectoprocta have the anus opening outside the tentacle circle and possess a well-developed coelomic body cavity.

In the Endoprocta the tentacles surround a small vestibule into which opens the mouth, anus, reproductive and excretory ducts. The digestive system is simple, there being an oesophagus, stomach and rectum. The animals filter-feed. There are a pair of simple protonephridia each ending in a flame cell. There is no well-developed body cavity, the space between the ectoderm and endoderm being filled with parenchyma and being the remnant of the blastocoelic space. Retraction of the polyp occurs by folding the tentacles into the vestibule and not by retraction of the external parts into a body cavity as seen in the Ectoprocta. The gonads are simple with ducts leading to the outside. *Pedicellina* is hermaphrodite with the sexes developing consecutively. *Loxosoma* is dioecious.

Development of the zygote by spiral or equal cleavage (depending on the species) leads to a trochosphere larva. This larva fixes itself by the oral surface and undergoes a complex metamorphosis. The alimentary canal rotates in a manner similar to that seen in the barnacles. The larva develops a stalk on the new aboral surface and takes on adult form.

Pedicellina (Fig. 173) is often found growing on algae and seaweeds in rock pools. It is a colonial form with a main stolon from which rise small individual polyps. From time to time the head of the polyp degenerates and a new head is formed. This may be a system of rejuvenation similar to that obtained by brown body formation in the Ectoprocta.

Loxosoma is a single polyp sometimes found commensal with sponges, crabs and ascidians. It reproduces asexually by forming small lateral buds which separate off to form small polyps.

Urnatella (Fig. 172) is a North American fresh-water form similar to *Loxosoma* except that usually two polyps remain attached to each other in a 'siamese twin' fashion.

CHAPTER VIII

THE PHYLUM ANNELIDA

DIAGNOSIS. Segmented worms in which the perivisceral cavity is coelomic; with a single preoral segment (prostomium); with a muscular body wall in which externally the elongated muscle cells are arranged with their longitudinal axes across the width of the worm (circular layer) while internally their axes are parallel to the length of the worm (longitudinal layer); with a central nervous system consisting of a pair of preoral ganglia connected by commissures with a pair of ventral cords which usually expand in each segment to form a pair of ganglia from which run nerves to all parts of the segment; with nephridia and coelomoducts; and the larva, if present, of the trochosphere type.

While the above definition is the only one that can be applied to all the annelids, typical representatives of the phylum can also be described as possessing a definite cuticle and bristles or *chaetae* composed of chitin, arranged segmentally, imbedded in and secreted by pits of the ectoderm (Fig. 175). The cuticle is thin and not composed of chitin, thus differing from that of the Arthropoda.

CLASSIFICATION

There are six classes of which the first three are the most important in the phylum.

CLASS 1. POLYCHAETA

Well-segmented annelids with chaetae and a spacious perivisceral coelom usually divided by intersegmental septa. The chaetae arise from special prominences of the body wall called parapodia. The animals usually have a distinct head which bears a number of appendages. They are nearly always dioecious and the gonads extend throughout the body. There is external fertilization and the fertilized egg gives rise to a free-swimming larva, the trochosphere. The group is mainly marine though there are some estuarine forms. Typical examples are *Arenicola, Sabella, Nereis*.

CLASS 2. OLIGOCHAETA

Well-segmented annelids with chaetae and a spacious perivisceral coelom usually divided by intersegmental septa. The chaetae are fewer in number than in the polychaetes and they are not situated on parapodia. There is usually a distinct prostomium in front of the mouth but it does not bear appendages. The animals are hermaphrodite, the male and female gonads being few in number (one or two pairs), the male always being anterior to the female. The special genital ducts, the coelomoducts, open by funnels into the coelom; spermathecae and a clitellum are present at sexual maturity. Reproduction

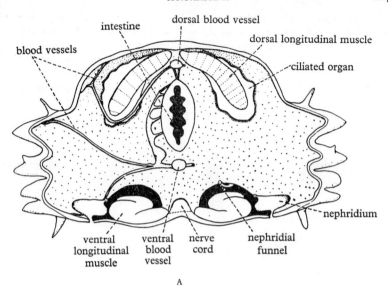

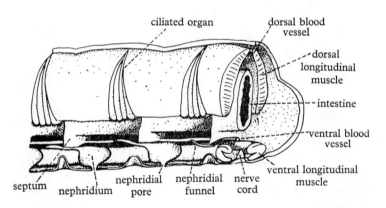

Fig. 174. Segmental structure of the annelids; diagram of *Nereis* to show the position of the segmental organs. A. Transverse section. B. Stereogram with the ectoderm of one side removed. (After Goodrich.)

is by copulation and cross-fertilization, eggs being laid in cocoons and developing directly without a larval stage. The group has terrestrial and freshwater forms. Typical examples are *Allolobophora*, *Tubifex*.

CLASS 3. HIRUDINEA

Annelids with a somewhat shortened body and a fixed number of segments. The segments are broken up externally into a number of rings or annuli. The animals are without chaetae or parapodia. The anterior and posterior ends of the body have suckers. The coelom is very much invaded by the growth of mesenchymatous tissue and is usually reduced to several longitudinal spaces (sinuses) with transverse communications. One primitive form *Acanthobdella* has chaetae and a spacious perivisceral coelom in its anterior segments. The animals are hermaphrodite and develop a clitellum at sexual maturity. Reproduction is by copulation and cross-fertilization; eggs being laid in a cocoon and developing directly without a larval stage. The group is represented by a few genera on land and in the sea, most genera are fresh-water forms. Typical examples are *Hirudo, Glossiphonia*.

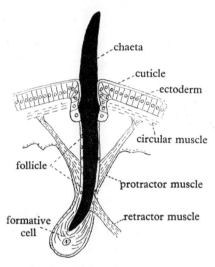

Fig. 175. Chaeta of body wall of *Lumbricus*. (Altered from Stephenson.)

CLASS 4. ARCHIANNELIDA

A small group of worms with a ciliated epidermis, little or no evidence of segmentation and only rarely possessing chaetae as in *Saccocirrus*. The animals are mostly marine. Typical examples are *Protodrilus, Polygordius, Saccocirrus*.

CLASS 5. ECHIUROIDEA

Marine burrowing annelids that have lost almost all traces of segmentation in the adult though the trochosphere larva shows mesoblastic somites or ganglionic rudiments. The chaetae are lost except in a few forms. The animals have a well-developed preoral lobe and a spacious coelom. Typical examples are *Echiurus, Bonellia*.

CLASS 6. SIPUNCULOIDEA

Annelids of doubtful affinities. They have a well-developed spacious coelom and a single pair of nephridia but the adult shows no sign of segmentation. The larva, however, shows three pairs of somites that quickly disappear.

They are sessile sand dwellers and they have a coiled intestine so that the anus is dorsal and anterior. Typical examples are *Sipunculus, Phascolosoma*.

POLYCHAETA, OLIGOCHAETA AND HIRUDINEA. The linking of these classes into subphyla presents several interesting problems. The classes are closely related, with the Oligochaeta possessing characters intermediate between those of the Polychaeta and the Hirudinea. One classification groups the Polychaeta and the Oligochaeta together as a class Chaetopoda with the common characters of chaetae and a well-developed coelom. On the other hand it is possible to link the Oligochaeta and the Hirudinea together to form the group Clitellata; both possess a clitellum during sexual maturity, are hermaphrodite, and show many close resemblances during embryonic development. The nervous system of the Polychaeta develops from two distinct centres whilst those of the Oligochaeta and Hirudinea develop from one centre. The mesoderm in the Polychaeta arises from the 4*d* cells whilst in the Oligochaeta and Hirudinea the mesoderm develops from 3D, 4D or 3*d*, but not 4*d*.

Two suggested groupings of the major classes of the annelids:

(1) **Chaetopoda** — Polychaeta, Oligochaeta

(2) **Hirudinea**

(1) **Clitellata** — Oligochaeta, Hirudinea

(2) **Polychaeta**

The remaining groups, Archiannelida, Echiuroidea and Sipunculoidea present a greater problem in deciding their affinities. The archiannelids are probably simplified polychaetes though through their simplification they show many apparently primitive characters. There is no clear indication that the different genera of the archiannelids are closely related; it is more probable that their similarities are due to convergence and not due to close phylogenetic relationship.

The Echiuroidea and Sipunculoidea used to be grouped together as the Gephyrea mainly on the fact that they were both burrowing animals that had lost their segmentation and showed certain common nephridial characters, all of which are features of doubtful phylogenetic significance. The Echiuroidea have more annelid affinities than do the Sipunculoidea though both show their annelid affinities more closely during their embryonic phase. The two groups have on occasions each been placed in a separate distinct phylum.

CLASS 1. POLYCHAETA

DIAGNOSIS. Well-segmented Annelida, with a spacious perivisceral coelom, usually divided by intersegmental septa; marine; numerous chaetae arising from special prominences of the body wall called parapodia; usually with a distinct head which bears a number of appendages; nearly always dioecious, with gonads extending throughout the body and external fertilization; with a free-swimming larva, the *trochosphere*.

GENERAL ACCOUNT

In a typical polychaete there is a distinct preoral region or *prostomium* and a postoral body composed of many segments. Each segment owes its distinctness to the development in the larva of a pair of mesoblastic somites which join round the gut, the cavities which develop in them becoming the perivisceral cavity of the adult segment. At the same time the larval ectoderm (epiblast) develops segmentally repeated organs: the *ganglia*, swellings in the continuous ventral nerve cords, the *nephridia* or excretory organs and the *chaetae*. The chaetae are borne in groups upon processes known as *parapodia*, whose projection from the body wall is due to the development of special muscles for moving the chaetae.

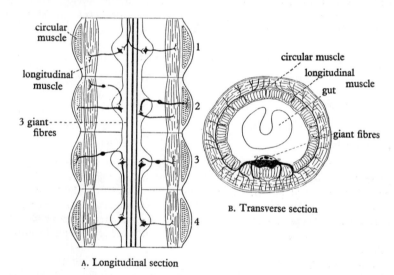

Fig. 176. Longitudinal and transverse section of *Lumbricus* to show the musculature and its innervation. In A, segment 2 shows neurones constituting an intrasegmental neurone arc; segments 3 and 4 show those which make up an intersegmental arc; B shows the distribution in the body wall of two segmental nerves and their branches.

NERVOUS SYSTEM. The chief feature of the nervous organization is that the musculature of all parts of the body is co-ordinated by metamerically repeated intra- and intersegmental reflexes (Fig. 176 A). In each segment there is, for example, a correlation of the circular and longitudinal muscles by the segmental nerves which acts so that contraction of one brings about automatically relaxation of the other. Then there are nervous connexions between adjacent segments which act so that excitation of a muscle layer in one segment leads to excitation of the same layer in the other segment. By the working together of the inter- and intrasegmental reflexes the normal peristaltic movement of the body is brought about.

There may also be a system of giant fibres running along the whole length of the ventral nerve cord (Fig. 176B). These are responsible for the reactions which require immediate co-ordination of the whole body in response to excitation of the higher centres, the supra- and subpharyngeal ganglia. The rapid contraction of the whole of the longitudinal musculature in response to a noxious stimulus is an example of this kind of reaction. On p. 197 it was shown that the primary function of the primitive central nervous system is that of a sensory relay. In the annelids there is added the second great function, that of inhibition. A nereid, which has had the suprapharyngeal ganglia removed, moves about ceaselessly, showing that a function of the ganglia in the normal animal is the inhibition of movement. If the supra- and subpharyngeal ganglia are both removed then the animal is permanently quiescent, a condition like that of a polyclad turbellarian when the cerebral ganglia are removed.

The head and accompanying sense organs may be well developed, for instance, in some of the pelagic Polychaeta where the eyes are remarkably complex. In such cases the brain (prostomial ganglia) may attain a structure almost as complicated as in the higher arthropods. The head processes (tentacles, palps) vary greatly. While they may be very complicated in the errant Polychaeta, they are frequently absent in burrowing members of that group.

COELOM. The coelom is bounded by an epithelial layer, the *peritoneum*, which gives rise to the *gonads* (which are usually developed in most of the segments), to the *yellow cells*, which play a part in the work of nitrogenous excretion, and to the *coelomoducts* by which the eggs and sperm pass from the coelom to the exterior. In most of the polychaetes the eggs are fertilized externally, forming a *trochosphere* larva, the method of reproduction thus conforming to that of other marine groups.

In some forms the coelom may be very spacious and have a ciliary circulation, as in *Aphrodite* (Fig. 181) where it develops at the expense of the blood system.

SKELETAL SYSTEMS. The annelids like the coelenterates, platyhelminthes, and molluscs have a hydraulic skeleton; the longitudinal muscles being extended due to an increase in the hydraulic pressure when the circular muscles contract. The longitudinal muscles are attached to a lattice-work of inextensible collagen fibres in the basement membrane. In *Arenicola*, where there are few septa dividing the coelomic cavity, the importance of the coelomic fluid during locomotion is shown by the following experiment. *Arenicola* normally takes about three minutes to burrow into the sand. If 0·38 ml. of coelomic fluid are removed with a hypodermic syringe, then the animal takes eight minutes to burrow. This indicates that the reduced coelomic pressure makes it more difficult for the circular and longitudinal muscles to co-ordinate. On the other hand if *Arenicola* is cut in half it takes four to five minutes to burrow. This is due to the contraction of the circular muscles at the cut surface which takes place and prevents the loss of the coelomic fluid.

In *Lumbricus* there are septa between each segment and each septum has a foramen through which the nerve cord passes. Injection of dyes and X-ray opaque substances show that there is no passage of coelomic fluid from one segment to the next during locomotion. The circular muscles of the septum around the nerve cord contract and so make each segment a closed discrete unit. It is possible that the development of small hydraulic units increases the efficiency of the locomotor system in the annelids.

BLOOD SYSTEMS. The blood system varies greatly. In small forms it is absent altogether. Typically it consists of a dorsal vessel in which the blood moves forward, and a ventral vessel in which it moves backward and from which the skin is supplied with venous blood. The whole of the dorsal vessel (Fig. 204 A) is usually contractile: there may also be vertical segmental contractile vessels which are usually called 'hearts'. In some forms, for example *Pomatoceros* (Fig. 179 C), there are no separate dorsal and ventral vessels but a *sinus* round the gut: the peristalsis of the latter brings about the movements of the blood. While the whole of the skin is sometimes richly supplied with blood vessels and usually performs an important part in the aeration of the blood there are often branched segmented processes which may rightly be called *gills* (*Arenicola*, Fig. 186): the alimentary canal is probably a respiratory organ too. While haemoglobin is often present in the blood, usually in solution, a related pigment, chlorocruorin, which is green, occurs in many tubicolous polychaetes. The variable state of the mechanism of respiration is shown by the fact that one species of a genus (the polychaete, *Polycirrus*) may possess haemoglobin while another has no respiratory pigment.

It is not clear to what extent the haemoglobin in annelids has the same function as haemoglobin in mammals. In many annelids the haemoglobin under normal conditions remains in an oxygenated condition and so does not help in oxygen transport. An alternative suggestion is that the haemoglobin acts as an oxygen store in emergencies such as prolonged periods of tidal exposure. In *Lumbricus* there does appear reasonable evidence that the pigment plays a part in oxygen conduction.

NEPHRIDIA. The nephridia are essentially tubes developed from the ectoderm which push their way inwards so that they project into the body cavity. In some polychaetes they end blindly—this is the primitive condition. In most annelids they have acquired an opening (nephrostome) into the body cavity itself. In some cases there is a partial fusion with a mesodermal element, the coelomoduct, so that a compound tube consisting mainly of ectoderm but partly of mesoderm exists (*nephromixium*). Nephromixia may take on the functions of coelomoducts where these do not exist independently. All types of tubes are termed here *segmental organs*.

DETAILED ACCOUNT

The structure of the Polychaeta is very variable and dependent on the habit of life, both externally (especially the head appendages and parapodia) and internally (especially the segmental organs). The variation in methods of reproduction is also very characteristic. For these reasons an account will first be given of some of the very large number of families into which the Polychaeta are divided, in which a rough ecological grouping is adopted. A summary of the variation in segmental organs and reproductive habits follows at the end.

1. Errant Polychaeta. With unmodified head and armed eversible pharynx (proboscis), fitted for an active life but often living in tubes; sometimes modified in structure and behaviour during the reproductive season. Euni-

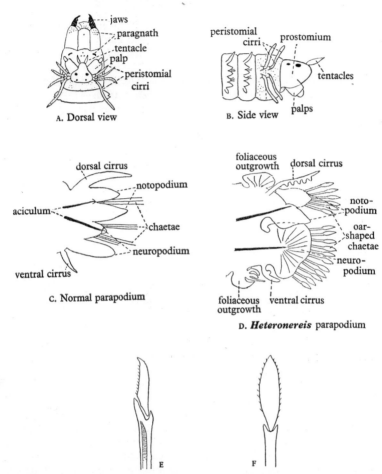

Fig. 177. *Nereis*; head, pharynx and parapodia. E, unmodified compound chaeta. F, oar-shaped compound chaeta of *Heteronereis*.

cidae: *Eunice, Leodice* (the palolo worm); Nereidae: *Nereis*; Syllidae: *Syllis, Myrianida*; Phyllodocidae: *Eulalia, Asterope*; Polynoidae: *Aphrodite, Lepidonotus, Panthalis.*

2. True tubicolous Polychaeta. Much modified for the collection of microscopic food; the anterior part of the gut is not eversible and there are no jaws; inhabiting tubes which they rarely or never leave. Chaetopteridae:

Chaetopterus; Terebellidae: *Terebella, Amphitrite*; Serpulidae: *Pomatoceros, Filigrana*; Sabellidae: *Sabella, Spirographis*.

3. **Burrowing Polychaeta.** With reduced head, usually having a proboscis. Arenicolidae: *Arenicola* (without jaws).

Errant Polychaeta

The external structure is known to the elementary student through the type *Nereis* (Figs. 177, 193). The prostomium bears two kinds of filiform, tactile appendages, the *tentacles* which are dorsal and the *palps* which are lateral; there are also one or two pairs of eyes upon it. The anterior part of the gut (*pharynx*) is eversible and serves for grasping food; its lining may be chitinized in places to form the jaws and paragnaths of *Nereis* or teeth as in *Syllis*. These are not necessarily the sign of a carnivorous habit but may be used for cutting up pieces of seaweed or boring in sponges.

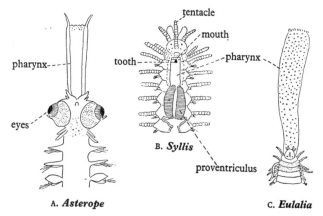

Fig. 178. Anterior regions of *Asterope*, *Syllis* and *Eulalia*. The peristomal segments are stippled to show extent of cephalization.

The ordinary *trunk segment* has a double parapodium consisting of a dorsal *notopodium* and a ventral *neuropodium*, usually with rather different types of chaetae. A *dorsal cirrus* and a *ventral cirrus* are nearly always present; they are filiform structures but may be modified to form pectinate gills (*Eunice*) or plate-like elytra (Polynoidae). From the conical noto- and neuropodia spring a bundle of chaetae; the chaetal sacs project into the coelom and each bundle is supported by an enlarged and wholly internal chaeta—the *aciculum*, which also forms the point of origin of the parapodial muscles. The chaetae are of two kinds, simple and compound.

The segment (or segments) just behind the mouth, forming the *peristomium*, is, however, much modified. There are no notopodia or neuropodia (except in occasional species, which retain chaeta-bearing processes as a primitive feature). But the cirri remain as the *peristomial cirri* in pairs consisting of a dorsal and ventral member. In *Nereis* there are two pairs of peristomial cirri on each side, indicating the fusion of two segments to form the peristomium.

In some families (Syllidae, Fig. 178 B) this is constituted by a single segment, but usually two or more have been pressed forward towards the mouth and modified. This is the first indication of the process of *cephalization* carried much farther in the arthropods and vertebrates.

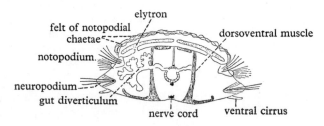

A. *Aphrodite*

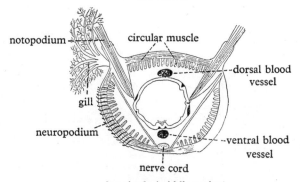

B. *Arenicola* (middle region)

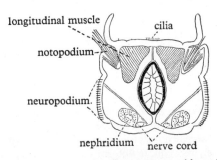

C. *Pomatoceros* (thorax)

Fig. 179. Transverse sections through various polychaetes.

The worms in this group used to be definitely classed as the Errantia or free-swimming forms, but a great number of them (e.g. the nereids) do live in tubes which, however, they can leave and reconstruct anew. The most beautiful example of tube-building in the Polychaeta is furnished by *Panthalis*, a polynoid. In this the chaetal pits of the notopodium produce not stiff bristles but plastic threads which are woven by the comb-like ventral chaetae and the shuttle-like action of the anterior parapodia into a continuous fabric which forms the lining of the mud-covered tube. *Aphrodite*, the sea mouse (Fig. 179 A), is a short, broad form which burrows in mud, and though it does not form a separate tube it covers its back with a blanket made from interwoven chaetal threads similarly formed from the notopodium. Between this blanket and the back is a space into which water is drawn by a pumping action of the dorsal body wall, being filtered through the matted chaetae. In this there are special plate-like modifications of the dorsal cirri—the *elytra*—round which circulates the water from which they possibly obtain dissolved

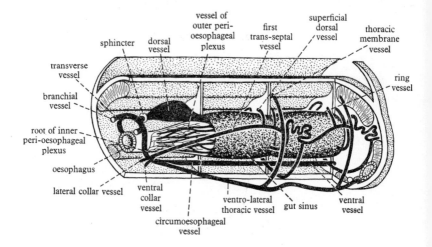

Fig. 180. Blood system of *Pomatoceros* showing the large blood sinuses. Lateral view of peristomial and 2nd and 3rd thoracic segments, with half of body wall removed. (From Hanson.)

oxygen. In other polynoids (e.g. *Lepidonotus*, which lives under stones but does not burrow) the elytra can have no respiratory function but are probably protective, spreading over the whole or greater part of the back (sometimes bits of sand or shell are attached to special papillae). Not all the dorsal cirri are modified to form elytra: typical filiform cirri are placed on alternate segments. *Aphrodite* has remarkable segmental caeca of the alimentary canal in which takes place digestion of the fine food particles which pass a sieve at the junction with the intestine.

The diagnostic features of *Nereis* and other genera mentioned in the classification are given below.

Nereis (Fig. 174). Two tentacles, two palps; pharynx with two jaws and twelve groups of paragnaths; noto- and neuropodium each double; chaetae all compound; most species have a special sexual form (*Heteronereis*).

Eunice. Five tentacles, two palps; pharyngeal armature well developed; a single peristomial segment; gills in many segments; chaetae simple and compound.

Eulalia (Fig. 178 C). Five tentacles, no palps; pharynx very long with soft papillae only; three peristomial segments; dorsal and ventral cirri leaf-like; chaetae all compound.

Asterope (Fig. 178 A). Similar to *Eulalia* but a pelagic polychaete with transparent body and enormous eyes of complicated structure.

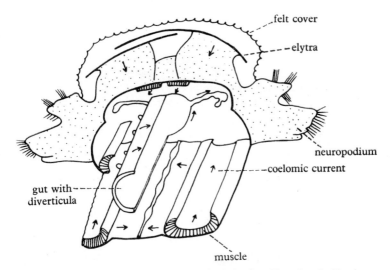

Fig. 181. Diagram of the coelomic circulation in *Aphrodite*. Note that the blood system is reduced and the main circulating system is coelomic. (After Kükenthal.)

Syllis (Fig. 192 A). Three tentacles, two fused palps; pharynx enclosed in a pharynx sheath with a single conical tooth and a muscular *proventriculus* which functions as a pump; no notopodium.

Autolytus (Fig. 192 C). Like *Syllis* but pharynx long, with a circle of teeth; no ventral cirrus. *Myrianida* has similar characters.

Myzostomum (Fig. 190). External parasite of crinoids. Disc-shaped, with ten pairs of cirri around the perimeter. Ciliated in patches. Has five pairs of parapodia between which are four pairs of suckers. Gut has lateral caeca. The coelom is obliterated and filled with connective tissue. Hermaphrodite; the egg gives rise to a typical trochosphere larva which in turn develops into a nereid-like larva.

Histriobdella (Fig. 190) is a parasite on the eggs of the lobster, having no chaetae but two pairs of 'feet' by which it executes acrobatic movements. It resembles *Dinophilus* in its reduced coelom and musculature but it has jaws the structure of which are very similar to those of the Eunicidae.

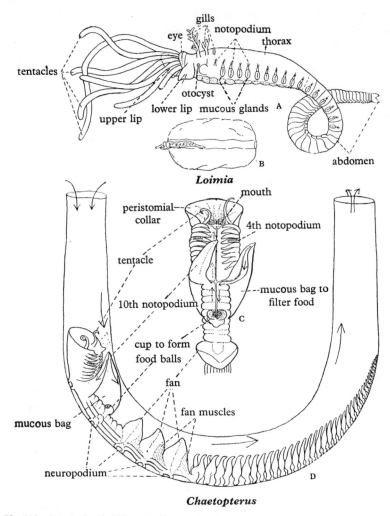

Fig. 182. *Loimia* (terebellid) and *Chaetopterus pergamentaceus*. In D, the direction of water is indicated by arrows.

True tubicolous Polychaeta

Here the prostomium has become much smaller and its appendages enormously modified and increased. The peristomium may be produced into a collar which in some forms grows round the prostomium and encloses a funnel-like cavity at the bottom of which lies the mouth. The food consists of small animals or plants or organic debris and it is collected by ciliary mechanisms. In the terebellids, serpulids and sabellids (Figs. 182–185), the appendages of the head, which probably correspond to tentacles, are very numerous.

Each tentacle has a ciliated groove running from the tip to the mouth and along this minute particles may be seen to travel. In the terebellids these tentacles are extensible and capable of independent movement when separated from the body. In the serpulids and sabellids, they are rather stiff branched structures, which can, however, curl up when withdrawn into the tube; they sometimes bear eyes and sometimes are wonderfully pigmented.

Besides the food-collecting tentacles there are gills in the terebellids. These are branched processes, usually three pairs, situated just behind the head, full of circulating blood. In the serpulids and sabellids, there are no special respiratory organs but the whole surface of the body serves for the exchange of gases.

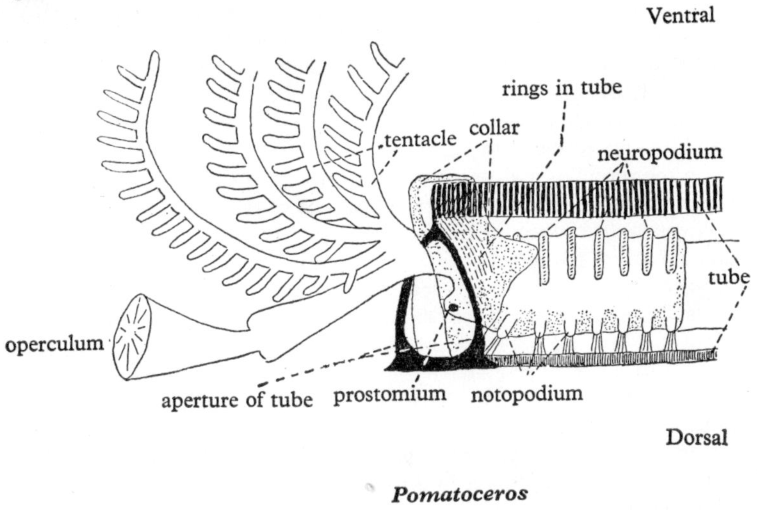

Fig. 183. Diagram of *Pomatoceros* in its tube. The sides of the tube are not represented.

In the terebellids the tubes are composed of a soft cementing substance mixed with mud or a parchment-like material to which adhere sand grains, sponge spicules, foraminifera or fish-bones. It is usually porous (so that change of water can take place through it) and the animal occasionally leaves its shelter; there are at least two openings to the exterior. The tube of the chaetopterids is parchment-like but in the serpulids there is a groundwork of mucin in which carbonate of lime is laid down. In the latter family there is only one opening from which the crown of tentacles emerges but never any more of the body. The tentacles are violently withdrawn in obedience to any such stimulus as touch or change of illumination.

In all the types except *Chaetopterus* the body is divided into two regions, an anterior *thorax* and a posterior *abdomen*. The thorax is composed of segments in which the notopodium is a conical structure with capillary chaetae while the neuropodium is a vertical ridge in which are imbedded short-toothed chaetae called *uncini*, which only just project from the body wall. It is suggested that the notopodium assists movement up and down the tube while the neuropodia are braced against the tube and maintain the worm in position. In the abdomen the arrangement of the parapodia is different, and in the serpulids and sabellids the uncini become dorsal and the simple chaetae ventral (introversion).

In the serpulids (Fig. 183) the peristomium is similar to the other thoracic segments but it is produced into a *collar* which folds back over the ventral surface and sides and secretes successive hoop-shaped rings which are added to the tube. Other features are the *thoracic membrane*, a lateral frill possibly respiratory, and the *operculum*, a much enlarged and stopper-like branch of a tentacle which exactly closes the mouth of the tube when the animal is retracted.

The renewal of water round the body is of the utmost importance in respiration. It is brought about by undulatory movements of the abdomen and sometimes by sharp rhythmic contractions and expansions of the body which pump the body in and out of the tube. The great development of the dorsal bands of longitudinal muscle seen in a transverse section of a serpulid (Fig. 179c) is characteristic of the tubicolous worm. Another typical modification seen in the serpulids and sabellids is the *median ciliated groove*, which starts from the anus, runs along the ventral surface of the abdomen, turning on to the dorsal surface when the thorax is reached. It serves to conduct the faeces to the mouth of the tube.

Chaetopterus (Fig. 182) is probably the most modified of all tubicolous worms. It lives in a parchment-like tube which is U-shaped with at least two apertures. There is a peristomial collar as in other tubicolous worms, but the tentacles are a pair of rudimentary processes. A very complicated mechanism exists for obtaining food, which can be observed by taking a live *Chaetopterus* from its tube and replacing it within a glass tube of the same calibre in an aquarium. The worm fits very loosely in its tube and there is plenty of room for a current of water to sweep through from end to end. Such a current is maintained by the rhythmical oscillation of the *fans* (fused notopodia) of the middle region. Food particles contained in the current are entangled in the mucus bag shown in Figs. 182 C, D. The food forms a bolus in the small cup at the end of the bag and periodically the cilia carry the bolus along a dorsal groove to the mouth.

FEEDING IN SABELLA. *Sabella* is a tubicolous polychaete living in the littoral region. In the feeding condition it has its tentacles extended into the water. A current of water is drawn over the tentacles by the activity of the lateral cilia. Food particles are trapped on the tentacles and are led to a small groove on the inside of the tentacle (Fig. 184). The food particles are carried down this tentacular groove towards the mouth, at the base of the tentacle they run into a deep groove which acts as a sorting device. The smallest particles drop to the bottom of the groove and are led to the mouth, the medium-sized particles fill the groove and are led to build the tube, whilst the largest particles settle on top of the groove and are led to the rejection tracts.

Sabella has its body specialized for tubicolous living. The body is divided into three regions:

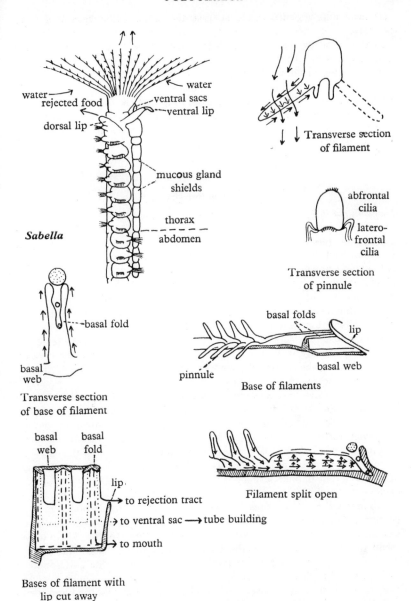

Fig. 184. Filter feeding in *Sabella*. The arrows indicate the direction of water currents. (After E. A. T. Nicol.)

(1) head and attached branchia, together with a collar region of three segments;

(2) thorax of five segments; the dorsal lobes of the parapodia are hair-like whilst the ventral lobes are hooked;

(3) abdomen of about three hundred segments. The dorsal lobes of the parapodia are hooked (uncigerous) whilst the ventral lobes are hair-like (setigerous). This arrangement is opposite to that of the thorax.

If the head is amputated a new head is formed at the cut surface and a few of the abdominal segments reorganize their parapodia into the thoracic pattern. If the head is amputated on one side only then reorganization of the abdominal parapodia takes place on that side only.

Burrowing Polychaeta

Arenicola marina (Fig. 186) is the type of a burrowing polychaete and it has a rounded cross-section like an earthworm. In its division of the body into regions, the modification of the parapodia, and the internal anatomy it resembles the tubicolous worms. The prostomium is much reduced, however, without any appendages and there is an eversible pharynx, covered with minute papillae, which is the organ for locomotion through the sand as well as for feeding. In general form it thus resembles an earthworm: the chief obvious difference is the presence of gills and parapodia. It is divided into three regions: the anterior, consisting of the peristomium, an achaetous segment, and six segments which have a notopodium with capillary setae and a neuropodial ridge with chaetae resembling uncini (crotchets); the median, the segments of which have gills in addition; and the posterior, in which parapodia and chaetae are entirely lost.

The body wall consists of the typical circular and longitudinal muscle layers as in *Lumbricus*, and by their alternating contraction and expansion the peristaltic movements which are characteristic of the earthworm and other burrowing forms are carried out. In *Nereis* and other surface-living forms progression takes place in two ways. (1) By alternate flexing of the two sides swimming movements are brought about. The longitudinal muscles, which are arranged in four bundles, are much more important than the circular and are capable of rapid contraction. (2) By successive movement of the parapodia crawling movements occur (as in a centipede), the special parapodial muscles coming into action. In tubicolous forms peristalsis occurs, but the longitudinal muscles are even more important than in *Nereis* for the violent movements of contraction which withdraw the animal into its tube. They form a bulky dorsal mass and resemble the columella muscle of the gasteropod in their action (Fig. 179 B).

Arenicola is the most convenient polychaete type for dissection and therefore the following details of internal anatomy are given (Fig. 186). In several prominent features it differs from *Lumbricus* and also from *Nereis* or *Eunice*. The body cavity is spacious, it is not encroached upon by the longitudinal musculature, and the vertical septa which primitively separate the body cavities of the segments have nearly all disappeared. Only the three anterior septa and an indefinite number of the most posterior are preserved. In the greater part of the body the coelom is thus uninterrupted. In its general development the alimentary canal resembles that of the earthworm. The muscular pharynx, however, is not well developed, the oesophagus is a thin-walled tube with no such development as the gizzard of the earthworm and

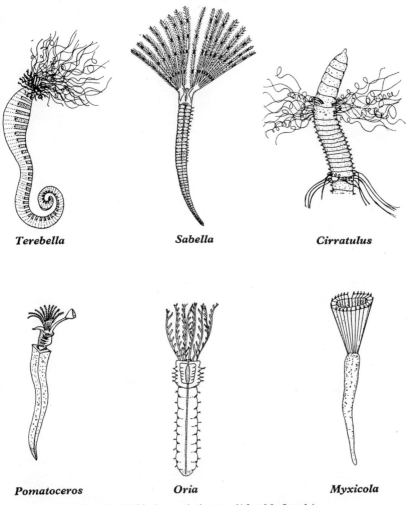

Fig. 185. Tubicolous polychaetes. (After MacIntosh.)

it bears only a single pair of caeca. The intestine is the longest part of the gut, the seat of digestion and absorption, and it is invested by a layer of yellow cells. The blood system, which also contains haemoglobin in solution in the plasma, differs slightly from that of *Lumbricus*: there is a single pair of large *hearts*, each divided into a *ventricle* and *auricle* which connect the important lateral intestinal vessels from which the branches supplying the gills are derived with the ventral vessel.

The circulation for that region just behind the heart may be expressed as follows: lateral vessels→auricle→ventricle→ventral vessel→afferent vessel to body wall and gill→efferent vessel to subintestinal vessel→intestinal plexus

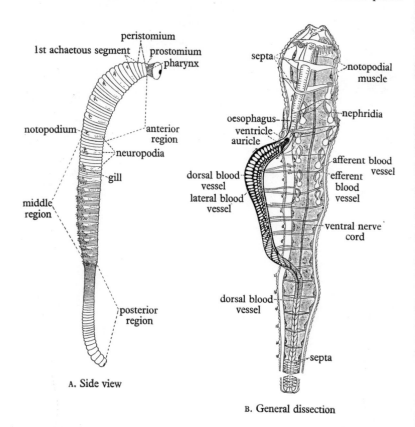

Arenicola marina

Fig. 186. *Arenicola marina*; side view and general dissection. In B the direction of blood flow is indicated by arrows. (After Ashworth.)

→dorsal vessel or lateral vessel. The dorsal vessel does not communicate directly with the heart.

The segmental organs are, like the gills, only found in the middle region. They are prominent organs lying beneath the oblique muscles, remarkable for the large size of the nephrostome, the dark secretory bag-like portion, the cells of which contain insoluble excreta, and the small gonad which lies just behind it. In *Arenicola* as in *Lumbricus* the gonads are restricted to a small number of segments, but the reproductive cells are shed into the body cavity at maturity and completely fill it.

In *Glycera* the prostomium is narrow and conical, the tentacles being very small. It possesses a very large proboscis armed with four sharp teeth. The parapodia are reduced in size and bear compound chaetae, and in its internal structure too *Glycera* comes nearer to the errant worms than does *Arenicola* (Fig. 190).

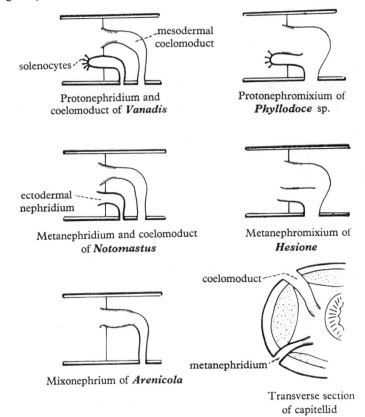

Fig. 187. Types of excretory and reproductive ducts in various polychaetes. The different combinations of nephridia (ectodermal) and coelomoducts (mesodermal) are shown. (After Goodrich.)

THE EXCRETORY AND REPRODUCTIVE ORGANS. Now that a survey of the chief types of polychaetes has been made, a brief description will be given of the segmental organs. These are tubes repeated in successive segments which serve to convey the excretory and reproductive products from the coelom to the exterior. There have been two views concerning the function and origin of the coelom. One maintains that the coelom is primarily a space for the storage of the gonad products. The other view holds that the coelom is a space devised for the storage of excretory products. The former view has been strongly supported by Goodrich who has suggested that the nemertean worms might demonstrate an intermediate condition between the coelomate and the acoelomate condition.

The ducts from the coelom to the exterior are of two types: *nephridia* derived from ectodermal rudiments and concerned primarily with excretion, and *coelomoducts* derived from mesoderm and concerned primarily with transport of

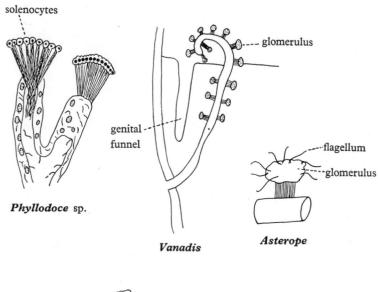

Phyllodoce sp.

Fig. 188. Nephridia and coelomoducts as found in various polychaetes. (From Goodrich.)

gonad products. It will be seen that their function is not restricted to the primary one and that many variations occur in the relationship of the nephridium and the coelomoduct (Figs. 187–189).

Protonephridium. This is a nephridium that terminates in the coelom as a blind tube. At the closed end of the tube are many solenocytes, cellular organs similar to the flame cells of the platyhelminthes and rotifers. The solenocytes may be grouped together in little packets or glomeruli. On the outside of these glomeruli are cilia which keep the coelomic fluid constantly in circulation over the solenocytes. The solenocytes are usually well supplied with blood vessels; this facilitates filtration and absorption from the blood. Protonephridia are found in *Vanadis* (Figs 187, 188).

Metanephridia. In these the nephridium opens into the coelom. The open end is ciliated and is called the nephrostome. These are found in the Capitellidae and *Notomastus* (Fig. 187).

The coelomoducts and nephridia show various degrees of association. The combination of the two into one organ may be such that they share only the same external opening or the fusion may be more intimate so that they share most of the same duct.

Protonephromixium. The coelomoduct is grafted into a protonephridium; a good example is seen in *Phyllodoce*. It can conduct both gonad products and excreta.

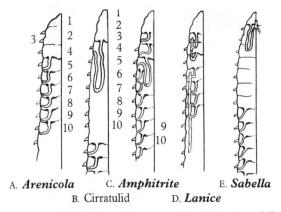

A. *Arenicola*   C. *Amphitrite*   E. *Sabella*
B. Cirratulid   D. *Lanice*

Fig. 189. Nephridia in the tubicolous polychaetes. These show a specialization of the anterior nephridia associated with excretion, whilst the posterior nephridia have mainly a reproductive function. (From Goodrich.)

Metanephromixium. The coelomoduct is attached to a metanephridium; *Hesione* (Fig. 187).

Mixonephrium. The coelomoduct is so closely associated with the nephridium that together they form an apparently simple funnelled organ (Fig. 187). There is no hard-and-fast boundary between a metanephromixium and a mixonephrium and there are many intermediate forms.

Ciliated organ. In some forms part of the coelomoduct becomes separated off from the metanephromixium and becomes attached to the dorso-lateral muscles as a ciliated organ. This keeps the coelomic fluid in circulation. A fine example is found in *Nereis* (Fig. 174).

There may be a specialization of the nephridia in different parts of the same worm. In the Serpulidae, Terebellidae and other families of tubicolous worms there are one to three pairs of long nephridia situated anteriorly. In most of the segments behind there are short open nephromixia which serve for the escape of the eggs and sperm. There is thus a division of function between the segmental organs in tubicolous worms; the anterior being specialized for excretion, the posterior ones for genital discharge (Fig. 189).

288 ANNELIDA

REPRODUCTION. The gonads in the polychaetes are usually patches of the peritoneal epithelium, repeated in most of the segments, proliferating until a great number of the germ cells have been detached into the body cavity which they almost entirely fill and where they undergo maturation (Fig. 199A). When ripe they reach the exterior usually through the segmental organs, but occasionally the body wall ruptures and so opens a way of escape.

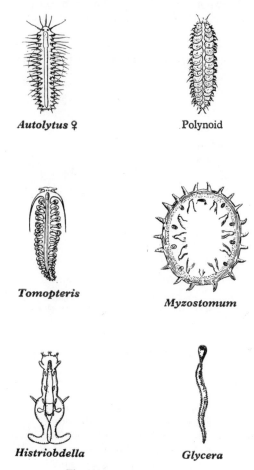

Fig. 190. Errant polychaetes.

Like so many other marine animals the polychaetes thus liberate eggs and sperm freely into the sea, fertilization taking place externally. This habit is associated in many forms with the phenomenon of *swarming* in which a worm, usually crawling or burrowing on the sea bottom, when sexually mature rises to the surface and swims vigorously, eventually discharging its genital products and sinking to the bottom as suddenly as it rose. In most nereids this occurs irregularly through the summer months, but in at least two forms (*Leodice viridis*, the 'palolo' of the reefs of the southern Pacific, and *Leodice fucata* of the West Indian reefs) the phenomenon (Fig. 191) has acquired the strictest periodicity. As the day of the last quarter of the October–November moon dawns in the Pacific, the palolo breaks off the posterior half of its body, already protruding from the mouth of its burrow in the coral rock, and these fragments rise to the surface in such quantities that the water writhes with worms and is later milky with the eggs and sperm discharged. Immediately afterwards the remaining anterior end begins to regenerate the missing portion, but a whole year elapses before the gametes are again ripe—even two days before spawning occurs fertilization cannot be brought about artificially. In the West Indian species the phenomenon is similar but takes place in the third quarter of the June–July moon.

In the syllids the phenomena of swarming are vastly more varied. The whole animal may produce germ cells and swarm. Usually, however, the gonads are confined to the posterior part of the body which is detached as a free-swimming unit; this often develops a head but never jaws and pharynx. It can live for some time but not feed. In the majority of forms a single bud is produced, but in *Autolytus* (Fig. 192 C) and *Myrianida* a *proliferating region* is established at the end of the original body and from this a chain of sexual individuals is budded off, the oldest being situated most posteriorly. The whole chain may be found swimming at the surface, the original worm dragging after it the chain of sexual individuals which one by one detach and lead a short independent existence. In some species of *Trypanosyllis* (Fig. 192 D) the zone of proliferation is in the form of a cushion of tissue on the ventral surface of the last two segments and this produces not a linear series of buds but successive transverse rows, amounting to more than a hundred—the fully formed sexual individual possesses a head but no vestige of an alimentary canal. The extraordinary branching form, *Syllis ramosa* (Fig. 192 B) shows remarkable capacity for heteromorphic growth in the production of sterile side branches from the stock and reproductive buds.

In the syllid there is usually no notopodium during asexual life but during the maturation of the gonads the parapodium is reconstructed, a notopodium being formed from which spring bundles of long capillary swimming chaetae, while a corresponding development of new muscles takes place. Even greater is the change in the parapodia of the maturing nereids. The muscles of the asexual period break down and the fragments are digested by leucocytes before the new muscles are formed. The parapodium of the sexual form, the *heteronereis*, is produced into membranous frills and contains a new type of oar-shaped chaeta (Fig. 177F). The eyes become immensely larger and the animal itself very sensitive to light. The heteronereis does in fact resemble those members of the Phyllodocidae and Alciopidae which have become permanently pelagic. The increase in the surface of the parapodia may be useful in swimming and floating; it has without doubt some connexion with the increased gas exchange associated with an active life.

It is easy to see in the swarming habit an adaptation for securing fertilization of the greatest possible number of eggs. There are remarkable cases in the syllids (*Odontosyllis*) where the meeting of the sexes is facilitated by the ex-

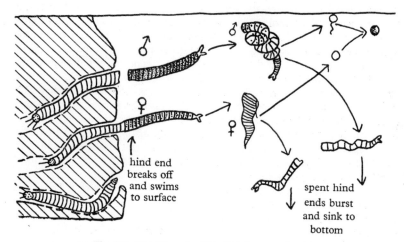

Fig. 191. The life cycle of the Palolo worm *Leodice*.

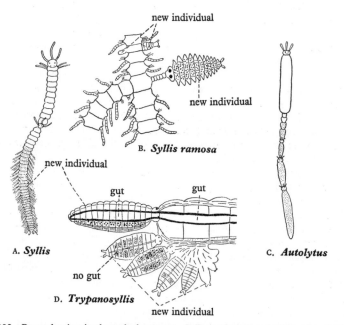

Fig. 192. Reproduction in the polychaetes: A, *Syllis*, with the posterior region forming reproductive individuals. B, *Syllis ramosa*, showing branching of the asexual stock and budding of reproductive individuals. C, *Autolytus*, with a chain of reproductive individuals, budded off successively from the proliferating region. D, *Trypanosyllis gemmipara*; longitudinal section through the end of budding stock showing two types of individual, one type containing alimentary canal, the other without.

change of light signals, and in the nereids the discharge of sperm may only be brought about by the influence of a secretion from the swarming female. Discharge of the gametes is nearly always followed by the death of the sexual individual.

The fertilized egg gives rise to an unsegmented larva, the trochosphere, which is described in the next section.

Development of the Polychaeta

The cleavage of the egg in the Polychaeta and the Archiannelida, the polyclad Turbellaria, the Nemertea and the Mollusca follows almost exactly the same plan. Division occurs rhythmically, affecting the whole or greater part of the blastomeres at the same time. The first two divisions are equal, producing four cells (Figs. 194–196) lying in the same plane, which are called A, B, C, D; each cell in its further cleavage resembles the others and gives rise to one of the *quadrants* of the embryo. D tends to be larger than the others and becomes the dorsal surface of the embryo, while B is ventral, A and C lateral. The next divisions (third, fourth and fifth) are unequal and at right angles to the first two and result in three *quartets* of *micromeres* being divided off successively from the *macromeres* as A, B, C and D are then termed. The region in which the micromeres lie is the upper or *animal pole* of the embryo, while the macromeres form the *vegetative pole*. The micromeres are not directly over the macromeres from which they are formed but in one quartet they are all displaced to the right, while in the next they will be displaced to the left of the embryonic radius and the next to the right again. The cleavage is therefore said to be of *spiral* type and successive cleavage planes are at right angles. At a later period it is replaced by cleavage in which there is no alternation of the kind described above, and the result is that the embryo becomes bilaterally symmetrical.

The rest of the description is drawn from the Polychaeta but can be applied with slight modifications to the other groups. Cell classification and fate are shown (in Figs. 195, 196).

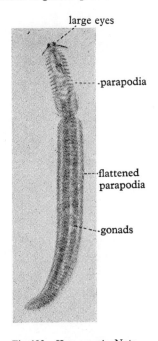

Fig. 193. *Heteronereis*. Note the large eyes and well-developed peristomial cirri, unmodified trunk region, and dark posterior region containing gonads.

The cells of the first three quartets give rise to the ectoderm of the larva and of the adult. The sixth division, however, results in the separation from the macromeres of a fourth quartet which is composed of cells differing notably in size and density from those of the first three. Of the fourth quartet $4d$ (Fig. 194D) alone produces the mesoderm, while the other three, $4a$, $4b$ and $4c$, reinforce the macromeres to form the endoderm. The mesoderm is, however, only in course of differentiation during larval life and a larval mesoderm

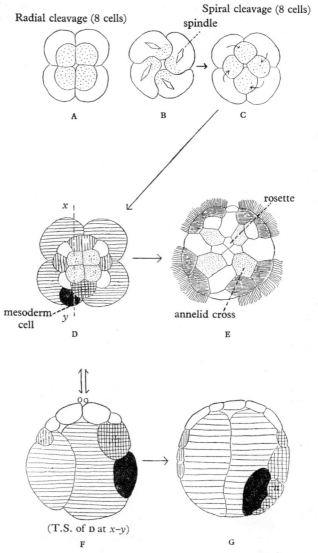

Fig. 194. Segmentation and development of the annelids. A, eight-celled stage in radial cleavage. B, C, eight-celled stage in spiral cleavage. D, segmenting eggs of *Nereis* seen from the animal pole, showing the macromeres (horizontal shading), the first three quartets of micromeres, and the mesoblast cell (black), in the fourth quartet. E, later stage also seen from the animal pole, to show the rosette, the annelid cross, and the prototroch cells (horizontal shading). F, vertical section through stage 4 at $x-y$. G, vertical section through later stage to show gastrulation. Cross-hatched cells are growing over the macromeres and the mesoderm cell is withdrawing into the interior.

or *mesenchyme* is produced from which particularly the musculature of the trochosphere is fashioned. The mesenchyme is derived from the inward projections of cells of the second and third quartets.

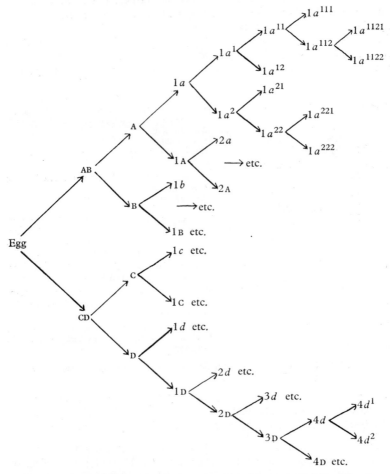

Fig. 195. Cell lineage in the polychaetes. Illustrating the method of classifying the different cells in the developing embryo. (From Costello.)

GASTRULATION (Fig. 194 G). The amount of yolk in the macromeres determines the character of the cleavage within certain limits and the type of gastrulation. In forms like *Polygordius* with very little yolk the micromeres and macromeres are nearly the same size and gastrulation takes place by invagination; in *Arenicola*, *Nereis* and nearly all Polychaeta and Mollusca the micromeres are much smaller than the macromeres, and as they divide to form the ectoderm they grow round the massive macromeres and an 'epibolic' gastrula is formed. The cells of the fourth (and fifth) quartets approach each other from the two sides. The mesoblast cell ($4d$) begins to withdraw from the surface into the blastocoele, and the blastopore, that is the uncovered surface of the macromeres, becomes much smaller and slit-like. Eventually as gastrulation is completed the lips of the blastopore join in the middle, the same cells meeting each other in every case, leaving an anterior opening which becomes the mouth and a posterior which closes, but in the neighbourhood of which the anus of the trochosphere arises later. The blastopore therefore represents the ventral surface of the larva. At the same time the macromeres withdraw

Cells	Fate
First quartet of micromeres ($1a$–$1d$)	Ectoderm of upper hemisphere, the annelid cross and the cerebral ganglion
$1a^{111}$–$1d^{111}$	Apical organ
$1a^2$–$1d^2$	Prototroch
$1c^{1122 1}$, $1d^{11221}$	Head kidneys
Second quartet of micromeres ($2a$–$2d$)	
$2a$–$2c$	Portions of circumoral ectoderm, ectoderm of lateral dorsal trunk region
$2a^2$–$2c^2$	Stomodaeum
$2a^{11}$–$2c^{11}$; $2a^{12}$–$2c^{12}$	'Post-trochal' cells
$2d$	The ventral plate, including ventral nerve cord, seta sacs, general ectoderm of post oral ventral and lateral regions. Portions of the nephridia? Ectoderm of the middle dorsal trunk region
Third quartet of micromeres ($3a$–$3d$)	Portions of the general ectoderm of the circumoral and circum-anal regions
$4d$ (The mesentomere)	Mesoderm bands, portions of the posterior part of the archenteron, some pigment cells of the anal region
Endomeres. ($3A$–$3C$; $4D$)	Endoderm, including a portion of the posterior archenteric wall, some pigment cells of the anal region

Fig. 196. Fate of different embryonic cells in polychaetes. (From Costello.)

into the interior to form a second cavity, the *archenteron*, bringing with them the cells of the fourth and fifth quartets ($4a$, $4b$, $4c$, $5a$, $5b$, $5c$). The *somatoblast* ($2d$) breaks up into a large number of cells to form the *ventral plate*.

TROCHOSPHERE. The change from gastrula to trochosphere (Fig. 197) follows quickly and with little further cell division. The first quartet of micromeres have by this time been differentiated (Fig. 194 E) into (1) the *apical rosette*, consisting at first of four small cells and becoming the *apical organ* of the trochosphere; (2) the cells of the so-called *annelid cross* which alternate with those of (1) and form the cerebral ganglia; (3) the *prototroch*, forming four groups of cells which constitute the preoral ciliated ring of the trochosphere; and (4) the intermediate girdle cells, forming most of the general ectoderm of the part in front of the prototroch, which is called the *umbrella*. The expansion of the *subumbrellar ectoderm*, i.e. that behind the prototroch, is due to the proliferation of a single cell in the second quartet of micromeres, $2d$ (the *somatoblast* (Fig. 194 F)). It forms a plate which spreads from its originally dorsal position round the sides, the two wings uniting behind the mouth to form the *ventral plate*, becoming the ventral body wall. The descendants of this single cell thus make up nearly the whole of the subumbrellar ectoderm. Its sisters 2*a*, 2*b*, 2*c* give rise to the *stomodaeum* and are tucked in at the mouth at the close of gastrulation. This marks the completion of the alimentary canal. The young trochosphere now possesses a very thin outer epithelium, thickened in the region of the apical disc and the equatorial ring of cilia, the prototroch, and in the region of the ventral plate, which is the rudiment of a large part of the trunk of the adult worm. It will form ventral nerve cord, chaetal sacs and the ventral and lateral ectoderm of the trunk. The larval gut opens by a *mouth* in the equatorial region and consists of an ectodermal oesophagus (stomodaeum) opening into the endodermal *stomach* and an ectodermal *hind gut* opening to the exterior by an *anus*. The cavity between the ectoderm and the gut (blastocoele) is spacious and traversed by

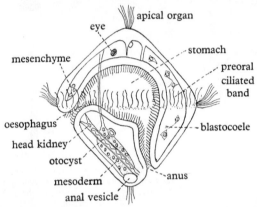

Fig. 197. Trochosphere larva of *Eupomatus*, side view. (After Shearer.)

the pseudopodium-like processes of the mesenchyme cells, larval muscles and nerves, and also contains the two *larval nephridia*, each of which is composed of two hollow cells placed end to end, one of which contains a 'flame' of cilia. They are descended from the first quartet of micromeres and sink in from the surface.

The trochosphere drifts hither and thither in the sea, swimming feebly by the action of the cilia of the prototroch and sometimes also by secondary postoral rings of cilia (e.g. metatroch formed from cells of the third quartet). During this pelagic existence the rudiments of the adult worm continue their development—which is best traced in *Polygordius*—the apical organ develops into the *prostomium* of the adult with brain, tentacles and eyes, while the trunk rudiment formed by the proliferation of the ventral plate and the mesoblast cell grows backwards as an ever-lengthening cylindrical process containing the end gut. In the ectoderm of this is developed ventrally the rudiment of the ventral nervous system, while to the sides of this and internally are the mesodermal strips (derived from the single cell 4*d*), which show at once *metameric segmentation* (Fig. 198), first as pairs of solid blocks, then with cavities, to form the *somites*. Each of these box-like mesodermal segments has then an inner wall which is applied to the gut (splanchnic mesoderm) and an outer (somatic mesoderm) lying under the ectoderm. The right and left rudiments meet in the middle lane and are only separated by the *dorsal* and *ventral* mesenteries which are formed by their apposed walls, while the anterior and posterior borders of each segment are *septa*. At the same time the adult nephridia develop from ectoderm rudiments and the blood vessels differentiate in the septa and mesenteries.

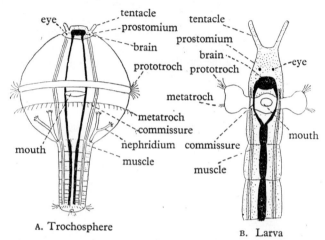

A. Trochosphere   B. Larva

Fig. 198. Development of *Polygordius*. A, trochosphere with rudiment of prostomium and trunk. B, metamorphosing larva, with prostomium and trunk brought close together by the contraction of longitudinal muscles; the umbrella of the trochosphere is shrivelled and about to be discarded. (After Wolterek.)

The advanced larva (Fig. 198 A) thus consists of two rudiments of the adult body, separated by the body of the larval trochosphere. They are joined by a pair of longitudinal muscles and of nerves, and in one species of *Polygordius* metamorphosis of the larva into the adult is brought about by the shrivelling up of the larval tissues and the drawing together and the union of the head and trunk assisted by the contraction of these muscles (Fig. 198 B). The larval mouth remains in the adult. After metamorphosis the animal sinks to the bottom and begins its adult life.

CLASS 2. OLIGOCHAETA

DIAGNOSIS. Annelids, with a comparatively small number of chaetae, not situated on parapodia, with prostomium distinct but usually without appendages; always hermaphrodite, the male and female gonads being few in number (one or two pairs), situated in fixed segments of the anterior region, the male always anterior to the female; with special genital ducts (coelomoducts) opening by funnels into the coelom, *spermathecae*, and a *clitellum* present at sexual maturity; with reproduction by copulation and cross-fertilization; eggs being laid in a cocoon, developing directly without a larval stage.

CLASSIFICATION

Order **1. Terricolae.** Oligochaetes living on land. *Lumbricus, Allolobophora, Eutyphoeus.*

Order **2. Limicolae.** Oligochaetes living in water. *Tubifex, Stylaria, Aeolosoma.*

GENERAL ACCOUNT

The pharynx is not eversible and pharyngeal teeth (such as frequently occur in the Polychaeta) are absent, except in one small family, the Branchiobdellidae, which have ectoparasitic habits similar to the leeches and resemble them in some particulars of structure.

Though the chaetae are not borne on parapodia they are usually divided into two bundles or groups on each side which roughly correspond to the noto- and neuropodia. They may be classified into hair chaetae which are long and fine (dorsal chaetae of *Stylaria*) and shorter chaetae which are rod-like (*Lumbricus*) or needle-like. The point of the needle is single- or double-pronged. There is not, however, the great variety found in the Polychaeta.

Certain main features of the reproductive system (Fig. 199) are the salient characters of the group. Its members are, without exception, hermaphrodite, and with a single possible exception cross-fertilization only is possible. The restriction of the gonads to a few segments occurs also in some sabellids among the Polychaeta and in some archiannelids. The sexual cells are shed into the coelom either into the general coelomic cavity as in the Polychaeta or into special parts of it divided off from the rest (*seminal vesicles* of *Lumbricus*) where they mature. Spermathecae are usually present to contain the spermatozoa received from another worm in copulation. The clitellum is a special glandular development of the epidermis whose principal function is the secretion of the substance of the cocoon and the albuminoid material which nourishes the embryo. It is a secondary sexual character which is only present in the reproductive season in most Oligochaeta, but the earthworms (*Lumbricus, Allolobophora*) used in zoological laboratories in this country always possess it. Both the clitellum and the cocoon produced by it are found in the Hirudinea. It may also be mentioned that many oligochaetes have special copulatory chaetae, sometimes hooked for grasping the other worm or with a sharp point for piercing it.

For the purposes of the elementary student it is probably best to recognize that the Oligochaeta contain two well-marked ecological types, the 'earthworm', a larger burrowing terrestrial form, and the aquatic oligochaete which is much smaller and simpler in structure. It is probable that the former type is the more primitive; the aquatic oligochaete shows many characters which resemble those of the archiannelids and are most likely due to a process of simplification. The reasons for the conclusion that the aquatic oligochaetes are not the oldest of these groups are given below.

Earthworms

These are divided into a number of families of which the most important are the Lumbricidae, containing *Lumbricus* and *Allolobophora*, and the Megascolecidae which is the largest of all.

The primitive forms in all families resemble *Lumbricus* in the following characters. There are a large number of segments and each one is furnished with eight chaetae arranged in pairs and all on the ventral side of the worm. A series of *dorsal pores* is found along the back in the intersegmental grooves. The alimentary canal is characterized by a large muscular *pharynx* by which the food is sucked in, with many glands, the secretion of which is used in external digestion. The oesophagus in one part of its length gives rise to one or more pairs of diverticula, the cells of which secrete calcium carbonate (*oesophageal pouches* and *glands*). At the end of the oesophagus or the beginning of the intestine there is a thick-walled *gizzard* in which the food is masticated with the aid of the soil particles. The intestine has a dorsal ridge, the *typhlosole*, to increase the absorptive surface. The nervous, muscular and circulatory systems exist throughout the earthworms with little variation from the condition in *Lumbricus*.

The variations which occur in more specialized members of all families are as follows. The chaetae may increase in number and come to be arranged in a complete ring round the body (*perichaetine*). The dorsal pores may disappear. The oesophagus may lose its calciferous glands and the gizzard may be absent or develop into several. The reproductive organs vary in small but important particulars. There are nearly always two pairs of *testes* in segments 10 and 11 and one pair of *ovaries* in segment 13, but the testes may be reduced to a single pair. There are usually two pairs of *spermathecae* but the number varies and occasionally they are absent altogether. The *prostate glands* (of unknown function) are nearly always present in earthworms except in the Lumbricidae.

REPRODUCTION. The reproductive system (Fig. 199 c) consists essentially of two pairs of *testes* in segments 10 to 11 and one pair of *ovaries* in segment 13, followed by ducts which open by large funnels just behind the gonads and discharge to the exterior in the next segment in the case of the oviduct, and several segments behind in the case of the sperm duct. The testes, at least, are enveloped by *sperm sacs* (vesiculae seminales) which are outgrowths of the septa, and in the cavity of these the sperm undergo development. In some earthworms there are no sperm sacs and this condition, resembling that in the Polychaeta, is probably the earliest in the group. There are two pairs of *spermathecae* in the region in front of the testes. In the neighbourhood of the male external aperture there are *spermiducal* (*prostate*) *glands* which do not actually open into the sperm duct. A single pair of segmental organs (open nephridia) is present in each segment.

The simplest method of copulation in earthworms is that found in *Eutyphoeus*, where the end of the sperm duct can be everted to form a *penis*. This is inserted into the spermathecal apertures and the spermatozoa thus pass directly from one worm to another. It is obvious that the mechanism of copulation is far more complicated in the Lumbricidae. Here the worms come into contact along their ventral surfaces and each becomes enveloped in a mucous sheath. Close adhesion is secured between the clitellum of one worm and the segments 9 and 10 of the other, partly by embracing movements of the clitellum and partly by the chaetae of the same region being thrust far into

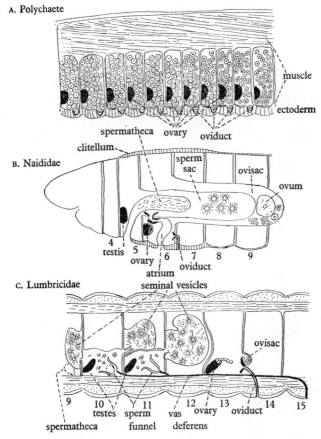

Fig. 199. Sexual reproductive organs: A, Polychaete; longitudinal section of *Serpula intestinalis* to one side of the mid line. B, Oligochaete, Naididae. (After Stephenson.) C, Oligochaete, *Lumbricus terrestris*. (After Hesse.)

the body wall of the partner. The sperm passes out of the male aperture and along the *seminal groove* to the clitellum; how it enters the spermathecae of the other worm has never been determined.

The cocoons are formed some time after copulation. The worm forms a mucous tube as in copulation. The cocoon is then secreted round the clitellum and finally the albuminous fluid which nourishes the embryo is formed between the cocoon and the body wall and the worm frees itself from the cocoon by a series of jerks. All three products, mucus, cocoon substance and albumen, are secreted by the clitellum and each probably by a distinct type of cell. The eggs are sometimes extruded and passed backwards into the cocoon while it is still in position on the clitellum but the spermathecae eject the spermatozoa when the cocoon passes over them.

The embryo of *Eisenia* is illustrated in Fig. 200. The prototroch is absent but the gut and stomodaeum are developed early to absorb the albumen in the cocoon. There are two mesoblast pole cells at the hinder end which bud

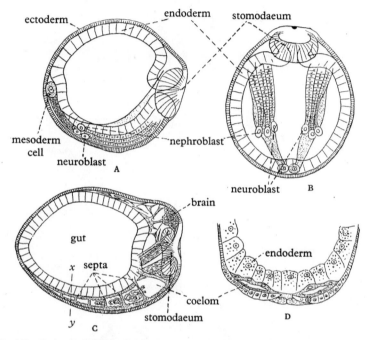

Fig. 200. Embryo of *Eisenia foetida*. A, lateral view of embryo in which the mesoblast is unsegmented. B, ventral view of small embryo. C, longitudinal section of the ventral part of a later embryo a little to one side. D, transverse section of same embryo along the line x—y in C. (After E. B. Wilson.)

off the mesodermal strips: there are three ectodermal pole cells on each side, the most ventral a *neuroblast* forming half the nerve cord and the two others *nephroblasts* giving rise to longitudinal rows of cells which divide up to form the nephridia.

In *Lumbricus* the larva goes through a type of metamorphosis within the cocoon. Larval excretory organs, musculature, and cilia around the mouth develop at an early stage only to be broken down at a later stage and replaced by the adult organs. A similar situation is seen in the development of some leeches. In fact there are many similarities in the development of the Oligochaeta and Hirudinea which mark them off from the Polychaeta. The Oligochaeta and Hirudinea like the Polychaeta show spiral cleavage. But whilst

OLIGOCHAETA

the mesoderm in the Polychaeta comes from the 4*d* cell, in the Oligochaeta and Hirudinea it may come from the 3D, 4D or 3*d*, but never 4*d*. Another difference is seen in the development of the nervous system. The central nervous system of the Polychaeta arises from two sites, one in the prostomium the other in the body. In the Oligochaeta and Hirudinea the central nervous system arises from only one site. These and other similarities link the Oligochaeta and Hirudinea into one group, the Clitellata.

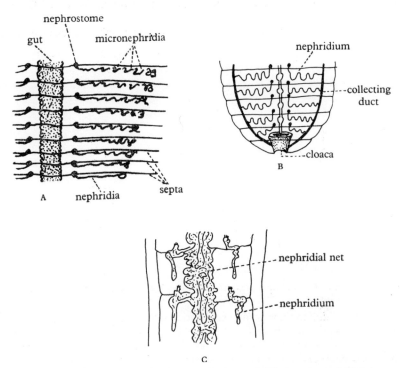

Fig. 201. Excretion. A, development of micronephridia (=meronephridia). B, *Allolobophora antipae* and C, *Lumbriculus*: arrangement of the nephridia.

SEGMENTAL ORGANS. The Oligochaeta like the Hirudinea but unlike the Polychaeta have their nephridia and coelomoducts separate. The nephridia are metanephridia and are usually present in each segment of the body whilst the coelomoducts are restricted to a few reproductive segments. The nephridia can either open to the outside of the body (exonephric) as in *Lumbricus* or they can open into the gut (enteronephric) as in *Pheretima*. When the original large pair of nephridia are still present in each segment they are called holonephridia. In other cases the nephridia may divide to form many small nephridia which are called meronephridia. In the development of the excretory system of *Megascolides* the segmental organs first appear as cords of cells, one pair in each segment. These holonephridia are later thrown into loops and each loop becomes separated off from the next to form a meronephridium (Fig. 201 A). *Lumbricus* has a pair of exonephric holonephridia in each segment. *Allolobophora antipae* has a pair of holonephridia in each segment but these open into a longitudinal duct that discharges into the hind end of the intestine. In the Indian earthworm *Pheretima*, there are three different types of nephridia.

(1) In segments 4, 5, and 6 there are many enteronephric meronephridia opening into the pharynx. These are called peptonephridia and may have a digestive function.

(2) Each segment posterior to segment 6 has a number of exonephric meronephridia.

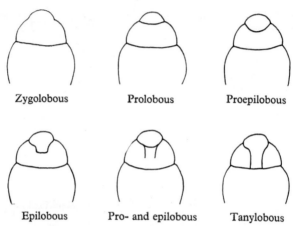

Fig. 202. Types of head present in the Oligochaeta. (From Stephenson.)

(3) In addition to these many exonephric meronephridia, segment 15 and all those posterior to it have 40–50 pairs of small meronephridia. These open into a pair of ducts that run along the dorsal wall of the intestine. These ducts have segmental openings into the intestine.

It has been suggested that the enteronephridia are of use in water conservation since *Pheretima* can survive drought conditions better than most earthworms.

CIRCULATION. There is a well-developed blood circulation. Blood flowing through the parietal and dorso-intestinal vessels of each segment is collected in the dorsal vessel. It is prevented from returning by an elaborate system of valves (Fig. 204 A). Waves of peristaltic contraction beginning at the hind end of the dorsal vessel and continued by the 'hearts' press it forwards and ventralwards into the ventral vessel which is the main distributing channel.

Aquatic Oligochaeta

As a type of these, *Stylaria*, belonging to the family Naididae, will be shortly described (Fig. 204 B). This is a transparent worm rather less than a centimetre long found crawling on water weed. The prostomium bears minute eyes and is produced into a long filiform process. In most of the segments there are two bundles of chaetae on each side, the dorsal consisting of hair chaetae and needle chaetae, while the ventral has only 'crotchets' with a double point. The first four segments have no dorsal bundles (incipient cephalization).

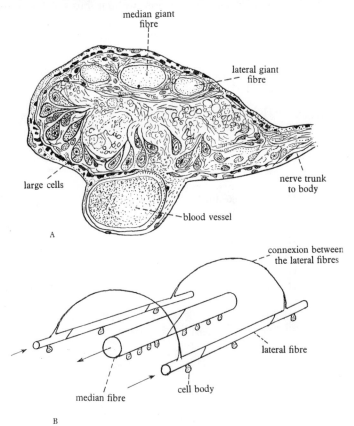

Fig. 203. Giant fibres in *Lumbricus*. A, transverse section of nerve cord. Showing large cells and giant fibres. B, stereodiagram of the arrangement of the three giant fibres. Note that the median one has four cell bodies to each unit of axon whilst the lateral ones have only one cell per axon unit. These cells conduct rapidly and facilitate the rapid contraction of the animal (After Stough.)

The alimentary canal is simpler in character than that of *Lumbricus*, a gizzard being absent. The intestine is ciliated and the action of the cilia brings in from the anus a current of water which probably assists respiration. The testes (Fig. 199 B) develop in segment 5 and the ovaries in segment 6, while a pair of spermathecae is found in the testis segment. The sexual cells develop in the seminal vesicle and the ovisac, which are unpaired backward pouchings of septa 5/6 and 6/7 respectively. The male ducts open by a funnel on septa 5/6 and discharge into an *atrium*, which is lined by the cells of the *prostate*. While sexual individuals are often met with and can be recognized at once by the appearance of the opaque clitellum in segments 5–7, individuals reproducing asexually are much commoner. Chains of worms attached to one another may be found, and the existence of one or more *zones of fission*, where new segments are being formed and separation of two individuals will take place, is easily observed under the microscope.

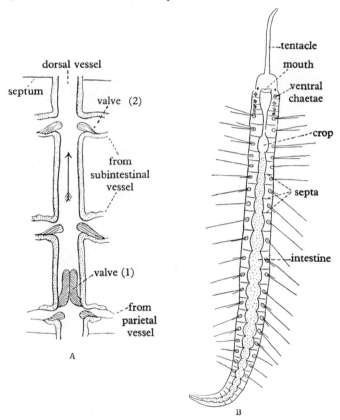

Fig. 204. A, dorsal vessel of *Lumbricus* to show connexions and valves. (After Rosa.) Valves 1, closed during the contraction of the dorsal vessel; valves 2, open during dilatation of the dorsal vessel. B, *Stylaria proboscidea*, dorsal view. The four anterior segments have hooked ventral chaetae only, the rest have long dorsal chaetae as well.

Stylaria is a delightful object of study. The operation of many of the organs can be easily observed with a low power and the results form a useful supplement to work with *Lumbricus* in understanding oligochaete organization.

From the above account it will be seen that *Stylaria* differs from *Lumbricus* not only in its small size and transparency but also in the number and appearance of the chaetae—which give it a certain resemblance to the Polychaeta. The reproductive organs, however, are entirely different from those of the latter group and it is in this system that the real contrast between polychaete and oligochaete lies.

The aquatic oligochaetes when they are of small size often show reduction of the vascular system, ciliation of the under-surface (in one form, *Aeolosoma*), and a nervous system of embryonic type. These are characters which may be primitive but, as in the archiannelids, so here, they are probably the results of simplification; it is generally agreed that the replacement of sexual by asexual reproduction is a secondary feature, and the frequency with which it is found in the aquatic Oligochaeta shows them to be, on the whole, specialized types.

Two common genera, *Tubifex* and *Lumbriculus*, are larger worms which in their appearance have more resemblance to earthworms. A brief description of them follows.

Tubifex. A small red worm with rather numerous chaetae in the dorsal and ventral bundles belonging to various types; without gizzard; testes and ovaries in segments 10 and 11 respectively.

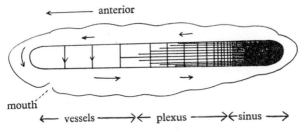

Fig. 205. Blood circulation in *Lumbriculus variegatus*. The animal has vessels anteriorly, capillaries medially, and a sinus system posteriorly. (After Haffner.)

It lives in the mud at the bottom of ponds and lakes with its head buried and its tail waving in the water; the latter movements are respiratory. They draw water from upper layers which contain more oxygen: when the oxygen content of the water in general falls a greater length of the worm is protruded and its movements become more vigorous. A great deal of detritus passes through its alimentary canal so that *Tubifex* plays the same sort of part in fresh water that the earthworms play on land.

Lumbriculus resembles *Tubifex* superficially but has only eight chaetae in a segment, placed as in *Lumbricus*; chaetae double-pointed; not often met with in sexual state but reproduces habitually by breaking up into pieces each of which regenerates the missing segments.

In this worm the primitive nature of the blood system is well seen (Fig. 205). At the posterior end there is a continuous *sinus* round the gut, in the middle region this becomes resolved into a dense plexus of capillaries and at the anterior end there is the beginning of a segmental arrangement.

CLASS 3. HIRUDINEA

DIAGNOSIS. Annelida with a somewhat shortened body and small, fixed number of segments, broken up into annuli and without chaetae (except in *Acanthobdella*) or parapodia; at the anterior and posterior ends several segments modified to form suckers; coelom very much encroached upon by the growth of mesenchymatous tissue and usually reduced to several longitudinal tubular spaces (sinuses) with transverse communications. Hermaphrodite, with clitellum. Embryo develops inside cocoon.

CLASSIFICATION. The Hirudinea may be divided as follows:

Acanthobdellidae. A group intermediate between the Oligochaeta and the Hirudinea, containing the single genus *Acanthobdella*.

Rhynchobdellidae. Marine and fresh-water forms, with colourless blood, protrusible proboscis and without jaws. *Glossiphonia*.

Gnathobdellidae. Fresh-water and terrestrial forms, with red blood and without a protrusible proboscis but usually with jaws. *Hirudo*.

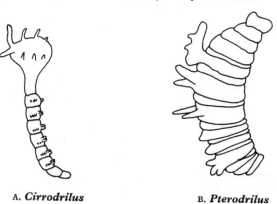

A. *Cirrodrilus*   B. *Pterodrilus*

Fig. 206. Branchiobdellidae; aberrant oligochaetes. A, *Cirrodrilus*; B, *Pterodrilus*.

GENERAL ACCOUNT

In the typical leeches the *constitution of the body* is remarkably constant. There is a prostomium and thirty-two body segments; an anterior sucker (in the centre of which is the mouth) is formed from the prostomium and the first two segments, and a posterior from the last seven. Both suckers are directed ventrally. The subpharyngeal 'ganglion' (Fig. 207 B) is composed of four single ganglia fused together and the posterior 'ganglion' of seven. Between them lie twenty-one free ganglia, and the number of segments is estimated by summation of all the ganglia. The number of annuli to a segment varies in different forms.

ALIMENTARY CANAL. The alimentary canal is highly characteristic and consists of the following parts. (1) A muscular *pharynx* with unicellular salivary glands. In the Gnathobdellidae, which includes *Hirudo*, there are three chitinous plates or jaws. In the Rhynchobdellidae (Fig. 208) there is a protrusible *proboscis* surrounded by a *proboscis sheath*. (2) A short *oesophagus* follows, leading into (3) the *mid gut* (crop) which is often provided with lateral caeca, varying in number, and is used for storing up the blood or other juices of the host. This is kept from coagulating by the ferment (anticoagulin) contained in the salivary secretion (*Hirudo*). In the mid gut a very slow digestion takes place, the blood appearing almost unchanged even after several months. (4) An *intestine*, which is also endodermal, and has, in *Hirudo*, a pair of diverticula. (5) A very short ectodermal *rectum* discharging by the anus, which is dorsal to the posterior sucker.

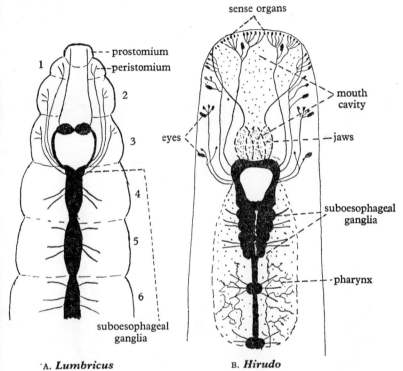

Fig. 207. Anterior part of nervous system: A, *Lumbricus*; B, *Hirudo*. The brain in both consists of a single pair of dorsal ganglia belonging to the prostomium. In *Lumbricus* the suboesophageal ganglia and the lower part of the circumoesophageal commissures give rise to nerves to segments 1, 2, and 3, and so belong to three segments. In *Hirudo* the suboesophageal ganglionic mass consists of four (or five) pairs of ganglia fused together. (A, after Borradaile; B, after Leydig.)

BODY WALL. The body wall consists of a single layer of ectodermal cells between which blood capillaries penetrate, a dermis with pigment cells and blood vessels, and an outer circular and inner longitudinal layer of muscles. The muscle fibres have a characteristic structure, consisting of a cortex of striated contractile substance and a medulla of unmodified protoplasm. Inside the musculature are masses of mesenchymatous tissue: in the Gnathobdellidae this is pigmented and forms the *botryoidal tissue*, the cells of which are arranged end to end and contain intracellular capillaries filled with a red fluid.

The mesenchyme almost completely occupies the space which is the perivisceral cavity in the earthworm. There are, however, longitudinal canals, constituting the *sinus system*, and these represent the remnants of the coelomic spaces; there are always dorsal and ventral and often (e.g. *Glossiphonia*, Fig. 209 B) two lateral sinuses, and there are numerous transverse canals in each segment. Into this reduced coelom the nephrostomes open and the gonads are found in it. The blood system consists of two contractile lateral vessels (and in the Rhynchobdellidae of dorsal and ventral vessels running

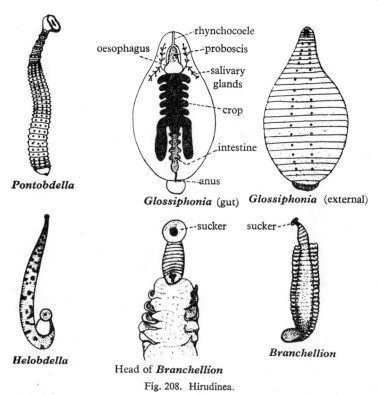

Fig. 208. Hirudinea.

inside the corresponding coelomic spaces). These vessels all communicate with one another. They also communicate with the sinuses of the coelom and with the capillaries of the botryoidal tissue, as has been shown by careful injection. This astonishing condition is unique, but a parallel may be drawn with the vertebrate in which the lymphatic system communicates both with the coelom and the blood system. The peculiar functions of the lymphatic system are not shared by the botryoidal vessels which have no particular connexion with the gut.

NERVOUS SYSTEM. The nervous system is of the usual annelidan type but characterized by the fusion of ganglia anteriorly (Fig. 207 B) and posteriorly.

HIRUDINEA

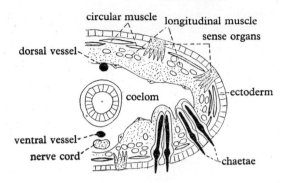

A. *Acanthobdella*

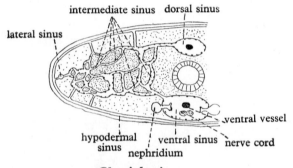

B. *Glossiphonia*

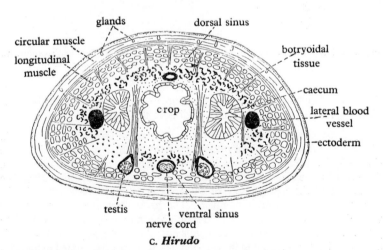

C. *Hirudo*

Fig. 209. Transverse sections of Hirudinea to show progressive restriction of the coelom. In *Acanthobdella* the coelom is continuous but encroached upon by the growth of parenchyma (stippled); in *Glossiphonia* it is broken up into a system of sinuses; in *Hirudo* the sinuses are reduced in size and there is no intermediate sinus.

There are segmental sense organs in the form of papillae, and on the head some of these are modified to form eyes and the so-called 'cup-shaped organs'.

NEPHRIDIA. The nephridia are much like the metanephridia of the Oligochaeta except that they are more specialized due to the reduction of the coelom and the masking of the primary segmentation. The nephrostome is the only ciliated part and projects into a ventral coelomic chamber (Fig. 210). The nephrostome leads into an expansion called the capsule. The capsule is usually filled with phagocytes and as a rule does not communicate with the following intracellular canal. This nephridial canal is much coiled and branched and in

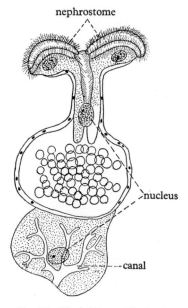

Fig. 210. Nephridium of the leech.

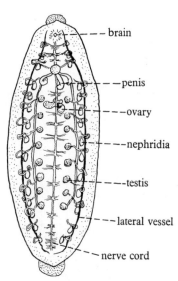

Fig. 211. Excretory and reproductive system of *Hirudo*.

the Ichthyobdellidae the branches link up to form a network. This is the plectonephric condition, the canals leading to the outside.

REPRODUCTION. The testes, of which there are often several pairs (nine in *Hirudo*), and the single pair of ovaries are also present as closed vesicles in the sinuses and are derived from the coelomic epithelium, but in distinction from the rest of the annelids they are continuous with their ducts (Fig. 211). The separation of the genital part of the coelom from the rest, begun in the Oligochaeta, here becomes complete. The testes discharge into a common vas deferens on each side; the two vasa unite anteriorly to form a median penis. Similarly the two oviducts join and the eggs pass through a single albumen gland and vagina to the exterior. The spermatozoa, united in bundles, are deposited on the body of another leech and appear to make their way through the skin to the ovaries where fertilization occurs. The eggs are laid in cocoons, the case of which is formed by clitellar glands in the same way as in *Lumbricus*.

Family Acanthobdellidae

Acanthobdella (Fig. 209 A), a parasite of salmon, is a link with the Oligochaeta. In it the specialized hirudinean characters are only partly developed. There is no anterior sucker but a well-developed posterior sucker formed from four segments. The total number of segments is twenty-nine compared with thirty-two in the rest of the group. There are dorsal and ventral pairs of chaetae in the first five body segments and the coelomic body cavity is a continuous perivisceral space, interrupted only by segmental septa as in the Oligochaeta. It is, however, restricted by the growth of mesenchyme in the body wall and split up into a dorsal and ventral part in the clitellar region. The so-called testes (really vesiculae seminales) are tubes running through several segments, filled with developing spermatozoa and their epithelial wall is continuous with that of the perivisceral coelom, another primitive feature. The vasa deferentia, moreover, open into the testes by typical sperm funnels.

It is interesting to find that in the Branchiobdellidae, a family of the Oligochaeta, parasitic on crayfish, there is the same sort of leech-like structure: a posterior sucker, annulated segments, absence of chaetae and presence of jaws. But the condition of the coelom, nephridia and generative organs is so like that of the Oligochaeta that the family must remain in that group.

Family Rhynchobdellidae

Pontobdella (Fig. 208), parasitic on elasmobranch fishes.

Glossiphonia (Fig. 208), a fresh-water leech feeding on molluscs such as *Limnaea* and *Planorbis* and on the larvae of *Chironomus*; body ovate and flattened; hind gut with four pairs of lateral caeca; eggs laid in the spring, the young when hatched attaching themselves to the ventral surface of the body of the mother.

Family Gnathobdellidae

Hirudo, the medicinal leech, at one time a common British species but now rare, jaws armed with sharp teeth.

Haemopis, the horseleech, common in streams and ponds, which it leaves to deposit its cocoons and in pursuit of prey; jaws armed with blunt teeth, which cannot pierce the human skin; a single pair of caeca in the mid gut.

This leech is carnivorous, devouring earthworms, aquatic larvae of insects, tadpoles and small fish. The land leeches of the tropics, of which *Haemadipsa* may serve as an example, live in forests and swamps and, mounted on leaves and branches, wait until a suitable mammalian prey presents itself.

CLASS 4. ARCHIANNELIDA

DIAGNOSIS. Small marine annelids with simplified structure, parapodia and chaetae being usually absent.

This group was founded to receive two genera, *Polygordius* and *Protodrilus*, which were formerly considered to be primitive forms from which the larger

312   ANNELIDA

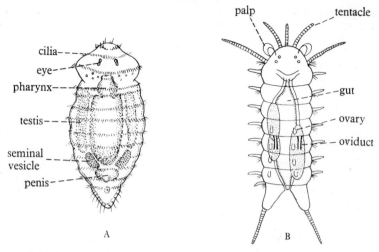

Fig. 212. A, *Dinophilus* and B, *Nerilla*. (A, after Harmer; B, after Goodrich.

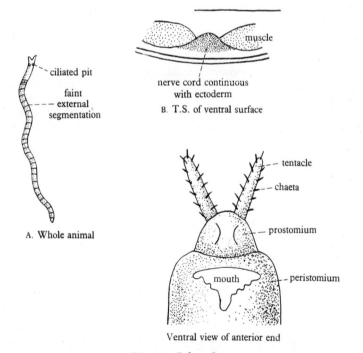

Fig. 213. *Polygordius.*

groups of annelids might be derived. From time to time other genera have been included which show some, but not all, of the characters which distinguish the original genera. The series of diagnoses of the best-known genera given below starts with *Polygordius* and works back to forms which come very close to the Chaetopoda. There can be little doubt that the Archiannelida are derived from this latter group by the loss of some of its distinctive features (e.g. parapodia and chaetae), and retention of juvenile characters (ciliation and connexion of nervous system with epidermis). These changes are also found within the limits of the Polychaeta, and if it were not that other characters link up its members the group might well be considered as a family of polychaetes. *Dinophilus* comes late in the series because, though evidently related, it does stand rather apart. It has a superficial resemblance to a small turbellarian enhanced by the great reduction of the coelom.

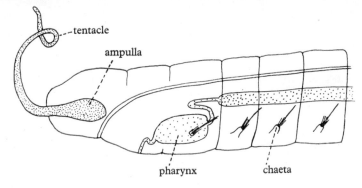

Saccocirrus

Fig. 214. Diagram of the anterior end of *Saccocirrus* showing the chaetae and segmentation. (After Goodrich.)

Polygordius (Fig. 213), with elongated cylindrical body, head with two tentacles and ciliated pits; without parapodia or chaetae; with segments of the coelom separated by septa with a pair of segmental organs opening into each by nephrostomes; with longitudinal muscles in four quadrants, the circular muscles being usually absent; with a reduced vascular system and nerve cords lying in the epidermis; with a trochosphere larva (Fig. 198).

Protodrilus. As in *Polygordius* but with segmentation marked externally by ciliated rings and with a longitudinal ciliated groove in the middle of the ventral surface; with a ventral muscular pharyngeal sac; hermaphrodite.

A single species, *P. chaetifer*, has been discovered with four short chaetae in each segment.

Saccocirrus (Fig. 214). As in *Protodrilus*, but with chaetae arranged in a single bundle on each side of each segment; with separate sexes, each with complicated genital apparatus, the females with spermathecae and males with a pair of protrusible penes in each segment behind the oesophagus.

Nerilla (Fig. 212B). As in *Protodrilus*, but with two bundles of chaetae separated by a single cirrus on each side of each segment; three prostomial tentacles and a pair of palps; with separate sexes and a reduced number of genital segments (three in male, one in female), three pairs of sperm ducts uniting at a common median genital aperture, and two oviducts with separate genital apertures.

Dinophilus (Fig. 212 A), with very short flattened body consisting of only five or six segments, a ciliated ventral surface and ciliated ring in every segment; without septa, dorsal and ventral mesenteries, or a vascular system; with greatly reduced coelom and longitudinal muscles; five pairs of 'closed' nephridia; separate sexes, male with median penis injecting spermatozoa into

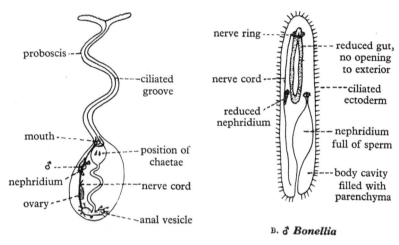

A. ♀ *Bonellia*

B. ♂ *Bonellia*

Fig. 215. *Bonellia*; A, female and B, male. (After Spengler.)

female through skin, female with eggs of two sizes, the smaller giving rise to males and the larger to females.

The value of the Archiannelida to the elementary student of zoology is that they illustrate an evolutionary process which may be called simplification or reduction (but not degeneration), and which is not unlike the changes which parasitic forms have undergone.

CLASS 5. ECHIUROIDEA

DIAGNOSIS. Annelids which show few signs of segmentation, with a spacious coelomic cavity, a well-developed prostomium, a terminal anus, a single pair of ventral chaetae, sometimes several pairs of segmental organs, and in *Echiurus* a trochosphere larva in the nervous system of which there appear to be as many as fifteen pairs of ganglionic swellings (Fig. 216).

Echiurus, with a spoon-shaped prostomium, two pairs of segmental organs and a trochosphere larva.

Bonellia (Fig. 215) with a prostomial proboscis bifurcated at the end, capable of enormous elongation and extremely mobile; a single segmental organ

ECHIUROIDEA

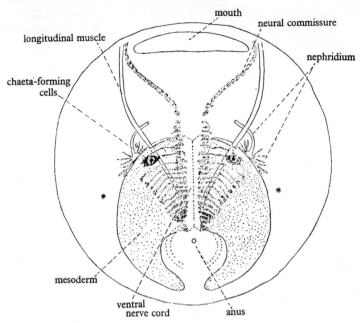

Fig. 216. Young *Echiurus*. Ventral view of larva showing segmentation of posterior end. (After Baltzer.)

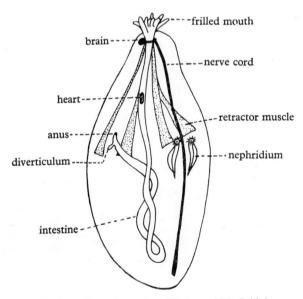

Fig. 217. *Sipunculus*. (From Shipley and MacBride.)

(brown tube); the female is the typical individual and the males are reduced to small ciliated organisms, like a turbellarian, which live in the segmental organ of the female.

It is now known that larvae of *Bonellia* carry the potentialities of both sexes. If they develop independently they become females. If they should come into contact with the body of the adult female, she exercises (probably through the action of some specific secretion) a largely repressive effect on further development, but a male gonad is formed.

Thalassema. The British representative, *T. neptuni*, has a single pair of segmental organs. In two Japanese species, *T. taenioides* and *T. misakiensis*, these have been greatly multiplied so that in the former there are 200 pairs rather irregularly arranged. From a consideration of these forms it appears that the multiplication of the segmental organs is a secondary phenomenon.

CLASS 6. SIPUNCULOIDEA

DIAGNOSIS. Annelids with a spacious uninterrupted coelomic cavity and few signs of segmentation: without prostomium in adult; chaetae always absent, anterior part of body invaginable into posterior part; anus dorsal and anterior; with a single pair of segmental organs (brown tubes); in *Phascolosoma* a trochosphere with three pairs of mesoblastic somites which soon disappear.

Sipunculus (Fig. 217) and *Phascolosoma* are British genera.

CHAPTER IX

THE PHYLUM ARTHROPODA

DEFINITION. Bilaterally symmetrical, segmented Metazoa; with, on some or all of the somites, paired limbs, of which at least one pair function as jaws; a chitinous cuticle, which usually is stout but at intervals upon the trunk and limbs flexible so as to provide joints; a nervous system upon the same plan as that of the Annelida; the coelom in the adult much reduced and replaced as a perivisceral space by enlargement of the haemocoele; without true nephridia, but with one or more pairs of coelomoducts as gonoducts and often as excretory organs; and (except in *Peripatus*) without cilia in any part of the body.

EFFECT OF THICK CUTICLE. The Arthropoda have much in common with the Annelida, and must be regarded as derived from the same stock as the Polychaeta in that phylum. The key to most of their peculiar features is an increase in the thickness of the cuticle. This brings with it the necessity for joints; and the stout, jointed limbs can now be adapted for various purposes to which those of polychaetes were not convertible. Always at least one pair of them become jaws; with this is usually associated the specialization for sensory functions of one or two pairs which have come to stand in front of the mouth, and thus the process of cephalization, begun in the polychaetes, proceeds farther here. Other limbs commonly become legs. In order to move the complex of hard pieces constituted by the jointed cuticle, the continuous muscular layer of the body wall of an annelid has become converted into a system of separate muscles; with this, and with the fact that turgescence of the body wall is no longer a factor in locomotion, is perhaps connected the replacement of the perivisceral coelom by a haemocoelic space. The loss of the nephridia which in annelids lie in the coelom is probably due to the reduction of that cavity. An interesting feature of difference between the Arthropoda and Annelida is the absence from the former phylum of the chaetae, imbedded in and secreted by pits of the skin, which characterize the annelids; though bristles, formed as hollow outgrowths of the cuticle, are common on arthropods. This difference, too, may be connected with the difference in the stoutness of the cuticle. Lastly, it is perhaps that thick covering, hindering the loss of water by evaporation from the surface of the body and providing the skeleton which the lack of support from the medium necessitates, which has enabled arthropods very successfully to invade the dry land.

Like those of all other phyla, their earliest known members, the trilobites, were aquatic. Of their surviving groups, only one, the Crustacea, remains predominantly of that habit. No other invertebrate phylum has so large a proportion of terrestrial members.

CLASSIFICATION. The classification of the arthropods is complex. A scheme of the main groups will be found in the list of contents to this book

Table 6. Somites and limbs of Arthropoda

Somite	Onychophora	Arachnida Scorpionida	Trilobita	Crustacea Malacostraca	Insecta	Chilopoda (Scolopendra)	Diplopoda (Iulidae)
1...*	Preantennae	Embryonic	?	Embryonic	Embryonic	Embryonic	?
2...	Jaws	Chelicerae	Antennae	Antennules	Antennae	Antennae	Antennae
3...	Oral papillae	Pedipalpi	1st biram. limbs	Antennae	Embryonic	Embryonic	Embryonic
4...	1st pair of legs	1st pair of legs	2nd ,, ,,	Mandibles	Mandibles	Mandibles	Mandibles
5...	2nd ,, ,,	2nd ,, ,,	3rd ,, ,,	Maxillulae	(1st) Maxillae**	1st Maxillae	Embryonic
6...		3rd ,, ,,	4th ,, ,,	Maxillae	Labium (2nd Maxillae)	2nd Maxillae	Maxillae
7...	Many (17 to 43) somites, each bearing a pair of legs	4th ,, ,,	5th ,, ,,	(1st) Maxillipeds	1st pair of legs	Maxillipeds †	Collum
8...		Embryonic††	Many somites, each bearing a pair of limbs	2nd Thoracic limb	2nd ,, ,,	1st pair of legs	1st pair of legs
9...		Genital operc. ♀ ♂		3rd ,,	3rd ,,	2nd ,,	2nd ,, ♀ ♂§
10...		Pectines		4th ,,	1st Abd. som.	3rd ,,	3rd ,,
11...		1st lung books		5th ,,	2nd ,,	4th ,,	4th ,,
12...		2nd ,, ,,		6th ,,	♀ 3rd ,,	5th ,,	5th ,,
13...		3rd ,, ,,		7th ,,	4th ,,	6th ,,	Many (20 to 100) double somites, each bearing two pairs of legs
14...		4th ,, ,,		8th ,,	5th ,,	7th ,,	
15...		No limbs		1st Abd. limb	6th ,,	8th ,,	
16...		1st som. Metasoma		2nd ,,	7th ,,	9th ,,	
17...		2nd ,,		3rd ,, ♀	8th ,,	10th ,,	
18...		3rd ,,		4th ,, ♂	9th ,,	11th ,,	
19...		4th ,,			(styles)	12th ,,	
20...		5th ,,		5th ,,	10th ,, som.	13th ,,	
21...		⋮		6th ,,	11th ,, (cerci)	14th ,,	
22...		⋮		⋮	⋮	15th ,,	
23...		⋮		⋮	⋮	16th ,,	
24...		⋮		⋮	⋮	17th ,,	
25...		⋮		⋮	⋮	18th ,,	
26...		⋮		⋮	⋮	19th ,,	
27...		⋮		⋮	⋮	20th ,,	
28...		⋮		⋮	⋮	21st ,, ‡	
29...		⋮		⋮	⋮	Genital limbs, ♀ ♂	
Postsegmental region	Last pair of legs ♀ ♂ Embryonic	⋮		⋮	Embryonic	⋯	Limbless somite
	Telson	Telson	Telson	Telson		Telson	Telson

* Eyes and frontal organs belong to a presegmental region which may have median mesoblast of its own, and may bear various ganglia which enter into the procerebrum.
† Terga fused in *Scolopendra*, free in *Lithobius*.
§ This somite appears to have no limbs, because the limbs of the 8th and 9th somites have each moved forward one somite.
** If the superlinguae be maxillules (see p. 433), the limbs behind them stand on somites 6, 7, etc.
†† Chilaria in *Limulus*.
‡ *Lithobius* has fifteen pairs of legs.

and fuller details will be found under each of the classes of the arthropods in the following pages. The main classes of the arthropods are as follows.

CLASS 1. ONYCHOPHORA

Arthropods with a thin cuticle and a soft muscular body wall. *Peripatus*.

CLASS 2. TRILOBITA

Fossil arthropods with body moulded into three lobes; one pair of antennae· *Olenus*.

CLASS 3. CRUSTACEA

Aquatic arthropods with two pairs of antennae. *Astacus*.

CLASS 4. MYRIAPODA

Terrestrial arthropods with one pair of antennae and many pairs of walking legs. *Iulus*.

CLASS 5. INSECTA

Terrestrial arthropods with one pair of antennae and three pairs of walking legs. *Blatta*.

CLASS 6. ARACHNIDA

Terrestrial and aquatic arthropods with the first appendage chelate not antennate. *Limulus*.

DETAILED ACCOUNT

A more detailed survey necessitates a brief exposition of the principal groups into which the phylum falls. One small section stands apart from the rest. The *Onychophora* have a thin cuticle, without joints; a continuous muscular body wall; eyes (p. 323) of annelid type; only one pair of jaws, which moreover are constructed on a different principle from those of other arthropods, biting with the tip and not with the base of the limb; and a long series of coelomoducts, of which the pair that are the oviducts are ciliated. Only in this group, too, does the first somite bear a pair of limbs: in all others that somite is an evanescent, embryonic structure without external representation in the adult. In all these respects the Onychophora show a lower degree of development of the peculiar features of arthropods than the rest of the phylum.

The remaining groups of the phylum fall into two sharply different sections; the *mandibulate* or *crustacean-insect-myriapod* section; and the *chelicerate* or *arachnid section*. These are sometimes referred to as the Mandibulata and the Chelicerata.

In the first of these sections, the first pair of limbs (those of the second somite) are antennae, the succeeding pair, if present, are also antennae, the third pair are mandibles, and behind these limbs are one or more pairs of additional jaws (maxillae). In the crustaceans and insects there is commonly a pair of compound eyes of a complex type peculiar to these animals. The trilobites belong to this section, but their appendages behind the first pair are undifferentiated. In the *arachnid* section none of the limbs have the form of antennae or mandibles, the first pair (chelicerae) being usually chelate, the second chelate, palp-like, or leg-like, and the third to sixth pairs leg-like, though often some of the postcheliceral limbs possess biting processes (gnathobases) on the first joint. The members of this section never possess true compound eyes of the crustacean-insect type.

The *Crustacea* differ from the Insecta and Myriapoda in possessing a second pair of antennae, and nearly always in being truly aquatic. The *Insecta* differ from the *Myriapoda* in possessing only three pairs of legs, and usually in the possession of wings.

TAGMATOSIS AND CEPHALIZATION. The series of somites which, with small pre- and postsegmental regions, constitutes the body of an arthropod is marked out, by differences in width, fusions of somites, or features of the limbs, into divisions known as *tagmata*. In the Onychophora, Crustacea, Insecta, and Myriapoda, the foremost tagma is a short division, known as the *head*, which carries the antennae and mouth parts, and the rest of the body, known as the *trunk*, is often divided into two sections called *thorax* and *abdomen*. In the Arachnida, the foremost tagma is the *prosoma* ('cephalothorax') and carries legs as well as the limbs used in feeding, while the divisions, if any, of the hinder part of the body (*opisthosoma* or 'abdomen') are known as the *mesosoma* and *metasoma*. It is important that the student should recognize that each of these divisions varies in size, and that consequently none of them comprises in all arthropods the same somites, so that, for instance, the thorax of an insect is a quite different entity from that of a crayfish. The most significant variation is that of the head, which, as the organization of its possessor becomes higher, increases in size, taking in behind somites whose appendages become jaws, while, by alteration in the position of the mouth, it adds others, whose limbs become antennae, to its preoral sensory complex. Thus, while the head of the Onychophora comprises only the first three somites, and only the first of these is preoral, in the Crustacea there are in the true head six somites (including the embryonic first somite), of which three are preoral, and thoracic somites, whose limbs (maxillipeds) function as jaws, are often united with the head (Fig. 218)

PAIRED LIMBS. The paired limbs of arthropods present an enormous variety of form, and attempts have been made to reduce them to a common type. Some of the evidence suggests an archetype with a nine-segmented axis bearing on the median side of the first segment a biting process (gnathobase) and on a more distal segment an outer branch (exopodite); but there are difficulties in the way of assuming this in all cases, and the problem is still far from solution.

CUTICLE. The arthropod cuticle has a thin, impermeable, non-chitinous external layer (*epicuticle*) and a thick, elastic, permeable, lamellar inner layer, largely composed of chitin (an amino-polysaccharide which resists most solvent agents), the outer lamellae usually hardened, often by salts of lime. From time to time during the growth of the animal, the hard outer layers of the cuticle are separated by solution of the inner layers by an enzyme, ruptured, and shed in a *moult* or *ecdysis*. A new cuticle which has formed under it then expands to accommodate the body.

ARTHROPODA

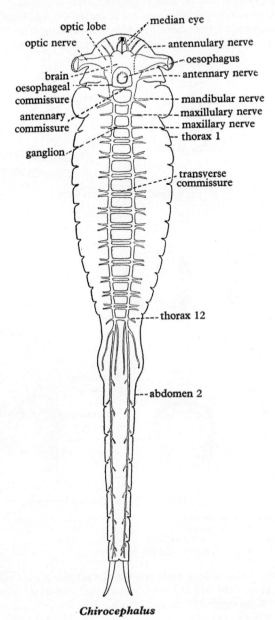

Chirocephalus

Fig. 218. A plan of the nervous system of *Chirocephalus*.

ARTHROPODA

NERVOUS SYSTEM. The nervous system of arthropods contains, in typical instances, on two longitudinal ventral cords and in a dorsal brain, a pair of ganglia for each somite, but where the somites are fused there is often a fusion of their ganglia, and where they bear no limbs their ganglia may be absent.

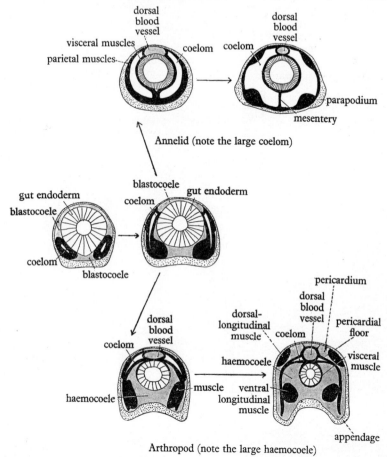

Fig. 219. Diagram to show the development of the perivisceral cavity, dorsal blood vessel and principal muscles in an annelid and arthropod. In annelids the main body cavity is coelomic, whilst in arthropods it is haemocoelic. The skin is dotted, the mesothelium black, the endoderm shown by radial lines, and the blastocoele (haemocoele) finely stippled.

The *brain* is a complex structure composed of the ganglia of the somites which have become preoral (though in a few crustacea the antennal ganglion remains postoral), of paired ganglia for certain primitively preoral presegmental sense organs (eyes, frontal organs), and sometimes also of a median anterior element (*archicerebrum*, in the strict sense). The ganglia of the first somite are known as the *protocerebrum*; with the ganglia anterior to them they constitute the *procerebrum* (*archicerebrum* of Lankester). The ganglia of the second somite

ARTHROPODA

are the *deutocerebrum* or *mesocerebrum*; those of the third somite are the *tritocerebrum* or *metacerebrum*. The identity of some of these ganglia may be lost, even in development. Concerning the functions of the central nervous system something is said on p. 458.

EYES. The eyes of the Onychophora are a pair of simple, closed vesicles, each with its hinder wall thickened and pigmented and its cavity occupied by a lens secreted by the wall. The eyes of all other arthropods (Fig. 220) consist of one or more units each of which is in essence a cup, or a vertical bundle, of cells, over which the cuticle of the body forms a lens. The cells which compose the bottom of each cup are (except in the median eye of the Crustacea) arranged in a sheaf or sheaves called *retinulae*; in the midst of each retinula is a vertical rod, known as the *rhabdome*, secreted by the cells of the sheaf in vertical sections which, when they are distinct, are known as *rhabdomeres*. Each bundle-unit has one such retinula. Sometimes in the cups the retinulae are surrounded by cells which bear on their free ends short *rods* of the same nature as the rhabdomeres. The retinula cells contain pigment and there is a ring of strongly pigmented cells around the cup. The eye units occur (1) as single cups each with several retinulae (ocelli of insects, Fig. 220 c″), (2) as groups of similar cups placed contiguously (eyes of myriapods), (3) as eyes composed of a number of small cups, each with a single retinula, united together (lateral eyes of *Limulus*), (4) as true compound eyes (Fig. 221) composed of a number of bundles of cells, each bundle (*ommatidium*) complex in structure and containing two or more refractive bodies, but each probably representing a narrowed and deepened cup. Compound eyes of this type are found in crustaceans and insects. They vary much in detail, but essentially the structure of an ommatidium is as follows (Fig. 220 D). At its outer end is a transparent portion of the general cuticle of the body, usually thickened to form for the ommatidium a biconvex lens. Under this lie the epidermal cells which secrete it (*corneagen cells*): the lens is one of the facets of the eye. Under the corneagen cells comes a bundle of two to five *vitrellae* or crystal cells, grouped around a refractive body, the *crystalline cone*, which they have secreted. The vitrellae taper inwards and their apex is clasped by a second bundle of cells, four to eight in number, which together form the *retinula*. Like the vitrellae the retinular cells secrete in the axis of the ommatidium a refractive body. This is the *rhabdome*, and is made up of *rhabdomeres*, one for each of the cells. Each retinular cell passes at its base into a nerve fibre which pierces the basement membrane of the eye and enters the optic ganglia. Around each ommatidium, separating it from its neighbours, there are usually pigmented cells, known as *iris cells*. The eyes of arachnids, other than the lateral eyes of *Limulus*, simulate the ocelli of insects, but are thought, from details of their structure, to have been formed by the degeneration of compound eyes resembling the lateral eyes of *Limulus*. The median eye of the Crustacea (Fig. 241 F) is composed of three cups, which may (some copepods) separate widely. The paired eyes probably do not, as has been suggested, represent a pair of appendages. The foremost, or preantennal, somite, to which they would in that case belong, possesses, in *Peripatus* and as a rudiment in embryonic stages of centipedes and certain insects, an appendage which coexists with the eye.

In most compound eyes, the pigment, both in retinular and in pigment cells, flows to and fro, being in dim light retracted towards the inner or outer ends of the cells so as to leave the sides of the ommatidia exposed, and in bright light extending so as to separate the ommatidia completely. In many diurnal insects it is permanently in the latter position. Vision takes place in two ways according to the situation of the pigment. When the latter is extended, in each ommatidium there falls on the retinula a narrow pencil of almost

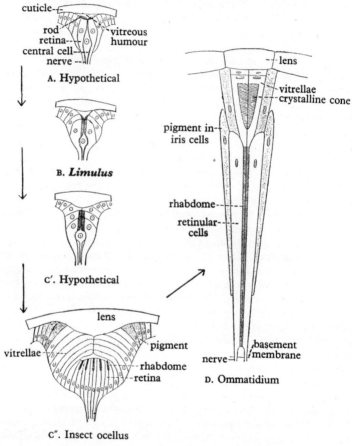

Fig. 220. Diagram of a series of eyes in arthropods.

parallel rays. There is then *mosaic vision*, an *apposition image*, composed of as many points of light as there are ommatidia, being formed on the whole retinal layer. When the pigment is retracted, each ommatidium throws a complete image of the greater part of the field of vision, and the images together form a *superposition image*, falling in such a way that their corresponding parts are superposed. Superposition images are less sharp than apposition images, but are formed with less loss of light. Compound eyes are especially adapted for perceiving the movements of objects, owing to the way in which such movements affect a series of ommatidia in succession.

THE ALIMENTARY CANAL. The alimentary canal of the Arthropoda possesses at its mouth and anus involutions of ectoderm, lined by cuticle, which are respectively the *stomodaeum* or *fore gut*, and *proctodaeum* or *hind gut*. These may be short, but in the higher Crustacea and Insecta form a considerable part, and sometimes nearly the whole, of the canal. The cuticular lining

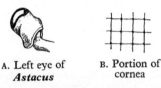

A. Left eye of *Astacus*

B. Portion of cornea

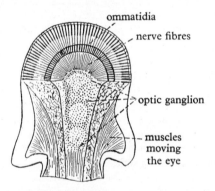

C. Longitudinal section of the eye

Fig. 221. The eye of *Astacus*.

of fore and hind gut is shed at moulting. The lining of the fore gut sometimes provides teeth for triturating or bristles for straining the food. Digestion is extracellular, save in certain acarina.

RESPIRATION. The respiration of aquatic arthropods, other than those which are but little modified from terrestrial ancestors, is sometimes, if the animal be small, effected only through the general integument of the body, but usually takes place by means of *gills* (branchiae). These are nearly always external processes, known as *epipodites*, which stand on the bases of the limbs, and are often branched or folded. Among terrestrial arthropods, some of the Arachnida possess *lung books*, which are generally held to have arisen by the enclosure of gill books, such as those on the limbs of *Limulus*, each within a cavity of the ventral side of the body. The remainder of the terrestrial Arthropoda breathe by means of *tracheae*, which are tubular involutions of the ectoderm and cuticle which convey air to the tissues. In some arachnids tracheae are present as well as lung books. Usually tracheae are branched, and strengthened by a spiral thickening of their chitinous lining. The study of the phylogeny of the Arthropoda leads to the conclusion that a tracheal system has arisen independently in the Onychophora, the Arachnida, and the Insecta and Myriapoda. Among the Crustacea, tufts of tubes which resemble tracheae are found in the abdominal appendages of woodlice.

VASCULAR SYSTEM. The vascular system is an 'open' one. That is, be the arteries long or short, they end by discharging their blood not into capillaries in the tissues from which veins conduct it to the heart, but into perivisceral cavities, known as *sinuses*, which bathe various organs. From these sinuses the blood collects into a *pericardial sinus* ('pericardium'), part of the haemocoelic system, which surrounds the heart. The latter is a longitudinal dorsal vessel, perforated by *ostia* by which it receives its blood from the pericardial sinus. Among the consequences of the structure of the vascular system are a low blood pressure and liability to severe bleeding from wounds. The latter danger is met, especially in the Crustacea, by very rapid clotting of the blood. Haemoglobin is present in the plasma of certain of the lower crustaceans and a few insects, haemocyanin in *Limulus*, scorpions, and some spiders.

COELOM. The coelom appears in the embryo as the cavities of a series of mesoderm segments ('mesoblastic somites', Fig. 359). It never assumes a perivisceral function, and in the adult is represented only by the cavities of the gonads and of certain excretory organs and occasional vestiges elsewhere.

EXCRETORY ORGANS. The excretory organs of arthropods are of very various kinds. True nephridia appear never to be present. *Coelomoducts* are present in a number of cases, though in the absence of perivisceral coelom they end internally each in a small coelomic vesicle or 'end sac'. These are found in the Onychophora in a long series of segmental pairs. In Crustacea there is either a pair of coelomoducts on the third (antennal) somite or a pair on the somite of the maxillae, or, rarely, both these pairs are present. In various crustaceans other glands, some ectodermal, some mesodermal, appear to have an excretory function, and sometimes replace both pairs of coelomoducts, which become vestigial. In arachnids, coelomoducts open on one or two of the pairs of legs. They are known as *coxal glands*, but are not homologous with the glands to which that name is applied in certain crustaceans. *Malpighian tubules* are tubular glands which open into the alimentary canal near the junction of mid and hind gut in the Arachnida, Insecta, and Myriapoda. In arachnids they are of endodermal origin, but in insects and myriapods they are part of the ectodermal hind gut. It is interesting that the subphyla differ in the nature of their nitrogenous excreta. In the Crustacea these are principally ammonia compounds and amines, in the Insecta they are urates, in the Arachnida guanine.

MUSCULAR SYSTEM. Nearly all the muscular tissue of arthropods is composed of striped fibres, but in *Peripatus* only the fibres of the jaw muscles are striped, and among the higher groups certain exceptions to the rule are known (some visceral muscles, etc.).

The innervation of the muscle system differs from that present in the vertebrates in three main respects.

(1) There are only a few nerve fibres running to the muscles. In the verte-

brates several hundred nerve fibres run from the ventral root to a single muscle. In insects and crustaceans only two to five nerves supply a muscle.

(2) In addition to motor and sensory nerves, the crustaceans at any rate, amongst the arthropods, possess an inhibitory nerve fibre running to the muscles. This nerve inhibits muscle contraction.

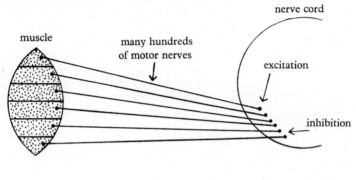

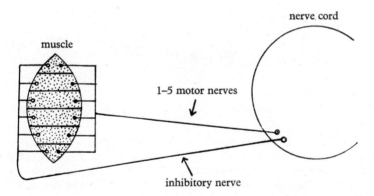

Fig. 222. Nerve-muscle systems. The arthropod system differs from that of the vertebrates in the number of nerves supplying the muscle, the possession of an inhibitory nerve, and the multiple innervation of each muscle fibre.

(3) In the vertebrates each single muscle fibre has only one motor end-plate; in the arthropods each muscle fibre has several motor end-plates which allow local contractures of the muscle fibre (Fig. 222).

GONADS. The gonads are always, owing to the reduction of the coelom, directly continuous with their ducts, which are probably coelomoducts.

328 ARTHROPODA

These have no constant position of opening in the phylum. In the Crustacea they nearly always open at the hinder end of the thorax. In the Arachnida their opening is similarly near the middle of the body. In the Onychophora, Insecta, and centipedes they open near the hinder end, but in the remaining groups of the Myriapoda their opening is not far behind the head.

EMBRYONIC DEVELOPMENT. Typical features of the embryonic development are shown in Figs. 223, 248. The ova are generally yolky, and their cleavage is typically of the kind known as 'centrolecithal', in which (Fig. 223 A) the products of division of the nucleus come to lie in a layer of protoplasm upon the surface of a mass of yolk which thus occupies the position of a

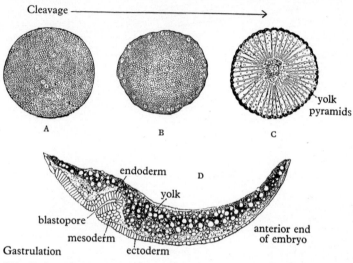

Fig. 223. Early stages in the development of *Astacus*. (After Morin and Reichenbach.

blastocoele. The mode of gastrulation varies from invagination (Fig. 223 D) to obscure processes of immigration and delamination. The formation of the mesoblast as a pair of ventral bands (Fig. 314), proliferated in primitive cases from behind, has already been mentioned (p. 127). As in annelids (p. 291), the mesoblast bands segment, and in most cases the segments ('mesoblastic somites') develop coelomic cavities (Fig. 219). The haemocoele arises by separation of the germ layers. The heart is formed by the dorsal ends of the mesoblast segments approximating. The nerve cords are proliferated from the ventral ectoderm (Fig. 359). In spite of the yolky eggs, there is a great variety of larval stages, though direct development is also frequent. The series of somites, which in the adult is often obscured by the loss, obsolescence, or fusion of some of its members, is usually more distinct in the embryo or larva, where the presence of a somite which it is difficult or impossible to recognize at a later stage is frequently indicated by one or more of three criteria: a pair of segments of mesoblast (mesoblastic somites), a pair of segmental ganglia, and a pair of limbs or limb rudiments.

CHAPTER X

THE CLASSES ONYCHOPHORA AND TRILOBITA

The two groups of animals with which this chapter deals both present in an apparently primitive condition features which are characteristic of the phylum Arthropoda. One at least of them existed in the Palaeozoic period. For these reasons, each of them has been regarded as giving indications concerning the ancestry of the Arthropoda. Whereas, however, the Trilobita are related rather closely to the Crustacea and more distantly to the other classes, the Onychophora are, as has been stated above, widely divergent from the rest of the Arthropoda. Some authorities, indeed, prefer to treat this group as an independent phylum. It must at least be regarded as representing a branch which parted at a very early date from the main arthropod stock. The Trilobita are indisputable arthropods, on the line of descent which gave rise to the Crustacea and perhaps the other classes.

CLASS 1. ONYCHOPHORA

DIAGNOSIS. Tracheate Arthropoda with soft thin cuticle and body wall consisting of layers of circular and longitudinal muscles; head not marked off from the body, consisting of three segments, one preoral, bearing preantennae, and two postoral bearing simple jaws and oral papillae respectively, also with eyes which are simple vesicles; the remaining segments all alike, the number varying according to the species, each bearing a pair of parapodia-like limbs which end in claws and contain a pair of excretory tubules; spiracles of the tracheal system scattered irregularly over the body; cilia present in genital organs; development direct.

CLASSIFICATION. There are two families.

1. **Peripatidae.** Equatorial animals with 22–43 pairs of limbs. *Peripatus, Oroperipatus, Typhloperipatus, Mesoperipatus.*

2. **Peripatopsidae.** Australasian animals with 14–25 pairs of limbs. *Peripatopsis, Peripatoides, Ooperipatus, Opisthopatus, Paraperipatus.*

BIOLOGY. The animals which constitute this very important group are few in number and uniform in structure. There are in all seventy different species which are distributed discontinuously over the warmer parts of the world; Trinidad, Jamaica, Haiti, Puerto Rico, African Congo, Malaya, Borneo, Sumatra, South Africa, Australia, New Zealand.

They are found in very retired positions which are permanently damp as, for instance, beneath the bark of dead trees and under stones. They have a superficial resemblance to other crawling animals which are found in the same places, like myriapods, slugs and earthworms, and until their anatomy was well known were classified, by different investigators, with all three of these. Certain of the characters of *Peripatus* such as the feebly developed sense organs, the simple structure of the jaws and feet and the soft skin may be linked with the environment in which they lurk away from light and enemies. Yet it can hardly be doubted that the Onychophora are a division of the Arthropoda which has preserved more simple features of an ancestral race than any other living form, terrestrial or aquatic.

PRIMITIVE FEATURES. Such features are in all probability the thin cuticle, the muscular body wall, the small number of head segments, the complete series of segmental excretory organs, the presence of cilia and possibly also the parapodia-like limbs (Fig. 224).

Fig. 224. *Peripatus capensis*, slightly enlarged. (From Sedgwick.)

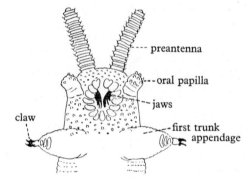

Fig. 225. *Peripatus capensis*, male; ventral view of anterior end. (After Sedgwick.)

EXTERNAL CHARACTERISTICS. The thinness of the cuticle is responsible for the absence of external segmentation (save for the repetition of the appendages). The head (Fig. 225) bears three pairs of appendages which are none of them very highly developed. While elsewhere in the arthropods the first segment is present in the embryo but disappears in the adult, here it persists and bears a pair of appendages which may be called *preantennae* to distinguish them from antennae. They are rather long and very mobile, but do not retract like the tentacles of the slug. The next segment bears the jaws, which are not unlike enlarged claws of the trunk appendages and so bite with the tip and not the side. They are, moreover, tucked in within the oral cavity, but they are borne on muscular papillae arising in the embryo and must without a doubt be regarded as appendages.

The trunk appendages are short, conical and hollow, bearing at their distal ends spinose pads and a retractile terminal foot with two recurved claws.

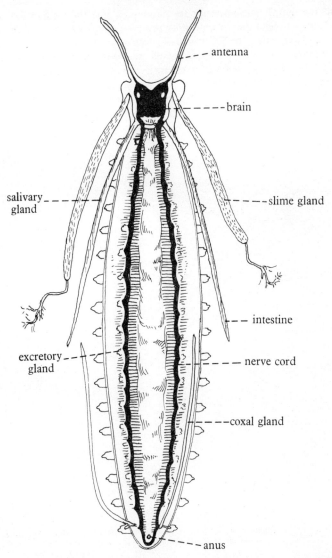

Fig. 226. *Peripatus capensis*, male; dissected to show internal organs. (After Balfour.)

The number of segments varies according to the species, in some of the Peripatidae the number is variable even in individuals of the same species. There is no tail.

BODY CAVITY. The adult body cavity is haemocoelic but the embryonic coelom is well developed. In the development of *Peripatus* just after the gastrula stage, the blastopore becomes elongated, the anterior part giving rise to the mouth, the posterior to the anus, whilst the median part closes (Fig. 229 B). Behind the blastopore is a primitive streak which forms the paired mesoblastic somites. The anterior pair move in front of the mouth and help to provide the mesoderm of the tentacular segment. None of the rest become preoral. In all segments the somites early acquire a cavity, the coelom, and later divide into two. Of these the ventral part migrates into the appendage as this is formed, and eventually becomes part of the segmental excretory

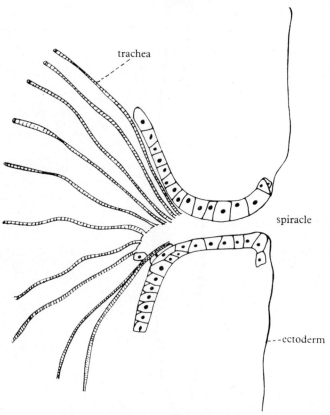

Fig. 227. *Peripatus capensis.* Section through tracheal pit and developing bundles of tracheal tubes. (After Balfour.)

organ. The other part approaches its fellow in a mid-dorsal position to form the heart lying between them (cf. Fig. 231 B) and while in the anterior region it mostly disappears, those of the posterior segments fuse longitudinally to form two tubes which become the gonads (Fig. 231 A, B, C).

At the same time the gaps between the organs become filled with blood and a dorsal part of the haemocoel so formed is marked off by a partition as the pericardium. This contains the heart, a long tube with a pair of ostia in nearly every segment. There are, however, no other blood vessels, so that the condition of the circulatory system is by no means so advanced as in the higher Crustacea and the more primitive Arachnida.

TRACHEA. The possession of the perivisceral haemocoel almost diagnoses the group as arthropods, but it was the discovery of the tracheae which led to the inclusion of *Peripatus* in that phylum. The spiracles are scattered over the surface of the body, most thickly on the sides and ventral surface, several occurring in each segment. Each spiracle leads into a pit, penetrating the muscle of the body wall, from which arise bundles of minute air-containing tubes which end in various organs of the body (Fig. 227). It can hardly be doubted that these tracheae are definitely arthropodan in type; their most significant difference from those of other forms is in their non-segmental character. Their irregular distribution is only possible because they originate as pits in soft skin; when once a cuticular exoskeleton has been established, tracheae can only be excavated in joints between segments. Probably then the Onychophora have never had a more definite cuticle than they possess at present; if they had, tracheae have been acquired since it was lost. The tracheae have no spiracle control hence they are unable to regulate the degree of air intake or of water loss through the tracheae. As a result of this the animals loose water twice as rapidly as an earthworm, forty times as rapidly as a caterpillar, and eighty times as rapidly as a cockroach, all under the same conditions. Thus though the skin is dry and less permeable than that of the earthworm, and though it has suitable breeding and feeding systems for living on land, and though it is uricotelic, it has been unable to depart from its cryptozoic life due to its lack of spiracular control.

DIGESTIVE SYSTEM. The alimentary canal consists of short ectodermal fore gut and hind gut and a very long endodermal mid gut, lined by a peritrophic membrane (p. 441) which is thrown off periodically. The fore gut consists of a buccal cavity into which open the large salivary glands, and a muscular suctorial pharynx. The mid gut possesses no separate glands.

EXCRETION. The excretory tubules (Fig. 228) are composed of a distal terminal bladder, a coiled secretory canal and a ciliated duct which opens into a much-reduced coelomic vesicle. The bladder is formed from ectoderm, the rest from mesoderm. It can perhaps be said then that the tubule is a modified coelomoduct which has attained its present condition by the tucking-in of ectoderm at its external opening. The tubules form a complete series but some of them have been converted into uses other than excretion. Thus the tubules corresponding to the oral papillae form the salivary glands and are much larger and more complex than in other segments. The anal glands and the gonoducts themselves have the same origin. Only the tubules corresponding to the jaws and the first three trunk segments disappear.

REPRODUCTIVE SYSTEM. The sexes are separate in *Peripatus* and the gonads paired, but the ducts unite to form a median passage opening just before the anus. In the male the filiform spermatozoa are bound up in spermatophores in the upper part of the vas deferens; the lower part is muscular and ejaculatory in function. Fertilization is usually internal but in *Peripatopsis capensis* the spermatophores have been seen deposited on the skin of the female, giving hypodermic impregnation similar to that observed in some leeches.

The ovaries are embraced by a funnel, the receptaculum ovorum, which communicates with an oval receptaculum seminis. The eggs are fertilized at the proximal end of the oviduct; they vary in size according to the species. In the larger, development takes place at the expense of the yolk and secre-

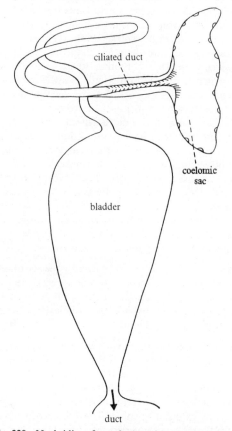

Fig. 228. Nephridium from the leg of *Peripatus capensis*.

tions of the uterine wall; but the embryos from the smaller eggs become attached to the uterine wall and a *placenta* is formed (Fig. 230). Cilia have been described in parts of the genital tract.

GLANDS. The crural glands are found on all the legs except the first and consist of a simple sac; these sacs are usually better developed in the male and for this reason are believed to play some role in reproduction. There is a single pair of slime glands which discharge from the oral papillae. They are made up of a much-branched secretory part and a large reservoir. The slime can be shot out to entangle an enemy. It is never used in obtaining food.

NERVOUS SYSTEM. The nervous system (Fig. 226) consists of a pair of supraoesophageal ganglia from which the preantennal nerves are given off, a pair of circumoesophageal commissures, and two ventral cords which are widely separated and connected by about ten transverse strands in each

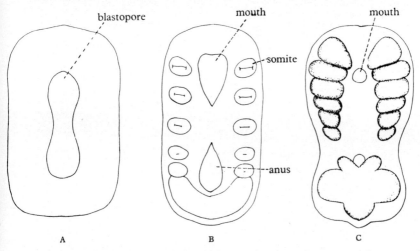

Fig. 229. Diagrams of the cleavage and early embryonic development of *Peripatus capensis*.

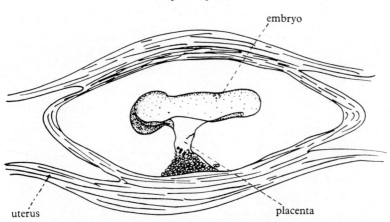

Fig. 230. Placenta of *Peripatus trinitatis*.

segment. There are slight enlargements in each segment which can be regarded as incipient ganglia, but the whole nervous system is primitive for an arthropod or even an annelid and can best be compared to that of *Polygordius* in the archiannelids or *Chiton* in the molluscs.

SENSE ORGANS. The animal possesses several series of papillae on the surface of the body which may act as tactile and chemoreceptors. The only

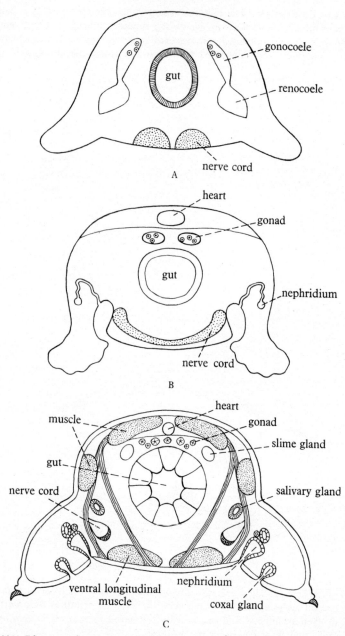

Fig. 231. Diagrams of transverse sections through developing *Peripatus capensis* to show the growth of the haemocoele and coelom.

specialized sense organs possessed by *Peripatus* are the pair of simple eyes. Each eye is a hollow vesicle, the cavity of which is filled by a cuticular lens. The eye is developed as an invagination of the embryonic brain, whilst the nervous system is still part of the ectoderm.

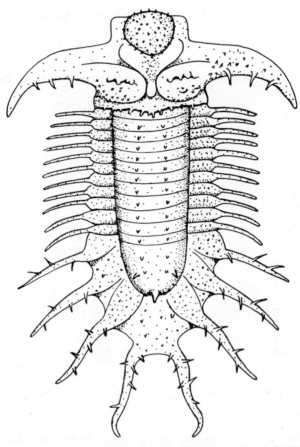

Fig. 232. *Terataspis*. From the Devonian of New York. Actual size 27 in. (From Shrock and Twenhofel, after Reimann.)

AFFINITIES. A few fossil forms have been described though some are only doubtful onychophorans. The most certain fossil is *Aysheaia* found in the Cambrian of North America. Though it has onychophoran features it is always found in association with marine fossils. If it is related to *Peripatus* it is a member of a line that gave rise to early Onychophora. In general the Onychophora came off from the main arthropodan stock at a very early stage when a typical haemocoel had been developed and cephalization had commenced, but when the epithelium had not finally specialized in the production of thick chitin (chitin is present in the annelids in thin layers) and was still ciliated in places.

CLASS 2. TRILOBITA

DEFINITION. Palaeozoic Arthropoda with the body moulded longitudinally into three lobes; one pair of antennae; and, on all the postantennal somites, appendages of a common type which has two rami and a gnathobase.

BODY FORM. The Trilobita were marine organisms and were very numerous in the Cambrian and Silurian but became extinct by the Secondary

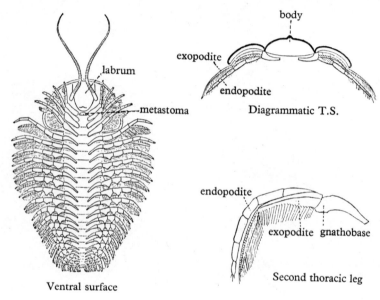

Fig. 233. *Triarthrus becki* from the Utica Slate (Ordovician) near Rome, New York. (After Beecher.)

period. The average size was quite small (2 inches) though one or two reached a length of 2 feet (*Terataspis*). Their body was oval and depressed, and consisted of a head and a segmented trunk, of which the anterior somites were movable on one another, but the hindermost, in varying number, were nearly always united to form a tagma known as the *pygidium*. The body could usually be rolled up like that of a woodlouse. Along its whole length longitudinal grooves divided lateral *pleural* portions from a middle region. In the head, this middle region is known as the *glabella* and transverse furrows usually mark out more or less distinctly five somites. The pleural portions of the head are known as the *cheeks*, and each bears in most species a sessile compound eye. On each cheek a longitudinal *facial suture* divides an outer from an inner area, passing immediately internal to the eye. The postero-

TRILOBITA

lateral angles of the cheeks are often produced backward as spines. Under the head a large *labrum* or *hypostoma* projects backward below the mouth, behind which is a small *metastoma*.

LIMBS. The antenna is uniramous and multiarticulate and is the only pre-oral appendage. Since it is the foremost of five head appendages it has the same position as the antennule of the Crustacea, with which it is probably homologous. In that case it would seem likely that a true first somite had already, as in modern crustacea, become merged in the anterior region of the head. Traces of a groove which exist in some species may perhaps indicate its existence.

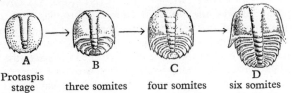

A
Protaspis stage

B
three somites

C
four somites

D
six somites

Fig. 234. Development of a trilobite, *Sao hirsuta*, Cambrian, Czechoslovakia. A–C, × 12; D, × 10. (After Barrande.)

The remaining limbs are all of one type, though there is a gradual progressive modification from one end of the series to the other. Each has two rami. Of these, one, usually held to be the outer ('exopodite'), bears a long fringe of bristles, while the other ('endopodite') is leg-like and divided into six joints. It is supposed by some authorities that the bristle fringe on the so-called exopodite was on the inner side of the limb, and was used for collecting food, like the fringes on the trunk limbs of branchiopoda (p. 371), but this surmise is not generally accepted. From the basal portion of the limb a process for the manipulation of food, the *gnathobase*, projects towards the middle line. The configuration of the basal portion (protopodite), and the relation of the rami to it, are obscure. The telson is without limbs.

DEVELOPMENT. The Trilobita hatched as a larva, the *Protaspis*, which was subcircular, and consisted, like the *Nauplius* larva of the Crustacea, principally of head. In its further development there appear, first the pygidium, and then free somites between the pygidium and the head, new somites being added in front of the telson while those at the front end of the pygidium become free.

It is probable that the majority of the trilobites lived upon the sea-bottom in shallow to moderately deep water, but others appear to have been adapted to burrowing, pelagic, and deep-sea conditions.

Examples are: *Terataspis, Olenellus, Triarthrus.*

CHAPTER XI

THE CLASS CRUSTACEA

DEFINITION. Arthropoda, for the most part of aquatic habit and mode of respiration; whose second and third somites bear antennae; and their fourth somite a pair of mandibles.

FACTORS IN THE EVOLUTION OF THE CRUSTACEA. (1) *Specialization of limbs*. The Crustacea are essentially aquatic arthropods. That fact alone makes it possible that in them the same appendages should combine the functions of locomotion (by swimming), feeding (by gathering particles from the water), respiration (by exposing a thinly covered surface to the medium), and the reception of sensory stimuli. There is perhaps no extant crustacean in which all four functions are thus combined—unless we may regard the trunk limbs of the Branchiopoda (see below) as sense organs in a minor degree—but not uncommonly three, and perhaps usually two, are performed by the same limb. In the lowest members of the class—the 'phyllopod' Branchiopoda (such creatures as the fairy shrimp, *Chirocephalus*, shown in Fig. 250)—a long series of somites of the trunk bear similar appendages which all function alike in swimming, respiration, and the gathering of food. Evolution within the crustacean group appears to have proceeded mainly by the specialization, for particular functions, of particular appendages of an ancestor which possessed along the whole length of the body a numerous series of limbs, of which all, except probably the first pair (antennules), were as much alike and capable of at least as many functions as those which the Branchiopoda now possess upon the trunk. Such a condition existed in the Trilobita, but in all modern crustacea the appendages of the head are already specialized for various uses, and in most members of the group the specialization has gone farther. Moreover, it has taken place in more than one way. Limbs which in one crustacean are adapted to some particular function are in others specialized for quite different services.

(2) *Shortening of the body*. Two other factors, added to, or perhaps consequent upon, the specialization of limbs, have taken part in bringing about the great variety of organization which exists in the Crustacea. One is a shortening of the body. As the efficiency of the limbs increases by specialization, there occurs a lessening of their number, and finally the reduction or loss of the somites whose limbs have thus disappeared. The reduction, which has occurred independently in every class, has taken place in the hinder part of the body, though as a rule the extreme hind end (telson) is relatively unaffected.

(3) *Development of carapace*. The other factor is the development, from the hinder part of the head, of a skin fold—the *carapace*—by which the important anterior region of the body is overhung and protected, and the setting up in the surrounding water of currents for purposes of respiration and feeding is facilitated. Not all crustaceans possess the carapace: in some

it has perhaps never existed, others have discarded it. In those which have it, its extent varies: in extreme cases it encloses the whole body.

The transformation of the external make-up of the body is of course reflected in the internal organization, which shows corresponding concentrations of function and differentiation of the contents of somites.

CLASSIFICATION

Subclass 1. BRANCHIOPODA. Free crustaceans with compound eyes; usually a carapace; at least four pairs of trunk limbs which are in most cases broad, lobed and fringed on the inner edge with bristles.

> *Order* 1. **Anostraca.** No carapace; stalked eyes; antenna of fair size but not biramous; trunk limbs numerous and all alike; caudal rami unjointed. *Chirocephalus, Artemia*
>
> *Order* 2. **Lipostraca.** Fossil order represented by *Lepidocaris*.
>
> *Order* 3. **Notostraca.** Carapace a broad shield above trunk; compound eyes sessile and close together; trunk limbs numerous, the first pairs differing considerably from the rest; multiarticulate caudal rami. *Apus, Lepiduris*
>
> *Order* 4. **Diplostraca.** Compressed carapace enclosing trunk and limbs; compound eyes sessile and apposed or fused; antenna large and biramous; 4–27 pairs of trunk limbs usually considerably differentiated.
>
>> *Suborder* 1. **Conchostraca.** 10–27 pairs of trunk limbs. *Estheria*
>>
>> *Suborder* 2. **Cladocera.** 4–6 pairs of trunk limbs.
>> *Daphnia, Sida, Leptodora, Polyphemus*

Subclass 2. OSTRACODA. Free crustaceans with a bivalve shell and an adductor muscle; and not more than two recognizable pairs of trunk limbs, these not being phyllopodia. *Cypris, Cypridina*

Subclass 3. COPEPODA. Free or parasitic crustaceans without compound eyes or carapace, typically six pairs of thoracic limbs of which the first is always and the sixth is often, uniramous, the rest biramous. No limbs situated on the abdomen.
Calanus, Chondracanthus, Lernaea, Cyclops

Subclass 4. BRANCHIURA. Crustacea temporarily parasitic on fishes, with compound eyes, a suctorial mouth, carapace-like expansions of the head, unsegmented limbless abdomen. *Argulus*

Subclass 5. CIRRIPEDIA. Fixed, for the most part hermaphrodite, crustaceans, without compound eyes in the adult, with a carapace (except in rare instances) as a mantle which encloses the trunk; usually with a mandibular palp and typically six pairs of thoracic limbs.

> *Order* 1. **Thoracica.** Cirripedia with alimentary canal, six pairs of thoracic limbs, no abdominal somites, permanently attached by the preoral region. *Lepas*

Order 2. **Acrothoracica**. Separate sexes; have an alimentary canal, less than six pairs of thoracic limbs, no abdominal somites. Permanently sessile. *Alcippe*

Order 3. **Apoda**. Hermaphrodite, no mantle, no thoracic limbs, no anus, the body divided by constrictions into rings. *Proteolepas*

Order 4. **Rhizocephala**. Parasitic on decapod crustaceans; never have an alimentary canal; adult has no appendages, develops fungus-like roots which penetrate into host.
Sacculina, Thompsonia

Order 5. **Ascothoracica**. Parasitic, with alimentary canal and six pairs of thoracic appendages. *Laura*

Subclass 6. **MALACOSTRACA**. Crustacea with compound eyes, usually stalked; typically a carapace covering the thorax; a thorax of eight somites; an abdomen of six somites (rarely seven); all except the rare seventh abdominal somite bear appendages.

Order 1. **Leptostraca**. Have seven abdominal somites; phyllopodia, large carapace not fused to any thoracic somite. *Nebalia*

Order 2. **Hoplocarida**. Shallow carapace fused to three thoracic somites. First five thoracic limbs subchelate. *Squilla*

Order 3. **Syncarida**. No carapace. *Anaspides, Bathynella*

Order 4. **Peracarida**. Carapace does not fuse with more than four thoracic segments. Have oostegites.

Suborder 1. **Mysidacea**. Carapace covers most of thoracic segments. *Mysis*

Suborder 2. **Cumacea**. Carapace covers only three or four segments. *Diastylis*

Suborder 3. **Tanaidacea**. Carapace small, covers only two thoracic segments. *Apseudes, Tanais*

Suborder 4. **Isopoda**. No carapace, body dorso-ventrally flattened.
Ligia, Armadillidium, Idotea

Suborder 5. **Amphipoda**. No carapace, body laterally flattened.
Gammarus, Caprella, Phronima

Order 5. **Eucarida**. Carapace fused to all the thoracic segments. No oostegites.

Suborder 1. **Euphausiacea**. Small scaphognathite, no statocyst.
Nyctiphanes

Suborder 2. **Decapoda**. Big scaphognathite, statocysts present.
  1. **Macrura**. Abdomen hard, long and extended. *Astacus*
  2. **Anomura**. Abdomen soft. *Eupagurus*
  3. **Brachyura**. Abdomen hard, short and folded beneath the body. *Carcinus*

GENERAL CLASSIFICATION OF CRUSTACEA. The specialization of the limbs, shortening of the body and development of the carapace has given rise to six subclasses of Crustacea. We must now briefly survey them.

(1) BRANCHIOPODA. In the Branchiopoda feeding is performed by the limbs of the trunk. In the 'phyllopod' groups of this subclass, mentioned above, it is only on the head that differentiation among the appendages has proceeded to any considerable extent. Of the head limbs each, as we have seen, is specialized for some particular function, such as the service of the senses or the manducation of food. On the trunk the limbs, which are numerous, are still similar and all subserve at least the functions of feeding and respiration. In the order *Anostraca*, to which *Chirocephalus* belongs, there is no carapace, and the trunk limbs, whose similarity is very strong, retain the function of swimming. In the order *Notostraca* (Fig. 251), also phyllopodous, there is a carapace but it is wide and shallow and does not enclose the trunk limbs, and they are still used for swimming. A certain degree of differentiation exists between these limbs, the anterior pairs for instance being capable of clasping objects. In both the foregoing orders limbs have been dispensed with on some of the hinder somites. The remaining phyllopod group, the *Conchostraca* (Fig. 252), are united with the non-phyllopod group Cladocera as the order *Diplostraca*. In the members of that order (except a few aberrant Cladocera) the carapace encloses the trunk limbs, which are not used for swimming, that function being taken over by the antennae. The Conchostraca alone among branchiopods retain limbs on all their trunk somites like the trilobites, but as in the Notostraca there is a certain degree of differentiation between the members of the series. In the *Cladocera* (Fig. 253), the highest group of the Branchiopoda, a compact and very efficient feeding apparatus is formed by some half-dozen pairs of limbs; the trunk is correspondingly shortened, and even so some of the hinder somites are limbless. In certain members of this group, such as the water-flea *Daphnia*, there is a high degree of differentiation between the trunk limbs (Fig. 254).

(2) OSTRACODA. A similar habit of body is even more strongly developed in the subclass Ostracoda (Fig. 258) which are very short-bodied and completely enclosed in a bivalve shell formed by the carapace. Whereas, however, in the Cladocera it is by trunk limbs that food is gathered, in the Ostracoda that function is performed by limbs of the head. The trunk limbs, which have lost the functions of swimming and respiration as well as that of feeding, serve relatively unimportant subsidiary purposes, and are reduced, at most, to two pairs. Some members of the class carry shortening to an extreme pitch by contriving to dispense with one or both of these pairs.

(3) The members of the subclass COPEPODA (Fig. 259) also feed by means of appendages on the head, though they use these differently from the Ostracoda. In contrast to that group they have no carapace, and they have retained a trunk of some ten somites, of which the first half-dozen bear limbs which are specialized organs of swimming. The hinder part of the trunk is without appendages, save a pair of styles on the telson, often shows coalescence of somites, and may become a mere stump. Some of those members of this class which are parasitic lose in the adult female the segmentation and most, or even all, of the appendages.

Table 7. Somites a...

Somite	*Apus* (Notostraca)		*Daphnia* (Cladocera)	*Cypris* (Ostracoda)	*Cyclops* (Copepoda)
1...	No limbs		No limbs	No limbs	No limbs
2...	Antennules		Antennules	Antennules	Antennules
3...	Antennae		Antennae	Antennae	Antennae
4...	Mandibles		Mandibles	Mandibles	Mandibles
5...	Maxillules		Maxillules	Maxillules	Maxillules
6...	Maxillae		Maxillae (ves.)	Maxillae	Maxillae
7...	Thor. limbs	1	Thor. limbs 1	Thor. limbs 1	Maxillipeds
8...	,,	2	,, 2	,, 2 ♂ ♀	Thor. limbs 2
9...	,,	3	,, 3		,, 3
10...	,,	4	,, 4		,, 4
11...	,,	5	—		,, 5
12...	,,	6	,, 5		,, 6 (c)
13...	,,	7	Thor. som. 7 ♀		a { Thor.som.7(d) ♂ ♀
14...	,,	8	,, 8		Abd. som. 1
15...	,,	9	,, 9		,, 2
16...	,,	10			,, 3
17...	,,	11 ♂ ♀	No true ab-	No abdominal	
18...	Abd. som.	1	dominal so-	somites or limbs	
19...	,,	2	mites. Last 3		
20...	,,	3	somites (13–15)		
21...	,,	4 — 2 to 5 pairs of limbs to each somite	are limbless and called abdomen		
34...	,,	17			
35...	,,	18			
36...	,,	19 — No limbs			
37...	,,	20			
38...	,,	21			
39...	,,	22			
	Telson with rami		Telson ♂ with rami	Telson with rami	Telson with rami

limbs of Crustacea

Somite	*Lepas* (Cirripedia)	*Nebalia* (Leptostraca)	*Gammarus* (Amphipoda)	*Astacus* (Decapoda)
1...	No limbs	No limbs	No limbs	No limbs
2...	Antennules	Antennules	Antennules	Antennules
3...	Lost in adult	Antennae	Antennae	Antennae
4...	Mandibles	Mandibles	Mandibles	Mandibles
5...	Maxillules	Maxillules	Maxillules	Maxillules
6...	Maxillae	Maxillae	Maxillae	Maxillae
7...	Thor. limbs 1 ♀	Thor. limbs 1	Maxillipeds	Maxillipeds I
8..	,, 2	,, 2	Legs I	,, II
9...	,, 3	,, 3	,, II	,, III
10...	,, 4	,, 4	,, III	Legs I
11...	,, 5	,, 5	,, IV	,, II
12...	,, 6 ♂	,, 6 ♀	,, V ♀	,, III ♀
13...		,, 7	,, VI	,, IV
14...		,, 8 ♂	,, VII ♂	,, V ♂
15...	No abdominal	Abd. limbs 1	Abd. limbs 1	Abd. limbs 1
16...	somites or limbs	,, 2	,, 2	,, 2
17...		,, 3	,, 3	,, 3
18...		,, 4	,, 4	,, 4
19...		,, 5	,, 5	,, 5
20...		,, 6	,, 6	,, 6
21...		Abd. som. 7		
	Telson with rami	Telson with rami	Telson	Telson

'Abdominal' somites are those between the last genital somite and the telson. ♂ indicates the position of the male opening, ♀ that of the female. *a*, joined but distinct. *b*, fused in female. *c*, uniramous and vestigial. *d*, genital operculum of female represents the thoracic limb.

346   CRUSTACEA

(4) In the small subclass of parasites known as BRANCHIURA (Fig. 263) which are sometimes placed in the Copepoda, but differ from that group in possessing compound eyes and in other important respects, there are carapace-like lobes at the sides of the head, but these do not enclose the trunk, and the general build of the body and the form and function of the thoracic limbs simulate those of a copepod. The abdomen is much reduced.

(5) The subclass CIRRIPEDIA or barnacles, which as larvae attach themselves by their antennules to some object upon which they henceforward lead a sedentary life under the protection of a large, mantle-like carapace, bear, upon the same trunk somites as do the Copepoda, limbs which, like those of the latter group, are biramous. These appendages, however, are used, not for swimming, but for gathering food-particles from the water; while of the head

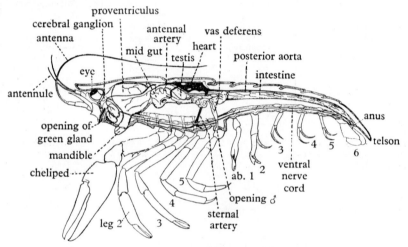

Fig. 235. *Astacus fluviatilis*, dissected from the left-hand side.
(From Shipley and MacBride.)

appendages the antennae are absent and the others are much reduced and not used in gathering food. The least specialized members of this subclass are, in respect of segmentation and appendages, on a par with the best-segmented of the Copepoda. Most cirripedes, however (the ordinary barnacles, Fig. 264), have lost the whole of the hinder (abdominal) region of the trunk. Others are deficient in the appendages of further somites, and the series ends with the sac-like parasites of the order *Rhizocephala* (Fig. 267).

(6) The subclass MALACOSTRACA (the highest crustaceans, including various 'shrimps', slaters, sandhoppers, crayfishes, etc.) obtain their food with the limbs on the anterior region (thorax) of the trunk, and, in primitive cases in which it is gathered as particles, strain it from the water with the last pair of appendages of the head (the maxillae). The thoracic limbs retain also the function of locomotion and normally are adapted for respiration by the presence upon them of gills, which are usually protected by a carapace of moderate size. Thus this region of the body of the Malacostraca is, in its own

ways, as many-functioned as the corresponding part of the trunk of *Chirocephalus*. The Malacostraca maintain in typical cases (Fig. 269) the swimming function of the limbs on the hinder portion (abdomen) of the trunk, and some of the subclass have found other uses (ovigerous, copulatory, etc.) for these appendages. Accordingly there is seldom any reduction in the fixed number of fourteen (or fifteen) trunk somites which, arranged always in a thorax of eight and an abdomen of six (or seven), characterizes the class. Nevertheless in all but one of the orders the abdomen has lost a somite, in the crabs (Fig. 278) and some others of the highest suborder (*Decapoda*) it is reduced, and in a few members of the subclass it is a limbless and unsegmented stump.

ENTOMOSTRACA. The name Entomostraca was formerly used in the classification of the class, to distinguish from the Malacostraca a division containing all the other subclasses. Since, however, these differ from one another as widely as each of them does from the Malacostraca, the name is no longer used in classification but is only a convenient designation for the lower crustacean subclasses as a whole.

GENERAL STRUCTURE

We must now proceed to review in more detail the common organization of the Crustacea and the variation which it presents throughout the group.

CUTICLE. The cuticle of a crustacean is, save for the joints, usually stout relative to the size of the animal, but is thinner and flexible in many parasitic genera. It is often strengthened by calcification, and in certain ostracods, barnacles and crabs this gives it a stony hardness. In each somite there may or may not be distinguishable the dorsal plate or *tergite* (*tergum*) and ventral *sternite* (*sternum*) usual in arthropods. The tergite may project at each side as a *pleuron* (Fig. 246).

SOMITES. There are embryological indications that the *body* should be regarded as containing, besides the *somites*, an anterior *presegmental region*, to which the eyes belong, corresponding to the prostomium of a worm, and a *postsegmental region* or *telson*, on which the anus opens. Each somite, except the first, which is purely embryonic, may bear a pair of *appendages*, though it is rarely that the appendages of all the somites are present at the same time. The somites never all remain distinct in the adult. Always some of them are fused together and with the presegmental region so as to form a head, and often there is also fusion of them elsewhere.

TAGMATIZATION: HEAD. Nearly always the somites are grouped into three tagmata, differentiated by peculiarities of their shape or appendages, and known as the head, thorax, and abdomen. These, however, are not morphologically equivalent in different groups.

The *head* always contains, besides the region of the eyes and the embryonic first somite, the somites of five pairs of appendages—two, the antennules and antennae, preoral; and three, the mandibles, maxillules, and maxillae, postoral. More somites are often included in the actual head, but as the additional appendages (maxillipeds) then usually show features of transition to those behind them, and as the fold of skin which forms the carapace first arises from the maxillary somite, the true head is held to consist only of the anterior portion of the body as far as that somite inclusive. There is evidence of an earlier head, carrying only the first three pairs of limbs which alone exist in the *Nauplius* larva, and still indicated in some cases (as in *Chirocephalus*, *Anaspides*, Fig. 269 B, and *Mysis*, Fig. 270 A), by a groove which crosses the cheek immediately behind the mandible. This *mandibular groove* is distinct from the

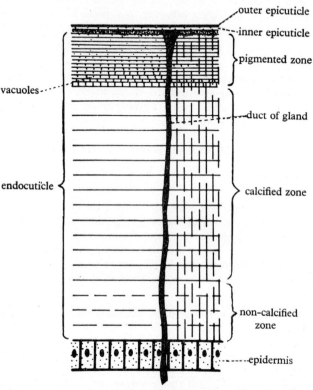

Fig. 236. Crustacean cuticle. In general it is very similar to the insect cuticle. (From Dennell.)

true *cervical groove* which often (as in *Astacus*, Fig. 277) marks the boundary between head and thorax: the two grooves may co-exist, as in *Apus* and in *Nephrops*. The Crustacea, indeed, admirably illustrate the way in which the process of 'cephalization' tends, in arthropods as in vertebrates, to extend backwards and to involve more and more segments. With it has gone a backward shifting of the mouth, which in the Crustacea now stands behind the third somite, with two pairs of appendages (antennules and antennae) in front of it. The commissure which unites the ganglia of the antennae still passes behind the mouth, and may usually be seen, as in *Astacus* (Fig. 240), crossing from one of the circumoesophageal commissures to the other. The head of the

Crustacea is unlike, and less specialized than, those of other arthropods in that its limbs are not entirely restricted to sensory and alimentary functions but often have also other uses, such as swimming, the setting up of currents, or prehension.

THORAX AND ABDOMEN. The head, though it varies in extent, is of the same nature throughout the group, being primarily, like the heads of other animals, the seat of the principal organs of special sense and of manducation. On the other hand, the two tagmata known as the *thorax* and *abdomen*, which usually can be recognized in, and together compose, the post-cephalic part of the body or *trunk*, vary much more in extent, and each of them has in the several groups no constant feature save its position relative to the other. The precise boundary between thorax and abdomen is sometimes difficult to fix. The names, as they are commonly used, are in this respect inconsistently applied, denoting in some groups limb-bearing and limbless regions, in others the sections of the trunk which lie before and behind the genital openings. For the sake of consistency we shall adopt the convention that the somite which bears the genital openings (or the hinder such somite when, as sometimes happens, the male opening is on a somite behind that of the oviduct) is always the last somite of the true thorax. In this sense, in certain cases (copepods, cladocera), somites which are commonly called abdominal are strictly to be reckoned as thoracic. In respect of segmentation the trunk varies from the condition of a limbless stump in certain ostracods to the possession of more than sixty somites in some of the Branchiopoda.

CARAPACE. A structure very commonly found in crustaceans is the *shell* or carapace, a dorsal fold of skin arising from the hinder border of the head and extending for a greater or less distance over the trunk. Its size varies greatly. In the Ostracoda (Fig. 258) and most conchostracans (Fig. 252) it encloses the whole body, extending forwards at the sides so as to shut in the head. In other cases (cirripedes, Fig. 264, most cladocera, Fig. 253), it only leaves part or the whole of the head uncovered. In typical malacostraca it covers the thorax (Fig. 277), but in some it is a short jacket, leaving several thoracic somites uncovered (Fig. 270D), and in some (the Syncarida, Isopoda, and Amphipoda, Figs. 271–275) it has disappeared. In the Anostraca (Fig. 250) and Copepoda (Fig. 259) it was perhaps never present. It may be a broad, flat shield over the back, as in *Apus* (Fig. 251), but is usually compressed, and in the Conchostraca and Ostracoda becomes truly bivalve, with a dorsal hinge. In the Cirripedia it is an enveloping mantle, usually strengthened by shelly plates (Fig. 265). In the Conchostraca, Ostracoda, Leptostraca, and Cirripedia it has an adductor muscle, but the adductors of these groups vary in position and are not homologous. The carapace may fuse with the dorsal side of some or all of the thoracic somites (the Cladocera, most of the Malacostraca): such somites are not on that account alone to be regarded as included in the head, though they may become so. The *chamber* enclosed by the carapace is known in various cases by various names as gill chamber, mantle cavity, etc., and performs important functions in sheltering gills or embryos, directing currents of water which subserve feeding or respiration, etc. In front, the carapace is continuous with the dorsal plate which represents the terga of the head, the cervical groove, if present, marking the boundary between them. We shall apply the term *dorsal shield* to the structure composed of the dorsal plate of the head with the carapace, if the latter be present. These terms have been used in various senses. In the usage here proposed, when there is no carapace fold, the dorsal shield is the dorsal plate of the head together with the terga of the somites that are fused with the head.

The dorsal plate of the head may be prolonged in front as a projection which is called the *rostrum* (Fig. 277).

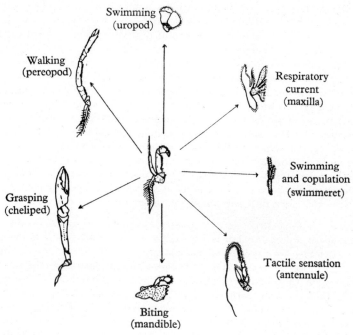

Fig. 237. Adaptive radiation of the crustacean appendage. The different limbs of *Astacus* are arranged according to their basic function. They may be regarded as derived from a central type such as the third maxilliped.

DORSAL ORGAN. A glandular patch or patches on the dorsal surface of the head, near its hinder limit, in many of the Branchiopoda, in *Anaspides*, and in the young stages of various other crustaceans, is known as the *dorsal organ* or *neck gland*. It is used by cladocera and conchostraca for temporary fixation. In other cases its function is not known. Possibly the organs to which this name is given are not all homologous. They must not be confused with the 'neck organ' of branchiopods (see p. 357).

APPENDAGES. Of the appendages or *limbs* of the Crustacea, the first, or antennule, is a structure *sui generis*, not comparable in detail with any of the others. Typically it is uniramous, and though in many of the Malacostraca it has two rami, these are probably not homologous with the rami, described below, of other appendages. The remaining limbs may all be reduced to one or other of two types—the 'biramous' limb usually so-called, to which most of them more or less clearly conform, and the *phyllopodium*, to which belong the trunk limbs of the Branchiopoda and some other appendages, chiefly maxillules and maxillae and notably the maxilla of the Decapoda. The name by which the first of these types is generally known refers to the fact that limbs which best represent it fork distally into two rami. Since, however, the phyllopodium possesses the same two rami, and bears them, though not as a distal fork, yet in the same way as a great number of limbs of the first type, it is well not to use a name which might imply that there is a constant difference in respect of the rami between the limbs of the two types. We shall therefore call the first type the *stenopodium*, referring to its usually slender form (Greek στενός, narrow).

STENOPODIUM. In the stenopodium (Fig. 238 A, H) the two rami—an inner *endopodite* and an outer *exopodite*—are set upon a common stem, the *protopodite*. In many cases the protopodite bears also, on its outer side, one or more processes known as *epipodites* (Fig. 238 A). In limbs in which the type is most perfectly developed the two rami are subequal and are borne distally upon the protopodite (Fig. 238 H) but in most cases the endopodite is the larger, and forms with the protopodite an axis, the *corm*, on which the exopodite stands laterally (Fig. 238 A, G, I). In a few instances the exopodite is the larger.

PHYLLOPODIUM. The phyllopodium (Figs. 238 B, C, D; 254 C), is a broader and flatter limb than the majority of stenopodia. Its cuticle is usually thin, and then the shape of the limb is maintained largely by the pressure of blood within it. In these cases the flexibility is such that no joints are needed. There is in this limb an axial portion or corm which bears on the median side a row of lobes known as *endites*, and on the outer side one or more lobes known as *exites*. Of the latter the more distal, standing usually opposite the third or fourth endite from the base and often known as the *flabellum*, is the homologue of the exopodite of the biramous limb. Exites proximal to this are epipodites. Of the *endites*, that which stands at the base of the limb is usually different in form from the rest and used in one way or another for manipulating the food. It is known as the *gnathobase*.

A limb of either type may vary by the lack of any of its parts. Notably the loss of the exopodite is liable to produce from either a *uniramous limb*. Moreover, though the two types are very distinct in cases in which they are perfectly developed, as in the swimmerets of *Astacus* (Fig. 238 H) and the trunk limbs of *Apus* (Fig. 238 B), there are many limbs which depart more or less from either type in the direction of the other—as, for instance, from the stenopodial type in the shape of the exopodite (Fig. 263 B), or, as stated above, in the relation of the latter to the rest of the limb, or from the phyllopodium in the proportions of the rami or the reduction of the endites.

PRIMITIVENESS OF THE PHYLLOPODIUM AND STENOPODIUM. The comparison just made between the phyllopodium and the stenopodium leaves untouched the question which of them is the more primitive, that is, more resembles the limbs of the ancestral crustacean. On this point there is

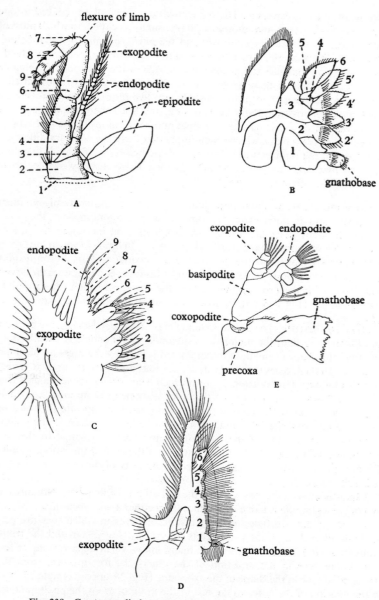

Fig. 238. Crustacean limbs:
A, second thoracic limb of *Anaspides* (Syncarida);
B, tenth thoracic limb of *Apus* (Notostraca);
C, maxilla of *Mysis* larva of *Penaeus* (Decapoda);
D, second trunk limb of female *Cyclestheria hislopi* (Conchostraca);
E, mandible of *Calanus* (Copepoda);

an old and as yet unsettled controversy. As proof of the primitiveness of the stenopodium it is pointed out (1) that this limb is more widespread than the phyllopodium, (2) that it occurs in the *Nauplius* larva (p. 366), the early phyllopod *Lepidocaris* (p. 373), and the trilobites, in all of which it is likely to

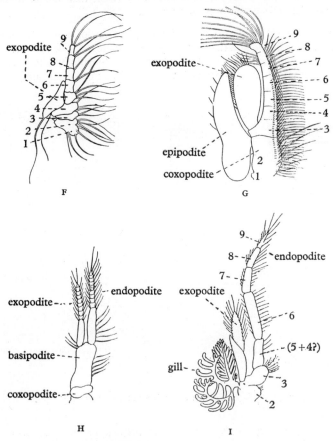

Fig. 238. Crustacean limbs:
F, maxilla of larva of *Sergestes arcticus* (Decapoda);
G, thoracic limb of *Nebalia* (Leptostraca);
H, swimmeret of a crayfish;
I, mid thoracic limb of an euphausid;
1–9 = endites or segments of the limb.

be primitive, (3) that it more nearly approaches the form of the majority of parapodia of the Annelida, from which the Crustacea are held to have taken origin. In demonstration of the ancestral nature of the phyllopodium it is urged (1) that typical stenopodia with subequal rami borne distally upon a protopodite are comparatively rare and usually occur in highly specialized crustaceans (Copepoda, Cirripedia; Malacostraca) (2); that the biramous limbs of the *Nauplius* and *Lepidocaris* are not primitive but adaptive, the relations of the rami of the limbs of trilobites are problematical, and the admittedly primitive Branchiopoda possess phyllopodia; (3) that the unjointed, turgid, lobed phyllopodium more nearly resembles the parapodia of certain annelids in which the neuropodium is axial, than the stenopodium resembles the normal biramous parapodium.

FUNCTIONS OF LIMBS. Concerning the functions of particular members of the series of limbs, and the corresponding modifications of their structure, little can be said that would hold good throughout the subphylum. There is an immense variety in these respects. The *antennules* and *antennae* are primarily

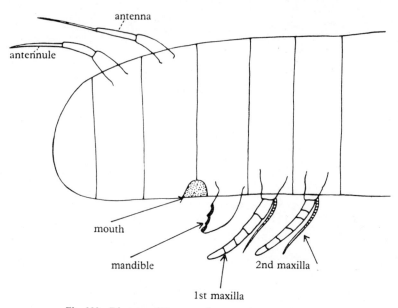

Fig. 239. Diagram of the appendages on the crustacean head.

sensory, and perhaps usually possess something of that function when they are also capable of swimming, prehension, attachment, etc. In the nauplius larva (Figs. 249 A, 262 A) the antennules are uniramous and the antennae biramous, and they normally retain these conditions in the adult. The *mandibles* always play, by means of their strong gnathobase, some part in preparing the food, whether by chewing or by piercing for suction, but the distal part of the limb (*palp*) may aid in locomotion or set up feeding currents. They generally lose in the adult the biramous condition which they have in the nauplius. The *maxillules* and *maxillae* tend to be phyllopodia. The maxillules have usually the function of passing food to the mouth but may serve other ends. The maxillae have various functions in connexion with feeding and respiration. The *limbs of the thorax* perform in various cases practically every function for which appendages are used. If a crustacean walks, it is usually by means of these limbs. Often in one or more of them the last joint can be opposed to the joint which precedes it, forming a *chela* (or a *subchela*), so that the appendage is adapted for grasping. Modification of the hinder thoracic or anterior abdominal limbs in connexion with reproduction is common. *Abdominal limbs* are lacking save in certain of the Branchiopoda and most of the Malacostraca. When they are present they are commonly used for swimming, for setting up currents of water, or for carrying eggs and young. (Fig. 237.)

FEEDING METHODS. When feeding is restricted to a few limbs it is often, though not always, accomplished in some other way than by the original habit of gathering food in small particles. Continuous and automatic straining-out of such particles, which is practised (though in different modes) by the most primitive members of all classes except the Branchiura, is superseded in various members of different classes by the intermittent seizure, by particular limbs, of particles of some size, and this by the grasping of larger objects, which may lead to a predatory habit. Finally, either of these modes of feeding may be replaced in parasites by suction or absorption, through organs which do not always represent appendages at all. (Parasites, however, are not known among the Branchiopoda or Ostracoda.) Needless to say, each change in the mode of obtaining nutriment has entrained numerous alterations in organs other than those by which the food is actually taken, as in the means of locomotion, sense organs, weapons of offence, etc. On the other hand, adaptations to mere differences of habitat, in the Crustacea, as in other arthropods, are, as a rule, strikingly small. There is, for instance, remarkably little difference between a land crustacean and its nearest marine relatives. Pelagic genera, however, are sometimes considerably modified.

ACCESSORY APPENDAGES. Three elements of minor importance complete the external make-up of the Crustacea. In front of the mouth is a *labrum* or upper lip; behind the mandibles is a lower lip or *metastoma*, usually cleft into a pair of lobes known as *paragnatha*; and on the telson usually (but in no adult malacostracan except the Leptostraca, Fig. 269 A) is a pair of *caudal rami* forming the caudal *furca*.

Appendages which are lost are *regenerated* at subsequent moults; and the highest members of the group possess an elaborate mechanism for *autotomy*—the breaking-off of limbs which have been injured or which have been seized by enemies.

INTERNAL SKELETON. An internal skeleton is usually present in the form of ingrowths of the cuticle, known as *apodemes*, which serve for the insertion of muscles. Sometimes (notably in the Decapoda) they unite to form a framework, the *endophragmal skeleton*. In the Notostraca, a mesodermal tendinous plate, the *endosternite*, lies under the anterior part of the alimentary canal.

NERVOUS SYSTEM. The nervous systems of crustacea exhibit a very complete series of stages from the ideal arthropod condition (see p. 322) to the extremest concentration. That of the Branchiopoda (Fig. 218) is in a very primitive state, having the antennal ganglia behind the mouth as the first pair of the ventral ladder, distinct ganglia for the following somites, and widely

356 CRUSTACEA

separated ventral cords. In the lower members of the Malacostraca (*Nebalia*, some mysids, etc.), the antennal ganglia have joined the brain and the ventral cords are closer together, but otherwise the primitive condition is retained. In other crustaceans various degrees of concentration of the ventral ladder

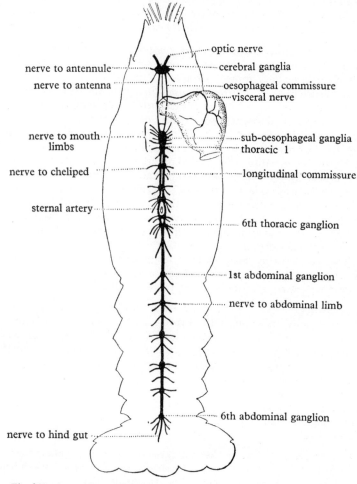

Fig. 240. A semidiagrammatic view of the central nervous system of *Astacus*. (From Borradaile.)

are found, beginning with the establishment of a suboesophageal ganglion for the somites of the mouth parts (Fig. 240), and ending in the formation, in the crabs (Fig. 282) and some other forms, of a single ventral ganglionic mass. In the Rhizocephala one ganglion (Fig. 267 F) supplies the whole body. The brain contains ganglia for the eyes (*optic lobes*), for the first or preantennulary somite (*protocerebrum*), and for the antennules (*deuto-* or *mesocerebrum*).

As in other arthropods, the name *procerebrum* is given to the anterior part of the brain, composed of the protocerebrum, the optic lobes, and sometimes other ganglia which are not connected with paired limbs. Except in the Branchiopoda it also contains the antennal ganglia (*trito-* or *metacerebrum*). A *visceral* ('sympathetic') system is present. In its main features the functioning of the nervous system resembles that of insects (p. 458).

SENSE ORGANS. Sense organs are well developed in the free members of the group. *Eyes* are of two kinds, the *compound* eyes, of which a pair is usually present except in the Copepoda and adult cirripedes, and the *median* eye. Details of the structure of the compound eyes have been given above (p. 323). They may be sessile or stalked, and the latter condition has given rise to a theory that they represent a pair of appendages. Since, however, there are no somites corresponding to their ganglia and since at their first appearance in the embryo they are sessile, this view is not generally accepted (see also p. 323). The median eye (Fig. 241 F) is the eye of the *Nauplius* larva, and it persists in most adults, though it is generally vestigial in the Malacostraca. It consists of three pigmented cups, one median and two lateral, each of which is filled with retinal cells whose outer ends are continued as nerve fibres. Thus the sense cells are inverted, as in the eyes of vertebrata. Sometimes each cup has a lens. In some of the Copepoda the lateral cups are removed from the median one and developed as a pair of lateral eyes. Senses other than sight are subserved by various modifications of the bristles which exist on the surface of the body and contain nerve fibrils in their protoplasmic contents. Most of these bristles are branched in various ways and have *tactile* functions, including that of appreciating the resistance of the water to movements. In the Decapoda and Syncarida on the basal joint of the antennule (Fig. 241 A) and in the Mysidae on the endopodite of the sixth abdominal appendage there is a pit whose wall bears such hairs while the hollow usually contains sand grains (most decapods) or a calcareous body formed by the animal (Mysidae). These organs are *statocysts* for the sense of balance. *Olfactory hairs* or *aesthetascs* (Fig. 241 E) with delicate cuticle stand on most antennules and on many antennae. A pair of groups of cells, sometimes surmounted by setae, standing on the front of the head and known as *frontal organs*, are found in many crustaceans and are supposed to be sensory. They are present as two papillae in the *Nauplius* larva (Fig. 267 A). The *nuchal sense organ* or 'neck organ' of many branchiopods is a group of cells on the upper side of the head containing refractive bodies and connected to the brain by a special nerve. Its function is unknown.

PIGMENTATION. As is well known, most crustaceans are pigmented. The pigments are of various colours—red, orange, yellow, violet, green, blue, brown, black, etc., though not all are found in any one species. The majority of them are lipochromes, though the brown and black are melanins. For the most part they are contained in branched cells (chromatophores), but some of the blue, and perhaps certain others, are diffused in the tissues. The chromatophores may lie in the epidermal layer, in the dermis, or in the connective tissue of deeper organs. Their behaviour has been studied in

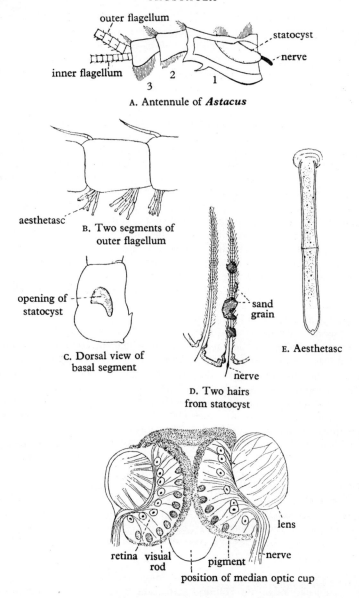

Fig. 241. Crustacean sense organs. (B, after Huxley; F, after Claus.)

various malacostracans. The pigment is often caused to expand or contract, which it does by flowing into and out of their processes. In this it is affected by light, responding both to intensity of illumination and to the nature of the background, but only rarely to colour (wave-length). In light of high intensity or on a light-absorbing (e.g. dull black) background it expands; in light of low intensity or on a light-dispersing (e.g. dull white) background it contracts. Different pigments are affected to different degrees, and thus both the degree and the pattern of the coloration of a sensitive species (notably, for instance, of many prawns), changes with its surroundings—usually, in nature, in such a way as to render the animal inconspicuous. The response to intensity of illumination is due to direct action of the light upon the chromatophores and will thus take place even in blinded animals; the response to background

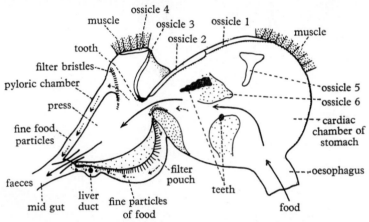

Fig. 242. Fore gut and mid gut of *Astacus*. Only the small particles of food are able to make their way through the filtering hairs to the digestive gland. (After Yonge.)

depends upon the eyes. The eyes, however, do not act through nerves to the chromatophores, but by causing certain endocrine glands to pass hormones into the blood.

ALIMENTARY CANAL. The alimentary canal (Figs. 235, 250A, 253, 274, 281) is with very rare exceptions straight, save at its anterior end, where it ascends from the ventral mouth. The *fore gut* and *hind gut* (stomodaeum and proctodaeum), lined with cuticle inturned at the mouth and anus, leave a varying length of *mid gut* (mesenteron) between them. The intrinsic musculature, sometimes supplemented by extrinsic muscles running to the body wall, is strongest in the fore gut, whose lining sometimes develops teeth or hairs. In the Malacostraca (Fig. 242) these elements become a more complex proventriculus ('stomach'), with a 'gastric mill' and a filtering apparatus of bristles which strains particles from the juices of the food, the mill and filter being often in separate 'cardiac' and 'pyloric' chambers. The mid gut usually bears near its anterior end one or more pairs of diverticula ('hepatic caeca') which serve for secretion and absorption and may branch to form a 'liver'.

This gland, however, unlike the liver of vertebrates, forms all the enzymes necessary for the digestion of the food and absorbs from its lumen the products of digestion. It stores the reserves in the form of glycogen and fat. Occasionally there is an anterior median dorsal caecum. Caeca are also sometimes found at the hinder end of the mid gut: these are more often

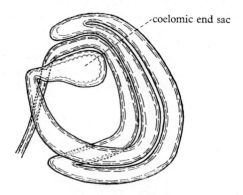

A. Maxillary gland of *Estheria*

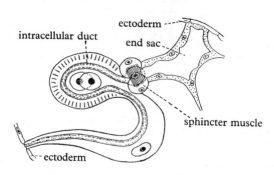

B. Antennal gland of early metanauplius of *Estheria*

Fig. 243. Excretory organs of *Estheria* larva and adult.

median. In a few cases the hind gut is absent and the mesenteron ends blindly. In the Rhizocephala and the monstrillid copepods (pp. 386, 394) the alimentary canal is absent throughout life, for these animals absorb through the skin during the parasitic period enough nutriment to last through an entire life history.

DIGESTION is extracellular. The fore gut is frequently the seat of mechanical processes, and sometimes of chemical action by juices secreted by the mid gut diverticula, but never of absorption. The latter process as well as most of the chemical work is performed by the mid gut, including the hepatic diverticula. In the hind gut the faeces are passed to the anus, being in some entomostraca sheathed in a so-called 'peritrophic membrane' composed of a mucoid substance secreted by certain cells of the epithelium.

EXCRETION. The principal excretory organs of the Crustacea are two pairs of glands, known as the *antennal* and *maxillary glands*, which open (Fig. 243) at the bases of the appendages from which they take their names. They are very rarely (Lophogastridae) both well developed at the same stage in the

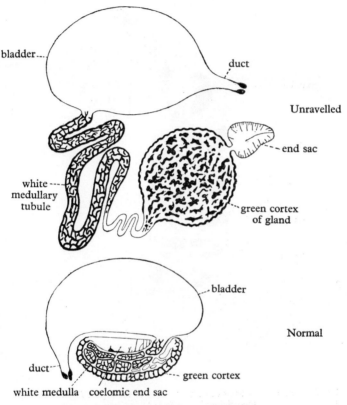

Fig. 244. Diagrams of the green gland of *Astacus*.
(From Parker and Haswell, after Marchal.)

same species, but one may succeed the other as a functional organ in the course of the life history: the antennal gland, for instance, is the larval excretory organ of the Branchiopoda, but the maxillary gland is that of the adult; and the Decapoda, whose adult kidney is the antennal gland, sometimes use as larvae the maxillary gland instead. The maxillary gland is the more widespread as an adult organ, the antennary gland being functional in the adult only in certain of the Malacostraca. In the Ostracoda and Leptostraca both are vestigial in the adult. Each of these glands (Figs. 243, 244) has an *end sac* and a *duct* leading from the end sac to the exterior. The end sac is always mesodermal and doubtless represents a vestige of the coelom. The duct is sometimes (in the Malacostraca probably always) a multicellular, mesodermal structure, and sometimes intracellular and of ectodermal origin. At the junction of end sac and duct there is often a sphincter. The antennal gland of the Decapoda is usually very complicated. That of the crayfish lacks extensions of the bladder which lie among the viscera in many other genera

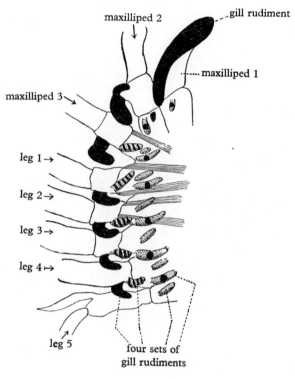

Origin of gills

Fig. 245. Part of the left side of a late larva of the prawn *Penaeus* to show the origin of the gills. (After Claus.)

as in crabs. All the parts of the organs are excretory, and the function of the sphincter of the end sac is perhaps to prevent the passage back into that vesicle of poisonous products excreted in the duct.

These glands are probably the remaining members of a series of segmental excretory organs. Their mesodermal portions are no doubt coelomoducts, homologous with those of the Annelida; their ectodermal portions probably are not the homologues of nephridia but represent ectodermal glands such as are common in the Crustacea. Various other glands, mostly of doubtful morphological significance, which occur in different crustaceans have been shown, or are suspected, to have an excretory function. Thus, in *Nebalia*, eight pairs of ectodermal glands at the bases of the thoracic limbs are excretory, while in ostracods a pair of rather complex glands, also of ectodermal origin, which lie between the folds of the shell in the antennal region, may have a similar function. Excretion appears also sometimes to be performed by caeca of the mid gut—as by some of those of the barnacles and by the posterior pair of amphipods—or by cells of the epithelium of the mid gut itself.

RESPIRATION in many of the smaller crustaceans, notably in the Copepoda, takes place through the general surface of the body. In forms with stouter cuticle or more bulky bodies this is supplemented or replaced by the use of special organs upon which the cuticle remains thin. The most important

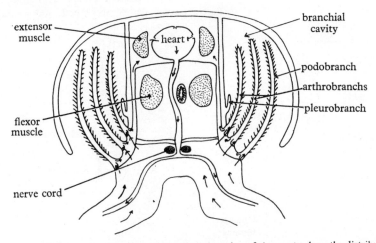

Fig. 246. Diagram of section through the thoracic region of *Astacus* to show the distribution of the gills on the limbs. The circulation of the blood is indicated by arrows.

of such organs are the lining of the carapace, if that structure be present, and certain epipodites which are known as gills and in many of the Malacostraca have their surface increased by branching or folding (Figs. 238I, 280). In the Decapoda incorporation of the precoxa with the flank of the body has brought it about that some of the gills (proepipodites, Fig. 250c) stand in that position and not upon the actual limbs (Figs. 245, 246). Such gills are known as 'pleurobranchiae'. In the Isopoda respiration is effected by the broad rami of the abdominal limbs. Renewal of the water upon the respiratory surfaces may be brought about by the movements of the limbs upon which they are located, but often certain appendages bear special lobes adapted to set up a current under the carapace and thus to flush the chamber in which the gills and the carapace lining are situated.

Some land crustaceans have no special adaptations for respiration in air. In others the gill chamber is adapted, by the presence of vascular tufts of the lining of the carapace, for use as a lung. The woodlice, which are terrestrial members of the Isopoda, are remarkable in approaching in their respiration the principle employed by normally terrestrial arthropods, for the integument of their abdominal limbs is invaginated to form branching tubes which resemble tracheae.

VASCULAR SYSTEM. The vascular system is seen in its most primitive condition in the Branchiopoda Anostraca (*Chirocephalus*, Fig. 250A). Here the *heart* runs the whole length of the trunk, situated above the gut in a blood sinus known as the *pericardium*, with which it communicates by a pair of ostia in each somite except the last. In front it is continued into the only *artery*, a short aorta, from which the blood flows direct into the *sinuses* of the head and thence through those of the trunk to the pericardium, eddies from a main ventral sinus supplying the limbs. In all other crustacea, except the Stomatopoda, the heart, if it be present, is in some degree shortened, and in the

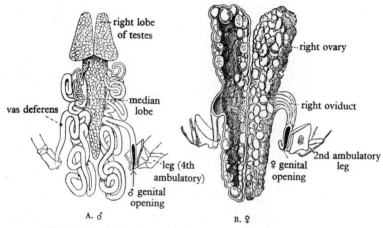

Fig. 247. Reproductive organs of *Astacus fluviatilis*. (From Howe.)

Malacostraca (Fig. 235) a system of arteries interposes between the heart and the sinuses, leaving the former by several vessels, which conduct the blood to the organs. In the Eucarida (Euphausiacea and Decapoda) the heart is shortened to a compact shape and has three pairs of ostia; in most of the Cladocera it is a sac (Fig. 253) with only one pair. In the Cirripedia and many of the Copepoda and Ostracoda the heart is absent and the blood is kept in movement only by the movements of the body and alimentary canal. In the parasitic copepod *Lernanthropus* and some related genera there is a remarkable system of closed blood vessels without a heart.

The *blood* is a pale fluid, which bears leucocytes except in ostracods and most copepods. It contains in the Malacostraca the copper-containing respiratory pigment *haemocyanin* (p. 130). In various entomostraca, notably in *Lernanthropus*, just mentioned, haemoglobin has been found.

REPRODUCTIVE SYSTEM. As is usual with animals that are free and active, the *sexes* are separate in the great majority of the Crustacea, though the Cirripedia, which are sessile, certain of the parasitic Isopoda, and a few exceptional species in other groups, are hermaphrodite. Parthenogenesis takes place in many of the Branchiopoda and Ostracoda, and in these it is often only at more or less fixed intervals that sexual reproduction occurs. The male is usually smaller than the female and in some parasites is minute and attached to her body. He has often clasping-organs for holding his partner, and these may be formed from almost any of the appendages. He may also possess organs for the transference of sperm: these may be modified

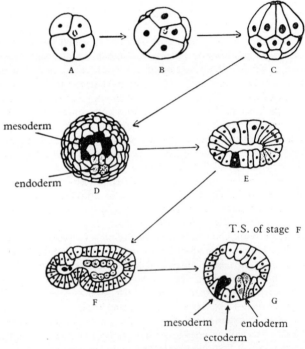

Fig. 248. Crustacean embryology. Development of *Polyphemus*. (After Kuhn.) For the development of a yolky egg, see Fig. 223.

appendages or protrusible terminal portions of the vasa deferentia. The *gonads* of both sexes (Fig. 247) are hollow organs from which ducts lead directly to the exterior. Primarily there is one gonad on each side, but they often unite more or less completely above the alimentary canal. The ducts usually open near the middle of the body, though the male openings of Cirripedia and some Cladocera are almost terminal and the female opening of Cirripedia is on the first thoracic somite. Save in the Cirripedia, the Malacostraca, and some of the Cladocera, the ducts of the two sexes open upon the same somite.

SPERM AND OVA. The spermatozoa are very varied in form and often of complex structure; usually, but not always, they are immobile. They are transferred to the female, often in packets (*spermatophores*). The *ova* have usually much yolk, and meroblastic, centrolecithal cleavage (Fig. 223 A, B, C), but sometimes are less yolky and undergo total cleavage. Gastrulation may be by invagination (Fig. 223 D), or by immigration. Occasionally the eggs are set free at laying, but in the great majority of cases they are retained for a time by the mother, either in some kind of brood pouch or adhering in some way to her body or appendages. *Development* is not infrequently direct, but in most cases involves a larval stage or stages.

LARVAL FORMS. Typically, the crustacean hatches as a *nauplius* larva (Fig. 249 A), a minute creature, egg-shaped with the broad end in front, unsegmented, but provided with three pairs of appendages—the antennules, which are uniramous, and the antennae and mandibles, which are biramous and should each bear a gnathobasic process or spine directed towards the mouth, though those of the mandibles are often not developed at first. The antennal ganglia are as yet postoral (see p. 355). The median eye is the only organ of vision. A pair of frontal organs (p. 357) are present as papillae or filaments. There is a large labrum. Fore, mid and hind guts can be recognized in the alimentary canal. Antennal glands may be present. This larva is found in some members of every class of the Crustacea, though among the Malacostraca only certain primitive genera possess it, and in the Ostracoda it is modified by having already at hatching a precociously developed bivalved carapace. In every class, however, it is also often passed over, and becomes an embryonic stage within the egg membrane or in a brood pouch, the animal hatching at a later stage, such as the *metanauplius* and *zoaea* mentioned below, or even almost as an adult (Fig. 249 G).

In the Branchiopoda and Ostracoda the nauplius is transformed gradually into the adult, adding somite after somite in order from before backwards by budding in front of the telson, much as somites are added to the trochosphere in the development of annelids, while by degrees the other features of the adult develop. The early stages of this process, which possess more somites than the nauplius, but have not yet the adult form, are known as metanauplii. The carapace is often foreshadowed quite early by a dorsal shield, which later grows out behind and at the sides to assume the form which it has in the adult, and the appendages, at first mere buds, gradually take on their final shapes.

VARIATIONS IN LIFE CYCLE. In most cases, however, the process just described is modified. (1) It makes a sudden great advance at one moult. In the Cirripedia the late nauplius passes with a leap to the so-called *cypris* larva, which has many of the features of the adult: a similar leap takes the copepod metanauplius to the first *cyclops* stage (p. 385) and those of Malacostraca to the *zoaea*. (2) Certain structures may be precociously developed. In those of the Malacostraca which have nauplii, the metanauplius is followed by stages, known as *zoaeae*, in which the abdomen is well developed, while the thorax, though it already possesses in front a few pairs of biramous appendages, is still rudimentary in its hinder part. In these larvae also the last pair of abdominal limbs usually appears, or comes to functional development, before the others. Zoaeae, however, most often are not preceded by a

CRUSTACEA

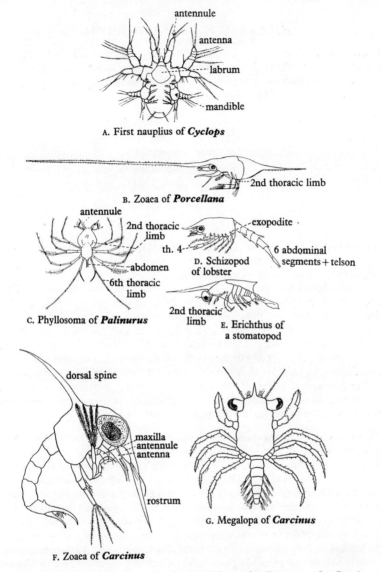

Fig. 249. Larval crustaceans. (A, after Dietrich; F, after Faxon; G, after Bate.)

free nauplius but appear as the first free stage (Fig. 249 F). (3) Temporary retrogression of certain organs takes place during the development of some of the Malacostraca: this affects some of the thoracic limbs in certain Stomatopoda and the prawn *Sergestes*, abdominal swimmerets and the antennule in the prawn *Penaeus*.

SUBCLASS 1. BRANCHIOPODA

DIAGNOSIS. Free-living Crustacea with compound eyes; usually a carapace; the mandibular palp very rarely present and then as a minute vestige; and at least four, usually more, pairs of trunk limbs, which are in most cases broad, lobed, and fringed on the inner edge with bristles.

CLASSIFICATION

1. *Order* Anostraca
2. *Order* Lipostraca
3. *Order* Notostraca
4. *Order* Diplostraca
   Suborder 1. Conchostraca
   Suborder 2. Cladocera

PRIMITIVE AND SPECIALIZED CHARACTERS. The Branchiopoda are, on the whole, the most primitive group of the Crustacea. This is seen in the varying and usually large number of their somites, the usually small amount of differentiation in the series of limbs on the trunk, the vascular system of the lower members of the group (p. 364), and the nervous system of all (p. 355). Their mouth parts, on the other hand, are small and simple in structure, a condition in which they are not primitive but exhibit reduction. Nearly all of them are, like sundry other archaic animals, of fresh-water habitat, and their characteristic mode of feeding is the taking, by means of setae upon their trunk limbs, of particles of detritus or plankton from suspension in the water.

The primary divisions of the subclass have been mentioned on p. 343. The most conspicuous differences between them are in the carapace, the compound eyes, the antennae, the trunk limbs, and the telson.

CARAPACE. The carapace is very variously developed. In the Anostraca it is not present. The Notostraca have it as a broad, shallow cover over the back. In the normal Cladocera ('Calyptomera') it bends down at the sides to enclose the trunk as a shell which forms a brood pouch over the back. In the two groups of aberrant Cladocera which (though they are probably not closely related) are together known as 'Gymnomera' this shell has shrunken to a dorsal brood pouch leaving the trunk partly or wholly uncovered. In the Conchostraca it forms in the same way as in the Cladocera a shell, but here the head is usually enclosed as well as the trunk, and there is a distinct dorsal hinge of thin cuticle separating two valves which can be closed by an adductor muscle situated in the maxillulary somite. Usually the carapace leaves the trunk free within it, but in the Cladocera it fuses with two—in *Leptodora* (p. 381) with all—of the thoracic somites.

ANTENNAE. The antennae, which in the nauplius are biramous and natatory, retain this condition in the adult of those forms (Diplostraca) in which the enclosing carapace has deprived the trunk limbs of the swimming function, and also in the extinct *Lepidocaris* (Lipostraca). In the Recent Anostraca the antennae are stout but uniramous and not natatory; in the male they are adapted to clasping the female. In the Notostraca, which apply the head to the ground in feeding, they are reduced to uniramous vestiges.

TRUNK LIMBS. The trunk limbs (except in the aberrant Cladocera which constitute the Gymnomera) are phyllopodia (p. 351) which bear on the median side endites furnished with feathered bristles and on the outer side, besides the exopodite or flabellum, a thin-walled branchia and often also one or two proepipodites.

FILTER FEEDING. With these appendages the Anostraca and Notostraca swim, and all members of the class breathe and gather food. Beating rhythmically forward and backward with a movement which each pair starts a little earlier than the pair in front of it, they cause, by a pumping action which shall be described presently (p. 371), a flow of water into the median gully whose sides are formed by the two rows of limbs, thence outwards into the spaces between each limb and its neighbours in front and behind, and then backwards. This current brings with it the particles which serve for food, bathes the branchiae, and causes, in the Anostraca and Notostraca, forward movement of the body. As the water passes outwards, the food particles are, by the bristles on the endites, strained off and retained in the median gully. The apparatus varies in detail with the nature of the food. In the Notostraca, which feed mainly by stirring up, with the tips of their thoracic limbs, detritus on the bottom and then filtering it, the bristles on the endites are not adapted to straining out fine particles (which therefore escape with the outgoing current) but detain coarser particles. This is perhaps the primitive mode of feeding of the Branchiopoda, and may even be inherited from the trilobites. In the Anostraca and Diplostraca there is a special apparatus for filtering off fine particles. This consists of a close set row of long, finely feathered setae, placed on the edge of the endites and so disposed as to cover the opening from the median gully to the space between the limb and its neighbour behind. Members of these orders which derive part or all of their food from detritus have various kinds of apparatus, composed of bristles, for removing the coarse particles and passing them backwards to be either swept away with the outgoing stream or broken up for food by the hinder members of the series of limbs. Finally, the material gathered is passed forwards to the mouth in a median 'food groove' along the belly by a current whose causation is a matter of dispute. The feeding apparatus whose principles have just been described differs greatly in detail in different branchiopods, and reaches its highest complication in the tribe of cladocera known as Anomopoda, to which the common water-flea *Daphnia* belongs. Examples of it are described more fully below.

The Gymnomera have slender, mobile, jointed trunk limbs with which they manipulate the relatively large organisms which serve them for food.

TELSON. The telson is in the Anostraca subcylindrical, with the caudal rami as elongate plates or styles; in the Notostraca it has the rami long and many-jointed, and is in *Lepidurus* produced backwards on the dorsal side as a plate. In the typical Diplostraca it is flexed ventrally and produced backwards laterally into a pair of strong, curved, toothed claws, and can be brought forward ventrally to clear the gully between the limbs. In the Gymnomera it has re-straightened.

COMPOUND EYES. The compound eyes are in the Anostraca stalked (in *Lepidocaris* they appear to have been absent). In the remainder of the class they are sessile and covered by an invagination of the outer cuticle, which forms a shallow chamber over them.

HABITAT. *Artemia salina* (p. 373) and a few marine cladocera are the only members of the class whose habitat is not in fresh water.

Throughout the group, thick-shelled eggs capable of resisting drought or freezing are produced by sexual reproduction. Often there is also parthenogenesis, the eggs of which are usually thinner shelled than those that are sexually produced (see p. 379).

'PHYLLOPODA'. The name Phyllopoda, which is applied sometimes to the whole class and sometimes to its members exclusive of the Cladocera, is on account of this ambiguity best not employed in systematic nomenclature.

1. *Order* Anostraca

DIAGNOSIS. Branchiopoda without carapace; with stalked eyes, with antennae of a fair size but not biramous; with the trunk limbs numerous and all alike; and with the caudal rami unjointed, and flat or subcylindrical.

CHIROCEPHALUS DIAPHANUS. We may take as an example of this group, *Chirocephalus diaphanus* (Fig. 250), one of its two British representatives. This creature turns up from time to time in temporary pools of water in various districts. It is about half an inch in length, transparent, and almost colourless, save for the reddened tips of most of the appendages and of the abdomen, the black eyes, and often a green mass of algae in the gut. It is incessantly in motion, swimming on its back. Its delicate appearance, and the iridescent gleaming of the bristles on its appendages as they are moved, have earned it the name of the fairy shrimp. The body is long, subcylindrical, and enlarged anteriorly to form the *head*, upon which the mandibular groove (p. 348) is conspicuous. The head has in front a *median eye* and a *neck organ* (p. 357), and bears at the sides: (1) the large, stalked *compound eyes*; (2) the *antennules*, slender, unjointed, and ending in a tuft of sense-hairs; (3) the stout *antennae*, triangular in the female but in the male (Fig. 250B) elongate, two-jointed, and carrying on the inside at the base a complicated, lobed 'frontal appendage' which comes into play when the limb is used for clasping the female; (4) the *mandibles*, whose bases are prominent at the sides of the head, while the remaining part of each of them is directed towards the mouth as a process with a blunt, roughened end. Below, the head bears (*a*) the large *labrum* which is directed backwards under the mouth; (*b*) the *paragnatha*, a pair of small, hairy lobes behind the mouth; (*c*) the *maxillules*, a pair of small triangular plates fringed by long bristles; (*d*) the *maxillae*, which are microscopic vestiges, each bearing three spines.

THORACIC LIMBS. Behind the head come eleven thoracic somites which each bear a pair of phyllopodia. Fig. 250 c shows that these possess all the typical features of such limbs but are remarkable for the distal position of the exopodite and for the very long basal endite, which may be simply the gnathobase (p. 351) but probably represents also the second endite. The fringe of long bristles on the median border is, in life, directed backwards, roughly at right angles to the main plane of the limb. The twelfth thoracic somite, upon which are the genital openings, is fused ventrally with the first abdominal. In the male, it bears a pair of ventrolateral processes in each of which is the terminal portion of a vas deferens, with a protrusible penis which probably represents an appendage. In the female there is here a median, ventral, projecting egg pouch, which, like the penes, is held to represent a pair of limbs. The *abdomen* consists of seven simple, limbless somites and a telson which bears a pair of caudal rami as narrow, pointed plates, fringed with bristles.

ALIMENTARY CANAL. The alimentary canal begins with a short, vertical fore gut, or oesophagus. This leads to a mid gut which continues as far as the telson, where it is succeeded by the hind gut or rectum. The mid gut is somewhat wider in the head, where it is known as the *stomach*, than in the trunk, where it is called the *intestine*. From the stomach proceeds a pair of sacculated diverticula ('liver'). The *food* consists partly of coarse detritus gathered by the trunk limbs from the bottom of the pool, and partly of small organic particles, especially unicellular algae, which are strained off from the water by the trunk limbs in the following manner (Fig. 250 D, E, F).

FILTER FEEDING. The space which exists between each limb and that behind it is enlarged at the forward stroke, which finishes with the limbs vertical, and narrowed at the back stroke, which ends with them roughly horizontal, lying against the body. During the forward stroke the enlarging of this space exerts a suction. The proepipodites, exopodite, and large distal endite are drawn back by the suction and pressed back by the resistance of the water, till they reach the limb behind and so convert the space just mentioned into a chamber which is closed except on the median side, where it is separated only by the backwardly directed bristle fringe from the median gully between the limbs of the right and left sides. From this gully, therefore, water is drawn into the chambers at the sides as they enlarge, particles which it contains being strained off by the bristles and remaining in the gully. The latter is of course replenished by the entrance of water from the ventral side. During the back stroke, the chambers, as they become smaller and the pressure of the water in them rises, open owing to this pressure lifting the structures which had closed them; and the water they contain is driven out and backward in two ventrolateral streams, the animal being driven forwards. Thus the same movement of the limbs serves both for the gathering of food and for swimming. The particles which are retained in the median gully are drawn dorsalwards because the suction of the side chambers is greatest where they enlarge most, at the bases of the limbs, and so get into a median food groove of the ventral surface. There they are carried forward to the mouth by a minor stream, which is said to be caused by the escape forwards at the bases of the limbs of some of the water contained in the lateral chambers at a certain phase of the movement. The food is agglutinated by a sticky secretion

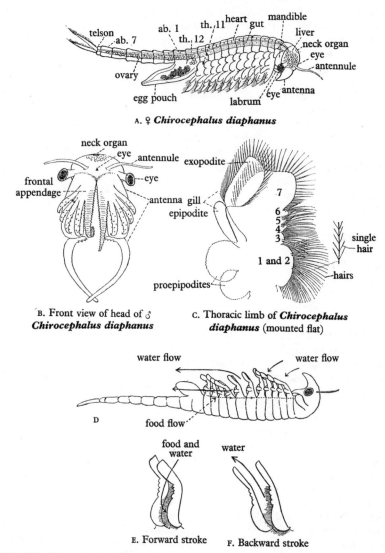

Fig. 250. *Chirocephalus*. The animal normally swims upside down; D, E and F illustrate the feeding mechanism and are fully described in the text. th. 11 and th. 12 = 11th and 12th thoracic phyllopods; ab. 1–ab. 7 = abdominal somites.

produced by glands in the labrum, and pushed by the maxillules between the mandibles, which pound it and pass it into the mouth.

INTERNAL ORGANS. The *organs of excretion* are a pair of maxillary glands (p. 361), situated in the hinder part of the head and the first thoracic somite. They are wholly of mesodermal origin. The *nervous system* (Fig. 218) and the *vascular system* have been described above (pp. 355 and 364). The *gonads* are a pair of tubes lying one on each side of the alimentary canal in the abdomen, and are continuous in front, each with a short duct. The vasa deferentia lead to the penes, the oviducts to a median uterus in the egg pouch. The eggs are enclosed in stout shells and will remain alive in dry mud for many months. The larva at hatching is a late nauplius in which, though there are no appendages behind the mandibles, the trunk is already distinct from the head.

Artemia salina, the other British species of anostracan, occurs in various parts of Europe in salt lakes and marshes and in pans in which brine is being concentrated. It can endure a very high concentration of salt, and some of its minor features change with the degree of the concentration, so that is has been described under different specific names. It differs from *Chirocephalus* in having only six abdominal somites and in the form of the antennae of the male.

2. Order **Lipostraca**

Lepidocaris, a minute, blind, fresh-water form from the Middle Devonian, was closely related to the Anostraca which survive (*Euanostraca*), but differed from them in the following, among other respects. It had biramous antennae which recall those of the Cladocera; a clasping organ on the maxillule of the male, instead of on the antenna; and the trunk limbs without branchiae and differentiated into two sets—the first three pairs adapted for gathering food, with gnathobase and with the last endite directed inwards and the exopodite lateral, and the remaining pairs adapted for swimming, with the last endite and the exopodite directed distally side by side at the end of the limb.

3. Order **Notostraca**

DIAGNOSIS. Branchiopoda with a carapace in the form of a broad shield above the trunk; the compound eyes sessile and close together; the antennules and antennae much reduced; the trunk limbs numerous, the first two pairs of them differing considerably from the rest; and slender, multi-articulate caudal rami.

Apus and Lepidurus. This order contains only the genera *Apus* and *Lepidurus*, which differ in but minor features. *Apus cancriformis* (Fig. 251) is British, but is now very rarely found in these islands. The head is broad and depressed, flat below and arched above, and forms with the carapace a horseshoe-shaped structure, which bears the eyes above and the small antennules and antennae beneath, at some distance from the sharp front edge. There is a dorsal organ, which is not used for fixation, but no nuchal sense organ. From under the carapace the hinder part of the trunk projects backwards, ending in two long, jointed caudal rami. The genital opening is on the 11th of the trunk somites. Each of these bears a pair of limbs until the 13th (second of the abdomen) is reached, after which there are two to five pairs to a somite as far as the 28th somite. Five limbless somites separate this from the telson. The first thoracic limb is a modified phyllopodium, with the endites slender and many-jointed, very long in *Apus* though shorter in *Lepidurus* (Fig. 252A). The second thoracic

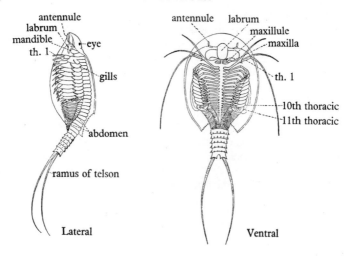

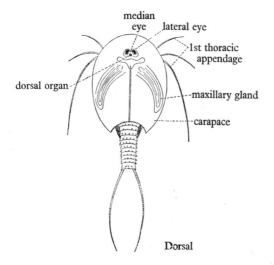

Apus cancriformis

Fig. 251.

limb is less modified in the same direction, the endites being shorter and unjointed. The remaining trunk limbs (Fig. 238 B) are normal phyllopodia: they decrease in size from before backwards, and those of the thorax have the endites well chitinized and mobile. Feeding is most often upon detritus (see p. 369), the flat underside of the head being applied to the bottom during the process, but the animals also devour the dead or living bodies of organisms, clasping them with their strong thoracic limbs and rasping fragments from them with the endites. The Notostraca swim well, but can also crawl with their thoracic limbs or clamber with the anterior pairs.

The limbs of the genital somite are in the female modified for carrying eggs, the flabellum fitting as a lid over a cup formed by the distal part of the axis. Males are rare, reproduction being normally by parthenogenesis.

4. Order Diplostraca

Branchiopoda with a compressed carapace which usually encloses the trunk and its limbs; the compound eyes sessile and apposed or fused; the antennae large and biramous; four to twenty-seven pairs of trunk limbs, often considerably differentiated; and the telson usually ending in a pair of curved claws.

Suborder 1. Conchostraca

DIAGNOSIS. Diplostraca with 10–27 pairs of trunk limbs; the carapace provided with adductor muscle in the maxillulary somite and with hinge, not fused with thoracic somites, and usually enclosing the head; and nearly always a nauplius larva.

No member of this order is British.

The animals haunt the bottom and are mainly or exclusively detritus feeders, dealing differently with fine and coarse particles (p. 369).

Estheria (Fig. 252) is a common European genus. A thoracic limb of a related but exotic form is shown in Fig. 238 D.

Suborder 2. Cladocera

DIAGNOSIS. Diplostraca with 4–6 pairs of trunk limbs; the carapace without hinge or adductor muscle, fused with two or more thoracic somites, and not covering the head; and without nauplius larva (save in *Leptodora*).

I. CTENOPODA. The members of this suborder are the water fleas. They fall into four tribes. Of these, the first, known as Ctenopoda, shows affinities with the lower Branchiopoda in that the trunk limbs, of which there are six pairs, are all alike and all strain food from the water, the gnathobase projects, and the heart is elongate. The shell is well developed and covers the trunk limbs. *Sida*, which may be taken among weeds in pools in various parts of Britain, is one of the Ctenopoda. *Penilia*, one of the few marine Cladocera, is another.

II. ANOMOPODA. The second tribe of the Cladocera, known as Anomopoda, contains most of the genera of the suborder. Its members retain a well-developed shell, but the trunk limbs, of which there are often only five, and sometimes only four, pairs, are highly differentiated for various parts of the process of feeding, only some of them doing the actual filtering off of the food particles. The gnathobases of the filtering limbs do not project but are enlarged to bear most of the filter fringe. The heart is a short sac in the first two trunk somites.

Daphnia and *Simocephalus*, common British forms, found swimming in ponds and ditches, are examples of this tribe. *Simocephalus* (Fig. 253) differs from *Daphnia* in possessing a cervical groove (p. 348), and in lacking a median dorsal spine which in *Daphnia* stands on the hinder edge of the carapace. The

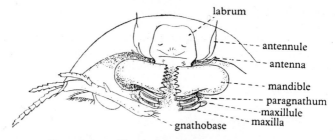

A. Ventral view of head of **Lepidurus glacialis**

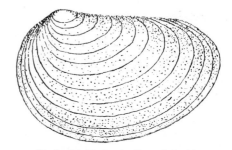

B. Shell of ♀ **Estheria** (from left side)

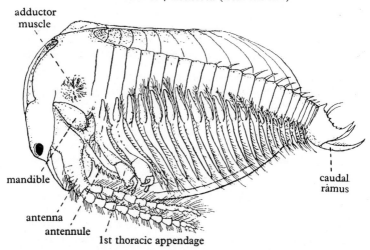

C. **Estheria** ♂, after removal of left valve of shell

Fig. 252. *Lepidurus glacialis* and *Estheria obliqua*. (From Calman.)

following description applies to both genera. The *head* is bent downwards, so that the median eye and the small antennules are ventral to the antennae. A large, sessile compound eye, formed by the fusion of a pair, stands in front. Above it is a nuchal sense organ. Of the rami of the antennae one has four joints and the other three, and both bear long, feathered setae. The mouth parts are much like those of *Chirocephalus* (p. 370). The segmentation of the *trunk* is obscure. The first two somites are fused with the head, as is shown by the position of their appendages. Behind these are three fairly distinct limb-bearing somites (so that there are in all five pairs of trunk limbs), and then three that are limbless and hardly distinguishable, and a telson, which is compressed and produced on each side of the anus into a toothed plate, bear-

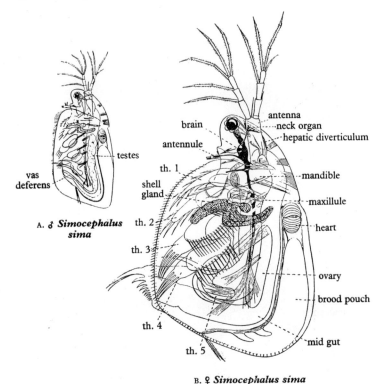

Fig. 253. *Simocephalus*; th. 1–th. 5 = thoracic limbs. (A, from Shipley and MacBride; B, from Cunningham.)

ing terminally a spine that may represent a furcal ramus. The third free somite is longer than the others and bears its limbs in the hinder part, which suggests that it is the fifth of the six pairs of *Sida* which is missing here. The limbless region is commonly known as the 'abdomen'. Two strong dorsal processes on it close the brood chamber behind.

The structure of the *trunk limbs* is shown in Fig. 254. Together they form a food-gathering mechanism which is very efficient because, instead of all working in the same way as those of the Anostraca, they are differentiated in adaptation to different parts of the task. The third and fourth pairs form a pumping and straining apparatus (Fig. 254–255) which in principle is the same as those formed by the limbs of *Chirocephalus*, but has for side walls the carapace, against which the proepipodites play, and is closed behind by a barrier formed by the fifth pair. The broad exopodites of the third and fourth pairs open and close the ventral side of the apparatus as they flap to and fro under

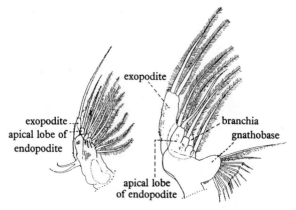

A. First thoracic limb   B. Second thoracic limb

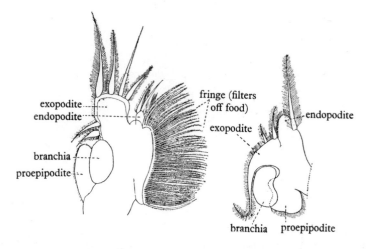

C. Third thoracic limb   D. Fifth thoracic limb

Fig. 254. Thoracic limbs of *Daphnia*. (After Lilljeborg.)

the pressure of the water. The long, feathered bristles of the first and of the distal part of the second pair guard the ventral opening of the median gully and keep too large particles from being drawn into it. The complex set of bristles upon the large endite or 'gnathobase' (which corresponds both to the first and to the second endite of the ideal series) in this limb play some part—

exactly what is disputed—in bringing the food to the mouth. Glands in the labrum produce a sticky secretion as in *Chirocephalus*.

The *alimentary canal* resembles that of *Chirocephalus* (p. 371), but the caeca are unbranched. The food on being swallowed passes direct to the middle part of the mesenteron, where it is digested, and then forwards to the anterior region and the caeca, where the digested products are absorbed and the indigestible residue sent backwards to be formed into faecal pellets in the hinder part of the mid gut. The *maxillary gland* lies in the carapace.

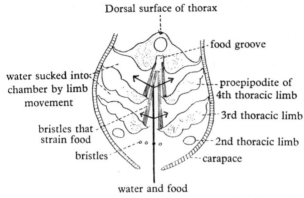

Fig. 255. Diagram of a transverse section through the thorax of *Daphnia* indicating mode of filter feeding. The dilatation and contraction of the chambers between the thoracic limbs set up a pumping movement by which water is made to flow through the bristle fringes from the median gully. (After Storch.)

The *gonads* are simple, elongated sacs lying in the trunk and continuous with their ducts, which open in the male on the telson, in the female dorsally behind the last limb. The eggs are yolky. They are of two kinds, 'summer' eggs which have relatively little yolk and develop rapidly by parthenogenesis in the brood pouch of the mother, and 'winter' eggs with much yolk which need fertilization and develop slowly. The winter eggs are fertilized in the brood pouch, but then the cuticle of the carapace, which has thickened, is thrown off as a case—the *ephippium*—in which they are contained. They go through the early stages of segmentation within a short time, but after this a period of quiescence sets in, during which they may be dried or frozen without injury. Sexual reproduction takes place at certain times only, normally twice a year. After the winter eggs develop in spring, there are for some half-dozen generations no males, and reproduction proceeds by parthenogenesis. Then, about May, a generation appears in which males are present. In this sexual and asexual reproduction go on side by side. The same thing occurs again in autumn or at other times when, in unfavourable circumstances, such as cold or starvation, males appear. It is interesting to note that, since parthenogenesis is never suspended by all the females, there is nothing to show that a sexual phase in the life cycle is necessary.

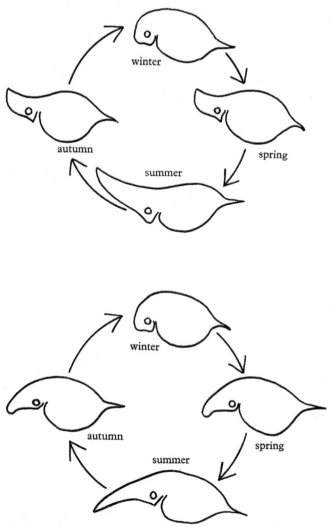

Fig. 256. Cyclomorphosis in *Daphnia*. The animals often show seasonal variation in form. Two different examples are shown here. The cycle need not be quite so simple and regular. (From Coker.)

III. ONYCHOPODA AND IV. HAPLOPODA. (Gymnomera.) The normal cladocerans which compose the tribes Ctenopoda and Anomopoda are often united under the name Calyptomera in contrast to the remaining two tribes, which are known as Gymnomera. These are aberrant forms whose food consists of planktonic organisms relatively much larger than the particles upon which *Daphnia* feeds. Their carapace has shrunk till it forms only the brood pouch and leaves free the comparatively slender, prehensile trunk limbs with which the food is handled, and their eyes are prominent and adapted to sighting moving objects. They are often bizarre in form.

Polyphemus, a British fresh-water genus, is an example of the tribe *Onychopoda*. It has a long telson, but the head and 'abdomen' are not elongate and the carapace does not fuse with the hinder part of the 'thorax'. The trunk limbs have gnathobases. In *Evadne* and *Podon*, marine members of the tribe, the telson is not elongate.

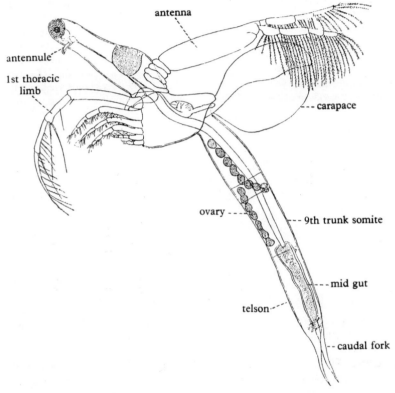

Fig. 257. A female *Leptodora kindti*. (After Lilljeborg.)

Leptodora (Fig. 257), the only member of the tribe *Haplopoda*, is a pelagic inhabitant of certain fresh waters in Britain and elsewhere. The body is long and slender owing to elongation of the head and of the 'abdomen', in which the segmentation is distinct. The fore part of the trunk bears six pairs of slender, jointed, uniramous limbs, without gnathobases. The carapace has fused with all the somites of this region and projects behind it as a brood pouch. The winter egg gives rise to a nauplius, the only instance of a larva in the Cladocera.

SUBCLASS 2. OSTRACODA

DIAGNOSIS. Free Crustacea, with or without compound eyes; with a bivalve carapace and an adductor muscle; a mandibular palp, usually biramous; and not more than two recognizable pairs of trunk limbs, these not being phyllopodia.

CHARACTERISTICS. The small crustaceans which compose this class differ little in the general form of the body but show very great variety in that of their appendages. All their cephalic limbs are well developed and complex; the trunk limbs are uniramous and one or both pairs may be lost. The adductor is in the maxillulary somite. There is often a gastric mill and usually a pair or more of hepatic caeca: the latter and the gonads may (*Cypris*) extend into the shell valves. Both antennal and maxillary glands are present, both have ectodermal ducts, and both are without opening in the adult. Other

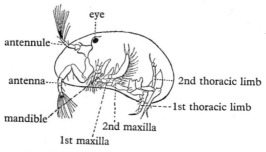

Fig. 258. Lateral view of *Cypris*. (After Zenker.)

glands may be excretory. The nauplius, if present, has a bivalve shell. There are among the ostracods fresh-water and marine, pelagic and bottom-living forms. Parthenogenesis is common among them, and in some males have never been found.

Cypris (Fig. 258) is a common British fresh-water genus. It swims well, by means of its antennae, but is not pelagic. It is omnivorous, feeding on algae, small animals, detritus, etc., and taking its food in various ways. Large objects are pushed into the shell by the antennae or pulled in by the mandibles, finer particles drawn in by the action of the epipodites of the maxillules (whose fan of setae is conspicuous in the figure), gathered by long bristles on the palps of the mandibles, and passed towards the mouth by the endites and endopodites of the maxillules, assisted by the gnathobase of the maxillae. The first trunk limb is used in crawling, and the second in cleaning. *Cypris* lacks the compound eyes and the heart which are found in some other members of the class—for instance in the marine *Cypridina*, which is also characterized by a large antennal exopodite, turned outwards in a notch of the shell for rowing.

SUBCLASS 3. COPEPODA

DIAGNOSIS. Free or parasitic Crustacea, without compound eyes or carapace; with biramous or uniramous palp, or with none, on the mandible; and typically with six pairs of trunk limbs, of which the first is always and the sixth often uniramous, the rest biramous, and none are situated behind the genital aperture (i.e. on the abdomen).

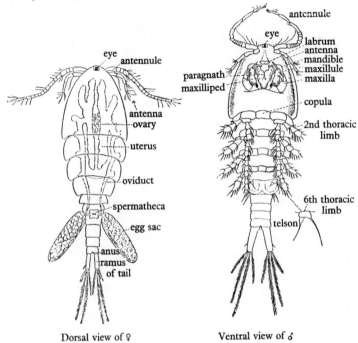

Fig. 259. *Cyclops*. (Partly after Hartog.)

CHARACTERISTICS. The *form of the body* varies greatly in the members of this subclass, from the pear-shaped or club-shaped free-swimming genera to the distorted, unsegmented, and sometimes even limbless adults of some of the parasites. In all cases in which the segmentation is complete the number of somites is the same—sixteen, including a preantennulary somite but not the telson—throughout the group, but the actual tagmata, which do not conform to the limits of the head, thorax, and abdomen, are not uniform in all members of the class.

Cyclops. We shall take as an example of the group the little fresh-water crustacean *Cyclops* (Fig. 259) which, though it is not one of the most primitive members of the Copepoda, is well segmented and can be obtained everywhere in ponds and ditches. The *shape* of this animal is that of a slender pear with a stalk. The front part of the pear is unsegmented; this is a compound head or 'cephalothorax', composed of the true head and the first two thoracic somites: beneath, in front, it bears a blunt projection, the rostrum. The rest of the broad part of the body contains three somites, the third to fifth of the thorax. The cephalothorax and these free thoracic somites are produced at the sides into low pleural folds. The stalk begins with a short somite which is united to, but distinguishable from, that which succeeds it. The next somite bears the genital openings and is therefore, on the convention we have adopted (p. 349), the last somite of the true thorax, but is usually reckoned as the first of the abdomen; in the female it is fused with the somite which succeeds it. Two free abdominal somites and a telson, which bears two styliform, setose

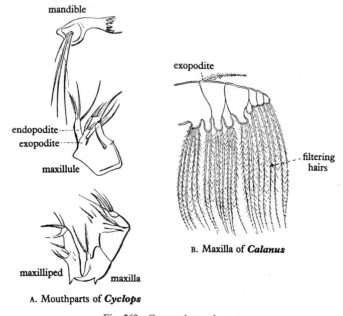

Fig. 260. Copepod mouth parts.

caudal rami, complete the body. The somites of the thorax bear limbs, which will be described presently. The limbs of the somite of the genital opening are present in the female only, and in her are reduced to the condition of small valves over the openings of the oviducts. The abdominal somites are without limbs in either sex. It will be seen that the actual *tagmata* of *Cyclops* are not the head, thorax, and abdomen, however the limit between thorax and abdomen be fixed, but are a cephalothorax of eight somites (including the preantennulary), a mid body (sometimes, but unsuitably, named the 'metasome') of three somites, and a hind body or 'urosome' of five somites and the telson.

On the head, the median *eye* is well developed. The *antennules* are long, uniramous, provided with sensory hairs, divided into seventeen segments, and in the male bent as hooks to hold the female. The *antennae* are shorter, slender, uniramous, and four-jointed. The *mandibles* (Fig. 260A) have a toothed blade (gnathobase) projecting towards the mouth and a papilla, bearing a tuft of bristles, which represents the palp. The *maxillules* have a larger gnathobase and small endopodite and exopodite. The *maxillae* are uniramous. The *maxillipeds* (first pair of thoracic limbs) are also uniramous; they stand immediately internal to the maxillae. The *2nd to 5th thoracic limbs*, of which the 2nd stands on the head, are biramous, with broad, flat, spiny rami (Fig. 259). The protopodites of each pair are united by a transverse plate or 'copula' so that they move together in swimming. The *thoracic appendages of the 6th pair* are small and uniramous.

The *swimming* of *Cyclops* is of two kinds—a slow propulsion by the antennae and antennules, and a swifter progression brought about by the use of the swimming limbs (2nd to 5th pairs) of the thorax. In the more primitive, pelagic copepods (*Calanus*, etc.) which have biramous antennae and biramous palps on the mandibles, the antennules do not take part in swimming. Such copepods *feed* by an automatic straining of particles from the water, though their apparatus for this purpose (see below) is very different from that of the Branchiopoda. *Cyclops*, on the other hand, in a manner of which the details are not understood, seizes its food particles from time to time.

The *alimentary canal* is of much the same nature as that of *Chirocephalus* but without mid gut diverticula. It possesses well-developed extrinsic muscles, of which those that run from its anterior region to the adjoining body wall produce rhythmical displacements of the canal and so cause a movement of the blood, while the dilators of the rectum draw in water which is believed to subserve respiration. Special organs for *circulation* and *respiration* are wanting in *Cyclops*, though other copepods have a saccular heart. *Maxillary glands* are present—probably entirely mesodermal. The ventral cords of the *nervous system* are concentrated into a single ganglionic mass. The *gonads* are single median structures which lie above the gut in the first two thoracic somites. The ducts are paired. In the female a large, branched uterus adjoins the ovary on each side, communicating with the lateral opening on the urosome by an oviduct which at its termination receives a duct from the spermatheca. The latter is median, in the same segment as the oviducal openings, with a median entrance of its own. The male transfers his spermatozoa to the female in a spermatophore. The eggs when laid are cemented into a packet (egg 'sac') which hangs from the opening of the oviduct, and are thus carried until they hatch. The possession of a pair of such packets gives a characteristic appearance to the females of *Cyclops*, as to those of many other copepods. In some genera, however, there is a single median packet, and in a very few the eggs are laid into the water.

The larva hatches as a typical nauplius (Fig. 249A). This is succeeded by several *metanauplius* stages, and then suddenly at a moult takes on the *first cyclops* stage, which has the general form of the adult but lacks appendages behind the 3rd pair of swimming limbs and also the somites of the urosome. In five successive cyclops stages the missing somites appear, the tale of limbs being meanwhile completed.

Calanus, which is marine and pelagic in all parts of the world, often occurring in enormous shoals which are an important item of food for fishes and whales, is in several respects more primitive than *Cyclops*, having the antennae and mandibular palps (Fig. 238 E) biramous, well-developed and biramous limbs on the 6th thoracic somite, and only one postcephalic somite in the cephalothorax. The 6th thoracic somite is included in the mid body, not in the urosome. The primitive custom of feeding by the automatic straining of food particles from the water is retained: the feeding current eddies from the swimming current which the antennae, mandibles, and maxillae set up, and is strained through a fringe of bristles on the maxillae (Fig. 260 B).

PARASITIC COPEPODS. The parasitic habit has been adopted by members of very different families of copepods, and to very various degrees even by members of a single family. Every stage may be found between normal, free-

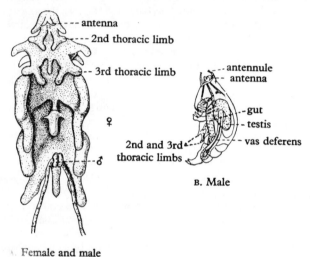

Female and male

Fig. 261. *Chondracanthus gibbosus*, an ectoparasitic copepod. (After Claus.)

living forms and the most degenerate parasites. Parasitic forms often have a suctorial proboscis, which is formed by the upper and lower lips enclosing mandibles adapted to piercing. Such a proboscis is not necessarily accompanied by a high degree of degeneration. The life histories of parasites are often complicated, and may involve remarkable changes of habit. Degenerate forms usually reach one of the cyclops stages and may pass through them all before they begin to degrade. Often the male is less degenerate than the female: he may be free-swimming while she is sedentary, or may be much smaller and cling to her body. It is only possible here to mention a few of the numerous genera of these interesting parasites.

Notodelphys, commensal in the pharynx of ascidians, is clumsy bodied, and has a large dorsal egg pouch on the 5th and 6th thoracic somites, but can swim and is sometimes captured outside the host.

Monstrilla has a very remarkable life history. The adults of both sexes are free-swimming, as are the newly hatched nauplii, but the intermediate stages are parasitic in various polychaetes, where they absorb nourishment by means of a pair of long, flexible processes which represent the antennae. In this stage they lay up a food supply for the entire life cycle, throughout which the animals are without functional mouth parts or alimentary canal.

Chondracanthus (Fig. 261), which infests the gills of various marine fishes, has in the adult stage a large female, whose body is produced into irregular, paired lobes and her appendages degenerate, though the mouth has not a proboscis but is flanked by the three pairs of minute, sickle-shaped jaws. The males are small, retain more of the copepod organization than the female, and cling by hook-like antennae to her body.

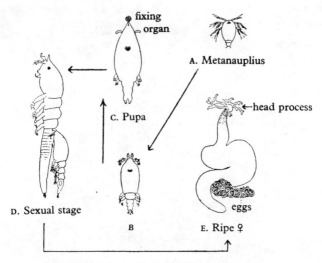

Fig. 262. Stages in the life history of *Lernaea*.

Caligus, ectoparasitic, mainly in the gill chambers of fishes, is clumsily built and has a suctorial proboscis, but retains the power of swimming. Its sexes do not differ greatly.

Lernaea (Fig. 262) hatches as a nauplius and at the first cyclops stage becomes parasitic on the gills of a flat fish, deriving nourishment from its host by means of suctorial mouth parts. Here it passes into a 'pupal' stage in which the power of movement is lost and retrogressive changes have taken place. Presently it regains the power of swimming and leaves the host in an adult copepod stage. In this stage impregnation takes place. The male develops no further, but the female attaches herself to the gills of a fish of the cod family, where by a great development of the genital somite she becomes converted into a vermiform parasite, anchored into the host by processes that grow out from her head, and retaining only the now relatively minute appendages of the thorax.

In *Herpyllobius*, parasitic on annelids, the female is reduced to a mere sac, drawing nourishment from the host by rootlets and bearing minute males which are also sac-like.

Xenocoeloma, also parasitic on annelids, is represented in the host's body only by the gonads, which are hermaphrodite, and some muscles, enclosed in a cylindrical outgrowth of the host's epithelium which forms a body wall for the vestiges of the parasite and contains a gut-like prolongation of the host's coelom.

SUBCLASS 4. BRANCHIURA

DIAGNOSIS. Crustacea, temporarily parasitic on fishes; which possess compound eyes; a suctorial mouth; carapace-like lateral expansions of the head which are fused to the sides of the first thoracic somite; an unsegmented, limbless, bilobed abdomen with a minute caudal furca; and four pairs of thoracic limbs, which are biramous, with usually a proximal extension of the exopodite.

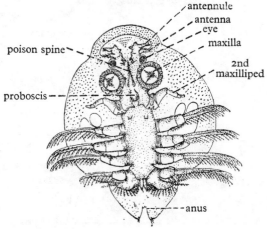

A. *Argulus americanus*, ♀

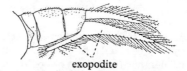

B. Second swimming leg
of *A. foliaceus*

Fig. 263. *Argulus*. (A, from Calman, after Wilson; B, after Hensen.)

CHARACTERISTICS. The members of this group in many respects superficially resemble the Copepoda, with which they are generally placed, but differ from that class in certain important features, notably in the possession of compound eyes, the lateral head-lobes, the opening of the genital ducts between the fourth pair of thoracic limbs, and the phyllopod-like proximal overhang of some of the thoracic exopodites (Fig. 263 B).

The carp-lice, as the Branchiura are called, are found both on fresh-water and marine fishes. They are good swimmers. The females deposit their eggs on stones and other objects. The larvae differ little from the adult.

Argulus (Fig. 263), the principal genus, has a pair of suckers on the maxillae and a poison spine in front of the proboscis. *A. foliaceus* is common on fresh-water fishes in Britain and the Continent.

SUBCLASS 5. CIRRIPEDIA

DIAGNOSIS. Fixed and for the most part hermaphrodite Crustacea; without compound eyes in the adult; with a carapace (except in rare instances) as a mantle which encloses the trunk; with usually a mandibular palp, which is never biramous; and typically with six pairs of biramous thoracic limbs.

CLASSIFICATION
1. *Order* Thoracica
2. *Order* Acrothoracica
3. *Order* Apoda
4. *Order* Rhizocephala
5. *Order* Ascothoracica

CHARACTERISTICS. The great majority of the Cirripedia are extremely unlike the rest of the class and would not be recognized as crustaceans at all by the layman. The familiar members of the subclass are the ordinary barnacles (Thoracica). Besides these, however, it contains several groups of related organisms, of which the parasitic barnacles (Rhizocephala) are the best known. The Ascothoracica link the subclass to other crustaceans.

1. *Order* **Thoracica**

DIAGNOSIS. Cirripedia with an alimentary canal; six pairs of biramous thoracic limbs; no abdominal somites; and permanent attachment by the preoral region.

LEPAS. We shall take as an example of this group the common goose barnacle, *Lepas* (Figs. 264, 265 A), found all the world over on floating objects in the sea. It hangs by a stalk or *peduncle* which, as we shall see, represents the foremost part of the head, greatly elongated but still bearing at its far end the vestiges of the antennules, imbedded in a cement by which it is held fast. The glands which produce the cement are contained in the peduncle, and open on the antennules.

The rest of the body is known as the *capitulum*, and is completely enclosed in the carapace or *mantle*, a fleshy structure strengthened by five calcified plates—a median dorsal *carina*, and on each side two known as the *scutum* and *tergum*. The scuta are anterior to the terga, that is, nearer to the peduncle. The mantle cavity opens by a long slit on the ventral side. Within the mantle cavity lies the body, turned over on its back with the appendages upwards (or downwards, as the animal hangs) and connected with the peduncle and mantle only at the extreme anterior end, where there is a preoral *adductor muscle* by which the sides (*valves*) of the mantle can be drawn together and so the opening closed. The *antennae*, which should be somewhere in this region, are absent. The prominent *mouth* is overhung by a large *labrum*. At its sides stand the *mandibles*, which have a flat, toothed process towards the mouth and a large, uniramous, foliaceous *palp*, and the *maxillules*, simple structures with a fringe of strong bristles on the notched median edge. A pair of simple, hairy lobes, united by a median fold, which shut in the mouth and its appen-

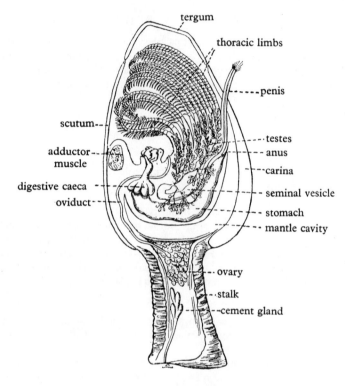

Lepas anatifera

Fig. 264. A view of *Lepas anatifera* cut open longitudinally to show the disposition of the organs. (From Leuckart and Nitsche, partly after Claus.)

dages from behind, represent the *maxillae*. The six pairs of thoracic limbs or *cirri* have each two long, many-jointed, hairy rami, curled towards the mouth. They are successively longer from before backwards. A couple of filamentous epipodites ('gills') stand on the protopodite of the first pair. Behind the cirri stands a long median ventral *penis*, and behind this again is the *anus*, with a pair of vestigial *caudal rami*.

FEEDING. The animal feeds by thrusting out the cirri through the mantle opening and withdrawing them with a grasping motion, whereby particles are gathered from the water by the setae upon the limbs. If it be molested the motion ceases and the valves are drawn to. The *alimentary canal* has an oesophagus (stomodaeum) directed forwards from the mouth to the long wide stomach which bears several caeca around its commencement and tapers behind into an intestine. Complicated *maxillary glands* open on the maxillae.

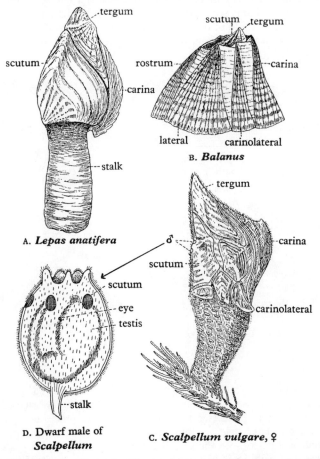

Fig. 265. Barnacles (Cirripedia; Thoracica). (A–C after Darwin; D, after Smith.)

There is no *heart* or system of blood vessels. The *nervous system* has a sub-oesophageal ganglion, and a separate ganglion for each pair of cirri behind the first.

REPRODUCTION. *Lepas* is hermaphrodite. The *ovaries* lie in the peduncle and the *oviducts* open on the bases of the first pair of thoracic limbs, much farther forwards than is usual in crustacea. The *testes* are branched tubes which lie at the sides of the alimentary canal and in the basal parts of the cirri. Each *vas deferens* enlarges into a *vesicula seminalis* whose duct joins that of its fellow in the penis. Impregnation takes place by the penis depositing a mass of spermatozoa on either side of the mantle cavity of a neighbouring individual, near the opening of the oviduct. It is possible that isolated individuals may be self-fertilized. The ova undergo their early development within the mantle cavity of the mother attached in a flat mass, the *ovarian lamella*, by a glutinous secretion manufactured by the terminal enlargement of the oviduct, to a fold of the mantle which projects on each side from near the junction with the body and is known as an *ovigerous frenum*.

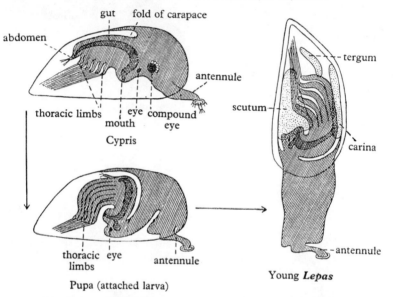

Fig. 266. Diagrams of three stages in the metamorphosis of *Lepas*. (After Korschelt and Heider.)

LIFE CYCLE. The young are set free as nauplii, characterized, as are those of nearly all cirripedes, by a pair of lateral *frontal horns*, on each of which opens a unicellular gland (see Fig. 267A). These are processes of a dorsal shield which in later stages acquires other spines. After several moults the larva suddenly passes into the so-called *cypris stage*. It is now enclosed in a bivalve shell with an adductor muscle, and possesses a pair of compound eyes. The antennules of this stage possess near their ends a disc on which opens the cement gland. The antennae have disappeared. There are six pairs of biramous thoracic limbs and a small abdomen of four somites. The cypris larva becomes fixed by the discs on its antennules, and its body rotates within the shell, so that the ventral surface is directed backwards (Fig. 266). Now the shell and body are rotated upwards on the antennae so that the adult position is assumed (Fig. 266); meanwhile the shell plates appear, the preoral region elongates to form the peduncle, and the abdomen disappears.

Scalpellum (Fig. 265 C, D) attaches itself to fixed objects, usually in deep waters. It differs from *Lepas* in possessing a number of additional plates on the capitulum, and scales of a similar nature on the peduncle. It is more remarkable in possessing what are known as *complemental males*. A few species of the genus are composed entirely of hermaphrodites as *Lepas* is. In most, however, some individuals are without female organs. These individuals are always smaller than those which possess ovaries, and live within, or at the opening of, the mantle cavity of the latter. In some species they almost perfectly resemble these in organization, but usually they are more or less degenerate, being sometimes even without an alimentary canal. As a rule the more degenerate live within the mantle cavity of the partner, the less degenerate on its mantle edge. In certain species, which have very degenerate males, the large individuals are without testes, so that the sexes are separate. The function of the complemental males is probably the effecting of cross-fertilization, for the species which possess them are of solitary habit. The phenomenon perhaps arose from the settling of young hermaphrodite individuals on the stalks of old ones, which is common in stalked barnacles.

Balanus (Fig. 265 B), the common acorn barnacle, differs from *Lepas* in the lack of a stalk, and in having an outer wall of skeletal plates homologous with some of the extra pieces on the capitulum of *Scalpellum*.

2. Order **Acrothoracica**

DIAGNOSIS. Cirripedia of separate sexes; with an alimentary canal; fewer than six pairs of thoracic limbs; and no abdominal somites; permanently sessile on the preoral region, in which the antennules are absent and the cement glands much reduced.

These are minute creatures whose females live in hollows which they excavate in the shells of molluscs, while the males are degenerate and have the same relation to the female as have those of the species of *Scalpellum* in which the sexes are separate.

Alcippe, British, lives in the columella of whelks, etc.

3. Order **Apoda**

DIAGNOSIS. Hermaphrodite Cirripedia; without mantle, thoracic limbs or anus; whose body is divided by constrictions into rings.

Proteolepas, the only known member of the order, is a small, maggot-like animal found by Darwin in the mantle cavity of the stalked barnacle *Alepas*. The antennules, by which it is attached, and the mouth parts, are those of a cirripede. Since the mouth is terminal, at least some of the more anterior of the eleven rings cannot represent somites.

4. Order **Rhizocephala**

DIAGNOSIS. Cirripedia which are parasitic, almost exclusively on decapod crustacea; have at no time an alimentary canal; and in the adult neither appendages nor segmentation; make attachment in the larva by an antennule; and are in the adult fastened to the host by a stalk from which roots proceed into the host's tissues.

Sacculina (Fig. 267), parasitic on crabs, is the best-known example of this group. Its life history is a very remarkable one. It starts life as a nauplius (Fig. 267 A), with the characteristic frontal horns of cirripede nauplii but without mouth or alimentary canal. The cypris larva (Fig. 267B) clings to a seta of a crab by one of its antennules. The whole trunk, with its muscles and appendages, is now thrown off and a new cuticle formed under the old one, with a dart-like organ which is thrust through the antennule and the thin cuticle at the base of the seta of the crab into the body of the latter. Through the dart the remnant of the larva, a mass of undifferentiated cells surrounded

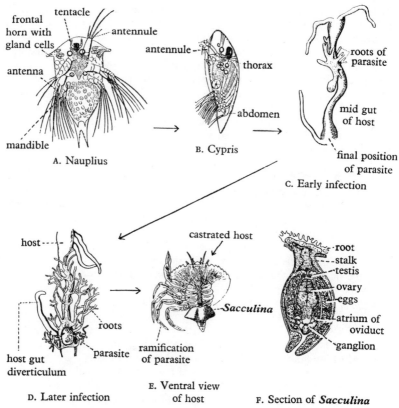

Fig. 267. Stages in the development of *Sacculina*. (After G. Smith.)

by a layer of ectoderm, passes into the host's body cavity. Carried by the blood it becomes attached to the underside of the intestine (Fig. 267C). There rootlets begin to grow out from it and eventually permeate the body of the crab to the extremities of the limbs. Meanwhile a knob also grows from the mass; forms within itself a mantle cavity surrounding an internal 'visceral mass' which contains the rudiments of genital organs and a ganglion; presses upon the ventral integument of the abdomen of the host, whose cuticle is thus hindered from forming at that spot; and consequently at the next moult of the crab comes to project freely under the abdomen, where it may be found in the adult condition.

The phenomenon known as *parasitic castration* is exhibited by crabs attacked by *Sacculina*. The moult at which the parasite becomes external produces a change in the secondary sexual characters in the new cuticle. The male crabs have a much broader abdomen, reduced copulatory styles (these may disappear altogether), and abdominal swimmerets (which carry the eggs in the female, and are absent in the normal male). There is, in short, a marked

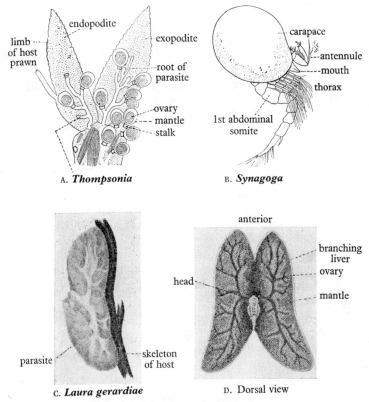

Fig. 268. *Thompsonia, Synagoga, Laura*, parasitic cirripedia. (A, after Potts; B, after Norman; C–D, after Lacaze-Duthiers.)

tendency to the female type. In the female crabs there is also a change, but this is held to be not towards the male but towards the juvenile type. The gonads disappear, but cases have been observed in which the parasite has been killed and months afterwards what was probably an originally male crab has regenerated a hermaphrodite gonad. Parasitic castration is the most evident expression of a remarkable and at present ill-understood interference by the parasite with the general metabolism and hormonic balance of its host.

Thompsonia (Fig. 268A), parasitic on crabs, hermit crabs, etc., is an extraordinary case of extreme reduction by parasitism, in which an arthropod is degraded to the level of a fungus. The rootlets of the parasite are widely diffused through the host. Their branches in the limbs give off sacs which become external at a moult of the host. These sacs contain neither ganglion, generative ducts, nor testes, but only a number of ova in a space of doubtful nature. When they are ripe the ova have become (probably by parthenogenesis) cypris larvae, which are set free by the formation of an opening. There is no parasitic castration of the host.

5. Order Ascothoracica

DIAGNOSIS. Parasitic cirripedia, which have an alimentary canal from which diverticula extend into the mantle; six pairs of thoracic appendages; and a segmented or unsegmented abdomen; and are not attached by the preoral region.

CHARACTERISTICS. These animals are parasitic and often imbedded in the tissues of their hosts. They are an early branch of the cirripede stock which has retained the abdomen, in some cases well segmented and provided with movable caudal rami, and has not the characteristic mode of fixation by the antennules, or frontal horns in the nauplius.

Laura (Fig. 268 C, D), imbedded in the tissues of the antipatharian *Gerardia*, has the mantle in the form of a very spacious sac with a narrow opening. Its abdomen has two somites and a telson.

Synagoga (Fig. 268 B), external parasite on *Antipathes*, has a bivalve mantle, from which usually protrudes the long abdomen of four somites and a telson. It is possible that this is an immature stage of an animal which is more retrograde when it is adult.

SUBCLASS 6. MALACOSTRACA

DIAGNOSIS. Crustacea with compound eyes, which in typical members of the group are stalked; typically a carapace which covers the thorax; the mandibular palp, if present, uniramous; a thorax of eight somites and abdomen of six (rarely seven), all (except the 7th abdominal) bearing appendages; and a complex proventriculus.

CLASSIFICATION

    1. *Order* Leptostraca
    2. *Order* Hoplocarida
    3. *Order* Syncarida
    4. *Order* Peracarida
      Suborder **1.** Mysidacea
      Suborder **2.** Cumacea
      Suborder **3.** Tanaidacea
      Suborder **4.** Isopoda
      Suborder **5.** Amphipoda
    5. *Order* Eucarida
      Suborder **1.** Euphausiacea
      Suborder **2.** Decapoda

The Malacostraca fall into two large groups and three smaller ones. Of the latter, the *Leptostraca* retain, in the hinder end of the abdomen, a primitive condition which has been lost in the other groups. The *Hoplocarida* (Stomatopoda) stand alone in possessing two free pseudosomites in the anterior part of the head, certain peculiarities of the thoracic limbs, and peculiar gills on the abdominal appendages. The *Syncarida* unite certain features which are characteristic of other groups. The large groups *Peracarida* and *Eucarida* contain most of the members of the subclass. The former of these two divisions is characterized by possessing a brood pouch, formed by plates (*oostegites*) upon the thoracic limbs, in which the young undergo a direct development, and by the freedom of some or all of the thoracic somites from the carapace. The Eucarida do not possess a brood pouch and usually have larval stages, their heart is a short chamber in the thorax, and their carapace fuses with the dorsal side of each thoracic somite. Independently in each of these two groups the caridoid facies has been lost to various degrees, so that the members of each can be roughly arranged in a series which, starting with prawn-like 'schizopods', ends in the Peracarida with the woodlice and in the Eucarida with the crabs.

CHARACTERISTICS. The Malacostraca contain a very large number of species, which exhibit great diversity. Nevertheless they are capable of reference to a common type in respect of more features than the members of any other group, though the Copepoda approach them in this. The *ideal malacostracan* has twenty somites, including the preantennulary and excluding the telson. Of these, six belong to the head (p. 347), eight constitute the thorax, and six the abdomen. This number is only departed from in the Leptostraca, which have an additional somite at the end of the abdomen. (In the embryos of Mysidacea such an additional somite is present, but in the adult it has fused with that which precedes it.) The female openings are always on the 6th thoracic somite, and the male on the 8th. A carapace encloses the thorax at the sides. The median eye is vestigial in the adult, and the compound eyes stalked. The antennules are biramous, as they are in no crustacean of any other group. The antennae have a scale-like exopodite by extending which the animal keeps its body level in the water. The mandibles have uniramous palps and the part which projects towards the mouth is cleft into 'incisor' and 'molar' processes. The maxillules have two endites (on the first and third joints) and the maxillae four, grouped in twos. The thoracic limbs have a cylindrical, five-jointed endopodite (p. 351), used when the animal has occasion to walk or to grasp large particles of food, a natatory exopodite, and two respiratory epipodites. The abdominal appendages are biramous; those of the first five pairs (*pleopods*) slender and fringed and used in swimming, those of the last pair (uropods) broad, turned backwards, and forming with the telson a tail-fan, used in rapid backward movement. There are no caudal rami. (The Leptostraca are the only members of the subclass which possess these rami in the adult.) Food is chiefly collected as particles in a stream which is set up by the action of the maxillae and which passes forwards through a filtering fringe of bristles upon the median margins of those appendages.

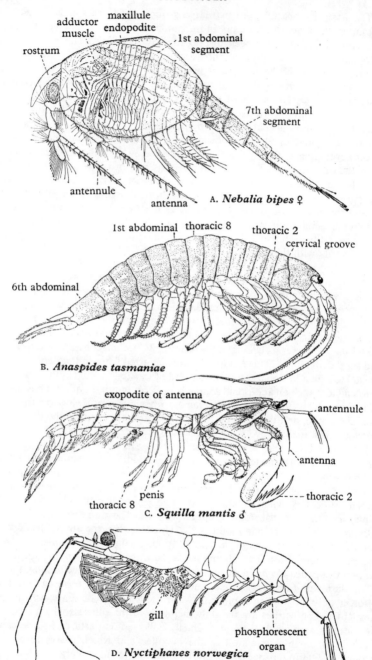

Fig. 269. Malacostraca. (A, from Calman, after Claus; B, from Woodward; C, from Calman; D, from Watase.)

This type is said to possess the *caridoid facies*. It is adapted primarily to swimming and is best exhibited in the small, prawn-like, pelagic forms, formerly classed together as *Schizopoda* but now distributed, as the suborders Mysidacea (Fig. 270A) and Euphausiacea, to the two main orders of the Malacostraca (see below). Departures from it are many and important, and most of its features have disappeared more than once independently. Thus the carapace, the inner ramus of the antennule, the scale of the antenna, the mandibular palp, exopodites of thoracic limbs, etc., have been lost in various branches of the malacostracan tree Only the number of the somites and the size of the tagmata are constant, save in the case of the Leptostraca already mentioned and in certain parasitic isopods. Departure from the caridoid facies is associated with the abandonment of the swimming habit for crawling or burrowing, and when that happens the animal ceases to gather food by filtration and adopts other modes of feeding, for which its limbs, and particularly the thoracic endopodites, become variously modified—as, for instance, by the development of chelae.

DEVELOPMENT. An exceptionally large number of members of this subclass have direct development. Of those which possess larvae only a few (*Euphausiacea*, a few of the Decapoda) hatch in the nauplius stage. A special characteristic of the larval development of the Malacostraca is the occurrence of a *zoaeal* stage (p. 366), in which the carapace and tagmata are present, the abdomen is better developed than the hinder part of the thorax, and the animal swims by biramous maxillipeds. In crabs, hermit crabs, and some related families the *zoaea* is succeeded by a *metazoaea*, which differs from it in having uniramous rudiments of thoracic limbs behind the maxillipeds. In other forms with larval development there is at this stage a prawn-like *schizopod larva* (*mysis* stage), with biramous limbs on all the thoracic somites, which is not always preceded by a zoaea.

1. *Order* Leptostraca

DIAGNOSIS. Malacostraca with a large carapace provided with an adductor muscle and not fused with any of the thoracic somites; stalked eyes; the thoracic limbs all alike, without oostegites, biramous, and usually foliaceous; seven abdominal somites, of which the last bears no appendages; and caudal rami on the telson.

Nebalia (Fig. 269A) is the commonest and typical genus of this group. *N. bipes*, the British species, may be found between tidemarks, under stones, especially in spots which are foul with organic remains. *Nebalia* has a rostrum, which is jointed to the head. The antennae have no scale, while the antennules are unique in possessing one. The carapace has an adductor in the region of the maxilla and encloses the four anterior abdominal somites. The thorax is short. Its limbs (Fig. 238G) are flat. Their endopodite is narrow and possesses five indistinct joints. Sometimes the long basipodite is divided and its distal region added to the endopodite as a preischium. The exopodite is broad and there is a very large epipodite, which serves as a gill. (The related *Paranebalia*, however, has a slender exopodite with a flagellum, and a small epipodite.) The first four pairs of abdominal limbs are large and biramous, the fifth and sixth small and uniramous.

The *alimentary canal* possesses a proventriculus of relatively simple type, several pairs of simple mid-gut caeca, and an unpaired posterior dorsal caecum. The *heart* is long, reaching from the head to the 4th abdominal somite. The *nervous system* is of primitive type (p. 355). The *excretory organs* have been alluded to on pp. 361, 362.

The animal *feeds* by straining particles from the water by means of an elaborate arrangement of setae of different kinds on the thoracic limbs, the necessary currents being set up by a pumping action of the same limbs. These work upon a principle similar to that employed by the Branchiopoda, the exopodites and epipodites acting as valves for pumping chambers between the limbs, but it is the *backward* stroke that enlarges the chambers, and they are closed by the *forward* flapping of their valves. *Development* is direct, the embryos being carried between the thoracic limbs of the mother, held in by the long setae on the limbs, but not glued to them like the eggs of the crayfish.

2. *Order* Hoplocarida (Stomatopoda)

DIAGNOSIS. Malacostraca with a shallow carapace which is fused with three thoracic somites and leaves four uncovered; two free pseudosomites on the head; stalked eyes; the first five thoracic limbs subchelate and the last three biramous; no oostegites; a large abdomen whose first five pairs of limbs bear gills on the exopodites, while the sixth forms with the telson a tail fan; and a large, branched 'liver'.

CHARACTERISTICS. The *second thoracic limb* bears a large, raptorial subchela. The *alimentary canal* has a rather simple proventriculus and a large branched 'liver'; the latter and the gonads extend along the large abdomen. In the *nervous system* eight pairs of ganglia are fused as the suboesophageal ganglion. The *heart* is very long, reaching from the head to the fifth abdominal somite. The *excretory organs* are maxillary glands. The *larvae* are pelagic and of the same general type as the zoaea but with a peculiar facies of their own (Fig. 249 E).

The members of the order are all marine, and for the most part live in burrows.

Squilla (Fig. 269 C) occurs in British waters.

3. *Order* Syncarida

DIAGNOSIS. Malacostraca without carapace; with eyes stalked, sessile or absent; most of the thoracic limbs provided with exopodites and none of them chelate or subchelate; no oostegites; a tail fan; and simple caeca on the mid gut.

CHARACTERISTICS. A small group of fresh-water malacostracans with a combination of features which forbids their inclusion in either of the other orders. In typical genera, they possess most of the features of the caridoid facies except the carapace; and the relatively slight differentiation of thorax from abdomen is a primitive character possessed by no other member of the subclass.

Anaspides (Fig. 269 B), from pools at 4000 ft. in Tasmania, is a normal member of the group.

MALACOSTRACA 401

Bathynella, from subterranean waters in Central Europe and England, small, degenerate, and eyeless, has various limbs reduced or absent and the first thoracic segment free.

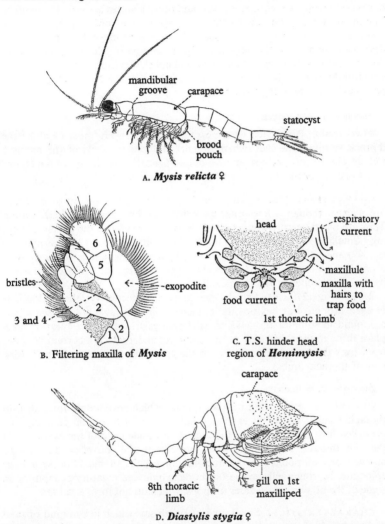

Fig. 270. Peracarida. (A, after Sars; C, after Cannon and Manton; D, after Sars.)

4. *Order* **Peracarida**

DIAGNOSIS. Malacostraca whose carapace, if present, does not fuse with more than four thoracic somites; whose eyes may be stalked or sessile; and which possess oostegites; a more or less elongate heart; and a few simple caeca on the mid gut.

CHARACTERISTICS. A large order containing several suborders, which range from the prawn-like Mysidacea, in which the caridoid facies (pp. 397, 399) is practically intact, to the Isopoda and Amphipoda (slaters and sandhoppers) in which the carapace is lost and other features are greatly modified. The important common characters which all these suborders possess are the presence of oostegites and the retention of the young, which are directly developed, in a brood pouch formed by those organs. Certain peculiarities, however, of the mandibles, which bear behind the incisor process a moveable structure known as the *lacinia mobilis*, of the thoracic limbs, etc., are also possessed in common by the Peracarida.

Suborder 1. Mysidacea

DIAGNOSIS. Peracarida with a carapace which covers most or all of the thoracic somites; the eyes (when present), stalked; the scale of the antenna well developed; exopodites on most or all of the thoracic limbs, of which one or two pairs are maxillipeds; and a well-formed tail fan.

CHARACTERISTICS. Small, usually pelagic crustaceans, most of which are marine, though a few occur as 'relicts' or immigrants in fresh waters. They are mostly carnivorous, but take vegetable matter in the course of feeding. Small food particles are obtained in a current set up by the maxillae (p. 399) and when there are no gills also by a whirling action of the thoracic exopodites, and are strained off by the maxillae: large food masses are seized by the endopodites of the thoracic limbs.

Mysis (Fig. 270A, B), British, possesses a statocyst on the endopodite of each uropod, but has not the branched gills (thoracic epipodites) which are found in some of the Mysidacea (Lophogastridae). Its respiration takes place through the thin lining of the carapace, under which a current is drawn from over the back by the action of the epipodites of the maxillipeds (first pair of thoracic limbs).

Suborder 2. Cumacea

DIAGNOSIS. Peracarida with a carapace which covers only three or four thoracic somites but is on each side inflated into a branchial chamber and produced in front of the head to lodge the expanded end of the exopodite of the first thoracic limb; eyes (when present) sessile; no exopodite on the antenna or endopodite on the maxilla; three pairs of maxillipeds; a large epipodite, bearing a gill, on the 1st thoracic limb and natatory exopodites on some of the others; and slender uropods, which do not form a tail fan.

CHARACTERISTICS. Small, marine organisms which live in mud or sand and are highly specialized, especially in their respiratory mechanism, for that habitat. The first thoracic exopodites form a valved exhalant siphon with the carapace lobes which lodge them.

Diastylis (Fig. 270D) is a British genus.

Suborder 3. Tanaidacea

DIAGNOSIS. Peracarida with a very small carapace, covering only two thoracic somites, with which it fuses; eyes (if present) on short, immovable stalks; a small scale, or none, on the antenna; thoracic exopodites absent or vestigial, a branchial epipodite on the maxilliped; and slender uropods, which do not form a tail fan.

CHARACTERISTICS. Small, marine crustaceans, usually inhabiting burrows or tubes, which are in an intermediate condition between the Cumacea and Isopoda in respect of the loss of the caridoid facies.

Apseudes (Fig. 276 A), and *Tanais*, which differs from it in having short, uniramous antennules and uropods and no antennal scale, and lives in a mass of fibres it secretes, are British genera.

Suborder 4. Isopoda

DIAGNOSIS. Peracarida without carapace; with sessile eyes; the body usually depressed; the antennal exopodite absent or minute, the thoracic limbs without exopodites, the first pair modified as maxillipeds, the remainder usually alike; the pleopods modified for respiration, and the uropods usually not forming a tail fan. (Any of these features may be absent in the adults of parasitic forms.)

CHARACTERISTICS. The Isopoda are a large group and exhibit much variety. We will study as an example *Ligia*, the shore slater (Fig. 271 A), found just above tidemarks in Britain and most parts of the world. This creature has a depressed, oval *body*, the cephalothorax, formed by fusion of the 1st thoracic somite with the true head, lying in a notch on the anterior edge of the 2nd somite of the thorax. Two large, sessile compound eyes take up the sides of the head. The abdomen continues the outline of the thorax, and its 6th somite is fused with the telson. The *antennules*, which are usually short in isopods, are here minute. The *antennae* are of a good length, which is due to the elongation of the two joints which precede the flagellum. The *mandibles*, unlike those of most isopods, lack the palp, but otherwise they are complicated, having between the incisor and molar processes a row of spines and the movable structure known as the *lacinia mobilis* (Fig. 272B) which is characteristic of the Peracarida. The maxillules and maxillae are less well developed than those of most isopods. The *maxillipeds* are broad and close the mouth region from behind. The rest of the *thoracic limbs* are uniramous and leg-like. Their coxopodites are fused with the body, so that the brood pouch plates (oostegites) of the female, which are epipodites of the legs, seem to arise from the sterna. The first five pairs of *abdominal limbs* are broad, with plate-like, respiratory endopodite and exopodite. The endopodite of the second pair of the male is produced into a copulatory style. The uropods have slender, styliform rami. The *alimentary canal* has an elaborate proventriculus, adapted, not to chew the food, but to press the juices out of it and to strain off solid particles from them; and there are three pairs of mid-gut caeca. The *heart* lies in the hinder part of the thorax and in the abdomen, where blood returns from the respiratory limbs to the pericardium. The *nervous system* has a concentration of ganglia in the abdomen as well as one for the mouth parts. The *gonads* are paired, and the testes bear three follicles, characteristic of the Isopoda (see Fig. 273). The young when set free from the brood pouch resemble the adult but lack the last pair of legs. *Ligia* is omnivorous, but chiefly eats *Fucus*. It gnaws with its mandibles, feeding hurriedly at low tide

Armadillidium, the common woodlouse, is more completely terrestrial in its habits than *Ligia*. Its antennae and uropods are short and thus permit the body to roll up into a ball in the familiar manner. The air tubes on the abdominal limbs have been alluded to on p. 363.

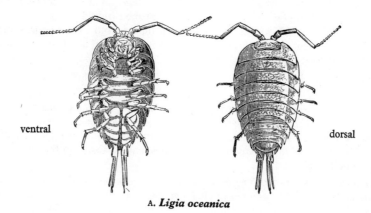

Fig. 271. Isopoda. (A, B, from *Cambridge Natural History*; C, after Fraisse.)

Asellus (Fig. 273), the hog slater, is a common fresh-water crustacean. It differs from *Ligia*, among other ways, in having all the abdominal somites fused, a flagellum on the antennule, a palp on the mandible, and free coxopodites on the legs.

Idotea, common among weeds, etc., on the British coast, differs from *Ligia* in having the last four abdominal somites fused with the telson and the uropods turned inwards as valves to cover the pleopods.

PARASITIC ISOPODS. Many of the Isopoda are parasitic. Among these there is found every grade from well-organized temporary parasites to some which are as adults mere sacs of eggs. *Aega* (Fig. 272A), a fish louse, has the ordinary isopod form, though heavily built, and with piercing mouth parts and some of the legs hooked. Its broad uropods form a tail fan. *Bopyrus* (Fig. 271 B), in the gill chamber of prawns, with dwarf males, is more degenerate but still recognizable as an isopod. *Cryptoniscus* (Fig. 271 C), a 'hyperparasite' on members of the Rhizocephala and a protandrous hermaphrodite, is extremely degenerate. Many of these parasites produce parasitic castration (see p. 395).

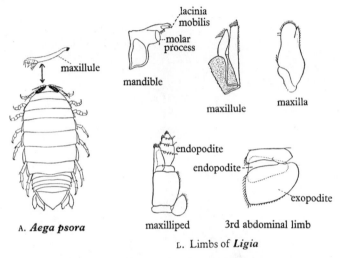

A. *Aega psora*

L. Limbs of *Ligia*

Fig. 272. Isopoda. (A, from Sars.)

Suborder 5. Amphipoda

DIAGNOSIS. Peracarida without carapace; with sessile eyes; the body usually compressed; no antennal exopodite; the thoracic limbs without exopodites, the first pair modified as maxillipeds, the remainder of more than one form, the second and third usually prehensile; the pleopods when fully developed divided into two sets, the first three pairs with multiarticulate rami, the last two resembling the uropods, which do not form a tail fan.

GAMMARUS. We will take as an example of this suborder *Gammarus* (Figs. 274, 275), of which closely related species occur in Britain in fresh waters and between tidemarks in the sea. The *body* of this animal is compressed and elongated, with the 1st thoracic somite fused to the head and no sharp distinction between the thorax and abdomen, which are of nearly equal length. At the sides of the head are pleural plates. The pleura of the thorax are short; but large, hinged coxal plates on the legs take their place. All the segments of the abdomen are free. The telson is deeply cleft. The antennules have two flagella; the uniramous antennae are much like those of *Ligia*. The mandibles have the same parts as those of *Ligia*, with a palp. The maxillules, maxillae, and maxillipeds are shown in Fig. 275A. The maxillipeds are united by the fusion of their coxopodites. The first two pairs of legs are subchelate, the third and fourth pairs are turned forwards and help the subchelae in feeding, the last three pairs are turned backwards and used when the animal crawls on

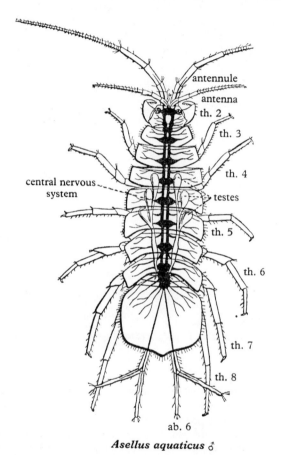

Asellus aquaticus ♂

Fig. 273. *Asellus aquaticus*. Dorsal view; th. 2–th. 8 = thoracic limbs. (From Leuckart and Nitsche.)

its side. The first three pairs of abdominal limbs are used in swimming and to direct water towards the gills, the last three pairs are used together to kick the ground in jumping. Simple *gills* (epipodites) are found on the coxopodites of the legs, and oostegites on those of the third to fifth pairs in the female (Fig. 274). The *alimentary canal* has a single-chambered but complex proventriculus, two pairs of 'hepatic' caeca, and a pair of caeca at the hinder end of the mid gut which have been supposed to be *excretory*. The principal

MALACOSTRACA 407

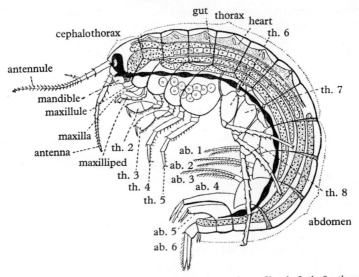

Fig. 274. *Gammarus neglectus*; female bearing eggs, seen in profile; th. 2–th. 8 = thoracic limbs; ab. 4–ab. 6 = posterior abdominal limbs. (From Leuckart and Nitsche, after Sars.)

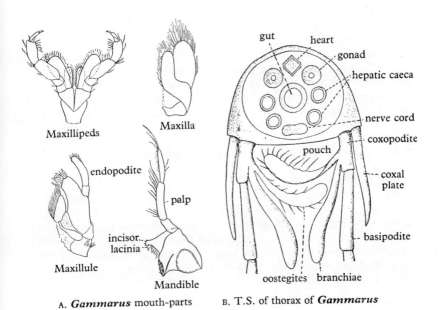

A. ***Gammarus*** mouth-parts   B. T.S. of thorax of ***Gammarus***

Fig. 275. Anatomy of *Gammarus*.

organs of excretion are antenna glands. The *heart* extends from the 7th to the 1st thoracic somite. The *young* are born with all their legs. The females with young are carried by males. After they have parted with the young they moult and are immediately re-impregnated. When the cuticle has set they are liberated.

Caprella (Fig. 276D), slender-bodied and living upon seaweeds, hydroids, etc., has two thoracic somites in the cephalothorax, no legs on the 4th and 5th thoracic somites, all the remaining legs subchelate, and the abdomen reduced to a minute stump.

Cyamus, the whale louse (Fig. 276B), is a *Caprella* with a short, wide body, adapted to its habit and habitat.

Phronima (Fig. 276C), marine and pelagic, often inhabiting pelagic tunicates, jellyfish, etc., is transparent and has a large head with immense eyes.

5. Order Eucarida

DIAGNOSIS. Malacostraca with a carapace which is fused with all the thoracic somites; stalked eyes; no oostegites; a short heart situated in the thorax; and a large, branched 'liver'.

The differences between the two orders which compose this subclass are not great. The small, prawn-like Euphausiacea are not far from the lower genera of the true prawns, members of the Decapoda.

Suborder 1. Euphausiacea

DIAGNOSIS. Eucarida in which the exopodite of the maxilla is small; none of the thoracic limbs are maxillipeds; there is a single series of gills, and these stand upon the coxopodites of thoracic limbs; and there is no statocyst.

The Euphausiacea are marine and pelagic, and at times form an important part of the food of whales. Like many pelagic animals they possess (in nearly all species), phosphorescent organs, which in this case are complex and situated on various parts of the body. They are filter feeders. Most (perhaps not all) are hatched as nauplii, and subsequently pass through stages of the zoaea type.

Nyctiphanes (Fig. 269D) is a British example of the group.

Suborder 2. Decapoda

DIAGNOSIS. Eucarida in which the exopodite (scaphognathite) of the maxilla is large; three pairs of thoracic limbs are more or less modified as maxillipeds, and five are 'legs'; there is usually more than one series of gills, of which some (*podobranchiae*) stand upon the coxopodites of thoracic limbs, others (*arthrobranchiae*) upon the joint-membranes at the bases of the limbs, and others (*pleurobranchiae*) upon the sides of the thorax; and a statocyst is usually present in the proximal joint of each antennule.

CHARACTERISTICS. The Decapoda owe their name to the condition of the hinder five pairs of thoracic limbs, which are adapted for locomotion, typically by walking but sometimes by swimming. Often, however, as in the crayfish, one of these pairs bears large chelae and is incapable of the locomotory function: others may also be incapacitated for it, as, for instance, the

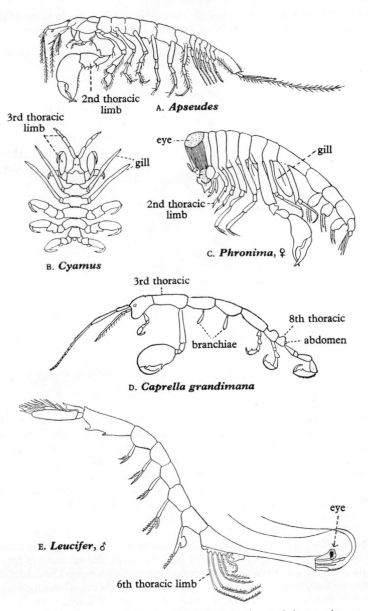

Fig. 276. Malacostraca. The numbers indicate somites or their appendages.

two small hinder pairs of the hermit crabs (Fig. 284A). Only in some of the lower genera is there any vestige of the exopodite upon these five pairs.

MACRUROUS TYPES. This order contains the most highly organized crustaceans. Among its members there is great diversity in the habit of body and in the form of the appendages, but two principal types can be observed. In the first or macrurous type the caridoid facies is in the main retained, the body is long and subcylindrical or somewhat compressed, the abdomen is long and ends in a tail fan, the appendages are usually slender, and any of the legs may be chelate. An example of this type, the common crayfish, *Astacus* (Figs. 223, 235, 237, 240, 242, 244, 246, 247, 277), is described in most textbooks of elementary zoology.

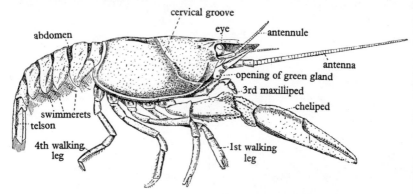

Fig. 277. The common crayfish, *Astacus fluviatilis*, seen from the side. (From Shipley and MacBride.)

BRACHYUROUS TYPES. The second or brachyurous type—which is not confined to the Brachyura *sensu stricto* but occurs independently in various members of certain groups, known collectively as the Anomura, that are intermediate between the macrurous divisions of the order and the Brachyura —has the cephalothorax greatly expanded laterally and more or less depressed, while the abdomen is reduced and folded underneath the cephalothorax. In it the appendages are as a rule shorter and stouter than in macrurous forms, and only the first pair of legs has a true chela.

The families *Penaeidea* (primitive prawns), *Caridea* (prawns and shrimps), *Astacura* (crayfishes and lobsters), and *Palinura* (crawfishes and bear-crabs) are macrurous. They are for the most part swimmers, though some of them, as the Astacura and Palinura, do more walking than swimming. The suborders *Anomura* and *Brachyura* are walkers, though some of the crabs have their own ways of swimming by means of flattened legs. The Brachyura proper are distinguished from other brachyurous forms by the occurrence in nearly all the latter of well-formed uropods, which the true crabs do not possess, and by a fusion of the edge of the carapace with the epistome, a sternal plate which lies in front of the mouth (see p. 412).

CARCINUS MAENAS. As an example of the Brachyura we shall describe

Carcinus maenas, the common shore crab of Britain (Figs. 278–282). The depression to which is due the difference in shape between the cephalothorax of this typical crab and that of a crayfish or prawn has brought it about that in a transverse section (Fig. 280 B) the carapace has at the sides (where, as the branchiostegite, it covers the gills) not an arched profile but runs out almost horizontally and is then bent in, at a sharp angle which is more acute in the anterior part of the body than in the hinder part, to end against the flank above the coxopodites of the legs. At the angle, the branchiostegite, viewed from above, describes the lateral part of the outline of the body. That outline begins between the eyes, where in the crayfish the rostrum

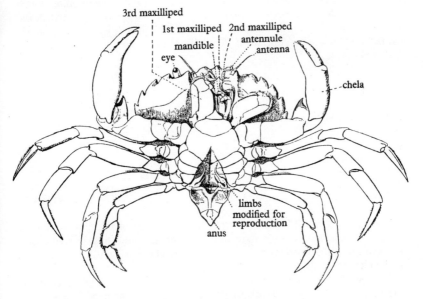

Fig. 278. The shore crab, *Carcinus maenas*, ventral aspect.
(From Shipley and MacBride.)

stands, with the *front*, a low, three-toothed lobe. On each side of this is the *orbit*, an excavation of the surface of the head for the reception of the eye. From the orbit the notched *anterolateral edge* curves outwards and backwards as a crest on the branchiostegite, forming with its fellow and the front a semicircle. From each end of the semicircle a slightly concave *posterolateral edge* carries the outline slanting inwards to the short, transverse *posterior edge* of the carapace.

To return to the transverse section: the thin inner layer of the fold which makes the branchiostegite is not so much drawn out as the stout outer layer, so that a considerable space is left between them. In the hinder region the two layers are not very widely separated, and there are in this space only blood channels and connective tissue, but anteriorly branches of the liver and gonad intrude there. The edge of the branchiostegite fits close against the flank of the thorax and the exopodites of the maxillipeds, leaving however the following openings: (1) small slits, one above each leg, (2) a large opening in front of the coxopodite of the chela, (3) a still larger opening in front of the mouth. These openings lead to and from the *gill chamber*. In the flattening of the body, the lateral wall of the thorax has come to face in great part upwards, so that the gills, instead of being directed vertically from their attachments, are directed more or less horizontally inwards over the convex, mound-like inner wall of the gill chamber. The *gills* are of the kind known as *phyllobranchiae*. That is, the axis of each, instead of bearing filaments as in the gills of the crayfish (*trichobranchiae*), has on either side a row of plates, set close like the leaves of a book. The podobranchiae stand out from the base of an epipodite, which bears also a slender process known as a *mastigobranchia*. In the crayfish the gill lies along this and is fused with it. The first maxilliped has a mastigobranchia without a podobranchia. The gill series of *Carcinus* is shown in the following table:

Table 8

	Mxpd I	Mxpd II	Mxpd III	Leg I (Cheliped)	Leg II	Leg III	Leg IV	Leg V	Total
Podobranchiae	—	1	1	—	—	—	—	—	2
Anterior arthrobranchiae	—	1	1	1	—	—	—	—	3
Posterior arthrobranchiae	—	—	1	1	—	—	—	—	2
Pleurobranchiae	—	—	—	—	1	1	—	—	2
Mastigobranchiae	(1)	(1)	(1)	—	—	—	—	—	(3)
Total	(1)	2+(1)	3+(1)	2	1	1	—	—	9+(3)

The mastigobranchiae lie in the gill chamber, that of the first maxilliped in the *epibranchial space* above (external to) the gills and those of the second and third maxillipeds in the *hypobranchial space* below the gills. Their function is the cleaning of the gills.

In front, the gill chamber narrows to an *exhalant passage*, which contains the scaphognathite and leads to the large anterior opening. The scaphognathite, working to and fro, drives water out of this opening and so draws in a current through the other apertures. The opening in front of the chela can be closed by a flange on the coxopodite of the third maxilliped, and so the current can be regulated. The water which enters this opening is prevented from taking a short cut to the exhalant passage by a large expansion of the base of the mastigobranch of the first maxilliped, which directs it under the gills. The current from the openings over the legs also passes under the gills. All the water then passes upwards through the gills into the epibranchial space above them and so to the exhalant passage. Thus the gills are thoroughly bathed.

Owing to the width of the body the *sterna* are more easily distinguished than in the crayfish. Those of the maxillulary to second maxillipedal somites are fused into a triangular mass. In front of the mouth the plate known as the *epistome* represents the mandibular and antennal sterna. From this a ridge extends to the median rostral tooth, separating two sockets in which stand the antennules. A downward process from the front, abutting on the basal joint of the antenna, separates each of these sockets from the orbit of its side. The two-jointed *eyestalk* arises close to the median line and passes through a gap between the frontal process and the antennal base to enlarge within the orbit.

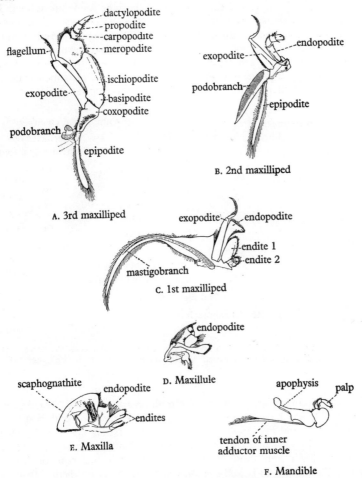

Fig. 279. Mouth parts of the right side of *Carcinus maenas*. (From Borradaile.)

The *abdomen* is reduced to a flap, turned forwards and closely applied to the sterna of the thorax. Its ventral (upper) cuticle is thin. It is broader in the female than in the male, in which its 3rd to 5th somites are fused. Two small knobs on the 5th thoracic sternum, fitting into sockets on the 6th somite of the abdomen, lock the two together as by a press button.

LIMBS. The *antennules* have short flagella and can fold back into the sockets mentioned above. The *antennae* also have a short flagellum. They have no exopodite (scale) and their coxopodite is represented by a small operculum over the opening of the antennary ('green') gland. The *mouth parts* are shown in Fig. 279. In the *mandibles*, the biting edge (incisor process) is toothless and the molar process reduced to a low mound behind the biting edge. The palp is stout and the first two of its three joints are united. The *maxillules* and *maxillae* have the usual endites well developed. The scaphognathite of the maxilla is shaped to fit the exhalant passage of the gill chamber. The *maxillipeds* have epipodites produced into long, narrow mastigobranchs, fringed with bristles which brush the gills. The flagella of their exopodites are turned inwards and the endopodite of the first of them is expanded at the end and helps to border the exhalant opening for the respiratory current. The

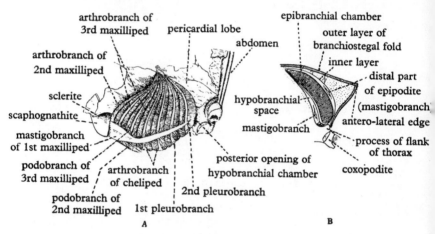

Fig. 280. Gill structure of *Carcinus*. A, dorsal view of the organs in the left branchial chamber. B, diagram of a transverse section through a branchial chamber. (From Borradaile.)

third pair are broad and enclose the mouth area from below. The *legs* lack an exopodite and have the usual joints (p. 351) in the stout endopodite, but the basipodite and ischiopodite are united. The first leg has a strong chela: concerning the physiology of its muscles something is said on pp. 140, 326. The others differ from those of the crayfish chiefly in that none of them are chelate. The animal, as is well known, walks sideways with them. *Abdominal limbs* are present in the female only on the 2nd to 5th somites. On a short, one-jointed protopodite they bear two long, equal, simple rami, covered with setae to which, as in other decapods, the eggs are attached by a covering secreted by dermal glands. In the male, the abdomen bears limbs only on its first two somites, and they are uniramous and adapted for transferring the sperm, the endopodite of the second working as a piston in a tube formed by that of the first.

FEEDING. In feeding the food is seized by the chelae, which place it between the mandibles. These do not chew it, but, unless it be soft enough for

MALACOSTRACA 415

them to sever a morsel when they close upon it, they hold it while the morsel is severed by the action of the hinder mouth limbs, The basal endites of the maxillules, the mandibular palps, and the pointed labrum push the food into the mouth.

ALIMENTARY CANAL. The alimentary canal resembles in general features that of the crayfish. Its fore gut has a similar apparatus for chewing, pressing, and filtering the food (Fig. 242). Its mid gut is short, and bears a pair of long

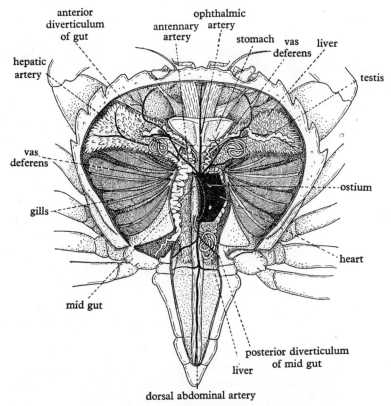

Fig. 281. The shore crab, *Carcinus maenas*, dissected from the dorsal side to show the viscera.

dorsal caeca which end, each in a coil, at the sides of the cardiac division of the proventriculus or 'stomach'. The hind gut, just before entering the abdomen, gives off dorsally a long tube coiled into a compact mass. The 'liver' is large and enters the carapace fold. In the *antennal glands* the whitish medullary portion found in the crayfish is lacking, and the bladder is prolonged into processes which lie among the other viscera.

In the *nervous system* (Fig. 282) the postoral ganglia are concentrated into a mass around the sternal artery. The *vascular system* (Fig. 281) is on the same plan as that of the crayfish. The *gonads* are in both sexes united across the middle line and prolonged laterally into the carapace fold. Each oviduct bears a spermatheca. The female opening is sternal; the male is on a flexible process of the coxa of the last leg.

LARVAE. The first larva is a typical zoaea (Fig. 249 F) with large compound eyes, carapace, rostrum, short unsegmented thorax, and long, strong abdomen with forked telson. Of its thoracic limbs only the first two pairs are present. After its first moult, which takes place almost at once, it has a median dorsal spine. The latter two features are characteristic of the zoaea of the crabs.

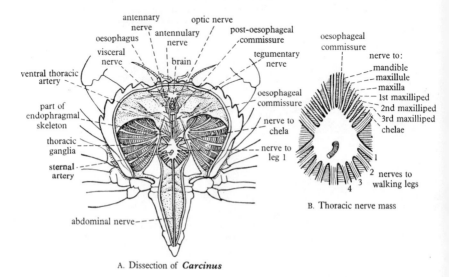

Fig. 282. A, *Carcinus maenas*, dissected from the dorsal side to show the nervous system. B, thoracic nerve mass enlarged.

A *megalopa* larva (Fig. 249 G), with the cephalothorax crab-like but the abdomen macrurous and carried at length, intervenes, as in other crabs, between the zoaea and the adult form.

Of the various examples of the suborder which are mentioned below, all except *Leucifer*, *Birgus*, and *Gecarcinus* occur in British waters.

The most aberrant member of the Decapoda is the minute, pelagic *Leucifer* (Fig. 276 E), which has a very slender, macrurous body with an extremely elongate head, long eyestalks, no limbs on the last two thoracic somites, no chelae, and no gills. Like the normally built prawn *Penaeus* and the rest of the group (Penaeidea) to which both belong, *Leucifer* starts life as a nauplius.

Leander, the common prawn, one of the Caridea, is macrurous like the crayfish, but built for swimming rather than walking, with phyllobranchiae, and with chelae only on the first two pairs of legs.

Crangon, the shrimp, is related to *Leander* but has a broader and flatter body, a very small rostrum, and the first leg subchelate.

Nephrops, the Norway lobster, one of the Astacura, differs from the crayfish in minor points, among others in having the podobranchs free from the mastigobranchs.

Homarus, the lobster, differs from *Nephrops* in size, form of chelae, etc.

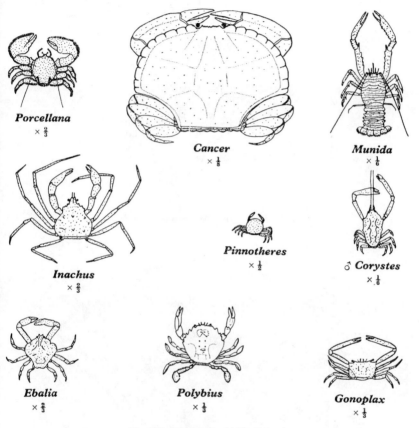

Fig. 283. Decapoda. (After Gosse.)

Palinurus, the crawfish or spiny lobster, one of the Palinura, differs from the crayfishes and lobsters in having a small spine in place of the rostrum, no antennal scale (exopodite), and no chela on any leg. It has a peculiar broad, thin schizopod larva, the *Phyllosoma* (Fig. 249 c).

Eupagurus (Fig. 284 A), the hermit crab, one of the Anomura, lives in the empty shells of gastropod molluscs. It has a large, soft abdomen, containing the liver and gonads, twisted to fit into the shell, and without appendages on the right side, save for the uropods, of which both are present, roughened,

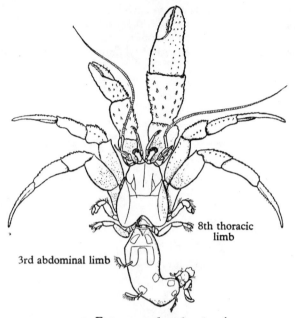

A. *Eupagurus bernhardus* ♂

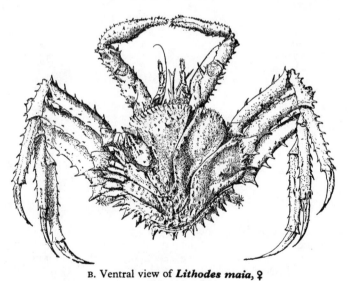

B. Ventral view of *Lithodes maia*, ♀

Fig. 284. Decapoda. (B, from the *Cambridge Natural History*.)

and serve to hold on the shell. The first three pairs of legs are as in a crab, the last two small and chelate.

Birgus, the robber crab, is a hermit crab which has grown too large to use the shells of molluscs, and has accordingly re-developed abdominal terga. It lives on land in the Indo-Pacific region, and is adapted to aerial respiration by the presence of vascular tufts on the lining of the gill chambers. Its zoaeae are marine.

Lithodes, the stone crab (Fig. 284 B), is by origin a hermit crab, but has lost the habit of living in shells and so thoroughly taken on the build of the true crabs that only some asymmetry of the abdomen and a few other minor points of structure betray its ancestry, even the uropods being absent.

Galathea, the plated lobster, another of the Anomura, is lobster-like but has the abdomen bent under the thorax, and the last leg small and slender and folded into the gill chamber.

Porcellana (Fig. 283), the china crab, related to *Galathea*, has a form of body resembling that of the true crabs, but possesses uropods. Fig. 249 N shows its remarkable zoaea.

Cancer (Fig. 283), the edible crab, is a member of the Brachyura, nearly related to *Carcinus* but more heavily built, without the slight powers of swimming possessed by the latter, and differing in other small points.

Gecarcinus, containing land crabs of the tropics, differs from *Carcinus* and *Cancer* in the shape of the third maxillipeds, which gape, the sternal position of the male opening, and the highly vascular lining of its swollen gill chambers. Its zoaeae are marine.

Maia, the spiny spider crab, is narrow in front, with bifid rostrum and feeble chelae, and a habit of decking itself with seaweed for concealment.

CHAPTER XII

THE CLASS MYRIAPODA

DIAGNOSIS. Land-living tracheate arthropods, usually elongated, with numerous leg-bearing segments; a distinct head with a single pair of antennae, a palpless mandible and at least one pair of maxillae; tracheal system with segmentally repeated stigmata, tracheae usually anastomosing; eyes, if present, clumps of ocelli; mid gut without special digestive glands, hind gut with Malpighian tubules; young hatching at a stage resembling the adult but possessing fewer than the adult complement of segments.

It has long been recognized that the group Myriapoda as defined above contains two chief divisions which are here treated as subclasses, one of which, the Chilopoda, is more closely related to the Insecta than the other, the Diplopoda. It is, however, convenient to retain the group, though the similarity of the chief members is probably more superficial than natural.

SUBCLASS 1. CHILOPODA

DIAGNOSIS. Carnivorous arthropods with the genital opening situated at the hind end of the body (opisthogoneate); body segments all similar (at least in the more primitive members of the division), body usually flattened dorsoventrally; ocelli present, head bears also antennae and three pairs of jaws (mandibles and two pairs of maxillae); the 1st body segment bears a pair of poison claws; the rest, each a single pair of ambulatory limbs, except the last two, which are legless; blood system consists of a dorsal heart and a ventral vessel connected by an anterior pair of aortic arches; tracheae typically branch and anastomose and have a spiral lining; gonads dorsal to gut.

LITHOBIUS. The type used for the study of this division is the centipede, *Lithobius* (Fig. 285 A), which is found under bark and stones, and is a much more active creature than the millipede, *Iulus*, which is found in the same situation. The chitinous exoskeleton is flexible and is moulted frequently. The body is flattened dorsoventrally and the legs in each pair are widely separated. The head consists of six segments all represented by coelomic sacs in the embryo which disappear in the adult, including a *preoral* and (between the antennae and the mandibles) an *intercalary*. All segments except the first are originally postoral but in development the mouth moves back and comes to lie between the mandibles. The number of head segments is the same as in the embryo insect and the crustacean, and a remarkable homology may be observed between the chilopod and insect head appendages. Thus the antennae are jointed mobile appendages varying in length; the mandibles are toothed plates without palps, the 1st maxillae consist of a basal portion bearing inner and outer lobes, while the 2nd maxillae are usually fused together to form a sort of labium and possess a palp-like jointed structure

(Fig. 286). The difference between the mouth parts of an orthopteran insect and a chilopod lies in the reduction in size of the two pairs of maxillae in the latter, which is possibly connected with the great development of the first pair of trunk appendages as maxillipeds, which are four-jointed, the distal joint being a sharp claw perforated by the opening of the poison gland, while the proximal joint is enlarged and meets its fellow in the middle line to form an additional lower lip.

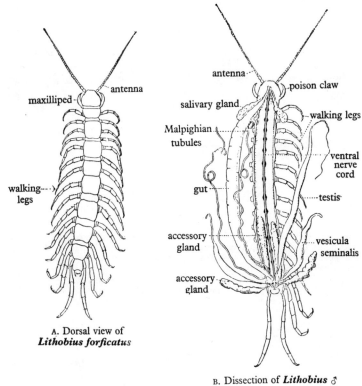

A. Dorsal view of *Lithobius forficatus*

B. Dissection of *Lithobius* ♂

Fig. 285. *Lithobius forficatus*. A, whole animal. B, male dissected to show internal organs. (From Shipley and MacBride.)

The body segments in *Lithobius* number eighteen. Of these, the 1st carries the poison claws (maxillipeds), the 17th the genital opening and usually a pair of modified appendages, the *gonopods*, and the last (telson), which is greatly reduced in size and not seen in Fig. 287, the anus, while the 2nd to 16th have each a pair of seven-jointed walking legs. Each segment has a broad tergum and sternum and between them a soft pleural region with a few small chitinous sclerites and the stigmata. In *Lithobius* and the group of chilopods to which it belongs, the terga are alternately long and short (Fig. 285). Only the segments which have long sterna have stigmata, but all have walking legs. In other centipedes, e.g. *Scolopendra* (see Table, p. 318), the terga appear equal

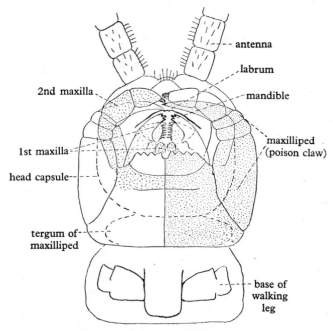

Fig. 286. *Lithobius forficatus*. Ventral view of head and two succeeding segments in a specimen boiled in potash and imbedded in Canada balsam. On the right of the observer, the maxilliped and the sternum belonging to it are lightly stippled; on the left the maxilla is more coarsely stippled. (Original.)

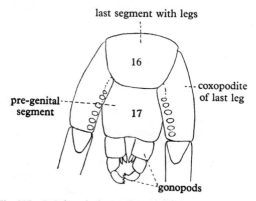

Fig. 287. *Lithobius forficatus*, female; hind end, ventral side.

throughout but very small tergites may be seen lying vestigial beween the large tergites. To this group belongs *Scutigera* (Fig. 288).

The alimentary canal consists of a short fore gut into which open two or three pairs of salivary glands, a very long mid gut without any associated glands, and a short hind gut into which open a pair of Malpighian tubules.

The vascular system is better developed than in insects. The heart runs the whole length of the body and possesses in each segment not only a pair of ostia but also lateral arteries. It ends anteriorly in a cephalic artery and a pair of arteries which run round the gut and join to form a supraneural vessel. The arteries branch and open into haemocoelic spaces. There is a pericardium and below it a horizontal membrane, perforated and provided with alary muscles as in insects. In the respiratory system the tracheae branch and anastomose and possess a spiral thickening, but in the remarkable form *Scutigera* the stigmata are unpaired and dorsomedian in position and the tracheae are unbranched and simple in structure.

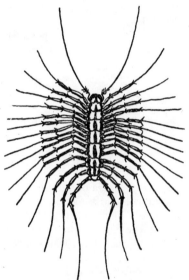

Fig. 288. *Scutigera nigrovittata.* (After Kükenthal.)

The reproductive organs (Fig. 285 B) consist of an unpaired ovary or testis, with a duct which divides into two and passes round the end gut to open by the median genital opening. There are two pairs of accessory glands and in the male two vesiculae seminales. Spermatophores are formed but it is doubtful whether copulation occurs. *Lithobius* lays its eggs singly and buries them in the earth. The young are hatched with seven pairs of legs.

The nervous system comprises a cerebral ganglion supplying the antennae and the eyes, a suboesophageal ganglion giving branches to the other head appendages and the maxillipeds, and a ventral chain with a pair of ganglia in each leg-bearing segment.

SUBCLASS 2. DIPLOPODA

DIAGNOSIS. Arthropods with the genital opening situated on the 3rd segment behind the head (progoneate); trunk segments arranged in an anterior region (*thorax*) of four single segments and a posterior region (*abdomen*) of double segments, each with two pairs of legs; body usually cylindrical; skeleton strengthened by a calcareous deposit; ocelli present, head bears also short club-shaped antennae, mandibles and a single pair of maxillae; vascular system well developed as in Chilopoda; tracheae arise in tufts from tracheal pouches, do not anastomose; gonads ventral to gut; young hatch usually in a stage with three pairs of legs and development takes place gradually.

Though the head of the adult millipede appears to have fewer segments than that of the Chilopoda a study of the embryo shows that there are really the same number. An intercalary segment exists between the antennal and mandibular segments and behind the mandibles a pair of rudimentary appendages appear but soon vanish. These are the first maxillae: the second maxillae (labium) persist in the adult.

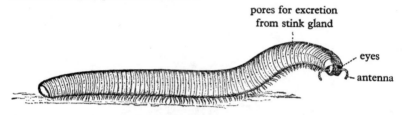

Fig. 289. *Iulus terrestris* (Diplopoda), × c. 3½. (From Koch.)

Iulus is one of the commonest genera of millipedes. It is vegetarian. It has an elongated body, consisting of a large number of segments (up to seventy), which can be rolled into a ball. The head (Fig. 290 A) carries a pair of short antennae with seven joints. The labrum is continuous with the front of the head and is a toothed plate; the mandibles, which have no palp, bear a movable tooth and a ridged and toothed plate; behind them is an organ known as the *gnathochilarium* (Fig. 290 C), which, in structure and position, recalls the labium of insects and, like it, is formed by the junction of paired appendages, the principal part of it by the appendages of the labial segment. Also a postlabial segment contributes to it forming the *basilar plate*. The tergite of this segment, however, forms what is apparently the first segment after the head. This is known as the *collum*; there are no stigmata and no separate appendages, though the first pair of legs appears to belong to it they are those of the second segment. The next three have a single pair of ambulatory legs apiece, a pair of ganglia and a pair of stigmata, and in the embryo a pair of coelomic sacs. These four segments may be said to constitute the *thorax*, though, as related above, the first takes part in the formation of a head structure. The genital openings are situated in the basal joint of the second pair of legs, which appear to arise from the second segment, but really belong to the third.

Behind this is the *abdomen* consisting of an indefinite number of double segments (up to a hundred in *Iulus*). The exoskeleton of a body segment consists of a tergum and two sterna. In the double segment of *Iulus* (Fig. 289) the sclerites of two segments are fused together to make a continuous ring. The sterna carry two pairs of stigmata and legs. In the embryo there are two pairs of coelomic sacs; there are two ostia in the heart and two pairs of ganglia. In *Iulus* the sterna are much shorter than the terga and also much narrower so that the legs come off very close together; also the terga are narrower in front so that they can be telescoped into the terga in front. The diagram here given (Fig. 290 D) shows that this relation occurs when the diplopod body is straightened out; when the animal rolls up the adjacent rings are completely disengaged.

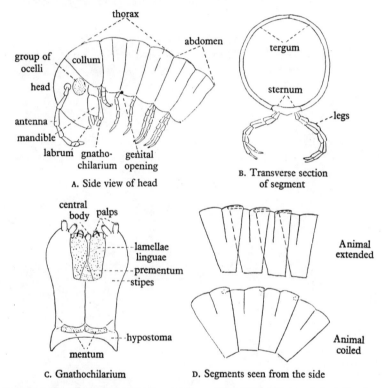

Fig. 290. *Iulus terrestris*; details of morphology. Note the anterior genital opening in A; stippled parts in C are believed to belong to the segment of the collum. (D, after Kükenthal.)

The stigmata are elongated slits, which can be closed by a valve, and they communicate with a tracheal pocket from which spring two thick bunches of unbranched tracheae. These are of two sorts; one long and slender, the other shorter and thicker with a spiral lining. In other millipedes (*Glomeris*) the tracheae may become much longer and branch but they never anastomose.

The circulatory system is in a stage of development higher than that of the insects. The alimentary canal bears a pair of long Malpighian tubules arising from the hind gut.

The legs consist of the same elements as in the insect, but the tarsus is divided into three joints, the last of which carries a claw. In the male the first leg is modified for copulation and in the 7th segment there is an auxiliary copulatory apparatus, consisting of processes used for transferring sperm into the vagina of the female. These processes may occur together with legs and so are not homologous with them. There are no similar organs in the female. The generative glands are unpaired with ducts opening on the 3rd body segment. The eggs are yolked and are laid after copulation in a nest made of hard earth. The mother keeps watch over them before hatching.

CHAPTER XIII

THE CLASS INSECTA (HEXAPODA)

DIAGNOSIS. Tracheate Arthropoda in which the body is divided into three distinct regions, the head, thorax and abdomen. The head consists of six segments and there is a single pair of antennae; the thorax consists of three segments with three pairs of legs and usually two pairs of wings; the abdomen has typically eleven segments and does not possess ambulatory appendages; genital apertures situated near the anus (Fig. 291).

CLASS 5. INSECTA (HEXAPODA)

SUBCLASS APTERYGOTA (AMETABOLA)
Super-order Entotropha
1. *Order* Collembola
2. *Order* Protura
3. *Order* Diplura

Super-order Ectotropha
4. *Order* Thysanura

SUBCLASS PTERYGOTA (METABOLA)

SECTION I. PALAEOPTERA (Exopterygota) (Hemimetabola)
Super-order Ephemeropteroidea
5. *Order* Ephemeroptera

Super-order Odonatopteroidea
6. *Order* Odonata

SECTION II. POLYNEOPTERA (Exopterygota) (Heterometabola)
Super-order Blattopteroidea
7. *Order* Dictyoptera
8. *Order* Isoptera
9. *Order* Zoraptera

Super-order Orthopteroidea
10. *Order* Plecoptera
11. *Order* Notoptera
12. *Order* Cheleutoptera
13. *Order* Orthoptera
14. *Order* Embioptera

Super-order Dermapteroidea
15. *Order* Dermaptera

SECTION III. OLIGONEOPTERA (Endopterygota)
(Holometabola)
Super-order Coleopteroidea
  16. *Order* Coleoptera
Super-order Neuropteroidea
  17. *Order* Megaloptera
  18. *Order* Raphidioptera
  19. *Order* Planipennia
Super-order Mecopteroidea
  20. *Order* Mecoptera
  21. *Order* Trichoptera
  22. *Order* Lepidoptera
  23. *Order* Diptera
Super-order Siphonapteroidea
  24. *Order* Siphonaptera
Super-order Hymenopteroidea
  25. *Order* Hymenoptera
  26. *Order* Strepsiptera

SECTION IV. PARANEOPTERA (Exopterygota)
(Heterometabola)
Super-order Psocopteroidea
  27. *Order* Psocoptera
  28. *Order* Mallophaga
  29. *Order* Anoplura
Super-order Thysanopteroidea
  30. *Order* Thysanoptera
Super-order Rhynchota
  31. *Order* Homoptera
  32. *Order* Heteroptera

INTEGUMENT. There can be no doubt that the properties of the integument of insects have greatly influenced the phenomena of growth in these animals and have also contributed largely to the success of insects in the world of living things and to their dominance among terrestrial invertebrates. The single epidermal cell-layer of which the integument consists, secretes on its surface, wherever it may be, a cuticular material of considerable complexity (Fig. 292). In its most generalized form this cuticle consists of an inner laminated layer, the endocuticle, which is chitinous and easily flexible. Outside this is a similar laminated zone into which sclerotized proteins have been incorporated, imparting to the cuticle as a whole both rigidity and elasticity. This layer, the exocuticle, is covered in turn by an epicuticle which,

INSECTA

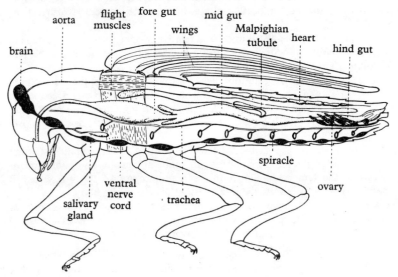

Fig. 291. Diagram of an insect to show its typical structure.

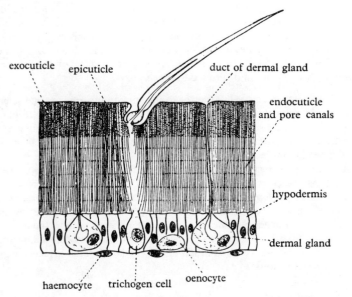

Fig. 292. A section of the cuticle and hypodermis of a typical insect. (After Wigglesworth.)

although usually extremely thin (often not thicker than 1μ), consists of several layers the most important of which is an outer cement layer overlying wax.

Modifications in the distribution and relative proportions of endocuticle to exocuticle enable some regions to be characterized by rigidity and others by flexibility. Thus, on the surface of wing-bearing segments where flight depends on elastic and rigid properties of the cuticular plates there is a plentiful covering of exocuticle. Intersegmentally, however, where freedom of movement is desirable, the exocuticle may be greatly reduced, or even absent. These are factors which have contributed greatly to the precision of movement of which insects are capable.

The epicuticle is responsible for the considerable impermeability to water of the cuticle of the terrestrial insect. Its removal by solution or abrasion, or by raising the temperature near to the melting-point of the wax which it contains, greatly increases the rate of water loss from the body.

We may conclude then that in the integument of the insect reside some of the most important qualities which determine the animal in its terrestrial environment where desiccation is a constant threat.

SKELETAL SYSTEM. The *head* is enclosed by an exoskeleton which consists of several plates or *sclerites*, both paired and unpaired, fused together and having no clear relation to the segmentation of the head. In front an unpaired *frons* lies between the antennal bases and this is surmounted by another sclerite, the *epicranium*. At the sides are the *parietal plates*, the lower parts of which are commonly termed the *genae*. From the lower border of the frons there depend in turn two median plates, the *clypeus* and the *labrum*, which, with the *hypopharynx* behind them, enclose the *cibarium*, a space in front of the true mouth (Fig. 293). The cibarial surface of the labrum is usually highly sensory and forms the *epipharynx*.

HEAD SEGMENTATION. Posterior to the epicranium and parietal sclerites there are, in some generalized insects, two further ones. These are narrow arched pieces, the *occipital* and the *postoccipital*. To the lower borders of these are attached the first maxillae and the labium. It is perhaps here alone that the head capsule has retained some small external sign of the segmental constitution of the head, the occipital and postoccipital sclerites representing its two most posterior segments. The latter of these two with the base of the labium bound the *foramen magnum* through which the internal organs pass between the head and the thorax. The segments of the head are indicated by the paired appendages, the ganglia of the nervous system (neuromeres) and, in generalized insects, the coelomic sacs (which can be demonstrated in sections of the embryo but which disappear later). This evidence can be summarized as follows:

Segment	Neuromere	Appendage
preantennal	protocerebrum	—
antennal	deutocerebrum	antenna
intercalary	tritocerebrum	embryonic
mandibular	mandibular ganglion	mandible
maxillary	maxillary ganglion	maxilla
labial	labial ganglion	labium

INSECTA   431

EYES. In addition to compound eyes there are simple eyes or *ocelli*, of two kinds. Lateral ocelli are usually the only type of eye in larval insects and represent the larval counterparts of the compound eyes which function in the adult. Dorsal ocelli on the vertex of the head of adult insects are structures distinct from the lateral ocelli and co-exist with compound eyes. Their structure is described later. The compound eyes (as described more fully in the

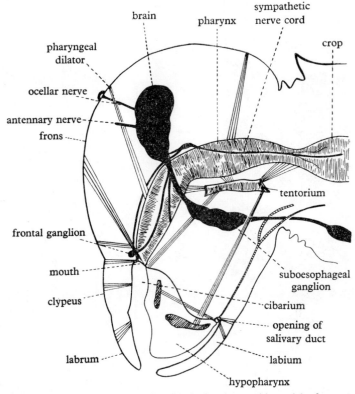

Fig. 293. Diagram of the anatomy of the head of an insect with special reference to its musculature and nervous system.

section on the Arthropoda) possess a cornea which is divided into a number of facets; corresponding to each facet is a group of visual cells, the *ommatidium*. The current theory of mosaic vision states that each ommatidium, isolated from its neighbours by a coat of pigment, conveys to the retinula at its base only such rays of light as travel parallel to the axis of the ommatidium. The total impression is that of a mosaic composed of as many separate pictures as there are ommatidia, every picture different from its neighbours, but all combining to form a single 'coherent' picture. The compound eye has probably the advantage that it can detect movements of very small amplitude. It gives, however, only a vague idea of the details of objects, for there is no focusing apparatus and only objects very close to the eye can be perceived clearly. In some insects the eye is divided into two parts: a dorsal with coarse facets which probably only serves to detect variation in illumination, and a ventral with finer facets which gives fairly definite images of objects. Possibly in some insects the first function in night vision, the second by day. It must also be mentioned that experiments show that many insects can distin-

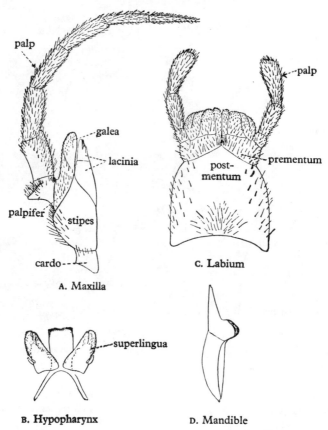

Fig. 294. Mouth parts of *Petrobius maritimus*. (After Imms.)

guish colours. The development of flower colour and pattern is generally supposed to have taken place simultaneously with that of the aesthetic senses of insects.

THE ANTENNAE. These are a pair of appendages consisting usually of many joints. They are sometimes filiform but may show complicated variations in structure. In all cases they carry sense hairs, and may serve olfactory, gustatory as well as tactile functions. In some insects sensillae occur on the antenna which are humidity detectors.

LIMBS. The first two pairs of mouth-parts, *mandibles* and first *maxillae*, lie at the sides of the mouth, while the second maxillae, invariably fused together, bound the mouth posteriorly, and are known as the *labium*. Such a fusion characterizes the maxillipeds of certain Crustacea. The primitive and generalized condition is, undoubtedly, that which is found in insects which feed on solid food, e.g. in the cockroach or in *Petrobius*. The mandible, which is rarely jointed, represents the toothed basal segment of an originally jointed limb, and corresponds in form and function to that of Crustacea, but never possesses a palp (Fig. 294). Each first maxilla is articulated to the head by a basal segment, the *cardo*. The succeeding segment, the *stipes*, carries an outer palp-bearing sclerite, and distally bears two lobes, the inner spiny *lacinia*, and the outer hood-like *galea*. In the labium the basal plates corresponding to cardo and stipes of the two sides are fused to form the *postmentum* and *prementum*. The more distal *prementum* bears a palp at either side and a number of lobes, typically four, between them, known collectively as the *ligula*. Where there are four as in the cockroach, the two median glossae are defined from the lateral paraglossae. In *Petrobius* (Fig. 294) further subdivision of the ligula lobes has occurred.

In the labium the distal prementum, to which the palps and the ligula are attached, represents the fused stipites of the first maxillae. Proximal to this sclerite is the postmentum, by means of which union with the head is effected. The postmentum, often divided further into proximal submentum and distal mentum, represents the fused cardines of the first maxillae. The hypopharynx, already referred to, bears the salivary aperture. A pair of sclerites, *superlinguae*, are normally fused to the sides of the hypopharynx.

Fig. 295 indicates the similarity between the insectan and crustacean mouth-parts. Such an attempt at a comparison is only possible with the more generalized mouth appendages of the Insecta.

With the evolution of different feeding habits, the structure of the mouth-parts, just described, has been departed from in a variety of ways. Comparative and embryological study, however, clearly reveals a uniformity of plan throughout, and the student must realize that the modifications to be met with in bugs, butterflies, bees and flies are all referable to the basal plan as exemplified in the mouth-parts of *Blatta* or *Petrobius*.

The *thorax* is separated from the head by a flexible neck region usually containing cervical sclerites, which, however, have no segmental significance. It consists of three segments—the *prothorax*, which carries a pair of legs but no wings, the *mesothorax*, and the *metathorax*, which each bear a pair of legs and, typically, wings. The legs are made up of five main segments, the *coxa* and *trochanter* (both of which are small), the *femur* and *tibia* (which form the greater part of the limb), and the *tarsus* (which is usually further subdivided by a number of joints, and ends in a pair of claws with a bifid cushion between them called the *pulvillus*). Of the many adaptations exhibited by the legs of insects the jumping type found in grasshoppers, the digging type in the mole-cricket *Gryllotalpa* and in the *Cicada* nymph, the swimming type in water-beetles like *Dytiscus* and *Gyrinus*, the prehensile type in the fore legs of the praying mantis may be mentioned, in addition to the ordinary running type as seen in a cockroach. Modifications for the production or reception of sound as in the Orthoptera and for the collection of food (the combs and pollen-baskets of bees) are also familiar (Fig. 296).

WINGS. The wings of an insect are thin folds of the skin flattened in a horizontal plane, arising from the region between the tergum and pleuron. A section of a wing bud shows two layers of hypodermis, the cells of which are greatly elongated (Fig. 317). Into the blood space between the layers grow tracheae, and when in a later stage the two layers of hypodermis come together and the basement membranes meet and fuse, spaces are left round

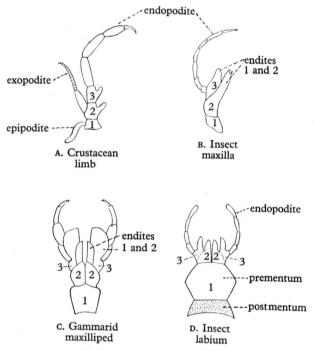

Fig. 295. Diagram to show the resemblance between the insect maxilla and labium and the biramous limb of the Crustacea. (From Imms, after Hansen.)

the tracheae which form the future longitudinal wing veins. These spaces contain blood, a trachea and a nerve fibre. The cuticle round the veins is much thicker than in the general wing membrane, so that the veins are actually a strengthening framework for the wing. The number and arrangement of the veins are highly characteristic of the different groups (Fig. 297). Though the majority of insects possess wings there are important orders which are wingless. Some, such as those to which the fleas and lice belong, are secondarily so, because of their parasitic habit. Others, however, constituting the large division *Apterygota*, are primitively wingless, and these, both on morphological and palaeontological evidence, must be regarded as the most ancient types known.

Among many orders of insects, there has developed a tendency for the two pairs of wings to act as one. This is accomplished by various devices which couple the fore and hind wings together, on each side. In the scorpion-flies, e.g. *Panorpa*, bristles project back from the posterior or jugal lobe of the fore wing to overlie the anterior border of the hind wing. Corresponding bristles to these, constituting the *frenulum*, project forwards from the anterior border

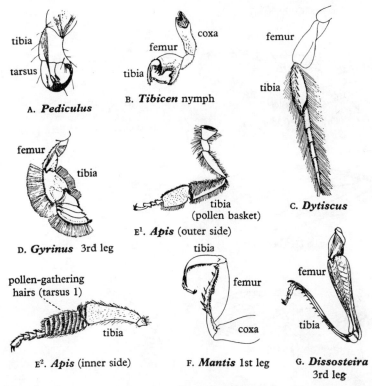

Fig. 296. Modifications of legs for performing various functions: A, for clinging to hair; *Pediculus*; B, for digging; nymph of *Tibicen* (Cicada); C, for swimming; water-beetle, *Dytiscus*; D, for surface swimming; whirligig-beetle, *Gyrinus*; E, for pollen gathering; hind leg of *Apis*; F, for catching prey; *Mantis* fore limb; G, for jumping; hind limb of locust, *Dissosteira*.

of the hind wing, and overlie the posterior border of the fore wing. In most Lepidoptera, frenular bristles of the hind wing are held in position by a group of curved setae forming a *retinaculum* on the fore wing. In the Hymenoptera, the two wings of a side are coupled by a row of hooks—the *hamuli*—on the anterior border of the hind wing, engaging in a fold of the posterior border of the fore wing.

In other orders, we find one pair of wings diverted to uses other than flight, the latter operation being then dependent on one pair of wings. The fore

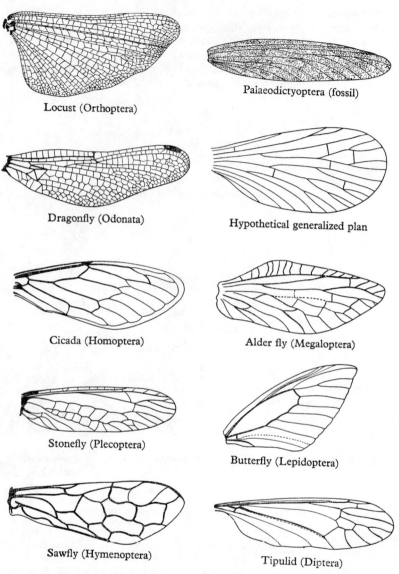

Fig. 297. Diagram illustrating the different types of wing venation.

wings, for instance, of Orthoptera and of Dermaptera, are protective to the more delicate folding flight wings behind them. The *elytra*, or fore wings of beetles, are similarly protective, and are held passively extended while the second pair of wings propel the animal through the air. In the males of *Strepsiptera*, the anterior, and in the male coccid bugs and all Diptera, the posterior wings are minute structures, flight being performed by the remaining pair, which are normally developed. Thus, either by linking two pairs of wings together, or by dispensing with one pair, flight is commonly brought about by one functional unit on each side of the body. The variations in form, consistency, and size of the wings are briefly dealt with under the different orders.

Simple up-and-down movements of the wings are sufficient to account for the elementary phenomena of insect flight. In moving through the air the anterior margin remains rigid, being strengthened by veins in this region, but the rest of the membrane, where the veins may be weaker or their distribution sparser, yields to the air pressure; so that when the wing moves downward it is bent upwards (cambered). As the wing moves upward, however, the membranous part is bent downwards; therefore, by becoming deflected in both upward and downward movement the wing encounters a certain amount of pressure from behind which is sufficient to propel it. The faster the wings vibrate the more they are cambered, the greater the pressure from behind, and the faster the flight. Smaller insects have as a rule a greater rate of wing beat. Thus a butterfly may make only nine strokes a second while a bee makes 190 and a house-fly 330. The wing muscles of insects thus contract immensely faster than those of any other animals, a condition brought about perhaps by each contracting fibre stimulating its antagonist to contraction, rapidly alternating contraction resulting. It is interesting to note that the intracellular respiratory pigment, cytochrome, occurs in high concentration in them.

To bring about wing movement, direct muscles attached to the small sclerites which lie in the membranes connecting the wing base with the tergum or pleuron and others called indirect inserted on the body wall are employed (Fig. 298). These two types of muscle differ both in appearance and in physiological properties. Direct muscles, controllers of flight attitude and direction, resemble leg muscles in appearance and contract slowly. Indirect muscles, the source of power, contract rapidly on the reception of low-frequency impulses. It appears that on the reception of an impulse their myofibrils are so excited as to respond by active contraction to the stimulus of stretching. Since dorso-ventral and longitudinal muscles are antagonistic to each other, the contraction of one of these sets of muscles causes the stretching of the other and hence, in turn, its contraction.

The extent to which direct and indirect muscles are present varies. In the Odonata a direct musculature is strongly developed, the muscles being attached to the intucked wing base. In the specialized orders Lepidoptera, Diptera and Hymenoptera, indirect muscle action is responsible for most of the movement and those muscles attached directly to the wing base serve for folding the wing to a position of rest as well as for flight purposes, rotating the wing on its base during upward and downward phases of movement.

Fig. 298 represents diagrammatically the condition in the winged aphids. The thorax is a box whose roof is capable of being arched and flattened by longitudinal and dorsoventral muscles respectively. Since the wing base has two points of attachment, (1) to the pleural plate, and (2) to the edge of the tergum, the wing operates as a lever of the second order. The arching of the tergum raises the wing base and depresses the wing, while a flattening of the tergum depresses the wing base and raises the wing. Fig. 299 shows diagrammatically the effect of contraction of direct muscles attached to small sclerites in the wing axis at its junction with the body. On the left, direct muscles are shown as inserted so as to depress the wing, and on the right, direct muscles are shown as inserted so as to elevate the wing.

ABDOMINAL APPENDAGES. The abdomen consists of a series of segments less differentiated than those of the head and thorax. The number is eleven, as seen to be present in the embryo insect (with the addition of a transient telson)

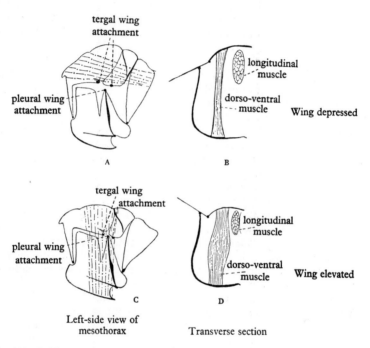

Fig. 298. To illustrate the mechanism of wing movement in an aphid. The effective muscles are shown in A and C by dotted lines. (After Weber.)

and in primitive groups (Thysanura and Odonata). In other groups the 11th segment is represented by the *podical plates* which bear the *cerci anales* (as for instance in the cockroach). In specialized insects the number of abdominal segments may be greatly reduced. In insect embryos rudiments of appendages are borne on each of the abdominal segments, but they disappear in the adult except in the Apterygota. In higher forms these appendages may be retained in a modified form as genitalia in the male (segment 9) and in the female (segments 8 and 9). The cerci of segment 11 are frequently retained. The arrangements of the genitalia are complicated; suffice it to say that in females they may be modified to form stings, saws, piercing ovipositors, etc., and in the two sexes to mould or to receive in some instances complicated spermatophores.

THE ALIMENTARY CANAL (Fig. 300) varies greatly in length; in many larvae it is no longer than the animal itself, but in certain types of insects like the Homoptera, which feed on plant juices, it is much coiled and may be several times the length of the body. It consists of an ectodermal *stomodaeum*

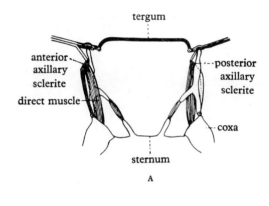

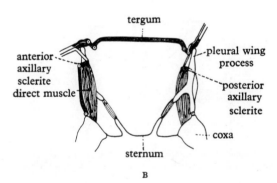

Fig. 299. To illustrate the effect of contraction of direct muscles attached to axillary sclerites which lie anterior and posterior to the pleural wing process. Contraction of the muscle attached to the anterior sclerite (left side of the figure) depresses the wing; contraction of the muscles attached to the posterior sclerite (right side of the figure) elevates the wing. In A the muscles are relaxed, in B they are contracted. (After Weber.)

or fore gut, an endodermal mid gut and an ectodermal *proctodaeum* or hind gut. The fore gut, in front of which is the cibarium, consists of (1) the buccal cavity succeeded by (2) the pharynx, which may be muscular and form a pumping organ (Fig. 293), and (3) the oesophagus, which has a posterior dilation, the crop. This latter functions as a food reservoir and may have a diverticulum enormously developed in sucking insects to store the liquid food.

Lastly there is (4) the *proventriculus* or gizzard, most typically developed in insects that eat hard foods, as in the Orthoptera. The cuticular lining of the fore gut is here greatly thickened and the sphincter muscles in this region control the passage of food between fore gut and mid gut. Into the buccal cavity discharge the salivary glands (Fig. 300), which may, as in the cockroach, have a very similar function to those of the mammal in producing enzymes for the digestion of carbohydrates. In other insects, however, they

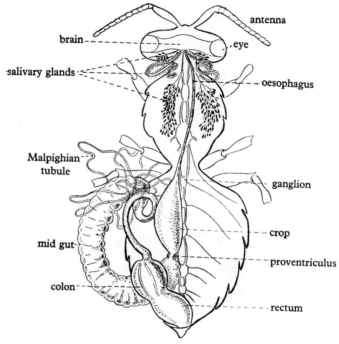

Fig. 300. General view of the internal organs of *Apis mellifica* as seen from above; musculature and tracheal system not shown. (From Carpenter.)

are specialized in ways which are mentioned later. Such glands are usually associated with the labium; in some insects, however, mandibular and maxillary glands are found.

The mid gut (Fig. 301) is lined by a layer of cells frequently all similar, which perform almost the whole task of digestion and absorption of all classes of foodstuffs. While secreting, the cells may break down and their contents are then discharged into the gut cavity. In the absorptive phase the border of the cells often has a striated appearance. The same cell may be capable of both absorption and secretion, but the epithelium as a whole often passes through rapid cycles which necessitate the constant supply of fresh cells. These are found (Fig. 301) in the troughs of folds or bottoms of pits into which the midgut epithelium is thrown. In many insects the surface is increased by the formation of long diverticula, the *pyloric caeca*, the cells of which are not in any way different from the rest of the epithelium. These vary greatly in number. Though the mid-gut epithelium has not an internal chitinous lining there is often a chitinous tube free in its cavity, the *peritrophic membrane*. This is produced either from special cells in the proventriculus, e.g. Diptera,

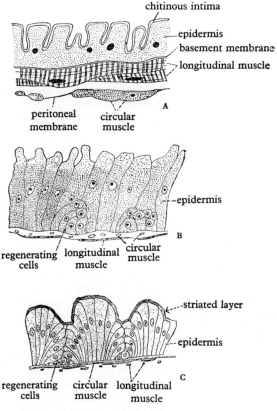

Fig. 301. Histology of the gut. A, longitudinal section of the wall of oesophagus of a termite. B, longitudinal section of the mid gut of a termite in secretory phase. C, transverse section of mid gut of *Blatta* in resting phase. (After Imms.)

or by delamination from the lining of the mid gut, e.g. Orthoptera and many Coleoptera. It consists of material similar to the endocuticle of the body wall and serves to protect the mid-gut cells from injury by food particles. Digestive enzymes readily pass through it as do the products of digestion for absorption by the mid-gut cells.

In certain cases, however, the mid gut is differentiated into functional regions. The first part of the mid gut of the *tsetse fly*, for instance, is concerned with water absorption which reduces the meal of blood to a viscid mass. Digestion of the food takes place in a region behind this, and in the lowest region of the mid gut absorption is effected. These functional regions are histologically distinct. In the cockroach in which no such histo-physiological distinction exists between the several parts of the mid gut, it appears that much digestion takes place in the crop to which place the enzymes from other parts may have to pass to meet the food before its further passage backwards. The so-called gizzard has been shown in this case to act not only as a triturating organ, but as a complicated sphincter guarding against the passage of any but the finest particles from the crop to the mid gut. After digestion has proceeded in the crop as the result of salivary and other secretory activity, the food passes through the gizzard, there to be triturated; and so on to the mid gut to meet the enzymes produced by the walls of this region. Resorption of the digested food takes place in the mid gut as well as in the hind gut.

The hind gut begins where the *Malpighian tubules* enter the alimentary canal and is usually divided into a small intestine or *ileum*, a large intestine or *colon*, in both of which the chitinous lining is sometimes folded and even produced into spines, and a short globular *rectum*. In most insects rectal glands in the form of thickened patches of epithelium occur. These have been shown to absorb water from the faeces and therefore play an important part in water conservation.

ENZYMES. Though the digestive enzymes of insects in the main belong to the same classes as those of mammals there are many significant differences. An omnivorous insect like the cockroach produces all the classes of enzymes except that represented by pepsin which is peculiar to vertebrates. Then also, the enzymes of insects appear to work in a rather more acid medium than do the enzymes of mammals. Finally the specialization in feeding habits in insects may be responsible for the absence of such free enzymes as are not needed and for either the acquisition of enzymes not generally found in the animal kingdom or the formation of a symbiotic partnership.

Thus when we compare the cockroach with such forms as the tsetse fly (*Glossina*) and the blow-fly *Calliphora*, we find the two latter deficient in certain enzyme classes, the former in carbohydrases, the latter in tryptases and peptidase. The evolution of the habit of feeding on blood (which consists so largely of proteins) involves the loss of the enzymes which digest carbohydrates and fats. Similarly the blow-fly, which exists on a diet in which carbohydrates are predominant, has to a certain extent lost its proteolytic and lipolytic enzymes.

This principle has an even wider application. In the leaf-mining caterpillars of the Lepidoptera, certain species are restricted to the upper, and others to the lower, parenchymatous layer of the leaf. If an egg of one species is accidentally deposited in the wrong layer of the leaf, death of the larva ensues owing to its inability to digest the proteins of that layer. Thus each species, it is said, has enzymes which are specialized in the narrowest degree for digestion not only of the proteins of a single plant but of those of a particular part of that plant, all others being unsuitable. Sucking forms, like *Aphis*, explore different regions of the plant tissue and it may perhaps be inferred that they have a wider range of enzymes than the leaf-miners.

Most interesting of all is the relation of phytophagous insects to cellulose, which is incapable of digestion by any vertebrate. Only a few wood-boring beetle larvae (Cerambycidae) have been shown to possess an enzyme that digests cellulose. The great majority of insects do not possess a cellulase and as all plant cell contents are contained within cellulose envelopes, it is clear that digestion can only follow when protoplasm is released by mechanical injury of the cell wall or when the enzymes are able to penetrate the cell wall and act upon the contained protoplasm. In lepidopterous caterpillars, which digest vegetable protoplasm with much greater success than do mammals, the latter explanation is shown to be true.

The insects which live on wood (excluding the Cerambycidae) can be divided into two classes: (1) those, like bark-beetles, which feed on fungi growing in their tunnels, and (2) those which harbour symbiotic organisms in special parts of their alimentary canal. In the latter class may be mentioned the wood-boring larvae of certain crane-flies and of death-watch beetles (e.g. *Xestobium*). In these cases the supposed symbiotic organism is the yeast, *Saccharomyces*. How it assists in the assimilation of wood is not known. On the other hand, those termites that eat wood in normal life always contain flagellates belonging to *Trichonympha* and other genera of the Hypermastigina (p. 70) living free in the intestine. The absolute dependence of certain termites on the flagellates is shown by the fact that when the flagellate fauna is removed (which can be done without harming the termite by heating to 40°C.) the termites will starve although they continue to eat their usual diet. It has been claimed that two-thirds of wood eaten is rendered assimilable by the digestive activity of the protozoa. This indicates a true symbiotic association, the termites receiving digested products from the Protozoa in turn for the anaerobic environment which they provide for them though the feeding of termites directly on the protozoa cannot be excluded.

The majority of so-called saprophagous insects are really phytophagous, in that they feed on yeasts and micro-organisms effecting the decomposition of the decaying matter. The house-fly is probably such a case. Blow-fly larvae feeding on decaying meat do, however, employ proteolytic enzymes, and to this extent are truly saprophagous, as is also the dung beetle *Geotrupes*. The flesh-fly *Lucilia*, though saprophagous in this way, still requires the microflora of the decaying food to complete a diet suitable for full development, these organisms supplying the vitamins necessary for growth.

The great range of environments occupied by insects as a whole is largely an expression of their diverse feeding habits, and few materials have escaped their attentions. In addition to the foods mentioned above may be noted keratin, which undergoes fermentative digestion in the larval gut of the clothes-moth *Tinea biselliella*. Silk can be utilized as the sole diet of the museum beetle *Anthrenus museorum*, the amino acids in this case supplanting both fats and carbohydrates.

The saliva of various insects shows great variety according to their habits; thus the larvae of the tiger-beetle (*Cicindela*), the flesh-eating larvae of flies, e.g. *Sarcophaga*, and the aquatic larvae of *Corethra*, pour their saliva, which contains a proteolytic enzyme, on their food and suck up the products of digestion (external digestion). Bees, with their reliance on pollen and honey as food, have four different kinds of salivary gland. These probably serve different purposes such as to invert sugars, to ensure preservation of food by adding formic acid, and to predigest pollen in the manufacture of 'bee bread' on which the young are fed. The proportion of carbohydrate to fat and protein in the food after the early stages of feeding as well as the absolute amount of food fed, determine whether a larval bee shall become a queen (fertile female) or a worker (sterile female). The former is fed throughout on a richer protein diet prepared from pharyngeal glands while the latter has its diet changed to pollen and nectar containing a higher carbohydrate content. In wood-boring larvae the secretion of a mandibular gland softens the wood and thus assists mastication, while in caterpillars, silk production is the main function of labial glands.

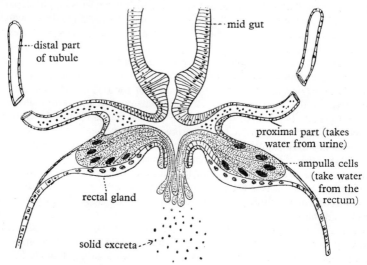

Fig. 302. Diagram of a longitudinal section through the gut of *Rhodnius prolixus* to show the entrance of the Malpighian tubules. Water and excretory material enter at the distal end, water is removed at the proximal end and by the protruding ends of the ampulla cells. (Modified after Wigglesworth.)

EXCRETION. The purpose of the excretory system is to maintain a constant internal environment. This is done by eliminating unwanted materials from the blood and by retaining those constituents which are needful to the organism. The principal excretory organs are the Malpighian tubules, ectodermal structures opening into the anterior end of the hind gut. They do not occur in Collembola or the Aphididae, while in the Thysanura, Protura and Strepsiptera they are doubtfully represented by papillae. In other orders they vary in number from four in the majority of Diptera to as many as a hundred or more in some Orthoptera. In *Periplaneta* there are sixty, equivalent to a surface area of 130,000 sq. mm.

Malpighian tubules are invariably blind tubes with an intercellular lumen opening proximally to the hind gut. The presence of uric acid crystals inside the cells in the lumen of the tube is proof of their function.

While their disposition varies in the different orders evidence is accumulating that there exist two types of cell, one for secretion of excretory matter into the tube and the other for reabsorption of water and ions back into the blood. The distribution of these two cell types may be random along the tube as in the fly *Ptychoptera*, the caterpillar of *Galleria* and the beetle *Dromius*. In other forms, e.g. *Rhodnius*, secretory cells with a close brush border and with dense granular cytoplasm form the distal half of the tube's length. Absorptive cells with a striated border composed of discrete filaments and with non-granular cytoplasm constitute the proximal or lower half of the tube.

It has been shown in *Rhodnius* that potassium urate and water pass into the tube in the distal part where the contents are alkaline while in the proximal part of each tube the contents are acid. Uric acid is the excretory product while potassium, probably as carbonate, is reabsorbed with water (Fig. 302). A circulation of water and base therefore exists within the system similar to that of the vertebrate kidney and serves for water conservation. A similar cytological specialization of the system is found in *Gryllotalpa* (Orthoptera), *Heptagenea* (Ephemeroptera), *Forficula* (Dermaptera) and in *Dytiscus* larvae (Coleoptera).

While the role of other organs supplementing the excretory functions of Malpighian tubules may have been exaggerated in the past, such additional mechanisms as the fat body, urate cells, the hypodermis, the cuticle, the gut wall and even nephrocytes should not be ignored. In these organs nitrogenous end-products seem to accumulate and thus assist the Malpighian tubules in their work, for instance the hollow wing scales of certain butterflies, e.g. *Pieridae*, are full of uric acid. It is possible, however, that in some of these cases, e.g. nephrocytes (cells found commonly associated with the fat body and the pericardium), their function is more concerned with intermediary metabolism of waste matters than with excretion.

It is interesting to notice that in certain aquatic larvae, e.g. the mosquito, excess water passing by osmosis into the body through the anal papillae is expelled to the gut through the agency of Malpighian tubules. These latter thus serve as osmo-regulators in addition to their other excretory functions.

Of non-nitrogenous excretory products may be mentioned the carbonates of calcium, potassium and magnesium. Calcium carbonate may be excreted in the integument, but in many cases it is eliminated by the Malpighian tubules, either gradually, e.g. *Drosophila*, or expelled *en masse* by way of the blood and the hypodermis during pupation, e.g. *Ascidia*, the celery-fly. In this latter example the recrystallization of the compound on the inner wall of the puparium (see p. 506) may serve to strengthen the weakness of the latter.

THE CIRCULATORY SYSTEM. There is a dorsally placed heart, primitively consisting of thirteen chambers, each corresponding to a segment, with a pair of ostia guarded by valves precluding outflow, at the base of each chamber. The blood is driven forward in these by muscular action of the heart wall, and passes into an anterior aorta which opens into the general body cavity in the head region. The haemocoelic body cavity is very spacious and the blood bathes all the organs. There is a dorsal horizontal diaphragm perforated by many holes which separates off the pericardium in which the heart lies. Fan-shaped muscles, having their origins laterally on the terga, are inserted into the pericardial diaphragm beneath the heart (Fig. 303), and by their contraction the passage of blood from the body cavity into the pericardium and so to the heart is facilitated. In some insects there is also a ventral diaphragm. Fig. 304 shows the general course of the circulation. Though the circulatory system is usually simple, accessory vessels are known which direct blood backwards along the nerve cord and upwards towards the pericardium in the metathorax (in the moth *Protoparce*). Further, in certain insects accessory hearts are present which assist in the circulation through special regions (in the thorax of the beetle *Dysticus* and in the bases of the legs of aphids where they propel blood through the wings and legs of these forms respectively). This much-reduced system is on the whole greatly in contrast with the complex arrangements of the decapod Crustacea and of such arachnids as *Limulus* and the scorpions where the respiratory pigment haemocyanin renders the blood of great importance in respiration.

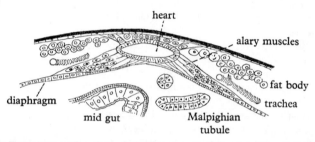

Fig. 303. Transverse section through the dorsal part of the abdomen of *Apis mellifica* to show attachment of heart to the body wall and to the diaphragm by the alary muscles. (After Snodgrass.)

The part played by blood in respiration introduces a topic which can only be adequately considered with the tracheal system next to be described. In anticipation of that account it may suffice to note the following points. The walls of the tracheae are freely permeable to gases and an exchange of gases between the blood and the air in the tracheae must therefore occur. In some insects the walls of air sacs within the tracheal system become intucked so as to form 'inverted tracheae' through which the blood circulates, thus giving rise to an organ which may act as a veritable lung, e.g. *Sphinx* and *Vespa*. Though these facts suggest a special oxygen-carrying function for the blood, it appears that its oxygen capacity is no greater than can be accounted for by physical solution. Haemocyanin does not occur and to this fact must be put down the rather vestigial nature of the circulatory system in insects. Haemoglobin occurs in a few, e.g. the larva of the midge *Chironomus*, the male apparatus of the water-bug *Macrocorixa* and in certain tracheal cells of the horse-fly *Gastrophilus*. This pigment may be derived from intracellular cytochrome and its occurrence may be of the nature of a chemical acciden of little functional significance. On the other hand it may serve, as it appears to do in *Chironomus*, as a means of enabling the animal to utilize oxygen when this occurs only at low tensions in the surrounding medium. The occurrence of chlorophyll invariably owes its origin to the food plant. Of the several kinds of blood cells which exist, perhaps those which play an important part in the histolysis of larval tissues during the pupation of holometabolous insects, e.g. the blow-fly *Calliphora*, are of most interest.

Having regard always to the restricted respiratory importance of blood, note should be taken of its other functions. Food substances are both transported and stored in blood. Hormones are carried about the body. Its water serves as a useful reserve enabling the animal to withstand considerable desiccation. It can also serve as a means of transferring pressure from one part of the body to another, thus assisting in moulting. Its cells, many of which are phagocytes, play an important part in protection against disease. Some of them also serve in the production of internal membranes.

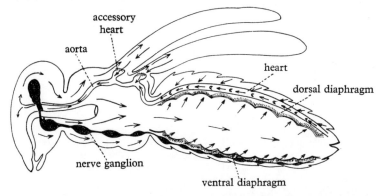

Fig. 304. Diagram to illustrate the general circulation of the blood in an insect.

OTHER ORGANS. Associated with the blood are the following cellular tissues, the *fat body*, the *nephrocytes*, the *oenocytes*, the *corpora allata*, and in various beetles, the *photogenic organs*. The *fat body* consists of closely adherent cells, in the vacuoles of which products of digestion are stored up. Fats, albuminoids and glycogen occur in this way. In addition are found urates showing that this organ serves for excretion. *Oenocytes* arise from metameric groups of ectodermal cells in the abdomen and are sometimes found in close association with the spiracles. They often exhibit a cycle of activity within the period of each instar. While much remains to be learned of their function, their inclusions of proteins, fats and the like, indicate their importance in intermediary metabolism. From their cyclical changes in relation to the histochemistry of the hypodermis, it appears likely that they produce the lipoproteins from which the cuticulin of the epicuticle is formed.

The *corpora allata* arise in the mandibular segment by invagination and subsequently come to lie above the oesophagus behind the brain. These small compact glands are now known to be hormonic in function, being responsible during growth for the retention of 'youthful' characteristics and after metamorphosis for the completion of the development of the eggs by the deposition of yolk.

Photogenic organs are probably derived from the fat body, some cells of which, typically, form a background of pigment in front of which are the photogenic cells covered by a transparent cuticle. They are richly tracheated, and by the oxidation of luciferin by the enzyme luciferase, energy in the form of light is produced. This may serve as a mating signal, the large-eyed male *Lampyris* beetle, for instance, being attracted by the light produced by the larviform female.

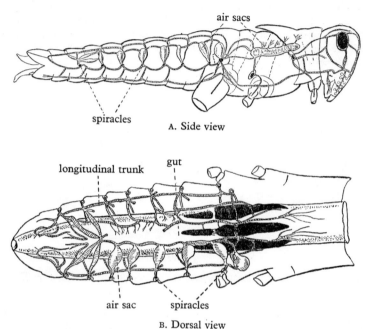

Fig. 305. Tracheal system of the locust, *Dissosteira carolina*. A, side view. B, dorsal view, the lower half to show air sacs, the upper half to show the tracheal supply to the alimentary canal. (Modified from Vinal.)

RESPIRATION. In the insects the tracheal system characteristic of terrestrial Arthropoda attains its most complete development. With rare exceptions (Thysanura and Protura and *Campodea* in the Diplura) the ectodermal tubes of the system form a network of which every part is in communication with every other part. Typically it communicates with the exterior by openings called *stigmata* or *spiracles* of which there are two pairs on the thorax and eight pairs on the abdomen (Fig. 305). The main branches leading from the stigmata not only divide into finer capillaries leading to the adjacent organs but also communicate by means of lateral trunks with each other. The capillaries or *tracheoles*, intracellular tubes of branching cells at the ends of tracheae, terminate either in or on the surface of cells and tissues, so that normally the oxygen is conveyed directly to the latter without the intervention of the blood. These end tubes, as may be seen in Fig. 306, are of the smallest calibre, being generally less than 1μ in diameter at their origin, diminishing to about 0.2μ at their endings. The cuticular lining, which in the main tracheae is strengthened, forming the spiral threads which prevent collapse of the tubes, is in the tracheoles thinned down so much that gaseous diffusion can take place easily between the cell fluid and the lumen of the tube.

During moulting the cuticular lining of existing tracheae is cast off and between moults new tracheae grow out from existing ones to become functional after the next succeeding moult. Tracheoles differ from tracheae in that their lining is not shed, and also in that once formed they persist throughout the life of the insect. To meet the varying respiratory demands of the various tissues of an insect, tracheoles can migrate actively to regions of oxygen deficiency and similarly regions deficient in oxygen stimulate the tracheal system into a richer branching to supply the needful regions after the next moult.

The stigmata are oval slits which, by mechanisms to be described later, can be closed and opened. By the timing of their opening and closing in relation to the muscular respiratory movements of the body there can be brought about a regular circulation of air in the main passages. Respiratory movements can easily be observed in such insects as wasps and grasshoppers. They are effected by the alternate contraction of the abdomen in its vertical axis by *tergosternal* muscles and recovery to the original form usually by the elasticity of the abdominal sclerites. Abdominal contraction with open spiracles results in expiration, but with closed spiracles the air already in the system will be forced into the finer capillaries where the oxygen pressure is thus increased.

In some Orthoptera it has been found that certain stigmata are normally inspiratory and others expiratory. Thus, in various grasshoppers (Fig. 305), the first four pairs are open at inspiration and closed in the expiratory phase, while the last six pairs are open in the expiratory phase, and closed at inspiration. In this way circulation of air through the main trunks is set up, aiding considerably in the diffusion of gas through the whole system. Air sacs (as mentioned above) in the form of thin-walled diverticula of the main tracheae occur in many insects (Fig. 305), particularly those, such as bees, migratory locusts and house-flies, with the power to fly for prolonged periods. These also assist considerably in the circulation of air through the tracheal system owing to the ease with which they can be compressed.

Thus to assist respiration in typical insects a neuro-muscular mechanism has been evolved which ensures some control of the ventilation of the tracheal system. Spiracular closing mechanisms and compressible air sacs are important in this process. The spiracular closing mechanism is also of considerable importance in controlling the rate of water loss from the body. The air in the tracheae is saturated with water and the insect with open spiracles must lose water at a rate directly proportional to the saturation deficiency of the atmosphere. The spiracles are therefore the principal site of water loss from the body, and their complete or partial closure under dry external conditions has a significant influence in reducing evaporation from the body. It follows from this that insects without closing mechanisms to their spiracles, e.g. some Collembola, must be restricted to life in a moisture-saturated atmosphere, otherwise they would suffer from desiccation. For these reasons it is deemed advisable to describe one or two types of closing mechanism.

The structure of spiracles varies considerably (Fig. 307). In the simplest instances, e.g. some Apterygota, they are simple openings of the tracheal system to the surface and devoid of any means of closing them or regulating

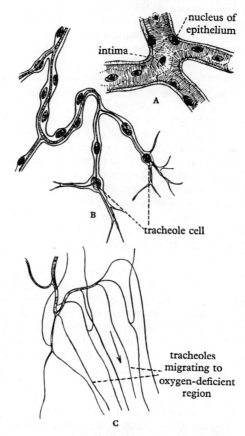

Fig. 306. Structure of trachea and tracheoles: A, diagram of trachea as seen by transparency. B, diagram of fine trachea ending in tracheoles. C, diagram of trachea with tracheolar endings migrating towards a region of low oxygen tension. (After Wigglesworth.)

their size. In most insects, however, the trachea opens at a spiracle situated at the bottom of an atrial depression in the cuticle, whose walls are often clothed with protective hairs. The closing apparatus of these latter spiracles may be either of two types. In the first type (mostly thoracic) the opening is provided with two external cuticular lips united by a ventral lobe and whose elasticity tends to maintain a spiracular opening between them. An occlusor muscle originating on the body wall and inserted on the aforesaid ventral lobe contracts so as to pull the two lips together (Fig. 307 C, D, E).

In abdominal spiracles there are no such external lips and the atrial wall is specialized into a fixed and a movable part. This latter is prolonged into a process to which two muscles are attached. The contraction of one of these (the occlusor) pulls the atrial walls towards each other so as to close the

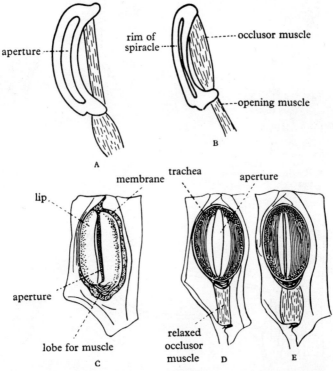

Fig. 307. Closing mechanism of spiracles. A, B, abdominal spiracles of *Apis* in open and closed condition respectively. Opening and closing muscles are present. C, external view of the metathoracic spiracle of a locust. D, internal view of same when open. E, internal view of same when almost closed. A single occlusor muscle closes the spiracle, the open condition being determined by elasticity of the surrounding cuticle. (A and B, after Weber; C, D, E, from Imms, after Snodgrass.)

passage between the atrium and the trachea. The contraction of the other muscle (the dilator) is antagonistic to the first-mentioned and opens the spiracle (Fig. 307 A, B).

Though a circulation of air certainly does take place in some insects, there are forms, such as lepidopterous larvae, which exhibit no respiratory movements and so, it may be inferred, possess no positive means of ventilating the air tubes. Forces of diffusion have been shown to be adequate to supply oxygen to the tissues of these forms. These same forces will also explain the transfer of oxygen from the wider to the narrower air-containing tracheae. In the flea, which has no ventilating mechanism, it has been observed that during spiracular closure the larger tracheae collapse, their contained air having been used up. When the spiracles open again the tracheae dilate again by

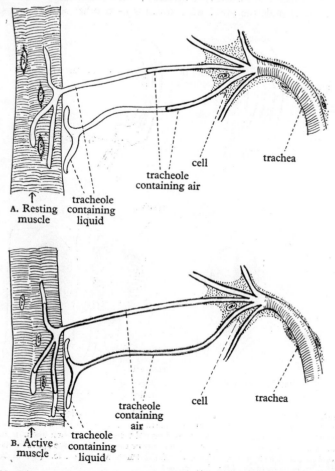

Fig. 308. Diagrams to show the regulation of tracheal respiration. In active muscle the osmotic pressure is higher and the fluid is withdrawn from the tracheoles, thus allowing air to get to the muscles more easily. (After Wigglesworth.)

their own elasticity. This implies a certain amount of ventilation, i.e. sucking in of air.

With regard to the ultimate problem concerning the way in which the air reaches the cell, modern theory on insect respiration assumes that the blind ends of the tracheolar tubes are bounded by a semipermeable membrane which is impermeable to lactic acid and such metabolites. Each tracheole contains a variable amount of watery fluid, the height of the column of which is determined in a state of equilibrium by hydrostatic pressure and capillarity on the one hand and by forces of osmotic pressure in the tissue fluids and of atmospheric pressure on the other. If now the osmotic pressure of the tissue fluids, for any reason, increases, the other forces remaining constant, water will then be absorbed from the tracheole tubes into the tissues and the column of air will be made to extend more deeply into the tissues. It has been shown that muscular activity of insects is associated with such withdrawal of water from the tracheoles. The evidence points to the conclusion that the chemical changes which accompany muscle contraction would provide the necessary osmotic changes to withdraw water from the tube and so bring the column of air to the tissues when and where their need is greatest (Fig. 308). From the air column, thus brought deeply into the tissues, the oxygen must diffuse into the surrounding tissue fluids.

We may note that since diffusion is so important in bringing oxygen to the tissues, the difficulty of gaseous diffusion in narrow tubes through any but the shortest distances at a rate adequate to supply wanting tissues, appears to have placed a limit to the size to which insects can grow. For instance, few insects grow to a size in which their diameter is greater than half an inch. The relatively large surface-bulk ratio which this smallness entails has placed a premium on the need of the insect for an impermeable cuticle to which reference has already been made. Diffusion in respiration, body size and cuticular structure are therefore all functionally related.

The control of respiratory movements by nerve centres is of interest. Though each nerve ganglion of the ventral chain serves as a centre for the respiratory movements of its own segment, there are certain regions of the nervous system which exercise a controlling influence over the respiratory activity of the insect as a whole. One example will serve to illustrate this point. The nymph of the dragon-fly *Libellula* pumps water for respiratory purposes into its rectum. In the natural state it responds to changes in the oxygen content of water quite readily, by increasing the rate of its respiratory movements when there is oxygen lack; reducing such movements in water saturated with oxygen. When, however, the prothoracic ganglion is destroyed, respiratory movements continue evenly, without reference to variations in the oxygen tension of the water.

There are thus primary respiratory centres, each responsible for movement in its own segment, and specially localized secondary centres, which can influence those movements in accordance with the demands for oxygen. The site of the secondary centre varies in different animals but never appears to lie in the head. Just as secondary centres respond to oxygen lack, so they have been shown to respond to the influence of carbon dioxide.

Though the above remarks would apply to the majority of insects, there are many stages of reduction of the tracheal system in the group, culminating in the Collembola, many of which have no tracheae at all, gaseous exchange taking place through the skin.

AQUATIC RESPIRATION. Aquatic insects fall into three main physiological groups. The first is distinguished by breathing air directly, at least one

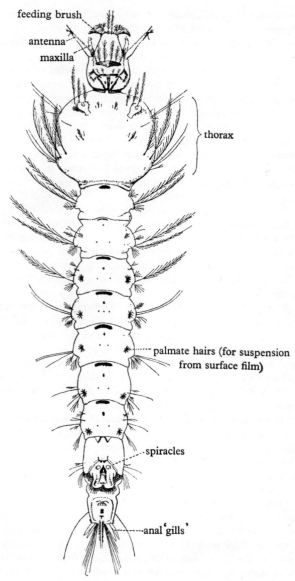

Fig. 309. Larva of *Anopheles maculipennis*. (After Nuttall and Shipley.)

pair of functional spiracles being retained. In the water-beetle *Dysticus* the abdominal spiracles communicate with a supply of air under the elytra which is renewed when the beetle comes to the surface. In the larva of the mosquito the spiracles are open to the air while the animal is suspended from the surface film (Fig. 309); the mosquito pupa, when surfacing, similarly presents open spiracles to the atmosphere (Fig. 310). In the direct air-breathers periodic visits are made to the surface to replenish the air in the tracheae, and to facilitate the making of this contact with the air devices involving hydrofuge hairs are common.

The second group includes the early stages of the Odonata, Plecoptera, Ephemeroptera and Trichoptera. These have no functional spiracles but breathe by means of tracheal gills—expansions of the body wall through whose thin walls respiratory exchange between the animal and the water is effected according to the laws of diffusion (Fig. 323). They are usually external

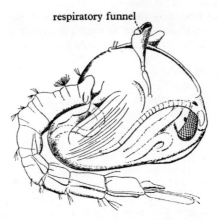

Fig. 310. Pupa of *Anopheles maculipennis*. (After Nuttall and Shipley.)

but in certain dragon-fly nymphs (*Aeschna* and *Libellula*) the rectal wall is raised into such gills and respiration is effected by pumping water in and out through the anus. Certain larvae show an even more complete adaptation to life in water in that, though they possess a tracheal system, this is entirely closed from the exterior and in their early stages is filled with fluid. Such forms respire of necessity by a process of simple diffusion through the general integument, e.g. *Chironomus* and *Simulium*.

The third adaptation to life in water is afforded by the phenomenon of plastron respiration. This is found in a number of aquatic adult insects, e.g. *Aphelocheirus* (Rhynchota), *Elmis* and *Haemonia* (Coleoptera), where open spiracles may coexist with the ability to respire under water so long as there is adequate dissolved oxygen in it. The body in the region of the spiracles is covered with a dense pile of exceedingly small hydrofuge hairs (as many as $2\frac{1}{2}$ million hairs per sq. mm. occur in *Aphelocheirus*). The gas trapped in this hair pile acts as a lung, oxygen passing from solution in the water through the air-water interface and so along an oxygen gradient to the tissues through the open tracheal system.

REPRODUCTION. The sexes of insects are separate, *Icerya purchasi*, a remarkable exception, being the only known self-fertilizing hermaphrodite in the class. The usual method of reproduction is by deposition of yolky eggs following copulation. The egg, except in many parasitic Hymenoptera, is richly supplied with yolk and invested with a vitelline membrane and further protected by a hard shell or *chorion*. The chorion exhibits different degrees of external sculpture and it is perforated at some point or points to allow of sperm penetration. A series of layers of cuticle very similar to those of the body wall form the bulk of the chorion. These are deposited over the oocyte by the follicular cells of the ovary. While the egg is still in the ovary the oocyte itself secretes a waterproof lining of wax to the chorion. Provision to prevent desiccation of the developing embryo is thus made. The spermatozoa, which are of the filiform type, may be transmitted to the female in the form of a spermatophore. Though insects are on the whole prolific creatures capable of producing large numbers of eggs, a few cases are met with where females lay only a few eggs in the course of their life. Thus, in the viviparous tsetse flies, a single egg is passed to the uterus about every nine or ten days. The larva is there nourished by special 'milk' glands till it is fully fed when it is passed out for immediate pupation. Viviparity and reduced egg production are here obviously associated with one another. In a large number of cases reproduction is effected without the intervention of the male. This phenomenon of *parthenogenesis* is best seen in the aphids or plant lice where several generations resulting in the production of parthenogenetic females are passed through. The racial advantage accruing from this greatly increased reproductive capacity is obvious.

Parthenogenesis is in certain cases, e.g. among the family Cecidomyidae of the order Diptera, found to occur in larval forms. In *Miastor*, for instance, a form living in decaying wood and under bark, reproduction in this manner (*paedogenesis*) occurs for the greater part of the year. These larvae contain prematurely developed ovaries from which as many as thirty larvae may grow. In summer, larvae occur which are morphologically different from the paedogenetic forms. These summer larvae pupate and the small midge-like flies which emerge lay four or five large eggs; from these a further series of paedogenetic larvae arises.

Among a few of the parasitic Hymenoptera, e.g. some Chalcididae, the phenomenon of *polyembryony* has been observed. This involves the development of more than one embryo from a single egg. In *Copidosoma gelechiae*, which parasitizes a caterpillar living on the golden rod *Solidago*, a hundred or more embryos may result from the deposition of a single egg.

ORGANS OF REPRODUCTION (Fig. 311). In the male the *testes* are usually small paired organs lying more or less freely in the body cavity. The extent to which they are divided into *follicles*, and the form of follicle, vary in different orders. Thus, in the Diptera, each testis is unifollicular, while in the Orthoptera a multifollicular condition prevails. Each follicle is divided into a *germarium* or formative zone, a zone of growth and maturation, and a zone in which spermatids are transformed into spermatozoa. In multifollicular testes the connexion between each follicle and the main duct is known as the *vas efferens* and each testis leads to the median *ejaculatory duct* by a *vas deferens* which is swollen at some point to form a *seminal vesicle*. The ejaculatory duct opens between the 9th and 10th abdominal sterna in association with the external genital plates (*gonapophyses*) of copulatory significance. Accessory glands of various kinds are usually found associated with the genital ducts. In some insects sperm-transfer to the female is by way of spermatophores. These are the product of the accessory glands. On transference to the female's *bursa* their walls are digested by enzymes produced by the bursa cells and the spermatozoa thus released.

The female organs (Fig. 311) consist of *ovaries, oviducts, spermathecae, colleterial glands* and a *bursa copulatrix*. Each ovary consists of a number of

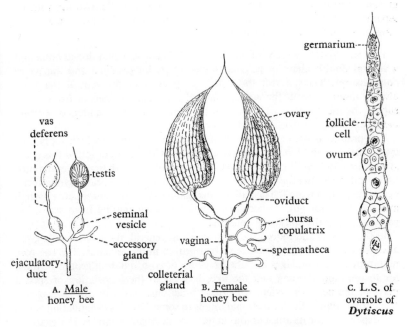

Fig. 311. Diagram of the reproductive organs.

ovarioles, corresponding to the testicular follicles of the male. Reduction of the ovary to a single ovariole occurs in such insects as *Glossina*, the tsetse fly, where the minimal number of eggs is produced.

Each *ovariole* (Fig. 311) is tubular and contains zones corresponding to those met with in the follicle of the testis. Nutritive cells are found associated with the developing ova. Such cells are concerned with the transference of yolk to the growing ova and they or other cells may entirely encircle the ova, round which they secrete the chorion or outer egg shell. The ovarioles forming an ovary are connected together anteriorly in the body cavity by their peritoneal coverings, in the form of terminal filaments, and these are attached either to the body wall or to the pericardial diaphragm, thereby maintaining the ovary in position.

The two *oviducts* leading from the ovaries become joined to a median ectodermal common oviduct leading to the hind border of the 8th sternum. The 9th sternum, by invagination, forms a vagina and receives the oviduct just mentioned.

Colleterial glands providing fluid for the formation of an *ootheca* (a case surrounding the eggs), or a sticky secretion for fastening eggs to surfaces, usually open into the vagina. The pouch for the reception of spermatozoa is the *spermatheca*. It is an ectodermal invagination, lined by cuticle and provided with a muscular coat. The spermatheca, which may be divided (e.g. Diptera), opens into the oviduct or into the bursa copulatrix, this being found in some forms (e.g. Lepidoptera) as an outgrowth of the oviduct with its separate and additional opening to the exterior on segment 8. The bursa receives the intromittent organ of the male.

THE NERVOUS SYSTEM of insects (Fig. 312) consists of a dorsal *brain* and a ventral double chain of ganglia connected by longitudinal and transverse commissures. The anterior three pairs of ganglia of the ventral chain are always fused to form the *suboesophageal ganglion*, the nerves from which supply the mouth parts. The suboesophageal ganglion is united by *circumoesophageal* connectives to the brain.

The brain consists of three pairs of closely fused ganglia which supply the eyes, antennae and labrum respectively (see p. 430). In addition to this is the *sympathetic system* (Figs. 293 and 312) which contains both sensory and motor fibres and innervates the anterior alimentary system, and the heart. Sympathetic nerve fibres pass from the last abdominal ganglion to the reproductive system and the posterior gut. Other sympathetic fibres emanating from segmental ganglia supply the spiracular muscles.

In the insects, and indeed the arthropods in general, there has been a great advance over the stage of nervous organization in the annelids. The complex nature of the appendages and the necessity of co-ordinating groups of these for locomotion, feeding and the like, has led to the association of special parts of the nervous system with these functions. We will call each part a 'functional unit'. Each functional unit is to some extent self-regulating and is not dependent for its autonomous action on the higher centres. For example a decapitated wasp can still walk and if a limb be removed from one side, compensatory movements of the remaining five legs enable the animal to walk in a straight line. But the working together of the functional units concerned, into different reactions, is controlled by the brain, and the inhibitory character of that control is shown when the ganglia are removed. A 'decerebrate' bee will try to fly, walk, feed and polish its abdomen all at the same time. This is because no inhibition is being exercised on the functional units, which themselves remain intact in spite of the removal of the higher centres.

SENSE ORGANS. There can be no doubt that insects perceive stimuli similar to those causing sensations in ourselves. They are sensitive to the waves of light and sound, to changes of temperature and humidity, to chemical stimuli by contact or at a distance, as in the sensations of taste and smell, and to tactile impressions. The sensory equipment is complicated, and the solution of the functional problem which many of its parts present is not made easier by the fact that though the principle of the reaction may be the same as in ourselves, insects often react to stimuli of an amplitude which is beyond our receptive capacity. For instance, they react to pitches of sound which the human ear cannot detect, and though they do not appreciate the full spectrum in colour vision, they can perceive ultra-violet rays.

No matter what the sense organ may be, the fundamental element is the *sensillum*. In the case of a simple sensory hair (*trichoid sensillum*) the following elements are present: a trichogenous cell which gives rise to the seta; a hair-membrane cell which produces the fine membrane at which the seta is

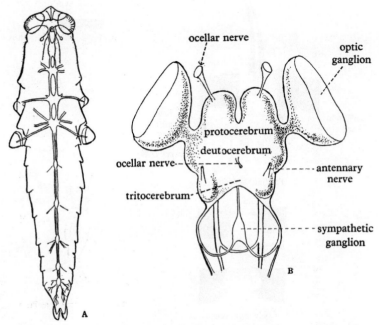

Fig. 312. Nervous system of a grasshopper. A, ventral chain. B, brain and associated nerves. (After Uvarov.)

articulated to the body wall, and a bipolar nerve cell which lies within the trichogenous cell (Fig. 313). Such sensilla are generally tactile though in certain cases olfactory, gustatory and heat-perceiving functions have been shown to rest in them. Olfactory sensilla commonly occur on the antennae and on the palps. These are generally *placoid* (with plate-like cuticle covering the sense cell) or *coeloconic* (where the covering plate is thin and sunk in a depression below the surface) (Fig. 313). But though the antennae are usually olfactory in function, this sense is also located elsewhere, since removal of the antennae may not entirely inhibit olfactory sensation.

The power of insects to diffuse scents from special glands is well known. These serve for defence, or to attract the sexes to each other, and their prevalence and wide distribution throughout the class postulate the existence of an olfactory sense. In moths the faculty possessed by males of discovering the exact position of unpaired females is of so astonishing a character that many observers have disbelieved the olfactory explanation, and resorted to theories of etheric wave-transmission. The production of a volatile chemical is clear, however, in those cases where male moths have assembled at an empty box in which a female had been recently housed.

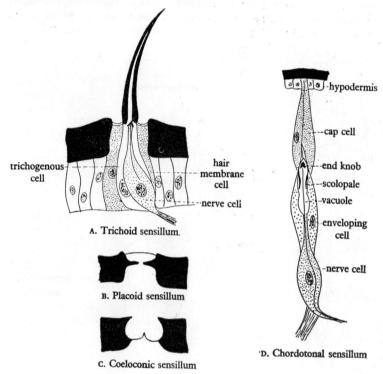

Fig. 313. Insect sensillae. (A, modified from Eltringham, after Snodgrass; B, modified from Imms, after Hess; C and D, from Imms.)

It is comparatively simple to demonstrate the existence of a taste sense in insects. Preferences for sugar to other substances in solution can readily be shown in a feeding butterfly. To find, however, that the taste organs lie in the feet, as, for example, in *Vanessa*, the red-admiral butterfly, is perhaps sufficient reason for using the term *chemo-tactile* for a sense which has no exact parallel in our own experience. Taste organs occur also in the mouth, and on the palps of the mouth parts.

Many insects, such as grasshoppers and cicadas, are provided with sound receptors known as tympanal organs, with which are incorporated *chordotonal* sensilla. Each of the latter consists of a sense cell, to one end of which is attached a nerve fibre. To the other end is connected a rod or *scolopale* which ends in an apical thickening or is free to vibrate in the fluid protoplasm of an enveloping cell. The whole structure is attached to the hypodermis by covering cells at one end and by a ligament at the other (Fig. 313 D).

Scolopale sensilla of this type may or may not be associated with a tympanum or ear-drum. When they are, as in cicadas and grasshoppers, there is clear evidence of response to sound waves set up by sound-producing organs possessed by themselves.

In the numerous cases in which no tympanum capable of responding to sound waves exists, a precise function is not clearly indicated. According to some, they may act as rythmometers, i.e. co-ordinators of the rhythmical movements of the insect's body. A more probable function is that of perceiving vibratory stimuli from without.

Though the organs of vision, with particular reference to compound eyes, have been dealt with in ch. IX, some account must be taken of other organs of light perception found in insects. Ocelli, both lateral as occurring in larvae, and dorsal as found in adults, present a considerable variety of structure. They have as their basis the ommatidium, which is itself a specialized sensillum of hypodermal origin. Lateral ocelli may occur in a group of some six or seven elements on each side of the head, or there may be only one such element at each side. Essentially there is a cuticular corneal lens beneath which may be found a crystalline lens of hypodermal origin. Beneath this lies the rhabdome, or sensory rod, the product of about seven sense cells from whose bases the nerve fibres pass away. The lateral ocelli not only have larval functions as organs of light perception, but also form the basis in association with which the adult compound eye undergoes its development. They vary as much in structure as they do in their perceptive capacity. They may serve merely as light perceptors in a general sense, or they may endow their possessors with a sense of both colour and form.

Dorsal ocelli are found in adult insects on the frontal region of the head. While they vary in structure, a common type is that in which a single lens overlies a group of sense cells associated with several rhabdomes. It appears likely that their function is more concerned with the perception of changes in light intensity than with the perception of images.

While sensilla respond to stimuli from without there are also those that serve as proprioceptors. For instance, the degree of flexion of joints of either the body or the limbs may be detected by suitably placed tactile bristles or by the tension put upon the dome-shaped covering membranes of *campaniform sensilla*. Campaniform sensilla on the wings are similarly proprioceptive.

That the stimulation of proprioceptive organs must have some importance in equilibrating the insect with reference to gravity seems clear. There are, however, in some insects, particularly in those that live in water, static organs containing air in contact with that of the tracheal system. In these organs sensory hairs suitably placed at the air/water interface enable the animal to detect differences of pressure at that interface and so serve to orientate the animal in respect to gravity. Being sensitive as they are in *Aphelocheirus* (Rhynchota) to uniform as well as to differential pressure changes, they can be said to direct the swimming movements of the animal upwards or downwards and to maintain the animal on an even keel.

The ability of such insects as bees to communicate to their fellows knowledge about food sources is a subject which falls outside the scope of this book. Yet these things, depending as they do on the sensory powers of the individual, cannot be passed without comment. Bees can recognize certain types of form difference and, while in some respects their colour sense is restricted, they are able to detect orange-yellow, blue-green and blue-violet from each other (but without discrimination within each of these wavebands, and without perception at the red end of the spectrum), and they are also sensitive to ultra-violet, a wave-length for which we ourselves have no perception.

By imparting the scents of visited flowers to their followers in a dance at the hive they can encourage other bees to exploit the same food source. They cannot, however, impart a similar knowledge of the form or colour of the flowers. Then, by means of dances (either in the hive or just outside it) in which they are followed by others, they can communicate information on the distance of the food source from the hive with considerable accuracy. They can also inform their fellows of the direction from the hive of the food source in relation to the position of the sun even when the sun is obscured by cloud, provided they receive light from a patch of blue sky. This has been shown to depend upon the power bees have to analyse the degree of polarization of the light emerging from the blue sky. This power in turn appears to rest in the eight visual cells surrounding the rhabdome of each ommatidium of the compound eye, which function as the analysers of polarized light. From these short remarks it seems clear that in some respects their sensory apparatus opens up to them knowledge of the environment which to us, except by special instruments, is closed.

EMBRYOLOGY. Though arthropod eggs vary in the amount of yolk contained within them they are for the most part yolky and are *centrolecithal* in type (p. 328). To this feature must be ascribed those distorting influences which make arthropod development so different from that of other invertebrates. Among insects it is only in the primitive Apterygota and in many parasitic Hymenoptera that are found small, comparatively yolkless eggs which undergo total cleavage. Though these may represent the primitive condition, they cannot be taken as typical of modern insects.

The typical yolky egg is provided with a vitelline membrane and a stout chorionic shell already described. After fertilization incomplete cleavage sets in, a process involving only the successive mitoses of nuclei. In this early stage, therefore, the egg is a syncytium of very yolky cytoplasm in which lie the cleavage nuclei. These wander to the peripheral cytoplasm, there to form an outer cellular layer or blastoderm (Fig. 314A, B). In this latter occurs a ventral thickening, thus differentiating embryonic from extra-embryonic blastoderm, and in its relation to the yolk the embryo now resembles an inverted chick embryo, but, as might be expected, its method of differentiation is highly different.

Gastrulation proceeds as follows. From the middle line of this embryo certain cells pass inwards towards the yolk by invagination, by proliferation or by the overgrowth of cells of the germ band lateral to them. This enclosed cell mass is mesoderm (together with endoderm in certain cases). The plate left outside constitutes the ectoderm (Fig. 314C, D). In such cases, where endoderm is not included in the enclosed mass as above, this layer rises from growth centres, anterior and posterior, at the places where the stomodaeum

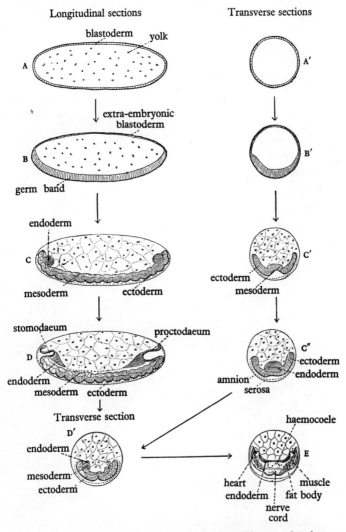

Fig. 314. Diagram to illustrate the main features of insect embryology. For details see text. (After Eastham.)

and proctodaeum will appear or have already differentiated. The result in any of these cases is a three-layered embryo relegated to the ventral side of the egg, i.e. beneath the yolk. It consists of a layer of outer ectoderm, within which is the mesoderm from which paired segmental somites develop.

Against the yolk lies the endoderm destined to form the mid gut. The mesoblastic somites give rise on their upper borders to the heart rudiments, and on their outer and inner borders to the muscles of body wall and gut respectively.

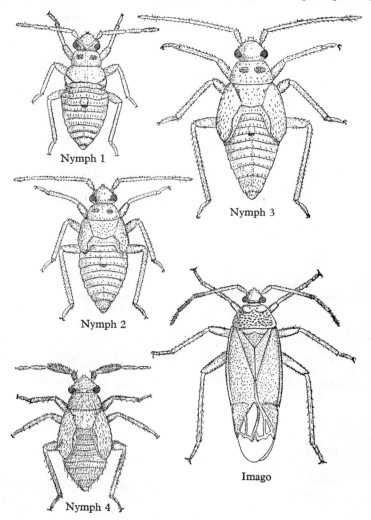

Fig. 315. Metamorphosis of a capsid bug, *Plesiocaris vagicollis*.
(After Petherbridge and Hussain.)

The lower border of each somite breaks down to form the fat-body. In so doing the coelomic cavity disappears, and, minute as it always was, becomes continuous with those spaces arising by separation of the germ layers from each other, namely, the haemocoele. This latter as in all arthropods constitutes the main body cavity (Fig. 314E).

GROWTH AND METAMORPHOSIS. Insects, like all other arthropods, attain their maximum size by undergoing a succession of moults or ecdyses. The number of moults which an insect passes through is fairly constant for the species, and the form assumed by the animal between any two ecdyses is termed an *instar*. The animal's existence is thereby made up of a succession of instars (growth) during which the insect is immature, followed by the attainment of the final adult instar (*metamorphosis*). In the simplest and most generalized insects the several instars are very similar to one another and only differ from their appropriate adults in the absence of wings and the incomplete development of the reproductive system. Where the adult is primitively wingless, as in silver-fish and springtails (Figs. 318, 320), the change from young to adult is so slight as to be ignored, and metamorphosis, involving only a development of the reproductive system, is conveniently regarded as being absent. The insect orders falling in this category are often grouped under the heading Ametabola.

In winged insects, however, the adult is in sharp contrast to the young stage not only because of its wings, but on account of the sexual appendages which, in the adult, have become prominent. Such forms are said to undergo a metamorphosis (Fig. 315). Here the degree of metamorphosis varies considerably, irrespective of wings, according as the young stages resemble their adults or not. A growth stage of a cockroach, for instance, possesses the general appearance of the adult. On the other hand the young stage of a house-fly is a grub and has no resemblance to the final stage with its wings, elaborate body form and mouth parts (Figs. 338, 342).

Metabolous insects—those passing through a distinct metamorphosis—manifest many degrees of this phenomenon. Classifications have often been based on this, but in the system presented later in this book the basis is largely palaeontological and structural. Be that as it may, groupings are possible as follows.

The Heterometabola are those insects whose young stages, known as nymphs, closely resemble the adult in body form and type of mouth parts. Such young stages possess compound eyes and, where wings develop, the growth of them is external and easily visible throughout most or all of the nymphal instars. Here are included the following orders: Dictyoptera, Isoptera, Zoraptera, Notoptera, Cheleutoptera, Orthoptera, Embioptera, Dermaptera, Homoptera, Heteroptera, Psocoptera, Mallophaga, Anoplura Thysanoptera, Ephemeroptera, Odonata and Plecoptera. The last three orders, having nymphs adapted for aquatic life, have sometimes been grouped accordingly as Hemimetabola.

The Holometabola, comprising the orders Coleoptera, Megaloptera, Raphidioptera, Planipennia, Mecoptera, Trichoptera, Lepidoptera, Diptera, Hymenoptera, Strepsiptera and Siphonaptera, have young stages known as larvae, which differ markedly from the adult in body form and mouth parts. These young stages do not possess compound eyes, but in their place are lateral ocelli. Further, their wings develop from within pockets of the hypodermis and are therefore not visible from the outside during growth. So great is the difference between the larva and the adult that an instar known as the *pupa* has been intercalated to bridge the gulf between them (Fig. 348).

This stage, one of apparent rest, is actually one of great physiological and developmental activity, and it is here that many larval tissues, e.g. the muscles and the alimentary canal, are broken down by phagocytic or other action and the new adult tissue is built up from many growth centres, generally known as *imaginal discs*. The change from larva to pupa is often accompanied by a period of inactivity at the end of the last larval instar.

A. Campodeiform larva of
Pterostichus (Carabidae)

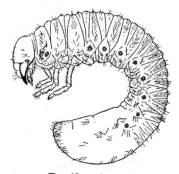

B. Eruciform larva of
Melolontha (Scarabeidae)

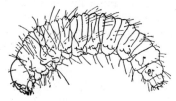

C. Legless larva of
Phyllobius (Curculionidae)

Fig. 316. Larval types of Coleoptera. (C, after Rymer Roberts.)

The origin of the phenomenon of *holometaboly* is obscure. That it is today associated with divergent specializations of larvae and adults in varying degrees is a matter of observation. It is not surprising, therefore, to find among the orders composing this group, as, for instance, in many Coleoptera, larvae which are rather nymph-like in that they are well cuticularized and possess well-developed legs, and mouth parts resembling those of the adults (Fig. 316A).

The forms of larvae vary considerably and indicate to a great extent the degree of metamorphosis passed through. A *campodeiform* larva (Fig. 316A) is one strongly resembling certain members of the ametabolous Thysanura and possesses well-developed legs, antennae, cerci and mouth parts, e.g. many Coleoptera. An *eruciform* larva (Fig. 316B) is fleshy and thin skinned, its legs are often in the form of supporting struts rather than organs of active locomotion, *prolegs* are often found on the abdomen, and there are no cerci, e.g. caterpillars of Lepidoptera and saw-flies (Fig. 335). A *grub* (Fig. 316C) is an apodous larva which in certain respects resembles the eruciform type, e.g. certain Diptera, Coleoptera and Hymenoptera.

Pupal modifications are also found; thus the *exarate* type, characteristic of the Hymenoptera, Coleoptera, Mecoptera and Neuroptera, is that in which the cases of the adult appendages lie are not fused along their length to the body (Fig. 348). In *obtect* pupae (Fig. 333) wing and leg cases are fused to the body wall, e.g. most Lepidoptera and Diptera. In the most specialized Diptera the last larval skin is retained as a barrel-shaped *puparium* over the pupa within. Such protected pupae are called *coarctate* (Fig. 342B).

In the Heterometabola the development of adult form is a gradual process and the appendages, including mouth parts, antennae and legs, grow directly into those of the adult. Wings in such forms develop gradually as external dorsolateral extensions of the meso- and metathoracic body wall (Fig. 315). All the Heterometabola have such a wing development and therefore the alternative name '*Exopterygota*' is often given to the group.

Larvae of the Holometabola on the other hand, as stated above, possess, for the most part, mouth parts having a form and mode of working different from that of their adults, their legs are reduced in size and complexity or even absent, and wing growth is hidden from view—hence the alternative name '*Endopterygota*'. It is in the pupal stage of Holometabola that adult appendages appear for the first time on the surface.

The development of adult appendages in the larva is only one of the many aspects of metamorphosis. The wings which suddenly appear in the pupa of the butterfly grow gradually through each of the five larval instars, but instead of growing externally as in the Heterometabola (Exopterygota) they arise as outgrowths from the bottom of intuckings of the body wall. In other words an accommodating fold of the body wall forming a sac, opening at the surface by a minute pore, hides the growing wing bud within it and this is the main difference between *endopterygote* and *exopterygote* development.

At pupation the sac carrying the wing disc or bud at its base becomes straightened out by contraction of its walls and the wing bud is thereby brought to view. Similar limb buds are to be found for the adult legs and mouth parts which always grow in association with the corresponding larval organs. Such buds are known collectively as *imaginal discs* and their existence characterizes all endopterygote insects (Fig. 317).

From this brief account we can see that two postembryonic processes occur in the lives of all insects. Both require the process of moulting, the shedding of the old and the production of the new cuticle. The first of these, growth, is the attribute of the young insect; the second, metamorphosis, is the attainment at the final moult of the condition of sexual maturity. From a wide range of sources it is now known that both these processes are under hormonic control from neurosecretory cells of the brain, the corpora allata and the prothoracic glands.

The interplay of these is complex, depending on a precisely timed activation. The result of this interplay is the suppression of adult characters from appearance during youth and the final attainment of sexual maturity when

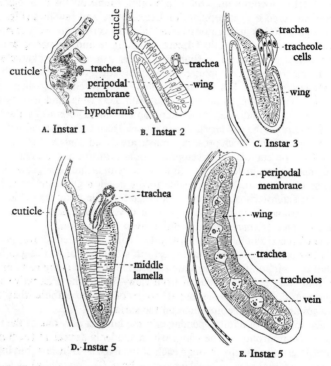

Fig. 317. The internal development of a wing in the larva of the butterfly, *Pieris rapae*, as seen in transverse sections.

the juvenile features are lost. The prothoracic gland is said to produce the hormones necessary for moulting but only when activated to do so by the neurosecretory cells of the brain.

The moulting thus stimulated is the route by which either further juvenile stages or the final adult stage can be attained. In the presence of the secretion of the corpora allata the juvenile state results. When these glands cease to function, normally at the end of the larval period, the moulting process, still activated by the prothoracic gland, gives rise to the adult.

FOSSIL RECORD. The first records of insects are to be found in rocks of the Devonian period. These take the form of fragmentary remains of insects similar to our present-day Collembola, and there seems little doubt that wingless insects of this kind abounded then, in a rich vegetation of pterydophytes, etc., under very wet conditions. It is possible, though not proved, that thysanuran insects not unlike our present-day silver fish, *Lepisma*, lived then.

Only in the Carboniferous do we find the burgeoning of winged insects, and in this period and the Permian which followed there lived insects which have been classified into orders which seem to be prophetic of the forms we have with us today. Some of these orders became extinct early. Thus, the Palaeodictyoptera, huge insects with a wing up to 50 cm., came into existence in the lower Carboniferous to disappear in the lower Permian. That they flew is certain, their wings having those arrangements of longitudinal veins indicative of use (Fig. 297). The retention of immobile flap-like extensions on the sides of the non-wing-bearing segments characterized some of these insects—a feature which may throw light on the origin of wings.

Following them there came into being a number of orders—primitive in their mouth parts and venation, and probably in wing growth and their simple metamorphosis. Such orders as Protoephemeroptera, Protodonata, Protohemiptera and Protohymenoptera have been established. These names need not indicate a direct ancestry to the modern orders whose names they bear. They do show clearly, however, that in the Permian there was occurring a steady deployment of types conferring much variation on the subphylum. The fact that they were all mandibulate and, as far as has been determined exopterygote in wing growth, lends strong support to the view that from these generalized characters were evolved later the more specialized features of many of the insects of today. Probably the strongest support of the view that insects with mandibulate mouth parts, external wing growth and with wings with more or less parallel longitudinal veins with a rich network of cross veins, are the primitive ones, rests in the persistence, with little change, of cockroaches, from the Carboniferous till today. The student making his first attempt at the intricacies of entomology by dissecting the cockroach may well keep in mind that he is dealing with a very ancient type—a real aristocrat among insect species.

Insects with internal wing growth and complex metamorphosis—those hitherto grouped as Endopterygota or Holometabola—came later. We thus find Diptera, Trichoptera, Lepidoptera, Neuroptera, Mecoptera and Hymenoptera in the Mesozoic and becoming abundant in the Tertiary period. The Coleoptera seem to be far older than these, extending back to the Upper Permian. In the Tertiary there had already occurred among them those evolutions in habit and form leading to water-beetles, weevils and the leaf-eating chrysomelids. This earlier establishment of the Coleoptera is not without interest since the Coleoptera, as we know them today, possess, particularly in their mouth parts, a number of features which place them in the generalized category.

Now if we consider the order of events briefly recounted here it will be seen that though the forerunners of our modern complex orders, Diptera, Hymenoptera, Lepidoptera, etc., may have existed in the Permian, this latter age, together with the Carboniferous, was essentially that of the generalized insect not yet equipped with feeding mechanisms for dealing with flowering plants. Be that as it may, the more interesting fact is that the main evolution of our specialized bees, flies and butterflies coincided in point of time with the evolution of the flowering plants, to which by their manner of feeding they are now on the whole so inseparably bound.

CLASSIFICATION. A natural classification of insects cannot be formed without reference to fossil history and comparative anatomy. The details of palaeontology and of much comparative anatomy, especially where it involves wing venation, are too complicated to be attempted in this book. The student wishing to go more deeply into these things is referred to the literature mentioned. We can, however, remark on some of the broader aspects of palaeontology in so far as they assist the forming of a classification.

Though the insects form a natural group—all the members being referable to some generalized form possessing, among other things, mouth parts efficient for chewing solid food, an eleven-segmented abdomen, three-segmented leg-bearing thorax and six-segmented head, two pairs of membranous wings carrying parallel longitudinal veins with a reticulum of cross veins between them—the orders are clearly defined. As now presented to us they are not easily linked by intermediate forms and much of the story of evolution within the class consists rather of disjointed sentences than a continuous theme.

A study of fossils from the Devonian onwards has, however, made possible some groupings of orders which are indicative of relationships. Thus, for example, among winged insects there are orders in which the wings are held unfolded when at rest and where the network of cross veins is prominent. These orders, Ephemeroptera and Odonata, have origins which are traceable from somewhat similar prototypes in the Upper Carboniferous and the Permian, and in consequence their union into a common section, Palaeoptera, appears justified.

Of the remaining winged insects, all of which are able to fold their wings over the body when at rest, it has been possible to form a section (Polyneuroptera) comprising those insects with rich wing venation, namely orthopteroids (crickets and grasshoppers), blattoids (cockroaches), dermapteroids (earwigs) and plecopteroids (stone-flies). Another section has been formed, the Oligoneoptera, in which fewer veins are found on the wing: the Coleoptera, Neuroptera, Trichoptera, Lepidoptera, Diptera, Hymenoptera, etc. The first of these sections, Polyneoptera, has its fossil roots also in the Upper Carboniferous and the Permian. The section Oligoneoptera, however, has more recent recognizable prototypes in the Permian and Mesozoic.

The Paraneoptera (psocids, bird-lice and body-lice, together with the rhynchotan bugs) are still more recent in origin: recognizable fossils relatable to our present-day forms have not been found with any certainty before the Mesozoic.

It is on evidence of this kind (only briefly mentioned here) that the groupings shown in the classification have been made. Since this classification is in many respects different from what has hitherto been presented in many text-books it has been deemed advisable to indicate in brackets how it stands in relation to the older system. Thus, among the Pterygota, Section I (Palaeoptera), Section II (Polyneoptera) and Section IV (Paraneoptera) include all those orders whose members show external wing growth (Exopterygota) and incomplete metamorphosis (Heterometabola). The remaining Section III, Oligoneoptera, combines within it all those higher orders of insects showing internal wing growth (Endopterygota) and complete metamorphosis (Holometabola).

With this explanation we may leave the matter, only remarking that the older divisions, Exopterygota and Endopterygota, Hemimetabola and Holometabola had about them a convenience born of expediency where some of the differences in metamorphosis, both in degree and kind, or of wing growth, were not fully taken into account.

Subclass 1. APTERYGOTA (AMETABOLA). Primitive wingless insects.

  Super-order 1. **Entotropha.** Mouth parts almost entirely hidden within the head.

    Order 1. **Collembola.** Six abdominal segments of which three carry appendages. Larva born with full number of adult segments. No metamorphosis. No Malpighian tubules. Springtails. *Podura*

    Order 2. **Protura.** Abdomen with twelve segments. Segments of body added progressively during post-embryonic life. Rudimentary Malpighian tubules. No antennae.
Acerentomum

    Order 3. **Diplura.** Abdomen of eleven segments. Larva born with full complement of adult segments. Segmented antennae. Cerci present. Malpighian tubules rudimentary, or absent.
Campodea

  Super-order 2. **Ectotropha.** Mouth parts visible.

    Order 4. **Thysanura.** Abdomen of eleven segments. Larva born with full complement of adult segments. Cerci and terminal filaments present. Abdominal segmental styles and coxal vesicles. Malpighian tubules present. 'Silverfish' and bristle-tails. *Lepisma*

Subclass 2. PTERYGOTA (METABOLA). Young stage born with full complement of adult segments. No abdominal locomotory appendages; simple metamorphosis. Malpighian tubules present. Mouth parts free.

  SECTION I. PALAEOPTERA (HEMIMETABOLA, EXOPTERYGOTA). Wings not folding. Hemimetabolous. Malpighian tubules numerous.

  Super-order 1. **Ephemeropteroidea.**

    Order 5. **Ephemeroptera.** Mouth parts regressed. Fore and hind wings unequal; raised in repose; rich cross-veining with intercalary longitudinal veins. Young stages aquatic. May-flies. *Chloeon*

Super-order 2. **Odonatopteroidea.**

Order 6. **Odonata.** Mouth parts for biting. Prothorax reduced. Wings with pterostigma, nodus and arculus; rich cross-veining. Young stages aquatic. Dragon-flies. *Aeschna*

SECTION II. POLYNEOPTERA (HETEROMETABOLA, EXOPTERYGOTA). Wings folded in repose. Heterometabolous. Malpighian tubules numerous.

Super-order 1. **Blattopteroidea.** Mouth parts for biting.

Order 7. **Dictyoptera.** Cerci multi-articulate. Eggs laid in ootheca. Tarsi pentamerous. Blattids and mantids. *Periplaneta*

Order 8. **Isoptera.** Mouth parts for biting. Anterior and posterior wings similar. No cross-veins. Termites. *Termes*

Order 9. **Zoraptera.** Mouth parts for biting. Small carnivorous insects with two jointed tarsi. No apparent copulatory organs. Females with wings and eyes. Males wingless and eyeless. *Zorotypus*

Super-order 2. **Orthopteroidea.**

Order 10. **Plecoptera.** Head free, prognathous;* mouth parts for biting. Long, multi-articulated cerci. Walking legs. No ovipositor. Stone-flies. *Perla*

Order 11. **Notoptera.** Head orthognathous,* mouth parts for biting. Cerci as in blattids. Tarsus pentamerous. Ovipositor present. No wings. *Grylloblatta*

Order 12. **Cheleutoptera.** Head prognathous. Mouth parts for biting. Short cerci. Walking legs. Eggs laid free. Rudimentary ovipositor. Phasmids, stick- and leaf-insects. *Dixippus*

Order 13. **Orthoptera.** Head orthognathous. Mouth parts for biting. Posterior legs enlarged for jumping. Typanum on the tibia. Stridulation organs. Locusts, grasshoppers and crickets. *Gryllus*

Order 14. **Embioptera.** Head prognathous. Mouth parts for biting. Antennae filiform. Tarsal silk glands. Winged males, wingless females. *Embia*

Super-order 3. **Dermapteroidea.**

Order 15. **Dermaptera.** Mouth parts for biting. Anterior wings as short elytra, posterior wings folded both transversely and longitudinally. Three-jointed tarsus; cerci as forceps. No ovipositor. Earwigs. *Forficula*

* The head is *prognathous* when its long axis is in line with that of the body; *orthonathous* when this axis is at right angles to the long axis of the body. When the mouth, thus directed downwards, is brought still farther back under the head the condition is described as *hypognathous*.

INSECTA 473

SECTION III. OLIGONEOPTERA (HOLOMETABOLA, ENDOPTERYGOTA). Wings folded in repose. Holometabolous. Endopterygote. Usually with few Malpighian tubules.

Super-order 1. Coleopteroidea.

Order 16. **Coleoptera.** Head prognathous. Mouth parts for biting. Prothorax free. Abdomen of nine segments. Fore wings as elytra, meeting along the mid-dorsal line in repose. Larva with an anal proleg on 9th segment. Beetles. *Carabus*

Super-order 2. Neuropteroidea.

Order 17. **Megaloptera.** Head prognathous. Mouth parts for biting. Prothorax short. Two pairs of membranous wings without pterostigma; rich venation. Pentamerous tarsus. Aquatic larva with paired segmental and segmented gills. Alder-flies. *Sialis*

Order 18. **Raphidioptera.** Head prognathous. Mouth parts for biting. Prothorax long. Two pairs of membranous wings with pterostigma. Tarsus pentamerous. Larva terrestrial. Snake-flies. *Raphidia*

Order 19. **Planipennia.** Head prognathous. Mouth parts for biting. Two pairs of long membranous wings, veins bifurcated at ends. Pentamerous tarsus. Terrestrial larva with long, curved, grooved mandibles for piercing and sucking. No abdominal legs. Malpighian tubules may secrete silk. Lace-wings and ant-lions. *Myrmeleon*

Super-order 3. Mecopteroidea.

Order 20. **Mecoptera.** Head orthognathous. Mouth parts for biting borne at the end of cephalic extension. Thorax little sclerotized. Pentamerous tarsus. Two pairs of membranous wings without pterostigma; veins simple at extremities. Larva terrestrial with prolegs. Scorpion-flies. *Panorpa*

Order 21. **Trichoptera.** Head orthognathous. Biting mouth parts weak. Antennae multi-articulate. Hairy membranous wings, roof-like when in repose. Tarsus pentamerous. Larva aquatic with tracheal gills, may construct case in which to live. Caddis-flies. *Phryganea*

Order 22. **Lepidoptera.** Mouth parts almost always forming a coilable proboscis. Scaly wings. Tarsus pentamerous. Larvae usually phytophagous, caterpillars with abdominal prolegs bearing hooks. Butterflies and moths. *Vanessa*

Order 23. **Diptera.** Mouth parts modified for piercing and sucking, or for sucking only. One pair of wings; the posterior wings are modified as balancers. Tarsus pentamerous. Larva legless with head often little differentiated. Flies. *Musca*

Super-order 4. **Siphonapteroidea.**

Order 24. **Siphonaptera.** Mouth parts for piercing and sucking. Wingless ectoparasites. Larva legless. Fleas. *Pulex*

Super-order 5. **Hymenopteroidea.**

Order 25. **Hymenoptera.** Head orthognathous. Mouth parts for biting and/or for fluid-feeding. Mandibles always present. First abdominal segment fused to thorax. Four membranous wings. Anterior wings linked to posterior wings by a groove and engaging hooks. Tarsus pentamerous. Female with ovipositor. Malpighian tubules numerous. Larva (with exceptions) legless, generally with a well-formed head capsule. Ants, bees, wasps. *Apis*

Order 26. **Strepsiptera.** Males winged. Anterior wings modified as balancer. Posterior wings folded at rest. Females endoparasitic and larviform. *Stylops*

SECTION IV. **PARANEOPTERA (HETEROMETABOLA, EXOPTERYGOTA).** Wings folded in repose. Malpighian tubules few in number. Exopterygote and heterometabolous.

Super-order 1. **Psocopteroidea.**

Order 27. **Psocoptera.** Head orthognathous. Mouth parts for biting. Antennae multi-articulate. Abdomen globular. Tarsus two- or three-jointed. Wings often absent. Book-lice. *Atropus*

Order 28. **Mallophaga.** Head orthognathous. Mouth parts mandibulate, reduced. Antennae short. Thoracic segments distinct. Tarsus one- or two-jointed. Wingless. Bird-lice. *Menopon*

Order 29. **Anoplura.** Head prognathous. Mouth parts for piercing and sucking. Rostrum labial. Antennae five-jointed. Thoracic segments not differentiated. Tarsus one- or two-jointed. Body-lice. *Pediculus*

Super-order 2. **Thysanopteroidea.**

Order 30. **Thysanoptera.** Head hypognathous.* Mouth parts for piercing. Rostrum labial. Antennae contiguous at base. Prothorax free. Two pairs of wings with setose margins; reduced venation. Wingless forms common. Abdomen of ten segments, the last being tubular. Tarsus one- or two-jointed with terminal adhesive vesicle. Thrips. *Kakothrips*

*See p. 472 n.

INSECTA 475

Super-order 3. **Rhynchota.**

Order 31. **Homoptera.** Head hypognathous, labial proboscis extending to anterior coxae. Wings; two pairs, membranous. Tarsus three-jointed. Cicadas and aphids. *Aphis*

Order 32. **Heteroptera.** Head ortho- or hypognathous, labial process not touching anterior coxae. Two pairs of wings, the anterior ones as hemi-elytra. Tarsus three-jointed. Water-bugs, capsids and bed-bugs. *Leptocoris*

SUBCLASS APTERYGOTA

Primitively wingless insects carrying on the abdomen a varying number of paired appendages other than the external genitalia and cerci. Metamorphosis slight or absent.

Super-order **Entotropha**

The tracheal system when present is usually of a simple type in which spiracles are without closing mechanisms and the tracheae associated with a spiracle remain unconnected with those of other spiracles.

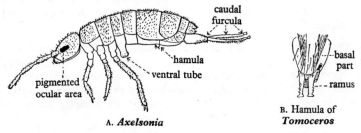

A. *Axelsonia*   B. Hamula of *Tomoceros*

Fig. 318. Collembola. (From Imms, after Carpenter and Folsom.)

1. *Order* **Collembola (Springtails)**

DIAGNOSIS. Small wingless insects with biting mouth parts deeply withdrawn into the head; compound eyes absent; a six-segmented abdomen which often carries three pairs of highly modified appendages serving the purposes of adhesion and jumping; a tracheal system is commonly absent and there are no Malpighian tubules; metamorphosis absent.

ACCOUNT. Four-jointed antennae, ocelli and, in some forms, post antennal sensory organs are characteristic features of the head.

There are no tarsi on the legs, claws being borne by the tibiae. The first abdominal segment carries a *ventral tube* which is moistened by a glandular secretion from behind the labium poured down a ventral groove running along the middle of the thorax. This ventral tube, regarded as adhesive, is formed by the fusion of the embryonic appendages of this segment. On the ventral side of the 3rd segment, the nearly complete fusion of a pair of appendages has resulted in the formation of the *hamula*, which engages the *furcula* prior to leaping. The latter is a forked structure representing a pair of limbs of the 4th segment (Fig. 318). By contraction of the extensor muscles of the furcula the latter is pulled down out of contact with the hamula and the animal is propelled forwards into the air.

Collembola are described as *protomorphic* in that they hatch with fewer segments than is characteristic of the class, this number being maintained into the adult stage.

The absence of tracheae is a secondary feature due to the small size of the animals rendering surface respiration sufficient for their mode of life. In such forms as are tracheate a spiracular closing mechanism is commonly absent. The single pair of spiracles is found in the neck region.

Collembola have a wide distribution. They are found along the seashore between tidemarks and submerged by each tide, e.g. *Anurida maritima*. Common aquatic forms are denizens of fresh waters, e.g. *Podura aquatica*. They have been reported to be so abundant at times in Arctic zones as almost to cover the snow, and in Europe sometimes to be present in such large numbers that the progress of railway trains is impeded owing to their having prevented the wheels from gripping the rails.

2. Order **Protura**

DIAGNOSIS. Minute insects without wings, eyes or antennae; with piercing mouth parts deeply inserted in the head capsule; with abdomen of twelve segments, the first three of which bear papillae (Fig. 319).

ACCOUNT. This is a small group of doubtful affinities. Its members are found in decaying organic matter. The fact that on hatching the abdomen is nine-segmented and that subsequent moults bring about the full number of segments (i.e. they are *anamorphic*) is regarded by some authorities as sufficient ground for their inclusion in a class distinct from the Insecta. An example is *Acerentomum doderoi* of Europe.

3. Order **Diplura**

DIAGNOSIS. These small insects of slender elongated form have multi-articulate antennae. The eleven-segmented abdomen terminates in a pair of jointed cerci which in some cases, e.g. *Japyx*, have become modified to forceps. There is no median caudal filament. Malpighian tubules are commonly absent. They live in the soil, under stones and in detritus. *Campodea* is a common British form (Fig. 319A).

Super-order Ectotropha

4. Order **Thysanura (Bristle-tails)**

DIAGNOSIS. Biting mouth parts (Fig. 294); antennae many-jointed; compound eyes present; abdomen of eleven segments, some or all of which bear styliform appendages which probably represent the coxites of limbs no longer present; anal cerci jointed.

ACCOUNT. *Lepisma saccharina* (Fig. 320) the common 'silver-fish' which inhabits dwellings of man, and (*Machilis*) *Petrobius maritimus*, found above high-tide mark along the seashore and estuaries, are common examples. In *Petrobius* (Figs. 294, 321) interesting features are presented by the well-

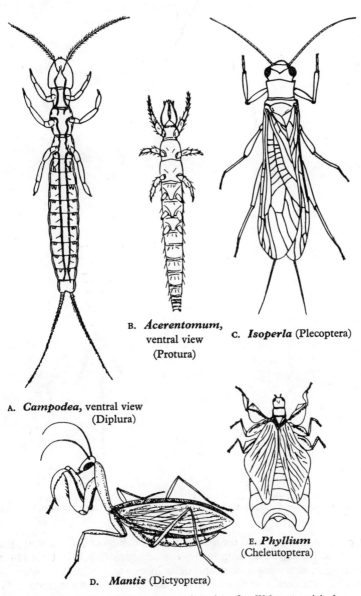

Fig. 319. Various primitive insects. (A and E, after Weber; B, original; C, after Despax; D, after Brehm.)

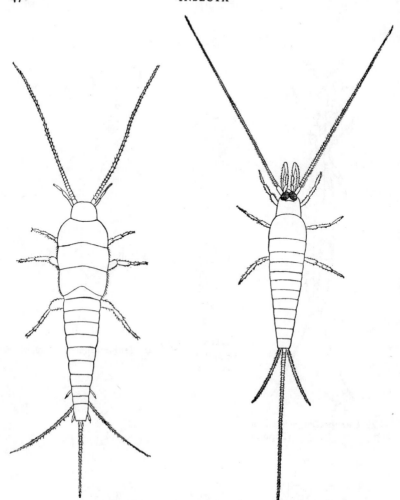

Fig. 320. *Lepisma saccharina*.
(From Imms, after Lubbock.)

Fig. 321. *Petrobius maritimus*.
(From Imms, after Lubbock.)

developed *superlinguae* and the jointed mandibles, both of which are primitive characters. The superlinguae in *Petrobius* are paired structures attached to the hypopharynx and possess inner and outer lobes and a palp-like process. In spite of a superficial resemblance to maxillae they have no segmental value. They are the homologues of the crustacean paragnaths.

SUBCLASS PTERYGOTA

SECTION I. PALAEOPTERA (Hemimetabola, Exopterygota)

Super-order **Ephemeropteroidea**

5. *Order* **Ephemeroptera (May-flies)**

DIAGNOSIS. Vestigial mouth parts reduced from the biting type; wings membranous with a reticulate venation; the hinder pair small; caudal filament and cerci very long (Fig. 322); metamorphosis hemimetabolous. The nymphs are aquatic and an active winged stage known as the subimago occurs before the last moult yields the adult.

Fig. 322. *Ephemera vulgata*. (From Imms.)

ACCOUNT. The eggs are laid in water, either scattered over the surface or attached to submerged stones, etc., by the female, which enters the water for the purpose.

The nymphs at first possess no gills but subsequent instars bear on the abdomen movable paired tracheal gills (Fig. 323), which may be branched or lamellate, exposed or protected in a branchial chamber. The body form varies with the habits. Thus inhabitants of fast-flowing streams have flattened bodies with legs provided with strong clinging claws, e.g. *Ecdyonurus*. Those which live in clear still water have a streamlined form for rapid movement, e.g. *Chloeon*, while burrowing types have fossorial legs, e.g. *Ephemera*, and are in some forms provided with protective gill opercula, e.g. *Caenis*. The mouth parts are of the biting type, and the two-jointed mandibles and well-developed superlinguae are features of importance. The nymphs are essentially herbivorous. Nymphal life is usually of long duration: as many as twenty-three instars may occur. In order to emerge, the fully fed nymph creeps out of the water

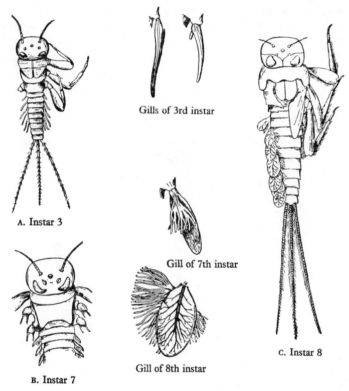

Fig. 323. Nymphal stages of *Heptagenia*. Showing the development of the gills. (After Imms.)

on to a plant stem. A moult gives rise to the winged subimago stage. This flies away and after a period which varies, according to the species, from a few minutes to about twenty-four hours, a final moult yields the adult which enjoys, as the name of the order implies, a similarly short life. In the adult the mouth parts are vestigial, no feeding is done, and the alimentary canal, full of air, serves no longer for digestion.

Economically these insects are of importance in so far as they constitute a proportion of the food of fresh-water fishes, the adults being caught by fish during their nuptial dance, and the nymphs being devoured by bottom-feeding fish.

Super-order **Odonatopteroidea**
6. Order **Odonata (Dragon-flies)**

DIAGNOSIS. Predacious insects with biting mouth parts; two similar pairs of wings with characteristic reticulate venation; prominent eyes and small antennae; elongated abdomen with accessory male genitalia on the 2nd and 3rd sterna; metamorphosis hemimetabolous; nymphs aquatic, possessing a modified labium known as the mask.

ACCOUNT. The members of this order are large insects, and in the Carboniferous period genera existed which had a wing expanse of two feet. They are strong and rapid fliers, catching their food, in the form of small insects, on the wing. The forwardly directed legs play an important part in catching the prey and holding it while it is masticated.

The thorax has a peculiar obliquity of form, the pleural sclerites being directed downwards and forwards at each side with the result that the leg bases are carried forwards towards the mouth and the wing bases backwards.

The wings (Fig. 324) have a complex venation of a reticular nature, characteristic features being a *stigma* or cuticular thickening of the wing membrane near the apex, a *nodus* or prominent cross-vein at right angles to the first two longitudinal veins, and a complex of veins near the wing base known as the *triangle* (Fig. 297). There is no coupling apparatus. All the mouth appendages are strongly toothed, maxillae and labium assisting the mandibles more efficiently in mastication than in most insects with biting mouth parts.

Though the male pore is on segment 9 of the abdomen, the copulatory apparatus is found in the sternal region of segments 2 and 3. Before copulation, spermatozoa are transferred to this apparatus. The male then grasps the female in the region of the prothorax by means of his posterior abdominal claspers. While in flight in this tandem position the female turns her abdomen down and forwards and receives sperm from the accessory copulatory apparatus of the male. Dragon-fly eggs are laid in water or on water-weeds. The nymphs breathe by means of tracheal gills and are of two kinds: (1) those with external gills in the positions of cerci anales and caudal filaments—*Zygoptera*, (2) those with gills on the walls of the rectum—*Anisoptera*. In the latter case water is pumped in and out through the anus, and this action may be made use of in locomotion—the sudden expulsion of water causing a rapid forward movement on the part of the nymph. The nymphs are, however, on the whole slow-moving creatures, lurking well camouflaged among water-weeds while in wait for their prey. The main difference between the mouth parts of the nymph and imago concerns the labium. In the adult this has normal proportions, but in the nymph the postmentum and prementum are elongated and capable of being shot out rapidly from the folded resting position, so impaling the prey, e.g. a tadpole, on the *labial hooks*.

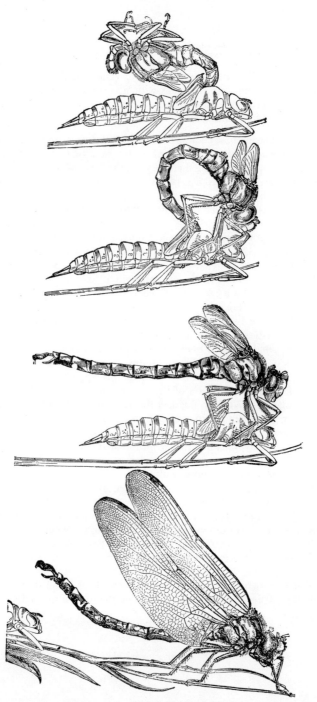

Fig. 324. The emergence of the dragon-fly, *Aeschna cyanea*. (After Latter.)

SECTION II. POLYNEOPTERA (Heterometabola, Exopterygota)
Super-order **Blattopteroidea**
7. *Order* **Dictyoptera (Cockroaches and Mantids)**

DIAGNOSIS. Insects with generalized biting mouth parts; tarsus five-jointed; anterior wings narrower and stouter than the posterior ones, which are more membranous and fan-folding; metamorphosis hemimetabolous. Jointed cerci. Styles in adult males only. Ovipositor small or lacking. Eggs laid in an ootheca produced by secretions of female accessory glands.

ACCOUNT. All these are insects of considerable size: included are the cockroaches and the praying mantids. The cockroaches (Blattidae) have dorso-ventrally flattened bodies, a pronotum which is large and shield-like, and strongly developed legs for rapid running—the coxae being broad so as to protect the lower surface of the body.

Though they are common in Britain, where they have accommodated themselves to life in dwellings where food is available and where an equable temperature prevails, they are in the first instance tropical or subtropical insects. Their mouth parts are indicative of their omnivorous habit, as also is their alimentary canal. Strongly cuticularized toothed mandibles are followed by prominent maxillae, each of which bears a five-jointed palp, a toothed setose lacinia and a sensory flexible galea. The labium has a four-lobed ligula consisting of a pair of small glossae flanked by larger paraglossae. The labial palps are three-jointed.

The alimentary canal is provided with a pair of salivary glands developed on the labial segment. They secrete amylase. A capacious thin-walled crop leads into a gizzard-like proventriculus, the inner lining of which is provided with prominent cuticular 'jaws' and spiny pads. These, worked by circular and longitudinal muscles, serve to break up the food into fine particles and to filter it in its passage to the mid gut.

The mid gut is the seat of the formation of a full complement of enzymes suitable to the mixed diet on which the animals feed. Examples are the cockroaches *Periplaneta americana* and *P. australasiae* and the less common German form, *Blattella germanica*.

The mantids also included here are predacious subtropical and tropical insects. Their chief character lies in the fore limbs which are raptorial (Figs. 296F, 319D). Here the femur bears a ventral longitudinal groove surmounted at its two edges by strong spines. Into this groove fits the blade-like tibia, its sharp toothed edge impaling the prey against the femur. The prothorax is long, and in conformity with the slenderness of the two posterior pairs of legs the insect moves only slowly and in an ungainly fashion. *Mantis religiosa* of south-western Europe is an example.

8. *Order* **Isoptera (Termites or White ants)**

DIAGNOSIS. Social and polymorphic insects with biting mouth parts; four-lobed ligula; wings very similar, elongate and membranous, capable of being broken off along a line at the base; cerci short; metamorphosis slight.

ACCOUNT. The animals of this order abound everywhere in the tropics. Like the true ants they have types of individuals (castes), specialized for the purpose of reproduction, labour and defence (Fig. 325). The termite community usually contains a de-alated *royal pair,* the king and queen, who are the founders of the colony, and also supplementary reproducing individuals of two kinds: (1) winged, which normally serve for the formation of new

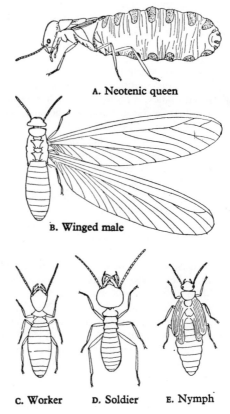

Fig. 325. Termites. *Hamitermes silvestri* Hill. Tropical Australia. (After Tillyard.)

colonies, and (2) wingless, which become capable of reproduction if occasion demands. There is usually a vast number of sterile wingless individuals belonging to two castes, the *workers* and *soldiers*. The termite nests may be merely series of burrows in trees, dry timber or in the ground, or they may be huge mounds made of earth cemented together with the saliva of the termites. Those living in the ground excavate the soil of the tropics, turning it over and enriching it just as earthworms do in temperate regions. Others may remove permanently from the soil much of its organic matter.

The winged sexual forms in several colonies usually swarm at the same time, so enabling intercrossing between members of different colonies to take place;

and of the countless numbers, a few individuals escape the attacks of birds and other animals and alight and cast their wings. A single pair forms a new colony first of all by making a small burrow, the *nuptial chamber*. The first-formed young are mostly workers and, having themselves been tended to maturity by their parents, take over the nursing of the young. The queen becomes enormous and helpless and is fed by the workers; she lays eggs at an incredible rate, up to a million eggs a year, it is said.

Their food consists chiefly of wood and other vegetable matter and many species are extremely harmful, e.g. *Neotermes*, which damages structural timbers, and *Calotermes militaris*, which bores into and does much harm to tea plants in Ceylon.

It is now known that digestion and growth of wood-eating termites can only go on when there is a protozoan fauna of trichonymphids (p. 70) and other flagellates in the hind gut. The fragments of wood are ingested by the protozoa and converted into sugars, being largely stored up in the form of glycogen. The termite needs the metabolic services of the protozoa to render the food available, and in return provides the anaerobic conditions which the protozoa are known to require.

Termites may forage by night for plant food, and members of the sub-family Microtermitinae cultivate in their nests *fungus gardens*. The fungus, which grows on a bed of chewed vegetable matter, serves as the food for the royal pair and the nymphs.

The workers and soldiers differ from the sexual individuals, not only in their sterility, but also in having more powerful mandibles. In the soldiers the head can produce a protective secretion and the mandibles are greatly specialized for defence (Fig. 325). Both these castes consist of males and females, though secondary sexual characters are not very marked. If, as is stated, slight caste differences are already apparent in the newly hatched young, caste formation cannot be a matter of nutrition.

9. *Order* **Zoraptera**

DIAGNOSIS. These are minute insects, winged or wingless, with nine-jointed moniliform antennae. Their mouth parts are of the biting type. Wings, of which there are two pairs, have a reduced venation, and are capable of being shed by basal fractures. They have a wide distribution in the warmer parts of the new and the old world. They live in colonies and some caste differentiation exists. An example is *Zorotypus* of West Africa.

Super-order **Orthopteroidea**

10. *Order* **Plecoptera (Stone-flies)**

DIAGNOSIS. This is an order of aquatic mandibulate insects with heterometabolous metamorphosis. Though in possession of two pairs of well-developed wings, they are weak fliers, and do not move far from their aquatic breeding grounds. Prominent, elongate antennae and cerci are characteristic features, as also are the three-jointed tarsi. According to some authorities the wing venation represents a primitive type (Figs. 297, 319). Much variation in venation is, however, found in the order.

ACCOUNT. The nymphs are always aquatic, for the most part inhabiting swift-flowing streams with stony beds. They possess the antennal and cercal features of the adult and breathe by means of gill tufts in various positions. In some cases gill vestiges are found on adults though these are not aquatic. Like most aquatic insects, they have a wide distribution, the most generalized families being found in southern, the most specialized in northern, regions. *Perla maxima* is a common species found in European streams.

11. Order **Notoptera**

This is a small order of aptercus insects comprising four species with generalized features lying intermediate in many respects between those of crickets and cockroaches, e.g. *Grylloblatta campodeiformis*.

12. Order **Cheleutoptera (Stick- and Leaf-insects)**

DIAGNOSIS. Mouth parts of the biting type. Legs usually long and slender, all alike. Short unjointed cerci. Prothorax reduced. Wings often lacking, when present the elytra-like fore wings (*tegmina*) are shorter than the hind wings. Ovipositor formed of six short valves.

ACCOUNT. These insects are all vegetable feeders—denizens of tropical and subtropical countries. They are notable for the many curious modifications, both of form and colour, which cause them to simulate sticks (stick-insects) and leaves (leaf-insects) (Fig. 319). Many of them are parthenogenetic and their eggs are often endowed with an extremely thick shell provided with an operculum which is detached at hatching. Examples are *Carausius morosus*, the stick-insect, and *Phyllium crurifolium*, a leaf-insect.

13. Order **Orthoptera (Locusts and Grasshoppers)**

DIAGNOSIS. Large insects, with mouth parts of the biting type; posterior legs with enlarged femora for jumping; fore wings as tegmina which overlap each other, cerci unjointed; pronotum with enlarged lobes hiding the pleural wall; ovipositor well developed; specialized stridulatory organs (Figs. 296, 297, 326).

ACCOUNT. This order includes grasshoppers, locusts and crickets. Among the grasshoppers are the *Tettigoniidae*, with large sword-like ovipositors consisting of three pairs of valves borne on the 8th and 9th abdominal segments. By means of these the eggs, not enclosed in an ootheca, can be deposited in plant tissues, on which these insects commonly feed. The antennae also are long, often extending backwards beyond the apex of the abdomen. Stridulation is brought about by rubbing a denticulous ridge on the left tegmen against a corresponding region of its right counterpart. This latter possessing a smooth tense membrane acts as a resonator when the tegmina are in motion and the noise, produced mostly at night, can be very loud. Auditory organs of some complexity are situated in each fore tibia. *Phasgonura viridissima* is found in England.

Other grasshoppers and locusts differ from these in their shorter antennae, which are seldom as long as the body, and in their less conspicuous ovipositor, the valves of which are short and curved. Stridulation in these 'short horned' grasshoppers and locusts (*Acridiidae*) is effected by rubbing the inner edges of the hind femora which are 'pegged' against the hardened veins on the tegmina. The latter in consequence vibrate and make a low buzzing sound. A prominent auditory organ is found on each side of the first segment of the abdomen.

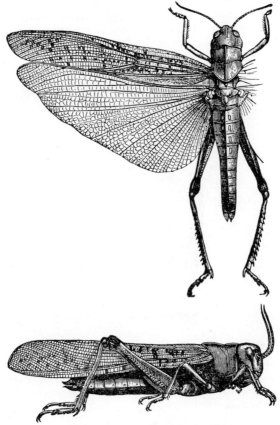

Fig. 326. Locusts, *Pachytylus migratorius*; natural size. (From Shipley and MacBride.)

Members of the Acridiidae are notable for their gregarious and migratory instincts. A species, such as *Locusta migratoria*, commonly leading a relatively harmless existence as a solitary grasshopper, may under certain conditions develop in countless numbers which, after travelling long distances, invade cultivated districts causing incalculable harm. Thus in the case of *Locusta migratoria*, when environmental conditions favour an increase in numbers, there is an inevitable trend towards the production of swarming migrants, i.e. the gregarious phase. The subsequent decline in numbers leads to the production of solitary non-migrants, i.e. the solitary phase. The two phases differ morphologically, biologically and in distribution so distinctly as to have been regarded as distinct species. Between them are transient individuals which form a series with no fixed characters, merging imperceptibly into the gregarious phase at one end and into the solitary phase at the other.

The crickets (*Gryllidae*) more closely resemble the long-horned grasshoppers, mentioned above, in their antennae, ovipositor and stridulatory apparatus, and appear to be directly related to them. *Gryllus domesticus*, the house-cricket, competes with the cockroaches for a place in domestic dwellings and leads there a similar life. *Gryllotalpa gryllotalpa*, the mole cricket, is subterranean in habit.

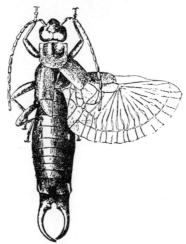

Fig. 327 Dermaptera: *Forficula auricularia* (male). (From Imms, after Chopard.)

14. *Order* Embioptera

DIAGNOSIS. Small tropical insects with elongated and flattened bodies; two pairs of similar wings with reduced venation; females apterous; cerci two-jointed, generally asymmetrical in male; metamorphosis absent in female, slight in male.

ACCOUNT. These insects are widely distributed in the warmer parts of the world. Many are gregarious, living in tunnels formed of silk produced by tarsal glands, e.g. *Embia major* from India.

Super-order Dermapteroidea

15. *Order* Dermaptera (Earwigs)

DIAGNOSIS. Insects with biting mouth parts; ligula two-lobed; fore wings modified to form short leathery tegmina; cerci unjointed, always modified into *forceps*; metamorphosis slight.

ACCOUNT. The common earwig, *Forficula auricularia* (Fig. 327) is a good example of this small but definite order, which comprises a number of small, usually nocturnal insects, omnivorous in diet. The female deposits the eggs in the soil, remains with them until they hatch, and even protects the nymphs afterwards. The hind wings have a characteristic venation and fold along transverse as well as longitudinal furrows, thus contrasting with the Orthoptera. When unfolded, the wing presents the appearance of a half wheel, the 'spokes' radiating backwards from the anterior border. The large posterior membranous portion corresponds to the anal wing area of Orthoptera, that part corresponding to the anterior area of the latter order being greatly strengthened by the coalescence of a number of longitudinal veins. The forceps are organs of defence and offence. In *Labidura* they are used for seizing the small animals on which this form lives.

SECTION III. OLIGONEOPTERA (Holometabola, Endopterygota)

Super-order **Coleopteroidea**

16. *Order* **Coleoptera (Beetles)**

DIAGNOSIS. Biting mouth parts; fore wings modified to form horny *elytra* which meet along the mid-dorsal line; hind wings membranous, folded beneath the elytra, often reduced or absent; prothorax large and mobile; mesothorax much reduced; metamorphosis complete, larvae (see p. 467) campodeiform, eruciform or, more rarely, apodous.

ACCOUNT. In the larvae the head is well developed (Fig. 316) and the mouth parts are of the biting type, resembling those of the adults. The most primitive larvae are those of the *campodeiform* type (found, for instance, among the *Cicindelidae* (tiger-beetles), *Carabidae* (ground-beetles) and the *Staphylinidae* (rove-beetles)). They are very active in movement and often predacious, with well-developed antennae and mouth parts, and cuticularized exoskeleton. In the *eruciform* type (Fig. 316B), found among plant-eating forms like the lamellicorn-beetles, the legs are shorter, and the animal much less active in its search for food, the body bulkier and cylindrical. Finally, there is the *apodous* type which is found in the *Curculionidae* (the weevils), in which not only are the thoracic legs lost but the antennae and mouth parts are reduced (Fig. 316C). The apodous larvae usually live inside the soft tissues of plants or beneath the soil attached to roots.

The relation which these larval forms bear to one another is perhaps indicated by the larval stages passed through in the life history of the oil-beetle, *Meloë*, the larvae of which are parasitic on solitary bees of the genus *Anthophora*. The first instar is known as the *triungulin*. This is an active campodeiform larva which attaches itself to its host after searching actively for it. The second instar, which is enclosed with an abundance of pollen and nectar in the cell of the bee, is intermediate in form between the campodeiform and eruciform types, legs being present, but very small. The third stage is a legless grub. From this series it may be inferred that the form of larva in Coleoptera is related to the ease or difficulty with which food is obtained. The pupa is exarate and usually thin-skinned. Obtect pupae, e.g. in certain Staphylinidae, do, however, occasionally occur.

In such a large order of insects it is to be expected that all manner of habits and food will be found. Beetles occur in large numbers in water, soil, and plant tissues. Circumscribed environments like dung, rotting vegetation, wood and fungi are never without prominent coleopteran associations. A large number, such as many coccinellids (lady-birds), carabids, e.g. *Carabus violaceus*, and staphylinids, e.g. *Ocypus olens*, are carnivorous and to this extent useful insects. On the other hand, among the phytophagous forms are to be found some of the most serious agricultural pests, the boll weevil, *Anthonomus grandis*, causing so much damage to the cotton crop in America that is has been seriously proposed to cease growing cotton for a period of

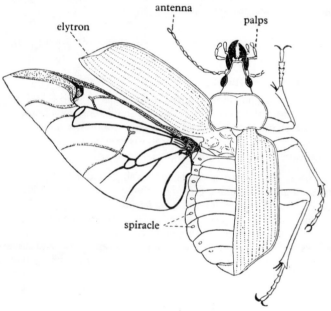

Fig. 328. External anatomy of *Calosoma semilaeve*, with left elytron and wing extended. (After Essig.)

time in order to eradicate this pest. Among the weevils one might mention also the several species of *Calandra* which do great injury to stored grains. A large number of beetles cause considerable damage to timber, probably the most notable being *Xestobium rufovillosum*, the death-watch beetle, destructive to structural timber. From all this we may expect a great variation in detail of life history, and for this reason no attempt is made here to deal with this aspect of their lives.

The order falls into two suborders, the *Adephaga* and the *Polyphaga*.

The *Adephaga*, for the most part carnivorous, are distinguished by filiform antennae, a five-jointed tarsus, and a larva of the campodeiform type, with a tarsus bearing two claws. To this group belong those families including the large water-beetle *Dytiscus*, the ground-beetles *Carabus* and *Calosoma* (Fig. 328), the tiger-beetle *Cicindela*, and the aquatic whirligig-beetles *Gyrinus*.

The second suborder, the *Polyphaga*, includes a large number of families grouped into several superfamilies, the members of which show much variation both in form and habit. There is a tendency towards reduction in the number of tarsal joints from five to three, and though some forms possess filiform antennae, clavate (clubbed), geniculate (elbowed), and lamellate (segments extended to form a 'book of closely arranged leaves or lamellae') antennae occur, as in the *Coccinellidae*, *Curculionidae* and *Scarabaeidae* respectively (Fig. 329). Larvae vary from the campodeiform to the legless grub, but where a tarsus is present it invariably carries only one claw.

The *Staphylinidae* range from carnivorous to phytophagous forms, and, as adults, are characterized by the short elytra which leave the abdomen exposed. The larvae are campodeiform, closely resembling those of ground-beetles, e.g. *Ocypus* (Fig. 329A). *Meloidae* or oil-beetles also have short elytra, but

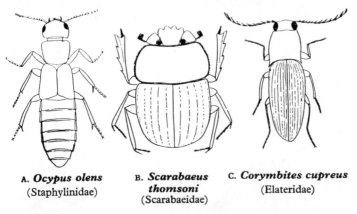

A. ***Ocypus olens***
(Staphylinidae)

B. ***Scarabaeus thomsoni***
(Scarabaeidae)

C. ***Corymbites cupreus***
(Elateridae)

Fig. 329. Three types of Coleoptera.

these being wider at the base than is the prothorax are readily distinguished from members of the staphilinid group. The interesting changes undergone by their larvae during metamorphosis have already been mentioned.

The *Chrysomelidae* or leaf beetles are exclusively phytophagous. Their bodies are rounded and smooth, and are often highly coloured with a metallic lustre. Antennae of these beetles are filiform and relatively short (e.g. *Phyllotreta*, the flea-beetle).

Weevils belonging to the family *Curculionidae* are easily distinguished by their greatly extended head, forming a rostrum at the end of which mouth parts are borne. *Anthonomus grandis*, the cotton-boll weevil of America, and *Ceuthorrhynchus*, the turnip-gall weevil of England, are typical examples. The larvae are apodous.

The chafer-beetles (*Scarabaeidae*) (Fig. 329B) have lamellate antennae. Their legs are often fossorial and bear four-jointed tarsi. Characteristic of these is the fat-bodied eruciform larva, almost incapable of movement, and which feeds on roots, e.g. *Melolontha* (Fig. 316B). *Aphodius* is a dung-beetle whose larva develops in the faecal matter of farm animals.

492   INSECTA

The family *Coccinellidae* (lady-birds) is of extreme importance, its members being carnivorous in young and adult stages, aphids and scale-insects figuring very largely in their diet. The beetle is smooth and rounded, with head concealed beneath the prothorax. The four-jointed tarsus appears to possess only

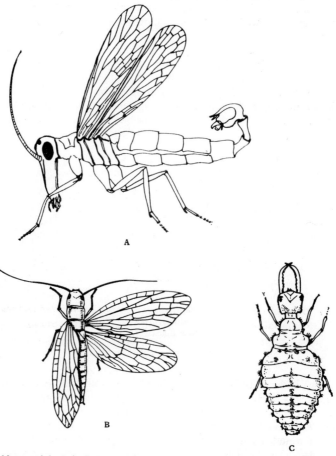

Fig. 330. A, adult male *Panorpa* (Mecoptera); B, *Sialis* (Megaloptera), and C, larva of *Myrmeleon*, ant-lion (Planipennia). (A, B, after Weber; C, after Grasse.)

three joints, owing to the small concealed third joint, e.g. *Coccinella* of Europe. *Novius cardinalis* is a classical example of a predatory insect being used in the biological control of the scale-insect, *Icerya purchasi*, of citrus trees.

Super-order **Neuropteroidea**

17. Order Megaloptera

The insects of this and the following two orders were formerly united in one, Neuroptera, in which they held subordinal rank. Similarity of mouth parts, venation and structure of wings certainly indicate close relationship.

DIAGNOSIS. Two pairs of large, broad wings, the posterior ones having a large posterior or anal field. The longitudinal veins branch freely and cross-veins are common, particularly behind the anterior border. A *pterostigma* or pigmented area placed laterally on the anterior border is either absent or ill-defined. At rest the wings are held over the back in a roof-like manner (Figs. 297, 330B).

ACCOUNT. Their carnivorous larvae (Fig. 331) are always aquatic and are remarkable in that they bear on the abdomen a series of pairs of gills which are jointed and which are moved by intrinsic muscles. *Sialis lutaria*, the alder-fly, is British. *Corydalus* of North and South America and *Archichauliodes* of Australia and New Zealand are other examples.

18. Order **Raphidioptera**

DIAGNOSIS. These are insects of small size. Two pairs of similar hyaline wings with freely branching longitudinal veins and a well-marked pterostigma are characteristic. The head projects forwards and is flattened above and narrowed towards the prothoracic junction. This feature, together with the elongated sub-cylindrical prothorax and the long-drawn-out 10th segment of the abdomen, accounts for their being named 'snake'-flies. The larva is terrestrial. *Raphidia* is British.

19. Order **Planipennia (Lace-wings and Ant-lions)**

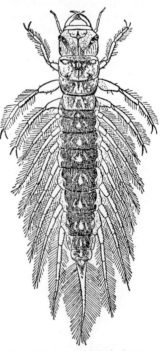

Fig. 331. Larva of *Sialis lutaria*. (From Imms, after Lestage.)

DIAGNOSIS. There is a wide range of form and size in this order, both very large and very small insects being included. There is also considerable variation in the wings since in some forms, e.g. *Ithone*, the two pairs of wings are alike while in others, e.g. *Nemoptera*, the posterior ones are drawn out in narrow strap-like structures several times larger than the body. However, all except the small *Coniopterygidae* agree in having a richly branching venation with many cross-veins. Metamorphosis complete.

ACCOUNT. The larvae of most of them are terrestrial, e.g. *Myrmeleon*, the ant-lion (Fig. 330c). Some, however, are aquatic, *Sisyra* being parasitic in the fresh-water sponge, *Spongilla*. Common, though not universal, features of the larvae of Planipennia are the forward-extended, curved and pointed mandibles and maxillae which are so arranged as to form between them at each side a food tube up which the blood fluid of their prey is sucked. Larvae are mostly predacious. *Hemerobius*, *Sisyra*, *Chrysopa*, among others, are British.

Super-order Mecopteroidea

20. Order Mecoptera (Scorpion-flies)

DIAGNOSIS. A small order of insects distinguished by their vertically directed and elongated head capsule carrying the biting mouth parts at its end; two pairs of similar wings with a simple venation in which a number of cross-veins divide the whole area into a number of nearly equal rhomboidal cells (Fig. 330A). Metamorphosis complete.

ACCOUNT. The male genitalia are prominent and the terminal segments of the abdomen carry them in a dorsally curved position in the manner of the scorpion's tail. The eruciform larvae are caterpillar-like and may possess prolegs on all segments of the abdomen. This feature, together with the presence of a large number of ocelli on the head (there may be twenty or more on each side), readily distinguishes these larvae from those of the Lepidoptera.

Panorpa communis, the common English scorpion-fly, lays eggs in crevices in the soil and the larvae hatching from these feed on decaying organic matter. Pupation occurs in an earthen cell and the life cycle is an annual one. Much information is still wanting on the life histories of the members of this order.

21. Order Trichoptera (Caddis-flies)

DIAGNOSIS. Medium-sized insects with bodies and wings well clothed with hairs; mandibles vestigial or absent; maxillary and labial palps well developed; two pairs of nearly similar membranous wings with few cross-veins and held in a roof-like manner when at rest (Fig. 332). Metamorphosis complete.

ACCOUNT. These obscurely coloured insects have considerable powers of flight and at sexual maturity may produce mating swarms. Oviposition may be directly into the water or the eggs may be deposited on plants above water where they will await immersion with the winter floods.

Larvae live in both stagnant and running waters. While most caddis larvae build cases of various materials (sand, small shells and particles of vegetation) in which they live, e.g. *Limnophilus*, there are others which either wander about freely without a case, e.g. *Hydropsyche*, or which lurk under stones in running water behind a silken web which they manufacture to catch small animals carried there in the water, e.g. *Plectrocnemius*. The case-builders are usually phytophagous, sluggish and eruciform. The net-spinners are commonly carnivorous, swift in movement and campodeiform. Eruciform larvae have segmental tufts of tracheal gill filaments on the abdomen. Campodeiform larvae do not possess these. In whatever degree these two larval types differ from each other they all possess the diagnostic feature of a pair of post-abdominal appendages bearing grappling hooks. Pupation takes place in a silken cocoon inside the case where this has been used by the larva. Water circulates through the cocoon making its entrance and exit through silken sieves at the two ends. The pupa possesses large mandibles by which it bites its way out of the cocoon and then by strong swimming movements of the mesothoracic legs it passes to the shore, there to emerge in the adult form after the final moult. Common examples of case-builders are *Phryganea*, *Odontocerum* and *Hydroptila*, and of net-spinners: *Plectrocnemius* and *Polycentropus*. *Rhyacophila* and *Hydropsyche*, common denizens of streams, are wandering carnivores.

22. Order **Lepidoptera (Butterflies and Moths)**

DIAGNOSIS. Mouth parts of the imago usually represented only by a sucking proboscis formed by the maxillae; two pairs of membranous wings, clothed with flattened scales, as also is the body; metamorphosis complete; larvae eruciform with masticating mouth parts, with three pairs of legs on the thorax and often five pairs of prolegs on the abdomen; pupa obtect, either enclosed in a cocoon or an earthen case, or free (Fig. 333).

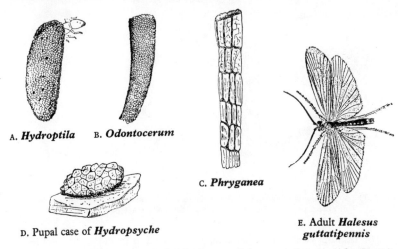

A. *Hydroptila* B. *Odontocerum*

C. *Phryganea*

D. Pupal case of *Hydropsyche*

E. Adult *Halesus guttatipennis*

Fig. 332. Trichoptera. A, B, C, D, Caddis cases. E, *Halesus guttatipennis*. (After Imms.)

ACCOUNT. The adults live on the nectar of flowers, and to absorb this a highly specialized proboscis has been formed from the greatly elongated galeae of the maxillae, each being grooved along its inner face and locked to its neighbour (Fig. 334). The laciniae are atrophied and the maxillary palp is usually much reduced. The mandibles are nearly always functionless, being immovably fused to the head capsule, and the labium is represented by a triangular plate and a pair of large, three-jointed palps.

Each half of the proboscis is a tube in itself into which passes blood from the head, and also a trachea and a nerve. Across the cavity of this tube there pass a number of oblique muscles. At rest the proboscis is tightly coiled like a clock spring under the head. When feeding the proboscis is extended and its tip placed in the food source. It is now recognized that the elastic properties of the cuticular wall of the proboscis account for the coiled condition when resting. Extension of the proboscis is brought about by the internal oblique muscle of each galea. These, working in conjunction with a stipital valve controlling the closure of the passage between cephalic and galea haemocoeles, cause the proboscis to develop a dorsal keel along its whole length. The attainment and retention of this new shape depends on the turgidity of the galea tube and the elasticity and flexibility of parts of the cuticular wall. For mechanical reasons it cannot in the keeled position be retained in the coiled state and extension of the proboscis results.

In feeding, a complex pharyngeal muscular apparatus causes the fluid food to be sucked into the mouth (Fig. 334B). The length of the proboscis in many cases corresponds to the depth of the corolla of the flower which the species frequents, and in the *Sphingidae* (hawk moths) may be greater than that of the body. Sometimes the organ is reduced or absent and the animal does not then feed in the adult stage at all.

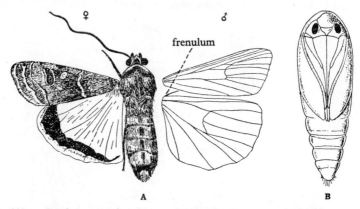

Fig. 333. A, *Tryphaena pronuba*, with venation and frenulum; male condition on the right. B, obtect pupa of *Platyhedra gossypiella*. (After Metcalf and Flint.)

The beginnings of the proboscis can be traced in primitive forms. In the *Micropterygidae* there are biting mandibles and maxillae of the type usually found in insects which masticate their food: in *Micropteryx* there is no proboscis, the animal feeding on pollen. In *Eriocrania* the mandibles are nondentate, the laciniae are lost and the galeae form a short proboscis.

The characteristic feature of the wings is the clothing of scales (Fig. 333). These latter are produced by enlarged hypodermal cells, and their main function appears to be the presentation of colour due either to striation of the surface causing interference colours, or in lesser degree to the pigment they contain (like the uric acid of the Pieridae). There also occur 'scent scales' which may have a sexual significance. Several methods of wing coupling have been developed independently in this order. In addition to the type already referred to on p. 435 and consisting of *frenulum* and *retinaculum*, there is the further method met with in the ghost-moths in which a jugal lobe from the fore wing engages the anterior border of the hind wing. In other forms there is neither frenulum nor jugum and the wings are coupled by a considerable overlap of the two wings of a side, e.g. the butterflies (*Papilionoidea*).

In the females of certain Lepidoptera, e.g. the winter-moth, *Cheimatobia brumata*, the wings are totally lost and the animals are confined to the food plant, fruit trees, on which they spend their larval life. The winged male is attracted to the female, under these circumstances, by scent.

Lepidopterous larvae (Fig. 335) have three thoracic and ten abdominal segments with nine pairs of spiracles situated on the prothorax and first eight abdominal segments. The mandibles are typically strong and dentate; the maxillae are stumpy and consist of a cardo, stipes and single maxillary lobe with a two- or three-jointed palp; the labium has a large mentum, a prementum bearing a median spinneret and small two-jointed palps.

The thorax bears three pairs of legs, and the abdomen five pairs of prolegs on segments 3–6 and 10. Such prolegs are different from the typical insect limbs, being conical and retractile and with hooks on the apex (Fig. 335). In many families there are less than five pairs of prolegs, but in *Micropteryx* there are eight pairs.

Lepidopterous larvae feed almost exclusively on flowering plants (exceptions being the lycaenid caterpillars which are carnivorous, feeding on aphids or entering ants' nests and devouring the larvae, and the micropterygid caterpillars whose diet consist almost exclusively of lichens and mosses). Their digestive enzymes are modified for dealing with plant tissues.

The pupa, which is disclosed after the last larval moult, is usually protected by a cocoon previously prepared by the larva. In the case of *Tortrix* moths the cocoon is largely composed of leaves drawn together by silk strands. In others, e.g. the silkworm moth, *Bombyx mori*, it is composed of silk and from it the silk of commerce is prepared. Agglutinated wood particles form a hard cocoon in the puss-moth, *Dicranura*. In *Pieris*, the pupa is naked and attached to the substratum by the hooked caudal extremity, the *cremaster*, and by a delicate girdle of silk about its middle. In the most primitive forms (e.g. Micropterygidae) the pupae are free, their segments are free to move and the appendages are not fused to the body. Obtect pupae, in which only few segments are movable and the appendages are fused to the sides of the body, are most common, e.g. *Platyhedra* (Fig. 333). Free or incompletely free pupae often emerge from the cocoon before the emergence of the adult. Emergence from the cocoon is often assisted by an armature of hooks and spines, though the most primitive moths, the *Micropterygidae*, have pupae with enlarged mandibles. By means of these they bite their way from the cocoon, recalling the Trichoptera in this respect.

Lepidoptera are almost invariably harmful in the larval stage, few plants being free from their attacks, and some of the world's most serious insect pests, such as the cotton-boll worm, *Platyhedra gossypiella*, and the gypsy-moth, *Porthetria dispar*, are included in this order. Some caterpillars have, however, been used as agents in the destruction of cacti in Australia.

CLASSIFICATION. The order is divided into two suborders. In the first of these, Homoneura, the fore and hind wings have venations which are almost identical. To this primitive feature may be added that of the included family Micropterygidae whose mouth parts are mandibulate and the structure of whose maxillae and labium are easily comparable with those of the cockroach. The ghost-moths or swifts (*Hepialidae*) are also included in this suborder. These nocturnal insects have vestigial mouth parts and short

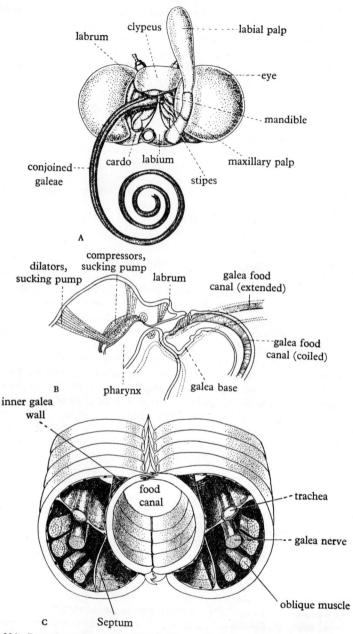

Fig. 334. Butterfly proboscis: A, head of butterfly seen from in front and ventrally, with proboscis turned to the animal's right side. B, diagram of sagittal section through the head of butterfly and proboscis—full line, when the proboscis is coiled (mouth closed), interrupted line, when the proboscis is extended (mouth open to food canal). C, tranverse section through a butterfly's probiscis at about the middle of its length. (All original.)

antennae. Their jugate type of wing coupling has already been described. In certain species, e.g. *Hepialus humuli*, the female searches for the male prior to mating. The larvae live in the ground and are white and hairless.

The second suborder, *Heteroneura*, is more specialized in that the venation of the hind wing has undergone reduction and so presents a venational pattern

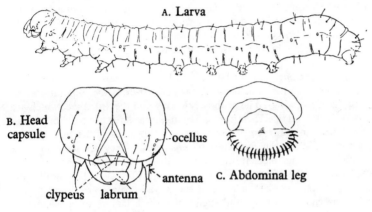

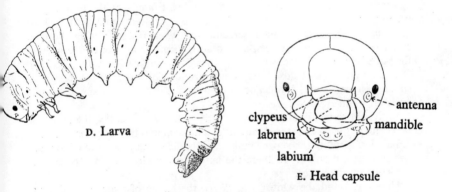

Fig. 335. Caterpillar of Lepidoptera and Hymenoptera.

very different from that of the Homoneura (Fig. 333). Here are included the vast majority of moths and butterflies. Since the families are distinguished largely on venational characters no attempt will be made to deal with them in a classificatory scheme.

Among the numerous families of this suborder may be mentioned the tineid moths—small species still retaining maxillary palpi and possessing narrow fringed wings, with a frenular bristle on the hind wing for coupling purposes. *Tineola biselliella* is one of the clothes-moths whose larvae can live on the keratin of woollen goods.

The goat-moths (*Cossidae*) are large moths without maxillary palps and with a frenular coupling apparatus. These are nocturnal, and lay their eggs on trees. Their larvae tunnel in timber, e.g. *Cossus*.

Ephestia, the flour-moth, and *Plodia*, the meal-moth, are most important as pests of stored products, while *Chilo* is a form whose larva bores into the shoots of the sugar-cane in India. *Galleria*, the wax-moth, able to digest wax after this has been broken down by bacteria, inhabits beehives in most parts of the world, having become artificially distributed. These belong to the family *Pyralidae*.

Hawk-moths (*Sphingidae*) are large, stoutly built moths whose fore wings are much larger than the hind ones. A further feature is the obliquity of the outer margin of the wings. The proboscis is long and the antennae, which are thick, end in a hooked tip. Their phytophagous larvae have five pairs of prolegs and usually bear an upturned spine or process on the back of the last segment.

Of slender build are the geometers (*Geometridae*). They are weak in flight and a wing-coupling mechanism is not always present. Some species, e.g. *Cheimatobia*, the winter-moth, are wingless as females. The family gets its name from the fact that in most of the larvae prolegs are borne by the 6th and 10th segments of the abdomen only. Such larvae, in consequence, walk by looping the body, bringing the hind segments near to the thoracic and so appear to be measuring distances along the surface walked upon.

The owl-moths or *Noctuidae* are the dominant family of the order. They usually fly at night and to this fact is related their sombre colouring which assimilates the insects to their surroundings when resting during the day. The larvae are almost hairless, and in such forms as pupate in the ground the pupa is naked. *Tryphaena pronuba*, the yellow-underwing moth (Fig. 333), is a common species whose larvae devour roots. The larvae of nearly related species, known as cut-worms and army-worms, rank among the worst insect pests of North America.

In the above-mentioned forms, collectively known as moths, the antennae taper to a point and the frenular coupling apparatus is common. The remainder, forming the super-family Papilionoidea, may be grouped for convenience as butterflies. In these the antennae are clubbed and there is no frenulum on the wings.

Here are found the whites, e.g. *Pieris*, the larvae of many of which are restricted to a cruciferous diet, and the 'blues' and 'coppers', in which the metallic colouring on the wings and the shape of the tapering larvae are distinguishing features. There are also the swallow-tails, e.g. *Papilio*, in which the hind wings are commonly extended into tail-like prolongations. Finally may be mentioned the 'skippers', so-called because of their erratic, darting flight quite distinct from the sustained flights of other forms.

23. *Order* **Diptera (Flies)**

DIAGNOSIS. Insects with a single pair of functional wings, the hind pair represented by highly sensory stumps (halteres) (Fig. 336); mouth parts suctorial and sometimes piercing or biting, elongated to form a proboscis;

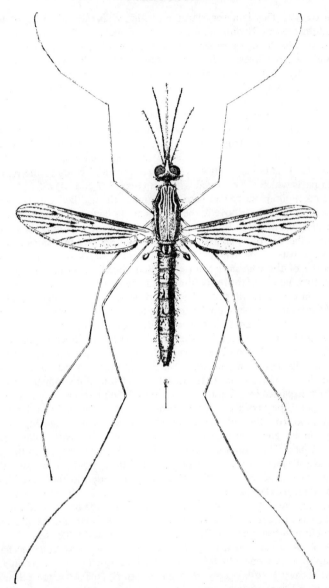

Fig. 336. *Anopheles maculipennis*, female. (After Nuttall and Shipley.)

prothorax and metathorax small and fused with the large mesothorax; metamorphosis complete, larvae often grub-like and always apodous, the head frequently being reduced and retracted; pupa either free or enclosed in the hardened larval skin (puparium).

ACCOUNT. This is a very large and highly specialized order of insects. The adults are mostly diurnal, feeding on the nectar of flowers, but a number are predacious, living on other insects (e.g. the robber-flies), while some, e.g. tachinids, are parasites. A further development which takes place in several families is the acquisition of blood-sucking habits. The representatives of this ecological class are of great importance because they harbour and transmit pathogenic organisms, causing such diseases as malaria, sleeping sickness, elephantiasis, yellow fever and some cattle fevers.

The several kinds of mouth parts which have been developed in the Diptera have departed widely from the primitive biting type. There is always a proboscis formed principally by the elongated labium, ending in a pair of lobes, the *labella*. This labium may serve as a support and guide to the remaining mouth parts which are enclosed within it (Figs. 337 and 341A).

The most complete system is to be found in the gadflies, e.g. *Tabanus* and *Chrysops*. Within the groove of the labium are to be found a pair of mandibles and a pair of maxillae. By lateral movements of the sword-like mandibles a wound in the skin of a mammal is made, to be deepened by the backward and forward thrusts of the maxillae. Into the wound so formed is inserted a tube composed of the *epipharynx*, an elongated chitinization of the roof of the mouth to which the labrum is fused, and the *hypopharynx*, a corresponding elongation of the mouth floor. The blood passes into this tube, being drawn up by the pharyngeal pump within the head. The hypopharynx carries a duct down which the salivary fluid is passed. Besides this, the proboscis of a gadfly can be used for taking up fluids exposed at surfaces. Such exposed fluid is drawn on the labellar surfaces into small channels, the *pseudo-tracheae* (later to be described), which converge to a central point on the underside of the labellar lobes. There it meets the distal end of the epi-hypopharyngeal tube, up which it passes by the pumping action of the pharynx.

The mouth parts of the female mosquito (Fig. 337A) differ in degree more than in kind from those described above. The labium is elongated, is grooved deeply on its upper surface and bears distally a median point and two labellar lobes. The paired maxillary and mandibular stylets lie in the groove as do also the median labrum-epipharynx and the hypopharynx. The labrum-epipharynx is grooved ventrally and its lateral borders are so curved inwards that by their overlapping they form the food tube. The hypopharynx carries the salivary duct. Fig. 337B shows how the food tube of the proboscis joins the pharynx in the head. In male mosquitoes mandibular and maxillary stylets are absent and feeding from fluid at exposed surfaces is the rule.

The blow-fly *Calliphora* (Fig. 338) has also lost its piercing mechanism. Mandibles are absent and the maxillae are represented by palps only. The labium is very broad, deeply grooved anteriorly to carry the labrum-epipharynx and hypopharynx within it and is so constituted that the whole of the proboscis can be folded up under the head when not in use. The labellar lobes are large and complex, making possible a variety of feeding attitudes according as the food is fluid or semi-solid.

The surface of each labellar lobe in its inner median part forms the pseudo-tracheal membrane. This is a flexible membrane which is interrupted by some thirty fine canals running transversely across it. These converge, either directly

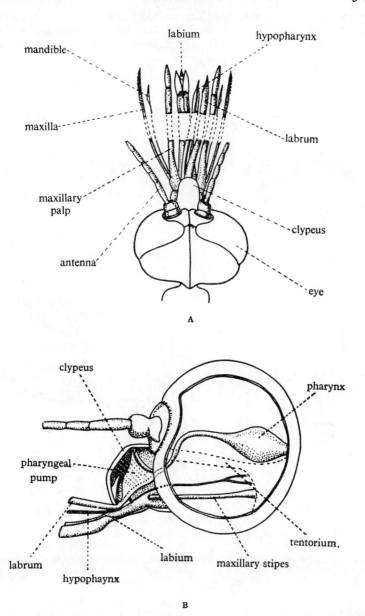

Fig. 337. Mosquito mouth parts. A, dorsal view of head and mouth parts of *Anopheles* female. B, sagittal sectional view of the head as seen from left side to show connexion between proboscis and alimentary canal. (After Robinson.)

or by union with a common canal, to the central region of the labellum towards which the epi-hypopharyngeal food tube is directed (Fig. 339). These fine canals are the pseudotracheae. Each forms a fine incomplete tube

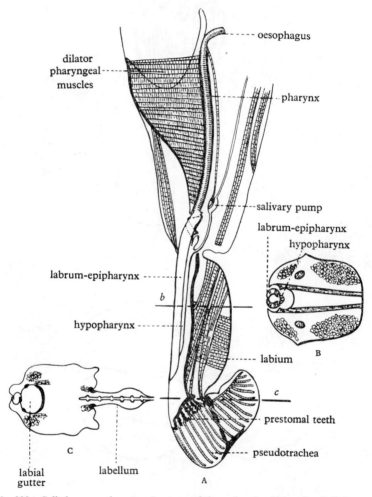

Fig. 338. *Calliphora* mouth parts. Anatomy of the proboscis of blow-fly, *Calliphora*, as seen in longitudinal section A, and in transverse sections B and C, taken at levels b and c respectively. (After Graham-Smith.)

imbedded in, and running parallel to, the labellar surface. The interior of each tube is in communication with the oral surface of the labellum through a very narrow zig-zag fissure, the lumen of the tube being kept open by means of incomplete cuticular rings running transversely round the tube—thus giving a superficial resemblance to a trachea. At the junction between the labellae and the body of the labium, to which place the pseudotracheal tubes converge, there is a complement of *prestomal* teeth (Figs. 338, 340) lying between the inner ends of the pseudotracheae.

The food tube is formed mostly by the conjoined labrum-epipharynx and the hypopharynx, and at the lower end of the labrum where the latter structures fail to reach the full distance the food tube is formed by the overlapping sides of the labial groove. In this way food collected at the median part of the pseudotracheal membrane to which the pseudotracheae converge is brought into contact with the food tube and so with the pharynx (Fig. 338).

Calliphora feeds largely on fluids, but in the presence of soluble solid food solution is effected by regurgitating alimentary fluid on to it. In other cases the prestomal teeth can abrade temporarily dried surfaces and so bring the regurgitated fluids more effectively into contact with the food material. Fig. 340 shows diagrammatically some of the different attitudes of the labellae when the insect is feeding on materials of different degrees of fluidity. When the labellae are completely retracted food enters the wide-open food tube irrespective of the pseudotracheae and the passage is then large enough for semisolid food or even the eggs of helminths to enter.

We may infer from all this that in the evolution of the Diptera there has been an evolution towards surface fluid feeding involving the loss of piercing stylets. The matter does not end there, however, since there are forms, closely related to the blow-fly type, in which mandibular and maxillary stylets are lacking, but which nevertheless can pierce the skins of mammals, using the labium for the pupose, e.g. the tsetse-fly *Glossina* and the stable-fly *Stomoxys*. In these cases (Fig. 341 B, C) the labium has become so rigid that it cannot be folded under the head. It thus, with its contained labrum-epipharynx and hypopharynx, projects stiffly forwards. The labellar apparatus is much reduced, consisting of three small, strong lobes bearing rasping teeth, some of which represent the prestomal teeth of *Calliphora*. Figure 341 B shows in transverse section how in *Glossina*

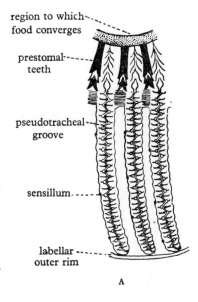

Fig. 339. Diagrams illustrating the structure of the pseudotracheal membrane: A, surface view of three adjacent pseudotracheal canals. B, chitinous supports to a pseudotracheal canal. (After Graham-Smith.)

the labium and the labrum-epipharynx fit together to form the food tube. It appears, then, that a second type of piercing apparatus has been evolved in the Diptera after the first piercing (stylets) mechanism had been lost.

The larvae of Diptera are among the most specialized in the class. Legs have been entirely lost, and the head and spiracular system have undergone varying degrees of reduction. Thus the most generalized larvae are at the same time *eucephalous*, i.e. with complete head capsule, and *peripneustic*, i.e. with lateral spiracles on the abdomen, e.g. *Bibio* (Fig. 342D). In the most specialized forms, on the other hand, we find the acephalous larva whose head capsule is entirely wanting, e.g. *Musca*. Such acephalous larvae may either

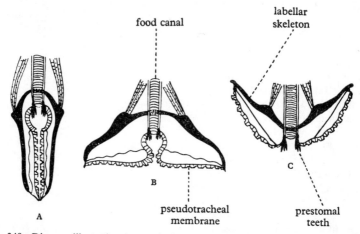

Fig. 340. Diagram illustrating three attitudes of the labellar lobes of *Calliphora*: A, when not in use; B, when taking surface food; C, when taking semi-solid food with prestomal teeth extended. (After Graham-Smith.)

be *amphipneustic*, with only prothoracic and posterior abdominal spiracles, or *metapneustic*, where only two spiracles are retained at the posterior end of the body. The first instar larva of *Musca* is metapneustic, subsequent instars being amphipneustic (Fig. 342A).

The eucephalous larva develops into an exarate pupa from which the adult emerges by a longitudinal slit on the thorax. The pupa resulting from the acephalous larva, on the other hand, is coarctate, the last larval skin being retained as a protective *puparium*, tracheal connexions maintaining contact between the pupa within and the larval skin outside it. Final emergence of the fly in this case clearly involves two processes, (1) the liberation of the fly from its pupal skin, and (2) the further liberation from the puparium. The latter splits transversely (Fig. 342B), the top being thrust away by an eversible head-sac, the *ptilinum*, which such flies possess. The features of metamorphosis just described are characteristic of many flies and, by defining one of the suborders, constitute an important basis of modern classifications (Fig. 342 A, B, C).

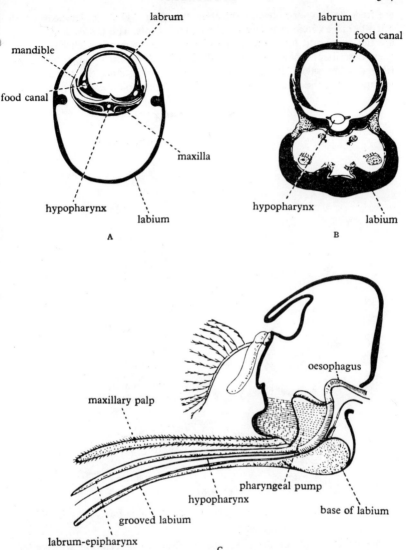

Fig. 341. A, transverse section through the proboscis of *Anopheles*. (After Robinson.) B, transverse section through the proboscis of *Glossina*. (After Weber.) C, mouth parts of *Glossina* in relation to the pharynx. (Original.)

CLASSIFICATION. *Suborder Orthorrhapha*. Included here are all those flies which are liberated by means of a longitudinal split in the mid-dorsal line of the pupal case. Such flies possess no ptilinum. Many of these, the Nematocera, have slender antennae and usually pendulous maxillary palpi. Their larvae are eucephalous with transversely biting mandibles and their pupae are free. To this series belong the crane-flies (Fig. 343 A), the larvae of which often damage cereal crops by devouring their roots. The *Culicidae* (Fig. 336) are the gnats and mosquitoes, the piercing proboscis of which has already been described. They are further distinguished by their wings which are fringed with scales. Both larvae and pupae are aquatic, the former being metapneustic, the latter *propneustic* (with anterior spiracles only). With the

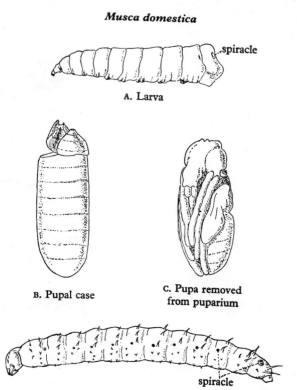

Fig. 342. Early stages of the Diptera. (A, B, and C, from Hewitt; D, original.)

blood-sucking habit of these flies has evolved an association with certain organisms which when, transmitted to man, cause disease. *Anopheles* is concerned with the transmission of malaria. *Stegomyia* transmits the causative organism of yellow fever, while *Culex fatigans*, a widely distributed tropical form, is a carrier of the thread-worm *Wuchereria bancrofti*, the cause of elephantiasis.

Nearly related to these are the *Chironomidae* (midges), the mouth parts of many of which are not adapted for piercing and sucking. A few of these, however, do suck blood, e.g. the midges of the genus *Forcipomyia*, whose larvae breed, some in water, others behind the bark of trees.

The *Cecidomyidae* (Fig. 343 C) are the gall-midges distinguished by their beaded antennae adorned with whorls of setae. The larvae of a few of these are parasitic, others are predacious, but the large majority are phytophagous, forming galls in plant tissues, e.g. of grasses. *Contarinia pyrivora* is the pear-midge, the larvae of which develop in the flowers of the pear so as to abort

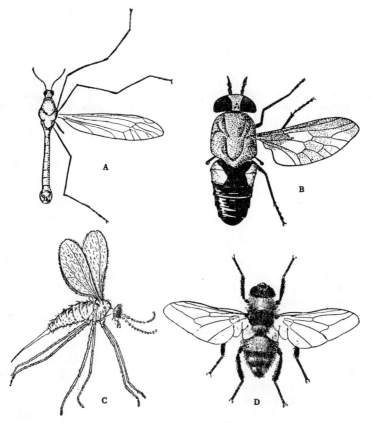

Fig. 343. Types of Diptera: A, *Tipula ochracea* (Tipulidae). B, *Chrysops caecutiens* (Tabanidae). C, *Contarinia nasturtii* (Cecidomyidae). D, *Hypoderma bovis* (Cyclorrhapha, Oestridae). (C, from Smith, after Taylor; D, from Smith, after Theobald.)

fruit production. *Miastor* lives behind tree bark in the larval state and, as mentioned above, is noteworthy for the phenomenon of paedogenetic parthenogenesis.

Another family of blood-sucking flies, known as the *Simuliidae*, consists of small flies with a hump-backed appearance and with broad wings. The spindle-shaped larvae live in running water and are characterized by the possession of prothoracic prolegs and an anal pad provided with setae by means of which they cling to rocks, etc., in the rapidly flowing water of their environment. Still included in the suborder Orthorrhapha are the flies with short antennae, the Brachycera. Though included in this scheme with the Orthorrhapha, their wing venational characters indicate a close relation with the Cyclorrhapha. In general, the basal joints of the antennae are larger than the terminal ones, these being reduced in number as compared with the nematocerous condition. The maxillary palpi are porrect (not pendulous). Their larvae are *hemi-cephalous* (head capsule incomplete posteriorly), with vertically biting mandibles; and the pupae are free and spiny.

From this vast assemblage of flies we may mention the *Tabanidae* or gad-flies (Fig. 343 B). These flies, to the mouth parts of which reference has already been made, are of stout build and possess large eyes occupying a great part of the head surface. Though a few transmit disease organisms (*Chrysops dimidiata*, as the vector of the nematode worm *Loa loa*, is responsible for calabar swelling in the natives of West Africa), the majority are harmful chiefly through the annoyance which their bites occasion. Tabanid eggs are usually laid on the leaves of plants overhanging water and their carnivorous larvae are either aquatic or ground-dwellers.

The robber-flies (*Asilidae*) are large bristly flies with a backwardly directed proboscis. They feed on all kinds of insects which they paralyse with their salivary fluid, and their legs, being strong and provided with powerful claws, are well adapted for grasping the prey. The *Empidae*, flies of more slender build, exhibit similar habits. Their larvae are terrestrial as are also those of the preceding family.

Suborder *Cyclorrhapha*. These flies emerge from a pupa which is enclosed in the last larval skin or puparium, and the commonly transverse or circular split in the latter, for release of the adult, gives the name to this suborder. It is therefore really a larval feature which establishes the position of these flies in the classification.

The antennae have three joints, the last of which is greatly enlarged, carrying a dorsal spine or *arista*. The maxillary palpi are one-jointed and porrect. A crescentic suture on the head lies above and encloses the bases of the antennae. This, known as the *frontal suture*, is a narrow slit along the margins of which the wall of the head is invaginated to form the ptilinal sac. The eversion of this enables the adult to emerge from the puparium. The extent to which the frontal suture is developed and the ptilinum persists, varies. The Syrphidae, for instance, have usually no persistent ptilinum and the frontal suture is not well developed. In the larva the true mouth parts have atrophied and the head capsule is lacking. There is a complex pharyngeal skeleton to which are attached mandibular sclerites which work in the vertical plane.

The *Syrphidae* (hover-flies) form an important family of brightly coloured flies, whose most obvious mark of distinction is the possession of a false longitudinal vein lying about the middle of the wing. Their larvae are amphipneustic leathery maggots, some of which (*Syrphus*) devour Aphidae, others live as saprophages in decaying material (*Eristalis*), others again are phytophagous (*Merodon*, the bulb-fly).

The remainder of the Cyclorrhapha may be considered under the heading of muscid flies. The frontal suture is prominent and the ptilinum persists.

Many families are included here, to some of which belong such serious agricultural pests as the frit-fly of oats, *Oscinus frit*, and the gout-fly of barley, *Chlorops taeniopus*. In such cases the larvae bore into the growing shoot, or into the stem. Larger and better known are the saprophagous house-fly, *Musca*, and the blow-fly, *Calliphora*. The larva of *Hypoderma lineatum* is parasitic in the bodies of cattle, causing 'warbles' on the backs of affected animals, while *Gastrophilus equi*, the bot-fly, is parasitic as a larva in the alimentary tract of horses.

The *Tachinidae* are important as parasites, chiefly of larval Lepidoptera, though larvae of Coleoptera, Orthoptera and Hemiptera frequently serve as hosts as do also more rarely myriapods and terrestrial isopods. The parasitic larvae become physically associated with their hosts in a variety of ways. This usually takes the form of an enclosing sheath produced by the host tissue enabling the parasite to breathe either from the outside air by a perforation in the body wall or from air in the tracheal system of the host. *Ptychomyia remota* is responsible for the very effective control of the Levuana moth, *Levuana iridescens*, of Fiji.

Blood-sucking muscids are important, e.g. *Glossina*, as the vector of trypanosomiasis causing sleeping sickness of man and cattle disease in Africa. The tsetse flies are pupiparous, larvae being nourished by special glands opening into the genital tract. The larvae are deposited as soon as fully grown and pupate immediately.

A number of members of this order present a greatly modified structure resulting from an ectoparasitic habit, some on mammals, others on birds. They are known as the Pupipara, being similar in their viviparity to *Glossina*. The following examples may be quoted: *Hippobosca* is a winged fly with body dorso-ventrally compressed, and is an ectoparasite of cattle. *Melophagus* is a wingless species, similarly associated with sheep, familiarly known as the sheep tick. *Nycteribia* is a wingless form parasitic on bats.

Super-order **Siphonapteroidea**

24. *Order* **Siphonaptera (Fleas)**

DIAGNOSIS. Wingless insects, ectoparasitic on warm-blooded animals; laterally compressed with short antennae reposing in grooves; piercing and sucking mouth parts, maxillary and labial palps present; coxae large; tarsus five-jointed; metamorphosis holometabolous; larva legless; pupa exarate, enclosed in a cocoon.

ACCOUNT. These insects are perfectly adapted to an ectoparasitic existence by their laterally compressed bodies, prominent tarsal claws, well-developed legs suitable for running between the hairs of their host, and for jumping, and by their mouth parts (Fig. 344). They exhibit only slight relationship to one other order, namely, Diptera, by features in their metamorphosis and to a less degree by their mouth parts.

The mouth parts (Fig. 345) consist of a pair of long serrated mandibles, a pair of short triangular maxillae with palps, and a reduced labium carrying palps. There is a short hypopharynx and a larger labrum-epipharynx reminiscent of the Diptera. The labial palps, held together, serve to support the other parts, a function which is performed by the labium in the Diptera. In piercing, the mandibles are most important and the blood is drawn up a channel formed by the two mandibles and the labrum-epipharynx. The thoracic segments are free and there are never any signs of wings. Though the eggs are laid on the host they soon fall off and are subsequently found in

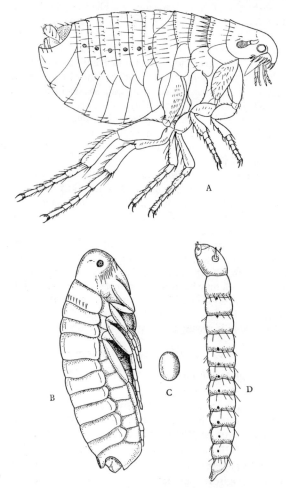

Fig. 344. The life history of the dog flea, *Ctenocephalides canis L.*: A, adult male; B, pupa; C, egg; D, larva.

little-disturbed parts of the haunts of the host. Thus in houses they come to lie in dusty carpets and unswept corners of rooms. In a few days the larvae hatch and feed on organic debris. The legless and eyeless larva possesses a well-developed head and a body of thirteen segments. At the end of the third larval instar a cocoon is spun and the creature turns to an exarate pupa from which the adult emerges, the whole life cycle occupying about a month in the case of *Pulex irritans*.

Pulex irritans is the common flea of European dwellings, but by far the most important economically is the oriental rat flea, *Xenopsylla cheopis*, which transmits *Bacillus pestis*, the bacillus of plague, from the rat to man. It appears that this bacillus lives in the gut of the flea and the faeces deposited on the skin of the host are rubbed into the wound by the scratching which

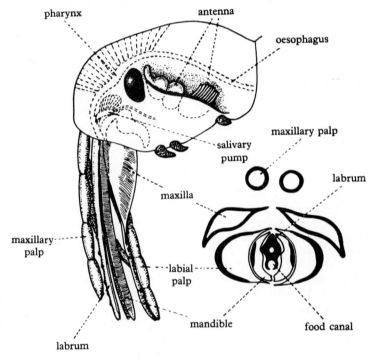

Fig. 345. Mouth parts of a flea, *Ctenocephalus*, as seen from the side and in transverse section. (After Weber.)

follows the irritation from the bite. *Ceratophyllus fasciatus*, the European rat flea, also transmits the plague organism as can also *Pulex irritans*, but since the latter does not live successfully on rats, it is never likely to prove a source of trouble.

Super-order **Hymenopteroidea**

25. Order **Hymenoptera**

DIAGNOSIS. Mouth parts adapted primarily for biting and often secondarily for sucking as well; two pairs of membranous wings coupled by hooklets on the anterior border of the hind wing engaging with a groove on the posterior border of the fore wing, hind wings smaller; first segment of the abdomen fused to the thorax, and a constriction behind this segment commonly found; an ovipositor always present, modified for piercing, sawing, or stinging; metamorphosis holometabolous; larvae generally legless, more rarely eruciform, with thoracic and abdominal legs; pupae exarate, protected generally by a cocoon.

ACCOUNT. This order is remarkable for the great specialization of structure exhibited by its members; for the varying degrees to which social life has developed, and for the highly evolved condition which parasitism has reached.

The mouth parts, though complex in some instances, seem seldom to have wholly lost the various parts recognizable in the generalized plan characteristic, for instance, of the Orthoptera. The climax of their evolutionary elaboration is seen in *Apis*, the honey-bee, and their least modified arrangements are found in sawflies.

In relation with this latter there have evolved certain features of the head capsule which are common to all but the more generalized Hymenoptera. Thus the head articulates with the thorax by a narrow neck and the *occipital foramen* which is therefore small is bounded below by a strong *hypostomal bridge* formed by the union of the post genae (Fig. 346B). Great mobility is thereby conferred on the head, the hypostomal bridge forming a strong base for the attachment of the maxillo-labial complex. In all Hymenoptera this complex is formed by the union in a common membrane of the maxillae and labium which are thus placed in a close working association with each other. The working of maxillae and labium as a functional unit is further ensured by their basal segments, cardo and stipes, submentum and prementum, being so arranged as to bend in a common plane. (Fig. 346E demonstrates this point in a sawfly.) Folding of the mouth parts under the head when at rest, as well as their forward extension when in use, is thus facilitated. In these several features, most of which are present in generalized forms, we have the foundations on which the structural evolution of the higher forms is based and without which presumably these latter might never have developed. Having regard always to these important facts we may still claim that among the sawflies are to be found the most generalized mouth parts. Wasps, too, are easily referable in these respects to the primitive omnivorous types with the additional feature of adaptation to licking of fluids by an extension of the bifid glossa and the setose maxillary galea. The mandibles here are well suited by their toothed form to feeding on solid food.

At the other end of the scale of specialization we have the elaborate elongated and extensible mouth parts of *Apis*, the honey-bee (Fig. 346A, B, C, D). The mandibles are large, smooth, spatulate structures articulated to the gena of the cranium. They are used for manipulation of wax and pollen within the hive and not for the gathering of food.

The labium has a short triangular postmentum, to the front border of which is articulated a long prementum. From this there projects forwards a long tongue, formed from fused glossae, and which is setose externally and grooved ventrally. At the base of the tongue are the short curved paraglossae, embracing it in such a way as to conduct fluid from the ventral glossal groove to the upper surface of the tongue base and so to the mouth which lies above.

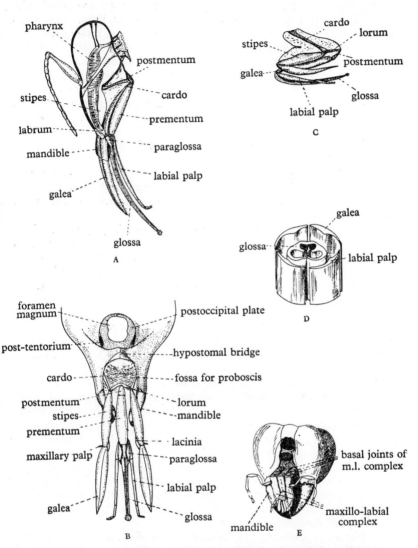

Fig. 346. Mouth parts of *Apis*: A, lateral view of head and attached mouth parts. B, posterior view of mouth parts in relation to the head showing the fossa into which the mouth parts are withdrawn when not in use, and the postgena separating this fossa from the foramen magnum. C, lateral view (left side) of the mouth parts folded up when not in use. D, transverse section of the proboscis, ventral surface facing the observer. E, posterior view of head of sawfly, *Tenthredo*, to show the maxillo-labial complex. (A and B, original; C, D, E, after Weber.)

Arising also from the distal end of the prementum are the labial palps consisting of several long segments whose inner surfaces, being concave, can partly encircle the bee's tongue ventrally for the whole of its length.

In line with the postmentum lies the maxillary cardo at each side. Basally each cardo is articulated to a cephalic apodeme projecting inwards to the head cavity. At its distal end it articulates both with the stipes and with a V-shaped sclerite, the *lorum*. This lies in the membrane which unites the labium with the maxillae and probably develops as a specialization of it. The lorum thus connects the two maxillae with each other, and into its apex fits the proximal angular border of the postmentum. The stipes of each maxilla lies at the side of the prementum and is of about the same length. Distally, on its outer side, lies the much-reduced maxillary palp, and on its inner side a similarly reduced lacinea. From between these two there projects the curved, blade-like, long galea. The two galeae have concave inner surfaces, like the labial palps, and with these latter complete the encirclement of the tongue dorsally (Fig. 346D).

Reference to Fig. 346A, B, C, D, shows, without lengthy explanations, how these several parts are spatially related to each other, both at rest and when extended. It is generally accepted that food can be drawn up the ventral groove of the tongue by capillary action. The food can also pass in larger quantities into the space surrounding the tongue enclosed by the galeae and the labial palps, passing within the folds of the paraglossae and being thereby directed to the mouth, which opens above this point.

Such a feeding mechanism is the climax in an evolutionary process which has involved in succession the fusion of the glossa lobes, as in the sawflies, the lengthening of the basal joints of the labium and maxilla as in *Colletes*, and finally the elongation of the glossa, e.g. *Apis* and *Bombus*.

The highly complex *social life* found in the bees, ants and wasps, in which caste development is a feature of prime importance, is foreshadowed in the interesting behaviour of solitary wasps and bees. The supply of food to the larva by *progressive feeding*, instead of *mass provisioning*, appears to enable the parent to become acquainted with its offspring, and this establishment of family life may be regarded as the forerunner of the complex social state of the higher forms. For instance, in English species of the wasp *Odynerus* the egg is laid in a cell and sufficient caterpillars stored to serve as food for the whole of the larval life (mass provisioning). Certain African species of this genus supply their growing larvae from day to day with fresh caterpillars (progressive feeding).

Another important feature in the development of social life has been that of *trophallaxis*. Among wasps, for instance, the worker taking food to a grub receives in turn a drop of saliva from the grub. This is eagerly looked for by the workers, and it is suggested that it is the mutual exchange of food between young and adult which engenders in the adult an interest in the welfare of the colony. That the exploitation of a particular form of abundant food has contributed to the development of the social system cannot be doubted. As examples we may quote pollen and honey for bees and dung as a basis for the simpler social life of some beetles. No feature determining cohesion of the bee colony seems to be of such paramount importance as the ability of the queen to satisfy the craving for a secretion produced by her (queen substance) which all members of the colony experience. The absence of a queen is quickly felt by the colony and its collective

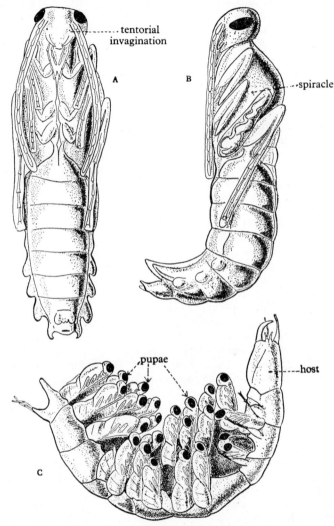

Fig. 347. A, B, exarate pupae of *Phaenoserphus viator*. C, pupae of same projecting from empty skin of host, the ground beetle, *Pterostichus*. (After Eastham.)

behaviour in consequence greatly disturbed. Ants and termites appear to be similarly dependent on the queen.

The admittedly complex environment in which a social insect lives has called into being a form of behaviour simulating intelligence. It is now known, for instance, that bees can with great effect acquaint each other of the presence of a food source. They can further inform each other by scent and dance of the position of the food source with considerable precision always provided the sky is not wholly overcast (see above, p. 462). The direction of the dance movement has reference to the position of the sun in the sky relative to the hive. Since this position can be determined by the bees in a sky in which the sun is not visible, though in which some blue sky is present, the conclusion has been drawn that in the ommatidia of the compound eye rests the power of the bee to analyse the degree of polarization of light emerging from a blue patch. The development of this important theme is outside the scope of this book. Perhaps the main conclusion we can here draw is that, concomitant with the social state's development, there has come about a complexity of behaviour which ultimately depends in turn on the enhanced sensitivity of the members of the colony.

The phenomenon of parasitism (Fig. 347) is highly developed in the Hymenoptera, ichneumons, chalcids and proctotrypids being almost entirely parasitic. Almost all orders of insects are affected by the activities of these very important insects, egg, larval, pupal and adult stages all being parasitized.

From the foregoing it will be seen that some of the most important insects are included in this order. The sawflies are important as agricultural pests. Flower-visiting bees are of great value in the pollination of flowers. Carnivorous wasps do good by devouring other insect pests such as aphids, while to a large extent the parasitic Hymenoptera are useful in checking the depredations of phytophagous insects.

Two main types of larvae are found in this order, the legged larva of the sawflies (Fig. 335D) and the legless form of bees, wasps and ants (Fig. 348). The sawfly larva has a superficial resemblance to the lepidopterous caterpillar, but is easily distinguished by its single pair of ocelli and the absence of crotchets or spines on the abdominal legs. The prolegs of the abdomen occur on different segments in the two forms under consideration as reference to Fig. 335 clearly shows.

CLASSIFICATION. The order falls naturally into two suborders, the *Symphyta* and the *Apocrita*.

Suborder **1,** the *Symphyta*, includes those species with the most generalized form, both as adults and as larvae. None of them show the highly specialized habits and instincts which characterize most of the other suborders, and with few exceptions they are phytophagous. The first abdominal segment is not perfectly fused to the metathorax nor is the fusion accompanied by the constricted waist so characteristic of the remaining Hymenoptera (Fig. 349D). The ovipositor is used in oviposition as a saw or drill for piercing plant tissues. The trochanter is two-jointed. Larvae are eruciform (Fig. 335D) and in addition to thoracic legs certain of the abdominal segments often carry prolegs devoid of distal spines or crotchets.

To this group belong the wood-wasps, the ovipositors of which are used as drills for perforating growing timber in which the eggs are laid. The six-legged, strong-headed larva bores through the wood (in the case of *Sirex gigas*, this stage lasts as long as two years), pupation occurring near the surface of the affected timber, from which the adult bites its way out. The sawflies (Fig. 349 D), with saw-like ovipositors, are most important as agricultural pests, and are distinguished from the wood-wasps by their softer bodies, their smaller size, and by the presence of two apical spurs on the anterior tibiae, e.g. *Nematus ribesii*, the gooseberry sawfly.

Suborder **2**, the *Apocrita*, includes all the remaining Hymenoptera. The second abdominal segment is invariably constricted to form a narrow waist or petiole, the first segment being firmly amalgamated with the thorax (Fig. 348). Larvae are apodous when full grown.

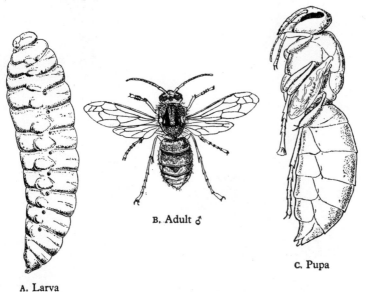

A. Larva B. Adult ♂ C. Pupa

Fig. 348. The hornet, *Vespa crabro*.

Ichneumon flies (Fig. 349 A) are distinguished by their slender curved antennae, and by the stigma on the wing. The ovipositor is often long and issues well forward from the tip of the abdomen. The larvae of Lepidoptera and of sawflies are their commonest hosts. *Rhyssa* parasitizes the larva of *Sirex*.

Cynipid flies have similarly slender antennae, but by the absence of the stigma on the wing and by their reduced venation are easily distinguished from the foregoing. Many of these form galls on plants, e.g. *Neuroterus* responsible for oak galls, and *Rhodites* for the pin-cushion galls of roses. Others, e.g. *Eucoila*, are parasitic on fly larvae.

Chalcid wasps (Fig. 349 B) also have the wing venation so reduced as to present no closed cells. The antennae are geniculate or elbowed. Though most of these small wasps are parasites, e.g. of lepidopterous and dipterous larvae, and of homopterous nymphs, a few feed on plant tissues. An example is *Harmolita* which produces galls on grasses.

In ichneumons, chalcids and cynipids the ovipositor issues beneath the abdomen well in front of its tip, and these insects differ in this feature from the *Proctotrypidae* in which the ovipositor is terminal. Dipterous larvae are

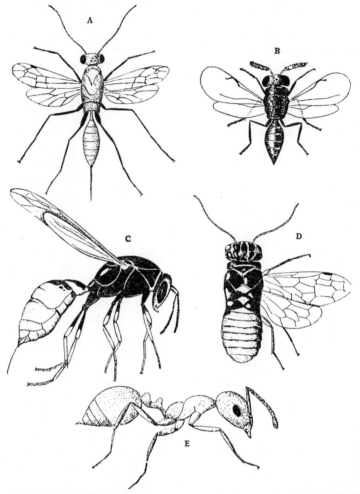

Fig. 349. Types of Hymenoptera: A, *Cryptus obscurus* (Ichneumonoidea). B, *Bruchophagus funebris* (Chalcidae). (After Howard.) C, *Polistes aurifer* (Vespidae). (After Essig.) D, *Pamphilus* sp. (Tenthredinidae). (Original.) E, *Monomorium minimum* (Formicidae). (After Essig.)

often parasitized by these insects, as are also the eggs of Orthoptera and Hemiptera. Many hyper-parasites, i.e. parasites of other parasites, occur in this family. *Phaenoserphus* is parasitic on carabid beetle larvae (Fig. 347) and *Inostemma* is an egg-parasite of dipterous gall midges.

Whereas parasitism is a characteristic, largely though not wholly, common to the foregoing families, the ants, wasps and bees next to be considered show a tendency, in varying degrees, towards the development of the social habit. The ants (*Formicoidea*) are social, polymorphic insects in which two segments are involved in the formation of the abdominal petiole. Further, this petiole is always characterized by the possession of one or two nodes (Fig. 349 E). The females are endowed with a well-developed sting, the modified ovipositor. Polymorphism reaches its highest degree of complexity in this group, as many as twenty-nine morphologically different castes having been recognized. Some of these are pathological phases due to infection by parasites, e.g. nematode worms, or other Hymenoptera. In such colonies as produce winged forms of both sexes, mating takes place during a nuptial flight in which several colonies in one neighbourhood indulge at the same time. This allows intercrossing between individuals from different colonies. The females then cast off their wings and start colonies in the ground, each one for itself. The workers are sterile females, whose power to lay eggs in certain circumstances may return. For instance, when a colony loses a queen several workers may, under the influence of a suitable diet, take her place. In addition to the environmental complexity which a social existence involves, the lives of ants are further complicated by association with other organisms. Some, e.g. certain myrmecine ants, have adopted an agricultural habit, living on fungi which they specially cultivate. Others gather seeds from which they destroy the radicle to prevent germination, special chambers or granaries in the nest being constructed for their storage. The pastoral habit characterizes others, a symbiotic relation being set up with such insects (e.g. aphids) as exude fluids which are palatable to the ants. In addition to associations of this kind there are numerous others of an indifferent or little-understood nature, but which may range from the symbiotic to the parasitic. Finally may be mentioned the slave-makers: *Formica sanguinea*, for instance, captures from the colonies of *F. fusca* pupae which on emergence serve as slaves in the colony which has adopted them.

The wasps of the super-family *Vespoidea* are both social and solitary in habit. In these, the abdominal petiole is smooth (Fig. 349 c) and, in species with a worker caste, this is always winged. The prothoracic tergum extends back towards the wing base. Wasps are essentially carnivorous. Only rarely have they resorted to plant food, e.g. some solitary masarine wasps that feed their larvae on pollen and honey. Among solitary species may be mentioned *Odynerus* which deposits caterpillars in its nest when its larvae are developing. Pompilid wasps are exclusively predatory on spiders. Certain forms have adopted the 'cuckoo' habit, laying their eggs in the nests prepared and provisioned by other species. Thus the ruby wasp, *Chrysis*, usurps the nest of *Odynerus*. *Mutilla* behaves similarly towards many solitary bees and wasps. Social wasps, e.g. *Vespa*, live in nests commonly constructed of paper obtained in the form of wood pulp by these insect architects. The larvae, living in closely arranged cells on horizontal combs, are fed on insect food gathered by the workers. In early summer, our common social wasps are useful in the control of such insects as plant-lice, etc. Later in the season, however, their liking for sweet fruits may make them a nuisance, both in the garden and in the home. In autumn the colony perishes, fertilized females being the only survivors. The inability to store animal food on which the larvae depend accounts for the disappearance of colonies in the autumn. Only in tropical regions where food is plentiful throughout the year do wasp colonies persist. *Vespa germanica* and *Vespa vulgaris* are common English wasps. *Vespa crabro* is the hornet (Fig. 348).

Closely resembling these are those wasps belonging to the super-family *Sphecoidea*, the distinctive feature of which is the possession of a prothoracic tergum which does not extend back as far as the wing bases. These are all solitary predacious forms, which sting their prey and so paralyse it before placing it in the larval cells which have been previously prepared, e.g. *Sphex*. A tendency towards the social habit is exhibited by *Bembex* which leaves its larval cells open and so can provision its young from day to day on small flies.

The super-family *Apoidea* includes the social and solitary bees. Distinctive of bees are the dilated hind tarsi (Fig. 296E) and the plumose hairs of the head and body to which the pollen adheres. Inner metatarsal spines of the posterior legs comb the hairs free of pollen, this being then transferred to the outer upturned spines (pollen basket) of the hind tibia of the opposite side. These legs are further adapted by possession of special spines for the manipulation of wax plates when being removed from the abdomen. The median glossa is also characteristic and in certain solitary forms, e.g. *Anthophora* and all the social bees, e.g. *Apis* and *Bombus*, is greatly elongated along with the parts other than the mandibles for gathering nectar from deep-seated nectaries of flowers. Larvae are fed exclusively on pollen, nectar and salivary fluids. *Megachile*, the leaf-cutter, is a solitary bee which makes cells of neatly cut leaf fragments. Each cell containing an egg is stored with honey and pollen. Such cells are commonly made in the walls of houses, the mortar being removed for this purpose. *Andrena* constructs burrows in the ground and, though solitary, is usually found in groups of individuals occupying a common terrain which may include a 'village' of several hundred nests. *Nomada* has adopted the 'cuckoo' habit.

Bombus enjoys a social existence similar to that of *Vespa* in that only impregnated females survive the winter. The colony of the honey-bee *Apis mellifera* has more permanence, only the males dying off in the autumn to leave the rest of the colony to hibernate. The nest is of wax, an exudation from abdominal glands of the worker (sterile female), and a material known as *propolis* of vegetable origin serves to fasten parts of the nest together and to render the whole weather-proof.

The workers of *Apis* are graded according to age into *nurses*, who see to the welfare of the larvae by incorporating salivary juices with their food, *ventilators* who, by wing-fanning, set up currents in the nest or hive to reduce the temperature and to evaporate the honey, *scavengers* or *cleaners*, and *foragers* who collect pollen and nectar. The changes from nursery work to house work and to field work are necessitated by changes in glandular capacity as age increases. Though the density of the population of the colony determines to some extent when a queen with a number of workers will depart from the hive as a swarm, it appears that this event is also dependent on other factors not as yet clear, one of which is the relative proportions of the above age-groups among the worker caste. The sexes are determined by a cytological mechanism. Thus, in bees, wasps and ants, haploid parthenogenesis results in the production of males. A fertilized (diploid) female has control over the fertilization of eggs which she lays. If an egg is fertilized by sperm from the spermatheca a female (diploid) offspring develops; if not, a male offspring (haploid) develops. Whether a young female becomes a worker (sterile) or a queen (capable of fertilization) depends on nutrition. Contrast this with diploid parthenogenesis in aphids (p. 533).

26. Order Strepsiptera

DIAGNOSIS. These interesting insects whose relationships are still a matter of controversy are characterized by a number of remarkable biological phenomena. Their parasitic activity is restricted to the growth stages and the adult female, the adult males being free-living. There is pronounced sexual dimorphism in that the females are *prothetelous*, i.e. they have a larval body form while being at the same time sexually mature.

The eggs, too, have been shown to be polyembryonic, as many as forty embryos developing from a single egg, e.g. *Halictoxenus* parasitic on *Halictus simplex* (Hymenoptera). They are usually described as having a type of hypermetamorphic life cycle in which two larval types exist according as the development is to be towards a male or female adult. The hosts to which their parasitic attentions are directed appear to be restricted to the Hymenoptera (Vespoidea and Apoidea), the Rhynchota (Homoptera), and, in one rare instance, the Orthoptera.

ACCOUNT. Among the hymenopterous hosts we may take as an example the solitary bee *Andrena* parasitized by *Stylops*. The female is totally endoparasitic in the host bee, and appears as a hernia-like extrusion from between the tergites of adjacent segments as in Polistes (Fig. 350B). The body of such a female is legless and wingless and composed of a subtriangular unsegmented cephalothorax and a clearly ten-segmented abdomen. It lies in the host so disposed that the cephalothorax is visible. Behind the mouth ventrally is a transverse slit which leads into a brood chamber found under the cuticle of the first five or six abdominal segments. Genital pores communicate between the internal genital system and the brood chamber (Fig. 350C). It is through the transverse slit on the cephalothorax that copulation with the male occurs. By way of it, too, are born the first-formed *triungulin* larvae, the female being viviparous.

The first-formed larvae are six-legged, segmented creatures of very small size (Fig. 350D, E). They leave the parent parasite and the host bee probably to await, on flowers visited by the latter, a new host to which they attach themselves. From there they are transported to the nest of the bee. In this position they seek out the larvae of *Andrena* into which they burrow to live as endoparasites. A moult leads to the appearance of a maggot which is said to nourish itself by absorption through the skin. Further instars occur during this growth stage though details of this are not fully known.

Should the resultant individual be a female a modification of the moulting process, during what must be regarded as the pupal period, occurs. In this process the last larval and pupal exuviae are not thrown off completely but

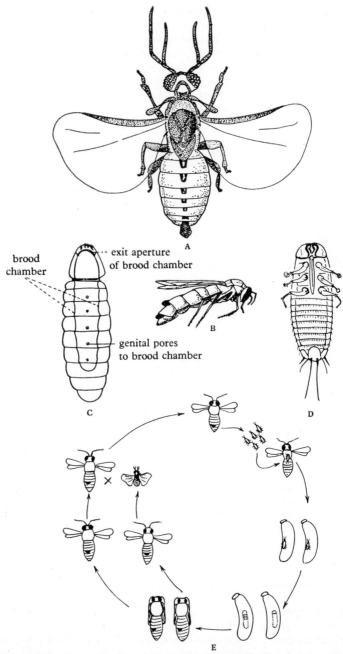

Fig. 350. General structure and life cycle of the Strepsiptera: A, male *Elenchinus*. B, a wasp (*Polistes*) bearing two female *Stylops* on the abdomen. C, ventral view of a female strepsipteran removed from a host bee. D, a first larva (triungulin) as released by a female to be transported by a bee to its larva in a nest. E, the life cycle of a typical strepsipteran parasite on a bee. (After Ulrich.)

become separated from the body ventrally to form a space—the brood-chamber—which by a transverse cleft in the anterior region makes contact with the exterior. The development of parasite and host proceeds together so that the adult phase of the two occurs simultaneously with the parasite's cephalothorax projecting in the manner described.

Should the resultant individual be a male, pupation occurs in the host after the head region has projected from between two abdominal segments of the bee. The casting of the pupal skin releases the male which flies away to seek a young endoparasitic female on another bee.

The male is characterized (Fig. 350A) by fore wings modified as small membranous balancers, the hind wings being expansive and fan-folded, with simple longitudinal veins but without cross-veins. Pro- and mesothorax have undergone considerable reduction. The legs vary in structure throughout the order and appear to serve more for attachment to the female at copulation than for locomotion.

The Strepsiptera are noted for the effects produced on their hosts by their presence. Such effects may lead to parasitic castration and concomitant effects on organization. For instance, parasitized bees may be deficient in their pollen-collecting apparatus and in many cases there are changes in colour and the adoption of secondary sexual characters belonging to the opposite sex.

The relationships of this order have been variously ascribed, by some to the Coleoptera, by others to the Hymenoptera. It is beyond the scope of this book to enter into the controversy and it has been decided here to accord ordinal rank to these insects.

SECTION IV. PARANEOPTERA (Heterometabola, Exopterygota)

Super-order **Psocopteroidea**

27. *Order* **Psocoptera (Book-lice)**

DIAGNOSIS. Small insects, either winged or wingless; with biting mouth parts; thoracic segments distinct; wings with reduced venation from which cross-veins are largely absent; metamorphosis slight.

ACCOUNT. These insects are to be found on bark and leaves of trees. They feed on lichens and dry vegetable matter. The eggs are laid on the bark or leaves and covered by a protecting sheath of silk by the female, e.g. *Peripsocus phaeopterus*.

Atropus pulsatoria, the book-louse, is found in damp, dark rooms and feeds on the paste of book bindings, wallpaper, etc.

28. *Order* **Mallophaga (Biting lice)**

DIAGNOSIS. These insects are ectoparasites of birds (less frequently of mammals). Their reduced eyes, flattened form and tarsal claws are features correlated with this mode of life. Unlike the Anoplura they have no piercing mechanism and devour with biting mouth parts small particles of feathers, hair, or other cuticular matter; metamorphosis wanting.

ACCOUNT. The common hen-louse, *Menopon pallidum* (Fig. 351), may be taken as an example. The head is semicircular in form and articulates with a prothorax which is freely movable on the rest of the body, a tagma formed by the fusion of the meso- and metathorax with the abdomen. The mouth is placed ventrally on the head and surrounded by biting mandibles and less prominent 1st and 2nd maxillae.

Eggs are laid separately on feathers or hairs and the life cycle is completed in about a month, the young instars resembling the adult in form and habit.

The various families of biting lice are strictly confined to particular groups of birds, indicating that evolution of the parasites has proceeded concurrently with that of their bird hosts.

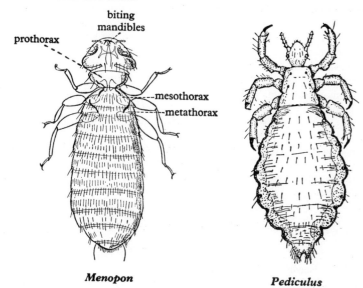

Menopon

Fig. 351. Hen-louse, *Menopon pallidum*. Dorsal view showing biting mandibles by transparency.

Pediculus

Fig. 352. Body-louse, *Pediculus humanus*. (After Imms.)

29. Order Anoplura (Sucking lice)

DIAGNOSIS. Ectoparasites of mammals, with mouth parts adapted for piercing the skin and sucking the blood of their hosts. The eyes are ill developed or absent. The single-jointed tarsus carries a large curved claw admirably adapted for clinging to the host (Fig. 296A). The thoracic segments are fused, and a flattened abdomen of nine segments possesses large pleural areas allowing the body to swell on feeding; metamorphosis wanting.

ACCOUNT. The minute mouth parts are accommodated at their bases in a stylet sac which is a diverticulum ventral to the pharynx. There are two stylets of which the dorsal is a paired structure, the halves of which maintain contact with each other distally to form a half-tube which is completed by the ventral stylet. This also consists of two elements. Between the dorsal and ventral stylets lies the salivary duct which appears distally to be a modification of the hypopharynx. The stylet complex can be sufficiently everted so as to make contact with the skin. Into the wound is poured the salivary fluid, and the mouth funnel is thrust in to enable the blood to be sucked up by the pharyngeal pump. Embryological evidence tells us that the 1st maxillae unite to form the dorsal stylet, the ventral being formed by the labium. A pair of mandibles also develops but these remain in a rudimentary condition.

Pediculus humanus, the body louse (Fig. 352), is associated with the spread of many diseases, such as typhus and relapsing fever. The disease known as trench fever, prevalent in all war areas during the first World War, has also been shown to be transmitted by this insect.

Eggs are laid attached to hairs of the body or clothing, and the three instars, passed through before attainment of the mature state, closely resemble the adult.

The louse has been found to lay about ten eggs daily, depositing in all about three hundred. Temperature plays a big part in controlling the development of these animals. Under average conditions the life cycle is completed in about three or four weeks.

Super-order **Thysanopteroidea**

30. Order **Thysanoptera (Thrips)**

DIAGNOSIS. Minute insects with asymmetrical piercing mouth parts with a short labial proboscis; prothorax large and free; tarsus two- or three-jointed with terminal protrusible vesicle; two pairs of similar wings, provided with a fringe of prominent long hairs, veins few or absent; metamorphosis slight, including an incipient pupal instar.

ACCOUNT. These insects are for the most part plant feeders, a few being carnivorous. They are regarded as serious pests in that they rob the plant of sap. They also often cause malformations and in some cases inhibit the development of fruit.

Parthenogenesis is of frequent occurrence. In the case of the pea thrips, *Kakothrips robustus*, the eggs are inserted in the stamen sheath of the flower and the nymphs emerging feed on the young fruit, inhibiting its growth. Later they feed on the soft tissues of pea pods, causing scar-like markings. The nymphs leave the plant and bury themselves deeply in the ground, where they remain till the following spring, when they pupate. Common thrips of importance are *Taeniothrips inconsequens* of pears and *Anaphothrips striatus* of grasses and cereals.

Super-order **Rhynchota (Bugs)**

DIAGNOSIS. The two orders Homoptera and Heteroptera into which this big group of insects is now divided comprise insects which differ from each other chiefly in the structure of their wings. In Homoptera both pairs of wings are uniformly membranous and transparent, and they are usually held at rest over the back in a roof-like manner. In Heteroptera the fore wings are hemi-elytra, the basal part being strongly cuticularized and pigmented. At rest they are placed flat over the back and overlap each other the one side over the other (Fig. 353).

528   INSECTA

Amid the welter of variation in size, form and habit which both orders present there are no differences more prominent and constant than these.

In both, the mouth parts adapted for piercing and sucking are morphologically identical, though the head in the Heteroptera is *prognathous* by virtue of a ventral head plate, the *gula* which thus carries the proboscis base to a forward position (Fig. 354 A). The absence of the gula in Homoptera leaves the head in a *hypognathous* position with the proboscis base lying in a backward-pointing position between the fore legs (Fig. 354 B).

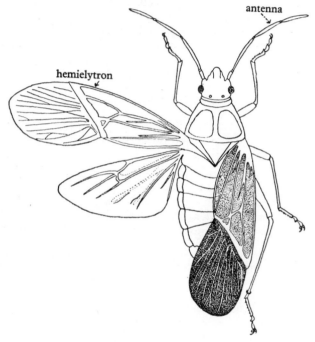

Fig. 353. External anatomy of *Leptocoris trivittatus* with wings spread on one side. (After Essig.)

In both orders the proboscis consists of a jointed dorsally grooved labium which receives two pairs of stylets, the mandibles and maxillae, so disposed as to form an effective piercing apparatus. Wing growth is external and the metamorphosis is gradual.

There is little doubt that though in the super-order there are many families whose members feed on animal juices, the group as a whole owes its existence to the feeding mechanisms which have made possible the exploitation of plant juices as a plentiful source of food. Mouth parts being fundamentally alike in the Homoptera and Heteroptera we will describe their main features and so make our account serve both orders (Fig. 355).

The labium typically is of some considerable length and may in some cases extend to between the coxae of the legs when at rest. It is jointed and except at its base it is grooved dorsally to receive the mandibular and maxillary stylets.

Basally the labial groove flattens out and the labrum here roofs over the stylet bases just before they enter the head.

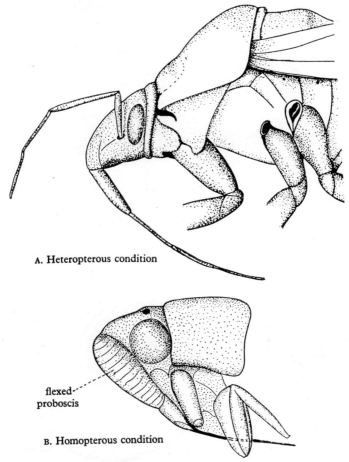

Fig. 354. Lateral views of proboscides of Rhynchota to illustrate the difference between the heteropterous and the homopterous conditions. A, *Deraecoris fasciolus* (Coreidae). (Modified after Knight.) B, *Zammara tympanum* (Cicadidae).

The mandibular and maxillary stylets are grooved along their median surfaces. The inner ones (maxillae) are so shaped as to fit together to form two enclosed tubes (Fig. 355C). The upper one of these is the food tube and the lower the salivary channel. In fitting together so as to form these two tubes the outer mandibles closely embrace the maxillae so that both contribute to form a proboscis of four closely fitting stylets. By means of the alternate action of protractor and retractor muscles inserted on them basally (Fig. 355B) the stylets can be thrust in and out of the tissue to which they have been applied. The toothed edges of the stylets distally make possible a purchase on the tissue pierced. In most forms the stylets lie wholly in the labial groove when at rest, being of the same length as the labium.

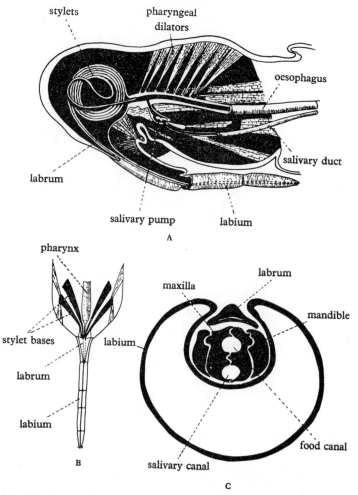

Fig. 355. Mouth parts of Rhynchota: A, sagittal section of head of *Aradus*. B, proboscis in relation to the head of a typical hemipteron. Protractor and retractor muscles are shown in relation to the bases of one pair of stylets. C, transverse section through the proboscis of *Aphanus*. (After Weber.)

There are many cases, however, where the stylets are much longer than the body. The stylet complex in that event is accommodated coiled up in a pocket in the head (Fig. 355A). These long stylets, capable of penetrating to great depths into host tissues, are 'paid out' from the basal coil, being assisted in this by the tip of the labium. The latter by a muscular mechanism can alternately grip and release the stylets as shown in Fig. 356 and so with the help of protractor muscles in the head pass the stylets ever deeper into the wound.

The saliva in blood-sucking forms carries an anti-coagulant but in plant-feeding bugs digestive enzymes cause the breakdown of cell tissues facilitating the insertion of the proboscis and at the same time causing much injury to the plant by their toxic properties. Indeed, it appears that injury to plants is greater from this cause than from the steady removal of plant sap.

That plant-sucking bugs also cause damage by the introduction of bacterial and virus infections, e.g. mosaic diseases of potatoes, may increase our respect for the Rhynchota as insects of very great economic importance.

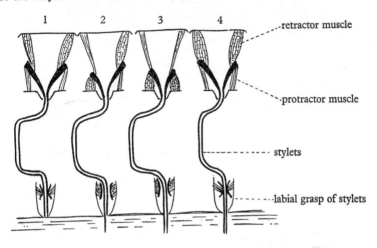

Psylla

Fig. 356. The penetration of the stylets of *Psylla* into plant tissue is effected by alternate action of basal protractor and retractor muscles and by the muscular action at the tip of the labium, by which the stylets are held in position prior to a retraction of the stylet base. The protractor muscles then thrust the stylets farther into the tissue after the stylets have been released from the labial grip. (After Weber.)

31. *Order* **Homoptera**

This order comprises a vast assemblage of forms ranging in size from the often microscopic Coccidae to the large tropical lantern bugs (Fulgoridae) and the cicadas which may be as long as 5 cm. with a wing expanse of 10 cm. With the cicadas we may join the leaf-hoppers, tree-hoppers and frog-hoppers, all active animals. In some instances the life cycle may be greatly prolonged, *Cicada septendecim*, for instance, having a seventeen-year cycle. The eggs are deposited in holes in twigs of trees, and the nymphs hatching from them fall to the ground into which they burrow to feed on the roots. After seventeen years of nymphal growth a stage resembling a pupa is passed through before emergence of the adult.

Scale insects (Coccidae) also belong to the Homoptera. *Pseudococcus* is the mealy bug, *Tachardia lacca* the lac-insect of commerce and *Aspidiotus perniciosus* the San José scale-insect of citrus trees.

Plant-lice (Aphididae) (Fig. 357 B, C), notable for their wide distribution and for their prolific reproduction, have transparent wings. The tarsus is two-jointed, that of the Coccidae being one-jointed. Wax-secreting cornicles are borne dorsally on the abdomen.

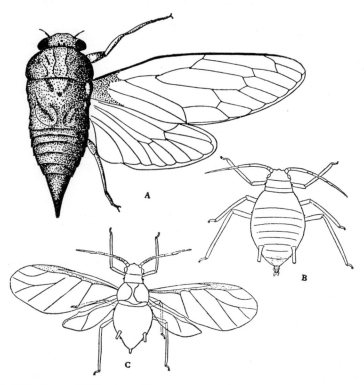

Fig. 357. Types of Rhynchota Homoptera. A, *Macrotrista angularis* (Homoptera, Cicadidae). B, *Aphis rumicis* (apterous viviparous female). C, winged viviparous female of the same. (B and C, after Davidson.)

In the last family the reproductive phenomena are of immense scientific and economic importance. A comparatively simple life cycle is that of *Aphis rumicis*. The winter is passed on the spindle tree *Euonymus* as eggs, which are laid in the autumn by fertilized females. In spring these eggs give rise to wingless viviparous parthenogenetic females. A variable number of these parthenogenetic generations is passed through in the summer, then winged parthenogenetic females appear which migrate to another host plant (the bean *Vicia faba* or other plants) and there reproduce, giving rise to generations of parthenogenetic females which eventually produce winged females which return to the primary host plant *Euonymus*. From these there now appear oviparous females to copulate with winged males, migrants from the secondary host plant, the bean.

The following summary will assist in the understanding of this life cycle:

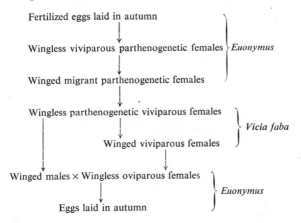

In other forms, such as *Phylloxera vastatrix*, the notorious pest of vineyards, the life cycle is immensely complicated and involves migrations between root and stem of the host plant. The reproductive capacity of these insects is most remarkable but is fortunately offset by the number of enemies they possess.

The cyclical reproductive phenomena in aphids as just described raise important problems relating to the intrinsic differences between sexual and parthenogenetic individuals, and to the environmental conditions governing the occurrrence of these phases in any life cycle.

Fertilized eggs produce only strictly parthenogenetic females. These multiply by diploid parthenogenesis, i.e. the eggs retain the full complement of chromosomes and are not capable of fertilization. Eventually come individuals capable of bearing sexual forms, *sexuparae*. The sexual forms arising from these produce haploid germ cells which have undergone normal reduction. It follows then that fertilization will restore diploid parthenogenesis. Sexual differences are indicated in the chromosomes; the female of *Aphis saliceti* possesses six, of which two are sex chromosomes; the male only five, one only being a sex chromosome. Sexual reproduction leads, however, only to the production of parthenogenetic females and not to males and females in equal numbers, as might be expected. This appears to be due to the fact that in the maturation of sperms, those with only two chromosomes die. Fertilization therefore is always between sperms and ova each with three chromosomes, of which in each case two are normal chromosomes (*autosomes*) and one is a sex chromosome X. The capacity of females with six chromosomes to produce male offspring with only five is due to the fact that in the maturation of male-producing parthenogenetic eggs, reduction in the number of chromosomes only affects the sex (X) chromosomes, one remaining in the egg, the other going to the polar body. In this way a parthenogenetic female with six chromosomes, i.e. $4 + XX$, gives rise to males with only five, i.e. $4 + X$.

A complete analysis of the environmental conditions governing the onset of sexual phases after a period of parthenogenetic reproduction is yet to be made. Food, temperature and light seem to be important, and of these a reduction of the last-mentioned factor seems to be associated with the production of sexual winged individuals.

32. Order **Heteroptera**

In addition to the distinguishing features already mentioned for this order may be mentioned the following whereby they differ from the Homoptera. The latter insects are always terrestrial and vegetarian. The Heteroptera on the other hand include aquatic as well as terrestrial forms, predators and blood-suckers as well as vegetarians.

They may conveniently be considered under two headings: the *Gymnocerata*, terrestrial forms with visible segmented antennae (Fig. 353), and aquatic forms whose antennae are short and hidden, the *Cryptocerata*.

In the first of these groups may be mentioned *Cimex*, the bed-bug, an ectoparasite with vestigial wings, flattened body and prominent claws. It inhabits human dwellings, and its retiring habits, coupled with its power to fast for long periods, make it a difficult creature to eradicate when once it is established. The shield-bugs (Pentatomidae) are phytophagous. The mesothoracic tergum is greatly enlarged to extend at least as far over the abdomen as the junction between the horny and membranous parts of the wing when these are at rest. The red bugs (Pyrrhocoridae) are also phytophagous. Certain species, e.g. of *Dysdercus*, are known as 'stainers' from their habit of feeding on cotton-bolls into which they inject a micro-organism responsible for the appearance of a red stain on the fibre. The Capsidae are almost exclusively phytophagous, some of their members being very serious pests of our English orchard trees and shrubs. *Plesiocoris*, until recent times restricted to such trees as willow, now attacks black-currant bushes, apple trees, etc. An exception to this phytophagous habit is found in *Cyrtorhinus mundulus* which sucks the eggs of the sugar-cane hopper, *Saccharicida*, thus effectively controlling this pest in Hawaii. In the family Reduviidae are many forms which transmit trypanosomiasis in the tropics, e.g. *Rhodnius prolixus*.

The Gymnocerata are notable for their numerous adaptations to aquatic life. They commonly lay their eggs in the tissues of submerged plants. Many, e.g. the water-boatman, *Corixa*, and back-swimmer, *Notonecta*, have powerful legs fringed with hairs which, by the simultaneous movement as members of pairs, propel the animal through the water as oars do a boat. They breathe air at the surface film, making use either of a terminal abdominal tube (*Nepa*) or of unwettable hairs between which air is trapped to enable the animal to breathe during its period of complete immersion (*Notonecta*). The aquatic Belostomatidae of subtropical (Ethiopian, Oriental and neotropical) distribution contain some of the largest known species, which may attain a length of 10 cm. We may also mention *Aphelocheirus* which, as already mentioned, breathes in water by a process known as plastron respiration.

CHAPTER XIV
THE CLASS ARACHNIDA

DIAGNOSIS. Arthropods with fully chitinized exoskeleton; the anterior part of the body (prosoma), never divided into head and thorax, consisting of six adult segments, the first (preoral) with prehensile appendages (chelicerae) usually three-jointed, the second (postoral) with appendages either sensory or prehensile (pedipalps) and the remaining four ambulatory; the posterior part (opisthosoma) consisting of thirteen segments and a telson in the most primitive forms but tending to become shortened, the first (pregenital) segment differing from the rest, the second bearing the genital opening; respiratory mechanisms of various types usually developed in the anterior part of the opisthosoma; coxal glands of coelomic origin in the 2nd and 5th prosomatic segments. Larval forms are absent in many Arachnida, important exceptions being *Limulus* (Xiphosura), *Linguatula* (Pentastomida) and the Acarina.

CLASSIFICATION

Order 1. **Scorpionidea**. Arachnids with the prosoma covered by a dorsal carapace, the opisthosoma divided into a mesosoma and a metasoma distinct from one another, containing twelve segments and a telson; chelicera and pedipalps both chelate; four pairs of walking legs; the first mesosomatic segment carries the genital operculum, the second the pectines, and the next four each a pair of lung books; the metasoma comprises segments reduced in size to form a flexible tail for wielding the terminal sting (telson) and bears no appendages. Viviparous. *Scorpio, Apistobuthus, Buthus, Hormurus.*

Order 2. **Pseudoscorpionidea**. Scorpion-like arachnids with no division of the opisthosoma into mesosoma and metasoma, i.e. no tail. *Microbisium, Chelifer.*

Order 3. **Eurypterida**. Extinct aquatic arachnids resembling scorpions with a meso- and a metasoma but showing greater variety in form. *Slimonia, Pterygotus, Hemiaspis.*

Order 4. **Xiphosura**. Aquatic arachnids with a broad prosoma divided by a hinge from the opisthosoma in which the first six segments are present and fused together dorsally. The caudal spine possibly represents the lost abdominal segments fused together. Five pairs of lung books present. *Limulus, Tachypleus, Carcinoscorpius.*

Order 5. **Araneida**. Arachnids with the prosoma covered by a single tergal shield but head marked off by a groove; opisthosoma (abdomen) separated by waist; soft, rarely having any sign of external segmentation; two to four pairs of spinning glands. *Epeira, Atypus.*

Order 6. **Palpigrada**. Small arachnids with elongated well-segmented body and long tail. *Koenenia.*

Order **7. Solifuga.** Arachnids with prosoma and opisthosoma not divided by pedicel. Body very hairy. No tail. *Galeodes.*

Order **8. Acarina.** Arachnids with rounded body with no boundary between the prosoma and the opisthosoma; basal segments of the pedipalps united behind the mouth; no gnathobase to the four walking limbs. *Tyroglyphus, Argas, Ixodes, Hydrachna.*

Order **9. Phalangida.** Arachnids with prosoma covered by a single tergal shield and united to the opisthosoma by its whole breadth; opisthosoma always segmented. *Phalangium, Oligolophus.*

Order **10. Pantopoda (Pycnogonida).** Arachnida in which the opisthosoma has disappeared with the exception of the pregenital segment which bears legs on which the genital pore opens. *Nymphon, Pycnogonum.*

DOUBTFUL ARACHNIDS

Order **11. Tardigrada.** Minute arthropods with four pairs of stumpy legs ending in claws; with oral stylets and a suctorial pharynx, without definite circulatory or respiratory systems. *Macrobiotus, Hypsibus, Echiniscus.*

Order **12. Pentastomida.** Elongate vermiform parasites with a secondary annulation and two pairs of claws at the sides of the mouth; without respiratory or circulatory systems. *Linguatula.*

GENERAL ACCOUNT

As has been pointed out in the introduction to the Arthropoda, the Arachnida are distinctly marked off from the rest of the phylum by the character of their appendages and especially by their chelicerae which furnish so strong a contrast to the sensory antennae, elsewhere found in the phylum. Moreover, nowhere else (except perhaps in trilobites) are true jaws absent, the prolongation of the basal joint of the anterior limbs toward the mouth (gnathobases) serving the arachnids for mastication. In the divisions of the group is found the greatest diversity in form, for though by no means active creatures, arachnids have become adapted to many kinds of environment.

SEGMENTATION. Besides the segments enumerated in the preamble, there is in the embryo of most arachnids a *precheliceral segment* (Figs. 358, 359 B, C). The variation in the segments of the prosoma is confined to minor details, the chelicera preserving much the same characters throughout the group, only losing a joint in the Araneida, and being either chelate or subchelate; the pedipalp, however, varies according to its function, being chelate in the scorpions, which seize their prey by means of it; modified for purposes of fertilization in the spiders; and merely an ambulatory appendage in *Limulus*. In most forms the tergites of the segments are fused together, but in the Pedipalpi and the Solifugae the last two prosomatic segments are entirely free.

It is in the opisthosoma and its segments that the greatest amount of variation can be seen. The *pregenital segment* (Fig. 359c) is always developed in the embryo, but tends to disappear in the adult. Thus in the Palpigradi and Pseudoscorpionidea it forms a distinct segment; in *Limulus* it is represented by a pair of rudimentary appendages, the *chilaria*; it is entirely missing in the adult scorpions. In addition to this segment there is a maximum of twelve segments and a terminal appendage, the telson, which is attained only by the embryo scorpions and the eurypterids; the Palpigrada and Pseudoscorpionidea have one less. In all these cases, there is a differentiation of the segments into two regions, the meso- and metasoma. In *Limulus* there are six segments only, but in the related extinct genus, *Hemiaspis*, there are three more. The Solifuga show ten. In the spiders, mites and phalangids, the body is much shortened; the phalangids have the anterior segments united to the prosoma. Lastly, the telson may be a sting in the scorpions, a jointed sensory flagellum in the Palpigradi, a fin in some eurypterids or a digging stick in others and in *Limulus*.

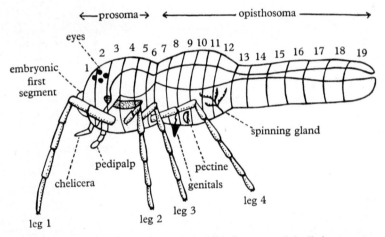

Fig. 358. Diagram of an arachnid to show the segments and the limbs present on each segment.

ALIMENTARY CANAL. A typical feature is the suctorial alimentary canal. The mouth is usually narrow and situated just behind the chelicerae; only in *Limulus* has it moved backwards, become enlarged and surrounded by the basal joints (gnathobases) of all the prosomatic appendages; in the scorpions the appendages of the 2nd–4th segments form gnathobases; the Palpigrada and Solifuga have no gnathobases. In all arachnids, except *Limulus*, the food is fluid and is drawn through a narrow oesophagus into a sucking stomach and thence into a straight mid gut, which is by far the longest part of the gut, and receives the openings of the digestive caeca; often, as in scorpions, there are several of these, segmentally repeated, very much branched and forming a compact, 'liver-like' organ. There may be important salivary glands entering the fore gut as in the scorpions. Posteriorly the mid gut, except in *Limulus*, gives off Malpighian tubules. The hind gut is short.

RESPIRATION. The respiratory organs of the Arachnida are distributed as follows. (1) 'Gill books' in the aquatic form, *Limulus*, and probably in the extinct eurypterids. (2) 'Lung books' in the terrestrial scorpions and Pedipalpa. (3) A combination of lung books and tracheae in the spiders. (4) Tracheae alone in the Solifuga, Pseudoscorpionidea, Phalangida and Acarina.

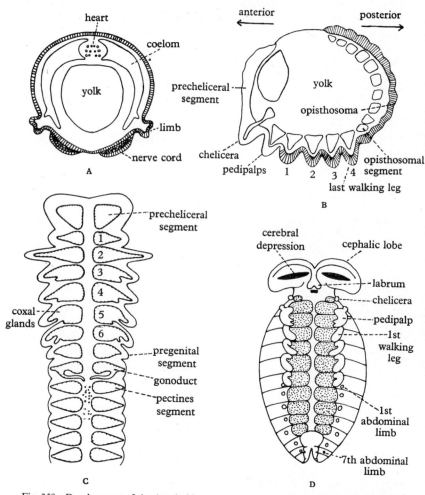

Fig. 359. Development of the Arachnida. A, transverse section of embryo spider (*Theridium*) (after Morin); B, longitudinal section of embryo spider (after Wallstabe); C, diagram of embryo scorpion (altered from Dawydoff); D, diagram of embryo scorpion (*Euscorpius carpathicus*) (after Brauer).

(5) Lastly, in the Palpigrada, smaller acarines and other forms, there are no special respiratory organs and exchange of gases takes place through the skin.

As the Arachnida apparently form a natural group, efforts have been made to derive these various methods of respiration one from the other. The gill books (Fig. 360A) are stated to be the most primitive respiratory organs. They

ARACHNIDA

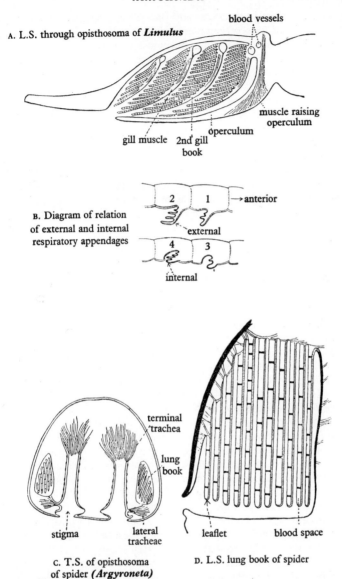

Fig 360. Respiratory organs in the arachnids. (A, from Shipley and MacBride; B, after Kingsley; C and D, after McLeod.)

are piles of leaflets, in which blood circulates, attached in each segment to the posterior face of freely oscillating plates, which are possibly appendages, resembling the abdominal appendages of the Isopoda which are also respiratory in function. There is a special muscular mechanism for opening and shutting the leaflets in the water and thus facilitating gaseous exchange. In the lung books of the scorpion there are also parallel leaflets, which are sunk into pits with a confined opening (pneumostome). The air circulates between these leaflets, but there is no evidence that air is actively pumped in and out of the lung. Gaseous exchange then appears to be entirely due to diffusion. In spiders, however, a complicated system of muscles has been described which brings about expiration by compressing the lung. Inspiration follows by the elasticity of the chitin lining.

It is generally supposed that the lung books of scorpions are derived from gill books by the withdrawal of the leaflets into special pouches, the lungs (Fig. 360B). The appendages or plates disappear or form the floor of the lung and the leaflets appear as folds of the lining. Lung books, according to this view, are organs which, originally intended for aquatic use, have been slightly adapted for terrestrial life, but while the scorpions in their long history have shown no capacity for further development, the rest of the Arachnida have developed the typical arthropod tracheal system. The spiders, at least, have passed through a primitive lung-book stage from which they have not all emerged. In fact they show all the stages of replacement of lung books by tracheae, which actually arise as diverticula of the lung itself. Thus we have the following stages in the spiders (Fig. 373):

(1) Two pairs of lung books and no tracheae in the families Atypidae, Liphistiidae and Aviculariidae.

(2) An anterior pair of lung books and a posterior pair of stigmata, opening into tracheae, seen in the majority of families (Fig. 360c).

(2a) An anterior pair of lung books, the posterior pair of stigmata and tracheae having entirely disappeared, in the family Pholcidae.

(3) Two pairs of stigmata, both opening into tracheae, in the family Caponiidae.

These form a complete series. The adherents of the theory that lung books have given rise to tracheae claim that, on the whole, those spiders which have two pairs of lung books are the most primitive in other respects. It may be pointed out, however, that there is also a connexion between the degree of development of tracheae in a family and the activity of its members. In inert forms, there may be reduction or even total loss of the tracheal system.

In all the forms in which lung books or gill books are present, there are processes in the embryo which can be identified as rudiments of appendages on the anterior abdominal segments (Fig. 359D). On the posterior border of these processes, leaflets develop at the same time as an invagination forms the lung cavity above them, so that the limb itself forms part of the floor of the cavity. On the whole then, embryology may be said to show the origin of lung books from gill books, and the comparative anatomy of spiders indicates that lung books have been replaced by tracheal systems. But there lie outside this series arachnid groups, like the Acarina, with tracheal systems of a different kind, which can only be derived with difficulty from the respiratory system of the other forms and may have had a separate origin.

MESODERM. In the arachnids, the mesoblast is formed as two lateral bands which segment into somites, just as does the same tissue in the annelids.

ARACHNIDA

The somites correspond with the external segmentation and in each one of them appears a coelomic cavity. This is best seen in the scorpions (Fig. 359c) and the spiders (Fig. 359b). They are formed near the ventral surface and extend on the one hand into the appendage and on the other towards the dorsal middle line, where the extensions from the two sides meet and form the heart between them. They also form diverticula varying in the different groups, which are the remains of a complete series of metamerically segmented coelomoducts. In the scorpions, the embryo (Fig. 359c) shows five pairs of these, in segments 3, 4, 5, 6 and 8. In only one case, that of segment 5, do the coelomoducts reach the external surface, and persist in the adult as a pair of excretory organs, the *coxal glands*. In segment 8 they grow towards the middle line and form the mesodermal part of the gonoducts. The other coelomoducts disappear and the coelomic sacs are resolved into mesenchyme which fills up the spaces of the body and forms the muscles, the blood and the fat body. In *Limulus* there are also a pair of coxal glands, which in development arise from the coelomic somites of no less than six segments, of which only segment 5 sends out a duct opening to the exterior.

1. Order **Scorpionidea**

DIAGNOSIS. Arachnids with the prosoma covered by a dorsal carapace; the opisthosoma divided into a mesosoma and metasoma distinct from one another, containing twelve segments and a telson; chelicerae and pedipalps both chelate; four pairs of walking legs; the first mesosomatic segment carries the genital operculum, the second the pectines, and the next four each a pair of lung books; the metasoma comprises segments reduced in size to form a flexible tail for wielding the terminal sting (the telson) and bears no appendages. Viviparous.

GENERAL STRUCTURE. The tergum of the prosoma bears a group of lateral eyes near the anterior border and a pair of median eyes, but some scorpions are blind. On the ventral surface there are inward projections from the basal joints of the pedipalps and the first two pairs of walking legs, which are masticatory in function (gnathobases). The walking legs are six-jointed and end in double claws. Between the basal joints of the last pair is a plate, the *metasternite*, which represents the fused sterna corresponding to these limbs; the sterna of the other prosomatic segments are not represented. At the beginning of the mesosoma there is in the embryo a pregenital segment with two limb rudiments. This disappears without leaving a trace in the adult. The two succeeding segments bear appendages: (1) the *genital operculum*, a small plate covering the openings of the genital ducts, which is formed by the union of two rudiments of appendages; (2) the *pectines*, flap-like structures attached by a narrow base with a distal border of chitinous spines like the teeth of a comb. They are tactile in function and derived from embryonic limb rudiments. There are no other exclusively sensory organs (except the eyes) on the body of the scorpion, but there are sense hairs scattered over the surface and more numerous on the pedipalps than elsewhere.

The lung books are found on segments 3–6 of the mesosoma. The 7th segment is without any external segmental organs. As has been already mentioned, there are, in the embryo, seven pairs of mesosomatic appendages (Fig. 359 D), those on the embryonic pregenital segment and on the six succeeding segments. Of these the 4th–7th never develop to more than papillae, but folds develop on their posterior surface and the skin behind is tucked in to form the lung sacs. When the sacs are complete, the folds become the leaves of the lung book. In the internal space of these folds, the blood circulates and is presumably aerated; it contains the respiratory pigment, haemocyanin. The circulatory system of the scorpion is remarkably complete (Fig. 362). The

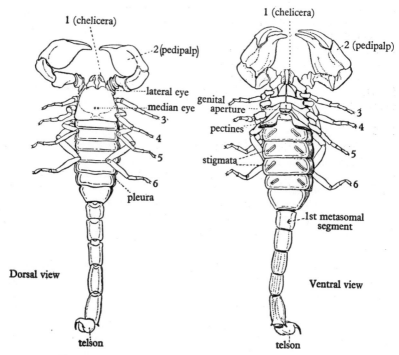

Fig. 361. *Scorpio swammerdami*, ×½; 3, 4, 5, 6 are walking legs.
(From Shipley and MacBride.)

heart consists of seven chambers (in the 7th–13th segments), into each of which a pair of ostia opens and from each there leave a pair of lateral arteries. In addition, there is an anterior and a posterior aorta, the former dividing into many branches in the prosoma, and one of these passes backwards as a supraneural artery. The arteries end in tiny vessels and many of these communicate with the special ventral sinus, which supplies blood to the lung books. Muscles run from the roof of this to the floor of the pericardium, and when they contract the ventral sinus enlarges and draws venous blood into it. When they relax, blood is forced into the lung books, whence it is returned to the pericardium by segmental vessels.

ARACHNIDA

A minute mouth opens into the pharynx which is suctorial, with elastic walls which can be drawn apart by muscles. A short oesophagus succeeds, and into this open the salivary glands. The endodermal mid gut is long and narrow and receives throughout its course several pairs of ducts which lead from the digestive glands. These together form a bulky mass, filling up the dorsal part of the mesosomatic body cavity. The food passes into the cavity of these to be digested. It consists mainly of insects, which are chewed by the gnathobases and the juices sucked up by the action of the pharynx. The beginning of the short hind gut is marked by the Malpighian tubules.

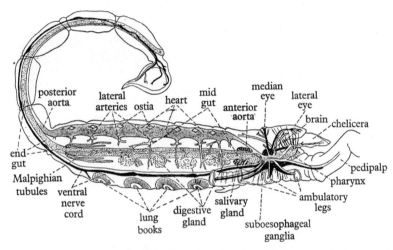

Fig. 362. View of the internal anatomy of *Buthus* showing digestive, circulatory and nervous systems. The nervous system is shown black, the circulatory system finely stippled, and the digestive system coarsely stippled. (Altered from Leuckart.)

The nervous system consists of a supraoesophageal ganglion which supplies the eyes, a large suboesophageal complex which gives branches to all the adult appendages, and two ventral cords which bear ganglia in the last seven segments.

The sexes are separate and the gonads constitute a network. The spermatozoa are filiform and fertilization is internal, being preceded by a courtship, described in lively fashion by Fabre as *danse à deux*. Scorpions are viviparous. Sometimes the eggs are rich in yolk and the young develop entirely at its expense; in *Scorpio* and other genera the eggs are small and yolk is entirely absent. In this case the young develop in lateral sacs of the uterus, attached to the mother by a kind of *placenta*. The young, when hatched, are sometimes carried on the mother's back.

The earliest scorpions are found in the Silurian, and it is of considerable interest that the first genus, *Palaeophonus*, was a marine animal. It closely resembles the terrestrial scorpions, except in its shorter and broader limbs without claws, and in the absence of stigmata.

2. Order **Pseudoscorpionidea**

DIAGNOSIS. Scorpion-like arachnids with no division of the opisthosoma into a meso- and a metasoma. No tail.

The chelicerae, pedipalps and walking legs are similar to those of the scorpions (Fig. 363). There are usually 2–4 eyes present. The animals respire by means of tracheae. There are two pairs of spiracles, one on the 3rd abdominal segment the other on the 4th. The mouth opens between the pedipalps and leads to a long thin oesophagus around which is condensed the central nervous system. There is a sucking pharynx followed by a broad mid intestine. The mid intestine has several glandular diverticula which fill the

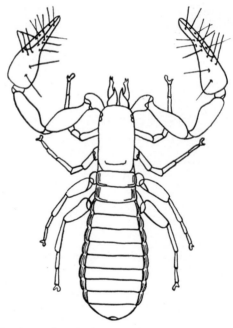

Fig. 363. Pseudoscorpion *Menthus rossi*. (From Grassé, after Chamberlin.)

general body cavity. The blood system is reduced, there being a single dorsal vessel forming the heart. The nervous system is condensed into a single mass in the anterior part of the cephalothorax. The sexes are separate.

The pseudoscorpions live under dead leaves, beneath moss (*Microbisium*) and in houses (*Chelifer*). They are only rarely found out on open ground (*Neobisium*).

3. Order **Eurypterida**

DIAGNOSIS. Extinct aquatic arachnids resembling the scorpions in the number and arrangement of the segments of the adult; the division of the abdomen into meso- and metasoma is not quite so marked; chelicerae short and three-jointed, chelate; the next four segments bear appendages which are

ARACHNIDA

often similar (but the pedipalps may be chelate); in the last (6th) prosomatic segment the appendages are always larger than the rest and are broad and paddle-shaped; first and second pairs of mesosomatic appendages unite to form the genital operculum; the first five mesosomatic segments bear indications of leaf-like branchiae; metasoma ends in a structure (telson) of variable form; mouth has moved backwards and is surrounded by gnathobases of all the limbs.

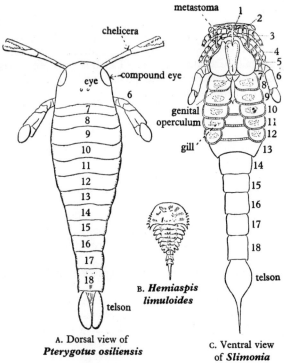

Fig. 364. Diagram of extinct Arachnida; all are Silurian forms. (A, after Schmidt; B, from Woods; C, from Laurie.)

GENERAL ACCOUNT. The great interest of this group lies in its similarity to the scorpions. There was, however, much more variety in external structure in these aquatic arachnids and they sometimes attained a length of 6 ft. Not only is there fundamental agreement in the segmentation and the division into meso- and metasoma, but also in characters like the shape and usually the size of the chelicerae, and the telson, which in primitive eurypterids has a recurved sting-like form. *Slimonia* (Fig. 364c) has a slightly modified telson. In one eurypterid (*Glyptoscorpius*) structures have been described which correspond to the pectines in position and structure. If this is substantiated, it constitutes a remarkable resemblance in detail.

A few special characters may be mentioned here. On the ventral surface a structure called the *metastoma* is seen which possibly represents the pregenital segment. Branchiae undoubtedly existed, but their exact nature is not known. Possibly the sterna of the segments which carried them were membranous and the branchiae were tucked in under them. There are five pairs and the first of these corresponds in position to the pectines of the scorpion (except possibly in *Glyptoscorpius*). Thus, when the ancestors of the scorpions became terrestrial, we may suppose that the first pair of respiratory appendages remained external and took on a sensory function, while the rest helped to form the lung books.

Minute forms with incompletely developed abdomen and enlarged eyes have been found which are thought to be the pelagic larvae of eurypterids. The adults were in all probability carnivorous forms, which crept and swam and sometimes burrowed at the bottom of shallow seas. In *Pterygotus* (Fig. 364A) and *Eurypterus* there were adaptive modifications of the telson for swimming and burrowing respectively.

4. Order Xiphosura

DIAGNOSIS. Aquatic arachnids with a broad prosoma divided by a hinge from the opisthosoma in which the first six segments are present and fused together dorsally; they bear six pairs of biramous appendages, of which the first form an operculum on which the genital apertures open and the remaining five carry the gill books; chelicerae of usual arachnid type, pedipalps not distinguished from the four pairs of ambulatory appendages which follow; mouth far back, surrounded by *gnathobases* of all the postoral limbs; caudal spine present, possibly representing the lost abdominal segments as well as the telson; pregenital segment represented by rudimentary appendages, the *chilaria*.

Limulus (Fig. 365), which is the sole living representative of the group, is evidently more affected by specialization than either the scorpions or eurypterids, and it is on this account that the attempts which have been made to indicate the king crab as an ancestral form of higher groups have usually been regarded as ingenious but illusory. It is essentially a shore-living, burrowing animal. Like a crab, its carapace is compact, dorsoventrally flattened and expanded laterally, so that the animal can shovel its way under sand and mud. Its legs are tucked under the carapace and the hinder pair kick out the sediment behind. To protect the gill books from this rough treatment, the operculum completely covers the appendages which bear them. But *Limulus* has not lost its tail, and an observer, watching the creature in an aquarium, will contrast it unfavourably for grace and efficiency with a crab. Its swimming movements, principally brought about by the flapping of the abdominal appendages, are slow and clumsy, and we can hardly consider it except as a sedentary animal.

The chelicerae are small, chelate and three-jointed, as is usual in arachnids. The succeeding four pairs of appendages are all alike in structure and function, consisting of six joints, the basal one being produced into a prominent spiny *gnathobase*; they are chelate (except the adult males, which are clawed). The last (6th) pair of appendages has four spines springing from the end of the last joint but one; while the four anterior legs are used for walking as well as masticating, this pair is particularly concerned with digging. They also possess an external spatulate process which is inserted under the operculum and cleans the gill books.

The chilaria, as has been stated, are the appendages of the pregenital segment. They are flattened processes without any function that has been discovered.

The appendages of the opisthosoma are shown in ventral view (Fig. 365), and vertical longitudinal section (Fig. 360A). They are all greatly flattened and expanded, consisting typically of a slender 'endopodite' and a broad plate which is the 'exopodite'. The anterior pair arise in the embryo as distinct

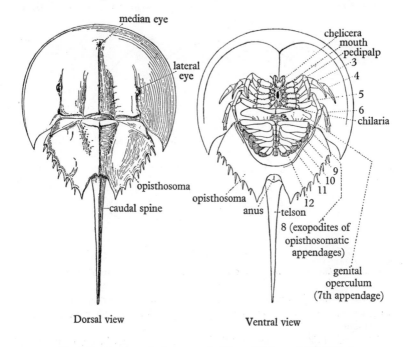

Fig. 365. *Limulus polyphemus*, × ⅓. (From Shipley and MacBride.)

rudiments, but fuse to form the genital operculum, on the under surface of which are the genital apertures. In all the others the appendages almost meet in the middle line, but remain distinct. From the posterior surface of the exopodite arise about 200 branchial leaflets. The appendages are provided with muscles by which the flapping movements are made which propel the animal in a leisurely way through the water and circulate water amongst the leaflets.

The mouth occupies a subcentral position under the carapace, surrounded by the gnathobases. Worms and small molluscs from the shore mud are seized by the chelae and, after mastication by the gnathobases, stuffed into the mouth, which leads to the fore gut consisting of an oesophagus and a chitin-lined 'stomach'; the mid gut is long and into it open two pairs of ducts from the digestive glands. These glands are very well developed and fill up much of the space inside the cephalothorax. There are no Malpighian tubules and no salivary glands in *Limulus*.

The circulatory system is very complete and like that of the scorpion in its main lines. A unique feature is the complete investment of the ventral nervous system by an arterial vessel which corresponds to the supraneural vessel of the scorpion.

The nervous system is of a very concentrated type. The supraoesophageal ganglia supply the eyes and are fused with the ganglia of all the succeeding segments as far as the opercular segment to form a ring round the oesophagus.

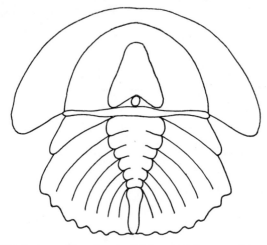

Fig. 366. *Limulus* larva. Note the superficial resemblance to a trilobite (Fig. 234).

From this a double ventral cord extends into the opisthosoma, swelling into ganglia in each of the 'gill-book' segments. Median and lateral eyes (p. 323) are present.

The coxal (brick red) glands arise from six segments in the embryo and open on the fifth pair of legs.

The reproductive organs consist of a network of tubules communicating with the exterior by paired ducts opening on the genital operculum. The eggs are laid far up on the shore at spring tides in holes dug for them by the mother, and the male, which comes ashore clinging to the carapace of the female, spreads the sperm over them, a method of fertilization very similar to that of the frog. The eggs are heavily yolked and the young hatch as a planktonic larva in a condition resembling the adult but with an opisthosoma showing separate segments and without the caudal spine. The larva, which swims by means of the abdominal appendages, as in the adult, has been called the 'trilobite' stage, because of an extremely superficial likeness to that group (Fig. 366).

While *Limulus* has existed since the Trias without any modification, it is of considerable interest that in the Palaeozoic very similar animals occur, in which there are three additional segments and a rather shorter caudal spine, indicating that the latter organ has been formed at the expense of the posterior opisthosomatic segments. These animals are *Hemiaspis* (Fig. 364B) and *Bunodes*.

5. Order **Araneida**

DIAGNOSIS. Arachnids with prosoma covered by a single tergal shield but head marked off by a groove; opisthosoma (abdomen) separated by a waist, soft, rarely having any trace of segmentation; two to four spinning glands: chelicerae two-jointed, subchelate: pedipalps modified in male for transmission of sperm.

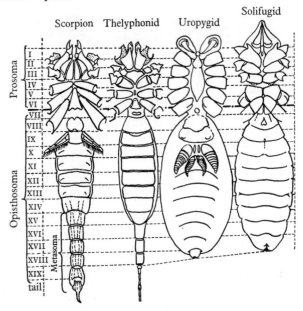

Fig. 367. Segmentation in different arachnids. Note the development of tagmata and loss of external segmentation. (From Grassé.)

GENERAL ACCOUNT

The Araneida differ from the other arachnids in that their abdomen is unsegmented and is joined to the cephalothorax by a narrow waist. One family of the Araneida, the Liphistiidae, still have the primitive character of a fully segmented abdomen, whilst *Tetrablemma* has a series of segmented plates on its abdomen. In the embryo spiders the segments of the opisthosoma are clearly indicated by the ten pairs of coelomic cavities (Fig. 359B). There are also five pairs of rudimentary abdominal appendages, the first of which disappears, the next two assist in forming the lung books, whilst the 4th and 5th become spinnerets. When more than two pairs of spinnerets are present, the additional ones are split off from pre-existing spinnerets. Embryological studies thus show that the existing forms with an apparently unsegmented abdomen are descended from ancestors with nearly the full number of segments typically found in the arachnids.

THE HEAD. The head bears eyes and mouth parts. The first pair of appendages are called the chelicerae; they are situated in front of and above the mouth and are of two segments each, a large basal segment and a terminal claw segment. The tip of the claw contains the opening from the poison gland on its convex side, thus when the claw is pressed into the prey the opening of the duct is not blocked and the poison is allowed to run freely into the open

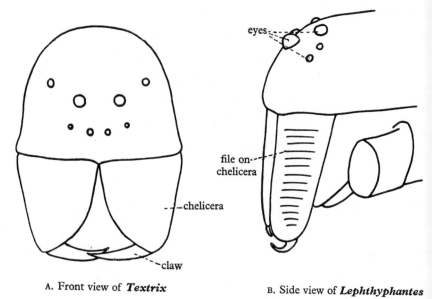

A. Front view of *Textrix*   B. Side view of *Lephthyphantes*

Fig. 368. View of araneid head showing the position of the chelicerae. (A, from Warburton; B, from Comstock.)

wound (Fig. 369A). The poison gland itself lies in the basal segment of the chelicera. In the burrowing tarantula the anterior edge of the basal segment has small teeth forming a rake against which the terminal claw presses. The anterior margin of the basal segment in most forms is covered with a dense mass of hairs forming the *scopula*. Though most chelicerae are subchelate, certain forms show a prolongation of the basal segment to form an opposable jaw and thus a chelate chelicera (Fig. 369B). In most spiders the chelicerae are held vertically but in the large bird-eating spiders and certain other primitive groups the chelicerae are held horizontally.

MOUTH. The upper lip lies between the chelicerae and the pedipalps and is called the rostrum. It is probably homologous with the labrum of the insects. The inner surface of the rostrum forms the roof of the epipharynx and leads to the oesophagus. The lower ventral part of this tube is formed by the labium or lower lip. This is not homologous with the lower lip of insects, where it is formed by the fusion of two appendages. The araneid mouth is adapted for the sucking of liquid food.

PEDIPALPS. The second pair of head appendages are the pedipalps. Each has six segments—a coxa at the base, followed by a trochanter, femur, patella, tibia, and tarsus. The pedipalp differs from a normal limb in that there is only one segment to the tarsus, the metatarsus being absent. The tarantulas have the coxa resembling that of other walking legs but in the true spiders the coxa of the pedipalp is swollen to form a crushing base or endite. This crush-

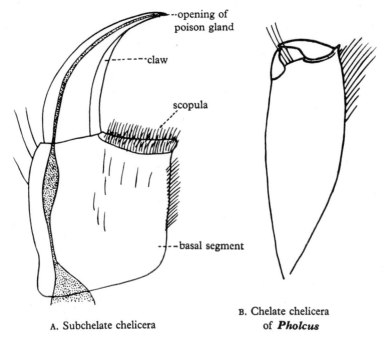

A. Subchelate chelicera

B. Chelate chelicera of *Pholcus*

Fig. 369. Chelicerae. (From Comstock.)

ing base may have a scopula and also a series of fine teeth, the *serrula*, which helps to crush the prey. The coxa and the endite form one part of the pedipalp whilst the other branch is made up of a series of joints, forming the palpus. In the females the palpus is very much like that of any walking limb except that it may have only one claw. In the males the palpus show considerable differentiation and specialization, it being used as an intromittent organ. Some tarantulas and the Linyphiidae use the palpus as part of a stridulating system.

WALKING LEGS. The thorax bears four pairs of walking legs. The dorsal part of the thorax, the tergum, is often grooved by furrows, these being lines along which the body muscles are attached. The sternum too may show certain

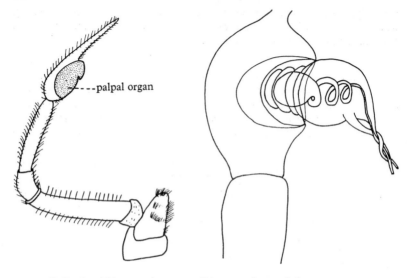

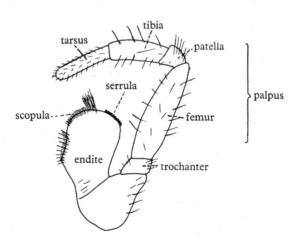

Fig. 370. Pedipalps. In the male the pedipalp is used as an intromittent organ.

grooves. The legs have seven segments: coxa, trochanta, femur, patella, tibia, tarsus and metatarsus. Those spiders having long flexible legs usually develop secondary joints. The tarsus usually has two claws. When the tarsus projects between the two claws at its end it is called an *empodium*, and the empodium

ARACHNIDA

may be claw-shaped thus forming three claws at the end of the foot. In other instances the empodium may be a soft pad used in climbing. Extra claws may develop from hairs on the tarsus but these differ from true claws in being developed not from a trichogen but from a spine-like process of the body wall formed from the hypodermis. Special hairs called *tenent hairs* (Fig. 371) are found on the tarsus; these are dilated at their extremity into a series of ridges which help the animal to obtain a purchase on smooth objects. An adhesive fluid is secreted through a small cavity into the space between the hairs. When the lower surface of the tarsus is covered with hairs it is called a scopula.

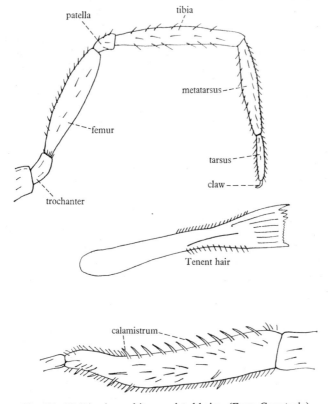

Fig. 371. Walking leg and its associated hairs. (From Comstock.)

CALAMISTRUM AND CRIBELLUM. Spiders that weave a hackled band of thread, i.e. a band containing both a warp and woof, develop a special series of curved spines on the upper margin of the tarsus of the hind legs. These spines form the calamistrum (Fig. 371). The calamistrum is used in conjunction with a special cover over the spinnerets called the cribellum (Fig. 374A). These two units take part in weaving the thread. Male cribellate spiders lose their cribellum at maturity but can still be recognized by the wide separation of the fore spinnerets. At the distal extremity of each segmen except the tarsus is a sense organ, the lyriform organ (Fig. 372A, B). It is probably used either as stretch receptor like the insect campaniform sensilla, or it may function as a hearing organ.

ABDOMEN. The abdomen is joined to the cephalothorax by a thin pedicel which is often concealed and overhung by the abdomen. In the ant-like spiders the thorax may form part of the pedicel. The dorsal wall of the pedicel has a sclerite called the *lorum* forming a centre for muscular attachment. The muscles of the abdomen are attached to the body wall and are often discernible externally by the indentations of the muscular wall of the abdomen. The abdomen often shows peculiar markings. Some primitive spiders show signs of segmentation in the adult abdomen either by the presence of distinct somites or by the presence of segmental sclerites. The abdomen possesses the external openings of the respiratory organs, the lung book and tracheae. The precise number of lung books and tracheae varies according to the group of spiders examined but the main patterns are shown in Fig. 373. The leaves of the book are seen to be thin plates with an internal space for the circulation of the blood. They are dotted with short chitinous spines and fused with the walls of the lungs. The cavity of the lung communicates by a narrow opening with the outside air and respiratory movements for the renewal of pulmonary air have been claimed by some observers. The tracheae arise as a series of parallel invaginations into the body. They do not branch as in insects but instead remain straight. They have typical tracheal structure, strengthened with spiral ridges of chitinous thickening. *Argyroneta* shows a richly developed tracheal system but in other forms, particularly spiders with slow movements, the number of tracheae is very much reduced till some have only a single pair of tracheae from each stigma. Recently it has been shown in *Lycosa amentata* both the lung books and tracheae are sites of oxygen diffusion into the body but the lung books are responsible for over 95 per cent of the respiratory exchange. The anterior part of the abdomen is usually more concave than the rest and is called the epigastrum. It is separated from the rest of the abdomen by the epigastric furrow.

REPRODUCTIVE ORGANS. The reproductive organs open into the mid line of the epigastric furrow. The male has a simple opening, the pedipalps forming the separate male intromittent organ, but the system is more complex in the female. The two internal ovaries open into a common duct which is usually, but not always, covered by a complex structure called the epigynum. The epigynum functions as a guide to the male intromittent organ and has guide grooves. The epigynum arises as a development of the anterior edge of the oviduct aperture. Usually there are paired spermathecae on either side of the oviduct and these are connected to the grooves in the epigynum (Fig. 375). The epigynum often develops into an organ of considerable complexity much used in systematic work on the araneids. It is in its simplest form as a simple plate in *Pirata* though it is usually more complex. In *Aranea angulata* there is in addition a flexible appendage which may be used in oviposition.

SPINNERETS. There are at the most three pairs of spinning organs situated on the posterior part of the abdomen. The spinnerets are referred to as the

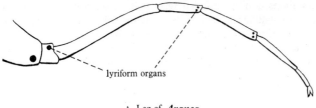

A. Leg of *Aranea*

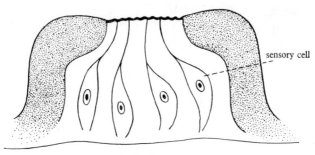

B. T.S. of lyriform organ

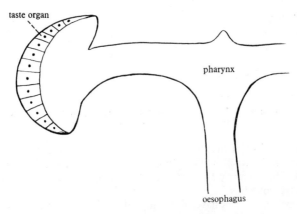

C. Taste buds in *Scytodes*

Fig. 372. Araneid sense organs. (A, after Vogel; B, after Kaston; C, transverse section of pharynx, after Millot.)

fore, median and hind spinnerets. The base of the spinnerets is firm but the tip is membranous and surrounded with hairs and barbs. These form the spinning field, the hairs in some way assisting the spinning of the silk threads. The spinneret is often surrounded by a chitinous ring to which is attached a tendon of a flexor muscle (Fig. 374B). This and embryological evidence indicates that they are most probably modified 5th and 6th abdominal legs.

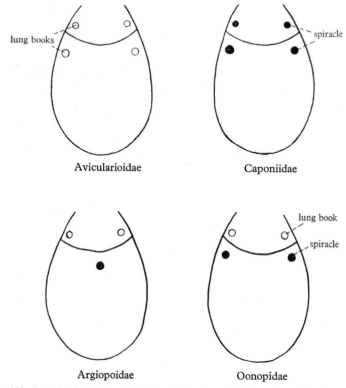

Fig. 373. Lung books and trachea (spiracles) in Araneida. Several combinations of these found in the spiders are shown here. (From Comstock.)

Over the surface of the spinning field are many small spinning tubes from which the silk is expelled. There may be as many as 100 spinning tubes per spinneret. The spinning tubes are of different shape, each associated with a different type of silk.

The cribellate spiders have, in front of the spinneret, an additional organ, the cribellum. This is a plate that runs transversely across the body and is covered with very small tubes, up to 10,000 in some species. The cribellum is used with the calamistrum to weave the bands of silk. Other spiders may have a slender organ, the colliculus, hanging in front of the spinnerets. It is absent in those families that have the cribellum and it is possibly homologous with the cribellum. Its function is not known.

FEEDING. Spiders have developed to an extreme the tendency found in other arachnids towards a carnivorous diet. Whilst most spiders on account of their size can only feed on insects, others are able to attack larger animals; birds in the case of *Mygale*. Digestion is largely external. The spider first of all injects a poison into the prey from its poison gland in the chelicerae. Proteolytic enzymes are secreted by the salivary glands in the under-lip. Some spiders inject enzymes and then wait till the inside of the prey is dissolved, whereupon they suck out the juices (Filistatidae); others crunch up the prey first of all with their mandibles and chelicerae (Argiopidae). A fly caught in

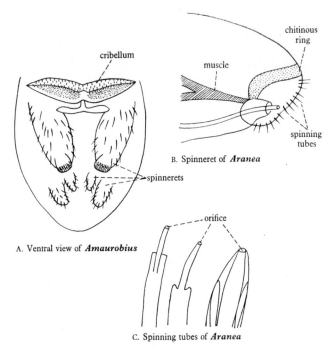

Fig. 374. Cribellum and spinnerets. (From Comstock.)

the web is pressed by the gnathobases of the pedipalps and droplets of liquid emerge from time to time. In a couple of hours the fly has been completely digested and the resulting fluid sucked into the alimentary canal by pulsations of the stomach. The chitinous skeleton of the fly remains as an empty husk. In most arthropods the area of the digestive system is increased by coiling and lengthening of the gut. In the spiders a different system is adopted, the increased gut area being achieved by the development of many diverticula (a similar system being found in the leeches and *Aphrodite*). There are three normal regions of the arthropod gut, fore gut, mid gut and hind gut. The former and the latter are lined with ectoderm, the median region is the only one lined with endoderm.

The fore gut consists of the pharynx, oesophagus, and sucking stomach. The entrance to the mouth is formed by the epipharynx. The lateral walls of

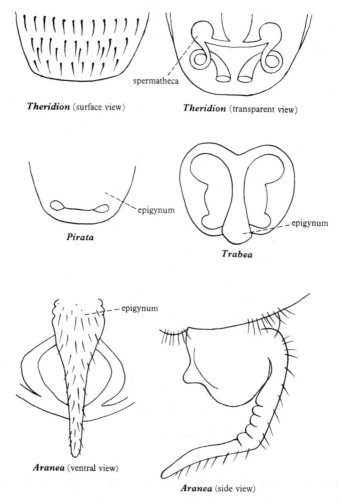

Fig. 375. Epigyna of various spiders. (From Comstock.)

mouth parts are lined with small hairs which filter off small particles and so prevent them from blocking the fine ducts in the intestine. The oesophagus has a small diverticulum that contains a chemoreceptor to taste the food (Fig. 372c). Sometimes a spider may be seen to rush up to a nasty-tasting insect, take a bite, then start to exude fluid and rush away and wipe its mouth on a leaf. The oesophagus has many fine grooves and when the walls of the oesophagus are pressed together the fluid can still pass through to the deeper regions, the coarser particles being restrained. The hind end of the oesophagus is surrounded by the circumoesophageal nervous system.

The sucking stomach is an enlargement of the posterior end of the oesophagus. The chitin in this region is much thicker and has many powerful muscles attached to it (Fig. 377). These muscles dilate the stomach, being opposed by a series of sphincter muscles that contract the stomach. The mid gut arises from the sucking stomach. It gives off two diverticula that run forwards and send a branch towards each leg whilst the main diverticulum passes back to the abdomen. In the abdomen the mid gut gives rise to several

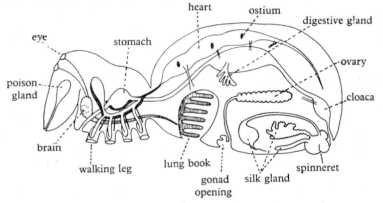

Fig. 376. Diagram showing the arrangement of the internal organs of a spider.

highly branching diverticula called the digestive gland or digestive diverticula. These tend to fill the abdominal cavity. The cells in the diverticula are of two types, secretory cells and absorptive cells. The secretory cells produce an albuminous secretion which mingles with the liquid in the gut and is then absorbed by special cells which break down the fats, proteins and amyloids into their more soluble forms. Certain of the absorbing cells take on a storage function. Other cells, as digestion proceeds and the soluble products diffuse away, become filled with indiffusible and insoluble materials. These may be transferred to the apical part of the cell and then passed into the lumen of the gut, or the whole cell may migrate into the gut and be cast away.

The hind gut is short and into it open the Malpighian tubules. In its posterior part there is a cloacal diverticulum in which faeces accumulate. The hind portion of the gut has sphincter muscles which allow the anus to be opened or closed.

EXCRETION. Excretion in the araneids proceeds by several different methods. (1) The adsorptive cells just described in digestion play a role in excretion. (2) There are certain intestinal cells situated near the hypodermis which become full of guanates. The guanine cells from time to time liberate their contents in liquid form into the lumen of the gut. The liquid passes to the hind gut cloacal pocket where it once more becomes crystalline due to the secretion of some acid by the hind gut. (3) The cephalothorax may contain nephrocytes, cells that take up carmine particles from solution and which would appear to play some role in excretion. (4) The main excretory organs are the Malpighian tubules. These are two sets of tubules arising from the cloacal pocket. The tubes are pseudociliated and a current caused by waves of muscular contraction that start at the tip of the tube and pass towards the cloaca, forces the excretory material in the tubes down towards the cloacal pocket. The tubules branch dichotomously into many branches. Each tube is syncytial with an internal lumen of about 100μ. The tubules secrete a slightly acid solution which contains guanates. Though precise experimental evidence on the excretory ability of these cells is not yet available some excretory function is indicated by the fact that they concentrate coloured dyes experimentally injected into the body.

NERVOUS SYSTEM. In the spiders the nervous system shows considerable condensation. In the cephalothorax, embryological studies have shown that the abdominal ganglia are included in the circumoesophageal mass. There are five pairs of ganglia associated with the chelicerae whilst in the suboesophageal ganglia there are many commissures between the two halves of the nervous system. In the liphistiomorphs there are about seventeen such ganglia, five being truly cephalothoracic, the other twelve coming from the abdomen (Fig. 378). The more anterior part of the brain receives tracts from the eight eyes and is thus dominated by the visual tracts. The more posterior and ventral parts of the central nervous system receive indirect visual fibres and also many correlating fibres. The association bodies are of two types, the median central body and the pedunculate bodies. The median body is believed to play a part in the co-ordination of visual patterns and also perhaps in controlling instinctive movements. In sessile spiders the development of this part is possibly related to the development of the spinning web. The corpora pedunculata are pairs of nodules joined together, the whole forming an H-pattern. These too are believed to play some part in visual co-ordination. The sedentary spiders have small corpora pedunculata. Little is known about the functions of different parts of the spider brain (Fig. 379).

Little or nothing is known about the peripheral innervation of the muscles in arachnids. It is most likely the same as that existing in the Crustacea and Insecta, i.e. a multiple innervation of the muscle fibres by a few motor nerves, each nerve fibre branching to every muscle fibre in a given muscle.

The araneids have well-developed eyes though in all cases they are simple ocelli and never compound eyes. The lens present in the eyes is not capable of accommodation. There are two types of ocelli. (1) The apical end of the nerve cell in the retina points towards the light. These are the direct eyes. In some of the jumping spiders (Salticidae) the eyes may have a complex series of extrinsic muscles that move the eyes and point them in the direction of jumping. (2) The nerves are bent over so that the apical end no longer points in the direction of the light. These are called the indirect or lateral eyes.

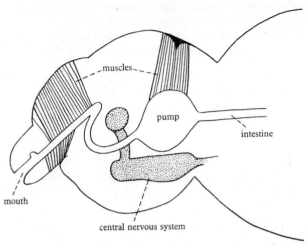

Fig. 377. Diagram to show arrangement of muscles in the sucking stomach of a spider. (After Millot.)

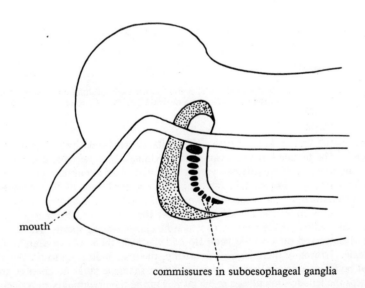

Fig. 378. Diagram to show the remains of segmentation (number of commissures) in the suboesophageal ganglion. The number varies according to the species, in the liphistiomorphs there are seventeen. (From Bristowe and Millot.)

The number of sensory cells, rhabdomes, in each ocellus varies according to the species of the animal and the position of the eye concerned. In the lycosids there may be 100 rhabdomes in the anterior eyes, 250 in the anterior median, 4,000 in the posterior laterals, and 4,500 in the posterior medians. The eyes have very short focal length but in the jumping spiders the eyes have differing focal lengths.

Other sense organs are present in the araneids. The gustatory and lyriform organs have already been mentioned. The males possess a chemoreceptor on their feet and they can detect whether or not a female has passed over a given spot. This sense is no longer present if the distal part of the foot is removed.

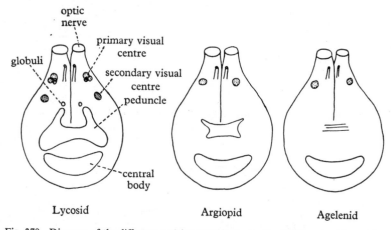

Fig. 379. Diagram of the different nuclei present in the brain of different spiders. Note the loss of the peduncle in the agelenids. (After Hanstrom.)

REPRODUCTIVE BEHAVIOUR. The genital system is relatively simple in the spiders. The female has two ovaries, two oviducts, one uterus and a vagina. The number of spermathecae varies. The male has a number of tubular testicles arranged in parallel rows lying on the ventral floor of the abdomen. The external copulatory organ in the male is the pedipalp. Many spiders show sexual dimorphism, the female being larger than the male. Prior to mating the male builds a little web on to which he ejaculates the sperm. The sperm are then sucked off into the pedipalp and the male then goes in search of a female. Normally there is some courtship, the male indicating to the female that he is not a normal article of food. Often the male takes no chances and grasps the female's mandibles in his own. During the preliminary movements the female becomes quiescent and copulation follows. In a minority of cases the male may be eaten by the female.

Some spiders wrap the fertilized eggs in a small cocoon. In *Stegodyphus lineatus* the cocoon is carried on the hind legs and turned over and over so that the sun equally warms the different sides of the cocoon. Other females brood over the cocoon and will not leave it or feed till the young have hatched out. In the lycosids when the young emerge they climb on to the back of the mother and are carried about for a few days.

GLANDS. Spiders are well supplied with glands. The main ones are the digestive glands, the poison glands and the silk glands. The poison glands are paired and lie in the anterior of the cephalothorax. They are formed embryologically as an invagination of the chelicerae. They are small in the mygalids but in the majority of spiders they are well developed. In *Sicarius* they are completely folded, rather like a cauliflower. The glandular tissue is surrounded by connective tissue and a muscular coat. There is very little poison in the gland and the store soon becomes depleted. The type of poison varies according to the species. In the mygalids the poison attacks the nervous system whilst in *Lycosa* it is necrotic and sets up gangrenous areas. Large spiders can kill birds and mice though it takes longer for the poison to act than it does when an insect is bitten. Very few spiders are dangerous to man and the few cases where death has occurred have usually been due to previous low health on the part of the victim.

The spinning glands are shown in Figs. 374, 376, in the ventral part of the abdomen. In the web spiders such as *Epeira* there are five types of gland of diverse structure and function, all opening by minute pores on the spinnerets. Thus the ampulliform glands supply the radial lines of the web. The spiral lines are made by the aggregate glands which also furnish the viscid fluid which covers them. The egg cocoon is formed by the tubuliform glands which are absent from the males. The aciniform glands manufacture the cords which are wrapped around the prey caught in the web. The pyriform glands make the attachment disc by which the silk threads are attached to the ground. Such spiders are well adapted to a sedentary web life. They have legs of great length compared with the body and on the ground they move slowly and uncertainly. The legs end in claws by means of which they can cling to the elastic threads of the web and which they also use to weave the threads of the web together.

The web-spinners are one of the most specialized branches of the araneids. There are other forms such as the wolf spiders (Lycosidae) and the jumping spiders (Salticidae) which are just as specialized as the Epeiridae. They run swiftly after their prey and may suddenly leap on to it. They usually possess two ampulliform glands which secrete a drag line which they leave behind them as they move. The web-spinner relies almost entirely on its sense of touch, and the vibration of the lines of the web affecting the tactile hairs on its legs acts as a guide to the entangled prey. But the hunting spiders find their victims by sight and have a remarkable range of vision compared with other spiders. This is used not only in the pursuit of food but also in the elaborate courtships that are characteristic of these two families, the males executing complex dances as they approach the female.

WEB-BUILDING. Though the stages leading to the development of the complex orb web are not known, there have been several explanations of the way in which it might have evolved. One of these, suggested by Bristowe, is that the web was originally a series of silk lines spun to protect the egg case.

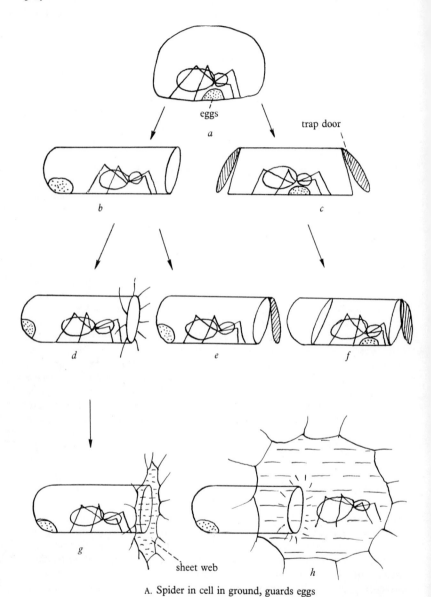

A. Spider in cell in ground, guards eggs

Fig. 380. A, diagram to show possible mode of evolution of web from a subterranean nest. (From Bristowe.)

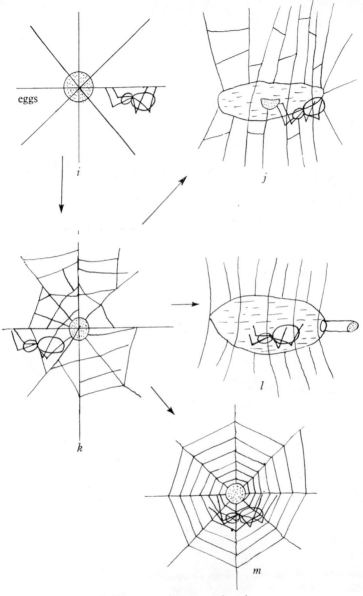

B. Spider suspends eggs on threads

Fig. 380. B, diagram to show possible mode of evolution of web from a series of strands protecting the eggs. (From Bristowe.)

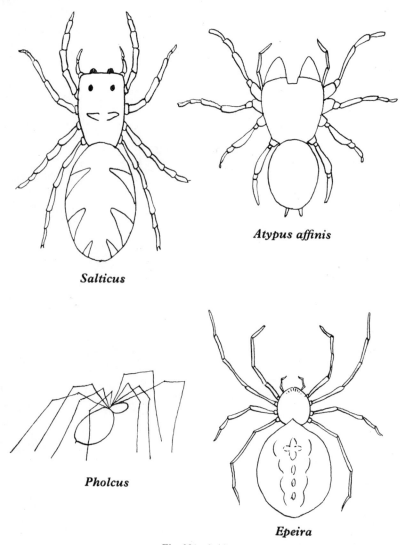

Atypus affinis

Salticus

Pholcus

Epeira

Fig. 381. Spiders.

There were two distinct patterns of evolution, one from the tube-building spiders, the other from the sheet-web spiders (Fig. 380 A, B). *Atypus affinis* spins a tube 7–10 in. long, three-quarters of which is submerged below the ground. The egg case is laid at the blind end of the tube. Any insect walking over the open end of the tube is seized by the spider and dragged inside. *Liphistius* has a few threads radiating from the mouth of the tube, this increasing the dangerous area for any passing insect. *Tegenaria* has gone farther in the development of these threads by having a flat sheet extending from the tube. The spider runs out on top of this web to attack any intruding insect. This series together with some more intermediate forms is shown in Fig. 380 A. It should be emphasized that these are merely illustrative stages in the development of the web; the theory in no way suggests that *Atypus* gave rise to *Liphistius*.

The second line of evolution came from spiders that hang their egg cases from shrubs and trees. Increase in the number of lines that suspend the egg sac has given rise to a simple snare which later gave rise to a simple sheet as seen in *Pholcus*. The snare could be made more complex as in *Linyphia*. This sheet web differs from the tube type of sheet web in that the egg sac is placed in the middle of the web (*Theridion*) and the spider always runs on the underside of the web. It is but a stage from this type of sheet web to the orb web. Some of the Oolboridae and the Argyopidae place their egg sacs in the centre of the orb webs, whilst it is quite likely that originally the orb webs were horizontal instead of vertical (Tetragnatha). It would appear that the orb type of web has at least two independent developments.

EXAMPLES OF ARANEIDA

Aranea, the familiar orb-webbed spider, found in houses and gardens; abdomen large and often patterned. Legs short and stumpy.

Pholcus, a house spider found in the south of England. Long thin legs and small body; web very loosely woven, the spider hanging below the web (Fig. 381).

Salticus, the jumping spider; cephalothorax longer than broad; anterior eyes well developed. Often found running on tarred fences (Fig. 381).

Atypus, the only British example of the Avicularoidea which have their chelicerae in a vertical plane instead of a horizontal plane. Small with stumpy cephalothorax bigger than the abdomen. Well-developed chelicerae.

Eurypelma. 'Tarantula' spider, up to 2 in. in length. Short stumpy hairy legs, large body; a bird-eating spider.

6. Order **Palpigrada**

Very small arachnids with a long, well-segmented body and an elongated tail. Found living under stones. *Koenenia* (Fig. 382).

7. Order **Solifuga**

Arachnids with body divided into prosoma and opisthosoma, no pedicel. The body and limbs are very hairy. There is no tail. The pedipalps have a small sucker. There is a well-developed tracheal system. The animals are tropical and subtropical. *Galeodes* (Fig. 383) and *Rhagodes*.

8. Order **Acarina (Mites and Ticks)**

DIAGNOSIS. Arachnids with a false head or capitulum set apart from the rest of the body and which carries the mouth parts. External segmentation is very much reduced or absent. Larval stages normally have three pairs of legs; nymphal and adult stages have four pairs of legs.

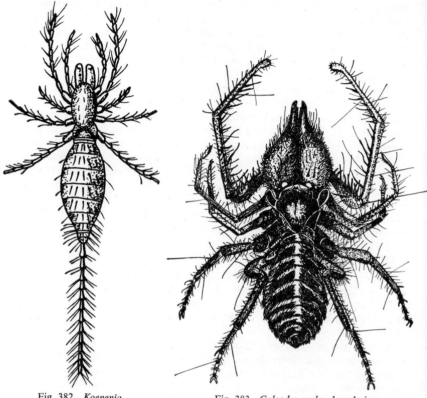

Fig. 382. *Koenenia*. (From Grassé.)

Fig. 383. *Galeodes arabs*, dorsal view. (From Grassé.)

CLASSIFICATION

Suborder 1. **Onchopalpida**. Acarina with typical ambulatory claws on the pedipalps and more than one pair of body stigmata. *Holothyrus*.

Suborder 2. **Mesostigmata**. Acarina with a single pair of stigmata lateral to the legs and associated with an elongated peritreme (chitinous tube associated with the tracheae and stigmata) or, if absent, degenerate parasites of the respiratory tracts of vertebrates. Haller's organ (a pit and seta on the first tarsus used in olfaction) is absent. Hypostome not developed for piercing. *Macrocheles*.

Suborder 3. **Ixodides**. Acarina with a pair of stigmata posterior or lateral to the coxae, associated with a stigmatal plate rather than an elongated peritreme. Haller's organ present. Hypostome modified as a piercing organ and provided with recurved teeth. *Ixodes*.

Suborder 4. **Trombidiformes**. Acarina with a pair of stigmata on or near the gnathostome, or absent. Palps usually free and highly developed.

Chelicerae modified for piercing. Anal sucker never present. *Oxypleurites, Hydrachna.*

Suborder 5. **Sarcoptiformes.** Acarina without stigmata or with a system of tracheae opening through stigmata and porous areas on various parts of the body. Coxae forming apodemes beneath the skin on the ventral side of the body. Mouth parts for chewing, strong chelae. A few parasitic forms with specialized chelicerae. Palpi simple. Anal suckers often present. *Tyroglyphus, Sarcoptes.*

GENERAL ACCOUNT. These forms are usually minute except in the case of the parasitic ticks. They are, variously, scavengers, ectoparasites on all sorts of plants and 'hangers-on' of all sorts of animals, but in the last case they become, by the modification of the chelicerae and pedipalps, blood-sucking parasites.

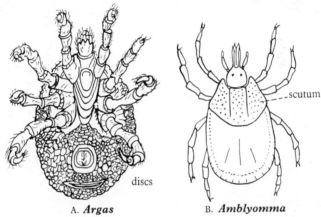

A. *Argas*   B. *Amblyomma*

Fig. 384. A, *Argas* (Argasidae) and B, *Amblyomma* female (Ixodidae). The Argasidae are distinguished by a leathery skin diversified by discs which mark the insertions of the muscles; the Ixodidae are distinguished by the hard scutum which covers the whole body of the male and the anterior part of the female. (From Nuttal and Warburton.)

In the most free-living of them, like the aquatic and predatory Hydrachnidae, the chelicerae are clawed piercing weapons and the pedipalps leg-like with sensory hairs. The chelate condition of the chelicerae may be seen in the cheese mite, *Tyroglyphus* (Fig. 386A), which is a typical saprophyte living on cheese only when it has begun to decay. The pedipalps are here no longer leg-like.

In a tick like *Argas* (Figs. 384A, 386C, 387) the pedipalps are sensory, but the chelicerae and the median *hypostome* are elongated and converted into serrated cutting tools; a sucking channel is formed between these. The mouth is usually minute and leads into a sucking pharynx and then into an endodermal stomach which gives rise to caeca in the ticks, where there are also salivary glands of large size opening into the pharynx. The saliva is said to contain an anticoagulin, as in leeches, and this renders easier the gradual digestion of the blood which is taken into the stomach. A remarkable phenomenon without parallel in the Arthropoda is the occurrence of intracellular digestion in some acarines. The cells of the stomach put out pseudopodia and the blood plasma is taken into vacuoles where it is digested.

The circulation is extremely degenerate. No heart has been observed with certainty and the blood system is lacunar in mites, but in the tick, *Argas*, there is a single-chambered pulsating vessel with a pair of ostia and an aorta running forward to a periganglionic sinus. The respiratory organs are tracheae, long and convoluted. These open by stigmata, the position of which varies in the main divisions of the group.

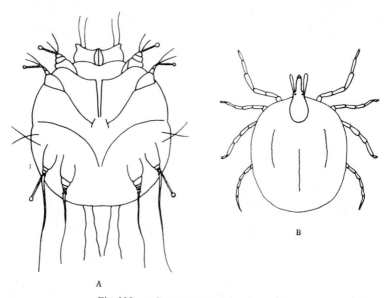

Fig. 385. A, *Sarcoptes* and B, *Ixodes ricinus*.

The life history of the parasitic forms is of great interest, especially in the case of the ticks. These are divided into the Ixodidae and the Argasidae. The ixodid *Boophilus annulatus* lives throughout its life attached to the cow, only interrupted by the necessity of moulting and reproduction. Though compelled to withdraw its mouth parts when its skin is cast, it plunges them back into the same place as soon as possible after the moult has occurred. Most Ixodidae, however, have three different hosts, one for each of the stages, larval, nymphal and adult. After each moult the argasid falls off and has to find a different host. The argasids are quicker feeders than the Ixodidae and whereas the Ixodidae have to distend a preformed cuticle in order to get a comparatively small blood meal, the argasids can engorge quite fully in fifteen to thirty minutes, the female sometimes increasing to five times its prefeeding weight. Young argasids can go without food for several months whilst the adult can go for over a year. It is necessary for the animal to have a meal before it can moult. Thus the six-legged larva feeds and then moults into an eight-legged nymph which feeds before it can become a sexually mature adult. In some cases the male does not feed after reaching sexual maturity but, after copulating with the female, dies. In the Ixodidae the female dies after laying the eggs but some Argasidae can take six or more meals in the adult stage, laying a new batch of eggs after each meal.

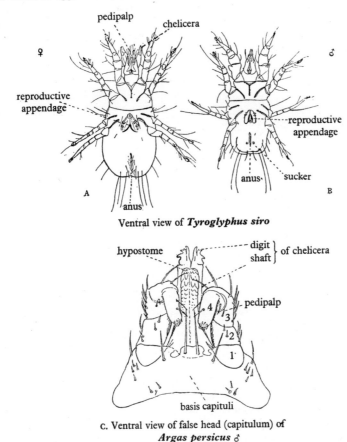

Fig. 386. A, B, *Tyroglyphus siro*, ♀ and ♂. 3, 4, 5, 6 = walking legs. (From Leuckart and Nitsche.) C, capitulum of *Argas persicus*, ♂. (From Nuttall.)

Many kinds of ticks carry disease, e.g. Texas fever of cattle is caused by *Babesia* (*Piroplasma*) transmitted by *Boophilus annulatus*; relapsing fever in man is caused by a spirochaete transmitted by *Ornithodorus moubata*.

Ixodes ricinus (Fig. 385B), the sheep tick, transmits diseases from one sheep to another. The female takes a meal of blood and drops to the ground. After some weeks she lays eggs at the roots of adjacent grasses. The eggs hatch out into larvae possessing three pairs of legs. The larvae climb to the top of the grass and stretch out their legs. They cling on to any passing host and find a suitable spot to sink their jaws into the flesh and imbibe blood. They feed for three to four days and then drop off the host. They hide in a dark crevice in the ground, moult, and develop into an eight-legged nymph. They climb up the adjacent vegetation and attach themselves to a new host for about five days during which they feed. They then drop off and moult into the adult form. They then seek another host. If the tick is a male it does not feed but seeks a female; if it is a female it feeds first before copulating. The larvae can survive starvation for fifteen months, the nymphs for thirteen months, and the adults for twenty-one months.

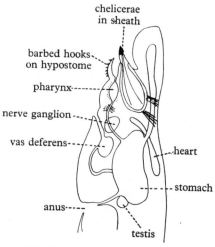

Fig. 387. *Argas persicus*. Median longitudinal section showing the proboscis, alimentary canal, and reproductive system. The chelicerae are within the cheliceral sheath, they are thrust forward by the contraction of the dorsoventral body muscles and cut their way into the host. The barbed hooks on the hypostome are then thrust into the wound and keep the tick in place. (Altered from Nuttall.)

Tyroglyphus farinae (Fig. 386A, B) lives in flour. Frequent contact with infected flour can lead to dermatitis in flour-mill workers. The adult lays twenty to thirty eggs which hatch into larvae with three pairs of legs. These feed for a few days and then moult to an eight-legged nymph. These can moult to a second nymphal stage and if conditions are unfavourable the second nymph can moult into a *hypopus* stage. There are two types of hypopus. One is a small motile animal with reduced legs but with many suckers on the ventral surface of the body. These help it to become attached to rodents or insects infecting the flour. The second type of hypopus is a small, light, non-motile animal with very reduced appendages. It is distributed by wind. These hypopus stages give rise to the adult.

Sarcoptes scabeii (Fig. 385A) is a small mite that burrows under the skin of mammals and causes severe irritation leading to mange or scabies. This may be followed by secondary infections leading to impetigo or eczema. The egg hatches out on the host and gives rise to a small six-legged larva. This burrows into the host's skin or descends down a hair follicle, moults and gives rise to the first nymphal stage with eight legs. The nymphal stage feeds and moults and gives rise, if the animal is a male, to the adult; in the case of the female, it has to make another moult before it can become an adult female. Thus the female life cycle has one more moult than the male and takes longer—seventeen days instead of nine to eleven days. The female is about twice the size of the male. It is considered that about two per cent of the British population is infected with these mites. They infect the wrist and hands, the mites taking twenty-five minutes to an hour to burrow into the skin.

Hydracarina. Small oval-bodied mites, 2–8 mm. in length, usually brightly coloured, aquatic. Between the palps they have a capitulum enclosing sucking mouth parts. The legs have small claws. Carnivorous. *Hydrachna* (Fig. 388).

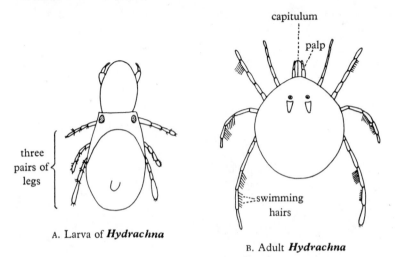

Fig. 388. A, six-legged larva, and B, the eight-legged adult of *Hydrachna*.

9. Order Phalangida

Arachnids with prosoma covered by a single tergal shield and united to the opisthosoma by its whole breadth; opisthosoma always segmented; chelicerae three-jointed and chelate; pedipalps leg-like; two simple eyes.

These creatures, with their enormous elongated legs (up to 160 mm. in the South American Gonyleptidae which have a body length of only 6 mm.), are familiar objects in the summer; the active predacious forms are supposed to live for a single season only, but some representatives are slow-moving and live longer. They feed on insects and other arthropods and suck their juices. The walking legs have the same number of joints as spiders, but the tarsus is multiarticulate. The opisthosoma contains at least ten segments. The animal breathes by tracheae and there are two stigmata on the first sternum of the opisthosoma, opening on each side of the reproductive aperture from which emerges a long protrusible process, which is an *ovipositor* in the female, a *penis* in the male. Small accessory stigmata are found in the tibia of the feet of the adults. These possibly help the respiration of the muscles in the extremity of the long legs.

Oligolophus (Fig. 389), *Phalangium* (harvest-men).

10. *Order* **Pantopoda (Pycnogonida)**

Arachnida, in which the opisthosoma has disappeared, with the exception of the pregenital segment which bears legs on which the genital pore opens.

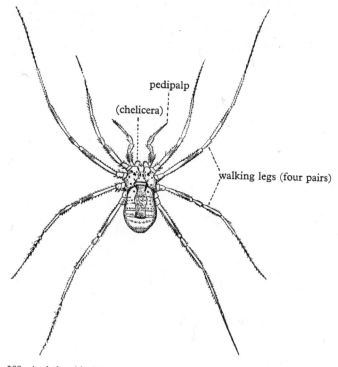

Fig. 389. A phalangid, *Oligolophus spinosus*, ♂, ×2. (From Shipley and MacBride.)

These extraordinary animals, e.g. *Nymphon* (Fig. 390A), are all marine and semisedentary, crawling slowly over seaweed and sedentary animals. They consist of the following regions: (1) the *proboscis*, a prolongation of the prosoma with the mouth at the tip; (2) four segments fused together bearing four eyes, the chelicerae, the pedipalps, the *ovigerous legs* which are present in both sexes and the first pair of walking legs; (3) three free segments bearing the remaining pairs of walking legs. The body is usually very small while the legs are enormously elongated. They have eight joints. The proboscis contains a sucking pharynx preceded by a filter of chitinous hairs which prevents any but fluid food from proceeding farther. The small stomach gives off digestive caeca which extend into the legs and other appendages. The common British form, *Pycnogonum littorale*, is found firmly attached by the terminal claws of the legs to the sides of sea anemones into which it inserts the proboscis and sucks the juices. There is a dorsal heart with three pairs of ostia; respiration is cutaneous. The nervous system consists of supraoesophageal ganglia and a ventral chain with suboesophageal and three or four other ganglia.

The sexes are separate and the males carry the eggs on the ovigerous legs. The gonads, like the alimentary canal, are branched and open on the 4th segment of the legs (the last pair of legs in *Pycnogonum* or all four pairs in

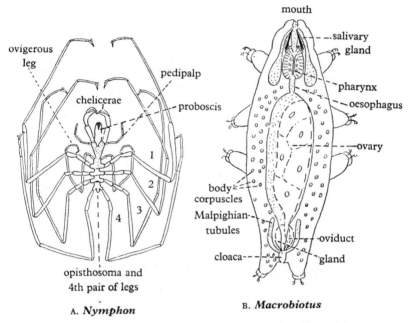

Fig. 390. A, *Nymphon*, an example of the Pantopoda; 1—4 = walking legs; note the ovigerous legs. (After Mobius.) B, *Macrobiotus* female, dorsal view. (Modified from Greef.)

Phoxichilidium femoratum). In the latter species the larvae are hatched as six-legged creatures, which form cysts in the polyps of the gymnoblast hydroid, *Coryne*.

The two small classes following have been associated with the arachnids but no sufficient reason can be advanced for this. They both exhibit simplicity of structure; in the case of the Pentastomida this is due to parasitism, but in the Tardigrada some of the traits of primitive arthropods may be preserved. In some ways the Tardigrada resemble *Peripatus* and their development is said to be of a very primitive type. But the size and specialized habitat incline the author to regard this as a case of 'simplification' such as is met with in the Archiannelida (p. 311).

11. *Order* **Tardigrada**

Minute arthropods with four pairs of stumpy legs ending in claws, with oral stylets and a suctorial pharynx, without definite circulatory or respiratory systems.

Representatives of this group, e.g. *Macrobiotus* (Fig. 390B), are found, for instance, in moss and in the sediment of rain gutters. They are minute and often very transparent animals, with a thin and flexible cuticle. The body is usually short and flattened; the tardigrades have been compared to the tor-

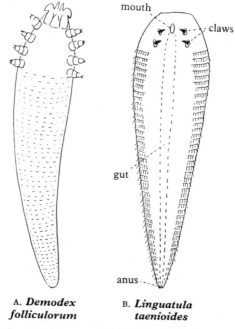

A. *Demodex folliculorum*   B. *Linguatula taenioides*

Fig. 391. A, *Demodex folliculorum*, a mite living in the hair follicles of man and domestic animals; ventral view. (After Blanchard.) B, *Linguatula taenioides*; ventral view of the stage at which it is eaten by the second host. (After Leuckart.)

toises among the vertebrates, from their slow and awkward gait. The mouth opens into a tube in which work the two chitinous stylets; a suctorial pharynx, the wall of which is composed of radiating muscular fibres, follows. Into the pharynx opens a pair of salivary glands. The animals pierce the wall of plant cells with the stylets and suck the sap by the action of the pharynx. Then comes a narrow oesophagus leading into a capacious stomach, and lastly the rectum, which is joined by two short tubes which probably represent Malpighian tubules, and by the duct of the gonad.

The perivisceral cavity contains no connective tissue cells but is crowded with numerous rounded corpuscles and traversed by bands of longitudinal muscle. The nature of the cavity is not known but the existing account of the embryology describes pairs of coelomic pouches arising as outgrowths of the archenteron, as in the echinoderms.

The legs resemble the appendages of *Peripatus* and each is terminated by two forked claws. The last pair are terminal and the anus opens between them. The nervous system consists of suprapharyngeal, subpharyngeal and four pairs of trunk ganglia, the latter corresponding to the appendages.

Physiologically the Tardigrada are interesting in their capacity for resisting desiccation. Like the rotifers and nematodes with which they are associated in habitat they shrivel up with loss of water, absorbing it again and returning to life at the next rain.

12. Order **Pentastomida**

Elongated vermiform parasites with a secondary annulation and two pairs of claws at the sides of the mouth; without respiratory or circulatory systems.

The commonest example, *Linguatula taenioides*, lives in the nasal passages of carnivorous mammals; the larvae, in which the claws of the adult are borne on prominences which may be called limbs, live in other mammals, chiefly herbivores. The eggs are passed out of the host, the larvae climb on to plants and are eaten by hares or rabbits; they traverse the wall of the gut and encyst in other tissues, often the liver. After a period of growth they wander once more through the body; they may at this stage be eaten by the second host and after wandering through the body reach the nasal passages. The larvae resemble certain parasitic mites (Fig. 391) and for that reason the group has been classed with the arachnids.

CHAPTER XV

THE PHYLUM MOLLUSCA

DIAGNOSIS. Unsegmented coelomate animals with a *head* (usually wel developed), a ventral muscular *foot* and a dorsal *visceral hump*; with soft skin, that part covering the visceral hump (the *mantle*) often secreting a shell which is largely calcareous, and produced into a free flap or flaps to enclose partially a *mantle cavity* into which open the anus and the mesoblastic kidneys (usually a single pair); a pair of *ctenidia* (organs composed of an axis with a row of leaf-like branches on each side, contained in the mantle cavity, originally used for breathing); having an alimentary canal usually with a buccal mass, radula and salivary glands, and always a stomach into which opens a *digestive gland*

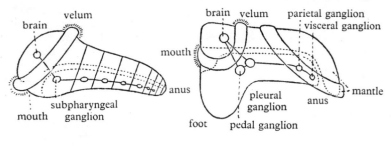

Annelid   Mollusc

Fig. 392. Comparison between annelid and molluscan organization. A, post-trochosphere larva with segmenting trunk. B, veliger larva of *Paludina* before torsion. (After Naef.)

or *hepatopancreas*; with a blood system consisting of a *heart*, a median ventricle and two lateral auricles, arterial system and venous system often expanding into a more or less extensive haemocoele, with haemocyanin as respiratory pigment; a nervous system consisting of a circumoesophageal ring, often concentrated into cerebral and pleural ganglia, pedal cords or ganglia and visceral loops; coelom, varying in development, but always represented by the *pericardium*, the cavity of the kidneys (which communicates with the pericardium), and the cavity of the gonads; often with larvae of the *trochosphere* type.

CLASSIFICATION

There are five classes in the Mollusca.

Class 1. **AMPHINEURA**. Molluscs with an elongated bilaterally symmetrical body, without tentacles or eyes; nervous system without ganglia.

   Order 1. **Polyplacophora**. Well-developed flat foot, shell made up from many units.   *Chiton*

Order 2. **Monoplacophora.** Well-developed flat foot, shell made from one unit, have an internal metamerism. *Neopilina*

Order 3. **Aplacophora.** Foot reduced. *Neomenia*

Class 2. GASTEROPODA. Molluscs possessing a head, tentacles, and at some stage of their development show torsion.

Order 1. **Prosobranchiata.** Gasteropods in which the adult shows torsion; the visceral loop is in a figure of eight, the gills are anterior to the heart. *Haliotis, Patella, Buccinum, Pterotrachea*

Order 2. **Opisthobranchiata.** Gasteropods in which the adults show detorsion by a process of untwisting. *Aplysia, Doris*

Order 3. **Pulmonata.** Gasteropods in which the adult's nervous system becomes symmetrical following torsion by a process of shortening of the abdominal commissures. *Limnaea, Helix*

Class 3. SCAPHOPODA. Bilaterally symmetrical molluscs living in a tubular shell open at both ends, reduced foot, tentaculate, no gills. *Dentalium, Cadulus*

Class 4. LAMELLIBRANCHIATA. Molluscs with a bilaterally symmetrical body laterally compressed and enclosed by a shell that develops as two valves. The head is rudimentary, tentacles are absent.

Order 1. **Protobranchiata.** Lamellibranchs with flat, non-reflected gill filaments. *Nucula, Yoldia*

Order 2. **Filibranchiata.** Lamellibranchs with reflected gill filaments, the filaments being joined by ciliary junctions. *Mytilus, Anomia*

Order 3. **Eulamellibranchiata.** Lamellibranchs with reflected filaments, the filaments being connected by vascular tissue. *Anodonta, Cardium*

Order 4. **Septibranchiata.** Lamellibranchs with gills no longer respiratory but, instead, forming a muscular septum. *Poromya, Cuspidaria*

Class 5. CEPHALOPODA. Bilaterally symmetrical molluscs; the head is well developed and surrounded by a crown of tentacles representing the foot; develops a siphon, typically have a chambered shell; nervous system greatly centralized, eyes usually well developed.

Order 1. **Dibranchiata.** Cephalopods with a single pair of gills and kidneys; shell internal. *Sepia, Loligo, Octopus*

Order 2. **Tetrabranchiata.** Cephalopods with two pairs of gills and kidneys; shell external. *Nautilus, Baculites*

DETAILED ACCOUNT

BASIC MOLLUSCAN PATTERN. While we do not know exactly what the ancestral molluscs looked like, we can make a very shrewd guess at their structure. They possessed the molluscan characters given in the definition above and they resembled the diagrammatic creature shown in side view in Fig. 393A. They had a head with tentacles, a flat creeping foot, a conical visceral hump covered by a mantle which possibly contained numerous

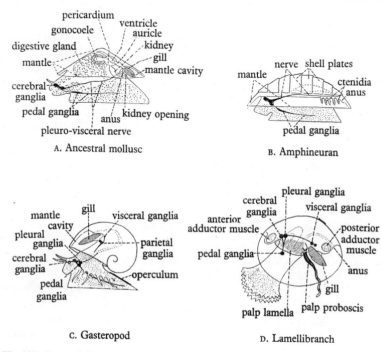

Fig. 393. Types of Mollusca; side view. The headfoot is stippled to contrast with the visceral hump and mantle. In A the mantle cavity has its original posterior position. In C it has become anterior, whilst in B and D it has extended forwards on both sides. (Partly after Naef.)

calcareous spicules and not a complete shell, and a posterior mantle cavity into which opened the median terminal anus and the common apertures of the kidneys and the gonads, and which also contained the ctenidia. In the alimentary canal the fore gut formed a muscular body, the buccal mass, and a radula (p. 608) and the mid gut an oesophagus, stomach and digestive glands and intestine. The heart had a median ventricle and a pair of auricles. The perivisceral coelom reduced by the development of an extensive haemocoele (p. 607) is represented by the pericardium with which communicates in front the cavity of the gonads and at the sides the two coelomoducts ('kidneys'). In the nervous system there were, as in annelids and arthropods, a circumoesophageal commissure or brain which may or may not have been ganglionated, ventral pedal cords, a visceral commissure coming from the pleural part of the brain, and a pallial commissure in the mantle edge. From this beginning diverged the different groups which we know today.

MAIN VARIANTS ON PATTERN. The chitons (Amphineura), which have departed least from the ancestral structure, became elongated but limpet-like forms (Fig. 393 B), their visceral hump being protected by eight shell plates, their mantle cavity extended all round the foot, while instead of a single pair

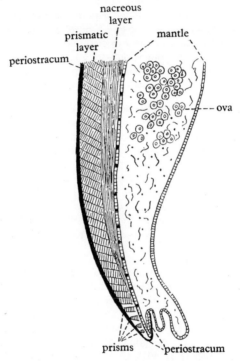

Fig. 394. Vertical section through the edge of the mantle of *Mytilus*. (After Field.)

of ctenidia many such pairs arose. The Gasteropoda remained as short creeping forms (Fig. 393 C); they are characterized by the growth of the visceral hump dorsally, but unequally so that it has coiled in a spiral (which is covered by a single shell). This caused a readjustment of the visceral hump which has revolved (usually to the right) on the rest of the body through 180° (torsion) and the mantle cavity is now anterior. The Lamellibranchiata (Fig. 393 D) are flattened from side to side, the whole body being covered by two mantle lobes secreting two shell valves united by a median hinge. The ctenidia inside the greatly enlarged mantle cavity have developed into huge organs of automatic food collection and so the head, rendered unnecessary and withdrawn into the mantle cavity, has become vestigial. Similarly the foot has lost its flat sole and has to be extended out between the valves to move the animal.

In the Cephalopoda, though there is an unequal growth of the visceral hump relative to the rest of the body, as in gasteropods, it is coiled in a plane spiral, but there is no torsion, the mantle cavity remaining posterior. The primitive forms in the group (Fig. 433 A) have an external shell which is divided into chambers, and those behind the body chamber contain gas. This has had a great effect on the development of the group, for by diminishing the specific gravity of the animals it has enabled them to become more or less free-swimming. They have tended, with the loss of the shell, to become more and more efficient swimmers, and this is associated with the development of their predatory habits. The anterior regions shows a kind of transformation new to the molluscs in its partial modification into circum-oral prehensile tentacles for seizing food. Lastly, and in connexion with all these changes, the brain and sense organs have become enormously developed and the cephalopods are seen to be one of the most progressive groups of invertebrates.

THE SHELL. Characteristically the ectodermal epithelium of the mantle secretes a *shell* in the Mollusca and in most of them the method of secretion is the same. The original shell is laid down by the mantle of the veliger larva (Fig. 399 B), but all extension takes place by secretion at its edge (Fig. 394). The outer shell layer, *periostracum*, formed of horny conchiolin, is first produced in a groove and then the *prismatic layer*, largely consisting of calcite or arragonite, is secreted underneath it by the cells of the thickened edge. The innermost *nacreous layer* (also mostly calcium carbonate) is, however, formed by the cells of the whole of the mantle, and under such conditions as occur in the formation of pearls this general epithelium is capable of secreting any of the three shell layers.

THE COELOM. The coelom is primitively represented as three pairs of cavities, the renocoele, the gonocoele and the pericardium, though they do *not* develop in a manner suggestive of mesodermal segmentation. Originally these three cavities intercommunicated and the reproductive cells discharged through the renocoele. In the more advanced molluscs there is a progressive separation of the renal and the gonadial products till they each discharge through their own duct (Fig. 402). A similar separation of the renal and gonadial systems is seen in the development of the vertebrate urino-genital system.

GILLS. The molluscs are mainly an aquatic group respiring by means of gills. These gills are usually supplied with cilia which bring a current of water flowing over the respiratory surface. Oxygen will diffuse most efficiently into a gill if the direction of water flowing over the gill is opposite to the direction of blood flowing inside the gill. This is referred to as 'the principle of counterflow' and it ensures efficient oxygenation of the blood. This has been of some importance in the evolution of the gill structure and their position in the mantle cavity. The arterial blood vessel always lies upstream to the venous vessel. In addition there is a special development of supporting tissue to hold the gill in position. The position of the gills in different molluscs is shown in Fig. 401.

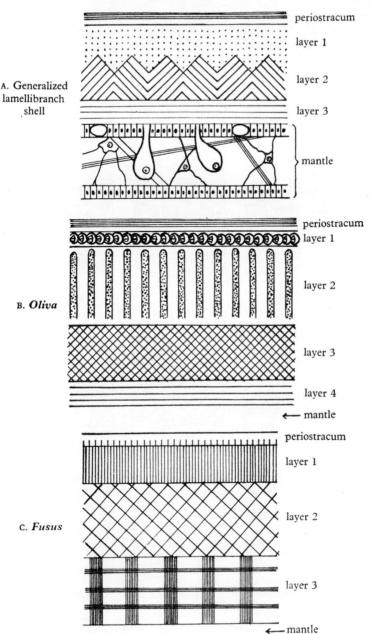

Fig. 395. Diagram of transverse sections of different molluscan shells showing the different layers present in the shell.

Fig. 396. Sculpturing found on the surface of lamellibranch shells. The name in parentheses is that of a genus which has the particular type of sculpture developed in some of its species. (From Turner, in Shrock and Twenhofel.)

THE MOLLUSCAN STOMACH. The stomach in the molluscs shows considerable variation in form and function correlated with the different feeding habits. The primitive molluscs were microphagous feeders, animals feeding on minute particles which they scraped off rocks by means of a well-developed radula. They also secreted a large amount of mucus; food particles and mucus were sent down to the stomach which had a complex series of folds that sorted out the food particles and prevented them from blocking the ducts leading to the digestive and absorbing gland (Fig. 397). The stomach has three distinct areas. (1) A series of ciliary sorting areas that filter-off the different food particles. (2) A cuticular lined gastric shield which protects the stomach against the action of sharp particles and also helps to crush large particles. (3) A small sac with strong cilia. These cilia seize the mucus string coming in

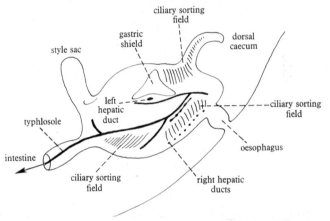

Fig. 397. Diagram of the molluscan stomach showing the grooves that guide the food through the stomach. (After Graham and Morton.)

from the mouth and rotate it and its contained food towards the intestine. The sac thus acts as a capstan. In the primitive animals the sac becomes full of stiff mucus imbedded with faeces; this is called the protostyle. The coarse foods go straight through the stomach to the intestine; the smaller particles are separated out by the ciliary fields and carried to the openings of the absorptive and digestive gland. This type of stomach is seen in *Diodora*.

Other primitive gasteropods such as *Trochus* still have the faecal style sac but the mucus strings that come from the oesophagus, instead of going to the sac, go to a specially developed caecum. The style sac still acts as a capstan and receives the mucus after it has been to the caecum and passes it on to the intestine.

The more intensive herbivores show an interesting modification of the style sac. Though it still retains some of its capstan-like action, it becomes filled with a crystalline material which contains an amylase which helps to digest the starches present in the food. In addition the style contains many spirochaetes which may secrete a cellulase also found in the style. The crystalline style and its associated enzymes have been evolved independently several times in the molluscs. It is found in the advanced herbivorous gasteropods such as *Crepidula, Cavolinia, Lambis,* as well as in several lamellibranchs.

In the carnivorous gasteropods the gut becomes muscular and loses its mechanical sorting areas, style sac and gastric shield. Instead there is a complex development of a muscular buccal mass and gut so that a well-

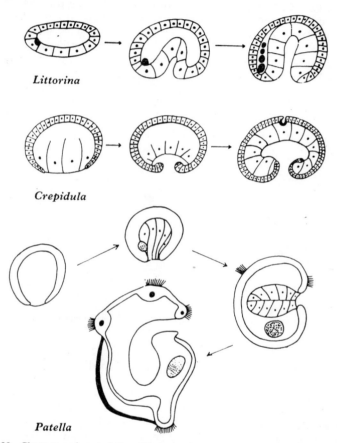

Fig. 398. Cleavage and gastrulation. There are three types of gastrulation shown here: invagination in *Littorina,* epiboly in *Crepidula,* and ingression in *Patella.*

developed peristaltic wave can occur. One usually imagines that a carnivorous animal can eat worms and other molluscs, but it should be remembered that animals feeding on coelenterates and polyzoans are also carnivores.

Some molluscs have very specialized feeding methods. *Vermetus gigas,* a gasteropod, traps its prey by releasing a series of mucus threads from the pedal gland. Small animals become trapped in these threads and at intervals the threads are swallowed and the animals passed to the stomach. Other gasteropods feed on molluscs. *Natica* bores holes in mollusc shells by applying its mouth to the shell and releasing sulphuric acid from a special gland on the snout. The acid dissolves a hole in the shell and *Natica* then inserts its radula into the hole and scoops out the animal's soft entrails. *Purpura* uses a mechanical method of boring into a shell, the radula slowly scraping away successive layers of the shell. *Philene* swallows small lamellibranchs whole and crushes them up in a muscular gizzard. *Chrysallida*, an opisthobranch pyramellid, is a parasite on sedentary polychaetes. It has no true radula but has developed a special hollow tooth in a proboscis that is also furnished

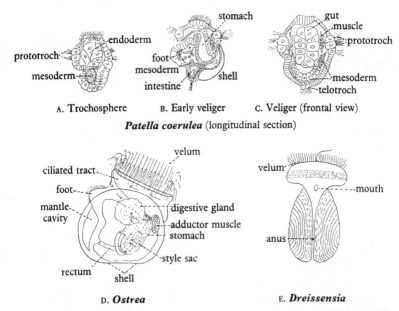

Fig. 399. Molluscan larval forms. A–C, stages in the development of *Patella coerulea*. (After Patten.) D, side view of veliger larva of *Ostrea edulis*; ciliary currents shown by arrows. (After Yonge.) E, ventral view of veliger larva of *Dreissensia*. (After Meisenheimer.)

with a sucker. The animal approaches a polychaete, extends the proboscis, grasps its prey with the sucker and then suddenly inserts the tooth through the body wall. The blood is then sucked out of the host by means of a muscular suction pump. *Calma*, a small opisthobranch living on the yolk of fish eggs, has a very reduced stomach, no gizzard, and no anus. The faecal material accumulates in the liver throughout the whole of the animal's life.

EMBRYOLOGY. In general the embryology of the molluscs closely resembles that of the polychaetes. To simplify matters the account given here will be restricted to the Gasteropoda amongst the molluscs. Most of the gasteropods are oviparous though viviparity is fairly common, being found in *Littorina rudis*, several species of *Helix*, and in *Paludina* to mention a few species. Those that lay eggs may either do so singly as in the case of *Haliotis* and *Acmea* or they may lay them in cocoons. The mortality in the cocoons is high though most of the cells that are laid in the cocoon are not true eggs; many function as nurse cells and provide the developing embryos with food. In *Buccinum* the development of the nurse cells has been particularly studied. The fertilized eggs start to cleave. Cleavage is total and spiral and is to some extent affected by the amount of yolk in the egg. The spiral cleavage is very much the same as that described for the polychaetes (Fig. 194). The first two divisions bring about the ABCD cells which then divide to form the four micromeres which sit in between the large macromeres. Further left and right

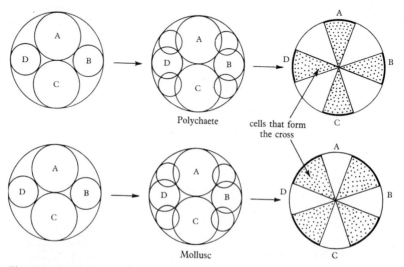

Fig. 400. Comparison of early development in molluscs and annelids. Though both groups show similar spiral cleavage there are certain differences, i.e. the position of the 'cross cells' relative to the plane of the first two cleavages. (After Dawydoff.)

division leads to the development of the four quartets of cells. During these cleavages the cells form the two structures noticed in the polychaetes, the apical cross and the rosette. It is important to note that though these two structures are found in both polychaetes and molluscs there are certain definite differences in the formation of the two groups. In the polychaetes the branches of the cross lie in line with the cells of the first macromeres A–D. In the gasteropods the branches of the cross lie between the cells of the first macromeres (Fig. 400). Another difference is that in the gasteropods the cells $1a^{112}$–$1d^{112}$ are not incorporated in the cross though they are in the annelid cross. The gasteropods on the other hand have in the cross the cells $2a$–$2d$, which do not take part in the annelid cross.

In general the fate of the molluscan quartets is very similar to that of the annelid cells (Figs. 194, 196). The main supply of the mesoderm arises from the $4d$ cells though some of the mesoderm arises from mesenchyme cells that

MOLLUSCA 589

migrate into the embryo. Gastrulation can occur by several methods, three of which are shown in Fig. 398. *Littorina* gastrulates by simple invagination, *Crepidula* by epiboly and *Patella* by ingression. The 4d cell migrates internally and forms the mesoderm rudiment, the coelom arising as a separation of the mesoderm cells. The mesoderm does not show segmentation, though at times there may be a very regular disposition of the mesoderm cells as a band on either side of the gut.

LARVAE. In the Mollusca the development of the trochosphere takes place in a fashion identical with that described for the annelid. In the diagram given here for *Patella*, we see the completion of gastrulation and the appearance of the ciliated rings of the trochosphere (Fig. 399A); also the single large cell which gives rise to the mesoderm. Then in Fig. 399B we see the early *veliger* with an internal organization similar to the annelid, with apical organ, larval nephridia and prototroch. The figure shows, however, organs which are not present in the annelid. On the dorsal side between the prototroch and the anus the larval ectodermal epithelium forms the rudiment of the *mantle* and even at this early age secretes the first *shell*. On the ventral side, there is a prominence which is the *foot* (formed by the union of two rudiments). The single mesoderm cell gives rise first of all to two regular mesoderm bands; and by the development of a cavity in each of these, right and left coelomic sacs are formed; then instead of segmenting as in the annelid, these largely break up into single cells, some elongating and becoming muscle cells (Fig. 399c). It is because there is never any commencement of segmentation in the embryonic mesoderm in molluscs that we have the strongest grounds for believing that molluscs never had segmented ancestors. The trochosphere is followed by a second free-swimming stage, the veliger, in which the prototroch develops into an organ, the *velum*, of increased importance, which serves not only for locomotion but also for feeding, the cilia creating a current which brings particles into the mouth. In the veliger stage the foot increases in size and the shell often becomes coiled in the Gasteropoda.

CLASS 1. AMPHINEURA (LORICATA)

DIAGNOSIS. Mollusca with an elongated, bilaterally symmetrical body, the mouth and anus at opposite ends; with a head, without tentacles or eyes, tucked under the mantle, which occupies the whole of the dorsal surface, and contains various kinds of calcareous spicules imbedded in cuticle, sometimes united to form continuous shells; a flattened foot sometimes reduced; a nervous system (Fig. 428A) without definite ganglia, the ganglion cells being evenly distributed along the length of the nerve cords, and composed of a circumoesophageal commissure and two pairs of longitudinal cords (*pedal* and *pallioviscoral*), each pair united by a posterior commissure dorsal to the rectum; a radula; usually a trochosphere larva.

CLASSIFICATION

1. Order **Polyplacophora**

Shore-living Amphineura with flat foot which occupies the whole ventral face of the body; mantle containing eight transverse calcareous plates as well

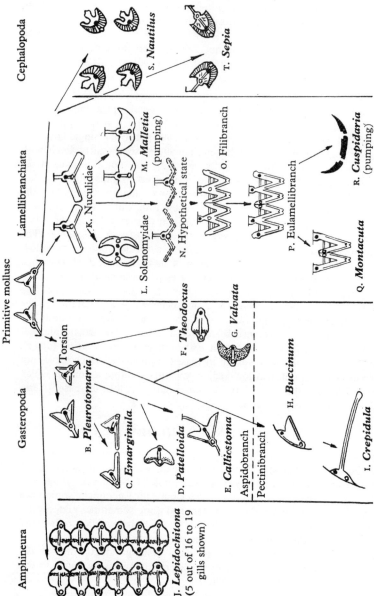

Fig. 401. Diagrams illustrating the probable mode of evolution of the gills throughout the Mollusca. Afferent blood vessel shown black throughout, efferent vessel as open circle, chitinous supporting rods indicated by thick black lines, where absent (D, F, G, J) the extent of the lateral cilia shown; supporting rods on afferent side of filaments in Cephalopoda (S, T) also shown. (From Yonge.)

AMPHINEURA

as spicules; in the mantle groove which runs entirely round the body there is a more or less complete row of ctenidium-like gills on each side. *Chiton* (Figs. 403, 404), *Craspedochilus*.

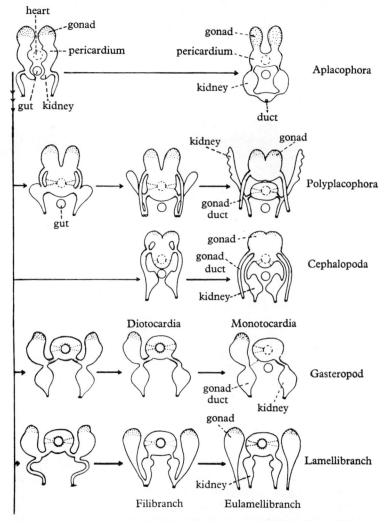

Fig. 402. Development and relationship of coelomic cavities in the molluscs. The coelom is represented by the gonad cavity (gonocoele), the kidney cavity (renocoele), and the pericardium. Note the tendency for the renocoele and gonocoele to separate. (After Goodrich.)

ACCOUNT. *Craspedochilus* is a small mollusc found underneath stones between tidemarks. It looks like an elongated limpet and has exactly the same habits, browsing on small algae and returning after excursions to a centrally situated home. In dorsal view there are seen the eight shell plates which articulate with one another and allow the animal to roll up like a wood-louse. Each plate is composed of two layers, the upper or *tegmentum* and lower or *articulamentum*. Both are calcareous, but the tegmentum is traversed by parallel canals containing ectodermal tissue which end on the surface in remarkable sense organs; some of these have the structure of eyes (the *aesthetes*). Young individuals, which possess a full equipment of aesthetes, are negatively phototropic. As, however, the valves become corroded and

Fig. 403. *Chiton.* Dorsal view

covered with encrusting organisms they become indifferent to light. The part of the mantle which surrounds the shells is called the *girdle* and this contains the spicules which are characteristic of the Amphineura as a whole.

On the ventral surface is seen the *head*, which does not project from under the shelter of the mantle. It bears no eyes and no tentacles, and is separated from the foot by a narrow groove. The mantle groove is shallow, running completely round the animal and containing a varying number of branchial organs, each of which resembles a *ctenidium*. There may be only six on each side crowded together at the posterior end, or they may occupy the whole groove from the head to the anus. It is probable that the forms with a small number of branchiae are the most primitive, and from the fact that the branchiae are graded in size it seems likely that one of them (the largest) is the original one and the others are derived from it. At any rate the repetition of the branchiae does not mean that the chitons were once metamerically segmented animals. There is no trace of any segmentation of the mesoblast in the larva and there is no correspondence between the numbers of the shell plates and of the branchiae.

The mantle groove also contains the *anus* in the middle line posteriorly, on each side, the renal apertures just in front of it, and the genital apertures a little farther forward. In this entire symmetry of the various apertures the chitons differ from any living gasteropods.

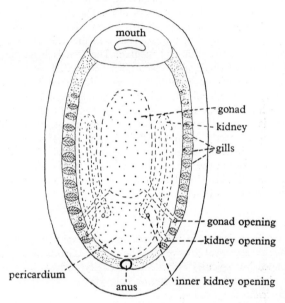

Fig. 404. Ventral view of *Chiton* to show external and internal bilateral symmetry. Mantle cavity finely stippled, the division of the coelom shown above the foot coarsely stippled.

The internal anatomy presents the features attributed above to the ancestral molluscs. Another feature which is probably primitive is the uniform distribution of nerve cells in the nerve cords and the consequent absence of ganglionic enlargements. The cords are connected by many commissures which form a nerve plexus (Fig. 428 A).

A point of great interest is the palaeontological antiquity of the group, forms with eight shell valves occurring in the Ordovician.

2. Order **Monoplacophora**

Almost bilaterally symmetrical Amphineura with internal metamerism. A single piece of shell covers the pallium which extends all over the dorsum. The anus is postero-median. Coelomic cavities well developed. The metamerically arranged, paired auricles deliver the blood to the two symmetrical long ventricles, one on either side of the intestine. Metamerically arranged nephridia arise from the coelomic sacs to open on the surface in the pallial furrow. Gonads symmetrically arranged, possibly metameric, opening through the nephridia. Nervous system primitively orthoneurous. *Neopilina, Tryblidium* (Fig. 405 A).

ACCOUNT. Recently ten specimens of a living mollusc were taken from a depth of 3600 metres near the west coast of Mexico. This animal, *Neopilina galatheae*, is bilaterally symmetrical with an external shell much like that of a limpet. One specimen had kept a dextrally coiled larval shell. There is a well-developed radula of sixteen rows of teeth. The stomach is situated close to the apex of the shell and gives off two symmetrical branches of the liver. There is a crystalline style in the stomach and the animal feeds on radiolarians. The anus is posterior and median. The foot is broad and situated between five pairs of small gills. Each gill is associated with a nephridium. There are five pairs of dorsoventral muscles which serve the foot, and three more muscles near the head. There is a well-developed coelomic cavity. The nephridia open at the base of the gills and the gonad products are passed to the exterior via some of the nephridia. The sexes are separate. The nervous system is almost exactly as in the Polyplacophora. The animals show a strongly marked metamerism and it is problematical whether this resembles the metamerism seen in the arm of the starfish or the coelomic segmentation seen in the annelids. It would be of great interest to have details of the coelomic development of *Neopilina*.

3. Order **Aplacophora**

Worm-like Amphineura in which the foot is absent or represented by a median ridge in a ventral groove and the mantle correspondingly enlarged. No shell plates but spicules only. Mantle cavity perhaps represented by a small cloacal chamber at the posterior end, gills present (*Chaetoderma*) or absent (*Neomenia*) (Fig. 405 B).

ACCOUNT. The Aplacophora are simplified forms of molluscs that have many worm-like characteristics but may be differentiated from the annelids in that they show no trace of segmentation and they have a molluscan type of coelom (pericardium, gonocoele and renocoele). In addition they sometimes possess a radula (Fig. 405B).

In *Neomenia* the mouth is ventrally placed and leads into a straight pharynx. The pharynx is followed by a stomach which is uncoiled and sends a dorsal diverticulum forward into the head region. The surface of the stomach is lined with secretory epithelium that corresponds to the liver of other molluscs. The intestine is small and leads to the cloacal chamber into which also open the kidneys. The circulation is open, though haemoglobin corpuscles may be present. The Aplacophora that possess gills have them as folds in the epithelium in open connexion with the haemocoele. There are a pair of renal tubes that run from the pericardium to the cloacal chamber. The animals are hermaphrodite and the gonads discharge into the pericardium from whence the products are voided through the excretory ducts.

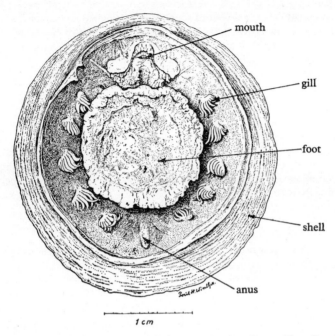

Fig. 405 A. *Neopilina galatheae*, ventral view, a monoplacophoran (From Lemche.)

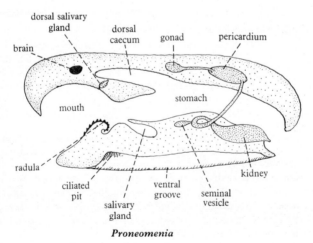

Fig. 405 B. Diagram of structure of *Proneomenia*, an aplacophoran.

There is a large supraoesophageal or cerebral ganglion which gives off circumoesophageal commissures and two sets of longitudinal nerve tracks; a dorsal pallial pair and a ventral pedal pair. The pallial and pedal chains are linked together by cross-connexions at different regions of the body.

Cryptoplax amongst the Polyplacophora has lost its shell, a fact that indicates a way in which the Aplacophora might have arisen.

Examples of the Aplacophora are *Proneomenia, Pararhopalia,* and *Chaetoderma.*

CLASS 2. GASTEROPODA

DIAGNOSIS. Mollusca with a distinct head bearing tentacles and eyes, a flattened foot, and a visceral hump which exhibits the phenomenon of torsion in various degrees and is often coiled; always exhibiting bilateral asymmetry to a certain extent; typically with a shell secreted in a single piece; nervous system with cerebral, pleural, visceral and usually pedal ganglia and a visceral loop; a radula; often a trochosphere larva.

CLASSIFICATION

Order 1. **Prosobranchiata (Streptoneura).** Gasteropods which exhibit torsion, nearly always with a shell and an operculum, with a visceral loop twisted in the form of a figure eight; the mantle cavity opening anteriorly; the ctenidia in front of the heart; separate sexes.

   Suborder 1. **Diotocardia (Aspidobranchiata).** Prosobranchs always with two auricles and sometimes two ctenidia. The ctenidia with two rows of gill leaflets (aspidobranch), the genital products discharging through the right kidney.

Haliotis, Diodora, Patella, Acmaea

   Suborder 2. **Monotocardia (Pectinibranchiata).** Prosobranchs with single auricle and ctenidium, the ctenidium always with one row of gill leaflets (pectinibranch). The gonads have separate ducts opening far forwards in the mantle cavity, the male has a well-developed penis.

  1. Adults pelagic. Heteropoda.   *Pterotrachea, Carinaria, Oxygyrus*
  2. Adults not pelagic.   *Buccinum, Purpura, Nassa, Littorina, Strombus*

Order 2. **Opisthobranchiata (Euthyneura A).** Hermaphrodite gasteropods descended from prosobranchs which have undergone torsion but themselves show a reversal of the process (detorsion), with the mantle cavity tending to occupy the posterior position again, the shell tends to become smaller, internal or entirely absent; the single ctenidium tends to disappear and be replaced by accessory respiratory organs, or the whole external surface may become a respiratory organ.

   Suborder 1. **Tectibranchiata.** Opisthobranchs which usually have a shell and nearly always a mantle cavity and ctenidium.

Acteon, Bulla, Aplysia, Cavolinia

Suborder 2. **Nudibranchiata.** Opisthobranchs usually of a slug-like nature which have neither shell nor mantle cavity, nor ctenidium. *Eolis, Doris*
(Pelagic opisthobranchs are called *Pteropods*.)

Order 3. **Pulmonata (Euthyneura** *B*). Hermaphrodite gasteropods which exhibit torsion and have a shell but no operculum, the nervous system is symmetrical due to shortening of the visceral connectives and a concentration of the ganglia into the circum-oesophageal mass; with a mantle cavity that has become a lung, without a ctenidium but with a vascular roof, and a small aperture, the pneumostome; a single kidney; development direct without a larva, from an egg richly supplied with albumen.

Suborder 1. **Basommatophora.** Pulmonates with eyes at the base of the posterior tentacles. *Limnaea, Planorbis*

Suborder 2. **Stylommatophora.** Pulmonates with the eyes at the tip of the posterior tentacles. *Helix, Arion, Testacella*

EVOLUTIONARY TENDENCIES. Two distinct evolutionary tendencies are found in the gasteropods, (1) coiling of the shell due to growth of the liver, (2) torsion.

GENERAL CHARACTERS. We can safely say that the Gasteropoda are descended from symmetrical unsegmented ancestors (p. 580), and that the most prominent differences among their present-day representatives are due to the varying degrees in which they exhibit the phenomenon of torsion. The ancestors of the Gasteropoda had not been affected by torsion. They possessed a symmetrical body with a straight alimentary canal ending in a posterior anus. On each side of this was a ctenidium, that is, a breathing organ composed of an axis with a row of leaf-like branches on each side. The ctenidia may have been free on the surface when they first arose, but they were soon contained in the posterior mantle cavity which developed with the visceral hump.

PRIMITIVE CHARACTERS. Many characters belonging to the primitive mollusc are still preserved in the gasteropods, the head with tentacles, the nervous system with cerebral, pleural, and pedal ganglia, the radula, the ventricle with two auricles and the two kidneys. Lastly, there is a flat creeping foot and a visceral loop formed by a connective from each pleural ganglion uniting with its fellow in the neighbourhood of the ctenidia.

VISCERAL HUMP. In the alimentary canal of molluscs there is a tendency for digestion and resorption to be confined to a dorsal diverticulum of the alimentary canal which develops into the digestive gland (liver). The growth of this causes the formation of a projection, the *visceral hump*, and a looping of the alimentary canal. This projection grows until it falls over, and this is the first step in the coiling of the visceral hump which is such a characteristic feature of the gasteropods. Growth proceeds until, in the snail, for instance, the visceral hump would, if uncoiled, be longer than the whole of the body. Owing, however, to the fact that one side of the hump grows faster during development than the other, the whole organ is twisted into a compact spiral which can be arranged so as not to interfere with the balance of the animal while crawling.

TORSION. In all gasteropods with coiled shells the mantle cavity is anterior, the opening directed forward and the coiling of the visceral hump is directed posteriorly. But in the development of these forms from the larva (Fig. 406) the mantle cavity first makes its appearance behind the visceral hump, and at a particular stage the visceral hump rotates in a counter-clockwise direction through an angle of 180° on the rest of the body (Fig. 406 D). This is what is

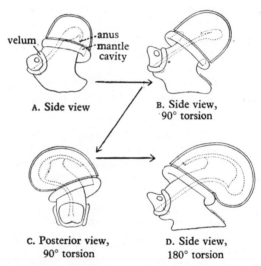

Fig. 406. To show torsion in *Paludina vivipara*. A, almost symmetrical stage, with mantle cavity behind, but anus twisted to the right. B, stage showing 90° of torsion anti-clockwise, mantle cavity and anus to the right. C, posterior view of B; animal with 90° of torsion. D, adult condition with 180° of torsion. (After Naef.)

known as *torsion*, and as shown above it is entirely distinct from the coiling of the visceral hump which precedes it, though it may have been necessitated by the antecendent phenomenon. Only the narrow neck of tissue (and the organs which pass through it), between the visceral hump and the rest of the body, is actually twisted; but the orientation of the mantle cavity and its organs is changed (Fig. 407). Before torsion the ctenidia and the anus point backwards, the auricles are behind the ventricle. After torsion the ctenidia project forward, the auricles are in front of the ventricle; the mantle cavity opens just behind the head. The uncoiled visceral loop has been caught in the twisting and one connective laid over the other, one passing over the intestine and the other underneath, but both coming together near the anus and completing a figure of eight. The whole process takes only two or three minutes in *Acmaea* so that it can hardly be brought about by differential growth. Muscular contractions must play their part.

The advantages of torsion are not clear, though several theories have been put forward. One, suggested by Garstang, states that torsion is an adaptation

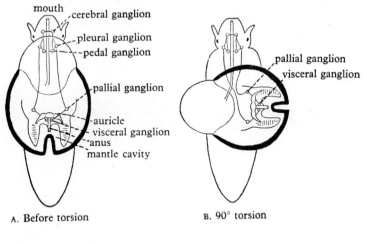

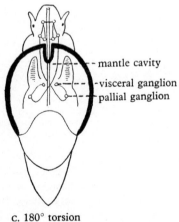

Fig. 407. Diagram to illustrate torsion, seen from above. (After Naef.)

to pelagic life during the larval stage. The untwisted prototroch larva swimming in the sea fell an easy prey to its predators, but the twisted larva with its mantle cavity curled above its head was able to retract its head and prototroch into the mantle cavity, stop the prototroch from effective beating and so fall to the sea bottom. In this way it could avoid its predators swimming in the water.

Though this theory is attractive there is much against it. In the first place there are many pelagic larvae, such as those of the lamellibranchs, that are not twisted but still survive pelagic larval life. Secondly, it has been shown that the cilia on some gasteropod larvae are under nervous control and so could be stopped by simpler means than forcibly withdrawing them into the mantle cavity. A third and more serious objection is found in the case of *Haliotis*. This animal rotates its shell in two stages, the first stage being through 90° whilst the second stage completes the torsion and brings the shell round the full 180°. But the animal is only pelagic during the first stage whilst the mantle cavity is still 90° to the head and so the head is unable to be retracted into the mantle cavity. The animal does not complete its torsion till it has settled on the bottom.

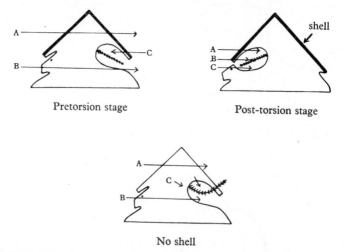

Fig. 408. Possible advantage of torsion. A, current in the sea; B, current due to animal's locomotion; C, respiratory current. In the twisted animal A and B augment C and do not oppose C as they do in the pretorsion condition. In the untwisted animals, the shell is lost and the gills are once more exposed to A and B as well as C.

An alternative explanation of torsion is as follows. Torsion is an advantage to the adult animal. The primitive gasteropod was not twisted and had its gills posteriorly on the body tucked inside the mantle cavity. The cilia on the gills had to draw the respiratory current in from behind the animal. If the animal moved it set up a current of water in a direction opposite to that in which it was moving and this stream would oppose the direction of the respiratory stream. It is possible that there were also weak currents flowing on the bed of the sea and that the primitive mollusc like most animals orientated upstream. This also would provide a stream in opposition to the respiratory current. Once the animal twists, however, all three streams would flow in the same direction and thus aid the flushing of the mantle cavity with fresh clean water (Fig. 408). In this way torsion would be an adaptation present in the adult to increase the ventilation of the mantle cavity.

The twist, however, brings the anus anteriorly and the faeces are discharged over the oral region. There is still some chance of interaction between the faecal material and the respiratory currents (though of course such interaction also could occur in the pretorted animals). To overcome mixing of the

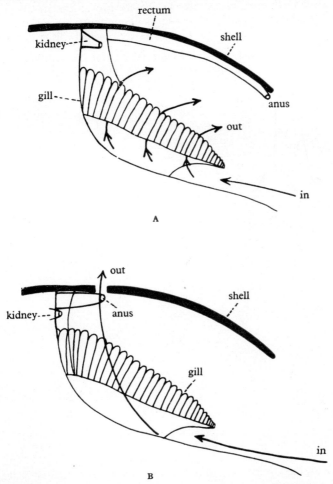

Fig. 409. Diagram of mantle cavity. A, primitive condition in which faeces may get on to the gills. B, condition in *Haliotis* and *Diodora* where there is a single current through the mantle cavity carrying the faeces away from the gills. (After Yonge.)

faecal and respiratory currents at least three different adaptations are found in the coiled gasteropods.

(1) The shell develops a hole or series of holes; the anus is retracted and the respiratory current sweeps over the gills and as it passes out of the mantle cavity carries with it the faecal products. This system is found in *Haliotis* and *Diodora* (Fig. 409).

(2) One of the gills and its corresponding auricle is lost. The filaments on the remaining gill develop only on one side of the main axis and the respiratory current sweeps laterally through the mantle cavity (Fig. 410). It is interesting to note that some of the Diotocardia such as *Trochus* are in an intermediate position in that the right gill has been lost but the two auricles are still present.

(3) The gills are reduced or lost and the respiratory surface is the mantle cavity which in some cases, as in *Patella*, develops pallial gills (Fig. 411A).

DETORSION is always associated with the loss of the shell and thus the liberation of the gills from their enclosing case. Once the shell is lost the gills

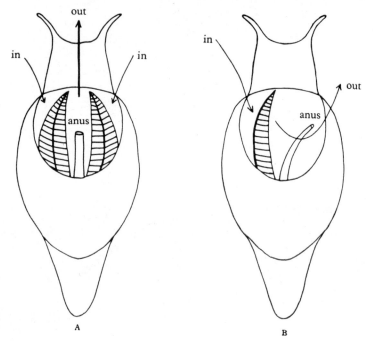

Fig. 410. Currents in mantle cavity. A, aspidobranch; B, pectinibranch. Note that the through-current in B carries the faeces away from the gills. (After Yonge.)

become exposed to the external currents and the anterior position of the gills is no longer so great an advantage. The degree of detorsion can be correlated with the extent to which the shell is reduced (tectibranchs and Pteropoda).

RADIATION OF GASTEROPODS. The large majority of gasteropods belong to the order which exhibits torsion in full development. It is called Prosobranchiata, because of the anterior position of the gills, or Streptoneura, because of the coiled visceral loop. The periwinkles, whelks and limpets of our shores, the fresh-water *Paludina*, and many others, belong to it. The order may, however, be divided into two groups, a primitive one in which the two ctenidia and consequently the two auricles are preserved (*Diotocardia* represented by *Patella, Diodora* and *Haliotis*) (Fig. 411 A, B, C) and a more specialized one in which the right (primitive left) gill, its auricle and even the right kidney have disappeared (*Monotocardia*, represented by *Littorina*, the periwinkle, and *Buccinum*, the whelk) (Fig. 411 D). Some of the Diotocardia, like *Trochus*, are in an intermediate state in which, though the right gill has disappeared, there is still a rudiment of the corresponding auricle. Besides this fundamental difference, there are others. For example, in the Monotocardia,

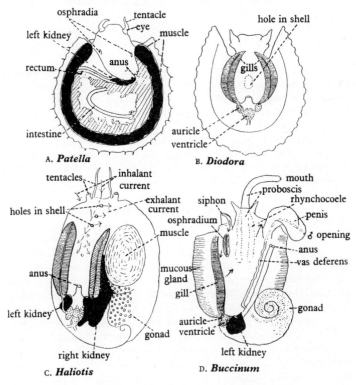

Fig. 411. Mantle cavities of streptoneurous gasteropods.

special generative ducts are developed (cp. also the penis of the male *Buccinum*), while in the Diotocardia, the generative organs open to the exterior through the right kidney (Fig. 402).

It is possible that the disappearance of the organs of one side is to be regarded as the consequence of the processes concerned in torsion and that in the Diotocardia the phenomenon cannot be regarded as having reached its climax. On the other hand, there is a large division of gasteropods called the Opisthobranchiata which show that the changes occurring in torsion are to a certain extent reversible. They have the ctenidium pointing backwards, the auricle behind the ventricle and the visceral loop untwisted and symmetrical. There are some forms (Bullomorpha, Fig. 417 D) included in the Opisthobranchiata which possess a complete coiled shell, but show only 90° of torsion, so that the anus and the ctenidium point laterally instead of anteriorly. The visceral loop also shows untwisting and the forms in this division are thus supposed to show partial reversion of torsion or *detorsion*. Forms like this pass into the typical opisthobranchs with complete detorsion, in which the shell is reduced or lost, the ctenidium directed posteriorly and the visceral loop is completely untwisted (*Aplysia*, Fig. 418 A). The Opisthobranchiata, it is plainly seen, are derived from the Monotocardia amongst the Streptoneura, since they have only a single ctenidium, a single auricle and a single kidney. They have not attained to complete bilateral symmetry, because the mantle cavity is still on the right side where yet present (tectibranchs), and the anus and genital aperture both open there.

The disappearance of the shell and the consequent uncoiling of the visceral hump, if not the cause of detorsion, is a constant accompaniment of the phenomenon. When it is complete, the mantle cavity and even the ctenidium may disappear and we arrive at the group known as the Nudibranchiata. In forms like *Eolis* (Fig. 418 C) their descent is shown by the fact that they possess a veliger larva with a coiled visceral hump which undergoes torsion (which reverses later). The adult shows evidence of streptoneurous ancestry in the presence of the anus at the right-hand side. In *Doris* (Fig. 418 B) the anus and renal aperture are median, but the genital aperture is still situated on the right side.

The last division of the Gasteropoda is the Pulmonata, which is usually united with the Opisthobranchiata to form the group Euthyneura. But 'euthyneury' or symmetry of the nervous system (more particularly the 'visceral' part of it) is arrived at in different ways in the two divisions. In the Opisthobranchiata, as shown above, it is by detorsion. In the Pulmonata, however, the shell is retained and the visceral hump coiled in typical members of the group (land snails). But the visceral loop is shortened and untwisted at the same time (Fig. 417 A, B), and finally it is incorporated with its ganglia into the circumoesophageal nerve collar, so that the nervous system becomes symmetrical. The most primitive members of the Pulmonata still show a twisted visceral loop which is beginning to shorten. All the group have lost the ctenidium but they retain the single auricle which shows them to be derived from the Monotocardia. This was brought about by a chain of circumstances involving migration from sea to shore.

Anatomy of Helix

The type of the Gasteropoda which is usually given for dissection is *Helix* (either *H. aspersa*, the common English garden snail, or *H. pomatia*, the edible snail). It possesses many features which are common to the whole of the Gasteropoda, but as has been seen above, the order Pulmonata to which *Helix* belongs is the most specialized and probably the latest developed division. *Helix* is a terrestrial animal breathing by a kind of lung, while the majority of gasteropods are marine animals breathing by gills, and besides the complications which this involves, the reproductive system is hermaphrodite with the most elaborate provision of glands and ducts which serve to produce eggs well stored with nourishment and are arranged so as to assure cross-fertilization. In the account of *Helix* which follows an attempt is made to distinguish clearly between the purely gasteropod features and the adaptive features which belong to the Pulmonata.

FOOT. The body of a snail is composed of three regions, the head, foot and visceral hump. The visceral hump is all that part which is covered by the shell when the animal is expanded, while the head and the foot make up the remainder outside the shell. There is no boundary between the latter two regions. The German zoologists refer to the whole as the 'Kopffuss' (the

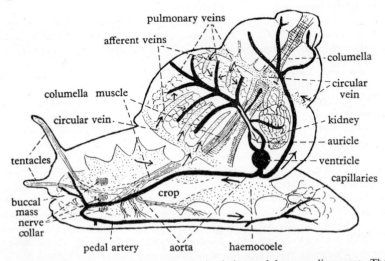

Fig. 412. *Helix pomatia.* Diagram of the circulation and haemocoelic spaces. The pulmonary veins, ventricle and arteries are shown in black; the veins and haemocoelic spaces are indicated by stippling. Only a few of the arteries and a small portion of the arterial capillary network in the posterior region of the foot are shown. The direction of the blood flow is given by arrows. The course of the columella muscle and its branches is indicated.

'head foot'), and this can be retracted as a whole within the shell by the action of the *columella muscle* (Fig. 412). The foot is particularly characteristic of the Gasteropoda. It possesses a flat ventral surface underlain by longitudinal muscle fibres. If a snail is observed crawling up a pane of glass, a series of rippling waves of contraction of very small amplitude are seen to pass regularly over the surface of the foot. They are co-ordinated by the action of a nervous network, such as occurs in the lower invertebrates (Fig. 85 A). The gliding movement of a snail indeed resembles that of a turbellarian, and we actually find that in some marine gasteropods the surface of the foot is clothed with cilia, which beat in unison, though they are perhaps capable of inhibition by the central nervous system. In most water snails, however, the foot moves by muscular contraction. To fit this kind of movement for passing over a hard dry surface, there is in the snail a copious secretion of slime from a *pedal* (*mucous*) *gland* which runs dorsal to the foot and opens just ventral to the mouth. As soon as the slime emerges it is spread out as a smooth bed of lubricating fluid along which the snail moves.

BODY. There are two pairs of *tentacles* on the head of the snail. The first are shorter and are supposed to be the seat of the sense of smell; the second bear a pair of simple eyes (Fig. 438 B) at their tip. Both are hollow and have attached to the inside of the tip a muscle whose contraction turns them outside in. The *mouth* is a transverse slit just ventral to the first pair of tentacles. On the right side of the body not far below and behind the second pair of tentacles is the *reproductive aperture*. On removing the shell, the junction of the visceral hump with the rest of the body is seen anteriorly as a thickened collar which is the edge of the mantle and the seat of secretion of the principal layers of the shell. It is fused to the head of the snail except for a round hole on the right side which is the *aperture of the mantle cavity* or *pneumostome*.

RESPIRATION. In the marine gasteropods the mantle cavity has a wide opening to the exterior, though a part of the mantle border (*siphon*) is modified to form a special channel by which fresh water for breathing may be drawn in by the action of the cilia clothing the gill. But in the air-breathing pulmonates where the cavity is converted into a lung, the injury of delicate respiratory tissues by evaporation must be avoided, and a pumping mechanism for renewal of air established. The restriction of the respiratory aperture is one of the necessary modifications. If a section is drawn across the lung of a snail it will be seen that the mantle forms the roof of the cavity and is covered with ridges in which run pulmonary veins converging towards the auricle. The floor of the cavity is arched and has a layer of muscles, which contract rhythmically. When they contract, the arch flattens and air is drawn in and at the limit of contraction a valve slides across the pneumostome. When the muscles relax, the cavity decreases in size and exchange of gases with the blood in the roof vessels is facilitated by the increase of pressure of the contained air. Then the pneumostome opens and air is expelled; the subsequent contraction of the floor muscles brings in a fresh supply. This 'breathing' is not so regular or so frequent as in a vertebrate; moreover, it may cease altogether in the winter when the snail hibernates.

COELOM. In dissection, a cut is made underneath the collar and another under the rectum and the roof of the mantle cavity turned back so as to show the pericardium enclosing the ventricle and single auricle, and the kidney, which is a yellow organ consisting of a number of folds covered by cells containing uric acid. The ureter is a thin-walled tube which runs along the right border of the mantle cavity parallel to the rectum and opens just behind the pneumostome and above the anus. Here again is a difference from the marine gasteropods in which the anus and kidney aperture discharge inside the mantle cavity, faeces and urine being swept away by the respiratory current. The pericardium and the kidney represent the coelom in the snail and, as is usual in Mollusca, their common derivation is shown by the connexion of the cavities by the *renopericardial canal*. The coelom, though thus represented, does not constitute the perivisceral cavity. On cutting the floor of the mantle cavity and continuing the cut forward towards the mouth a large body cavity (the haemocoele) is revealed which contains the anterior part of the alimentary canal and the greater part of the reproductive organs.

BLOOD SYSTEM. This is a haemocoele almost as well developed as that of arthropods. Its connexion with the rest of the blood system and the general

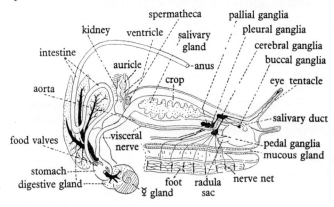

A. Digestive and nervous system of *Helix pomatia*

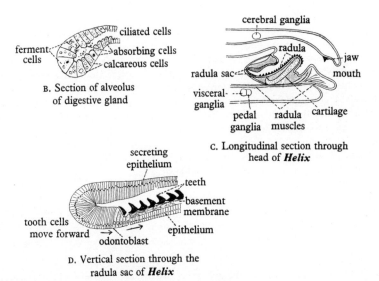

B. Section of alveolus of digestive gland

C. Longitudinal section through head of *Helix*

D. Vertical section through the radula sac of *Helix*

Fig. 413. *Helix pomatia*. (After Meisenheimer.)

course of the circulation may be briefly described here as follows: the ventricle pumps arterial blood through a single aorta which soon divides into an anterior aorta supplying the buccal cavity and a posterior which supplies the visceral hump. The terminal branches of these arteries eventually communicate with the general haemocoele (stippled in Fig. 412) and this discharges into the *circulus venosus* leading to the lung and heart.

DIGESTIVE SYSTEM. The alimentary canal (Fig. 413 A) commences with the *buccal mass*. On the roof of the mouth is a small transverse bar, the *jaw*, and in conjunction with this works the *radula*, which is a strip of horny basement membrane on which are fastened many rows of minute recurved teeth. It is formed in a ventral diverticulum of the buccal cavity called the *radula sac* (Fig. 413 D) in which proliferating tissue is constantly producing transverse rows of cells called *odontoblasts*, each of which helps to form a tooth, and other cells which secrete the basement membrane. The whole radula is pressed forward by the new growth so that fresh surfaces are constantly coming into use as the old part is worn away. The radula is supported by masses of tissue, resembling cartilage, which also serves for the attachment of muscles, and the whole forms the rounded organ which is the *buccal mass*.

The buccal cavity is succeeded by the *oesophagus*, which widens out into the *crop*, which in life contains a brown liquid secreted by the 'liver'. On the side of the crop are the branching white *salivary glands*, which empty their secretion by two ducts running forward into the buccal cavity. The secretion is partly mucus, partly digestive fluid containing an enzyme acting on starch. The crop is succeeded by the *stomach*; this is imbedded in the *digestive gland* (liver), which occupies most of the visceral hump. The 'liver', though apparently solid, is composed of a number of tubes and the end portion (*alveolus*) of each tube is glandular; the rest is ciliated and serves to introduce small fragments of food into the active alveolus. The alveoli contain cells of three kinds, secretory, resorptive and lime-containing (Fig. 413 B). The secretory cells produce the brown fluid found in the crop; this contains a ferment which dissolves the cellulose of plant cell walls and liberates the protoplasmic contents, no portion of which is digested in the crop or stomach. But these contents in the form of small granules are actually introduced into the alveoli of the liver and there taken up and digested by the resorptive cells which possess intracellular proteolytic enzymes. In carnivorous gasteropods digestion follows a different course. The glands of the alimentary canal secrete proteolytic enzymes and the digestion of protein takes place in the stomach and not in the cells of the digestive gland (see *Murex*, p. 613). A combination of extra- and intracellular digestion is highly characteristic of Mollusca, but in the possession of a cellulose-dissolving ferment *Helix* stands almost alone in the Animal Kingdom, and may be indeed said to be physiologically adapted to a plant diet (cf. *Teredo*, p. 635). The intestine runs from the stomach, within the liver, and then as the rectum in the roof of the mantle cavity.

REPRODUCTION. The reproductive organs are extremely complicated (Fig. 414 A) but a function has been assigned to each part of what appears to the elementary student as an unmeaning tangle of tubes. Eggs and sperm are produced in the same follicle of the *ovotestis*, a small white gland in the apex of the visceral hump. But while ripe sperm is found throughout a large part of the year, mature eggs only occur for a very short space indeed. Both eggs and sperm pass from the ovotestis to the albumen gland through the *herma-*

phrodite duct, the terminal portion of which is a pouch (*receptaculum seminis*) where sperm is stored and fertilization is said to occur. After fertilization, the eggs enveloped in albumen from the gland enter the rather voluminous female duct, which runs almost straight to the exterior. They then receive a calcareous shell secreted by the epithelium of the duct. The terminal portion of the duct is the thick-walled muscular *vagina*, into which open the *mucous glands*, the *dart sac* and the *spermathecal duct*.

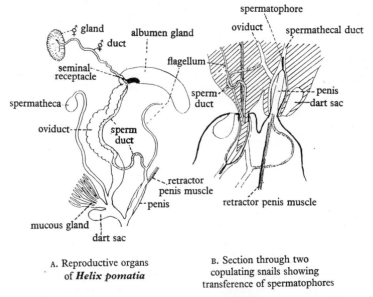

A. Reproductive organs of **Helix pomatia**

B. Section through two copulating snails showing transference of spermatophores

Fig. 414. *Helix pomatia*: A, reproductive organs. B, section through two copulating snails showing the transfer of spermatophores. (After Meisenheimer.)

The sperm, on the other hand, passes down a male duct which is at first only partly separate from the female duct, the cavity of both ducts being in communication until the male duct leaves the company of the female duct altogether, slips under a muscle, and joins the *penis* at its junction with the slender *flagellum*. In this latter the spermatozoa are compacted together and enclosed in its secretion to form *spermatophores*. The penis is muscular and has a special *retractor penis* muscle also attached to it. Both vagina and penis open into a common *genital atrium*, with an opening to the exterior far forward on the right side.

COPULATION. Cross-fertilization is the rule in nearly all species of *Helix* but cases of self-fertilization have been known. Usually, however, there is reciprocal fertilization, preceded by a remarkable preparatory event in which two snails approach each other and evert the genital atrium so that the male and female apertures appear externally. The dart sac mentioned above contains a calcareous sculptured weapon, the dart, which can be secreted anew very quickly by the epithelium of the sac. This is propelled by the muscles of the sac out of the female aperture when the other snail is almost in contact—in fact the two darts are launched almost simultaneously, with such force that they pierce the body wall, traverse the cavity and are found imbedded in various internal organs. Some time after this drastic stimulation, the two snails approach each other again and reciprocal fertilization takes place, the penis of each individual being inserted in the vagina of the other (Fig. 414 B).

The following account of further events has been given and shows, as in the earthworm, the remarkable complexity of the arrangements which are made to prevent self-fertilization in such common hermaphrodites. The foreign spermatophores find their way up the spermathecal duct to the terminal *spermatheca*, where the chitinous covering of the spermatophore is dissolved, and the spermatozoa set free. These now retrace their path to the junction with the female duct and then move up that duct to the fertilization pouch. Fertilization takes place in May or June but the eggs are not laid till July. It is said that the foreign sperm remains in the pouch during this time, and that immediately before ovulation the sperm produced by the individual itself degenerates within the hermaphrodite duct so that the eggs pass down the duct without any danger of being self-fertilized and meet the foreign sperm at the end.

After fertilization, the egg cell passes down the oviduct where it is enveloped with such quantities of albumen that the diameter of the albumen envelope is 20–30 times that of the egg cell itself. In the outer layer of albumen a skin appears, and in this crystals of calcium salts are laid down which aggregate to form a definite shell. The eggs are laid in July and August in small holes in the earth and hatch after about twenty-five days of development.

HIBERNATION. In the autumn the snail loses its appetite and hides, often in company with large numbers of its fellows, under leaves, making a small hole in the ground with its foot and shell in which it lies with the aperture upwards. The head and foot are withdrawn into the shell and the edges of the mantle approximate to form an almost complete disc filling up the aperture, leaving only a small hole for breathing. They secrete a membrane (*epiphragma*) mostly composed of $Ca_3(PO_4)_2$. Several such membranes may be found behind each other. In this winter sleep the snail remains for about six months; respiratory movements are carried on slowly and the heart beats sink from about 10–13 to 4–6 per minute. The rate of heart beat is closely dependent on the temperature, and at a temperature of 30° C. is from 50 to 60 beats per minute.

1. *Order* **Prosobranchiata (Streptoneura)**

DIAGNOSIS. Gasteropoda which exhibit torsion, nearly always with a shell and an operculum, with a visceral loop twisted in the form of a figure of eight, the mantle cavity opening anteriorly, the ctenidia in front of the heart, and separate sexes.

CLASSIFICATION. As mentioned on p. 596 the prosobranchs are subdivided into two suborders, the Diotocardia and the Monotocardia. There is another classification of the prosobranchs that has received considerable following amongst malacologists. The main difference is in the way that the

GASTEROPODA 611

Monotocardia are subdivided. It is fairly clear that they are not a natural monophyletic group. This classification divides the prosobranchs into three groups: order Archeogasteropoda, which corresponds to the Aspidobranchiata and contains such animals as *Acmaea* and *Neritis*; order Mesogasteropoda which contains the majority of the Monotocardia including the Heteropoda; order Neogasteropoda in which are *Buccinum*, *Murex*, *Terebra* and *Voluta*, animals possessing a well-developed penis, a good osphradium and a concentrated nervous system.

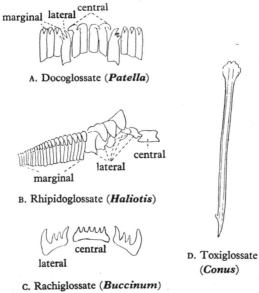

Fig. 415. Radulae of various types. A, used for rasping encrusted layers of algae off rocks. Radula of relatively enormous length; the teeth are quickly worn away. B, lateral and central teeth as in *Patella*, used in browsing on algae growing on stones. The marginals, of which only about half are shown, are probably used as a sieve to prevent fragments of food of too great size from entering the oesophagus. C, teeth of carnivorous type with sharp cusps. D, specialization of carnivorous type in which only two teeth (laterals) remain in each row, are hollow, and are used as poisoned daggers, carrying the secretion of the salivary glands.

In the account given here the old classification is retained and the Monotocardia and Diotocardia will be considered as the two suborders of the prosobranchs.

Suborder 1. **Diotocardia**

Haliotis, the ormer (Fig. 411 C), is a greatly flattened gasteropod which lives between tide-marks, as far north as the Channel Islands, browsing on seaweed and eating all kinds of dead organic material. It can move with considerable speed (5–6 yards a minute), but adheres very firmly to stones. The mantle cavity is very spacious and contains two ctenidia, the left being rather the larger, each with two rows of filaments. The mantle has a slit which

39-2

runs in the roof of the mantle cavity, its position being shown by a row of holes in the shell which serve for the escape of the exhalant current. The anus opens at the posterior end of the mantle cavity and the two kidneys on each side of the anus. There is a well-marked visceral loop and the pedal nerve centres have the form of long cords in which ganglion cells are evenly distributed. The gonad has no ducts but the genital cells are discharged into the right kidney. The radula has numerous marginal teeth arranged in a fan-like manner (rhipidoglossate type).

Diodora, the keyhole limpet (Fig. 411 B), is so-called because of the hole which perforates the mantle and the apex of the shell. It possesses two equal ctenidia. The visceral hump and shell are completely uncoiled, but in other respects it resembles *Haliotis* and possesses the same type of radula.

Patella, the limpet (Fig. 411 A), represents a type of complete adaptation to life on an exposed coast between tide-marks. Its conical shell only shows coiling in its early stages and offers the minimum of resistance to the waves. As in the above forms there is no operculum, but the mollusc cannot be detached from rocks without using great force, owing to the enormous power of the pallial muscles which press the shell against the rock. The mantle cavity is restricted anteriorly and the ctenidia have disappeared, though the *osphradia* connected with them are present as minute yellow specks. But a secondary mantle cavity extends all round between the foot and the mantle and contains a series of folds which are known as *pallial gills*. In the related Acmaeidae there are various stages of the loss of the ctenidia and their replacement by pallial gills. The enormously elongated radula is composed of very strong teeth and there are a small number of marginals (docoglossate type). This type of radula is suited to the feeding habits of the limpet, which scrapes the crust of minute algae off the surface of rocks. Limpets have a remarkable 'homing' sense, returning after excursions for food to the same spot, which may be marked by a depression in the rock.

Suborder 2. **Monotocardia**

Buccinum, the whelk (Fig. 411 D), lives between low-water mark and 100 fathoms. It is active and carnivorous, feeding on living and dead animals, which it grasps by means of its foot. It has a remarkable and highly developed proboscis which can be retracted within a proboscis sheath. The true mouth is situated at the end of the proboscis. The radula (of the rachiglossate type) is used for rasping away flesh, but it can even bore holes in the carapace of Crustacea.

There is only a single ctenidium with a single row of filaments. This is the primitive right member of the pair, though situated on the left of the mantle cavity. A very prominent organ is the bipectinate *osphradium*, which is easily mistaken for a ctenidium. There is a single kidney which is not used for the passage of the genital products. The gonads have separate ducts and in the male there is a penis.

The eggs are laid in capsules which usually contain several hundred and the capsules are attached to each other forming the sponge-like masses so often flung up by the tide.

Murex is nearly related to *Buccinum* and is also carnivorous. It has been recently shown that the salivary glands and the 'liver' all contain the same proteolytic enzymes. These have been separated by adsorption and found to comprise a proteinase, a carboxy-polypeptidase, an aminopolypeptidase and a dipeptidase. These are just such enzymes as occur in the vertebrates and the higher Crustacea; but, in contrast to vertebrates, in *Murex* there is no division of labour amongst the digestive organs.

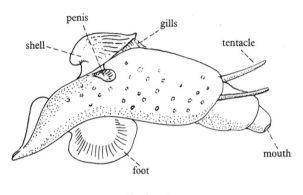

Carinaria

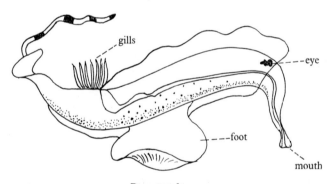

Pterotrachea

Fig. 416. Heteropods. These are pelagic monotocardians. Both animals swim upside down.

Littorina, the periwinkle, is interesting because it exhibits tendencies toward a terrestrial habit which is reflected in its structure. In certain species the filaments of the ctenidium are extended over the roof of the mantle cavity to form a kind of vascular network not unlike that in *Helix* and other pulmonates. *Littorina rudis* lives almost at high-water mark and spends more of its life in air than in water.

The structure of this form is very similar to *Buccinum* but it has no proboscis and is not carnivorous.

Paludina, on the other hand, is a fresh-water form of common occurrence in this country which still preserves the ctenidium and so must be regarded as a direct immigrant from sea water into fresh water. It possesses a kind of uterus in which embryos of relatively enormous size are developed.

Pterotrachea (Heteropoda) (Fig. 416) is an inhabitant of the open sea with many adaptations to pelagic life. It is laterally compressed; the tissues are transparent except for the digestive gland and pericardium compressed into a

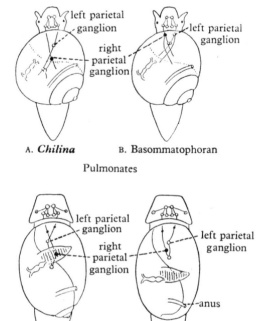

Fig. 417. To illustrate the origin of the uncrossed nerve tracts in the Pulmonata (shortening) and Opisthobranchiata (uncrossing). In A the left pleural connective has shortened and pulled the left parietal ganglion forward. In B both pleural connectives have shortened and pulled both parietal ganglia forward and hence uncrossed them. In C and D there is an untwisting and hence an uncrossing. (After Naef.)

small visceral hump. The animal swims ventral surface uppermost, using its foot as a fin. The *sucker* is a rudiment of the crawling surface. It is predacious, seizing worms and other animals with its radula and swallowing them whole.

2. Order **Opisthobranchiata**

DIAGNOSIS. Hermaphrodite gasteropods which are descended from prosobranchs which have undergone torsion but themselves show a reversal of torsion (detorsion); with the mantle cavity, where present, tending to occupy a posterior position again, the shell to become smaller, internal or entirely absent and the single ctenidium to disappear and be replaced by accessory respiratory organs or by the whole external surface becoming a respiratory organ.

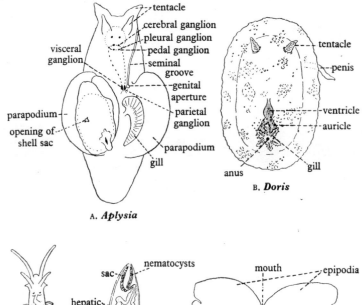

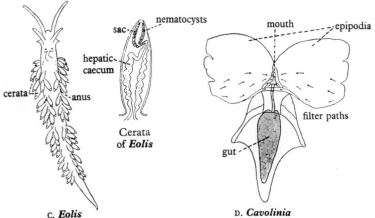

Fig. 418. Opisthobranchiate molluscs in dorsal view. D, the direction of the ciliated currents on the epipodia is indicated by arrows. (C, after Alder and Hancock; D, after Yonge.)

CLASSIFICATION

Suborder **1. Tectibranchiata**. Opisthobranchiata which often have a shell and nearly always a mantle cavity and ctenidium. *Actaeon, Bulla, Aplysia, Cavolinia.*

Suborder **2. Nudibranchiata**. Opisthobranchiata usually of slug-like habit which have neither a shell, nor a mantle cavity, nor a ctenidium. *Eolis, Doris.*

Aplysia (Fig. 418 A), the sea hare, is found crawling on seaweeds which form its food. The younger forms occur in rather deeper water and are red in colour, matching the red algae on which they occur, while the larger individuals, between tide-marks, devour green seaweeds such as *Ulva* and are olive-green. The head possesses two pairs of tentacles, the anterior being large and ear-like (hence the animal's name), while those of the second pair are olfactory in function and have each a simple eye at their base. From the sides of the foot in the posterior region rise two upwardly directed flaps, the *parapodia*: by using these the animal can swim. The mantle is reflected over the shell so as to cover all except a small area and the mantle cavity lies to the right of this with the ctenidium pointing backwards, while the anus is at the posterior end. In the walls of the mantle cavity are unicellular glands which secrete the purple pigment ejected by the animal when it is molested. There is a single generative aperture and a single duct for the sperm and ova but a *seminal groove* runs forward from the aperture to the head and reciprocal fertilization is impossible. The only internal characters which need be mentioned are the nervous system, with its well-developed but perfectly symmetrical visceral loop, and the alimentary canal which, in front of the stomach, is dilated into a *crop*, lined with horny plates, in which the seaweed is masticated before digestion.

Cavolinia (Fig. 418 D) is an example of the *Pteropoda* (sea butterflies), a special group of the Opisthobranchiata which are modified for pelagic life. They usually have a transparent uncoiled shell in the form of a quiver or a vase, from the aperture of which projects the foot in the form of two fins, the epipodia. By the slow flapping movement of these the pteropods progress through the water (Fig. 419). There are ciliated tracts on the fins, and by the action of the cilia on these, small organisms are sifted from the water and collected in the mouth, the radula assisting in swallowing. *Limacina* is a pteropod with a coiled shell.

Eolis (Fig. 418 C) is a nudibranch which possesses a series of dorsal processes (the *cerata*), which contain diverticula of the digestive gland, each of which opens to the exterior at the tip of the process. The animal feeds on hydroids or sea anemones, and while most of the food is digested or passes out of the anus, the nematocysts are collected in terminal sacs in the cerata and when the animal is irritated they are ejected and everted. This is a unique example of the use in defence by one animal of the offensive weapons of another. The cerata are often brilliantly coloured and experiments with fish show that sea slugs are avoided on account of their 'warning' patterns.

Hermaea is another nudibranch with similar cerata, which have not, however, openings to the exterior. The animal feeds on green algae (Siphonales). The radula, in each row of which there is only a single sharp tooth, forms a saw by which the cell wall of the alga is opened. Then by dilatation of the buccal cavity the fluid protoplasm is sucked out.

Doris (Fig. 418 B), the sea lemon, is a short flattened nudibranch, sluggish in movement, which feeds on incrusting organisms like sponges. There is a tough mantle, which is usually pigmented and often resembles the feeding ground, and is reinforced by calcareous spicules. Anteriorly there is a single pair of short tentacles and posteriorly a median anus surrounded by a tuft of accessory gills. In front of the anus is the median kidney aperture. The

Limacina retroversa

Akera bullata

Fig. 419. Swimming of opisthobranchs. The flexible foot sweeps the animal along. *Limacina* is a Pteropod; *Akera* is an Aplysioid. (After Morton; Morton and Holme.)

nervous system is centralized round the oesophagus, and the generative aperture occurring on the right side is the only external organ which is asymmetrical.

3. Order **Pulmonata**

DIAGNOSIS. Hermaphrodite gasteropods, most of which exhibit torsion and have a shell (but no operculum), but which have a symmetrical nervous system, the symmetry being due to the shortening of the visceral connectives and the concentration of the ganglia in the circumoesophageal mass; with a mantle cavity which has become a lung, without a ctenidium, but with a vascular roof and a small aperture (pneumostome); with a single kidney; without a larva, development being direct from an egg richly supplied with albumen.

CLASSIFICATION

Suborder 1. **Basommatophora**. Pulmonata with eyes at the base of the posterior tentacles. *Limnaea, Planorbis.*

Suborder 2. **Stylommatophora**. Pulmonata with eyes at the tip of the posterior tentacles. *Helix, Arion, Testacella.*

A few members of the Basommatophora are marine, but these are all shore forms and breathe air. The group, like the Opisthobranchiata, must have been derived from the Streptoneura Monotocardia, as they possess a single kidney. While they are usually united with the Opisthobranchiata to form the Euthyneura, which includes all forms in which the visceral loop is untwisted, there is no real justification for the establishment of the group, for the 'euthyneurous' condition is one which has been arrived at in two different ways, by detorsion in the Opisthobranchiata and by shortening of the visceral commissures in the Pulmonata. The important characters of the Pulmonata are those associated with the assumption of the terrestrial habit, namely the existence of the lung and the physiological characters correlated therewith. So strongly impressed are these that in almost all the forms which have secondarily returned to water (to fresh water as a rule), the lung continues to function as such and never contains water. *Limnaea,* for example, may be observed in an aquarium to approach the surface of the water at frequent intervals, expel a bubble of air from the lung and protrude the pneumostome through the surface film for a fresh supply. There are, however, a few species (*Limnaea abyssalis*) which live at great depths in lakes, and here the mantle cavity is full of water.

The other general characters of a pulmonate have been given at the beginning of the chapter in the description of *Helix*. They include the concentrated nervous system (it will be seen in Fig. 420 B that the visceral loop of *Limnaea* is not so much shortened as that of *Helix*; in other respects also it is a more primitive form), the complicated reproductive system, with its adaptations for cross-fertilization, and the digestive tract, specialized for the consumption of vegetable food. *Helix*, as has been seen, is thoroughly adapted for this purpose, but in the case of some of the slugs there is an exception to the general rule in the development of the carnivorous habit. This culminates in such a form as the predacious *Testacella*, which pursues earthworms under-

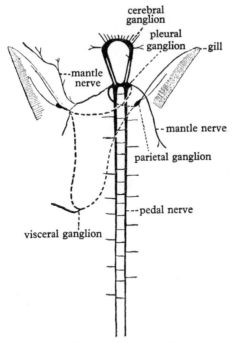

A. *Haliotis tuberculata*

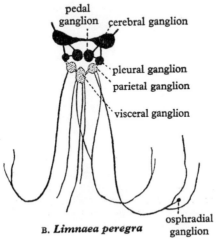

B. *Limnaea peregra*

Fig. 420. Gasteropod nervous system. (From Shipley and MacBride.)

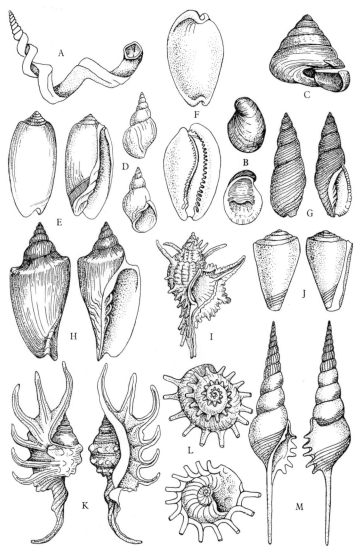

Fig. 421. Gasteropod shells: A, *Vermicularia*; B, *Crepidula*; C, *Pleurotomaria*; D, *Limnaea*; E, *Oliva*; F, *Cypraea*; G, *Mitra*; H, *Voluta*; I, *Murex*; J, *Conus*; K, *Lambis*; L, *Astrea*; M, *Rostellaria*. All shells are about one-third natural size. (From Turner, in Shrock and Twenhofel.)

ground and seizes them with the aid of the strong recurved teeth of the radula which can be thrust out of the mouth, the everted buccal cavity forming a huge proboscis. When the worm is swallowed it is digested in a large crop by the action of the juices of the digestive gland.

The reduction of the shell is shown in the slugs, some of which, like *Testacella*, have a small cap-like shell, which cannot possibly contain the visceral hump, while others have an internal horny disc like the shell of *Aplysia* and still others none at all. The mantle cavity of slugs opens by a pneumostome but there are no respiratory movements as in *Helix*. In other respects the organization of the slugs is very similar to that of snails.

The details of reproduction and development are uniform throughout the group, but in some snails, like *Bulimus*, the amount of albumen added as food for the developing embryo is so great that the egg is the size of a bantam's egg.

CLASS 3. SCAPHOPODA

DIAGNOSIS. Bilaterally symmetrical Mollusca with a tubular shell open at both ends, a reduced foot used for burrowing, a head with many prehensile processes, a radula, separate cerebral and pleural ganglia; ctenidia absent and circulatory system rudimentary; and a trochosphere larva.

This is a small group of molluscs which in some ways stands between the Gasteropoda and the Lamellibranchiata. They are greatly specialized for burrowing. Thus the shell is tubular and perforated at the apex. The foot emerges from the wider opening, while the apex remains above the surface of the sand when the animal is burrowing, and serves alike for the entrance of water into and its exit from the mantle cavity. The head is proboscis-like in form and has none of the usual sense organs, but in *Dentalium* (Fig. 422), the one common genus, there are extensible filaments, the *captacula*, with suckerlike ends, which arise from the dorsal side of the head and serve partly as sense organs and partly for seizing the food. The foot is conical and can be protruded for use as a digging organ.

There is a well-developed radula, a mantle, which in the larva is produced into two lobes (which fuse later), a nervous system with separate cerebral and pleural ganglia and a symmetrical visceral loop. The kidneys are paired; they do not have an opening into the perivisceral coelom. These characters, with the exception of the first and last, bring the Scaphopoda near to the primitive lamellibranch. In the two following morphological features the group is so specialized that it stands apart from any other division of the Mollusca.

There are no ctenidia, respiration taking place by means of the mantle. The circulatory system is remarkably simplified and there is no distinct heart.

The gonad discharges into the right kidney as in the Diotocardia among Gasteropoda (Fig. 402).

CLASS 4. LAMELLIBRANCHIATA (PELECYPODA)

DIAGNOSIS. Mollusca in which typically the body is bilaterally symmetrical, much compressed from side to side and completely enveloped by the mantle which is divided into two equal lobes; each lobe secretes a shell valve, the two valves being joined dorsally by a *ligament* and *hinge* and closed ventrally by the contraction of one or two transverse *adductor muscles*; the head is rudimentary, eyes, tentacles and radula being absent; there is a pair of *labial palps*

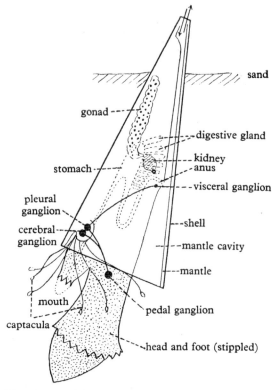

Fig. 422. Diagram of the structure of *Dentalium*. (Altered from Naef.)

with the mouth situated between them; the foot is ventral, without a crawling surface but usually wedge-shaped and adapted for progression in mud or sand; there are two ctenidia in the mantle cavity, often greatly enlarged and with a complicated structure; their cilia, together with those of the labial palps, form a mechanism for the collection of small food particles; the sexes are nearly always separate, and there is a trochosphere and a veliger larva in the marine forms.

CLASSIFICATION

Order 1. **Protobranchiata.** Lamellibranchs in which the gills have flat non-reflected filaments disposed on two rows on opposite sides of the branchial axis. *Solenomya, Nucula, Yoldia.*

Order 2. **Filibranchiata.** Lamellibranchs in which the gills form parallel ventrally directed and reflected filaments. The successive filaments are joined together by interciliary junctions. *Anomia, Arca, Mytilus, Pecten.*

Order 3. **Eulamellibranchiata.** Lamellibranchs in which the branchial filaments of the gills are united at regular intervals by vascular junctions which transform the linear filamentous spaces into a series of fenestrae. *Ostrea, Anodonta, Tellina, Cardium, Pholas.*

Order 4. **Septibranchiata.** Lamellibranchs in which the gills have disappeared as respiratory organs and now form a muscular septum which runs from the anterior adductor muscle to the point of separation of the siphons. *Poromya, Cuspidaria.*

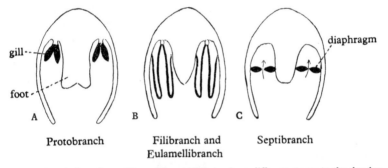

Fig. 423. Vertical sections of Lamellibranchiata to show different stages in the development of the ctenidia. The arrows in c show the direction of water flow through the 'diaphragm' when the latter moves downward. (After Sedgwick, from Lang.)

DETAILED ACCOUNT

The lamellibranchs are most conveniently classified by the structure of their ctenidia. We have firstly three groups which can be arranged in an evolutionary series, showing the ctenidia to become larger, more complex and solid organs. Lastly there is an isolated group, the Septibranchiata, in which the habit of life has completely changed and the ctenidia have practically disappeared (Fig. 423).

In Figs. 401, 423 A, B, the difference is seen between the Protobranchiata with their short and simple filaments and the next two groups in which each filament is greatly elongated and upturned so that descending and ascending limbs can be distinguished. The contrast between the Filibranchiata and the Eulamellibranchiata is expressed by Fig. 424, in which a transverse section through a 'gill' is shown, showing the component filaments separate in the first case, save for the ciliary junctions, united in the second. Lastly, in Fig. 423c, it is seen that in the Septibranchiata, the ctenidia are replaced by a horizontal muscular partition (which moves up and down like the piston of a pump) with apertures connecting the ventral and dorsal divisions of the mantle cavity.

The ciliation of the filaments is the same in all the first three divisions. Even in the Protobranchiata, the ciliary apparatus for food-collecting has been

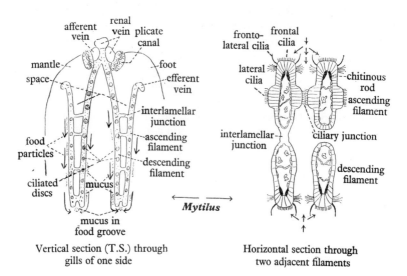

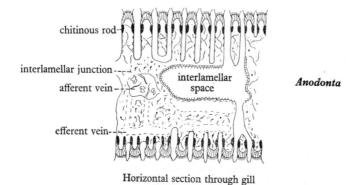

Fig. 424. The ctenidia of the Lamellibranchiata. The arrows indicate the direction of the food current and the path of the food particles it contains. *Mytilus* = Filibranch; *Anodonta* = Eulamellibranch.

developed as in the rest of the group, and it has been pointed out that there are ciliated discs, adjacent pairs of which act as ciliary junctions and hold the filaments together to form lamellae. There is, moreover, a subdivision of the mantle cavity into inhalant (ventral) and exhalant (dorsal) chambers in spite of the small size of the ctenidia.

GILL SYSTEM. The development of the ctenidia (Fig. 424) is the outstanding morphological and physiological character of the lamellibranchs. The arrangement of the shell valves, which allows the mantle cavity to extend the whole length of the body, also makes possible a great extension of the ctenidia. The axis increases in length and the branches on each side not only increase in length, becoming *filaments*, but also turn up at the ends so that there is a *descending* and an *ascending limb*. The limbs of adjacent filaments are connected together by *ciliary junctions* (*Mytilus*), or by growth of tissue (*Anodonta*), so that all the filaments are joined together to form *gill plates*, each gill plate consisting of two *lamellae* formed from all the ascending and all the descending limbs respectively. The lamellae are united by cords of tissue which constitute the *interlamellar concrescences*. The extent to which the gills are welded together to form continuous plates is the distinction between the three main groups of the Lamellibranchiata, the *Protobranchiata* (*Nucula*), the *Filibranchiata* (*Mytilus*) and the *Eulamellibranchiata* (*Anodonta*). But even in the last-named group there are left occasional holes through which water passes into the *interlamellar spaces* then into the *epibranchial space* dorsal to the gills.

Belonging to the same physiological system are the *labial palps*, two folds, one in front of the mouth and one behind, which are turned backwards and prolonged on each side of the visceral mass so as to form two pairs of richly ciliated triangular flaps, embracing the anterior end of the ctenidia, and enclosing a groove which leads to the mouth.

In the anterior part of the mantle cavity the axis of the gill is attached to the side of the animal dorsal to the foot, which here forms a vertical partition dividing the cavity into a right and left half. The mantle cavity continues behind the foot, however, and here the upturned ends of the inner rows of filaments of both ctenidia are united so that the mantle cavity is now divided by a horizontal partition into an upper or epibranchial cavity and a lower main cavity. The former opens at the *dorsal siphon*, the latter at the *ventral siphon*. A constant current of water is maintained during activity, entering by the ventral siphon, passing through the gill lamellae, and leaving by the dorsal.

FILTER FEEDING. From this the animal separates its food in the form of minute plants and fragments of organic debris. The current can easily be demonstrated by pipetting a suspension of carmine particles in the neighbourhood of the siphons, and the details of the process worked out by observing the motion of the coloured granules over the surfaces of the mantle cavity when one of the shells and its mantle lobe have been removed. In this way the direction of the ciliary currents of the ctenidia which transport the food particles can be demonstrated (Fig. 425). On entering the wide mantle cavity the velocity of the inhalant current is checked, and the heavier particles sink down and are taken up by the ciliary currents of the mantle which run towards the posterior region in the neighbourhood of the siphons. The main ingoing current with the smaller particles of carmine is drawn over the surface of the ctenidium and impinges against the individual filaments. Their structure and the distribution of the groups of cilia which all perform different functions is shown in the diagram of a transverse section through a ctenidium (Fig. 424). That the main current of water is drawn into the mantle cavity at all is the result of the activity of the *lateral cilia*. When the current which they have drawn to the ctenidium impinges on its surface the large *latero-frontal cilia* perform their task of deflecting the particles on to the face of the filaments

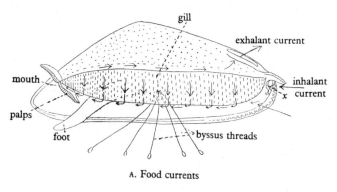

A. Food currents

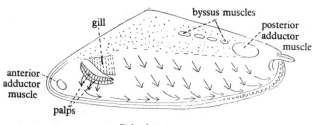

B. Rejection currents

Fig. 425. Diagrams to show the ciliary currents of *Mytilus*. (Adapted from Orton.) A, food currents; with the left lobe of mantle removed to show the outer lamella only of the left gill and the two palps of the left side separated and not embracing the front end of the gill as they normally do in life. The vertical arrows represent the currents caused by the frontal cilia, those at the bottom of the gill the main food current running to the mouth, and that at the top of the gill the exhalant current; x represents a curtain which prevents the inhalant current from directly impinging on the surface of the gill, an opportunity being thus afforded for a preliminary rejection of particles. B, rejection currents; *Mytilus* with foot and gills removed to show the interior of the right lobe of the mantle. The direction of the currents caused by the cilia is indicated by the arrows. The palps on the left side and the anterior end of the outer left gill remain and the rejection current marked by three parallel arrows is shown. The collector current runs along the groove under the mantle edge to the pouch.

where they come under the influence of the *frontal cilia*, which produce a constant stream down over the surface of the ctenidium towards its ventral edge. During the passage the particles in the stream become entangled in mucus, and on reaching the edge the string-like masses of food and mucus are directed by other cilia along the edge in the direction of the mouth, travelling partly in the '*food groove*'. When the labial palps are reached the collected material may, according to its nature, either be swept straight into the mouth or come under the influence of cilia working along rejection paths which direct it away from the mouth and toward the outgoing circulation on the mantle (Fig. 425).

This complicated but well co-ordinated ciliary mechanism is nearly always working when the lamellibranch is covered with water, and the amount of water which passes through the mantle cavity of a single mussel is surprisingly large. But it must be remembered that this current also serves the purpose of respiration, though the exchange of gases takes place through the medium of the mantle rather than the ctenidia. At low tide the animal must close its shell and carbon dioxide accumulates within the mantle cavity. This chemical change depresses ciliary activity and finally brings the cilia to rest, so that the store of oxygen in the tissues is conserved. When the tide rises, however, the cilia immediately resume activity.

Though the majority of the lamellibranchs have the power of movement it is thus seen that they feed in the manner of a sedentary organism, and it is not surprising that there are many fixed and burrowing forms among them.

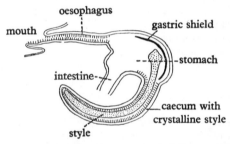

Fig. 426. Section of part of the alimentary canal of *Donax*. (After Barrois.)

DIGESTIVE SYSTEM. A short oesophagus leads directly into the *stomach*, which is a wide sac receiving on each side the ducts of the *digestive gland*, which is similar to that of *Helix* but contains only one kind of cell. This cell takes up the finely divided food which reaches the gland and digests it by intracellular ferments. The *intestine* runs into the foot and makes one or more loops, eventually returning to near the hind end of the stomach. It then passes through the pericardium where it is usually surrounded by the ventricle, ending as the *rectum*. The peculiarity of the digestive system is the presence of a diverticulum of the intestine, the cells of which secrete a *crystalline style* (Fig. 426); some cilia in the diverticulum rotate this and others move it forward so that at its free end, projecting into the stomach against a structure called the *gastric shield*, it is constantly dissolved and the style material mixed with the contents of the stomach. It is composed of protein to which is adsorbed an amylolytic ferment and it may be broken down and re-formed periodically. There is no doubt that this represents a special provision for the digestion of carbohydrates and it is also found in some gasteropods. For the rest, digestion of proteins and absorption take place in the digestive gland, the cells of which have a surprising power of taking up solid particles. In the oyster, it may be mentioned, there is an extraordinary abundance of leucocytes which wander here, there and everywhere, through the body. It has been shown that they enter the stomach and ingest diatoms and other food particles there, speedily digesting them and wandering over the body afterwards, so that they play a unique part in the transport of food.

BLOOD SYSTEM. The blood system of the lamellibranchs is best explained by reference to that of *Mytilus*, the common mussel (Fig. 427A). Here the heart, as in *Anodonta*, consists of a *ventricle* surrounding the rectum and two *auricles*, each of which opens into the ventricle by a narrow canal and is attached by a broad base to the wall of the pericardium over the insertion of the ctenidia into the mantle. A single vessel, the *anterior aorta* (a *posterior aorta* is also present in *Anodonta*), leaves the ventricle, dilates into an aortic bulb and then divides into many arteries. Of these, the most important are the *pallial arteries* going to the mantle and the arteries forming part of the visceral circulation (the *gastro-intestinal*, *hepatic* and *terminal* arteries, the last named supplying the most anterior part of the body, including the foot). The arteries break up into a network of vessels in all the tissues and these join to form veins and sinuses which are largely situated on the inner side of the mantle and the superficial parts of the body. The skin, being bathed in water and devoid of any cuticular covering which might hinder diffusion, is a general organ of respiration and the mantle is the most important part of it. Most of the blood from the *pallial circulation* is returned to the network of vessels in the kidney through the ribbon-like organs, known as *plicate canals*, which extend along the mantle just above the insertion of the ctenidium.

The visceral vessels likewise return blood to the kidney network so that practically the whole of the blood passes through the excretory organ and is purified. A part of the blood from the kidney network enters the *ctenidial circulation*, discharging into the longitudinal *afferent branchial* vein, which gives off to each filament a vessel which descends one side and ascends the other. The ascending vessels join to form a longitudinal efferent vessel, which discharges into the longitudinal vein of the kidney. Into this longitudinal vein is collected the blood from the kidney network in general and by this channel blood is returned to the auricle. It will be seen that the branchial circulation is not important in *Mytilus*; in *Anodonta* (Fig. 427B, C,) it is more developed.

FOOT CONTROL. In *Anodonta* (Fig. 427c) where the foot is larger than in *Mytilus* and movement more continuous the pedal artery is more important than the visceral arteries. The veins from the foot and the viscera join to form a *pedal sinus* and this opens into the *vena cava*. The junction of these is marked by a sphincter muscle (Keber's valve). This sphincter is closed when the foot is extended. The relaxation of the muscles and the pumping of the blood into the sinuses of the foot bring about the swelling of the foot. When the foot is retracted the blood is largely contained in spaces in the mantle. The *pallial* circulation is maintained during movement when the visceral circulation is interrupted as described above.

NERVOUS SYSTEM. While the Protobranchiata have a nervous system with four distinct pairs of ganglia (Fig. 393D), in the remainder of the class the

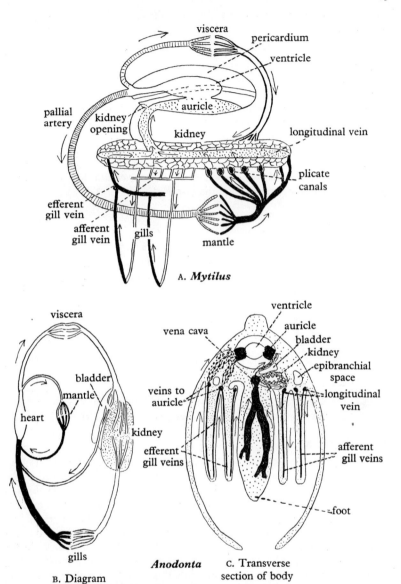

Fig. 427. Circulation in lamellibranchs. In *Mytilus* most of the blood from the viscera passes to the mantle for oxygenation. In *Anodonta* the gills are more important. (A, slightly altered from Field.)

number is reduced to three by the fusion of the cerebral and pleural ganglia (Fig. 428).

REPRODUCTION. The sexes are usually separate in the Lamellibranchiata, but some species of *Ostrea* and *Pecten* are always hermaphrodite, while this condition is frequent in *Anodonta*. In the Protobranchiata the gonad discharges into the kidney, but in most forms there is a separate generative aperture. While most marine forms and the fresh-water *Dreissensia* have trochosphere and veliger larvae, some lamellibranchs incubate the embryos within the ctenidia, and in the family Unionidae, which includes *Anodonta*, the larvae are much modified (*Glochidium*). When they are ripe the mother liberates them if a fish swims near her, and they attach themselves to the gills or fins and become encysted there. After a parasitic life which varies greatly in length they escape from the cyst as young mussels.

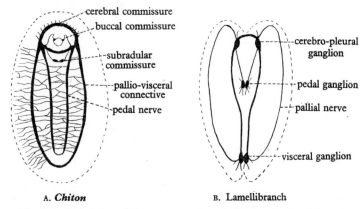

Fig. 428. Dorsal view of the nervous system of *Chiton* and a lamellibranch.

1. Order **Protobranchiata**

The best-known representative is *Nucula* (Fig. 393 D). It has a shell of very characteristic appearance with numerous teeth on the hinge-line and a foot which, when fully extended, has a flat ventral surface which has been compared with that of the gasteropod. But instead of creeping by means of it the animal uses it for burrowing; it is folded up (as is seen in the diagram), and thrust into the mud, then opened out and used as a holdfast, and the contraction of the retractor muscles draws the body below the surface. While the surface of the ctenidium is so small that the organ is of little use for feeding, the *labial palp* is enormous and divided into three parts. One of these is a kind of proboscis which is thrust out of the shell and collects food by ciliary currents. This is sorted and forwarded to the mouth by the other two parts without the intervention of the ctenidium.

The nervous system has distinct cerebral and pleural ganglia and the gonads have retained their original connexion with the kidneys. These and some less important characters show that *Nucula* and its relations are probably the most primitive of living lamellibranchs. The specialization of the labial palps has had as its consequence the partial suppression of the ctenidia, which remain in an undeveloped condition. In this respect the Protobranchiata can hardly be held to resemble the ancestral lamellibranch.

2. Order **Filibranchiata**

Mytilus (Fig. 425). While the majority of lamellibranchs are semi-sedentary, the sea mussel has developed the sedentary tendency and marks a half-way stage to the oyster which remains fixed through adult life. The mussel lives in association in *beds* between tide-marks where the conditions are favourable. The very extensible *foot* is tongue-like in shape with a groove on the ventral surface which is continuous with the *byssus* pit posteriorly. In this a viscous secretion is poured out which enters the groove and hardens gradually when it comes into contact with sea water. The tip of the foot is pressed against the surface to which the mussel attaches itself, and in a cup-like hollow which ends the groove the attachment plate is formed at the end of the byssal thread. When one byssal thread has been formed the foot changes its position and secretes another thread in another place. The byssus thus consists of a mass of diverging threads arising from the byssus pit and by means of it the animal is firmly attached to stones or other mussels. But mussels, particularly when young, creep about both by using the cup at the tip of the foot as a sucker and also by forming a path of threads along the surface of the substratum, as can be easily seen in the laboratory. While the development of the byssus is the most outstanding characteristic of the mussel, it may also be mentioned that a pair of simple eyes are developed, anterior to the inner ctenidial lamella; these are an inheritance from the larval mussel. The invasion of the mantle by the generative organs is another peculiar point. In the breeding season the aeration of blood in the mantle is reduced and the plicate canals (Fig. 427A) become the chief organ of respiration.

Pecten (Fig. 429). There are two common British species, *P. maximus* and *P. opercularis*, which are commonly known under the name of 'scallops'. The animal is found free and it moves not by the ordinary lamellibranch method but by swimming. The two valves are unequal, the right being larger and more convex, and the animal rests on this valve; in *P. opercularis* the valves are almost equal. In swimming the valves open and close very rapidly, forcing out the water between them. Usually the water is forced out dorsally on each side of the hinge line and the animal moves with the free ventral border forward; but on sudden stimulation the current passes out directly ventrally and the hinge-line becomes anterior. There is a single large adductor muscle: this is divided into two parts and the larger of these serves for the rapid contractions which cause swimming movements; the fibres are transversely striated; the smaller part has fibres which are capable only of strong, long-continued contraction and keep the valves closed. (Cf. ch. IV, p. 140.)

The foot is very much reduced, but it has nevertheless a distinct function, that of freeing the palps and gills from sharp and disagreeable foreign material; in the larva it is used actively in locomotion. The ctenidia, while resembling the typical filibranch gill of *Mytilus* in general, differ in the possession of two kinds of filaments and in the vertical folding of the gills. The larger *principal* filaments lie at the bottom of the troughs between successive folds and the

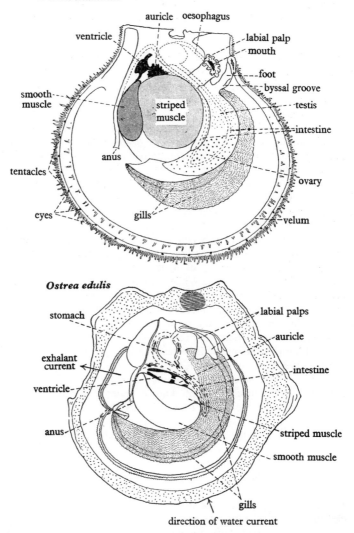

Fig. 429. *Pecten* and *Ostrea*, both with the right valve removed. (*Pecten*, after Dakin; *Ostrea*, after Yonge.) The arrows indicate the direction of the water currents.

descending and ascending limbs of each principal filament are connected by a sheet of tissue, the *interlamellar septum*. In one species, *Pecten tenuicostatus*, there are organic connexions between filaments instead of ciliary junctions only, and the existence of this condition is a valid criticism of the classification of the lamellibranchs by ctenidial structure.

Pecten is hermaphrodite. The ovary has a very vivid pink colour when the eggs are ripe. The testis lies behind it and is cream-coloured. The remaining feature to be noted is the presence of a large series of *stalked eyes* (Fig. 438 D), of a very complicated structure, at regular intervals all round the mantle.

3. Order Eulamellibranchiata

Anodonta (Figs. 424, 427 B, C). Many of the characters of this fresh-water genus are described above.

Ostrea (Fig. 429). In this form the adult is always fixed by the left (the larger) valve. As in *Pecten*, there is only one adductor muscle (the posterior) in the adult (but the spat possesses two equal muscles), and this is divided into

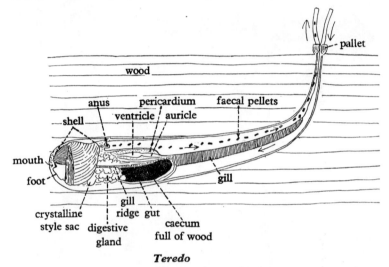

Fig. 430. *Teredo* boring into wood. Sawdust formed by the rotating movement of the shell enters the mouth, and faecal pellets of undigested wood are sent out via the exhalant chamber. The arrows indicate the direction of water currents. (Original.)

two parts, one with striated the other with non-striated fibres. The foot has disappeared entirely; the two auricles are fused together. Of great interest are the reproductive habits: it has been established that individuals of *O. edulis* function alternately as males and females. Spawning tends to take place at full moon as in some echinoderms. Another point of physiological importance is the great part which leucocytes play in digestion; the lumen of the alimentary canal is invaded and diatoms and similar bodies ingested, digested and transported by the leucocytes into the connective tissue.

A figure of the veliger larva of *Ostrea* is given (Fig. 399 D) to show the ciliary currents by which food is obtained, the crystalline style, which is revolved by the action of the cilia of the style sac, and the foot, which is lost in the adult.

Teredo (Fig. 430) is the most specialized of the boring lamellibranchs. While most lamellibranchs burrow in mud, others tend to work in consolidated

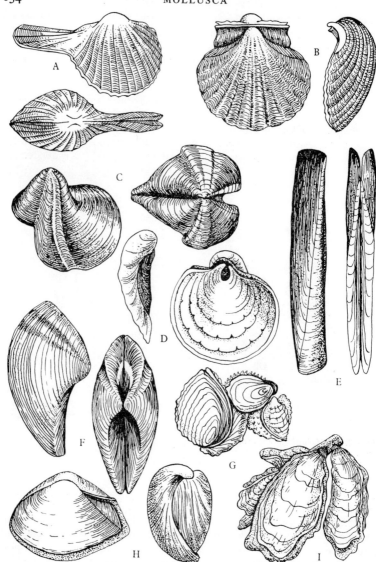

Fig. 431. Shells of lamellibranchs: A, *Cardiomya*; B, *Pecten*; C, *Xylophaga*; D, *Anomia*; E, *Ensis*; F, *Leda*; G, *Chama*; H, *Corbula*; I, *Ostrea*. All shells except that of *Ostrea* are about half natural size. (From Turner, in Shrock and Twenhofel.)

sediments such as *Pholas* in chalk and sandstone, and *Saxicava* in the hardest limestone. *Teredo* and *Xylophaga* bore in wood. The latter makes shallow pits, but *Teredo*, working with extraordinary speed, excavates long cylindrical tunnels (sometimes as much as a foot in a month or two). The wood is reduced to sawdust by the rotatory action of the two shell valves, in which the adductor muscle fibres maintain a rhythmical contraction. The sawdust is swallowed by the animal and is largely retained in a relatively enormous caecum of the stomach, but a great deal of the material passes into the cavity of the digestive gland and is there ingested by the epithelial cells. There is no doubt that *Teredo* has developed enzymes which are almost unique in the Animal Kingdom, which digest cellulose and hemicellulose. The structure of the animal is

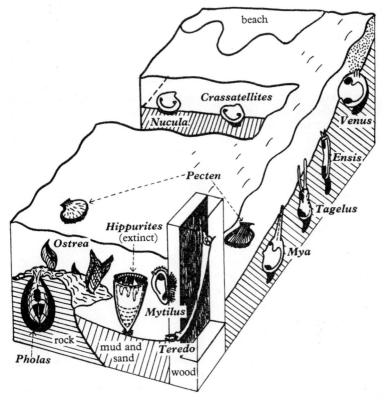

Fig. 432. The localization of lamellibranchs on the shore. (From Turner in Shrock and Twenhofel, after Berry.)

remarkable for the extraordinary long siphons and mantle cavity; while the mantle often lays down a calcareous lining to the tube and always a pair of calcareous valves, the pallets, which close the mouth of the tube when the siphons are retracted. The foot is very much reduced. A constant current into and out of the mantle cavity is maintained by ciliary action, and the ctenidia, though so greatly modified and elongated, constitute a collector mechanism; but it does not seem that diatoms obtained in this way form any part of the normal food of the creature, which exists almost entirely on the carbohydrates furnished by wood which also contains small quantities of proteins.

CLASS 5. CEPHALOPODA (SIPHONOPODA)

DIAGNOSIS. Bilaterally symmetrical Mollusca with a radula and a well-developed head which is surrounded by a crown of mobile and prehensile tentacles, sometimes held to be part of the foot, which certainly forms the *funnel* or *siphon*, a muscular organ, originally bilobed, used for the expulsion of water from the mantle cavity; one or two pairs of typical ctenidia; coelom sometimes exceedingly well developed, the genital part being continuous with the pericardium; typically a chambered shell in the last chamber of which the animal lives, though in most modern representatives it is reduced and internal or wholly absent; nervous system greatly centralized and eyes of great size and often complex type; eggs heavily yolked and development direct.

CLASSIFICATION. The Cephalopoda fall into two groups, in one of which (Tetrabranchiata) there are two pairs of ctenidia and a well-developed external shell, while the members of the other (Dibranchiata) have one pair of ctenidia and either one internal shell or none at all. Of the Tetrabranchiata *Nautilus* is the only living member; of the Dibranchiata, *Sepia*, a common form in the Mediterranean and elsewhere, is a convenient type. The organization of the group will best be understood from a description of these examples. As *Sepia* is the more easily obtained we shall describe it first and in more detail, though it is in many respects less primitive than *Nautilus*.

1. Order **Dibranchiata**

DIAGNOSIS. Cephalopoda with a single pair of ctenidia and kidneys; shell internal, enveloped by the mantle and in various degrees of reduction; 8–10 tentacles; the two halves of the funnel only seen in the embryo; chromatophores present; eyes of complex structure.

CLASSIFICATION

Suborder **1. Decapoda**

Dibranchs with ten tentacles and with a well-developed coelom. Internal shell consisting of phragmocone, rostrum and proostracum or very much simplified.

(1) *Tribe* Belemnoidea. Fossils from Mesozoic rocks which have given rise to the following tribes:

(2) *Tribe* Sepioidea. Decapoda with specially modified 4th pair of tentacles which can be retracted into pits; eyes with a cornea, internal shell sometimes with phragmocone bent ventrally: fins not united posteriorly; shore- and bottom-living forms. *Spirula, Sepia, Sepiola*.

(3) *Tribe* Oegopsida. Decapoda with anterior chamber of eye open; tentacles usually all alike; suckers often modified to form hooks; shell only represented by a horny gladius; strong swimmers. Includes many abyssal forms with phosphorescent organs; some gigantic forms, like *Architeuthis*, 60 feet long.

(4) *Tribe* Myopsida. Decapoda with a cornea in the eye, a simple gladius, specially elongated 4th pair of tentacles, not retractile into pits; fins united posteriorly; shore forms. *Loligo* (Fig. 440).

CEPHALOPODA 637

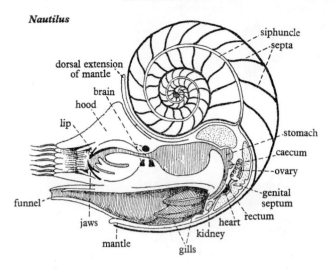

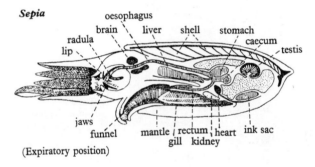

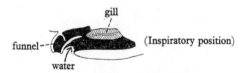

Fig. 433. Diagrammatic median section through *Nautilus* and *Sepia* for comparison of the organization of a tetrabranch and a dibranch respectively. (Altered from Naef.)

Suborder 2. **Octopoda**

Dibranchs with eight tentacles and a reduced coelom. *Octopus, Argonauta, Opisthoteuthis.*

DECAPOD MORPHOLOGY. *Sepia officinalis* is a shallow-water form, in which the shell has become internal. The general disposition of the organs remains much as it would be if the animal inhabited the last chamber of a shell like that of *Nautilus* (cf. Fig. 433). The whole body is cylindrical. At one end,

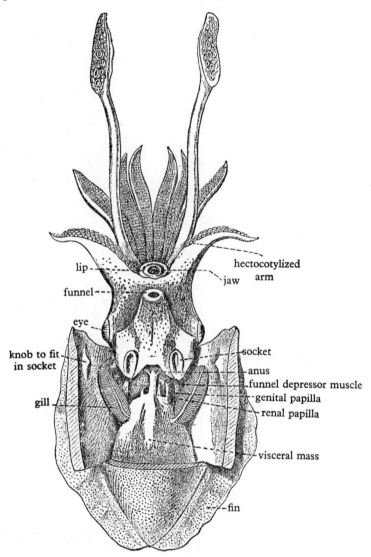

Fig. 434. Ventral view of male of *Sepia officinalis* with mantle cavity opened to expose its contents. (From Shipley and MacBride.)

which would have projected from the shell, is the head with the mouth in the centre and the two relatively enormous eyes at the sides. Round the mouth are the tentacles (arms) for seizing prey which are often considered to be part of the foot. Four pairs of these are short and stout and covered with suckers on their inner surface. The fourth pair (counting from the dorsal surface) are long and can be retracted into large pits at their base; there are suckers only at their free end. The left-hand member of the fifth pair in the male is slightly modified by suppression of the suckers. At one side, called posterior, is the mantle cavity, and protruding from its opening is the funnel, which is the remaining part of the foot. The visceral hump is the conical apex of the animal. Instead, then, of being protrusible like that of a lamellibranch or used for gliding like that of a gasteropod, the main part of the cephalopod foot is greatly modified for respiratory purposes. In view of the fact that there is no boundary between the head and the foot in molluscs, discussion as to whether the tentacles are part of the head or the foot is difficult and unimportant.

MANTLE. The shell has become internal and is a rather substantial plate which acts as an endoskeleton. The absence of a rigid envelope has made it possible for the mantle to become very mobile and to develop thick muscular layers, circular muscles running round the mantle cavity and longitudinal running towards the apex of the hump. When the latter contract and the former relax the mantle cavity enlarges and draws in water which circulates round the ctenidia; when the reverse action takes place the first effect of the contraction of the circular muscles is to draw the mantle lobe tight round the neck and then, when the contraction reaches its height, the water is expelled through the funnel. In rest these movements are gentle and rhythmic and only effect the change of water necessary for respiration. At the same time the animal is usually swimming slowly forward by the undulatory movement of the lateral fins. But if *Sepia* is alarmed or excited the muscles contract violently and the spasmodic ejection of water through the funnel causes the animal to dart quickly backwards. Equally by turning the funnel backward it can move quickly forward.

COLOUR CHANGE. Not only is the mantle highly muscular but the dermis contains large cells filled with pigment, the *chromatophores*, which can be dilated by the contraction of radiating muscle fibres attached to the cell wall. By alternate contraction and expansion of the chromatophores, waves of colour are made to pass rapidly over the surface of the animal. The colour change which is brought about in this way may be to a certain extent a response to the character of the background but it is also stated to be the expression of emotions.

LOCOMOTION. *Sepia* swims with the longest axis horizontal, the upper flattened surface is that under which the shell lies and the lower the mantle-cavity surface. These surfaces are *dorsal* and *ventral* respectively and the mouth and tentacles are anterior. All round the mantle in the horizontal plane rises a horizontal fin by which the gentler swimming movements are effected.

MANTLE CAVITY. When the mantle cavity is opened as shown in Fig. 434, the *funnel* is seen with its narrow external and wide internal openings, and at the base of it two sockets which fit corresponding knobs on the mantle. This locking arrangement ensures that the mantle fits tightly on the neck and so that all water is expelled by the funnel. At the anterior end of the visceral hump is seen the central *anus* at the end of a long papilla, so placed as to discharge the faeces directly into the cavity of the funnel, the shorter *renal papillae* immediately on each side, and on the left side only the *genital aperture*, also at the end of a long papilla. More posterior still are the large and typical *ctenidia*.

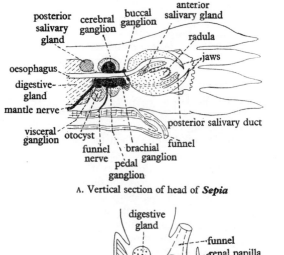

A. Vertical section of head of **Sepia**

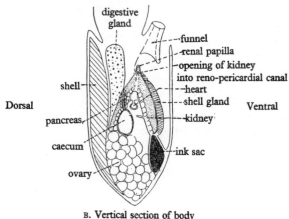

B. Vertical section of body

Fig. 435. Anatomy of *Sepia officinalis*. (B, after Grobben.)

On the face of the visceral hump in mature animals the accessory genital glands are seen through the skin; the chief of these are the shell-forming *nidamental glands* of the female which occupy a considerable area. Between these and in front of them are the *accessory nidamental glands*. Posterior to them is the *ink sac*, usually seen through the integument from which a narrow duct runs ventral to the rectum, opening into it a short distance behind the anus. The first step in dissection is to strip off the skin and then dissect out the gland and its duct as carefully as possible. It usually contains a large amount of the ink, which is composed of granules of melanin pigment formed by the oxidation of the amino acid tyrosin by the agency of an enzyme, tyrosinase.

This substance is ejected into the mantle cavity and through the funnel to form a 'smoke cloud' when the animal is attacked.

COELOM. The next stage in dissection is the opening up of the kidneys by cutting through the thin outside wall. It will at once be seen that the cavity of the organ contains a large amount of spongy excretory tissue, developed round the veins which run straight through the kidney. Just inside the renal papilla is a small rosette which carries the *renopericardial aperture*. This leads into the long narrow *renopericardial canal* running in the outer wall of the kidney and opening posteriorly into the *pericardium*, a wide space lying behind the kidneys which is only separated by an incomplete partition from the still more spacious *genital coelom* occupying the apex of the visceral hump (Fig. 435B).

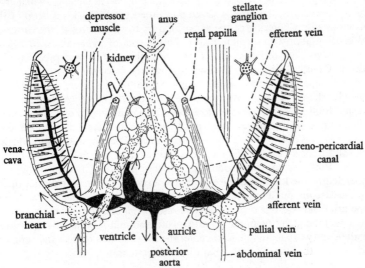

Fig. 436. *Sepia officinalis*; dissection from the ventral side to show kidneys and blood vessels. Arrows show the direction of flow of blood.

BLOOD SYSTEM. The median ventricle and the two lateral auricles are spindle-shaped bodies arranged in a line at right angles to the longitudinal axis of the body. Arterial blood is sent to the body from the ventricle by an *anterior aorta* running dorsal to the oesophagus towards the head and a *posterior aorta*; venous blood returns to the heart from the head by a very important vessel, the *vena cava*, which splits in the kidney region into two *branchial veins*, which run to the so-called *branchial hearts*, special muscular dilatations which pump blood through the capillaries of the ctenidia. The blood which is oxygenated there is sucked out of the ctenidium by the expansion of the auricle. Blood is also returned directly to the branchial heart from the mantle by the abdominal veins (and a smaller pair), and by the unpaired genital and ink sac veins which run first into the right branchial vein (Fig 436).

DIGESTION. In describing the alimentary system it must first be mentioned that *Sepia*, as a type of the Decapoda, possesses ten tentacles of which the fourth pair are longer than the others. These two tentacles have a slender stem and a swollen terminal portion to which the suckers are confined. Each tentacle can be rapidly extended and attached to the living prey, and with equal rapidity retracted into a pit near the mouth, thus bringing the food into the reach of the other tentacles, which hold it while it is being devoured. Round the mouth are frilled lips and just within it are the characteristic *beaks*, corresponding to the jaws of the gasteropod, which bite upon each other. The *buccal mass* is large and contains a well-developed radula and is traversed by the narrow oesophagus. Just within the buccal mass is the first pair of *salivary glands* and immediately in front of the digestive gland is the second pair, which produce not a digestive juice but a poison. In *Octopus*, which lives largely upon crabs, the prey is seized and bitten by the beaks, a drop of the poisonous saliva entering at the same time by the punctures in the carapace and causing almost immediate death. This is true of *Sepia* also which lives upon prawns and shrimps. The food—sometimes of considerable size—is bitten into pieces by the jaws and passed down the oesophagus (which though narrow is capable of considerable distension) to the muscular, non-glandular *stomach*. Here it is mixed with the secretion from the digestive gland and the digested food passes to the spiral *caecum*. This contains an elaborate ciliary mechanism which removes solid particles from the caecum, leaving only liquid products of digestion to be absorbed there. The digestive gland consists of a solid bilobed gland ('liver') and a more diffuse and spongy part ('pancreas'). Both are enzyme-producing, but the 'pancreas' (which in *Sepia* is suspended in the kidney sac) is also partly excretory. The single duct opens into the caecum, but a groove guides its secretion into the stomach. The 'liver' is the principal 'storage organ' for food reserves; it seems probable that these only reach the gland from the blood-stream, and that food is all absorbed in the alimentary canal, and does not enter the liver. In this respect the cephalopods appear to differ from the majority of invertebrates.

NERVOUS SYSTEM. The nervous system of *Sepia* is of great interest from the large size and intimate association of the ganglia round the oesophagus, which form a genuine 'brain' (Fig. 437B) in which special centres for the co-ordination of vital activities and for the simple reflex actions have alike been detected. In contrast to vertebrates there is a concentration of nerve cells in the brain, only a few outlying ganglia being present. For the protection of this large nervous mass a 'skull' has been developed composed of a tissue very similar to cartilage, which also forms the supports of the fins and tentacles. The nerve net found in the foot of gasteropods is absent.

The brain consists of the following ganglia: dorsally the *cerebral* or *supra-oesophageal*, ventrally (1) the *pedal*, divided into the brachial (the motor centre for the tentacles) in front and the *infundibular* (supplying the funnel) behind, and (2) the *visceral* supplying the mantle and the visceral hump. The cerebral ganglia are much more differentiated than any of the others. They can be divided into separate regions which co-ordinate the movements of organs for the performance of such complicated actions as feeding, swimming and creeping. In the visceral ganglia there are also two sharply defined centres which control the movements of the whole mantle in in-breathing and out-breathing respectively as well as numerous small centres, the stimulation of which causes contraction of small muscle patches in the mantle, while in the brachial ganglia there are separate centres for gripping by the suckers and for letting go.

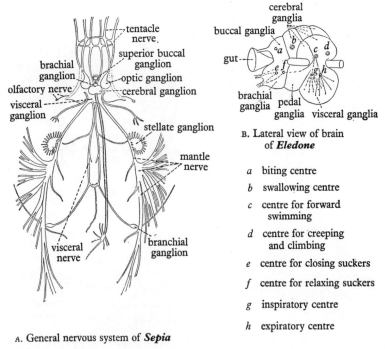

A. General nervous system of *Sepia*

B. Lateral view of brain of *Eledone*

a biting centre
b swallowing centre
c centre for forward swimming
d centre for creeping and climbing
e centre for closing suckers
f centre for relaxing suckers
g inspiratory centre
h expiratory centre

Fig. 437. Cephalopod central nervous system. (A, after Hillig; B, after Buddenbrock.)

From the cerebral ganglia there run forward a pair of nerves which end at the border of the buccal mass in a pair of *superior buccal* ganglia; a circum-oesophageal commissure links up these with the *inferior buccal*. From the visceral ganglia there is a pair of nerves running to the very prominent *stellate ganglia* in the mantle; there is also a visceral loop which sends off branches to the gills and a sympathetic loop ending in the *gastric ganglion* between the stomach and the caecum. The infundibular ganglion gives off a pair of nerves to the funnel and the brachial ganglia a separate nerve, which carries a ganglion on its course, to each arm.

In the dissection of the nervous system, a general view of the different parts of the brain is best obtained by making a longitudinal vertical section with a sharp scalpel. Such a section is shown in Fig. 435A. Afterwards the dissection of the nerves coming away from the brain can be carried out.

EYES. *Sepia* possesses very large eyes (Fig. 438 c), similar in their structure and development to those of a vertebrate. In the embryo, the eye originates as an ectodermal pit, the lining of which forms the retina and the contents of which become the vitreous humour. The pit closes up and at the point of closure the interior part of the lens is formed. Later appears a circular fold which forms the iris, limiting the pupil of the eye and forming an outer eye

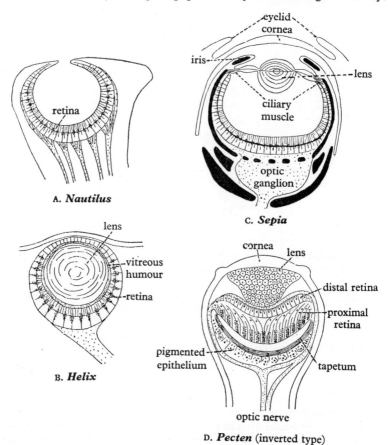

Fig. 438. Eyes of molluscs. In *Sepia* the cartilage is shown black.

chamber which is finally enclosed by the growth of a cornea. The external half of the lens is formed at the same time. A special ciliary muscle regulates the position of the lens. When it is relaxed the eye is focussed on the distance: when it contracts, increasing the pressure of the vitreous humour and so pushing the lens forward, the eye is focused on near objects.

REPRODUCTION. The ovaries and the testes are simply parts of the wall of the coelom. The ova are cells of large size; they are nourished by other peritoneal cells, the *follicle cells*, which surround the ova and pass on food from the special blood supply. The surface of contact between these cells and the egg is increased by folding. When ripe the ova escape into the genital coelom and pass into the genital duct. This has a terminal glandular enlargement and there are also the nidamental glands, unconnected with the genital ducts, which have already been mentioned. These secrete an elastic substance which forms the egg envelope.

The sperm pass similarly into the genital coelom and then by a very small aperture into the sperm duct which is modified to form in turn the seminal vesicle, the prostate gland and the terminal reservoir, called Needham's sac. All these play their part in the formation of the remarkable spermatophores, elastic tubes which by an elaborate arrangement burst and liberate the sperma-

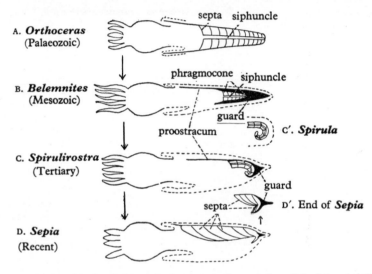

Fig. 439. A series of cephalopods to demonstrate the evolution of the internal shell. (After Naef.)

tozoa after copulation. The spermatophores are passed directly from the extended genital papilla into the funnel and then on to one of the arms (the *hectocotylus*) which is modified for the purpose of transferring the sperm to the female. In *Sepia*, the modification shows itself only by the suppression of some rows of suckers at the base of the arm, but in other forms it is profoundly modified. In *Octopus*, the end of the arm is spoon-shaped and the arm is extended so as to enter the mantle cavity of the female. In other octopods, a cyst, in which the spermatophores are stored, is formed at the end of the arm; from it a long filament is protruded. In *Philonexis* and *Argonauta* the modified arm is charged with spermatozoa, inserted into the mantle cavity of the female and then detached. This arm was described by early observers as a parasitic worm and named *Hectocotylus*.

OTHER DIBRANCHIATA. The members of this group are classified in two suborders, whose members respectively possess, like *Sepia*, ten arms (*Decapoda*), or, like *Octopus*, only eight (*Octopoda*). In no member of either division is there any known form in which the shell is external; in all cases the shell is more or less rudimentary or, in the case of the Octopoda, entirely absent. There is a well-known and extremely numerous fossil group, the

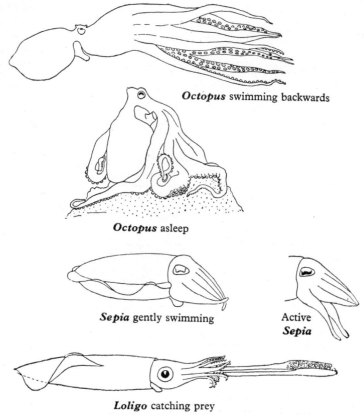

Fig. 440. External appearance of dibranchiates.

Belemnoidea (Fig. 439 B), in which impressions of the entire creature show the internal shell, the ink sac, and the ten arms beset with hooks. The shell consists of a chambered *phragmocone*, protected by a thickened *guard*, and with an anterior plate, the *proostracum*. It may well have been derived from a nautiloid form like *Orthoceras* (Fig. 439 A), as may be seen in the accompanying series of diagrams, in which the soft parts are of course partly conjectural. In a rare living form, *Spirula* (Fig. 439 c'), the chambered shell is reduced, but not quite so much as is the case in the belemnites. It is coiled and there is no guard or proostracum. Both are, however, present in the related fossil *Spirulirostra* (Fig. 439c). Finally, in *Sepia* (Fig. 439D) the guard is represented by the minute *rostrum* and, according to one interpretation, one side of the phragmocone has expanded to cover the surface of the proostracum, the septa forming the oblique calcareous partitions of the cuttle bone, while the other side forms a minute lip in which the septa are crowded together (Fig. 439D'). The siphuncle (p. 648) is a short wide funnel in between the two sides.

In *Loligo* there is only a horny *pen*, which represents the proostracum, while in the Octopoda there is no skeleton at all.

RADIATION OF DIBRANCHIATA. The Dibranchiata are specialized in two ways. The first is for a pelagic life; their bodies become elongated, fins develop and they become transparent. They may, exceptionally, develop such speed in the water that they take off from the surface and glide for considerable distances through the air, in the manner of the flying fish, aided by their spreading fins (*Todarodes sagittarius*). *Loligo* (Fig. 440) is a well-known example of the pelagic type and may be seen in aquaria swimming in troops, keeping their distances and turning with military precision.

The second mode of specialization is for a semisedentary life on the bottom. In this the body is short and the arms, which are much larger and more mobile than in the other type, are used for crawling. *Octopus* (Fig. 440) hides itself among stones and seeks its prey only at night. *Sepia* and *Sepiola*, though capable of active movement, spend long periods of rest half-covered with sand, assuming by means of chromatophore expansion brown ripple-marking on their mantles. The most sedentary form is the flattened *Opisthoteuthis*, which is almost radially symmetrical and has a remarkable resemblance to a starfish; the arms are all joined together and form a suctorial disc by which the animal applies itself to a rock.

2. Order **Tetrabranchiata**

DIAGNOSIS. Cephalopoda with well-developed calcareous shells. Living forms with two pairs of ctenidia and kidneys; arms very numerous, without suckers; eye simple; chromatophores absent; funnel in two halves.

CLASSIFICATION

Suborder 1. **Nautiloidea.** With membranous protoconch, central siphuncle and simple suture line, e.g. *Nautilus, Orthoceras*.

Suborder 2. **Ammonoidea.** With calcareous protoconch, marginal siphuncle and usually complicated suture line, e.g. *Phylloceras, Baculites*.

Nautilus. A brief description of *Nautilus*, the only surviving cephalopod with an external chambered shell, must be given here. The shell is coiled in a plane spiral; the earliest formed portion was membranous and is represented by a small central space. In the ammonoids there is a calcareous chamber, the *protoconch*, in this position. Succeeding this are the numerous *chambers*, separated from each other by the curved *septa*, each one marking a stage in the animal's growth. As the shell is added to, the animal moves forward and from time to time shuts off a space (the chamber) behind it by the secretion of a new septum. The terminal *living chamber* is much larger than the rest and is occupied by the body of the animal. All the others contain gas (which differs from air in its smaller proportion of oxygen); by means of this the heavy shell is buoyed up in the water and the animal can swim freely. The septa are perforated in the middle and traversed by the *siphuncle* which is a slender tubular prolongation of the visceral hump. It contains blood vessels and probably secretes gas into the chambers to maintain a constant pressure.

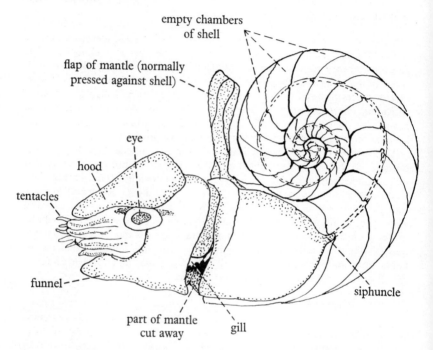

Fig. 441. *Nautilus*; animal partly extruded from the shell.

The relations of the different parts of the body in *Nautilus* are easily compared with those in *Sepia* (Fig. 433). The shell coils forward over the neck of the animal (exogastric); the mantle cavity is posterior as in all cephalopods. In other words differential growth of the visceral hump is not here associated with torsion. The *mantle* is thin and adheres to the shell; it cannot therefore be associated with the respiratory and locomotory movements. The 'head-foot' is produced into two circles of *arms* which are very numerous; they are retractile and adhesive but have no suckers. The anterior part of the region where it touches the shell is very much thickened to form the *hood*, and when the animal is retracted into the living chamber the hood acts as an operculum. The third region of the head-foot is the *funnel*, here composed of two separate lobes.

DIFFERENCES BETWEEN NAUTILUS AND DECAPODS.

The other principal points in which *Nautilus* differs from the rest of the living cephalopods are as follows:

(1) There are *four* ctenidia and *four* kidneys, without renopericardial apertures. The pericardium opens independently to the exterior by a pair of pores. The fact that in the most primitive cephalopod now existing there is a kind of segmentation of the body cavity and mantle organs has been taken to support the origin of the cephalopods from a metamerically segmented ancestor. This 'segmentation' may, however, be secondary. Certainly the

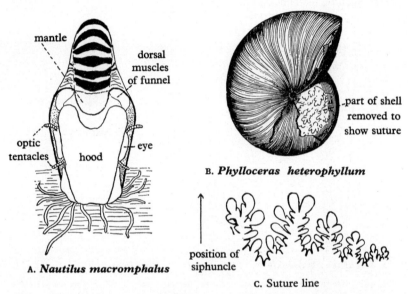

Fig. 442. A, *Nautilus macromphalos* adhering to the substratum in a vertical position by means of its tentacles. It usually lies horizontally. (After Willey.) B, *Phylloceras heterophyllum*, from the Lias. Part of the shell has been removed to expose the sutures. C, suture line of *Phylloceras*. (From Woodward.)

absence of a renopericardial connexion is not a primitive feature. There is nothing to prove that the fossil chambered-shell cephalopods had four ctenidia and four kidneys.

(2) There are very simple eyes (Fig. 438A) consisting of an open pit with no lens, the surface of the retina being bathed by sea water. This appears to be a primitive feature, but *Nautilus* is nocturnal and the eyes may have undergone reduction.

(3) There is no ink sac in *Nautilus*, nor apparently in the other forms grouped in the Tetrabranchiata.

Nautilus lives at moderate depths on some tropical coasts. It either swims near the bottom or crawls over the rocks, pulling itself along by its arms like *Octopus* (Fig. 440). The gentler swimming movements are caused by the contraction of the muscles of the funnel only; the more violent movements are probably caused by the animal suddenly withdrawing into the shell, thus expelling the water from the mantle cavity. It is nocturnal and gregarious and a ground feeder.

The chief interest of *Nautilus* lies in the fact that it is the sole living representative of a vast multitude of cephalopods with external chambered shells which flourished between the earliest Cambrian and the late Cretaceous period, a space of time embracing much the longest part of the history of life on the earth. After being the dominant type of marine invertebrate in the Mesozoic they suddenly became extinct, and the Cephalopoda are now mainly represented by the Dibranchiata with their internal shells.

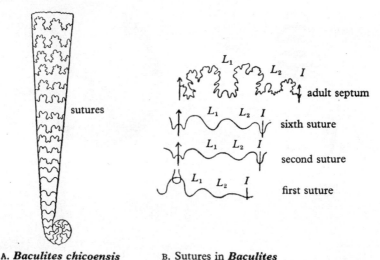

A. *Baculites chicoensis*   B. Sutures in *Baculites*

Fig. 443. Patterns of sutures in the ammonite, *Baculites* (from chalk). Note the variation in the pattern of the suture with age. L_1, L_2 = 1st and 2nd lateral lobes. (After Perrin-Smith.)

FOSSIL FORMS. The Tetrabranchiata are divided into two groups, the nautiloids and the ammonoids. The first of these contains *Nautilus* and other forms which agree with it in the position of the siphuncle and the shape of the septum. They reach their maximum development in the early Palaeozoic, where the dominant forms have straight shells like *Orthoceras* and *Actinoceras*, which were sometimes as much as 8 feet long. It is difficult to suppose that shelled animals of this size were anything other than sedentary organisms. There is a tendency for the shell to become coiled in other forms, exhibiting itself first in slightly curved forms like *Cyrtoceras*, then in loosely coiled forms like *Gyroceras* and finally in the closely coiled *Nautilus*. There is also the reverse tendency, and in *Lituites* the young shell is closely coiled but in adult life it straightens out completely.

The ammonoids appeared first of all in the middle of the Palaeozoic but reached their zenith in the Mesozoic. From the beginning of the Trias onward new families, genera and species are ceaselessly evolved. These are differentiated by the shape and sculpture of the shell whorls, but particularly by the patterns of the *suture line*, that is, the junction line of the septum and the outer shell (Fig. 442B). These patterns reach the greatest complexity. A great deal of interest attaches to the fact that in these characters the earlier formed chambers of an ammonoid individual usually differ from those of the adult shell (Figs. 442B, 443). There may, in fact, be several changes in the life of an individual and the succession of such changes has been recorded as evidence for tracing the descent of particular ammonoids. The most striking manifestation of the phenomenon is afforded by ammonoid stocks, particularly in the Cretaceous, in which the approach of extinction is heralded by 'uncoiling' in various stages. In *Scaphites* the shell is coiled in youth but later straightens out and finally hooks back. In *Baculites* (Fig. 443) only the very earliest chambers form a coiled shell; nearly the whole of the shell is straight. But the suture lines, though tending to become simplified, show the type of the family from which the uncoiled form is derived, and it is possible to show quite definitely that such genera as '*Scaphites*' and '*Baculites*' are not natural but polyphyletic; both scaphoid and baculoid forms occur in different lines of descent.

CHAPTER XVI

THE MINOR COELOMATE PHYLA

This chapter includes four phyla (1) Ectoprocta; (2) Brachiopoda; (3) Chaetognatha; (4) Phoronida. The phylogenetic position of these animals is not clear and they are grouped together in this chapter more as a matter of convenience than to indicate phylogenetic relationship. This chapter can be compared with the chapter on the minor acoelomate phyla, page 228.

PHYLUM ECTOPROCTA

DIAGNOSIS. Coelomate unsegmented animals, always sedentary and nearly always colonial; with a circumoral ring (*lophophore*) of ciliated tentacles, and a U-shaped alimentary canal; with anus outside the lophophore; with a coelomic body cavity and a lophophore retractile into a tentacle sheath; without definite excretory organs; usually with a ciliated free-swimming larva; asexual reproduction by budding.

The ordinary individuals in a colony of polyzoa at first sight resemble hydroid polyps—in their general shape, size and circle of tentacles. Closer inspection shows that they are triploblastic animals. In the majority of the Ectoprocta each individual consists of two distinct parts, the *zooecium* or body wall, and the *polypide*, consisting of the alimentary canal, the tentacles and the tentacle sheath (which contains the tentacles when contracted). The polypide can be entirely retracted within the zooecium and, as will be seen below, has a much shorter life than the latter.

CLASSIFICATION

Class 1. PHYLACTOLAEMATA. Fresh-water Ectoprocta with a horseshoe-shaped lophophore, an epistome and statoblasts.
Plumatella, Cristatella.

Class 2. GYMNOLAEMATA. Ectoprocta mostly marine, with a circular lophophore, without an epistome, with various types of trochosphere larva.

Order 1. CYCLOSTOMATA, with tubular zooecia, aperture without operculum, embryonic fission characteristic. *Crisia.*

Order 2. CHEILOSTOMATA, with aperture of zooecium closed by an operculum. *Bugula, Flustra, Membranipora.*

Order 3. CTENOSTOMATA, with aperture of zooecium closed by a folded membrane when the lophophore is retracted. *Alcyonidium.*

GENERAL ACCOUNT. The form chosen for illustration, *Plumatella* (Fig. 445), belongs to a class (Phylactolaemata) of the Ectoprocta in which the lophophore is not a simple circle, as is the case in the other class, the Gymnolaemata, but is horseshoe-shaped. A small ridge, the *epistome*, overhangs the mouth in this group but not in the Gymnolaemata. The mouth opens into the *oesophagus* which passes into a capacious *stomach* with a *caecum* which is attached by a strand of mesoderm, the *funiculus*, to the body wall. From the upper end of the stomach, the *intestine* runs to the anus which is situated just outside the lophophore. The food, consisting of small organisms like diatoms

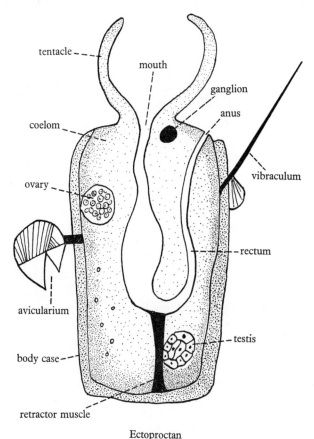

Ectoproctan

Fig. 444. Diagram of the structure of an ectoproct polyzoan.

and protozoa, is collected by the cilia of the lophophore and transported through the whole of the alimentary canal by cilia.

COELOM. The body cavity in all Ectoprocta is a true coelom containing a colourless fluid, and the cells which line it give rise to the germ cells. They are hermaphrodite; the testes are formed on the funiculus and the ovary on the body wall. When the germ cells are ripe the so-called *intertentacular organ* often appears; this is a tube which opens within the lophophore and serves for the escape of the genital products. Part of the coelom is shut off from the rest by an incomplete septum, as the *ring canal* which is prolonged into the tentacles. The intertentacular organ opens internally into this.

NERVOUS SYSTEM. The nervous system is represented by a single ganglion, situated between the mouth and the anus, and many nerves chiefly supplying the tentacles and gut. There are no special sense organs. No trace of a *vascular system* exists.

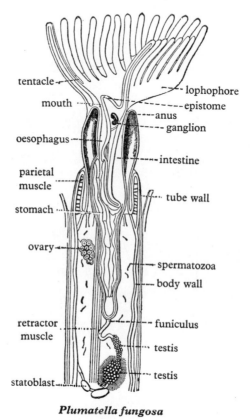

Plumatella fungosa

Fig. 445. View of the right half of *Plumatella fungosa*. (After Allman and Nitsche.)

BROWN BODY. A remarkable phenomenon very characteristic of the Ectoprocta is the formation of the brown body. Tentacles, gut, in fact the whole of the polypide, degenerates and forms a brown, compact mass. A new polypide is regenerated from the zooecium and the brown body often comes to lie in the new stomach and is evacuated through the anus. This periodical renewal of the individual is a normal process in most polyzoa (Fig. 450). It was formerly explained as due to the accumulation of excreta in the absence of specific excretory organs. It can, however, be hardly doubted that animals

ECTOPROCTA

so small and with so great an area of epithelium in contact with the water are able to rid themselves of excreta in a simpler fashion.

MUSCLES. As triploblastic metazoa with a centralized nervous system the Ectoprocta possess a more efficient contractile mechanism than the hydroids. The most prominent feature of this is the parietal system of muscles which circle round the body wall. By their contraction the internal pressure is raised and the polypide protruded. The retractor muscle which runs from the lophophore to the opposite end of the zooecium has an opposite action to the parietal system. The Ectoprocta are fascinating but exasperating objects under the microscope: they emerge with infinite caution from the zooecium and withdraw with incredible rapidity. With the lophophore a flexible part of the body wall is also invaginated and this is called the *tentacle sheath*.

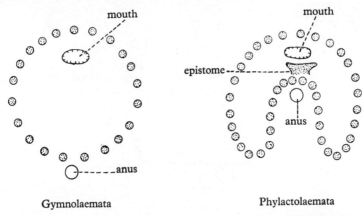

Fig. 446. Structure of the lophophore in the two main classes of the Ectoprocta.

The colonies differ greatly from those of hydrozoa in their habit and this is largely due to the absence of a connecting *coenosarc*. They are often incrusting like *Membranipora* and *Flustra* (hence the name of 'sea mats'), with all their zooecia packed closely together in a single layer; they may also be slender or massive; in the latter case they have a superficial resemblance to the actinozoan corals. While the outer layer of the body wall is often horny or flexible it frequently becomes incrusted with calcium carbonate and thus rendered rigid.

In the incrusting forms, especially the Cheilostomata, the zooecia are rigid boxes, in contact with one another along all four sides and with the substratum at the bottom. These are usually strongly calcified and only the top of the box, the *frontal surface*, is flexible (Fig. 447A).

The parietal muscles, which in primitive ectoprocts formed a continuous layer of circular muscles as in Chaetopoda, here form detached groups running from the side walls through the coelom to the frontal surface. When the muscles contract the latter is depressed and the lophophore is protruded. The process of calcification may extend to the frontal membrane and the mechanism of protrusion has then to be altered. In one large group of the

Cheilostomata there is a membranous diverticulum of the ectoderm under the calcareous frontal surface. This is called the *compensation sac* (Fig. 447B); to its lower surface the parietal muscles are attached. When they contract and the tentacles are extruded the sac fills with water, and when they relax the sac empties.

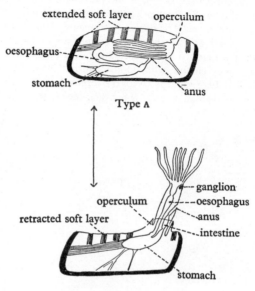

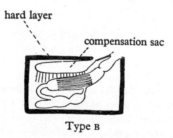

Fig. 447. Protrusion of the polyp of two types of cheilostomatous Ectoprocta. When there is a hard case to the polyp B the polyp can only protrude if water flows into a compensation sac. (After Harmer.)

POLYMORPHISM. Polymorphism is a feature of ectoproctan as it is of hydrozoan colonies. Perhaps the most remarkable modifications are to be seen in the individuals known as *vibracula* and *avicularia* of such forms as *Bugula* (Fig. 448). The vibracula are nothing more than long bristles which are capable of movement and often act in concert throughout a part of the colony, sweeping backwards and forwards over the surface, preventing larvae and noxious material from settling on the colony. The avicularia resemble the head of a bird, possessing a movable mandible which is homologous with the operculum of an unmodified polyp, and this is provided with powerful muscles. The avicularia suddenly snap their jaws and enclose as in a vice small

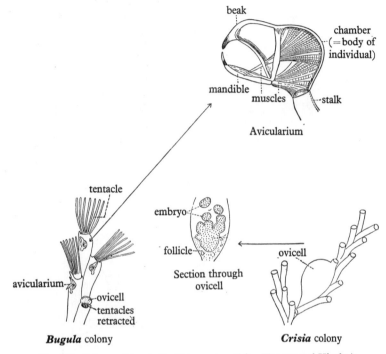

Fig. 448. Polymorphism in the Ectoprocta. (After Harmer and Hincks.)

roving animals which touch them, particularly the larvae of incrusting animals. In the most primitive cases, an avicularium is found in the same position in the colony as an ordinary zooecium and may even possess a functional polypide (Fig. 449). Further evolution led to displacement of the avicularia so that they became appendages of other zooecia, situated near the orifice. The two kinds of modified individuals thus perform tasks which in the Hydrozoa are allotted to the dactylozooids and in the Echinodermata to the pedicellariae. There are other modified zooecia as in the ovicell and the rooting cells.

Most of the Ectoprocta are marine and are amongst the most familiar objects of the beach. A complete division, the Phylactolaemata, are fresh-water. The marine forms possess a variety of free-swimming larvae, which are of the trochosphere type. In the Phylactolaemata, certain internal buds called *statoblasts* are formed from lens-shaped masses of cells on the funiculus and are enclosed by chitinous shells. The polypides die down during the winter and in the spring the statoblasts germinate and produce new colonies.

The free-swimming larvae may be assigned to the 'trochosphere' type. In most cases they are much modified and only the *Cyphonautes* larvae among the Ectoprocta possess an alimentary canal and are able to feed. The Cyphonautes is, then, the typical form (Fig. 452). It possesses a bivalve shell, each

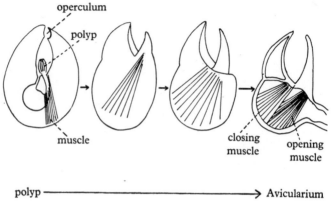

Fig. 449. Development of the avicularium from the normal type of polyp.

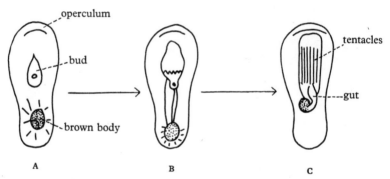

Fig. 450. Budding accompanied by brown body formation.

valve being triangular. The *apical organ* and *ciliated ring* (corresponding to the prototroch) can be seen projecting from between the valves, and in addition there are various characteristic organs, such as the *internal sac*, by which attachment is effected, prior to metamorphosis, and the *pyriform organ* of unknown function. On attachment the alimentary canal degenerates and the first individual of the colony is formed from a polypide bud consisting of an internal layer of ectoderm and an external of mesoderm. The ectoderm gives rise to the tentacles and tentacle sheath, the ganglion and the alimentary canal of the new polypide. A polypide bud which develops in exactly the same way is formed in the course of regeneration after the formation of a brown body.

ECTOPROCTA

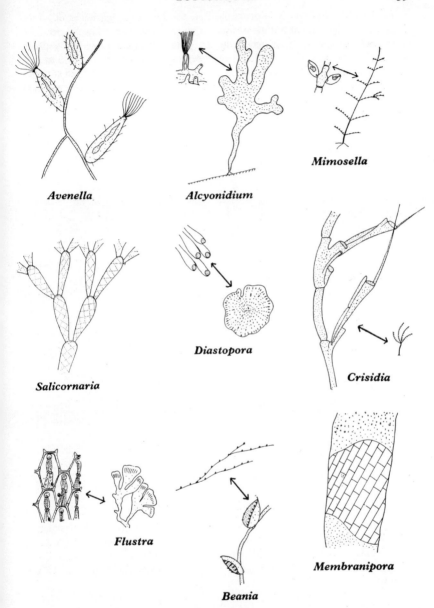

Fig. 451. Ectoprocta.

In the Cyclostomata it is probable that the fertilized egg never develops into a single individual but always into a large number by what is known as *embryonic fission*, such as occurs in the parasitic Hymenoptera. A much modified zooecium, the so-called *ovicell*, serves as a brood pouch and in that the *primary embryo* is formed and attached to follicular tissue which supplies nourishment. Masses of cells are nipped off to form the *secondary embryos* each of which becomes a free-swimming larva (Fig. 448).

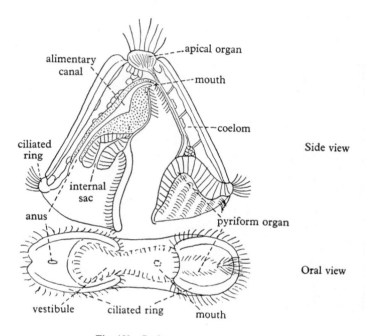

Fig. 452. Cyphonautes larva.

PHYLUM BRACHIOPODA

DIAGNOSIS. Coelomate unsegmented animals with a bivalve shell which is always attached, the valves being respectively dorsal and ventral in position; a complex ciliated circumoral organ, the *lophophore*, which maintains a circulation of water in the mantle cavity and leads food currents to the mouth.

CLASSIFICATION

Class 1. ECARDINES. Brachiopoda having shells with no hinge, no internal skeleton, and alimentary canal with an anus.   *Lingula, Crania.*

Class 2. TESTICARDINES. Brachiopoda having shells with hinge and internal skeleton, without anus.   *Terebratula, Waldheimia.*

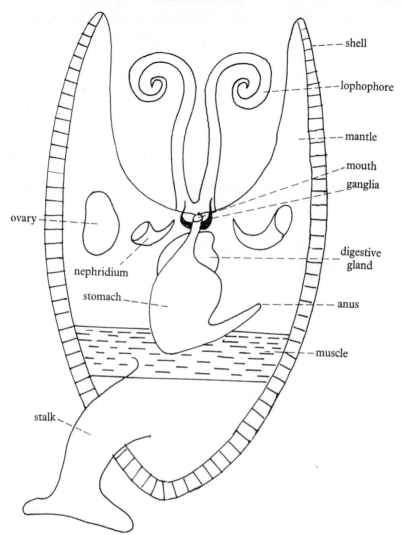

Fig. 453. Diagram of brachiopod organization.

GENERAL ACCOUNT. The group contains only marine animals with a strong but superficial resemblance to the lamellibranchs among the Mollusca. In the Palaeozoic and Mesozoic it was more abundantly represented than the Mollusca, but at the present day it contains but few genera and species. Of the former *Terebratula* and *Waldheimia* (in which the valves meet to form a hinge and which belong to the Testicardines) are found in deep water off our own coasts. Examples of hingeless forms (Ecardines) are *Crania*, which occurs abundantly in shallow water in the West of Ireland, and *Lingula*, which is not found in Britain, but in the tropics is sometimes exceedingly abundant in mud between tide-marks.

In such forms as *Waldheimia* and *Terebratula* (Figs. 454, 456), the ventral shell valve is larger than the dorsal and has a posterior beak or *umbo* perforated by a round aperture through which passes the *stalk* for attachment to a stone or rock. Each valve is secreted by a corresponding *mantle flap*, but in a way which differs from the corresponding process in the Mollusca. The mantle epithelium is produced into minute papillae which traverse the substance of the shell. The cells, of which the papillae are composed, are often of a minutely branching type which resemble the bone corpuscles of vertebrates. It must be supposed that the papillae are concerned with the secretion and growth of the shell. Each valve (Fig. 455) is composed of an outer layer of organic material (*periostracum*), under which is a thin layer of pure calcium carbonate

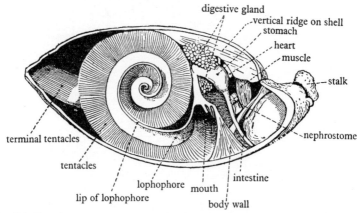

Fig. 454. Longitudinal section of *Magellania* (*Waldheimia*), slightly to the right of the mid line. (After J. J. Lister.)

and a thick inner *prismatic layer* composed mainly of calcareous but partly of organic material. The shell valves are opened and closed by a muscle system which is much more complicated than that of the lamellibranchs.

The hinge line is posterior and the mantle cavity is thus anterior. On opening the shells it is seen to be largely occupied by a complicated organ known as the *lophophore* of which a description follows. The *mouth* is placed in a transverse groove which is bounded dorsally by a continuous lip and ventrally by a row of tentacles. The groove is enormously extended and its boundaries drawn out laterally into two *arms* which are often coiled spirally in these and other members of the phylum. The tentacles are long and may be protruded from the shell opening. The cilia on the tentacles and on the mantle surfaces produce two ingoing currents of water at the sides opposite the two arms of the lophophore; the outgoing current is central, between the two arms (Fig. 457). This ciliary mechanism is similar to that of the lamellibranch ctenidium. On each side the current of water is directed between the tentacles of the lophophore, and the smaller and lighter particles suspended in it are sieved away and pass into the ciliated buccal groove and so towards the mouth. Heavier particles drop on to the ventral mantle lobe and are removed by outgoing ciliary currents and sudden clapping movements of the valves. When the ingoing currents have passed between the spirals of the lophophore they unite in the median dorsal part of the mantle cavity and become the outgoing current. The lophophore of Testicardines is supported by calcareous processes of the dorsal valve (the *brachial skeleton*) which assumes diverse and diagnostic forms in the different genera.

The mouth leads into a ciliated alimentary canal. There is a *stomach* into which opens the *digestive gland* composed of branching tubes in the cavity of which most of the digestion takes place. In *Waldheimia* the *intestine* ends blindly, but in *Lingula* and *Crania* there is an *anus*. The coelom is spacious and

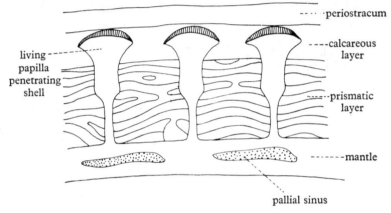

Fig. 455. Vertical section of the shell of *Magellania* (*Waldheimia*) *flavescens*. Note the mantle papillae running through the layers of the shell. (After King.)

divided into a right and left half by a dorsoventral mesentery; transverse mesenteries also exist. It is prolonged into the lophophore and tentacles and into the mantle as the *pallial sinus*. A pair of segmental organs, short tubes with large nephrostomes, which also function as generative ducts, are situated in the coelom; their external openings are at the sides of the mouth. The generative organs are developments of the coelomic epithelium and eggs and sperm alike dehisce into the body cavity. The sexes are usually separate in the brachiopods.

The blood system is very little developed and consists only of a longitudinal vessel in the dorsal mesentery, in one region of which a contractile vesicle may be distinguished as the *heart*, and a number of vessels running forward to the mouth and backward to the mantle and generative organs; all end blindly.

The nervous system mainly consists of a supraoesophageal and a suboesophageal ganglion in front of and behind the mouth respectively, connected by circumoesophageal connectives. A nerve runs to each tentacle but no special sense organs are known.

Lingula (Figs. 458, 459) is a persistent form, which has lived since the earliest period of which we have an organic record, the Cambrian, in precisely the same stage of development, if we can judge from the hard parts. It lives in mud or sand and has a very long contractile stalk by which it roots itself

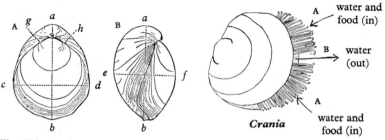

Fig. 456. *Terebratula semiglobosa*; Upper Chalk: A, posterior; B, anterior; *c–d*, breadth; *a–b*, length; *e–f*, thickness; *g–h*, hinge line; ×⅔. (From Woods.)

Fig. 457. *Crania* in the act of feeding with protruded tentacles. (After Orton.)

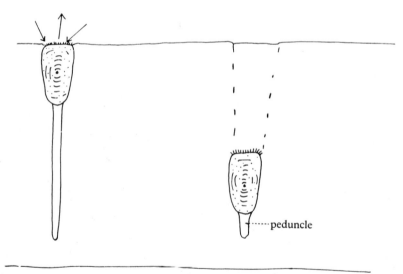

Fig. 458. *Lingula* in natural positions in mud. Arrows show directions of current.

and can withdraw from the surface. The opening of the shell is usually situated near the surface and the mantle secretes chaetae, like those of annelids, which project from the anterior border, and with the help of mucus and the mantle border form inhalant siphons at the side and an exhalant siphon in the middle. The shell valves are equal in size and horny in consistency, being composed of alternating layers of chitin and calcium phosphate.

Crania (Fig. 457) is a form without a stalk. The ventral valve is flat and fixed by its whole surface to a rock; the dorsal valve is conical. The tentacles of the lophophore are protruded from the shell margin.

The Brachiopoda have free-swimming larvae which are usually divided into three regions: an anterior like the preoral region of the trochosphere, a

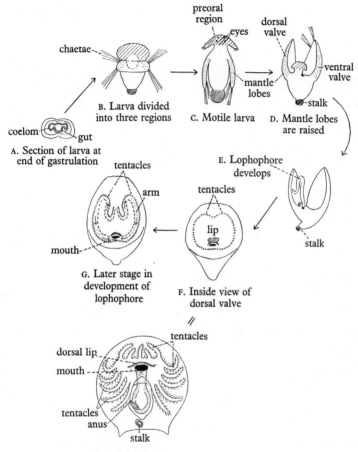

Fig. 459. Development of a brachiopod. The ciliated groove is indicated by stippling, and the movement of food to the mouth by arrows. (Altered from Delage and Herouard.)

median region in which the two lobes of the mantle are early produced, and a posterior one, hidden by the mantle lobe, which becomes the stalk (Fig. 459 B). The mantle lobes develop four bundles of chaetae (Fig. 459 C), and then turn forward to envelop the anterior region (Fig. 459 D). This now begins to develop the lophophore (Fig. 459 E, F, G) and shell valves form on the mantle lobes, while the posterior region grows into the stalk.

The coelom develops as a pair of pouches or a single pouch from the archenteron (Fig. 459A). Though the presence of mantle lobes, the presence of chaetae and the resemblance of the larva to a trochosphere relates the Brachiopoda to the annelid-mollusc stock, there is no evidence of segmentation and they cannot come very close to the Annelida; but possibly are nearer to the Mollusca. On the other hand the enterocoelic development of the body cavity suggests affinities to the echinoderms and chordates. Cleavage is usually radial though one species has shown spiral type of cleavage.

PHYLUM CHAETOGNATHA

DIAGNOSIS. Coelomate animals with an elongated body divided into three regions, head, trunk and tail, and with lateral and caudal fins; head with a pair of eyes and two groups of chitinous teeth and jaws; cerebral ganglion and ventral ganglion (in the trunk) connected by circumoesophageal commissures; body wall containing a layer of longitudinal muscle cells of peculiar type arranged in four quadrants; alimentary canal straight; no localized excretory or respiratory organs or vascular system; hermaphrodite and cross-fertilizing; free-swimming larva.

GENERAL ACCOUNT. The structure of an individual of this small and homogeneous group is shown in Fig. 460A. Very little need be added to the definition. The muscles are of a primitive type, each elongated cell consisting of a core of unmodified cytoplasm and an outer shell ring of contractile substance; they have thus some resemblance to those of the nematodes. The chaetognaths are, however, capable of executing very rapid movement by suddenly contracting these longitudinal muscles and are able to pounce upon and capture their food, which consists of diatoms, copepods and larvae of various kinds including fishes, in fact of most of their planktonic neighbours. These are seized by the hook-like jaws and swallowed whole.

The *coelom* is well developed with a distinct epithelial lining, and it is divided into two halves by a complete median and vertical mesentery, and also by two transverse septa into three chambers corresponding to the head, the trunk and the tail. Of these the head cavity is mainly occupied by the jaw muscles, while in the trunk and tail cavities are developed the ovaries and the testes respectively. The *ovaries* (Fig. 460B) are elongated solid organs attached laterally to the body wall. Traversing each ovary on its inner side is a duct with a blind anterior end (*oviduct*); this encloses a second duct (*sperm pouch*) also with a blind anterior end and with indefinite walls, containing sperm derived from another animal. Both ducts open into a small bulb-like *seminal receptacle* with an external aperture just in front of the second septum. The maturing egg is fertilized by a spermatozoon which passes into the ovary from the second duct and the zygote then passes through the wall of the oviduct and then to the exterior.

There is a solid *testis* in each half of the tail cavity and from these sperm mother cells are constantly budded off into the coelom, which is thus filled with sperm in all stages of development. The sperm passes into *vasa deferentia*, which are long tubes with a small internal opening behind the testes and a terminal dilatation, the *vesicula seminalis*, which opens to the exterior.

The eggs are laid in the sea and develop rapidly, passing through typical blastula and gastrula stages, after which the coelom is developed as a pair of anterolateral pouches of the archenteron (Fig. 460c). After gastrulation two cells become very prominent. These are the mother cells of the generative organs. The primary coelomic cavity is divided up first of all by the separation of the head cavity (Fig. 460 D) and at a later stage by a second septum between

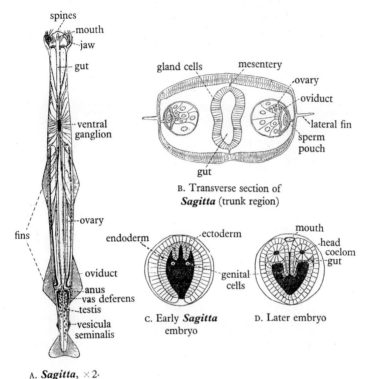

Fig. 460. *Sagitta*. In C the pouches still open into the archenteron. In D the pouches forming the head coelom have completely separated off from the archenteron, and the archenteric folds have grown back so as partly to separate off the second pair of pouches. (A, after Hertwig; B–D, after Burfield.)

trunk and tail, which divides the genital cells, which now number four, into an anterior pair, the mother cells of the ovaries, and a posterior pair, those of the testes.

Sagitta bipunctata is one of the most characteristic and cosmopolitan members of the plankton and is a typical pelagic organism with its glassy transparent body and its powers of vertical migration; off the coast of California it lives at a depth of 15–20 fathoms during the day and the greater part of the night, but at sunrise and sunset it rises to the surface, the light intensity and temperature there being at an optimum for the species at those times. Other living examples are *Spadella, Krohnia*.

The Chaetognatha are a very early offshoot of the coelomate stock and cannot very well be compared to any other phylum. While it is tempting to liken the tripartite division of the coelom in Chaetognatha with that in echinoderms and protochordates, it must be realized that in *Sagitta* the two transverse septa arise at different times and for different reasons. There is, however, a true tail here which is elsewhere found only in the Chordata and the development of the body cavity is enterocoelic.

The fossil, *Amiskwia*, occurring in the Cambrian, has been assigned to this group, but it appears to differ from the living forms in the absence of a septum between trunk and tail and in the presence of tentacles on the head.

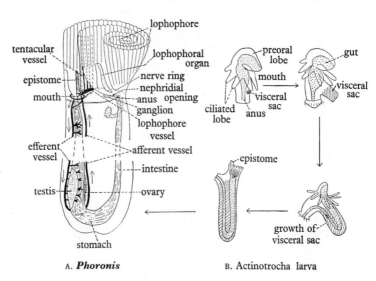

A. *Phoronis*   B. Actinotrocha larva

Fig. 461. *Phoronis*. In A the arrows indicate the direction of blood flow.
(A, altered from Delage and Herouard.)

PHYLUM PHORONIDEA

Coelomate unsegmented animals, sedentary, hermaphrodite and tubicolous, with a horseshoe-shaped lophophore, an epistome, a vascular system with haemoglobin, and two excretory organs.

This is a very small group: the genus *Phoronis* (Fig. 461) includes most of the species. They are all marine animals, usually of inconsiderable size, and like all sedentary forms they have a free-swimming larva; this is called an *actinotrocha* and it can be referred to the trochosphere type. It passes into the adult by a remarkable metamorphosis which is illustrated in Fig. 461 B.

Phoronis has a strong resemblance to a polyzoan like *Plumatella* but it differs from such a form in the presence of a vascular system and in other respects.

CHAPTER XVII

THE PHYLUM ECHINODERMATA

DIAGNOSIS. Coelomate animals; bilaterally symmetrical as larvae, radially symmetrical as adults; whose dermis contains calcareous ossicles; whose coelom in the larva consists of three segments, and in the adult forms a perivisceral cavity and several intricate systems of spaces, one of the latter being a water vascular system which pushes out the surface of the body as a series of delicate tentacles, the podia or tube feet; whose vascular system is represented by strands of lacunar tissue; whose principal nervous system remains in contact with the ectoderm from which it arose (though it may be invaginated with the latter); which have no nephridia; and whose gonads discharge direct to the exterior by special ducts.

CLASSIFICATION. The group includes the animals familiarly known as starfishes (Asteroidea), brittle stars (Ophiuroidea), sea-urchins (Echinoidea), sea-cucumbers or trepangs (Holothuroidea), and sea-lilies (Crinoidea) (Fig. 462).

There are two subphyla (not including all the fossil forms).

Subphylum 1. **ELEUTHEROZOA**. Free-living echinoderms without a stalk.

Class 1. **ASTEROIDEA**. Star-shaped or pentagonal echinoderms; whose arms contain caeca of the alimentary canal and are not sharply marked off from the disc; which have an aboral madreporite; open ambulacral grooves; usually have suckers on the tube feet and pedicellariae. *Asterias, Astropecten*

Class 2. **OPHIUROIDEA**. Star-shaped echinoderms whose arms are sharply marked off from the disc; arms do not contain caeca of the alimentary canal; with madreporite on the oral side; ambulacral groove covered; tube feet without suckers; and no pedicellariae. *Ophiura, Ophiocomina, Ophiothrix*

Class 3. **ECHINOIDEA**. Globular cushion-shaped or discoidal echinoderms; without arms; with small adambulacral areas in which lie the madreporite; ambulacral grooves covered; tube feet end in suckers; numerous long spines and pedicellariae.
Echinus, Spatangus, Echinocardium

Class 4. **HOLOTHUROIDEA**. Sausage-shaped echinoderms without arms, without recognizable adambulacral area, usually without external madreporite in the adult; with the ambulacral grooves covered; some of the tube feet are modified into tentacles around the mouth, and some or all of the tube feet if present are provided with suckers; a muscular body wall containing very small ossicles; no spines; no pedicellariae.
Holothuria, Psychropotes, Labidoplax

Subphylum 2. **PELMATOZOA.** Sessile echinoderms, usually stalked.

Class 1. **CRINOIDEA.** Echinoderms possessing branched arms. The oral surface is directed uppermost and the animals are attached throughout all or part of their life by a stalk which springs from the aboral apex. The tube feet have no suckers. The ambulacral grooves are open. There is no madreporite and no spines or pedicellariae.

Antedon, Rhizocrinus, Botryocrinus

The fossil echinoderms are discussed on p. 703.

GENERAL ACCOUNT

The great unlikeness between these animals and all other coelomata is chiefly due to the radial symmetry which they assume at metamorphosis and which distorts all their systems of organs to its own mould. The radii, which are nearly always five in number, diverge from the mouth. The surface of the body upon which the mouth lies is known as the *oral* or *ambulacral*, the opposite surface as the *aboral* or *abambulacral*. The terms 'ventral' and 'dorsal' should not be applied to these surfaces, for they correspond not to the ventral and dorsal but to the left and right sides of the larva. The anus, if present, lies usually on the aboral side, but in the Crinoidea it lies on the oral side. The alimentary canal runs a straight or devious course from mouth to anus. The other systems consist each of a ring around the axis which passes through the mouth and the middle of the aboral side, and a tube or cord along each radius. The *radii* are constituted by the presence of the radial members of the various systems. The areas between the radii are known as *interradii*. Most of the systems lie close under the ambulacral surface, and the tube feet project from it, forming radial bands known as the *ambulacra*. In the Asteroidea and Crinoidea the tube feet of each ambulacrum stand on either side of an *ambulacral groove* at the bottom of which lies the highly nervous strip of epithelium which is the radial 'nerve cord'. In the other classes the ambulacral groove is roofed in, forming an *epineural canal* over the nerve cord. In the Asteroidea, Ophiuroidea and Crinoidea the body is prolonged as *arms* in the direction of the radii, and the ambulacral and abambulacral surfaces are subequal in extent. On the other hand, in the spherical or cushion-shaped Echinoidea and the sausage-shaped Holothuroidea, the body is compact, and the ambulacral surface extends over most of it, leaving only in the Echinoidea a small, and in the Holothuroidea a minute, aboral area opposite to the mouth (Fig. 462). Externally and internally the symmetry is never quite perfect. At best the presence of the madreporite (see below), or of the anus, or of a genital opening, differentiates one of the interradii, and in some echinoids and holothurians a new and conspicuous bilateral symmetry has developed, and affects a number of organs.

In life, the Crinoidea are fastened to the ground by a stalk which arises from the middle of their aboral surface, and, though a few of them break free when they are adult, the mouth is directed upwards by them all. The other existing groups (Eleutherozoa) are free. In the Asteroidea, Ophiuroidea, and Echinoidea the mouth is directed downwards. The Holothuroidea apply one side of the long body to the ground, so that the mouth is directed horizontally (Fig. 462).

TUBE FEET. The tube feet (*podia*), whose function was perhaps originally a sensory or food-collecting one, are (or some of them are) in the Asteroidea,

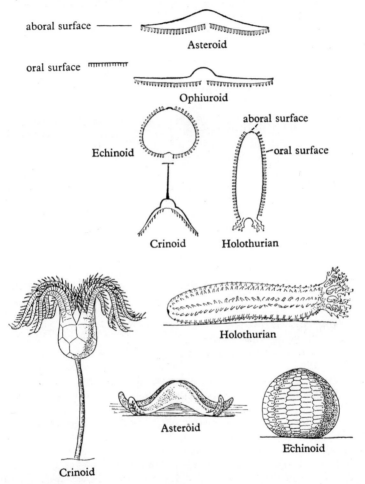

Fig. 462. Diagrams to show the relative extent of the oral and aboral surface and to compare the form of body in the several classes of echinoderm. The upper figures show the animals in the same morphological position, the lower figures show the animals in their natural position.

Echinoidea, and Holothuroidea adapted, by the presence of suckers at their ends, to walking. Probably they always subserve respiration, and in the 'irregular' echinoids some of them are modified for this function. They may also be modified for seizing food. They are protruded and retracted by alterations of the pressure of the fluid within them.

EPIDERMIS. The epidermis is usually ciliated, but not in ophiuroids or, except in the ambulacral groove, in crinoids. Usually, also, it contains gland cells and sense cells, the latter with their bases prolonged into fibrils which enter a plexus, formed by them and by branched nerve cells, among the bases of the epithelial cells—a nerve-net. The characteristic *ossicles* of the dermis may be scattered, so as merely to impart a leathery consistency to the skin, or united by muscles as a skeleton, or firmly apposed so as to constitute an armour. Some of them usually project as *spines*, over which the epidermis may presently wear away. *Pedicellariae* (Fig. 483) are sets of two or three spines arranged to bite together as pincers. They are of various types, often complicated, but only occur in asteroids and echinoids.

ALIMENTARY CANAL. The alimentary canal differs greatly in the several groups. It is axial in the Asteroidea and Ophiuroidea, coiled in the other classes. It possesses various diverticula in different cases, but not large glands like those which are common in other phyla. The anus is lacking in the Ophiuroidea and a few asteroids, and when present is more or less excentric except in the Holothuroidea.

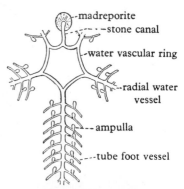

Fig. 463. A diagram of the water vascular system of a starfish. (From Borradaile.)

COELOM. The coelom of the adult is present as several distinct systems of spaces, the following being the most important:

(1) The large *perivisceral cavity* in which lie all the principal viscera.

(2) The *perihaemal system* consists of a radial vessel (in asteroids divided longitudinally by a vertical septum in which lies the principal 'blood' vessel) along the radius of each arm and a ring vessel around the mouth.

(3) The *aboral sinus system* enclosing the genital rachis and the gonads (see below).

(4) The *water vascular system* (Fig. 463) which lies above the perihaemal system and consists of a ring around the mouth, a tube known as the stone canal because it is frequently calcified, leading to an opening, the madreporite; a radial vessel along each radius with lateral branches from the radius to the tube feet which, when used for walking, possess swellings called ampullae, the contraction of which extends the tube feet.

(5) The *madreporic vesicle*, an inconspicuous cavity of morphological importance.

(6) The *axial sinus* is a space which varies greatly in development. It is inconspicuous in the asteroids, small in the echinoids and ophiuroids, very small in the holothuroids and merged with the perivisceral cavity in the crinoids. It communicates with the exterior (or, as in most holothurians, with the coelom) by a pore or set of pores situated in one of the interradii. This opening forms the *madreporite*.

ECHINODERMATA

The stone canal opens into the axial sinus just below the madreporite, and so the latter serves as the opening of the stone canal. In the Asteroidea and Echinoidea the madreporite is a conspicuous structure on the aboral side, pierced by many pores. In the Ophiuroidea it is on the oral side, and has one pore, or only a few pores. In most of the Holothuroidea it becomes detached, in the course of development, with its tiny axial sinus, from the body wall, and hangs into the perivisceral cavity, with which, instead of with the exterior,

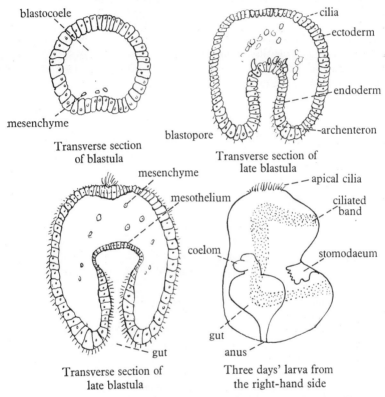

Fig. 464. Stages in the development of *Asterias vulgaris*. (After Field.)

it now makes communication, by a number of pores. In this group, by meristic repetition, there may be several or many stone canals, each with an 'internal madreporite'. In the Crinoidea, the stone canals, of which there are several, end each by a single opening into the perivisceral cavity, and the latter communicates by a number of pores with the exterior.

In the bilateral larva (*Dipleurula*), the coelom (Fig. 465) is present as three pairs of sacs, of which the first is preoral. The second pair is connected by a passage with the first: the third is independent. In outline, the relation between these sacs or segments of the larval coelom and the coelomic spaces of the adult is as follows: the perivisceral cavity of the adult is formed by the fusion of the main portions of the hinder pair; the aboral sinus system becomes separated from the perivisceral cavity; the perihaemal system arises as outgrowths from the left hinder cavity (in some cases it receives a component also from the left anterior cavity); the water vascular system ('hydrocoele') is formed by the transformation of the left second cavity (the right second cavity disappearing); the axial sinus is the persistent left anterior cavity, its madreporite being derived from a 'water pore' which puts that cavity into communication with the exterior. The opening of the stone canal into the axial sinus is the remains of the connexion between the left anterior cavity and the left second cavity, which latter, as we have seen, becomes the water vascular system. The madreporic vesicle is budded off from the right anterior cavity (the rest of which disappears); in the larva this vesicle pulsates; it probably represents the pericardium of the Hemichorda, which retains its contractile function in the adult (pp. 711, 713).

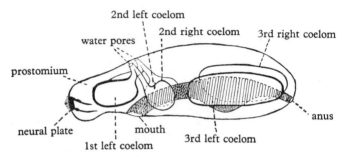

Fig. 465. Diagram of the arrangement of the coelom in the ideal dipleurula larva. (After Sedgwick.)

All echinoderms except the Holothuroidea possess a peculiar structure known as the *axial organ*, composed of connective and lacunar ('vascular') tissue, with cells derived from the genital rudiment, known as the *genital stolon*. The axial organ adjoins the axial sinus where the latter is present; in the Crinoidea it lies in the axis of the body. Its function is unknown; it has been regarded as a heart for the lacunar system on account of contractions which it is said to perform, and as an organ of excretion because in echinoids it takes up carmine injected into the body cavity.

EXCRETION. Of excretion in the echinoderms little is known. It appears to be performed by the wandering out, through the walls of the gills, of amoeboid cells laden with granules of excreta, by the organs of respiration, and by the intestine, but no constant and conspicuous organs subserve it alone. There are no nephridia. The nitrogenous excreta consist largely of ammonia compounds and contain practically no urates.

RESPIRATION. Respiration is performed through a variety of structures, some of which expose the coelomic fluid to the external water, while others carry the water into the body and expose it to the fluid in the coelom. To the first class belong the podia, and the 'gills' of asteroids and echinoids; to the second belong the 'genital bursae' of ophiuroids and the respiratory trees of holothurians.

VASCULAR SYSTEM. The vascular system of other animals is represented in the Echinodermata by a system of strands of a peculiar *lacunar tissue*, containing intercommunicating spaces which have no epithelial lining. Ultimately, this system is of the same nature as the blood vessels (haemocoele) of other animals, since both are systems of spaces derived from the blastocoele and filled by a fluid matrix containing free cells; but in appearance, and probably in the mode of its functioning, it is very different. A ring of lacunar tissue surrounds the mouth, lying in or immediately above the perihaemal ring and giving off in each radius a strand or 'vessel' which similarly lies above the radial perihaemal canal. Another portion of the system lies in the axial organ and connects the oral ring with an aboral ring, which accompanies the genital rachis (see below) and sends strands to the gonads. In the Echinoidea and Holothuroidea two strong 'dorsal' and 'ventral' vessels from the oral ring accompany the alimentary canal, running on opposite sides of that organ and giving off a plexus of branches which ramify on it, and in holothurians also in a perforated fold of the peritoneum. A 'vascular' plexus is also present on the alimentary canals of other groups. Contractions are said to have been observed in parts of the system, but it is very doubtful whether anything in the nature of a regular circulation takes place in it, though it probably maintains communication by diffusion between various parts of the body.

The main circulatory system in the echinoderms is a ciliary one. Materials are conducted through the body by means of specialized ciliary tracts leading to and coming from the more important organs of the body (Fig. 471).

REPRODUCTION. With rare exceptions, the sexes of echinoderms are separate. The *genital organs* are remarkable for their simplicity. They possess neither organs of copulation, nor accessory glands, nor receptacles for the retention of ova, nor a reservoir for the storage of sperm in either sex, and they discharge direct to the exterior and not, as is usual in coelomate animals, through the coelom or through ducts proper to that cavity. Nevertheless they arise in ontogeny from the coelomic wall. The genital system consists, except in the Holothuroidea, of the *genital stolon*, a collection of cells in the axial organ; the *genital rachis*, a ring connected with the stolon (Fig. 469) (aborally in the Asteroidea, Ophiuroidea, and Echinoidea, orally in the Crinoidea); the *gonads* proper, which are sacs or tubes, often branched, borne upon long or short branches of the rachis and varying in number in the different groups; and the short *ducts*, lacking in the Crinoidea. In the Holothuroidea there is only one gonad, which lies in the 'dorsal' interradius and has a duct in the dorsal mesentery and a vestigial stolon lying upon the duct, but no rachis.

THE NERVOUS SYSTEM. The nervous system may be classified into two systems: the sensory association system and the motor system. The sensory association system is made up of networks of the nerve axons and cells lying subepidermally in the ectoderm. This system is specialized and thicker in two regions where the nerve axons become grouped together to form distinct nerve tracts. These regions are, (1) the radial nerve cord, along the ventral surface of each arm, (2) the nerve ring, around the mouth. The radial nerve cords and the circumoral nerve ring, during the development of most echinoderms except the asteroids and crinoids, become invaginated and removed from the surface of the body. (Figs. 477, 484, 491.)

The sensory association nerves form a repeating pattern of nerve tracts running from the dorsal sheath and the tube feet to the radial nerve cord.

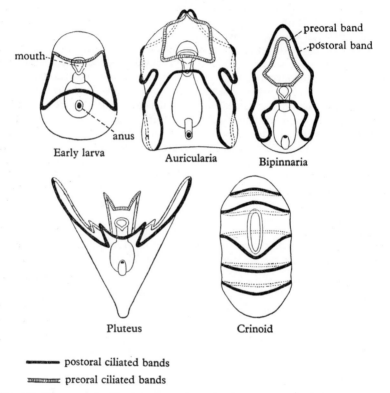

Fig. 466. Diagrams of echinoderm larvae. The postoral part of the early ciliated band is drawn heavily (except where remote), the preoral part is cross-hatched.

The motor nerves in the subepidermal plexus innervate the muscles in the arm, pedicellariae and tube feet. In addition to these there are a series of nerve cells and tracts that lie within the radial nerve cord underlying the mesodermal epithelium of the perihaemal canal. These motor nerves are sometimes called the deep oral system. They are poorly developed in the echinoids but well developed in the asteroids where they form metameric units innervating the tube feet and the dorsal sheath.

A third motor system, the aboral or apical system, is developed in the crinoids from the peritoneum of the body wall. In the adult it is removed from the general peritoneum and is enclosed in ossicles forming a nerve running along each arm and joining in the body. This system is not found in the Holothuroidea and though it was believed to be present in the asteroids it has recently been shown that the asteroid system is really a lateral continuation of the deep oral system.

The asteroid nervous system possesses many ganglia, groups of cells and axons making synaptic connexion with other dendrites. The starfish has these ganglia metamerically arranged, there being a ganglion lateral to each foot, a ganglion inside each tube foot, and another associated ganglion in the deep oral system.

SENSE ORGANS. The Echinodermata are poorly provided with *sense organs*. There is a general sensitiveness of the epithelium of the body, at least to tactile stimuli, which is heightened in the podia and in the *terminal tentacle* which stands at the end of each radial water vessel in the Asteroidea, Ophiuroidea, and Echinoidea. The olfactory sense is perhaps also located in the podia or in some of them, especially in those that are situated around the mouth and in the Holothuroidea are developed into tentacles. An eye-spot is situated at the base of each terminal tentacle in the Asteroidea, and certain holothurians possess statocysts in the skin.

All echinoderms are marine in habitat. Few of them are pelagic: none are parasitic. Only the Crinoidea are fixed, and some of these are only temporarily so.

LARVAE. The majority of members of the phylum have free, pelagic larvae; though some, as *Asterina*, pass a considerable time in the egg membrane and have larvae which are not pelagic; and a few, chiefly polar or deep-sea species, keep the young in brood pouches until they have the adult form. The eggs of the species which possess pelagic larvae are small; the others larger and more yolky in proportion to the lateness of the stage at which they are set free. Fertilization takes place in the sea or in brood pouches. Cleavage (radial, Fig. 194A) is total and forms a hollow, one-layered blastula (Fig. 464). This, by invagination or unipolar ingrowth, forms a gastrula with a wide blastocoele into which typical mesenchyme cells wander from the wall of the archenteron. The blastopore becomes the anus, and the mouth is formed by the breaking through of a stomodaeum. Meanwhile the archenteron has budded off, at the anterior end, a vesicle which, by processes that differ in detail in different cases, will give rise to the three segments of the coelom described above (p. 673). The future ventral side of the larva becomes concave. The larva is now known as the *Dipleurula*. The cilia which uniformly covered the blastula become sparse over most of the body but, except in the Crinoidea, grow stronger and more numerous in a *longitudinal band* around the ventral concavity. This band is the organ of locomotion. Growing more rapidly than the rest of the ectoderm, it becomes thrown into folds, the *larval arms* (which have nothing to do with the arms of adult echinoderms), whose length and arrangement differ so as to characterize a special type of larva in each class (Fig. 466).

In the *Auricularia* larva of the Holothuroidea the body is elongate and the band lengthens fore and aft and outlines a strong *preoral lobe*. The *Bipinnaria* of the Asteroidea resembles the Auricularia in general features, but in it the border of the preoral lobe separates completely from the rest of the longitudinal band. In the *Plutei* of the Ophiuroidea and Echinoidea the band

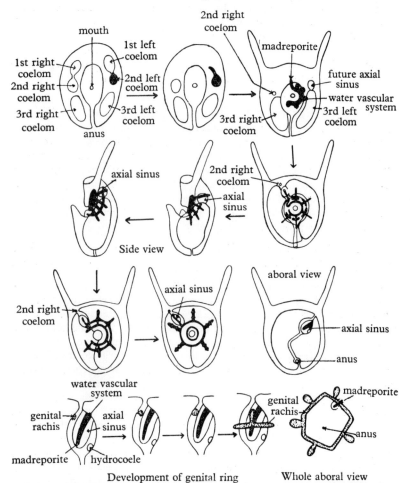

Fig. 467. Development of the adult starfish from the larval form. It will be seen that the adult rudiment of the water vascular system arises from the second left coelomic cavity which surrounds the mouth and forms the centre for the development of the adult features. The figures on the bottom line show the way in which the genital ring arises relative to the axial sinus. (After Delage and Herouard.)

remains continuous, but forms only a small preoral lobe, and the postanal region of the body develops greatly, while the slender arms are supported by calcareous rods. The *Pluteus* of the Ophiuroidea (*Ophiopluteus*) has a different appearance from that of the sea urchins (*Echinopluteus*), owing to the fact that the former of these larvae has fewer arms than the latter and that in it the arms known as the 'posterolateral arms' are the largest and are directed forwards, whereas these arms, if they are present in the *Echinopluteus*, are there small and directed outwards or backwards. The *larva of the Crinoidea* has no longi-

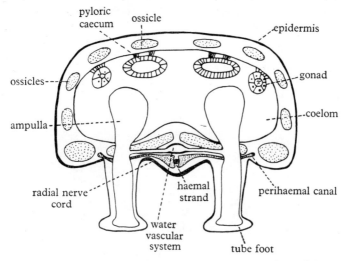

Fig. 468. A diagram of a transverse section of the arm of a starfish.

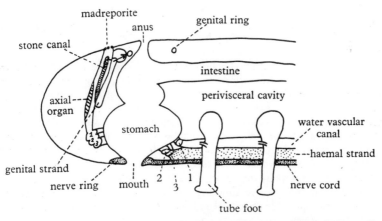

Fig. 469. A diagram of a section of a starfish passing through the madreporic interradius and along the opposite arm.

tudinal band, but five rings of strong cilia around the body. In the development of the Holothuroidea the *Auricularia* is succeeded by a stage known as the *pupa*, in which the longitudinal band breaks up and rearranges itself into a series of five transverse rings somewhat resembling those of the crinoid larva. The *Bipinnaria* of the Asteroidea is succeeded by a *Brachiolaria* which differs from it in possessing in the preoral region three processes by which the larva can hold fast to objects.

The larvae become transformed into adults by a *metamorphosis* which differs in the several classes. In all it involves an alteration of the position of the mouth, which in groups other than the Crinoidea is removed to the left side, and in the Crinoidea to the posterior end, taking with it the coelomic cavities of the left side. The fate of the several divisions of the larval coelom has been described above (p. 673). In the Crinoidea and Asteroidea the larva becomes fixed by the preoral lobe at the time of metamorphosis, a *fixation disc* developing for the purpose. In crinoids the fixation persists, at least until the adult is completely formed. In starfishes it is only temporary.

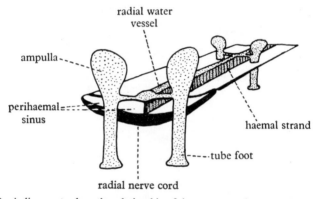

Fig. 470. A diagram to show the relationship of the water vascular canal, the perihaemal system and the haemal strand in the arm.

SESSILE ANCESTRY. The fixation of the sea lilies, and the fact that starfishes are fixed when the bilateral symmetry of the larva changes to the radial symmetry of the adult, are interesting facts in view of the fixation which is general in the other great group of radially symmetrical animals, the Coelenterata. Radial symmetry is essentially the symmetry of a sessile animal, which is in the same relation with its surroundings on all sides, whereas bilateral symmetry is that of a travelling animal, which needs differentiation not only of the upper side from that which faces the ground, but also of the fore from the hind end. It is likely that at one time all echinoderms were fixed, and that those which are now free retain the radial symmetry of their sessile ancestors. This supposition is supported by the fact that the earliest known fossil members of the phylum were fixed.

For the rest, the Dipleurula and its metamorphosis suggest that the early sessile echinoderms were descended from a free, bilateral ancestor; and the close resemblance between the Auricularia and the *Tornaria* larva of *Balanoglossus*, together with the history of the coelom (see p. 707), and the nature of the nervous system, indicate an affinity between that ancestor and the Enteropneusta.

SUBPHYLUM 1. ELEUTHEROZOA
CLASS 1. ASTEROIDEA

DIAGNOSIS. Star-shaped or pentagonal Echinodermata; whose arms contain caeca of the alimentary canal, and are usually not sharply marked off from the disc; which have an aboral madreporite; open ambulacral grooves; and usually both suckers on the tube feet, and pedicellariae.

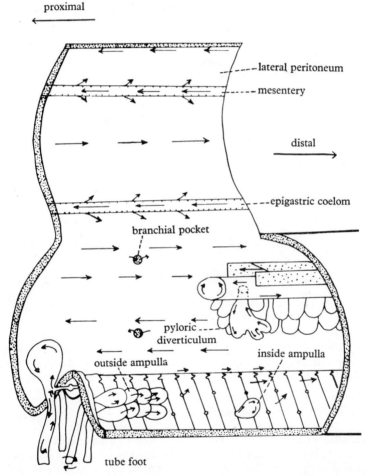

Fig. 471. Diagram of the ciliary currents in the arm of the starfish. (From Budington.)

THE OSSICLES (Fig. 468) of the body wall of a starfish may, as in the familiar *Asterias*, constitute a toughening mesh-work, or may have the form of more closely set plates, but are not united to form a continuous shell. Along the sides of the arms run two rows of strong pieces, the *supero-* and *infero-marginal ossicles*, which are hidden in *Asterias* but in many genera appear on the surface. The ossicles bear *spines*, which vary much in size and shape and arrangement, being often longer than the stumpy structures on the back of *Asterias*. Around and between the spines are usually *pedicellariae* of various kinds, the most perfect of which is the *forcipulate*, found in *Asterias*, which has a basal ossicle: its jaws may be *straight* or *crossed*. Over interspaces between the ossicles arise delicate, hollow outgrowths, the *gills*, into which the perivisceral cavity is prolonged. Above each ambulacral groove runs a double row of large, transversely placed, *ambulacral ossicles*, movable upon one another by muscles. Each has a smaller *adambulacral ossicle* at its outer

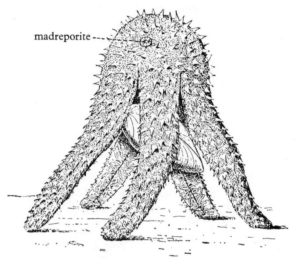

Fig. 472. *Echinaster sentus* in the act of devouring a mussel.
(From Shipley and MacBride.)

end. *Adambulacral spines* stand on the adambulacral ossicles. In *Asterias* they are long, and bear groups of large pedicellariae of the kind with uncrossed jaws. They can be turned inwards to protect the ambulacral grooves.

THE MOUTH leads through a short *oesophagus* (Fig. 469) into a large sac-like *stomach*, with two retractor muscles in each arm. Above is a five-sided *pyloric sac*, from each angle of which, separately or, as in *Asterias*, by a short common duct, arises a pair of branched *pyloric caeca*, which are slung, each by a double mesentery, from the roof of an arm: the epithelium of these secretes the digestive ferments and stores nutriment. From the pyloric sac a short, conical *rectum*, bearing in *Asterias* two glandular *rectal caeca*, rises to the anus, which is slightly excentric, in the interradius which is next, clockwise, after that of the madreporite. Animals of any kind that can be seized serve for *food*, and usually the stomach can be extruded to envelop and digest prey which are too large to be swallowed. Some species clasp bivalves with their arms

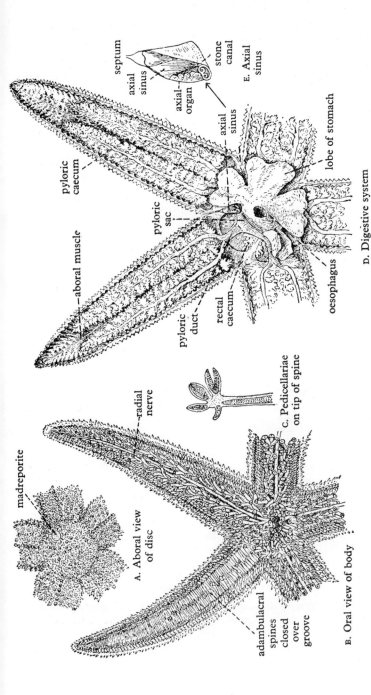

Fig. 473. Morphology of the starfish *Asterias rubens*. A, an aboral view of the disc showing the madreporite. B, oral view of the body. C, tip of an ambulacral spine showing pedicellariae. D, aboral view with dorsal sheath and part of the digestive system removed. One lobe of the stomach has been cut away, another has been turned back. E, enlarged view of the axial sinus and associated structures. (D, E, after Borradaile.)

(Fig. 472) and pull them open with the tube feet so that the everted stomach can be applied to the soft parts of the mollusc.

CAVITIES. In each interradius a stiff septum projects into the perivisceral cavity between the arms. To the septum in the interradius of the madreporite is attached a sac, the axial sinus, and into this, so as to appear to lie in it, project the axial organ and the stone canal, whose wall is calcified and infolded so as to increase its surface. Orally, the stone canal joins the water vascular ring, which bears nine small *Tiedemann's bodies*, of gland-like structure, and often, but not in *Asterias*, several stalked sacs, the *Polian vesicles*. The radial water vessel of each arm lies under the ambulacral ossicles, and between them and the radial nerve is the perihaemal sinus, divided by a

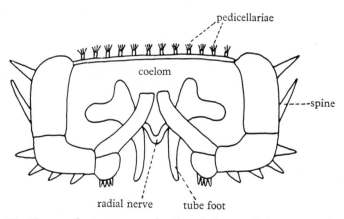

Fig. 474. Diagram of a transverse section through the arm of *Astropecten*. There is a greater degree of ossification in *Astropecten* than in *Asterias*; note also the pedicellariae grouped on the dorsal surface.

septum in which runs the 'blood vessel' (Fig. 470). The *gonads* are ten in number, shaped like bunches of grapes and varying in size with the season. They are attached to the body wall by their ducts, which open one on each side at the base of each arm.

Asterias (Figs. 468, 469, 471, 473). A typical member of the class. Its principal features have been mentioned above. British.

Astropecten (Figs. 474, 479). Without anus; without suckers on the tube feet; and with conspicuous marginal ossicles. Lives on a bottom of hard sand, into which it burrows, and upon which its tube feet are adapted to walk. British.

Asterina. With the arms short and wide, so that the body is pentagonal; and without pedicellariae. Has a shortened development, with a larva which is not a *Bipinnaria*. British: between tide-marks.

Brisinga (Fig. 479). With numerous, long, slender arms, sharply distinct from the disc, which is small. A deep-sea genus.

CLASS 2. OPHIUROIDEA

DIAGNOSIS. Star-shaped Echinodermata; whose arms are sharply marked off from the disc and do not contain caeca of the alimentary canal; with madreporite on the oral side; ambulacral groove covered; tube feet without suckers; and no pedicellariae.

THE SPECIAL FEATURES of the organization of a brittle star are connected with the fact that the animal moves, not by means of its tube feet, but by pushing and pulling upon surrounding objects with its arms. In adaptation to

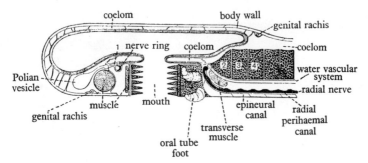

Fig. 475. Diagram of a section from an ophiuroid passing through an interradius and part of the opposite arm. 1, 2, 3, 4, are the ambulacral ossicles.

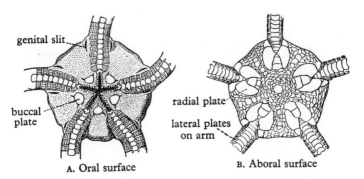

Fig. 476. A, *Ophiura*, oral surface of disc and part of the arms. B, *Ophiura*, aboral surface. (From Woods.)

this the arms are sharply distinct from and freely movable upon the disc, on the underside of which they are inserted. They are armoured by *skeletal plates* (Figs. 476, 477), in an upper, two lateral, and an under series. The epidermis is vestigial; there is a strong cuticle; spines on the side plates give grip; and the under plates, covering in the ambulacral groove, which is thus converted into an *epineural canal*, protect the nerve cord during the movements of the arm. The ambulacral ossicles of each pair fuse to form one of a series of vertebrae,

686   ECHINODERMATA

which articulate by an arrangement of knobs and sockets and can be moved upon one another in various directions by four muscles. The large vertebrae reduce the perivisceral cavity in the arm to a canal, in which there is no room for caeca of the alimentary canal. The *nerve cord* bears ganglia corresponding

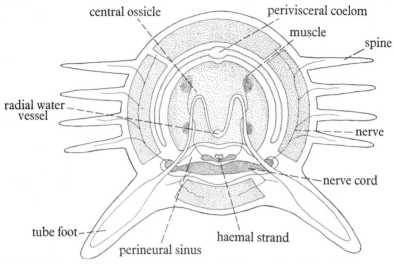

Fig. 477. Diagram of a transverse section through the arm of an ophiuroid. Note the large central ossicle and the internal nerve cord.

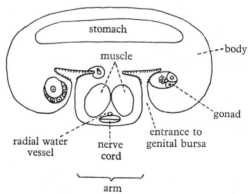

Fig. 478. Diagram of a transverse section through the disc of an ophiuroid. The section does not pass through the centre of the disc and includes a section of one of the arms.

to the muscles between the vertebrae and formed by increase of the coelomic (deep oral) component of the cord. The perihaemal vessel is shallow and not divided by a septum. The *tube feet* have no suckers and no ampullae and are often provided with warts of sense cells.

The *alimentary canal* (Figs. 475, 478) is a mere bag, not protrusible through the mouth, which is armed with an arrangement of spines serving as teeth.

OPHIUROIDEA

The food of some species consists of animals captured by the arms: others shovel mud into the mouth with the adjacent tube feet and digest the food it contains. There is no anus. The *madreporite*, aboral in the young, becomes oral in the adult because the disc, growing independently of the arms, and faster aborally, comes to overhang in the interradii. In coming over, the madreporite brings with it the axial sinus, stone canal, axial organ and madreporic vesicle, which are all orally placed. The gonads open, not directly

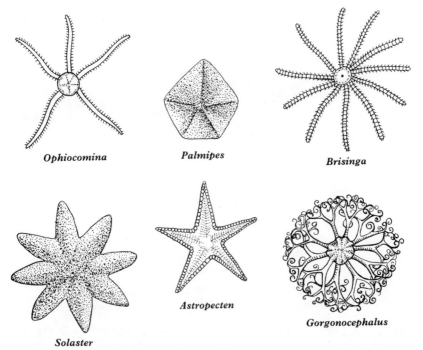

Fig. 479. Body form in asteroids and ophiuroids. Asteroids: *Solaster, Astropecten, Palmipes, Brisinga*. Ophiuroids: *Ophiocomina, Gorgonocephalus*.

to the exterior, but into *genital bursae*, of which one opens on each side of the base of each arm (Figs. 476 A, 478). The ectoderm lining the bursae retains its cilia and causes currents which subserve respiration.

Ophiura (Fig. 476), *Ophiocomina, Ophiothrix, Amphiura*. British genera, separated by relatively unimportant differences, which are chiefly evident in the ossicles and spines. *Amphiura* is hermaphrodite and viviparous.

CLASS 3. ECHINOIDEA

DIAGNOSIS. Globular, cushion-shaped, or discoidal Echinodermata, without arms; with small adambulacral area, in which lies the madreporite; ambulacral grooves covered; tube feet ending in suckers; numerous long spines; and pedicellariae.

MORPHOLOGY. The characteristic form of body of the Echinoidea is such as would result if the arms of a starfish were drawn up into the body by shrinkage of the aboral surface.

We shall describe the anatomy of this group by an account of a typical member of it—*Echinus esculentus*, a large species common in Britain. This animal (cf. Fig. 481 A) has the shape of a sphere with one side flattened, slightly polygonal in equatorial outline. In the middle of the flattened side is the mouth. Under the delicate, ciliated epidermis an armour, the shell or *corona*, composed of dermal plates firmly sutured together, encloses most of the

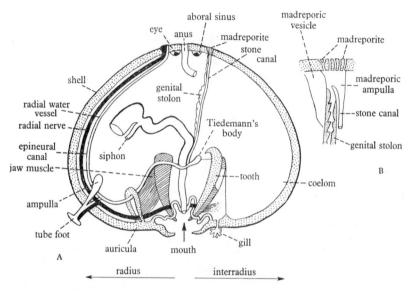

Fig. 480. A diagram of a vertical section of *Echinus* passing on the left through a radius and on the right through an interradius. Certain structures not immediately in the plane of the section are shown. A, whole section. B, region of the madreporite.

body, but at the two poles there are leathery areas, the *peristome* around the mouth, and the *periproct* in which the anus lies excentrically. The corona (Fig. 481 A) is composed of twenty meridional rows of plates, two in each radius (ambulacrum), and two in each interradius. The plates of the ambulacra are distinguished by the presence on them of the pores for the tube feet. These pores are in pairs, since each ampulla communicates with its tube feet by two canals. Thus water can circulate in and out of the tube feet and respiration is facilitated. At the aboral pole each radius ends in a single *ocular plate*, which bears the opening of the terminal tentacle, and each interradius in a *genital plate* which abuts upon the periproct and bears the opening of a gonoduct. One of the genital plates bears also the madreporite. All the plates are studded with bosses of various sizes, to which articulate the concave bases of the large and small spines and the pedicellariae.

ECHINOIDEA

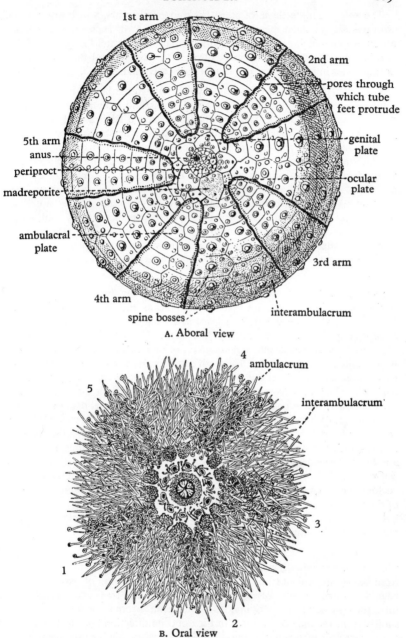

Fig. 481. *Echinus* structure. A, aboral view of the dried shell after the spines and pedicellariae have been removed. B, oral view. Note the pentamerous pattern of the tube feet in the five ambulacra.

The *spines*, unlike those of starfishes and brittle stars, which are moved with the ossicles under them, have muscles of their own. These are in two sets, an outer one which causes movements, and an inner 'catch' muscle (p. 140) which holds the spine firmly in position. On level ground the spines take part at times in locomotion, the animal using them like stilts.

THE PEDICELLARIAE (Fig. 483), which have three jaws, are of several kinds. *Gemmiform* pedicellariae have stiff stalks and globular heads with a poison gland in each jaw. The *tridactyle* kind have a flexible stalk and long jaws. The *ophiocephalous* kind are smaller and have a flexible stalk and broad, toothed jaws. The *trifoliate* pedicellariae are the smallest and have very flexible stalks and broad, blunt jaws. It is said that the gemmiform kind are weapons of defence against large foes, the tridactyle against small, the ophiocephalous seize small animals for food, and the trifoliate destroy debris. The peristomial edge of the corona is indented in each interradius by two notches, where stand the *gills*—delicate, branched outgrowths of the body wall, each containing a cavity which is continuous with the lantern coelom (see below). Ten little plates on the peristome around the mouth carry openings for the ten short, stout, sensory *buccal tube feet*, the proximal pair of podia in each radius.

LANTERN. The mouth, which is surrounded by five strong, slightly projecting, chisel-shaped, interradial teeth, leads into a relatively narrow oesophagus, whose lower part is enclosed in a framework, known as *Aristotle's lantern* (Fig. 482A), which supports the teeth. The lantern consists of five composite *jaws*, each clasping a *tooth*, and five radial pieces, known as *rotulae*, which unite the jaws aborally. The teeth can be moved outwards and inwards by muscles running from the jaws to radially placed arches, known as the *auriculae*, which arise from the inside of the corona near the lantern. Under each auricula, which perhaps represents a pair of ambulacral ossicles, runs a radial nerve, with its epineural canal, and the radial perihaemal canal, 'blood vessel', and water vessel. Within the lantern is a space, known as the *lantern coelom*, which is an enlarged perihaemal ring. Muscles, running from the auriculae to slender ossicles, known as *compasses*, which overlie the rotulae, can raise and depress the roof of the lantern, and thus pump the fluid of its coelom into and out of the gills. In some urchins, but not in *Echinus*, pouches of the lantern coelom project upwards into the perivisceral cavity. These are known as *Stewart's organs* or *internal gills* and when they are present external gills are often lacking.

THE OESOPHAGUS enlarges into a flattened tube, the *stomach* (Fig. 482B), which runs horizontally round the body in a clockwise direction as viewed from below, suspended from the shell in festoons by strings of tissue. At its beginning there is a short *caecum*, and it is accompanied by a small, cylindrical tube, the *siphon*, which opens into it at either end. From its distal end a tract similar to it, the *intestine*, returns in the opposite direction and then ascends as the narrower *rectum* to the anus. The food consists chiefly of seaweed.

THE WATER VASCULAR RING has five Tiedemann's bodies. It is situated above the lantern, and the radial vessels run downwards and outwards from

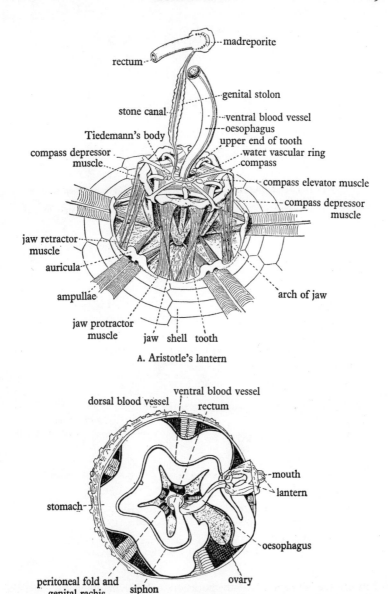

Fig. 482. Echinoid structure: A, Aristotle's lantern of *Echinus esculentus*. (From Shipley and MacBride.) B, oral view of a sea-urchin with part of the shell removed to show the course of the alimentary canal. Aristotle's lantern is deflected to the right. (After Cuvier.)

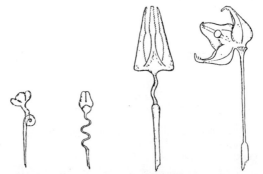

Trifoliate Ophiocephalous Tridactyle Gemmiform

Fig. 483. Pedicellariae of *Echinus miliaris*. Enlarged, but not accurately to scale.

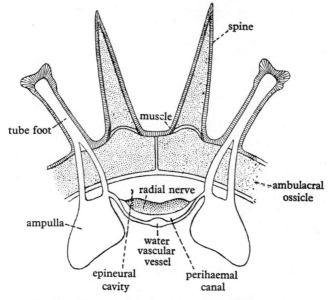

Fig. 484. A diagram of a section across the radius of *Echinus*.
(From Shipley and MacBride.)

it between the jaws and under the auriculae and then meridionally under the radial plates of the corona, to end each in the pigmented tentacle of an ocular plate. Each is accompanied in its course under the shell by the radial nerve cord, epineural canal, and perihaemal canal (Fig. 484). It is said that a small radial 'blood vessel' (not shown) runs between the perihaemal canal and water vessel. From the water vascular ring the stone canal, which is not calcified, ascends vertically to the madreporite, accompanying the axial organ, which surrounds the small *axial sinus*. Under the madreporite, however, the sinus is free and enlarged and forms an 'ampulla' into which the stone canal opens.

The oral ring of the *lacunar system* lies below the water vascular ring. The features of this system have been described on p. 675. The *gonads* are five large masses hanging into the perivisceral cavity from the region of the genital plates. The rachis is degenerate in the adult.

Echinus is an example of the *regular* sea-urchins (order *Endocyclica*). The class contains two other orders, *Clypeastroida* and *Spatangoida*, known collectively as the *irregular* urchins (*Exocyclica*), in which the anus, with its periproct, is displaced from the apical position which it occupies in the regular forms, and lies in an interradius, known as the *posterior interradius*, so that the body, which is considerably or very much flattened, has a marked bilateral symmetry. The madreporite remains in position and extends over the region vacated by the periproct. In most of the irregular urchins (though not in certain primitive forms known as Holectypoida or Protoclypeastroida) the aboral parts of the ambulacra are expanded to an oval shape (petaloid) and bear flattened, respiratory tube feet. These peculiarities are associated with the habit, possessed by typical members of both orders, of living partly or wholly buried in sand (see below).

1. Order **Endocyclica**

Echinoidea in which the mouth is central, the anus remains within the apical area, and the ambulacra are not petaloid.

Echinus (Figs. 480–484). A typical example, described above.

2. *Order* **Clypeastroida**

Echinoidea in which the mouth is central and furnished with a lantern, the anus outside the apical area, the dorsal parts of the ambulacra nearly always petaloid, and the body usually much flattened.

The members of this order live at or near the surface of the sand, and walk by means of the tube feet, which are very numerous. They extract food from the sand, which they shovel into the mouth by means of the teeth.

Clypeaster. A typical member of the group, of large size, widespread in tropical waters.

Echinocyamus. Small, oval, and not extremely flattened. *E. pusillus* is a British species.

3. *Order* **Spatangoida**

Echinoidea in which the anus and often also the mouth are excentric, the lantern has disappeared, the dorsal parts of the ambulacra are petaloid, and the body cushion-shaped or heart-shaped.

Typical members of this order live buried at some depth in the sand and move, not by means of their tube feet, but by ploughing their way with numerous, curved, flattened spines. In such forms the body has a heart-shape, owing to the depth of the anterior ambulacrum, which differs from the rest and has special tube feet, capable of great elongation and provided with fringed discs. These gather sand rich in food, which is then pushed into the mouth by stout buccal tube feet.

Spatangus and *Echinocardium* (Fig. 485) are typical members of the order, found in British waters. *Echinocardium* comes into shallower water than *Spatangus*, burrows deeper, and differs in respect of the arrangement of the spines.

CLASS 4. HOLOTHUROIDEA

DIAGNOSIS. Sausage-shaped Echinodermata, without arms; without recognizable abambulacral area; usually without external madreporite in the adult; with the ambulacral grooves covered; some of the tube feet modified into tentacles around the mouth, and some or all of the rest, if present, provided with suckers; a muscular body wall containing very small ossicles; no spines; and no pedicellariae.

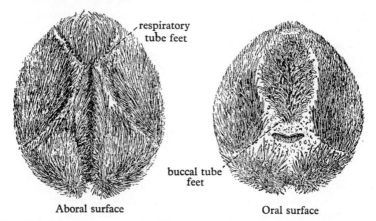

Fig. 485. *Echinocardium cordatum*. Note that the tube feet are very much contracted.

THE TYPICAL FORM OF BODY of the Holothuroidea is well seen in the members of the widely distributed genus *Holothuria* (Fig. 488), to which the familiar British 'cotton spinner' belongs. It is such as would result if in a regular echinoid the ossicles were reduced and the body drawn out in the oro-anal axis, the madreporite with the gonad which adjoins it being displaced along their interradius to a position not far from the mouth, and the other gonads lost. As has been explained on p. 673, the madreporite usually becomes internal. It is so in *Holothuria*.

Owing to the presence in one interradius of the primary madreporite and the gonad, the body always possesses a rudimentary *bilateral symmetry*. In many cases, as in *Holothuria*, this symmetry becomes conspicuous owing to the fact that the animal constantly applies to the ground the three radii of the side opposite to the madreporic interradius, and this side becomes differentiated as 'ventral' from the 'dorsal' side which contains the madreporic interradius with the two radii which adjoin it. The differentiation consists in a more or less marked flattening of the ventral side and the loss, or conversion

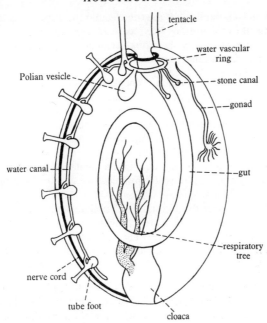

Fig. 486. A diagram of the structures in a holothurian. This diagram should be compared with the transverse section shown in Fig. 487. In both these diagrams the haemal system has been very much simplified.

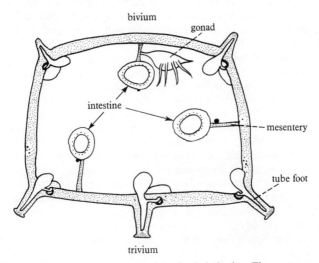

Fig. 487. A diagrammatic transverse section of a holothurian. The pentamerous symmetry is shown by the five rows of tube feet and their associated systems (radial nerves, water canal and haemal system).

into pointed sensory papillae, of the tube feet on the dorsal side. The tube feet may be confined to the radii or may, as in *Holothuria*, spread over the interradii, obliterating externally the radial arrangement, though internally the radial structures retain their position.

THE TENTACLES may be much branched (*dendritic*), provided with lateral branches only (*pinnate*), or, as in *Holothuria*, furnished only with a terminal circle of short branches, which themselves branch (*shield-shaped*). Shield-shaped tentacles are retractile owing to the presence of ampullae. Dendritic tentacles do not possess ampullae but are withdrawn by means of *retractor muscles* inserted into the radial pieces of the calcareous ring around the oesophagus (see below), which pull in the tentacular crown as a whole. Pinnate tentacles are withdrawn by retractor muscles or by ampullae or by both.

In the *body wall*, under the dermis, which contains minute ossicles of a form characteristic of the species, there are transverse muscles between the radii and longitudinal muscles under the radii. The radii contain the same structures as in the Echinoidea. Only one tube runs from the radial water vessel to each tube foot.

THE ALIMENTARY CANAL (Fig. 486) is slung to the body wall by a mesentery. It runs (except in *Labidoplax*) an S-shaped course, looping almost the whole length of the body—backwards in the mid-dorsal interradius, forwards in the left dorsal interradius, and finally backwards in the right ventral interradius to the anus. It starts as an *oesophagus*, enclosed in a *calcareous ring* of ten ossicles, five radial and five interradial, which has been thought to represent the auriculae and lantern of an echinoid. The oesophagus is followed by a short muscular region known as the *stomach*, this is succeeded by a thin-walled intestine which forms the greater part of the canal, and finally there is a short, wide *cloaca*. Into the latter usually open two long, branched *respiratory trees*, whose ramifications end in thin-walled ampullae through which water, when pumped in by contractions of the cloaca, passes into the body cavity, carrying oxygen to the coelomic fluid, and so to the organs. In *Holothuria* and a number of other genera the lower branches of the respiratory trees are converted into *Cuvierian organs*, tubes covered with a sticky substance which in sea water elongate and form a mass of sticky threads. When the animal is attacked or otherwise irritated a violent contraction of the muscles of the body wall sets up in the perivisceral cavity a pressure which ruptures the cloaca and drives out the Cuvierian organs (and often subsequently the rest of the alimentary canal). The enemy is entangled by the sticky threads. Except in the Dendrochirotae, the food is extracted from sand or mud which is shovelled into the mouth by the tentacles. Dendrochirotae entangle small organisms on their sticky tentacles and, putting the latter one by one into the mouth, contract upon them and, by drawing them out, strip off the catch.

THE AXIAL ORGAN is represented only by a cord-like genital stolon near the gonoduct. The *aboral sinus* and *vascular ring*, *genital rachis*, and *apical nervous system* are absent. The *lacunar system* consists of an oral ring, radial 'vessels', 'dorsal and ventral vessels' of the alimentary canal, and a plexus on the latter. In the middle part of the intestine of *Holothuria* and many other genera, the 'dorsal vessel' hangs from it on a perforated fold of the peritoneum, but remains connected with it by a plexus known as the *rete mirabile*. In perforations between the strands of this plexus the branches of the left respiratory tree are entangled. The condition of the *water vascular system* is that described on p. 672; for that of the genital system see p. 675.

The Holothuroidea are divided into six orders. Of these the Aspidochirotae and Dendrochirotae contain between them the bulk of the members of the class.

1. *Order* Aspidochirotae

Holothuroidea with shield-shaped tentacles; no retractor muscles, but tentacle ampullae; podia on the trunk; the madreporite internal; and respiratory trees.

Holothuria (Fig. 488).

2. *Order* Pelagothurida

Holothuroidea of pelagic habit; with shield-shaped tentacles; no retractor muscles, but large tentacle ampullae which push out the body wall; no podia on the trunk; the madreporite external; and no respiratory trees.

Pelagothuria (Fig. 488). The only pelagic holothurian. The animal swims by a webbed circle of projections caused by the enlargement of the tentacle ampullae.

3. *Order* Elasipoda

Deep-sea, benthic Holothuroidea with shield-shaped tentacles; no retractor muscles or tentacle ampullae; podia on the trunk, the madreporite external or internal; and no respiratory trees.

Deima (Fig. 488).

4. *Order* Dendrochirotae

Holothuroidea with dendritic tentacles; retractor muscles but no tentacle ampullae; podia on the trunk; the madreporite internal; and respiratory trees.

Cucumaria (Fig. 488). Body pentagonal, with two rows of tube feet on each radius and usually no other podia except the tentacles. British.

5. *Order* Molpadida

Holothuroidea of burrowing habit; with slightly pinnate or unbranched tentacles; tentacle ampullae, and sometimes also retractor muscles; without podia on the trunk; with respiratory trees; and with internal madreporite.

Trochostoma (Fig. 488).

6. *Order* Synaptida (Paractinopoda)

Holothuroidea of burrowing habit; with pinnate tentacles whose ampullae are vestigial; with retractor muscles; without radial water vessels, or podia on the trunk; or respiratory trees; and with internal madreporite.

Labidoplax (Fig. 488). Ossicles anchor-shaped. British.

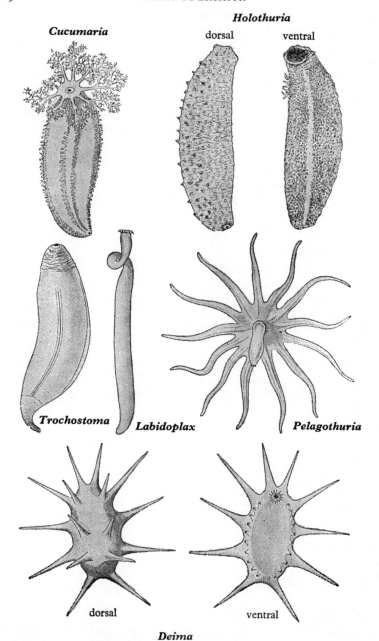

Fig. 488. Examples of the orders of the Holothuroidea: *Cucumaria* (Dendrochirotae); *Trochostoma* (Molpadida), *Labidoplax* (*Synapta*) (Synaptidae), *Pelagothuria* (Pelagothuridae), *Deima* (Elasipoda), *Holothura* (Aspidochirotae).

SUBPHYLUM 2. PELMATOZOA
CLASS 1. CRINOIDEA

DIAGNOSIS. Echinodermata with branched arms; the oral surface directed upwards; attachment during the whole or part of their life by a stalk which springs from the aboral apex; suckerless tube feet; and open ambulacral grooves; and without madreporite; or spines; or pedicellariae.

GENERAL STRUCTURE. The majority of the members of this class are extinct, and of those that survive the typical, stalked forms (Fig. 492 A) live

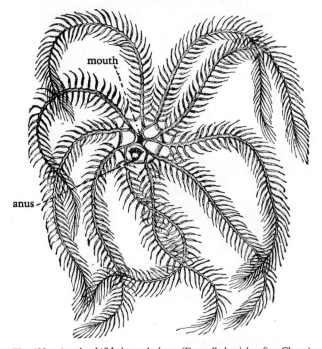

Fig. 489. *Antedon bifida* in oral view. (From Sedgwick, after Claus.)

in deep water and are less familiar than the shallow water feather stars (*Antedon* and *Actinometra*) which, when they are adult, break off from their stalks and swim by waving their arms. It will therefore be convenient to choose one of the latter to illustrate the anatomy of the group. *Antedon rosacea*, the common feather star, may be dredged in ten fathoms of water off the coast of England. Its body is composed of a small central region or *calyx* and five pairs of long, slender *arms*, each bearing a double row of alternate branchlets known as *pinnules*. On the convex aboral side, the calyx bears in the middle a knob, formed by the *centrodorsal ossicle*, which is the stump of the stalk and is fringed with numerous slender, jointed *cirri*, each ending in a small hooked claw and used for temporary attachment.

The flat top of the calyx is covered with a leathery *tegmen*, in the middle of which is the *mouth*, while the *anal papilla* stands in one of the interradii. Five *ambulacral grooves* start from the mouth, where they are separated by five low triangular flaps, the *oral valves*, and radiate across the tegmen, each running to one of the pairs of arms, to supply which it bifurcates. The groove on each arm gives a branch to each pinnule. Along their whole course the grooves bear a row of finger-shaped podia on each side, and they and the podia are

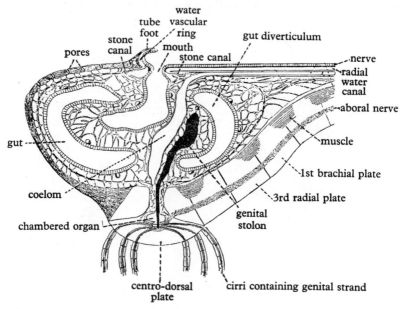

Fig. 490. A transverse section through the disc and base of an arm of *Antedon rosacea*. (After Ludwig.)

ciliated. Down this system a current set up by the cilia conveys particles gathered from the water to the mouth for food. The podia can only serve for respiration and to increase the ciliated surface; they are not prehensile, and it is said that only those around the mouth are sensory.

SKELETON. Everywhere except in the grooves, the ectoderm is vestigial and cuticulate like that of the Ophiuroidea. The dermis, which on the oral side is merely leathery, contains on the aboral side a *skeleton* of large ossicles. This consists of: (1) the *centrodorsal*; (2) the small *rosette* (formed by the fusion of five larval pieces known as basals), which is internal and roofs a cavity, presently to be described, in the centrodorsal; (3) in each radius, three *radials*, of which the first is usually not visible externally; (4) in each arm, a row of *brachials*; (5) in each pinnule, a row of *pinnularies*; (6) in each cirrus, a row of *cirrhals*, which are hollow. The ossicles of the appendages of the body are movable upon one another by muscles.

THE ALIMENTARY CANAL consists of a short, vertical *oesophagus*, a wide *stomach*, curved horizontally around the axis of the calyx and bearing two long diverticula and some low pouches, and a short *intestine*, which ascends to the anus.

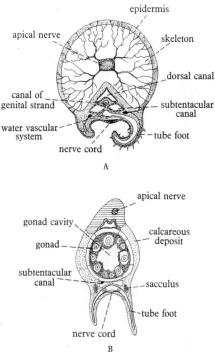

Fig. 491. A, diagram of a transverse section through the arm of a crinoid. B, diagram of a transverse section through the pinnules on the arm of a crinoid. (Both after Sedgwick.)

THE PERIVISCERAL COELOM (Figs. 490, 491) of the calyx is traversed by numerous calcified strands (*trabeculae*). In the arms there are present (1) a pair of *subtentacular canals*, (2) aboral to these, a *genital canal*, (3) aboral to this again, a *coeliac canal*, which is derived from the right posterior coelom of the larva. All these canals lead from the perivisceral cavity. It is said that there is a tiny perihaemal vessel in each arm but no oral perihaemal ring. There is no genital ('aboral') ring sinus. In the hollow of the centrodorsal ossicle lies what is known as the *chambered organ* (Fig. 490). This consists of five radial compartments, derived from the larval right posterior coelom; its wall is richly nervous and constitutes the centre of the *aboral* or *apical nervous system*. From the centre issue five interradial nerves, which branch and form a complicated plexus with a co-ordinating circular commissure, and from this plexus radial nerves supply the arms and pinnules. Nervous prolongations of the chambered organ also pass down the cirri. The whole of this system is enclosed in the ossicles of the adult, but it originates from the wall of the adjacent coelom. It controls the movements of the animal. If it be destroyed they cease; but the *ectoneural system* (which has the same arrangement as that of a starfish) can be cut away without affecting the movements.

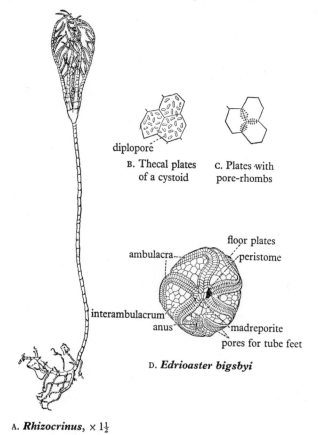

Fig. 492. Pelmatozoa. A, *Rhizocrinus*, a modern crinoid found in the abyssal depths of the sea. (From Sars.) D, *Edrioaster bigsbyi*. (After Bather.)

THE AXIAL ORGAN lies in the axis of the body. Starting as a slender strand in the centre of the chambered organ where the walls of the chambers meet, and enlarging in the perivisceral cavity, it narrows again orally, where it is continuous with a circular *genital rachis*. From this again *genital cords* pass down the arms in the genital canals and so reach the pinnules, where they enlarge into *gonads*. The genital cells are dehisced by rupture when ripe. The *lacunar system* has an oral ring and 'vessels' from this to a plexus on the stomach and to the lacunar tissue of the axial organ. It is doubtful whether radial vessels are present.

THE WATER VASCULAR RING closely surrounds the mouth, and from it numerous stone canals open without madreporites to the perivisceral cavity, which in turn communicates by many isolated pores, lined by cilia, with the exterior. This arrangement is due to the fusion of the larval axial sinus with the perivisceral cavity and subsequent multiplication of stone canals and pores (see p. 673). There are no ampullae, but the diameter of the canals can be varied by muscular strands which traverse them.

Two other recent crinoids may be mentioned here. *Pentacrinus* is a deep-water form with a long, jointed stalk, bearing whorls of cirri at intervals. The adult, like *Antedon*, breaks free and swims by waving its arms, but trails its stalk behind it. *Rhizocrinus* (Fig. 492A) has a jointed stalk without cirri except at the distal end, where some branching root-cirri are developed. By these the animal is permanently rooted. It is found at great depths in the Atlantic.

EXTINCT CLASSES

Echinoderms belonging to several groups now extinct are numerous as fossils in Palaeozoic rocks. They are all sessile by the aboral side, a fact whose significance has been mentioned above (p. 680). Their body wall contains an armour (*theca*) of plates, in which mouth, anus, and madreporite can often be identified. Some of these echinoderms are classified with the Crinoidea as *Pelmatozoa*, in contrast to the free members of the phylum, which constitute the *Eleutherozoa*.

The following are the groups referred to in the subsequent paragraphs:

Subphylum Pelmatozoa (attached forms)
   Class Cystoidea (M. Ord.–L. Perm.)
      Subclass Hydrophoridea (M. Ord.–M. Dev.)
      Subclass Blastoidea (M. Ord.–L. Perm.)
   Class Eocrinoidea (L. Camb.–M. Ord.)
   Class Paracrinoidea (M. Ord.)
   Class Crinoidea (L. Ord.–Recent)
   Class Edrioasteroidea (L. Camb.–Perm.)
   Class Carpoidea (M. Camb.–L. Dev.)
   Class Machaeridea (M. Camb.)
   Class Cyamoidea (M. Camb.)
   Class Cycloidea (M. Camb.)

Subphylum Eleutherozoa (free and vagrant forms)
   Class Asteroidea (M. Ord.–Recent)
   Class Ophiuroidea (M. Miss.–Recent)
   Class Auluroidea (L. Ord.–U. Miss.)
   Class Somasteroidea (L. Ord.)
   Class Echinoidea (M. Ord.–Recent)
   Class Holothuroidea (M. Camb.–Recent)

Class 1. **CYSTOIDEA.** These are sac- or vase-shaped echinoderms with or without a stem. The body has a variable number of hexagonal calcareous plates, and there is a trend seen in the group for the more primitive forms to have many irregularly shaped plates whilst the more advanced forms have a few plates of regular shape and pattern. The animals have a mouth, anus, genital pore and a madreporite (hydropore). The thecal plates are perforated by pores of unknown function and these pores are diagnostic of the Cystoidea. The thecae may contain food grooves or the grooves may lie on the oral side of the body.

    Subclass 1. **HYDROPHORIDEA.** Numerous irregular plates; the animals have no blastospires.

        Order 1. **Diploporita.** Thecal pores in pairs. Food groove on thecal surface (i.e. epithecal). *Aristocystites*

        Order 2. **Rhombifera.** Thecal pores in diamond-shaped pattern. Food grooves on special arms (brachiolar and extrathecal). *Lepadocrinus*

    Subclass 2. **BLASTOIDEA.** Highly organized cystoids with an ovoid-shaped stalked body. Certain plates of the theca (basal, radial, deltoids, and lancets) are uniform in arrangement throughout the whole of the group. The ambulacra are bordered by rows of pinnacle-like arms (brachioles) and at the side of the ambulacral grooves run long elongated pouches, the hydrospires, which open to the exterior at the oral end by spiracles. *Pentremites*.

Class 2. **EOCRINOIDEA.** A small group of fossils with characteristics intermediate between those of the cystoids and the crinoids. They have a ventro-lateral anus, there are no thecal pores. The pattern of the plates in the calyx is tricyclic whilst the arms show biseriate branching. The group is regarded by some authorities as the ancestral group of the Pelmatozoa. *Macrocystella*.

Class 3. **PARACRINOIDEA.** A problematic group with crinoid and cystoid features. The plates of the body are not arranged in cyclic patterns but they possess a system of subthecal pores. *Comarocystites*.

Class 4. **EDRIOASTEROIDEA.** Cushion-shaped animal without a stalk or arms but with five food grooves radiating over the body and converging at the mouth. The test is soft and flexible and made up of irregular plates. *Edrioaster*.

Class 5. **CARPOIDEA.** Stalked animals with a secondary bilateral symmetry due to lateral compression. Upon this plane of compression lie the mouth, hydrospire (madreporite) and anus. Two food grooves on the theca have been described. Sometimes there are two arm-like spines at the end of the oral ridge. It has been suggested that there is not really sufficient evidence to classify these animals as definite echinoderms. *Dendrocystis*.

Class 6. **MACHAERIDEA.** Bilateral multiplated animals of doubtful affinities. They have been classified as fossil cirripedes but are now placed in the echinoderms, mainly on the affinities of their plate structure with that seen in the other fossil echinoderms. *Turrilepas.*

Class 7. **CYAMOIDEA.** Small, bilaterally symmetrical animals with a theca of five plates. The plates are not composed of prismatic material. *Peridionites.*

Class 8. **CYCLOIDEA.** Small, bilaterally symmetrical animals with a theca composed of plates of prismatic material. *Cymbionites.*

CHAPTER XVIII

THE PROTOCHORDATA

DIAGNOSIS. Animals possessing chordate characteristics; gill slits, dorsal hollow central nervous system, enterocoelic coelom, notochord, postanal tail; but do not have a vertebral column and are therefore invertebrates (Fig. 493).

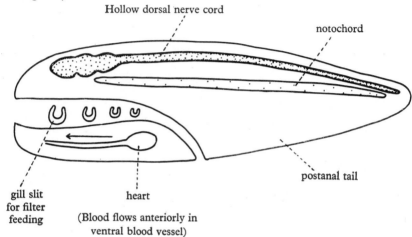

Fig. 493. Diagram of a generalized chordate.

CLASSIFICATION. This is based on the development of the notochord.

Subphylum 1. **HEMICHORDA**. Protochordates with nerve cord restricted to the neck region and a small notochord coming off from the pharynx. *Balanoglossus*.

Subphylum 2. **UROCHORDA**. Protochordates with hollow dorsal nerve cord and notochord only in the larval stage. *Ciona*.

Subphylum 3. **CEPHALOCHORDA**. Protochordates with notochord very well developed in the adult, extending from head to tail. *Amphioxus*.

CHORDATA. Most of the members of the phylum Chordata do not come within the scope of this book. But, though a position in that phylum is accorded by all authorities to the Tunicata and by many also to the Hemichorda, it is often convenient to treat of both these groups with the other invertebrate animals, and we shall take that course. It will be well, however, first to indicate what are the features which the groups in question share with the other chordate animals—the Vertebrata proper, and the Cephalochorda

(*Amphioxus*), which are usually studied with the Vertebrata. The Chordata are bilaterally symmetrical, coelomate metazoa which have in common certain fundamental features stated in the following paragraphs (Fig. 493).

(1) With the single exception of the minute, sessile *Rhabdopleura*, every member of the group possesses, at least in its early stages, lateral perforations from the pharynx to the exterior which are known as *gill clefts* or, by the name which is applied to those of them which in vertebrata do not bear gills, as *visceral clefts*. Moreover, the gill clefts of the Cephalochorda, the Hemichorda (except *Cephalodiscus*), and many tunicates, have the further resemblance that the perforations which originate them are subsequently divided by tongue bars, so that each gives rise to two of the definitive clefts. It is probable that the original function of the 'gill' clefts was the filtering off of food from water taken in through the mouth. This function they still retain in the lower members of the phylum (*Balanoglossus*, *Amphioxus*, tunicates, many fishes), though something of a respiratory function is perhaps always superadded to it.

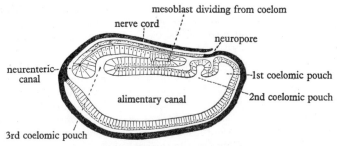

Fig. 494. Diagrammatic longitudinal section of an embryo of *Amphioxus*.
(From Shipley and MacBride.)

(2) In all the Chordata the *central nervous system* (*a*) arises from a median dorsal strip of ectodermal epithelium from which it never parts, (*b*) is, except in the trunk cord of *Balanoglossus* and the whole central nervous system (ganglion) of *Cephalodiscus* and *Rhabdopleura*, removed from the surface of the body, by invagination or by overgrowth of the epithelium at its sides, so as to form a tube, lined by its persistent epithelium. These features of the nervous system of the Chordata have analogies in that of the Echinodermata (pp. 670, 675).

(3) The common features of the *coelom* of the Chordata are more obscure. The coelom of the Hemichorda and the Cephalochorda arises, as in the Echinodermata (p. 673), by pouches of the archenteron forming three segments, of which the anterior (the *proboscis cavity* or *head cavity*) is at least in its beginning median and communicates on the left side (and in the Hemichorda often on both sides) by a pore with the exterior. In the Hemichorda the three segments retain their entity throughout life: the first does not divide into lateral halves, the second (collar cavities) acquires a pair of pores to the exterior, the third (trunk cavities) does not undergo transverse division. In the Cephalochorda, the first divides into two halves, of which the left, by

the opening out of its pore, becomes Hatschek's pit in the ectodermal depression known as the wheel organ, the second forms the first mesoderm segment (mesoblastic somite) and some cavities around the mouth, the third subdivides to form all the mesoderm segments except the first. In the Vertebrata the coelom forms as a split in a mass of mesoderm, though there are indications that the mesoderm rudiment should be regarded as a solid pouch arising in the same position as the hollow pouches of the Cephalochorda. The head cavity is represented by the premandibular segment of the embryo, a median structure with an opening to the exterior in the form of a communication with the ectodermal invagination for the pituitary body. Certain peculiarities of the mandibular segment indicate that it is the homologue of the first mesoderm segment in the Cephalochorda and so of the collar cavity. The remaining segments must represent those of the Cephalochorda and so the trunk cavity of the Hemichorda. In the Tunicata the mesoderm arises as a solid mass in the same position as the pouches of the Cephalochorda, but the coelom, except for certain doubtful vestiges, is non-existent.

(4) Except in the Hemichorda, the *notochord*, a skeletal rod which arises from the endoderm of the median dorsal line of the gut, runs the whole, or a considerable part, of the length of the body. In the Hemichorda the dorsal side of the gut at the anterior end, over the mouth, forms a skeletal outgrowth into the proboscis. This outgrowth has received the same name as the notochord, on the theory that it represents the anterior portion of that structure.

(5) With the exception of the Hemichorda, all the Chordata possess the *tail*, a postanal prolongation of the body in the direction of its main axis, without viscera, but containing extensions of the other principal organs—muscles, nerve cord, notochord, and, in the Vertebrata, backbone. A true tail is found only in the Chordata. In them it is a very important organ, used primarily in locomotion and maintaining position, though it may become an organ of prehension or a weapon.

SUBPHYLUM 1. HEMICHORDA

DIAGNOSIS. Chordata without tail, atrium, or bony tissue; with notochord restricted to the preoral region; central nervous system partly or wholly on the surface of the body; and three primary segments of the coelom retained in the adult in corresponding, externally visible regions of the body, the foremost of which is preoral.

CLASSIFICATION

Class 1. **ENTEROPNEUSTA**. Free-living worm-like hemichordates with many gill slits and a straight gut. *Balanoglossus, Ptychodera, Harrimania.*

Class 2. **PTEROBRANCHIATA**. Sessile tubicolous hemichordates with gill reduced or absent and the gut U-shaped. *Rhabdopleura, Cephalodiscus, Siboglinum.*

Class 3. **GRAPTOLITA**. Fossil forms known mainly from the structure of their tubes. *Dendrograptus, Dictyonema.*

HEMICHORDA

ACCOUNT. This small group contains the *Enteropneusta*, burrowing worms of the genus *Balanoglossus* and related, slightly different genera, and the *Pterobranchia*, the remarkable little organisms *Cephalodiscus* and *Rhabdopleura*, which live at considerable depth in the sea, in tubular houses which they secrete for themselves by their proboscis.

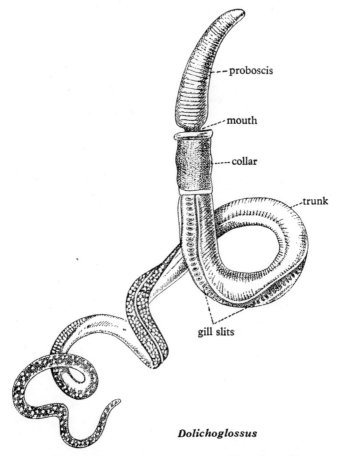

Fig. 495. *Dolichoglossus kowalewskii*, ×1. (From Spengel.)

CLASS 1. ENTEROPNEUSTA

Balanoglossus. The *body* of *Balanoglossus* (Fig. 495) has a conical preoral lobe, the *proboscis*, which behind, by a narrow stalk, joins the short, wide *collar* region. This overhangs in front the stalk of the proboscis and behind the beginning of the long *trunk*. Each of these regions contains one of the three segments of the coelom, the proboscis segment undivided, the collar and trunk segments each in two lateral halves. (The trunk cavities send forward into the collar a pair of 'perihaemal' prolongations at the sides of the dorsal blood vessel, mentioned below.) A pair, left and right, or a single, left, *proboscis pore* opens at the base of the organ. The mouth opens on the ventral side, between the overlap of the collar and the proboscis stalk. A pair of *collar pores* open backward from the collar cavities into the first gill pouch (see below). Dorsolaterally on the first part of the trunk is on each side a row of numerous small gill openings; these lead into deep pouches which communicate with the pharynx each by a tall opening virtually divided into two by a tongue bar, which hangs from the dorsal side but does not quite join the

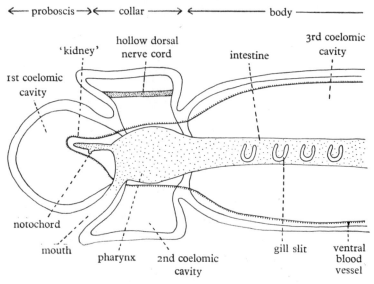

Fig. 496. Diagrammatic longitudinal section through a typical enteropneustan. Note the small notochord, hollow dorsal nerve cord, and the three coelomic segments. (After Delage and Herouard.)

ventral side as do the tongue bars in *Amphioxus*. In the region of the gills, and a little way behind it, the trunk is somewhat flattened above or has a pair of lateral ridges or folds, known as the *genital pleurae* because when they are present the gonads lie in them. Behind this *branchiogenital* region the trunk becomes more cylindrical and tapers gradually, as the *abdominal region*, to the *anus*, which is terminal.

BURROWING. The proboscis and collar are used in burrowing. They are distended by the taking in of water by the action of cilia in the tubes leading to the pores, and contracted by muscles in the body wall, the water being thus driven out. The proboscis first enters the mud and the collar follows and, by distending, gives a purchase for drawing forward the trunk.

NERVOUS SYSTEM. The body is covered by a ciliated *epithelium*, with gland cells and at its base a net of nerve fibrils to which processes of epithelial cells contribute. This net is thickened along the dorsal and ventral median lines of the trunk in the form of *nerve cords*, which are united by a ring thickening immediately behind the collar. The dorsal cord alone is continued into the collar, and here it is somewhat thicker and is invaginated to form a tube, by which arrangement it is protected during the movement of this prominent part of the body. On the stalk of the proboscis the cord communicates with the general net on that organ. There are no special sense organs. No *dermis* interposes between the epithelium and the muscles of the body wall.

ALIMENTARY CANAL. The alimentary canal (Fig. 496) is straight. From the *buccal cavity* in the collar, the hollow *notochord* (see p. 708) projects forward into the hinder part of the proboscis, strengthening the neck of that structure and supporting a group of organs (heart, pericardium, glomerulus), which

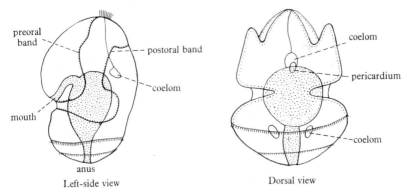

Fig. 497. Tornaria larvae. The larva is in the regressive stage and numerous secondary foldings of its ciliated rings have been lost. (From Sedgwick, after Metschnikoff.)

form with it the *proboscis complex*: at its root is a skeletal thickening of its basement membrane. Backwards, the buccal cavity leads into the *pharynx*, from which the gill slits open. In most species there is a ventral gutter, below the gill slits, leading to the intestine, which lies in the abdominal region. Along this gutter passes the mud which the animal swallows for food, excess of water leaving by the gill slits, which thus act as a straining apparatus.

BLOOD VESSELS. The blood vessels are for the most part mere crevices between the basement membranes of the ectoderm, endoderm, and mesoderm, which otherwise are everywhere in contact, having no mesenchymatous connective tissue between them. A *dorsal vessel* above the alimentary canal widens over the notochord in the hinder part of the proboscis into a space known as the *heart*. This is covered by a vesicle known as the *pericardium*, whose lower wall, contracting, communicates pulsations to the blood. From the heart the blood passes into a plexus contained in an organ known as the *glomerulus*, which is formed by a puckering of the hinder wall of the proboscis cavity around the end of the notochord. It is thought that this organ acts as a kidney, taking waste matters from the blood and throwing them into the proboscis cavity, whence they are expelled through the proboscis pores when the organ contracts. From the glomerulus the blood is gathered into two vessels which lead backwards one on each side to a *ventral vessel* below the gut. From this vessel is supplied a plexus in the wall of the alimentary canal, including the bars

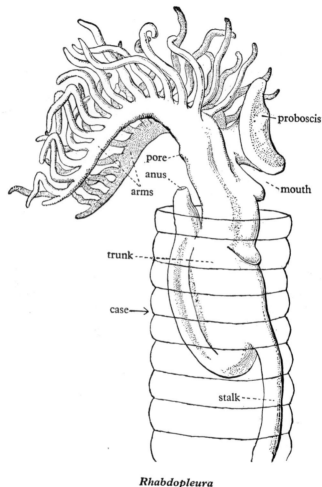

Rhabdopleura

Fig. 498. Zooid of *Rhabdopleura normani* seen from the right-hand side. (From Lang.)

between the gill openings. From this plexus blood passes to the dorsal vessel. The blood flows forwards in the dorsal vessel and backwards in the ventral.

REPRODUCTION. The sexes are separate. The gonads are mere sacs lying at the sides in the anterior region of the trunk. When they are ripe, openings break through from them to the exterior. Though they have no connexion with the coelom of the adult, they are developed from the coelomic wall.

ENTEROPNEUSTA

In most species the egg is small, and development passes through a pelagic larval stage known as the *Tornaria* (Fig. 497), which closely resembles the *Auricularia* larva of holothurians (Fig. 466), but differs in possessing a perianal band of cilia in addition to the longitudinal band, and in the presence of a

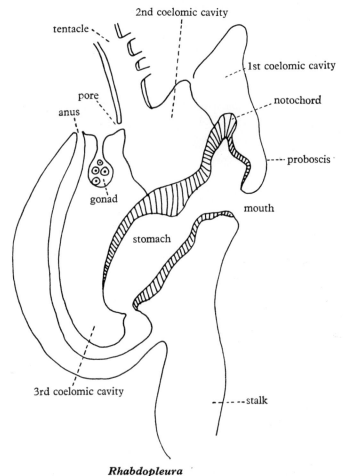

Rhabdopleura

Fig. 499. Diagrammatic longitudinal section of *Rhabdopleura* showing the three coelomic segments. Compare with Fig. 496. (After Sedgwick.)

couple of eyespots in the patch of epithelium which bears the apical tuft of cilia. The larva presently sinks to the bottom and undergoes a gradual transformation into the adult, retaining its original symmetry. The pulsating vesicle, which in echinoderms becomes the madreporic vesicle of the adult (pp. 673, 674), is in *Balanoglossus* the rudiment of the pericardium. In some species the egg is larger and there is no *Tornaria* stage. In all, however, cleavage of the ovum is complete and gastrulation is by invagination.

CLASS 2. PTEROBRANCHIATA

Cephalodiscus and *Rhabdopleura* (Figs. 498–501) are minute animals in which, owing to a protrusion of the ventral surface, the body is vase-shaped and the gut drawn down into a U, so that the anus opens upwards. The collar bears in *Rhabdopleura* two, and in *Cephalodiscus* several, hollow, branched *arms* which by means of cilia collect the food of the animal. On the forepart of the trunk are in *Cephalodiscus* the single pair of gill clefts and the pair of gonadial

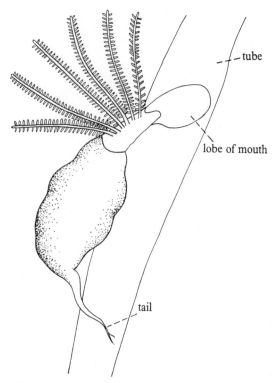

Fig. 500. Diagram of *Cephalodiscus* emerged from its tube. (From Dawydoff, after Anderson.)

openings, in *Rhabdopleura* only a gonadial opening on the right side. On the belly is a peduncle which bears buds. In *Cephalodiscus* these become free; in *Rhabdopleura* they remain in continuity with the parent so that a colony of zooids is formed. Both genera have all the characteristic features of *Balanoglossus*, save that *Rhabdopleura* has no gill clefts or glomerulus, and that in both the dorsal nerve patch of the collar is not invaginated.

Siboglinum caullery (Fig. 502). The Dutch Siboga Expedition to the Malay Archipelago brought back several specimens of this animal from the abyssal depths. It is a small transparent tubicolous worm-like animal with a glandular

PTEROBRANCHIATA

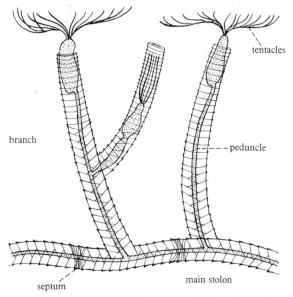

Fig. 501. Diagram of a colony of *Rhabdopleura*. (After Schepotieff.)

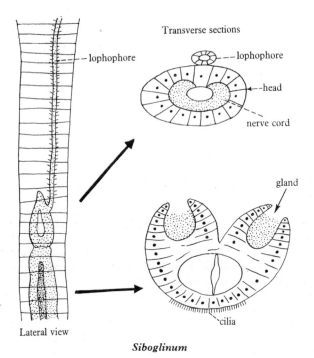

Siboglinum

Fig. 502. *Siboglinum*, a pterobranch. (From Caullery.)

head containing a hollow nerve-like organ. Attached to the head is a single tentacle or lophophore. Some regions of the body have small chitinous spike-bearing plates.

The phylogenetic position of this animal is not certain. The single lophophore, the hollow nervous system, and the structure of the tube are similar to those found in the Pterobranchiata. *Cephalodiscus siboga* has no gut and only two tentacles and so in some ways is similar to *Siboglinum*.

CLASS 3. GRAPTOLITA

The graptolites are extinct colonial hemichordates that resemble the pterobranchiates both in the structure of their tubular skeleton and in the mode of budding. They are known mainly from the structure of their external skeleton, little or nothing being known about their internal organization.

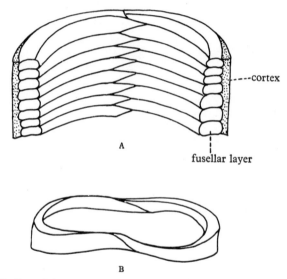

Fig. 503. Part of the tube of a graptolite. Note the resemblance to the tube of *Rhabdopleura*. (From Shrock and Twenhofel, after Kozlowski.)

Graptolites are found first of all in the Middle Cambrian and they achieved world-wide distribution in the Ordovician. They declined steadily through the Silurian and Devonian and became extinct in the Lower Carboniferous. Due to the wealth of the fossil record it has been possible to trace out in considerable detail the phylogeny of different groups.

The animals showed colonial habit similar to that seen in *Rhabdopleura*, there being a series of polyps housed in a case or theca, budded off from a main stem. The structure of the tubular skeleton is very much like that of *Rhabdopleura*, being made up of a series of concentric rings, each ring being in two halves (Fig. 503).

Rhabdopleura has two types of polyp, male and female; similarly many of the graptolites show polyp differentiation, there being three types of polyp:
(1) stolothecae, which bud off the other two types of thecae;
(2) Autothecae, thecae surrounding the female polyp;
(3) Bithecae, thecae surrounding the male polyps. (Fig. 504).

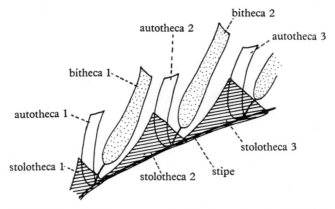

Fig. 504. Diagram of the arrangement of the thecae on a graptolite stipe.

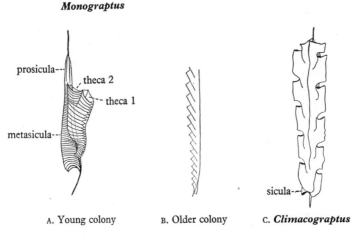

Fig. 505. Graptolites. (A, after Kraft; C, after Wimann.)

Some forms have but one type of theca, the autotheca, and in these cases the sexes are not separate.

There are certain definite trends seen in the evolution of the graptolites. First of all there may be variation in the arrangement of the thecae on the main stem or stipe. Thus they may be uniserate, with the thecae on one side of the stipe; biserate, with thecae on two sides; or tetraserate, with thecae on four sides of the stipe (Fig. 506 C). Another variation is seen in the orientation of the thecae along the stipe. In the early forms the thecal openings point upwards but in later forms the thecal openings point downwards. This may reflect a change in habit from one where the colony is attached to a piece of rock or weed on the sea bottom, to one in which they are attached to floating weed, or develop a pneumatophore, and thus adopt a pelagic life. A reduction may occur in the number of stipes that develop. In early forms there were many stipes whilst in the later forms there are only two or even one stipe.

CLASSIFICATION. There are five orders.

1. *Order* **Dendroidea**

The dendroids show a branching pattern of uniserate stipes (thecae on one side of the stipe only). Adjacent stipes may fuse together or be joined by short septa. Three types of thecae are present, autothecae (♀), bithecae (♂), and stolothecae (budding forms). A typical region of the stipe of *Dictyonema* is shown in Fig. 504.

During the development of *Dictyonema* the first individual of the colony secretes a small cone, the *sicula*, which is attached by a thread, the *nema*, to a floating mass of seaweed. The sicula is formed from two parts, the prosicula and the metasicula (Fig. 505A). From the side of the metasicula there develops the first theca, a stolotheca which buds off the first triad of thecae, autotheca (1), bitheca (1) and stolotheca (2). The stolotheca (2) then buds off another triad and in this way a uniserate stipe with three types of thecae is formed. A complication arises in that the colony often branches. In this case the triad develops into an autotheca, and two stolothecae, each of which continues budding. Examples of dendroids are: *Dictyonema, Acanthograptus, Callograptus, Dendrograptus*.

2. *Order* **Tuboidea**

These resemble the Dendroidea except that the thecae are not arranged so regularly and the branching is much more varied. There are the three types of thecae. Examples are: *Tubidendrum, Cyclograptus*.

3. *Order* **Camaroidea**

These graptolites have certain resemblances to the ectoproctans in their encrusting development. There are two types of thecae, bithecae (which are present only in some forms) and autothecae. The autothecae have a specialized shape with the bottom part expanded into a large chamber or camera. Example: *Cysticamera*.

4. *Order* **Stolonoidea**

These are very irregular branching forms, the branches being made up of complete rings and not half-rings as in the other Graptolites. Only autothecae have been found. Example: *Stolonodendrum*.

5. *Order* **Graptoloidea**

These animals consist of a series of uniserate, biserate, or even quadriserate stipes hanging from a common disc to which they are connected by a thread called a nema or virgula. The stipe has only autothecae (though one or two

cases of bithecae have been described). This is the best known of the five orders and shows considerable variation in form. The trends seen in the evolution of this order are:
(1) Change from the pendant to an erect position.
(2) Reduction in the number of stipes.
Examples: *Clonograptus*. An early form with many stipes.
Glossograptus. Has many biserate stipes attached to a central disc. This is believed to have been filled with air and acted as a float.
Phyllograptus. Has tetraserate stipe.
Tetragraptus. Has four pendant uniserate stipes (Fig. 506).
Bryograptus. Two pendant uniserate stipes.
Monograptus. One uniserate stipe.

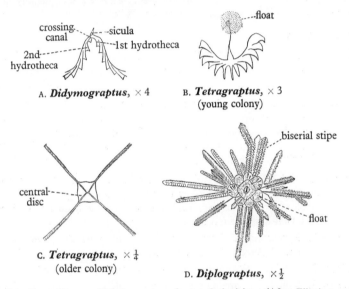

A. *Didymograptus*, × 4

B. *Tetragraptus*, × 3 (young colony)

C. *Tetragraptus*, × ¼ (older colony)

D. *Diplograptus*, × ½

Fig. 506. Graptolites. A, *Didymograptus v-fractus*, Ordovician. (After Elles.) B and C, from Ordovician rocks. D, *Diplograptus foliaceus* from the Utica Slate, New York. (After Ruedemann.)

Morphological details of the graptolites, other than information on the gross structure of the thecae and the types of branching of the stolon, have not been available till recently due to the inadequate state of preservation of the fossils. For this reason the phylogenetic position of the graptolites has been very obscure. Due to the presence of thecae and the incorrect inclusion of certain fossil coelenterates as graptolites, the groups have been classified by some authorities as Hydrozoan Coelenterates. Other investigators had been struck by the encrusting habit and growth pattern of certain graptolites and placed them in with the Ectoprocta. Recently certain siliceous deposits in Poland have yielded fossils that after treatment with hydrofluoric acid have shown fine details of the skeletal morphology. These show that the skeleton is made up of a series of half-rings (fusellar structure), a system very similar to that seen in the Pterobranchiate Hemichordates. In addition there is some similarity between the branching systems of the graptolites and the pterobranchiates. Though the information about the fine structure of the polyps is still very inadequate, the graptolites seem to be more like Pterobranchiata than coelenterates or Ectoprocta.

SUBPHYLUM 2. UROCHORDA (TUNICATA)

DIAGNOSIS. Chordata without coelom, segmentation, or bony tissue; with a dorsal atrium in the adult; notochord restricted to the tail, which is present in the larval organization only; the central nervous system removed from the surface of the body and in the adult degenerate; and a test, usually largely composed of a substance (tunicin) related to cellulose.

CLASSIFICATION

Class 1. **LARVACEA (APPENDICULARIA).** Tunicates in which the sexually mature form retains the organization of the larva.
Oikopleura, Fritillaria

Class 2. **ASCIDIACEA.** Tunicates in which the adult is sedentary and has no tail, a degenerate nervous system, an atrium which opens dorsally, a stolon (if any) that is of simple structure and several gill clefts which are nearly always divided into stigmata by external longitudinal bars.
Ciona, Botryllus

Class 3. **THALIACEA.** Pelagic tunicates in which the adult has no tail, a degenerate nervous system, an atrium which opens posteriorly, a stolon of complex structure, and clefts which are not divided by external longitudinal bars.

Order 1. **Pyrosomidae.** Feeble muscles develop in complete rings round body but at end of body only.
Pyrosoma

Order 2. **Salpidae.** Strong muscles but incomplete round body. *Salpa*

Order 3. **Doliolidae.** Strong muscles in complete rings round the body.
Doliolum

GENERAL CHARACTERS—*Ciona.* In the adult form, the members of this group are extraordinarily unlike the rest of the phylum. They have lost all the characteristic features of chordate animals except the gill clefts, and are rather shapeless objects which lead a sluggish existence by means of an organization of a low grade. Most of them are sessile, and there is no doubt that this habit has established the peculiarities of the group.

We shall describe the organization and life of the Tunicata by giving an account of a typical example, *Ciona intestinalis* of the British coasts, one of the simple 'ascidians'. The adult of this animal (Fig. 508A) is a subcylindrical sac, which reaches a height of several inches, sometimes nearly a foot, seated by the blind posterior end upon some solid object on the bottom, and at the other bearing two openings, a terminal *mouth* or *branchial opening* and an *atrial opening*, seated on a tubular projection a little way below the mouth.

This projection, which marks the dorsal side of the animal, is known as the *atrial siphon*. Beyond its origin the body narrows as the *oral siphon* towards the mouth. The latter is surrounded by eight small lobes with red pigment spots between them. The atrial opening has six lobes. Both apertures can be narrowed and virtually closed. When the animal is in water and has them open a current may be seen to set in at the mouth and out at the atrial opening. By sudden contractions of the body water may be forced out of both of them.

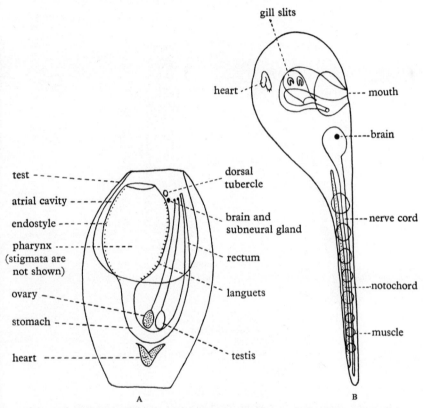

Fig. 507. Ascidian structure. Diagram of A, adult ascidian, and B, ascidian tadpole.

TEST. The body is covered by a tough, translucent test, remarkable for being composed largely of tunicin, a substance closely related to cellulose, and therefore not to be expected in an animal. The test is a cuticular secretion of the ectoderm, but contains cells of mesodermal origin which have wandered into it, and ramifying tubes in which blood circulates, which enter it at a point near the base of the sac. Below the test lies the true body wall or *mantle*, which contains numerous longitudinal and transverse strands of muscle by which the shape of the body can be altered, and sphincters around the openings, where test and mantle are tucked in for a short distance.

ALIMENTARY CANAL. The alimentary canal (Fig. 508 B) begins as a tube, the *stomodaeum* or *buccal cavity*, lined by the inturned test. This leads to the very large *pharynx*, a circlet of *tentacles* standing at the junction. A short *prebranchial zone* of the pharynx lies between the tentacle ring and the *peripharyngeal band*—a couple of ciliated ridges, which run round the pharynx, with a groove between them. In the dorsal middle line of the prebranchial zone stands the *dorsal tubercle*. This is the protuberant, horseshoe-shaped opening

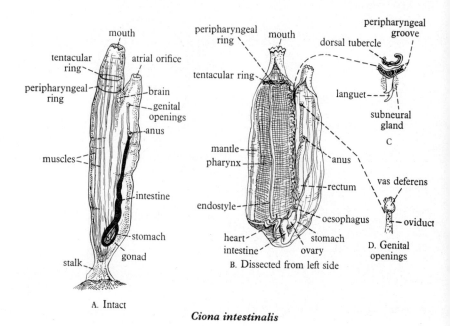

Fig. 508. *Ciona intestinalis*. (From Shipley and MacBride.)

of a *ciliated funnel* which receives the duct of a *subneural gland* that lies under the brain. The function of the gland is unknown. The funnel is innervated from the brain and is supposed to be sensory.

BRANCHIAL CHAMBER. The rest of the pharynx constitutes the spacious branchial chamber. The lateral walls of this chamber consist of a basket-work, formed by the subdivision of the original gill clefts. The openings of the basket-work (Fig. 509) are known as *stigmata*. They are longitudinally elongate and stand in transverse (dorsoventral) rows. Between them are *transverse* and *longitudinal bars*. The inner surface of this basket-work is crossed by *internal longitudinal bars*, slung from it and bearing papillae which project into the branchial cavity. All the bars of this apparatus are hollow and contain blood.

The epithelium which covers them is ciliated, the cilia being longer on the sides of the stigmata. Ventrally the basket-work walls are separated by a narrow, longitudinal, imperforate tract known as the *endostyle*. This is folded

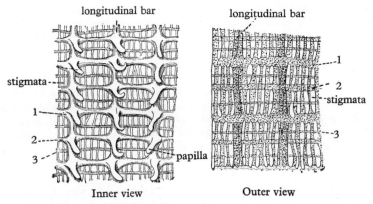

Fig. 509. Portions of the pharyngeal wall of *Ciona intestinalis*: 1, 2, 3 = transverse bars of the 1st, 2nd and 3rd orders. (From Sedgwick, after Vogt and Yung.)

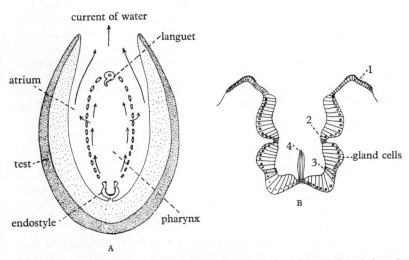

Fig. 510. A, a diagram of a transverse section of *Ciona*; and B, with an enlarged view of the endostyle. The arrows show the course of the main current and the food particles.

into a groove (Fig. 510) lined by an epithelium which is glandular and ciliated in alternate longitudinal strips, arranged in a manner similar to those in the endostyle of *Amphioxus*. To right and left a ciliated strip runs beside the groove. Posteriorly the groove passes into a caecum and the lateral ciliated strips curve up, as the *retropharyngeal band*, to the opening of the oesophagus, which is dorsal at the hinder end. Anteriorly, the same strips are continuous with the posterior peripharyngeal ridge, on each side of a gap in the latter. Dorsally, the lateral walls are separated by a *hyperpharyngeal band*, from which there hangs down into the branchial cavity a row of processes, the *languets* which are curved to one side.

The stigmata lead, not directly to the exterior, but into a cavity, the *atrium*, which opens externally at the atrial opening, is lined by ectoderm, and is placed dorsally like a saddle upon the branchial chamber, surrounding the latter completely except (1) in front, and (2) in the median line ventrally, posteriorly, and for a short distance from the front end dorsally. The atrium is crossed by vascular trabeculae from the branchial basket-work to the mantle. Its median dorsal part is known as the *cloaca*, the parts at the sides of the branchial chamber as the *peribranchial* cavities. One important difference between this atrium and that of *Amphioxus* should be noted. The atrium of the Cephalochorda is ventral: that of the Tunicata is dorsal (Fig. 511).

FEEDING. By the apparatus just described the animal feeds. The working of the cilia at the sides of the stigmata drives water through the latter, from the pharynx to the atrium, whence it passes out by the atrial opening as the current which has already been mentioned. This results in water being drawn in through the mouth to replenish the pharynx. Mucus secreted by the gland cells of the endostyle is by the cilia of that organ passed on to the inner face of the branchial basket-work, over which by further ciliary action it is passed to the dorsal languets. These receive it with their curved ends, slung in which it is worked backwards as a rope to the oesophagus. As it passes over the pharyngeal wall particles brought in with the current through the mouth are entangled in it, to be carried to the hinder part of the alimentary canal and there digested if they be fit for food. A similar function is performed by mucus which passes dorsalwards along the peripharyngeal band.

The short *oesophagus* leads backwards to a fusiform *stomach*. From this an *intestine*, whose ventral wall projects inwards as the typhlosole does on the dorsal side of the intestine of the earthworm, loops forwards to become the *rectum*, which runs a straight course half-way along the atrium, lying near the dorsal side of the pharynx. The epithelium of the digestive part of the alimentary canal is ciliated, and a gland ramifies in the wall of the stomach, into which it opens by a duct. The faeces are cast out by the outgoing current from the atrium.

The stomach and intestine lie in a section of the body known as the *abdomen*, which is behind (basal to) the region (called the *thorax*) in which the pharynx and atrium are situated.

EPICARDIA. The viscera just mentioned are enclosed in a perivisceral space, known as the *epicardial cavity*, which is of a very peculiar kind, since it is formed by two thin-walled outgrowths from the pharynx, one on each side of the retropharyngeal band. Epicardial diverticula of the pharynx are found in many tunicates (p. 734), but it is only in *Ciona* that they expand and form a perivisceral cavity, applying their walls as a peritoneum to the contained organs. In this cavity lies also the *heart*, a V-shaped tube placed in the intestinal loop, near the hinder end of the endostyle.

HEART. The heart has no proper wall but is formed by the folding-in of one side of the tubular *pericardium*, which on this side is muscular. The other *blood vessels* also have no walls of their own, but are mere vacuities between various structures. In these respects the Tunicata resemble the Hemichorda (see p. 711). Each end of the heart is continuous with a blood vessel. The vessel from one end runs forwards under the endostyle and communicates with the blood spaces in the branchial bars: these in turn join a vessel, in the

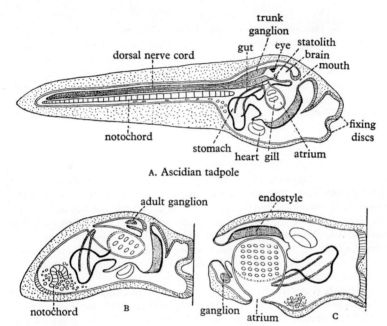

Fig. 511. Diagram of the metamorphosis of an ascidian tadpole larva. A, at time of fixation. B, midway in metamorphosis. C, metamorphosis complete.

hyperpharyngeal band, which gives off branches to the digestive organs, gonads, and body wall. To these same organs runs the vessel from the opposite end of the heart. The course of the circulation is remarkable. The heart for several beats drives the blood towards one end and then reverses its action. Thus the blood passes, at one time, like that of a fish, through the gills to the rest of the system, and at another in the opposite direction. The plasma is colourless, but contains nucleated corpuscles, some of which are of various colours owing, remarkably enough, to the presence of compounds of vanadium.

REPRODUCTION. The animal is a hermaphrodite. The *ovary* lies between the stomach and intestine as a compact mass; the *testis* ramifies over the stomach and intestine. The *genital ducts* run side by side along the rectum and at some distance beyond its end open into the cloaca. The vas deferens is the narrower, and has a patch of red excretory cells around its enlarged end, where it opens by a rosette of small pores. Fertilization takes place in the water, and the spermatozoa will not unite with ova from the same individual.

NERVOUS SYSTEM. The central nervous system is reduced to a single elongated solid ganglion (brain) on the dorsal side between mouth and atrial opening. From its ends nerves are given off. There are no organs of special sense unless the pigment spots between the lobes of the mouth be functional for the perception of light.

COELOM. Not the least striking feature of the remarkable organization which has just been described is the absence of any space that can with certainty be identified as coelom. Epicardium, pericardium, the cavities of the gonads, and even those of the closed excretory vesicles that lie around the intestine in many ascidians have been held to be of that nature, but there is no incontrovertible evidence on this point concerning any of them. Nephridia are also absent. *Excretion*, so far as is known, is performed only by the cells mentioned above, which store urates as solid concretions.

ASCIDIAN TADPOLE. So far, the student will have seen little ground for regarding *Ciona* as a chordate animal. When, however, we turn to consider its *life history*, no doubt remains upon this point. The *eggs* are small, though they contain some yolk; their cleavage is total and at first nearly equal. The *early stages* of development much resemble those of *Amphioxus*, but differ in that the cells which are to form the rudiments of various organs are very early recognizable (determinate cleavage), and that the mesoderm, which arises from the sides of the archenteron, does so, not as pouches, but as clumps of cells. Eventually there is formed a *larva*, about a quarter of an inch in length, which is known as the *Appendicularia* larva, and often as the 'ascidian tadpole'. This creature (Fig. 511) has a *tail* about four times as long as its *trunk*. In the tail are a notochord, a hollow dorsal nerve cord, a muscle band on each side, and a few mesenchyme cells. Dorsal and ventral median flaps of the test serve as fins, the tail being a swimming organ. In the trunk, notochord and muscle bands are lacking, and along with the alimentary canal the brain and pericardium are found. The mouth lies dorsally at some little distance from the front end. It leads through a short oesophagus into a large pharynx, in which the endostyle is already well developed. There is no branchial basket-work, but on each side a gill slit leads from the pharynx into an ectodermal pouch, which in turn opens dorsolaterally. Later the two pouches become united above the pharynx and thus the atrium comes into existence. Meanwhile the gill clefts increase in number by the breaking through from the pharynx of new clefts and the subdivision of existing clefts, in the course of which they pass through a U-stage with tongue bars. (The basket-work is ultimately established by the formation across each gill cleft of longitudinal bars which divide it into stigmata.) From the pharynx leads the rest of the alimentary canal, which early shows rudiments of oesophagus, stomach, and intestine, the latter curving dorsally and eventually opening into the left half of the atrium.

The dorsal nerve tube of the tail is in the trunk enlarged to form the brain. The hinder part of this is thick-walled and is known as the *trunk ganglion* (it does not become the 'ganglion' of the adult). The anterior part is larger than the trunk ganglion and for the most part thin-walled, and is known as the *cerebral vesicle*. Dorsally on the right it is differentiated to form the eye, a cup whose cavity is directed inwards, filled with pigment. On the floor a stalk projecting into the vesicle carries a concretion, the *statolith*, probably a sense organ for balance. Presently the vesicle acquires an opening into the dorsal side of the pharynx, near the mouth. The *pericardium* arises towards the end

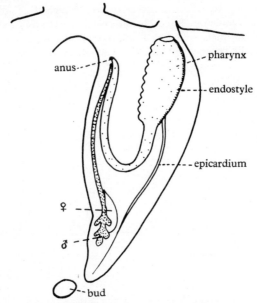

Fig. 512. Diagram of a zooid with a portion of one of its neighbours imbedded in an ascidian colony. Note the bud.

of larval life as an outgrowth from the ventral side of the pharynx. It does not form the heart until metamorphosis. The front end of the body is a prominent *chin*, and bears three *fixation papillae* of glandular cells. Except for the tips of these papillae, the animal is entirely covered with test, which even closes the mouth, so that feeding is impossible. After swimming for a short time, the larva fixes itself to some solid object by the papillae, and proceeds to undergo a metamorphosis (Fig. 511), by which it assumes the adult form. The tail is devoured from within by phagocytes. By growth in the region between the chin and the mouth, the latter and the atrial opening are shifted back until they point upwards from the region of fixation. Meanwhile, the central nervous system degenerates, save for certain portions of the cerebral vesicle, which forms from its hinder region the ganglion of the adult and from its ventral and anterior regions the subneural gland and the ciliated funnel;

the pharynx develops in the way described above; the heart is formed; the epicardial diverticula grow out from the pharynx; and the gonads arise from a mass of mesoderm.

Ciona is a solitary animal. Some other tunicates resemble it in this respect, but a large number establish by budding *colonies* of zooids, each zooid having the essential features of an individual of *Ciona*. In a few cases (*Perophora*, Fig. 518 A), the zooids are free from one another save at their bases, where they are united by the stolon from which they were formed. In most genera, however, the zooids of a colony are imbedded in a common test, with only the mouths and cloacal openings at the surface (Figs. 512, 518, 519). In such cases the original connexion between the zooids is lost, though their atrial openings usually join in a common cloaca. In the pelagic genera *Salpa* and *Doliolum* and their relatives buds are formed but, instead of remaining together as a colony, eventually become free and lead a solitary existence.

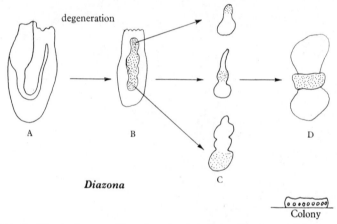

Fig. 513. Budding in *Diazona*. Note the regression of the polyp in B, and the formation of buds in C, and the new individual in D. (After Berrill.)

BUDDING IN ASCIDIANS (Figs. 513–515). The ascidians show considerable powers of regeneration and asexual reproduction. One of the simplest cases is seen in *Eudistoma* where the pharynx, oesophagus and stomach regress and turn into a shapeless mass of cells. This mass later reorganizes into a new animal, the epicardial cells playing an important role in the reorganization. In some ways this rejuvenation is similar to that seen in planarians following starvation (p. 196). In *Diazona* after the period of sexual reproduction in July–September, the thoracic region of the body starts to degenerate. The epidermal cells in different regions of the thorax start to grow and as they do so they constrict off small groups of the underlying tissues. These cell masses form buds each of which later develops into a new ascidian. In *Diazona* each of these buds occupies its original position in the new animal, i.e. if it had originally been the basal part of the oesophagus then it forms the basal part of the oesophagus in the new individual (Fig. 513). In *Morchellium*, on the other hand, the bud undergoes complete reorganization so that it loses all trace of its former organization and can form any part of the new individual.

In *Diplosoma* the epicardial tissue develops two small buds which lie midway along the oesophagus. The anterior bud develops into the posterior part of a new animal and this links up with the anterior part of the parent to form a new individual. In a similar way the posterior bud develops into the anterior part of a new individual and links with the posterior part of the parent to form a new individual. In this way the animal divides to form two complete

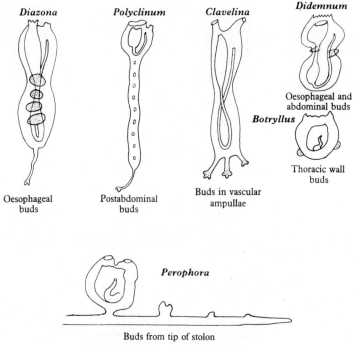

Fig. 514. Budding in ascidians. For details see the text. (After Berrill.)

animals. In *Clavelina* amoeboid cells migrate from the epicardial tissue into the basal region of the stolon. There the cells form small buds which do not develop until they are separated from the adult zooid. The adult inhibits the buds from developing. In *C. lepadiformis*, the buds separate off from one another and this leads to solitary individuals. In *C. picta* the buds remain connected together so that the animal forms colonies. Thus within closely related species both solitary and colonial forms can be seen (Fig. 514).

The active constriction of the bud is entirely due to the epidermal cells and all the inner structure such as the oesophagus, epicardium, and muscles are cut through by these constricting cells. The tissues that are represented in the bud show considerable variation from genus to genus. Six different variations are shown below.

730   PROTOCHORDATES

1. Epidermis, epicardium, reserve cells, part of the intestine. *Diazona*.
2. Epidermis, epicardium, reserve cells. *Euherdmania*.
3. Epidermis, endostylar growth, mesenchyme. *Salpa*.
4. Epidermis, septal mesenchyme. *Perophora*.
5. Epidermis, atrial epithelium. *Botryllus*.
6. Epidermis, atrial epithelium, reserve cells. *Distomus*.

Budding is accomplished in various ways. (1) It most often takes place from a *stolon*, which is a median ventral tubular outgrowth of the visceral

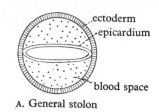

A. General stolon

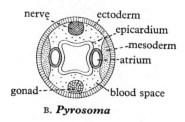

B. **Pyrosoma**

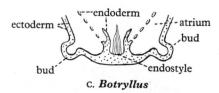

C. **Botryllus**

Fig. 515. Budding in tunicates, showing the different tissues present in the bud.

(abdominal) region of the parent, usually containing an inner tube that consists of the united distal portions of the two epicardial diverticula of the pharynx, and some mesenchyme cells in a blood space between this tube and the stolon wall. The epicardial tube will form the alimentary canal of the buds. Often it also forms the atrium and the nervous system. In the class Thaliacea, however (see below), the stolon is more complex and contains special tubes or strands of cells for the atrium and gonads and sometimes also for the nervous system. (2) In other cases (*Perophora* and *Clavelina*) the stolon contains, not an epicardial tube, but a longitudinal septum of mesoderm; and the internal organs of the bud are formed by the complication of a vesicle which arises by the hollowing out of a mass of mesoderm that comes into being in a swelling at the end of the stolon. (3) In *Botryllus* and its allies budding is effected in yet another way. These genera, which, unlike *Ciona* but like most solitary ascidians, possess no epicardium (epicardial diverticula), form their buds by paired outgrowths that are of quite a different kind from the stolon, for they arise from the atrial wall and each contains an inner vesicle which is a prolongation of the epithelium that lines the atrium of the parent: this vesicle forms the internal organs as well as the atrium of the bud. It should be noticed that in budding the origin of the organs takes place without regard to the germ layers from which they arise in the development of the ovum, for

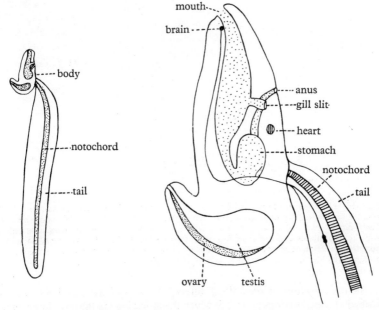

Fig. 516. *Oikopleura*. Note that the adult retains the structure of the larval ascidian but that the tail comes from the middle of the body.

the endodermal inner tube of ordinary stolonial budding often forms atrium and nervous system, which should be of ectodermal origin, and the ectodermal (atrial) inner vesicle of the 'pallial' budding of *Botryllus* forms the alimentary canal, which should be endodermal, while in *Perophora* and *Clavelina* all these organs arise from a mass of mesoderm (Fig. 515).

A zooid which arises by budding is known as a *blastozooid* (*blastozoite*): one which arises from an ovum is an *oozooid*. The oozooid, which in the Thaliacea differs considerably from the blastozooid, has always lost the power of sexual reproduction. In the Salpida and Doliolida the blastozooid has lost the power of budding, so that there is a regular alternation of generations.

CLASS 1. LARVACEA (APPENDICULARIA)

Tunicata in which the sexually mature form retains the organization of the larva.

The test is not composed of tunicin. It forms a remarkable 'house' that does not adhere to the animal, which from time to time it leaves and secretes a new one. The habitat is pelagic, and food is filtered from the water by an apparatus which forms part of the house and through which water is caused to flow by the movements of the tail.

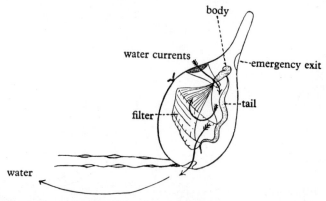

Fig. 517. *Oikopleura albicans* in its house. Movements of the tail cause water and food to enter the house, and impinge on the filter apparatus which removes the food particles. The food is sucked off the filter by the animal. When the water pressure rises sufficiently, a spring door opens at the tail end and the water passes out; this jet drives the house in the opposite direction. The animal can escape by pushing open a door (emergency exit). It does not re-enter but secretes a new house. (From Borradaile.)

The organization of the animal differs from that of the ascidian larva described above in various points, of which the following are the most important. Gonads are present in the hinder region of the body: nearly always they are hermaphrodite and protandrous. The tail is attached to the ventral side, near the hinder end of the body. The two simple gill clefts open ventrolaterally directly to the exterior. The intestine also opens directly to the exterior, ventrally or on the right-hand side. The brain is a compact fusiform ganglion, and the existence of a cavity in it or in the nerve cord is doubtful. There is no eye and a statocyst lies beside the brain on the left. In certain of these respects the animal resembles the larva of *Doliolum* (Fig. 523).

The Larvacea are now generally regarded as an instance of neoteny (p. 141).
Oikopleura (Figs. 516, 517). Common in British waters.

CLASS 2. ASCIDIACEA

Tunicata in which the adult is sedentary and has no tail; a degenerate nervous system; an atrium which opens dorsally; a stolon (if any) of simple structure and several gill clefts, which are nearly always divided into stigmata by external longitudinal bars.

The colonial members of this group are known as 'compound ascidians' and are sometimes classed together as *Ascidiae compositae*. But they are not of one origin; some of them have stolonial budding and dorsal languets and

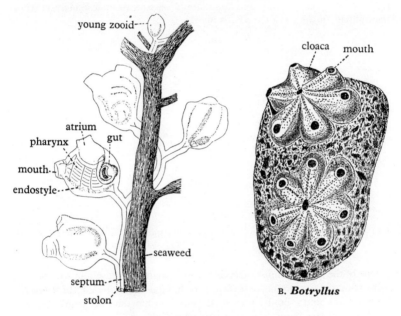

A. *Perophora*

Fig. 518. A, part of a colony of *Perophora*. (After Lister.) B, two groups of individuals of *Botryllus violaceus*. (Magnified, after Milne Edwards.)

are related to such solitary forms as *Ciona*; the others, with pallial budding and a continuous *dorsal lamina* in place of the languets, are related to solitary forms, such as *Ascidia*, which have no epicardium and possess a dorsal lamina.

Ciona (Fig. 508). Described on page 720.

Clavelina. Resembles *Ciona* in general features but has a stolon and forms clusters of individuals by the breaking off of buds from the ends of the stolon branches; the stolon has a mesodermal septum; the zooids and their tests are generally free from one another.

Polyclinum. As *Clavelina*; but the stolon contains an epicardial tube; and the zooids are imbedded in a common test with only the branchial and atrial openings at the surface.

Ascidia. Solitary; without epicardium; with the viscera at the side of the body, not in an abdomen; with dorsal lamina in place of languets. British.

Perophora (Fig. 518 A). Colonial; the zooids are free from one another but connected at their bases by a stolon with a mesodermal septum; and have dorsal lamina and viscera at the side as in *Ascidia*.

Botryllus (Figs. 518 B, 519). Colonial; the zooids imbedded as in *Polyclinum*; but with pallial budding; and with dorsal lamina, and viscera at the side.

ASCIDIAN CLASSIFICATION. This is of particular interest since it illustrates the differences between a natural classification designed to represent phylogenetic relationships, and a taxonomic classification designed to arrange the animals in the most expedient manner.

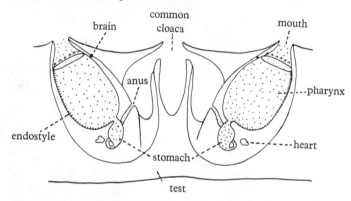

Fig. 519. Diagrammatic section through *Botryllus* showing two polyps opening into a common cloaca. (After Delage and Herouard.)

One of the first attempts to classify the ascidians was based on whether they were solitary or colonial. However, as has been mentioned for *Clavelina*, both solitary and colonial forms are found within the one genus. Another classification was based upon the structure of the gills.

(1) Aplousobranchia (Krikobranchia). Branchial sac has no folds, no inner vessels or bars.

(2) Phlebobranchia (Dictyobranchia). Branchial sac has no folds but does have internal vessels.

(3) Stolidobranchia (Ptychobranchia). Branchial sac has longitudinal folds and longitudinal vessels.

There are two slight handicaps in this classification. In the first place different authorities have given the groups quite different names which has led to considerable confusion. In the second place the actual structure of the gill varies considerably according to the size of the animal. Nevertheless, gill structure provides a quickly ascertainable taxonomic feature.

A more natural classification is provided by a study of the variation in the structure and development of the epicardium, though this can only be determined by embryological study. The epicardium develops as a pair of evaginations from the base of the pharynx. These epicardia probably correspond to the coelom since they arise from the enteron and are concerned in reproduction and excretion.

(1) Diplocoela. The two epicardia form distinct sacs opening into the branchial cavity. *Ciona*.

(2) Epicardiocoela. The two epicardia no longer open into the branchial cavity. They usually fuse together and descend into the stalk with the rest of the viscera.

  (*a*) Dicardia (heart V-shaped). *Diazona*.
  (*b*) Unicardia (heart straight). *Clavelina*.

(3) Acoela. No trace of the epicardia. *Perophora*.

(4) Nephrocoela. Epicardia take the form of vesicles which often become filled with excretory granules.

  (*a*) Enterogona. Gonads only on one side of the body. *Ascidia*.
  (*b*) Paragona. Gonads on each side of the body. *Styela*.

A comparison of the three classifications is shown below:

Lahille's classification	Seeliger's classification		Berrill's classification	
Stolidobranch	Ptychobranch	Family Molgulidae ⌉ Pyuridae Styelidae Botryllidae ⌋	├──Paragona ⌉ │ │ ├─Nephrocoela	
Phlebobranch	Dictyobranch	Ascidiidae ⌉ Rhodosomatidae ⌋──Enterogona Perophoridae ─────────Acoela Cionidae ──────────────Diplocoela		
Aplousobranch	Krikobranch	Diazonidae ⌉ Synoicidae │ Distomidae ⌋──── Dicardia ⌉── Epicardiocoela Didemnidae ⌉──── Unicardia ⌋ Clavelinidae ⌋		

CLASS 3. THALIACEA

Pelagic Tunicata in which the adult has no tail; a degenerate nervous system; an atrium which opens posteriorly; a stolon of complex structure; and gill clefts which are not divided by external longitudinal bars.

Thus defined, the group includes the Pyrosomatidae (Luciae), which are transitional from the Ascidiacea, with which they are usually placed, though by their essential peculiarities they belong here. The three orders of the class differ considerably, though two of them (Pyrosomatidae and Salpidae) are more nearly related to one another than either is to the third (Doliolidae).

In all, the *muscular strands* of the mantle are arranged as rings which encircle the barrel- or lemon-shaped body. These are complete, but feeble and present at the ends of the body only, in *Pyrosoma*, strong but usually incomplete ventrally and convergent dorsally in the Salpidae (Fig. 521); strong, complete and regular in the Doliolidae (Fig. 523). Their contractions cause (in *Pyrosoma*, assist) the locomotion of the animal by driving water from the atrial opening—in the Salpidae and Doliolidae direct to the exterior,

Pyrosoma

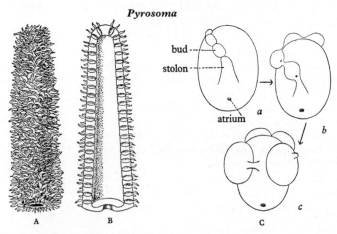

Fig. 520. *Pyrosoma*: A, colony. B, the same cut open longitudinally. C, diagram of the budding of the first individuals of a colony by the cyathozooid.

Salpa

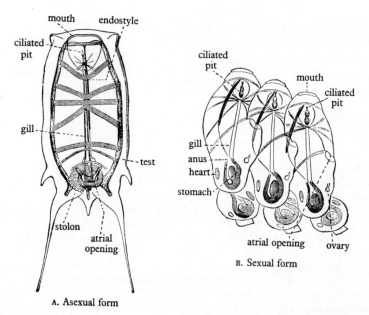

Fig. 521. *Salpa democratica-mucronata*: A, the asexual form (oozooid). B, the end of a stolon showing part of a chain of sexual individuals (blastozooids) about to be set free. (From Sedgwick, after Claus.)

in *Pyrosoma* (Fig. 520) into the lumen of a cylindrical colony and thence through the single external opening of the latter. The *gill clefts* are in *Pyrosoma* numerous (up to fifty), tall dorsoventrally, and crossed by internal, though not by external, longitudinal bars. In the Salpidae the first-formed cleft persists and becomes in the adult a single gigantic opening which occupies the entire side of the pharynx. The Doliolidae have a varying number (few in the oozooid, more numerous in the blastozooid) of short openings.

In *Pyrosoma* and the Salpidae the egg is retained long in the parent: in *Pyrosoma* it is yolky and meroblastic; in the Salpidae the embryo is nourished through a *placenta*. Development is direct, the tailed larval stage being omitted; and the buds formed by the oozooid on its stolon (which has a single epicardial tube) hang together for some time as a chain. In *Pyrosoma* this chain (of four zooids) coils into a circle around the body of the degenerate oozooid (*cyathozooid*, Fig. 520c), and its members then bud in such a way as

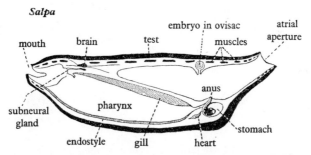

Fig. 522. A semidiagrammatic view of the left side of *Salpa*. (From Herdman.)

to form a cylindrical colony, closed at one end, composed of blastozooids. This is the form in which the animals pass their free existence, the oozooid never leaving the body of its parent. In the Salpidae oozooid and blastozooids are alike well developed and free swimming, and the blastozooids, of which there is a long chain, though they may coil into a circle (*Cyclosalpa*), are incapable of budding and eventually break away in groups (Fig. 521 B). In the Doliolidae there is a tailed larva, and the buds formed on the stolon (in which the epicardial tubes remain separate) break free one by one, though they subsequently make attachment to a dorsal process of the mother, by whom they are carried for some time.

1. *Order* **Pyrosomatidae (Luciae)**

Thaliacea which have no larval stage; whose oozooid is degenerate and retained within the parent; whose stolon contains a single epicardial tube; and whose blastozooids at first form a short chain, but subsequently by budding constitute a cylindrical colony of ascidian-like individuals.

Pyrosoma (Fig. 520). The only genus. The colonies vary in length from an inch or two to several feet, and are phosphorescent, from which fact the generic name is derived.

738 PROTOCHORDATES

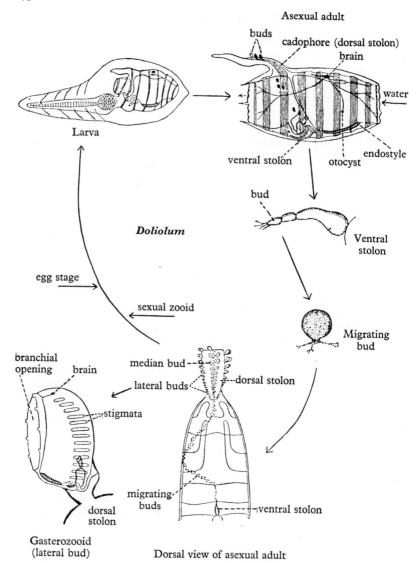

Fig. 523. *Doliolum* and its life history. See the text for details. (From the *Cambridge Natural History*, after Uljanin and Barrois.)

2. *Order* Salpidae (Hemimyaria)

Thaliacea which have no larval stage; whose oozooid is well formed and free; whose pharynx has no lateral walls, owing to enlargement of the primary pair of gill clefts; and whose blastozooids are incapable of budding, but adhere as a chain from which they eventually break free in groups.

All that is left of the walls of the branchial chamber (Fig. 522) is the endostyle and a dorsal (hyperpharyngeal) bar, known as the 'gill', which runs in a slanting direction along an immense internal cavity formed by the confluence of the branchial and atrial chambers through the absence of lateral branchial walls. The animal is as transparent as glass, save for a small, coloured 'nucleus' where the stomach and intestine are situated.

Salpa (Figs. 521, 522). The chain of blastozooids is band-like.

Cyclosalpa. The chain of blastozooids forms rings.

3. *Order* **Doliolidae (Cyclomyaria)**

Thaliacea which have a tailed larval stage; whose oozooid is well formed and free; whose pharynx has several stigmata on each side; and whose blastozooids break free one by one from the stolon as buds which subsequently make attachment to a dorsal process (cadophore) of the parent, by whom they are carried for some time.

The larva has the barrel-shaped body of the adult with a tail attached ventrally at the hinder end: it lies free in its test. Dorsally behind it has already the rudiment of the cadophore. In the adult oozooid the cadophore elongates, as the buds, wandering round the body from the ventral stolon, begin to settle down and develop. The bodies which break off from the stolon are known as *probuds*. They travel by the pseudopodial activity of certain of their ectoderm cells, and on arriving upon the cadophore divide several times to form the definitive buds. These fasten themselves in a lateral and a median row on each side. Eventually they grow into individuals of three kinds, which co-operate in a remarkable manner. Those of the lateral row become *gasterozooids* which gather food for the community, those of the median row *phorozooids* which act as nurses, and others *gonozooids* carried by the phorozooids. The latter presently break free with their charges, which are ultimately liberated to reproduce sexually.

Doliolum (Fig. 523).

CHAPTER XIX
LITERATURE

There is a vast amount of zoological literature and most of it is available to any student possessing patience and a ticket to a library that belongs to the inter-library lending service. It is, however, necessary for the time-pressed student to know the books that will repay study and also how to find the primary authority for any set of facts. This brief section of the book is to help the student obtain a preliminary orientation to the zoological literature. References have not been given in the body of the book, but instead a few references will be given here together with some indication of how to pursue the subject still further.

The student is advised to consult Roger C. Smith's book, *A Guide to the Literature of the Zoological Sciences* (5th ed. 1962), published by the Burgess Publishing Company, Minneapolis.

Though this book does not give lists of references to original papers, it does provide a very valuable explanation of the way in which zoological journals are classified, the names of the different scientific journals, where they are published and the sort of material that each contains. There is also a list of taxonomic works.

KEY WORKS

There are four works of paramount importance in searching for original publications:

Zoological Record, published by the Zoological Society of London, has come out every year since 1864. It is divided into sections each having a list of titles and authors of the papers published on that particular phylum over the course of the stated year. There is a very good subject index to these sections, with cross-references to morphology, ecology, physiology, and also to species. From the title of the paper and the journal, one can then look up the relevant journal and read the paper in detail.

Biological Abstracts (1926–), published at the University of Pennsylvania, covers the whole of biological literature. It consists of short summaries (10–200 words) of practically all the biological papers published in the previous year. There is a very exhaustive subject and author index to each year.

International Abstracts of Biological Sciences, published by Pergamon Press, for the Council of Biological and Medical Abstracts, is similar in many ways to *Biological Abstracts*. It concentrates on the experimental literature and its author and subject index comes out four times a year.

World List of Scientific Periodicals. Having found an interesting reference to a journal it is necessary to find the journal. Often if you apply to your librarian, the journal will be obtained for you. There is, however, a publication, the *World List of Scientific Periodicals* (3rd ed. 1952) that lists all the different journals and indicates which libraries in the British Isles possess them. For example if one reads a review or text-book and finds a reference

LITERATURE 741

to a paper in the Proceedings of the Zoological Society of Bengal which one is interested in reading, one can look up the *World List* to find the location of this journal.

There one finds: *Proceedings of the Zoological Society of Bengal* (1948–) L.BM^N.; Z.; BM.U.; BR.U.; E.AG.; R.; U.; MD.H. This means that the journal will be found in London at the British Museum library (Natural History), and at the Zoological Society library. It will also be found at the libraries of Birmingham University, and Bristol University, at the Animal Genetics Department, Edinburgh University, and also at the libraries of the Royal Society of Edinburgh and the University Library. It is also at the Commonwealth Bureau for Horticulture at Maidstone.

A somewhat similar publication is *Union List of Serials in the Libraries of the United States and Canada*, edited by W. Gregory (3rd ed. 1952), published by H. W. Wilson, New York. This provides information on the location of journals other than government publications. It is sometimes easier to get hold of a microfilm or photographed copy of a particular paper than to borrow the journal itself. Your librarian will be able to advise you on this matter.

Biological Bulletin published in 1944 a supplement to their normal issues which contained the titles of the 2258 serial publications that are taken by their library. Microfilm copies of papers are sometimes obtainable.

Though journals are very important and the above works give some guidance in searching for a particular reference, to use them straight off is much like examining a specimen with a high-power objective before one has made a low-power examination. The equivalent to using the low-power objective is to consult one of the major text-books and see if the information is available there first of all. It should, however, be remembered that text-books have the habit of making tentative hypotheses appear in the guise of concrete facts.

TEXT-BOOKS AND TREATISES
English

There is a scarcity of advanced zoological treatises in the English language. There is the *Cambridge Natural History* published in 1895 in ten volumes and there is also the *Treatise on Zoology* edited by Ray Lankester which unfortunately was never finished. Details of the different volumes in these works are presented below. Several of the volumes are still of very great use.

One of the best of the text-books for the general student is that written by Sedgwick. This was published in 1898–1908 and is in three volumes. It provides a balanced readable account of the morphology of all the major groups of invertebrates and vertebrates.

Recently Hyman has done a great service to English-reading zoologists by writing a series of volumes on the invertebrates; so far five volumes have appeared and others are in preparation.

A Treatise on Zoology (1900–), edited by Ray Lankester (Adam and Charles Black, London).
Part I. *Fascicle* 1. Introduction and Protozoa.
Fascicle 2. Introduction and Protozoa 2.

Part II. Porifera and Coelenterata.
Part III. Echinodermata.
Part IV. Platyhelminthes, Mesozoa and Nemertines.
Part V. Mollusca.
Part VIII (third fascicle only). Crustacea.
The Cambridge Natural History (1895–1909), edited by S. F. Harmer and A. E. Shipley (Macmillan, London; reprint available from Wheldon and Wesley, Codicote, Hitchin, Herts).
1. Protozoa, Porifera, Coelenterates, Ctenophora and Echinodermata.
2. Flatworms, Mesozoa, Nemertines, Nematodes, Rotifers, Polychaetes, Earthworms, Leeches, Gephyreans, Phoronis, Polyzoa.
3. Molluscs and Brachiopods.
4. Crustacea, Trilobites, Arachnids.
5. Peripatus, Myriapods, Insects (part 1).
6. Insects (part 2).
7. Hemichordates, Ascidians, Amphioxus and Fishes.

Student's Text-book of Zoology by A. Sedgwick (Sonnenschein, London).
Vol. 1. 1898 (but second edition in 1927). Protozoa, Platyhelminthes, Nematodes, Molluscs, Annelids, Chaetognatha.
Vol. 2. 1905. Vertebrata.
Vol. 3. 1908. Protochordates, Echinodermata, Arthropoda.

The Invertebrates (1940–), by L. H. Hyman (McGraw-Hill Book Company, New York and London).
Vol. 1. Protozoa through Ctenophora.
Vol. 2. Platyhelminthes and Rhynchocoela.
Vol. 3. Acanthocephala, Aschelminthes and Entoprocta.
Vol. 4. Echinodermata.
Vol. 5. Chaetognatha, Hemichordata, Pogonophora, Phoronida, Ectoprocta, Brachipoda, Sipunculida. *Retrospect.*
The volumes in preparation are Molluscs, Annelids, and Arthropods.

These volumes are valuable; for besides providing a detailed account of the morphology, physiology and classification of the groups, Dr Hyman discusses such problems as the relationship between the polyp and medusa in the coelenterates, the nature of segmentation and strobilation, and many other topics of interest that are discussed in no other easily available source. In volume 5, the *Retrospect* reviews the literature (1940–59) that appeared on the subject matter of the preceding volumes.

French

There are several excellent texts on the invertebrates written in French. A very good, though incomplete, series is *Traité de Zoologie Concrète* written by Delage and Hérouard. These volumes have excellent coloured illustrations as well as some interesting discussions of general zoological topics. The groups that are covered, are covered well.

In recent years (1948–) Grassé has edited a series of modern texts that provide a well-balanced, documented account of the different groups of animals. These include vertebrates and invertebrates.

Traité de Zoologie; Anatomie—Systématique—Biologie, edited by P.-P. Grassé (Masson et Cie, Paris).

Vol. 1. *Fascicle 1. Introduction, Protozoa (Flagellates).
  *Fascicle 2. Protozoa (Rhizopoda, Sporozoa, Cnidosporidia).
Vol. 2. Protozoa, Ciliata.
Vol. 3. Sponges, Cnidaria, Ctenophora.
Vol. 4. Fascicle 1. Platyhelminthes, Nemertines.
  Fascicle 2. Nemathelminthes, Rotifera.
Vol. 5. *Fascicle 1. Annelids, Sipunculids, Echiurids, Priapulids, Endoprocts, Phoronids.
  *Fascicle 2. Bryozoa, Brachiopods, Chaetognatha, Pogonophora, Mollusca; Aplacophora, Monoplacophora, Lamellibranchs.
  Fascicle 3. Mollusca, Gasteropods, Cephalopods.
Vol. 6. *Onychophora, Tardigrada, Arthropoda; Introduction, Trilobita, Chelicerata.
Vol. 7. Crustacea, Myriapoda.
Vol. 8. *Insecta, Introduction, Anatomy and Physiology.
Vol. 9. *Insecta, Palaeontology, Lower insects: Coleoptera.
Vol. 10. Insecta. Higher insects.
Vol. 11. *Echinoderms, Hemichorda, Protochorda.

The asterisk indicates that the volume has been published; the others are in preparation.

German

There are many extremely good German text-books. Two of the outstanding treatises are those edited by Kükenthal and Krumbach, and those by Bronn.

Bronn's *Klassen und Ordnungen des Thier-Reichs* is an exhaustive series of volumes that are a veritable treasure trove for the zoologist, often providing the last word on a subject. It was started in 1873 and not only covers the morphology of the different groups, but also goes into the details of their ecology and in some cases their physiology. The series was published by Akad. Verlag. Leipzig.

Handbuch der Zoologie—eine Naturgeschichte der Stämme des Tierreiches (1923–   ), edited by W. Kükenthal and T. Krumbach (W. de Gruyter, Berlin and Leipzig).

1. Protozoa, Porifera, Coelenterata, Mesozoa.
2.1. Nematomorpha, Platyhelminthes.
2.2. Annelida, Priapuloidea, Sipunculoidea, Echiuroidea.
3.1. Tardigrada, Pentastomida, Myzostomida, Arthropoda, Crustacea.
3.2.1. Chelicerata, Pantapoda.
3.2.2. Onychophora, Phoronida, Brachiopoda, Bryozoa, Rhabdopleura, Enteropneusta.
4.1. Progoneata, Chilopoda, Insecta (Apterygota).
4.2.1. Insecta (Hymenoptera-Lepidoptera).
4.2.2. Insecta (Diptera-Homoptera).
5.1. Mollusca, Echinodermata.
5.2. Tunicata.

EMBRYOLOGY

There are relatively few texts on invertebrate embryology. Dawydoff, in addition to his book of 1928, has written many of the embryological sections that are in Grassé's *Traité de Zoologie*. Older, but nevertheless valuable works, are those by Korschelt and Heider (there is an English translation of this work, by Bernard and Woodward), and the volume by MacBride.

Many of the great advances that have been made in the field of experimental embryology have been carried out on invertebrate material such as the eggs of *Echinus* and *Chaetopterus*; these experiments are discussed in the books by Needham, Willier, Brachet, and details of how to obtain material and handle the eggs, given by Costello.

Brachet, J. (1960). *The Biochemistry of Development* (Pergamon Press).
Costello, D. P., Davidson, M. E., Eggers, A., Fox, M. H. and Henley, C. (1957). *Methods for obtaining and Handling Marine Eggs and Embryos* (Marine Biol. Lab., Woods Hole, Mass.).
Dawydoff, C. (1928). *Traité d'embryologie comparée des invertébrés* (Masson et Cie).
Korschelt, E. and Heider, K. (1893–1909). *Lehrbuch der vergleichenden Entwicklungsgeschichte der wirbellosen Tiere* (Fischer, Jena). (2nd ed. 1936.)
MacBride, E. W. (1914). *Textbook of Embryology*. Vol. 1. *Invertebrates* (Macmillan, London).
Needham, J. (1942). *Biochemistry and Morphogenesis* (Cambridge University Press).
Willier, B. H., Weiss, P. and Hamburger, V. (1955). *Analysis of Development* (W. B. Saunders).

PRACTICAL ZOOLOGY

The invertebrates due to their vast number of 'types' are much neglected when it comes to practical books. There are many books on the preparation of slides from invertebrate material; Baker's book, *Cytological Technique*, discusses the principles involved in making such slides, and detailed methods are given by Peacock, by Pantin, and by McLung Jones. Casselman's book gives useful details for the histochemical localization of specific compounds in microscopical sections.

A very useful book is that edited by Galtsoff; in it are details and recipes for keeping invertebrates alive in the laboratory. Bullough provides dissecting instructions for some invertebrates, whilst excellent details on the anatomy of many invertebrates suitable for dissection will be found in Brown's *Selected Invertebrate Types*. There is also a series of monographs published by the Liverpool Marine Biology Committee.

For the experimentally minded student a very important book is that by Welsh and Smith. This provides a series of experiments together with full references to the experimental literature.

Baker, J. R. (1961). *Cytological Technique* (Methuen).
Brown, F., ed. (1950). *Selected Invertebrate Types* (Wiley).
Bullough, W. S. (1958). *Practical Invertebrate Anatomy* (Macmillan).
Casselman, W. Bruce (1959). *Histochemical Technique* (Methuen).
Galtsoff, P., ed. (1937). *Culture Methods for Invertebrate Animals* (Comstock, but reprinted by Dover).
Gatenby, J. B. (1950). *The Microtomist's Vade-Mecum* (Churchill).

Jones, R. McLung, ed. (1950). *Handbook of Microscopical Technique* (Cassell).
Pantin, C. F. A. (1946). *Notes on Microscopical Technique for Zoologists* (Cambridge University Press).
Peacock, H. A. (1940). *Elementary Microtechnique* (Arnold).
Rowett, H. G. Q. (1953). *Dissection Guides*, V. 'The Invertebrates' (John Murray).
Welsh, J. and Smith, R. (1960). *Laboratory Exercises in Invertebrate Physiology* (Burgess Publishing Co.).

Liverpool Marine Biological Committee's memoirs on typical British marine plants and animals, edited by W. A. Herdman and J. Johnstone, are a series of booklets each dealing with the detailed anatomy of one or possibly two animals. Though dissecting instructions are not given, the anatomy is described in sufficient detail to allow one to determine what the scalpel has revealed. The animals dealt with are as follows:

Alcyonium	Galathea	Buccinum	Ciona
Arenicola	Cancer	Haliotis	Ascidia
Tubifex	Lernaea	Cardium	Botryllus
Aphrodite	Eupagurus	Mytilus	Asterias
Pomatoceros	Ligia	Eledone	Echinus
Lineus		Sepia	Antedon
Polychaete larvae		Patella	Echinoderm larvae

PALAEONTOLOGY

An excellent account of the palaeontology of the different invertebrate groups will be found in the book by Shrock and Twenhofel. It is a book that all students of invertebrates should consult. There are, in addition, two treatises that deal with invertebrate fossils in great details. These are the volumes edited by Moore, and by Piveteau.

Moore, R. C., ed. (1952–). *Treatise on Invertebrate Paleontology* (The Geological Society of America). [* Means already published.]
   A. Introduction.
   B. Protista 1. (Chrysomonads, Coccolithophorids, Diatoms, etc.)
   C. Protista 2. (Foraminiferans.)
*D. Protista 3. (Radiolarians, Tintinines.)
*E. Archaeocytha, Porifera.
*F. Coelenterata.
*G. Bryozoa.
   I. Mollusca 1. (Chitons, Scaphopods, Gastropods.)
   J. Mollusca 2. (Gastropods.)
   K. Mollusca 3. (Nautiloid Cephalopods.)
*L. Mollusca 4. (Ammonoid Cephalopods.)
   M. Mollusca 5. (Dibranchiate Cephalopods.)
   N. Mollusca 6. (Pelecypods.)
   O. Arthropoda 1. (Trilobita.)
*P. Arthropoda 2. (Chelicerata, Pycnogonids.)
   Q. Arthropoda 3. (Ostracods.)
   R. Arthropoda 4. (Branchiopods, Cirripeds, Malacostracans, Myriapods, Insects.)
   S. Echinodermata 1. (Cystoids, Blastoids, Edrioasteroids.)
   T. Echinodermata 2. (Crinoids.)
   U. Echinodermata 3. (Echinozoans, Asterozoans.)

*V. Graptolithina.
W. Miscellanea (Worms, Conodonts).
X. Addenda (Index).
Piveteau, J., ed. (1952). *Traité de Paleontologie* (Masson et Cie). 3 volumes.
1. Protozoa, Sponges, Coelenterates, Bryozoa.
2. Brachiopoda, Chaetognatha, Annelida, Mollusca.
3. Onychophora, Arthropoda, Echinodermata, Stomochorda.
Shrock, R. and Twenhofel, W. (1953). *Principles of Invertebrate Paleontology* (McGraw-Hill).

PERSPECTIVE

It is often easy to become so immersed in the immense amount of detail involved in invertebrate zoology that 'the wood becomes obscured by the trees'. On such occasions it is useful to consult a work giving details of the history of the subject, since in this way the reader will appreciate the full significance of the various facts and realize that the knowledge of the present rests entirely on the information obtained by innumerable past research workers. The reader is advised to consult either Singer's or Nordenskiold's books for general details, and Dawes' book for more specific information about the development of ideas and factual knowledge over the past hundred years.

Dawes, B. (1952). *A Hundred Years of Biology* (Duckworth).
Nordenskiold, E. (1920). *The History of Biology* (Tudor Publ. Co.).
Singer, C. (1950). *A History of Biology* (Schuman).

PHYSIOLOGY

The accent of the present work has been mainly morphological. It is, of course, necessary for the student to have some knowledge of the physiology of the invertebrates and there are several texts that provide this. Some, such as Davson's and Scheer's *General Physiology*, provide an account of the general principles of physiology. Others, such as Carter's *General Zoology*, are concerned with the physiological processes found in the invertebrates. Scheer's *Comparative Physiology* discusses each phylum in turn, whilst Prosser deals with the different physiological function in order, i.e. nutrition, excretion, digestion, sense organs, etc. There are also many specialized works such as those of Krogh which give details of one specific physiological function. Finally there are the standard physiology text-books for medical students.

von Buddenbrock, W. (1953). *Vergleichende Physiologie*, 6 volumes arranged according to topics (Verlag Birkhauser, Basel).
Carter, G. S. (1951). *General Zoology of the Invertebrates* (Sidgwick and Jackson).
Davson, H. (1959). *Textbook of General Physiology* (Churchill).
Evans, C. Lovatt (1956). *Principles of Human Physiology* (Churchill).
Jenkins, P. L. (1961). *Animal Hormones* (Pergamon Press).
Krogh, A. (1939). *Osmotic Regulation in Aquatic Animals* (Cambridge University Press).
Krogh, A. (1941). *The Comparative Physiology of Respiratory Mechanisms* (University of Pennsylvania Press).
Lowenstein, O., ed. (1962). *Recent Advances in Comparative Physiology and Biochemistry* (Academic Press).

Nicol, J. A. C. (1960). *The Biology of Marine Animals* (Pitman).
Prosser, C. L., ed. (1950). *Comparative Animal Physiology* (Saunders).
Ramsay, J. A. (1952). *Physiological Approach to the Lower Animals* (Cambridge University Press).
Ruch, T. C. and Fulton, J. D. (1960). *Medical Physiology and Biophysics* (Saunders).
Scheer, B. T. (1948). *Comparative Physiology* (Wiley).
Scheer, B. T. (1953). *General Physiology* (Wiley).
Scheer, B. T., ed. (1957). *Recent Advances in Invertebrate Physiology* (University of Oregon Press).
Wright, S. (1961). *Applied Physiology* (Oxford University Press).

GEOGRAPHICAL DISTRIBUTION

Most zoologists at some time or another wonder how it is that some groups of animals are found in one part of the world but not in another. A very interesting case is that of *Peripatus* (p. 329). There are two extreme views concerning the geographical distribution of animals. The first suggests that there were land bridges joining the different land masses at different geological periods. The second view is that much of the fauna arrived on the separated islands either by their own flight (spiders carried on the web, spores, etc.) or else floating on rafts or logs. These two views are to be found in Gadow and Matthews.

Beaufort, L. F. (1951). *Zoogeography of the Land and Inland Waters* (Sedgwick and Jackson).
Beirne, B. P. (1952). *The Origin and History of the British Fauna* (Methuen).
Elton, C. (1958). *The Ecology of Invasions by Animals and Plants* (Methuen).
Gadow, H. F. (1913). *The Wanderings of Animals* (Cambridge University Press).
Lindroth, C. (1957). *The Faunal Connections between Europe and North America* (Wiley).
Matthews, W. T. (1915). Climate and evolution. *Ann. New York Acad. Sci.* **24**, 171–318.
Mayr, E., ed. (1952). The problem of land connections across the South Atlantic. *Bull. Amer. Mus. Nat. Hist.* **99**, 85–256.
Simpson, G. G. (1953). *Evolution and Geography*. Condon Lectures. 64 pp. (Oregon University Press).

ECOLOGY

A book that all invertebrate zoologists should read is that by C. M. Yonge, called *The Sea Shore*. This book provides a great deal of information about the life of the animals in the littoral regions and presents the different creatures as real live animals, not just as Latin names in an anatomical text (C. M. Yonge, *The Sea Shore*. Collins). A somewhat similar book is *Between Pacific Tides*, by E. F. Ricketts and J. Calvin, third edition revised by J. Hedgepeth (Stanford University Press). Oceanic ecology is dealt with in a fascinating way by A. C. Hardy in his book, *The Open Sea* (Collins). An important work is *Ecological Animal Geography*, by Hesse, Allee and Schmidt, 1951 (Wiley). This deals with the different habitats, i.e. lakes, ponds, streams, forests, alpine communities, etc., and the animals living in these specialized environments. There are many straightforward texts on animal ecology, such as *Animal Ecology*, by Charles Elton, 1927 (Sidgwick and Jackson); *Principles of Animal Ecology*, by Allee, Emerson, Park, Park and Schmidt, 1949 (Saunders); *Animal Ecology*, by MacFadyen, 1957 (Pitman); *Practical Animal*

Ecology, by Dowdeswell, 1959 (Methuen). Kevin has edited a very useful account of the modern work in terrestrial ecology, *Soil Zoology*, 1955 (Butterworth).

There are several journals that deal with articles on ecological subjects. *Ecology* and *Ecological Monographs* are published in the United States, whilst the *Journal of Animal Ecology* is published in England. In German there is the *Zoologische Jahrbücher*. Note that this periodical appears in three different 'streams'. One deals with ecology and systematics, the second deals with anatomy whilst the third deals with physiology. The ecological one is *Abteilung für Systematik, Ökologie und Geographie der Tiere*.

PARASITOLOGY

Many of the invertebrate animals discussed in the preceding pages are parasites. This mode of life, besides raising interesting physiological and ecological questions, also raises medical and veterinary problems. Many of the inhabitants of Africa and Asia are hosts to these parasites, and it must be admitted that not all Europeans and Americans are entirely free from nematodes, platyhelminthes, and protozoans! Parasitology, like most other branches of zoology, has become a subject in its own right. A few references will be given here to appropriate books; other references will be given in the sections on nematodes and platyhelminthes.

Baer, J. G. (1951). *Ecology of Animal Parasites* (University of Illinois Press).
Ball, G. H. (1943). Parasitism and evolution. *Amer. Nat.* **87**, 345.
Brand, T. von (1952). *Chemical Physiology of Endoparasitic Animals* (Academic Press).
Cameron, T. W. M. (1956). *Parasites and Parasitism* (Methuen).
Caullery, M. (1952). *Parasitism and Symbiosis*, translated by A. Lysaght (Sidgwick and Jackson).
Chandler, A. C. (1955). *Introduction to Parasitology* (Wiley).
Davenport, D. (1955). Specificity and behaviour in symbiosis. *Quart. Rev. Biol.* **30**, 29.
Lapage, G. (1951). *Parasitic Animals* (W. H. Heffer).
Lapage, G. (1957). *Animals Parasitic in Man* (Penguin Books).
Salt, G. (1941). The effects of hosts on their insect parasites. *Biol. Rev.* **16**, 239.
Salt, G. (1960). Experimental studies on insect parasitism. *Proc. Roy. Soc.* B, **151**, 446.
Yonge, C. M. (1944). Experimental analysis of the association between invertebrates and unicellular algae. *Biol. Rev.* **19**, 68.

There are several journals that publish research papers on parasitological subjects. *The Journal of Parasitology* and *Parasitology* publish papers over a wide field, whilst *Experimental Parasitology* is particularly concerned with the physiology of parasites. In addition there are various journals that specialize in groups of animals; *Journal of Helminthology, Journal of General Microbiology* (this includes papers on parasitic Protozoa).

BEHAVIOUR

There are many books that deal with the general behavioural reactions of animals. That of Maier and Schneirla discusses the different reactions of each phylum, whilst that of Fraenkel and Gunn gives more concern to the classification of the different behavioural patterns. Carthy's book describes the sensory behaviour of invertebrates, whilst the volumes of Warden, Jenkins and Warner are particularly valuable for their exhaustive bibliography. The two main approaches, the vitalistic approach, and the mechanistic approach to animal behaviour, are provided by the works of E. S. Russell and T. H. Savory respectively. The attitude of the ethologists is expressed by the works of Tinbergen and Baerends, and criticized by Eibl-Eibesfeldt.

Baerends, G. P. (1959). Ethological studies of insect behaviour. *Ann. Rev. Ent.* **4**, 207.
Carthy, J. D. (1958). *An Introduction to the Behaviour of Invertebrates* (Allen and Unwin).
Cloudsley-Thompson, J. L. (1960). *Animal Behaviour* (Oliver and Boyd).
Eibl-Eibesfeldt, I. and Kramer, S. (1958). Ethology, the comparative study of animal behaviour. *Quart. Rev. Biol.* **33**, 181.
Fraenkel, G. S. and Gunn, D. L. (1940). *The Orientation of Animals* (Oxford University Press).
Grindley, G. C. (1937). *The Intelligence of Animals* (Methuen).
Jennings, H. S. (1915). *Behaviour of the Lower Organisms* (Columbia University Press).
Maier, N. and Schneirla, T. C. (1935). *Principles of Animal Psychology* (McGraw-Hill).
Russell, E. S. (1938). *The Behaviour of Animals* (Arnold).
Russell, E. S. (1945). *The Directiveness of Organic Activities* (Cambridge University Press).
Savory, T. H. (1936). *Mechanistic Biology and Behaviour* (Watts).
Savory, T. H. (1960). *The World of Small Animals* (University of London Press).
Thorpe, W. H. (1956). *Learning and Instinct in Animals* (Methuen).
Tinbergen, N. (1951). *The Study of Instinct* (Oxford University Press).
Warden, C. J., Jenkins, T. N. and Warner, L. H. (1936–40). *Comparative Psychology* (Ronald Press). Vol. 1, *Principles and Methods*; Vol. 2, *Invertebrates*; Vol. 3, *Vertebrates*.

There are in addition two journals that deal with behavioural studies, *Behaviour* and *Animal Behaviour*. Various other journals deal with the more 'psychological' approach, e.g. *The Journal of Comparative Physiology and Psychology*, *Zeitschrift für Tierpsychologie*.

IDENTIFICATION

It is often necessary to identify an unknown animal by means of a key; but first of all one must know which key to use. A list of such keys will be found in Smart and Taylor's *Bibliography of Key Works for the Identification of the British Fauna and Flora*, 1953 (British Museum). American students can find a somewhat similar list in Smith's *Guide to the Literature in Zoological Sciences* (Burgess Publishing Company). The student will find the *Pocket Guide to the Sea Shore*, by J. M. Barrett and C. M. Yonge (Collins, 1958), a very useful book for the identification of shore animals. Another useful series is published by W. Brown, Iowa, entitled '*How to know the———*'. So far volumes have appeared on the Protozoa, insects, insect larvae, beetles and spiders. The Ray Society of London publish each year a monograph on one particular group, the work usually having a systematic bias. They have been doing this since 1845 and one of these is Darwin's work on the cirripedes.

There are several useful French books. *La Faune de La France*, by R. Perrier (Delagrave, Paris), is a series in ten volumes, full of keys and useful thumb-nail sketches of the different animals. The volumes are small and easily slip into the pocket. Then there is *Faune de France*, published in 1921 (Lechevalier, Paris). This is a large detailed series of over forty volumes, many of which are the authoritative work on the subject.

For identifying fresh-water animals there is the small book by Mellanby, *Animal Life in Fresh Water* (Methuen); the small book by Macan, *Guide to Fresh Water Invertebrates*, 1959 (Longmans), and also the valuable new edition of *Fresh Water Biology* by Ward and Whipple, edited by W. T. Edmonson, 1959 (Wiley). The Linnean Society is publishing a series of keys called *Synopses of the British Fauna* dealing with each group of animals in turn.

The Linnean Society of London, *Synopsis of the British Fauna*.
1. Opiliones (Arachnida). T. H. Savory.
2. Caprellidae (Amphipoda). J. R. Harrison.
3. Gammaridae. D. M. Reid.
4. Fresh Water Bivalves (*Corbicula, Sphaerium*, etc.). A. E. Ellis.
5. Fresh Water Bivalves (Unioacea). A. E. Ellis.
6. Lumbricidae. L. Cernoscitov and A. C. Evans.
7. Talitridae (Amphipoda). D. M. Reid.
8. Slugs. H. E. Quick.
9. British Woodlice. E. B. Edney.
10. British Pseudoscorpiones. G. O. Evans and E. Browning.
11. British Millipedes (Diplopoda). J. G. Blower.
12. British Echiurids, Sipunculids, and Priapulids. A. C. Stephen.

It is very necessary to have the animals in a good condition, or at least in a good state of preservation when one comes to identify them. Two books that help in preserving specimens are *Collecting, Preserving and Studying Insects*, by H. Oldroyd, 1958 (Macmillan, New York), and *Preservation of Natural History Specimens*, by R. Wagstaffe and J. H. Fidler (Witherby, London).

REVIEWS

There are several journals that contain reviews of specific topics in zoology. These reviews are extremely valuable since they provide an up-to-date and critical account of the work in a given field. The English journal, *Biological Reviews*, is particularly up-to-date since the authors are allowed to present an addendum on the material that has been published during the time that their article was in the press. The American journal, *Quarterly Review of Biology*, in addition to subject reviews, has extensive and often valuable book reviews. For example, the September 1960 number of 81 pages has 37 pages devoted to reviews of biological books. The German review journal is *Ergebnisse der Biologie*. For the strictly physiological literature there is *Physiological Reviews*. A very readable series of articles was published by Penguin Books as *New Biology*, there are a total of thirty-one volumes. The journal *Endeavour* usually has one article per number of biological interest, the illustrations being particularly well done. The same is true for *Scientific American*, a popular monthly publication.

In addition to the review publications there are yearly American publications such as *Annual Review of Biochemistry, Annual Review of Physiology, Annual Review of Entomology, Annual Review of Microbiology,* in which the preceding year's research is reviewed. The reviews are usually not very discursive, the accent is more on coverage of data than critical appraisal.

There are several symposia that are of interest. Two of these that specialize in biological subjects and appear every year are *Cold Spring Harbor Symposia,* and *Biological Symposia,* both being published in the United States. There is also a yearly series of volumes published by the Cambridge University Press for the Society for Experimental Biology. A specific topic is chosen and a series of authors write articles on this topic. Appearing at infrequent intervals is *Biological Progress,* edited by B. H. Glass. This has interesting and critical articles on different aspects of biology. There are also series called *Progress in Genetics* and *International Review of Cytology* that frequently contain papers of interest to a zoologist.

JOURNALS

There are many journals devoted to publishing the results of original research in specific fields. Usually each journal specializes in some section of zoology, though certain ones, such as *Nature* and *Science* which appear every week, have a wider field and are often used for preliminary announcements of researches as well as reviews of meetings and discussions.

For convenience some of the more strictly zoological journals are listed below under headings that give a *general* indication of their contents. It is helpful to develop the habit of 'browsing' through journals. It also helps if the student realizes that it is not necessarily the best thing to read a research paper through from start to finish; it is often easier to read them in the order—title, summary, introduction, methods—and then look carefully at the figures and tables to see if these support the conclusions indicated in the summary.

Morphology
    Quarterly Journal of Microscopical Science
    Journal of Morphology
    Zoologische Jahrbücher
    Proceedings of the Zoological Society (*London*)
    Bulletin of the American Museum of Natural History
    Systematic Zoology
    Bulletin of the British Museum (*Natural History*)
    Journal of Embryology and Experimental Morphology

Physiology
    Journal of Experimental Biology
    Physiological Zoology
    Journal of General Physiology
    Comparative Biochemistry and Physiology
    Zeitschrift für vergleichende Physiologie
    Journal of Cellular and Comparative Physiology
    Journal of Experimental Zoology

Marine Zoology
  Journal of the Marine Biological Association
  Biological Bulletin, Woods Hole
  Cahiers de Biologie Marine
  Archiv für Hydrobiologie
  Journal of Marine Research
  Pubblicazioni della Stazione zoologica di Napoli
Cytology and Methods in Histology
  Journal of Biophysical and Biochemical Cytology
  Stain Technology
  Journal of Histochemistry and Cytochemistry
  Journal of the Royal Microscopical Society

This list of journals is by no means comprehensive; there are many other journals such as *Evolution*, and *American Naturalist*, etc., that well repay reading. Other journals are mentioned in the reference lists on the appropriate subjects.

PHYLOGENY

Within recent years, interest has revived in determining the possible relationships of the various groups of animals. The new interest does not presume that any definite solution will be found to these problems, but instead considers that a re-evaluation of the known evidence allows the student to see things more clearly, and also allows him to tie the factual information together more easily. The journal *Systematic Zoology* is of especial interest in providing a sounding-board for new views and re-examination of older opinions.

Carter, G. S. (1954). On Hadzi's interpretation of phylogeny. *System. Zool.* **3**, 163.
Carter, G. S. (1957). Chordate phylogeny. *System. Zool.* **6**, 187.
Hadzi, J. (1953). An attempt to reconstruct the system of animal classification. *System. Zool.* **2**, 145.
Hanson, E. D. (1958). On the origin of the Eumetazoa. *System. Zool.* **7**, 16.
Jagersten, G. (1955). On the early phylogeny of the Metazoa. *Zool. Bidr. Uppsala*, **30**, 321.
Kerkut, G. A. (1960). *Implications of Evolution* (Pergamon Press).
Marcus, E. (1958). On the evolution of the animal phyla. *Quart. Rev. Biol.* **33**, 24.

PROTOZOA

These animals, with their relatively simple organization, present what at first sight seem to be ideal opportunities for research. Only too often the simplicity is superficial and involves the fundamental problems of the nature of cell membranes, protoplasm, the nucleus, etc. However, progress has been made and studies are available on the fine structure of the Protozoa (Pitelka, Grimstone, Picken), the relationship between the nucleus and the cytoplasm (Lorch and Danielli), the nature of amoeboid locomotion (de Bruyn, Goldacre, Noland), the nature of inheritance in protozoans (Beale, Sonneborn), and various other topics of interest.

Certain rearrangements of the status of the major protozoan groups will be found in most of the recent accounts of the Protozoa. The Flagellata and the Sarcodina are put together into a subphylum Rhizoflagellata, thus recognizing the ability of many flagellates to form pseudopodia, and of sarcodines to form flagella. The Sporozoa are also raised to subphylum rank, even though they are generally recognized as being a heterogeneous group of animals linked, in the main, by their parasitic habit. The Ciliophora too are given subphylum status so that the phylum Protozoa is then divided into three subphyla: Rhizoflagellata, Sporozoa, and Ciliophora. There is a discussion of this in Corliss (1959).

Corliss in his book, *The Ciliated Protozoa*, has presented a modified classification of the Ciliophora. This classification is shown in outline below together with the classification used in the present book. Certain differences will be seen. First, the Ciliophora are given the previously mentioned subphylum rank. Secondly, and of greater importance, the Suctoria, Peritricha, and Chonotricha are reduced in rank. The Suctoria were a subclass, now they are placed as an order within the Holotricha, as are also the Chonotricha and Peritricha. Previously the Chonotricha and Peritricha were of equal rank to the Holotricha. This new view is to some extent based on the work of Guilcher (1951), who showed that the young stages of the suctorians and chonotrichs had an infraciliature indicative of holotrich affinities. The reader is advised to consult Corliss for fuller details as well as an interesting and exhaustive bibliography of the ciliates.

Present scheme	Corliss's scheme
Class Ciliophora	*Subphylum* Ciliophora
Subclass Ciliata	*Class* Ciliata
Order 1. Holotricha	*Subclass* 1. Holotricha
Suborder 1. Astomata	Order 1. Gymnostomatida
Suborder 2. Gymnostomata	Order 2. Trichostomatida
Suborder 3. Hymenostomata	Order 3. Chonotrichida
Order 2. Spirotricha	Order 4. Suctorida
Suborder 1. Polytricha	Order 5. Apostomatida
Suborder 2. Oligotricha	Order 6. Astomatida
Suborder 3. Hypotricha	Order 7. Hymenostomatida
Order 3. Peritricha	Order 8. Thigmotrichida
Order 4. Chonotricha	Order 9. Peritrichida
Subclass Suctoria	*Subclass* 2. Spirotricha
	Order 1. Heterotrichida
	Order 2. Oligotrichida
	Order 3. Tintinnida
	Order 4. Entodiniomorpha
	Order 5. Odonstomatida
	Order 6. Hypotrichida

Abbott, B. C. and Ballantine, D. (1957). A toxin from *Gymnodinium veneficum*. *Jour. Mar. Biol. Ass.* **36**, 169.

Beale, G. H. (1954). *The Genetics of 'Paramecium aurelia'* (Cambridge University Press).

Bishop, A. (1959). Drug resistance in Protozoa. *Biol. Rev.* **34**, 445.

Bonner, J. T. (1954). The development of cirri and bristles during binary fission in the ciliate *Euplotes eurystomus*. *Jour. Morph.* **95**, 95.

Calkins, G. N. and Summers, F. M., eds. (1941). *Protozoa in Biological Research* (Columbia University Press).

Carter, L. (1957). Ionic regulation in the ciliate *Spirostomum ambiguum*. *Jour. Exp. Biol.* **34**, 71.

Corliss, J. O. (1955). The opalinid infusorians, flagellates or ciliates? *J. Protozool.* **2**, 107.

Corliss, J. O. (1956). Evolution and systematics of the ciliates. *System. Zool.* **5**, 68–91, 121–40.
Corliss, J. O. (1959). Comments on the systematics and phylogeny of the Protozoa. *System. Zool.* **8**, 169.
Corliss, J. O. (1961). *The Ciliated Protozoa* (Pergamon Press).
De Bruyn, P. (1947). Theories of amoeboid locomotion. *Quart. Rev. Biol.* **22**, 1.
Doflein, F. (revised by E. Reichenow, 1954). *Lehrbuch der Protozoenkunde* (6th ed., G. Fischer, Jena).
Doyle, W. (1943). The nutrition of the Protozoa. *Biol. Rev.* **18**, 119.
Elliott, A. M. (1959). Biology of *Tetrahymena*. *Ann. Rev. Microbiol.* **13**, 79.
Fritsch, F. E. (1929). Evolutionary sequence and affinities amongst the Protophyta. *Biol. Rev.* **4**, 103.
Garnham, P. C. C. (1954). Life history of malarial parasites. *Ann. Rev. Microbiol.* **8**, 153.
Goldacre, R. J. (1952). The folding and unfolding of protein chains as a basis for osmotic work. *Int. Rev. Cytol.* **1**, 135.
Grell, K. G. (1956). Protozoa and algae. *Ann. Rev. Microbiol.* **10**, 307.
Grimstone, A. V. (1961). Fine structure and morphogenesis in Protozoa. *Biol. Rev.* **36**, 97.
Guilcher, Y. (1951). Contribution à l'étude des cilies gemmipares chonotriches et tentaculafères. *Ann. Sci. nat. Zool.* (ser. 11), **13**, 33.
Hastings, J. W. (1959). Unicellular clocks. *Ann. Rev. Microbiol.* **13**, 297.
Hisada, M. (1957). Membrane resting and action potentials from a protozoan *Noctiluca scintillans*. *Jour. Cell. Comp. Physiol.* **50**, 57.
Huff, C. G. (1947). Life cycle of malarial parasites. *Ann. Rev. Microbiol.* **1**, 43.
Jahn, T. (1946). The euglenoid flagellates. *Quart. Rev. Biol.* **21**, 246.
Jepps, M. W. (1956). *The Protozoa Sarcodina* (Oliver and Boyd).
Kirby, H. (1950). *Materials and Methods in the Study of Protozoa* (University of California Press).
Kitching, J. A. (1938). The physiology of contractile vacuoles. *Biol. Rev.* **13**, 403.
Kitching, J. A. (1954). On suction in the Suctoria. *Recent Developments in Cell Physiology*, p. 197 (Butterworth).
Kitching, J. A. and Padfield, J. E. (1960). The physiology of contractile vacuoles. *Jour. Exp. Biol.* **37**, 73.
Klein, R. L. (1959). Transmembrane flux of K^{42} in *Acanthamoeba*. *Jour. Cell. Comp. Physiol.* **53**, 241.
Kudo, R. R. (1954). *Protozoology* (4th ed. C. C. Thomas).
Lackey, J. B. (1938). A study of some ecological factors affecting the distribution of Protozoa. *Ecol. Monog.* **8**, 503.
Levine, L. (1960). Cytochemical ATPase of *Vorticella* myonemes. *Science*, **131**, 1377.
Levine, R. P. and Ebersold, W. T. (1960). The genetics and cytology of *Chlamydomonas*. *Ann. Rev. Microbiol.* **14**, 197.
Lorch, I. J. and Danielli, J. F. (1954). Nuclear transplantation in amoeba. *Quart. Jour. Micr. Sci.* **94**, 445.
Lund, E. E. (1941). The feeding mechanisms of various ciliated Protozoa. *Jour. Morph.* **69**, 563.
Lwoff, A. (1950). *Problems of Morphogenesis in the Ciliates* (Wiley).
Lwoff, A. (1951–55) (editor of vols. 1 and 2). *The Biochemistry and Physiology of Protozoa* (Academic Press).
Margolin, P. (1956). The ciliary antigens of stock 172 *Paramecium aurelia*, variety 4. *Jour. Exp. Zool.* **133**, 345.
Noble, E. R. (1955). The morphology and life cycles of trypanosomes. *Quart. Rev. Biol.* **30**, 1.
Noland, L. (1957). Protoplasmic streaming. *Jour. Protozool.* **4**, 1.
Picken, L. E. R. (1960). *The Organization of Cells* (Oxford University Press).

Pitelka, D. (1962). *The Electron Microscopic Structure of Protozoa* (Pergamon Press).
Preer, J. R. (1959). Nuclear and cytoplasmic differentiation in Protozoa. *Growth Symp.* **16**, 3.
Pringsheim, E. G. (1941). The interrelationship of pigmented and colourless Protozoa. *Biol. Rev.* **16**, 191.
Sassuchin, D. (1934). Hyperparasitism in Protozoa. *Quart. Rev. Biol.* **9**, 215.
Sleigh, M. A. (1960). The form of beat of cilia of *Stentor* and *Opalina*. *Jour. Exp. Biol.* **37**, 1.
Sonneborn, T. M. (1947). Recent advances in the genetics of *Paramecium* and *Euplotes*. *Advanc. Genet.* **1**, 264.
Sonneborn, T. M. (1950). Methods in the general biology and genetics of *Paramecium aurelia*. *Jour. Exp. Zool.* **113**, 87.
Sonneborn, T. M. (1957). Breeding system, reproductive methods and the species problem in Protozoa, in *The Species Problem* (Amer. Ass. Adv. Sci.).
Sussman, M. (1956). The biology of the cellular slime moulds. *Ann. Rev. Microbiol.* **10**, 21.
Tartar, V. (1961). *The Biology of Stentor* (Pergamon Press).
Turner, J. P. (1933). The external fibrillar system of *Euplotes* with notes on the neuromotor apparatus. *Biol. Bull.* **64**, 53.
Wagendonk, W. J. (1955). Encystment and excystment of Protozoa. *Biochemistry and Physiology of Protozoa*, **2**, 85.
Wang, H. (1960). Experimental proteolysis in *Paramecium aurelia*. *Phys. Zool.* **33**, 271.
Watson, J. M. (1946). The bionomics of coprophilic Protozoa. *Biol. Rev.* **21**, 121.
Weiss, P. B. (1954). Morphogenesis in the Protozoa. *Quart. Rev. Biol.* **29**, 207.
Wichterman, R. (1955). *The Biology of 'Paramecium'* (Blakiston).

SPONGES

Interest in the sponges in the past has been concentrated on their regeneration (Galtsoff, Wilson), their spicules (Jones), and their water currents (Bidder). Willmer compares the tissue culture organization with that of sponges and his views are more generally expressed in his book. The ontogeny and systematics are discussed by Levi who also provides a detailed account of the embryology. Kerkut discusses the concept that sponges are the most primitive of the Metazoa.

Two new morphological points have arisen. One is the description of electron micrographs of the collar cells. Rasmont *et al.* have shown that these cells, instead of having a continuous protoplasmic collar, have a series of fine separate protoplasmic strands that form a ring around the flagellum. The other point is the descriptions by Pavans de Ceccatty of bipolar and multipolar cells that have many of the morphological features of nerve cells.

On the biochemical side, Bergman has isolated sterols from sponges and has found that many of them were hitherto unknown to occur in animal tissues.

Bergman, W. (1949). Comparative biochemical studies of the lipids of marine invertebrates with specific reference to the sterols. *Jour. Mar. Res.* **8**, 137.
Bidder, G. P. (1923). The relation of the form of a sponge to its currents. *Quart. Jour. Micr. Sci.* **67**, 293.
de Ceccatty, M. P. (1955). Le système nerveux des éponges calcaires et siliceuses. *Ann. Sci. Nat. Zool.* 11 sér. **17**, 203.

de Ceccatty, M. P. (1959). Les structures cellulaires de type nerveux chez *Hippospongia communis*. *Ann. Sci. Nat. Zool.* 12 sér. **1**, 105.
de Ceccatty, M. P. (1960). Les structures cellulaires de type nerveux et de type musculaire d'éponge siliceuses *Tethya lyncurium*. *C.R. Acad. Sci., Paris*, **251**, 1818.
Dendy, A. (1926). On the origin, growth and arrangement of sponge spicules. *Quart. Jour. Micr. Sci.* **70**, 1.
Dubosque, O. and Tuzet, O. (1937). L'ovogenèse, la fécondation et les premiers stades du développement des éponges calcaires. *Arch. Zool. exp. Gen.* **79**, 157.
Galtsoff, P. (1925). Regeneration after dissociation (an experimental study on sponges). I and II. *Jour. Exp. Zool.* **42**, 183–221, 223–55.
Hartmann, W. D. (1958). A re-examination of Bidder's classification of the Calcarea. *System. Zool.* **7**, 97.
Jagersten, G. (1955). On the early phylogeny of the Metazoa. *Zool. Bidr. Uppsala*, **30**, 321.
Jepps, M. W. (1947). Contribution to the study of sponges. *Proc. Roy. Soc. B*, **134**, 408.
Jones, W. C. (1959). Spicule growth rates in *Leucosolenia variabilis*. *Quart. Jour. Micr. Sci.* **100**, 557.
Kerkut, G. A. (1960). *Implications of Evolution* (Pergamon Press). (Porifera, the most primitive Metazoa, pp. 54–71.)
Levi, C. (1956). Embryologie et systématique des démosponges. *Arch. Zool. exp. Gen.* **93**, 1.
Levi, C. (1957). Ontogeny and systematics in sponges. *System. Zool.* **6**, 174.
Lufty, R. G. (1956). Relation between the dictyosome and contractile vacuoles in sponges. *J. Roy. Micr. Soc.* **76**, 141.
Parker, G. H. (1914). On the strength of water currents produced by sponges. *Jour. Exp. Zool.* **16**, 443.
Rasmont, R., Bouillon, J., Castiaux, P. and Vendermeerssche, G. (1958). Ultrastructure of choanocyte collar cells in fresh water sponges. *Nature, Lond.*, **181**, 58.
van Weel, P. B. (1948). On the physiology of the tropical fresh-water sponge, *Spongilla proliferens*. *Physiol. Oecol. Comp.* **1**, 110.
Willmer, E. N. (1945). Growth and form in tissue culture. *Essays on Growth and Form* (Oxford University Press).
Willmer, E. N. (1960). *Cytology and Evolution* (Academic Press).
Wilson, H. V. and Penny, J. T. (1930). The regeneration of sponges (*Microsciona*) from regenerating cells. *Jour. Exp. Zool.* **56**, 73.

COELENTERATA

Two recent developments in coelenterate studies are the application of electron microscope techniques to elucidate the structure of small forms such as *Hydra* (Wood, Slautterbuck, Hess), and the acceptance, by some authors, of the view that the coelenterates are not a primitive group, but that they are derived from acoelous platyhelminthes. This view was suggested by Hadzi (1953), and is supported by de Beer (1954). In Hadzi's view, the most primitive of the coelenterates are the Anthozoa which in turn gave rise to the Scyphozoa and Hydrozoa. Jagersten (1955) supports this view, and the evidence against is given by Hand (1959), Pantin (1960) and Kerkut (1960).

The coelenterates, being simple Metazoa, have roused the interest of physiologists who have analysed the reactions of the simple nerve net (Parker, Pantin, Horridge, Mackie, Gwilliam), tested the reactions of the independent effector organs—the nematocysts (Jones, Picken, Robson, Yanagita), and chopped polyps such as *Hydra* and *Tubularia* into pieces to watch them regenerate (Barth, Rose, Fulton). In addition much interesting work has been done on their life cycles (Berrill, 1949).

Barth, L. G. (1940). The process of regeneration in hydroids. *Biol. Rev.* **15**, 405.
Barth, L. G. (1944). The determination of the regenerating hydranth in *Tubularia*. *Phys. Zool.* **17**, 335.
Batham, E. J., Pantin, C. F. A. and Robson, E. A. (1960). The nerve net of the sea anemone, *Metridium senile*, the mesenteries and the collum. *Quart. Jour. Micr. Sci.* **101**, 487.
de Beer, G. R. (1954). The evolution of the Metazoa. In *Evolution as a Process* (Allen and Unwin).
Berrill, N. J. (1949). Developmental analysis of Scyphomedusae. *Biol. Rev.* **24**, 393.
Berrill, N. J. and Liu, L. K. (1948). Germ plasm, Weismann and Hydrozoa. *Quart. Rev. Biol.* **23**, 124.
Browne, E. T. (1897). On keeping medusae alive in an aquarium. *Jour. Mar. Biol. Ass.* **5**, 176.
Bryden, R. B. (1952). Ecology of *Pelmatohydra oligactis* in Kirkpatricks Lake, Tennessee. *Ecol. Monog.* **22**, 45.
Bullock, T. H. (1943). Neuromuscular facilitation in Scyphomedusae. *Jour. Cell. Comp. Physiol.* **22**, 251.
Busnett, A. L. (1959). Histophysiology of growth in *Hydra*. *Jour. Exp. Zool.* **140**, 281.
Busnett, A. L. and Garofalo, M. (1959). Growth patterns in the green Hydra, *Chlorohydra viridissima*. *Science*, **131**, 160.
Chapman, G. (1953). Studies on the mesogloea of coelenterates. *Quart. Jour. Micr. Sci.* **94**, 155; *Jour. Exp. Biol.* **30**, 440.
Chapman, G. (1958). The hydrostatic skeleton in the invertebrates. *Biol. Rev.* **33**, 338.
Crowell, S. (1957). Differential responses of growth zones to nutritive level, age, and temperature in the colonial hydroid *Campanularia*. *Jour. Exp. Zool.* **134**, 63.
Cuttress, C. E. (1955). Systematic study of Anthozoan nematocysts. *System. Zool.* **4**, 120.
Darwin, C. (1896). *The Structure and Distribution of Coral Reefs* (John Murray).
Davis, W. M. (1938). *The Coral Reef Problem* (Amer. Geograph. Soc.).
Fowler, G. H. (1891). Hermit crabs and anemones. *Jour. Mar. Biol. Ass.* **2**, 75.
Fulton, C. (1959). Re-examination of an inhibitor of regeneration in Tubularia. *Biol. Bull.* **116**, 232.
Garstang, W. (1946). The morphology and relations of the Siphonophora. *Quart. Jour. Micr. Sci.* **87**, 107.
Goreau, T. F. and Goreau, N. I. (1960). The physiology of skeleton formation in corals. *Biol. Bull.* **118**, 419; **119**, 416.
Gwilliam, G. F. (1960). Neuromuscular physiology of a sessile scyphozoan, *Haliclystus auricula*. *Biol. Bull.* **119**, 454.
Hadzi, J. (1953). An attempt to reconstruct the system of animal classification. *System Zool.* **2**, 145.
Hale, L. J. (1960). Contractility and hydroplasmic movement in the hydroid *Clytia johnstoni*. *Quart. Jour. Micr. Sci.* **101**, 339.
Hamilton, E. (1957). The last geographic frontier; the sea floor. *Sci. Mon., N.Y.* **85**.
Hammen, C. S. and Osborne, P. J. (1959). Carbon dioxide fixation in some marine invertebrates. *Science*, **130**, 1409.
Hand, C. (1959). On the origin and phylogeny of the Coelenterates. *System. Zool.* **8**, 163.
Hess, A., Cohen, A. I. and Robson, E. A. (1957). Observations on the structure of *Hydra* as seen with the electron and light microscope. *Quart. Jour. Micr. Sci.* **98**, 315.

Hickson, S. J. (1918). Evolution and symmetry in the order of Sea Pens. *Proc. Roy. Soc.* B, **90**, 108.

Hickson, S. J. (1924). *An Introduction to the Study of Recent Corals.* (University of Manchester Press).

Horridge, G. A. (1956). The coordination of the protective retraction of coral polyps. *Phil. Trans. Roy. Soc.* B, **240**, 495.

Horridge G. A. (1960). The nerves and muscles of medusae. *Jour. Exp. Biol.* **36**, 72; **32**, 555; **31**, 594.

Hoyle, G. (1960). Neuromuscular activity in the swimming sea anemone *Stomphia coccinea*. *Jour. Exp. Biol.* **37**, 671.

Hyman, L. H. (1940). Observations and experiments on the physiology of medusae. *Biol. Bull.* **79**, 282.

Jagersten, G. (1955). On the early phylogeny of the Metazoa. *Zool. Bidr. Uppsala*, **30**, 321.

Jones, C. S. (1941). The place of origin and the transportation of cnidoblasts in *Pelmatohydra oligactis*. *Jour. Exp. Zool.* **87**, 457.

Jones, C. S. (1947). The control and discharge of nematocysts in *Hydra*. *Jour. Exp. Zool.* **105**, 25.

Kerkut, G. A. (1960). *Implications of Evolution*, pp. 81, 94 (Pergamon Press).

Kramp, P. L. (1961). Synopsis of the Medusae of the world. *Jour. Mar. Biol. Ass.* **40**.

Laubenfels, M. W. (1955). Are the coelenterates degenerate or primitive? *System. Zool.* **4**, 43.

Lenhoff, H. M. and Schneiderman, H. A. (1959). The chemical control of feeding in the Portuguese man-of-war, *Physalia physalis*, and its bearing on the evolution of the Cnidaria. *Biol. Bull.* **116**, 452.

Lilly, S. J. (1955). Osmoregulation and ionic regulation in *Hydra*. *Jour. Exp. Biol.* **32**, 423.

Loomis, W. F. and Lenhoff, H. M. (1956). Growth and sexual differentiation of *Hydra* in mass culture. *Jour. Exp. Zool.* **132**, 555.

Mackie, G. O. (1960). Structure of the nervous system in *Velella*. *Quart. Jour. Micr. Sci.* **101**, 119.

Pantin, C. F. A. (1950). Behaviour patterns in the lower invertebrates. *Soc. Exp. Biol. Symp.* **4**, 175.

Pantin, C. F. A. (1952). The elementary nervous system. *Proc. Roy. Soc.* B, **140**, 147.

Pantin, C. F. A. (1960). Diploblastic animals. *Proc. Linn. Soc.* **171**, 1.

Parker, G. H. (1919). *The Elementary Nervous System* (Lippincott).

Picken, L. E. R. (1953). A note on the nematocysts of *Corynactis iridis*. *Quart. Jour. Micr. Sci.* **94**, 203.

Picken, L. E. R. (1957). Stinging capsules and designing Nature. *New Biol.* **22**, 56.

Ralph, P. M. (1959). A note on the pteromedusan genus *Tetraplatia*. *Jour. Mar. Biol. Ass.* **38**, 369.

Ralph, P. M. (1960). *Tetraplatia*, a coronate scyphomedusan. *Proc. Roy. Soc.* B, **152**, 263.

Rees, W. J. (1957). Evolutionary trends in the classification of capitate Hydroids and Medusae. *Bull. Brit. Mus. (Nat. Hist.). Zool. Ser.* **4**, 453.

Robson, E. A. (1953). The nematocysts of *Corynactis*. *Quart. Jour. Micr. Sci.* **94**, 229.

Robson, E. A. (1957). The structure and hydromechanics of the musculo-epithelium in *Metridium*. *Quart. Jour. Micr. Sci.* **98**, 265.

Romanes, J. G. (1885). *Jellyfish, Starfish and Sea Urchins* (Kegan Paul).

Rose, S. M. (1957). Polarized inhibitory effects during regeneration in *Tubularia*. *Jour. Morphol.* **100**, 187.

Rose, S. M. (1957). Cellular interaction during differentiation. *Biol. Rev.* **32**, 351.

Rose, S. M. and Rose, C. F. (1941). The role of the cut surface in *Tubularia* regeneration. *Phys. Zool.* **14**, 328.

Ross, D. M. (1960). The effects of ions and drugs on neuromuscular preparations of sea anemones. *Jour. Exp. Biol.* **37**, 732.
Russell, F. S. (1955). *Medusae of the British Isles* (Cambridge University Press).
Russell, F. S. and Rees, W. J. (1960). The viviparous scyphomedusa *Stygiomedusa fabulosa. Jour. Mar. Biol. Ass.* **39**, 303.
Slautterback, D. B. and Fawcett, D. W. (1959). Development of cnidoblasts in *Hydra*, an E.M. study of cell differentiation. *J. Biophys. Biochem. Cytol.* **6**, 441.
Southward, A. J. (1955). Observations on the ciliary currents of the jellyfish *Aurelia aurita* L. *Jour. Mar. Biol. Ass.* **34**, 201.
Spangenberg, D. B. and Ham, R. G. (1960). The epidermal nerve net of *Hydra. Jour. Exp. Zool.* **143**, 195.
Stephenson, T. A. (1928; 1935). *British Sea Anemones.* 2 vols. (The Ray Society.)
Weill, R. (1934). Contributions à l'étude des cnidaires et de leurs nématocystes. *Trav. stat. zool. Wimeraux*, **10–11**, 1–701.
Welsh, J. H. and Prock, P. B. (1958). Quarternary ammonium bases in the coelenterates. *Biol. Bull.* **115**, 551.
Wilson, D. P. (1947). The Portuguese man-of-war, *Physalia physalis*, in British and adjacent seas. *Jour. Mar. Biol. Ass.* **27**, 139.
Wittenberg, J. B. (1960). The source of CO in the float of the Portuguese man-of-war, *Physalia physalis. Jour. Exp. Biol.* **37**, 698.
Wood, R. L. (1959). Intercellular attachments in epithelium of *Hydra* as revealed by E.M. *J. Biophys. Biochem. Cytol.* **6**, 343.
Woodcock, A. H. (1956). Dimorphism in the Portuguese man-of-war. *Nature, Lond.*, **178**, 253.
Yanagita, T. M. (1959). Physiological mechanisms of nematocyst responses in sea anemones. *Jour. Exp. Biol.* **36**, 478.
Yonge, C. M. (1958). Ecology and physiology of reef-building corals. In *Perspectives in Marine Biology*, edited by A. Buzzati-Traverso (University of California Press).

MESOZOA

Dodson, E. O. (1956). A note on the systematic position of the Mesozoa. *System. Zool.* **5**, 37.
Stunkard, H. W. (1954). The life history and systematic relations of the Mesozoa. *Quart. Rev. Biol.* **29**, 230.

PLATYHELMINTHES

The Acolous turbellarians have been described by Hadzi as being the most primitive of the Metazoa; these views have already been mentioned in the section on Coelenterata. The regenerative ability of the planarians has been reviewed by Bronsted and also by Rose; the behaviour of the free-living planarians is discussed by Hovey and Olmsted and Ullyott; the possession of nematocysts and their use as weapons of defence is analysed by Kepner, whilst the physiology of tape-worms is reviewed by Smyth. There are also some interesting papers on the behaviour of the free-living larvae of the parasitic forms and the way in which they penetrate their hosts (Barlowe). The parasites often cause the host to set up a reaction which leads to immunity from further invasions. This is discussed by Taliaferro; Penfold; and Nigrelli. Some references are also given in the section of this chapter on Parasitology.

Barlowe, C. H. (1925). The life history of *Fasciolopsis buski. Amer. Jour. Hygiene Monog.* 4.
Baylis, H. A. (1938). Helminthes and evolution. *Evolution, Essays presented to Goodrich* (Oxford University Press).

Beuding, E. and Most, H. (1953). Helminthes, metabolism nutrition and chemotherapy. *Ann. Rev. Microbiol.* **7**, 295.
Brand, T. von (1957). Recent trends in parasite physiology. *Exp. Parasitol.* **6**, 233.
Bronsted, H. V. (1955). Planarian regeneration. *Biol. Rev.* **30**, 65.
Budington, R. (1924). The manner of copulation of a turbellarian worm, *Planaria maculata. Biol. Bull.* **47**, 298.
Cameron, T. W. (1934). *The Internal Parasites of Domestic Animals* (A. and C. Black).
Carter, G. S. (1954). On Hadzi's interpretation of animal phylogeny. *System. Zool.* **3**, 163.
Castle, W. A. (1928). An experimental and histological study of the life cycle of *Planaria velata. Jour. Exp. Zool.* **51**, 417.
Clark, R. B. and Cowey, J. B. (1958). Factors controlling the changes of shape of certain nemertean and turbellarian worms. *Jour. Exp. Biol.* **35**, 731.
Cort, W. W. (1922). Study of the escape of cercariae from their snail hosts. *Jour. Parasitol.* **8**, 177.
Curtis, W. C. (1906). The formation of proglottides in Cestoda. *Biol. Bull.* **11**, 202.
Dawes, B. (1956). *The Trematoda* (Cambridge University Press).
Erasmus, D. A. (1957). Phosphatase systems of cestodes. *Parasitology*, **47**, 70.
Hadzi, J. (1953). An attempt to reconstruct the system of animal classification. *System. Zool.* **2**, 145.
Hadzi, J. (1962). *The Relationship between the Turbellaria and Cnidaria* (Pergamon Press).
Hanson, E. D. (1958). On the origin of the Eumetazoa. *System. Zool.* **7**, 16.
Hovey, H. B. (1929). Associative hysteresis in marine flatworms. *Physiol. Zool.* **2**, 322.
Huff, C. G. (1956). Parasitism and parasitology. *Jour. Parasitol.* **42**, 1.
Hutchinson, W. M. (1959). Growth of the adult phase of *Taenia taeniaformis. Exp. Parasitol.* **8**, 557.
Jenkins, M. M. (1959). Effect of thiourea and some related compounds on regeneration in planarians. *Biol. Bull.* **116**, 106.
Jennings, J. B. (1959). Observations on the nutrition of the land planarian *Orthdemus terrestris. Biol. Bull.* **117**, 119.
Johnson, M. L. (1949). The tapeworm. *New Biol.* **7**, 113.
Keeble, F. and Gamble, F. (1907). Nature and origin of the green cells of *Convoluta roscoffensis. Quart. Jour. Micr. Sci.* **51**, 167.
Kepner, W. and Barker, J. F. (1924). Nematocysts of *Microstoma. Biol. Bull.* **47**, 239.
Lapage, G. (1957). *Animals Parasitic in Man* (Penguin Books).
Llewellyn, J. (1958). The adhesive mechanisms of monogenetic trematodes, the attachment of species of the Diclidophoridae to the gills of gadoid fishes. *Jour. Mar. Biol. Ass.* **37**, 67.
McCoy, O. R. (1935). The physiology of helminth parasites. *Physiol. Rev.* **15**, 221.
Moore, A. R. (1923). The function of the brain in locomotion of the polyclad worm, *Yungia aurantiaca. Jour. Gen. Physiol.* **6**, 73.
Nigrelli, R. F. and Breder, C. M. (1934). Susceptibility and immunity of certain fish to *Epibdella. Jour. Parasitol.* **20**, 259.
Olmsted, J. D. (1922). The role of the nervous system in the locomotion of certain marine polyclads. *Jour. Exp. Zool.* **36**, 57.
Pantin, C. F. A. (1931). The adaptation of *Gunda* to salinity changes. *Jour. Exp. Biol.* **8**, 63.
Pearse, A. S. and Wharton, G. (1938). The oyster 'leech' *Stylochus inimicus* (polyclad) associated with oyster beds along the coast of Florida. *Ecol. Monog.* **8**, 605.
Pederson, K. J. (1958). Morphogenetic activities during planarian regeneration as influenced by triethylene melamine. *Jour. Embryol. Exp. Morph.* **6**, 308.

Penfold, W., Penfold, H. and Phillips, M. (1937). *Taenia saginata*, its growth and propagation. *Jour. Helminth.* **15**, 41.
Pullen, E. W. (1957). A histological study of *Stenostomum virginianum* (Rhabdocoel). *Jour. Morph.* **101**, 597.
Reid, W. M. (1942). Certain nutritional requirements of the fowl cestode, *Raillietina cesticullus. Jour. Parasitol.* **28**, 319.
Reynoldson, T. B. (1958). The quantitative ecology of lake-dwelling Triclads in Northern Britain. *Oikos*, **9**, 94.
Rogers, W. P. (1955). The physiological basis of parasitism. *Rep. Aust. New-Zeal. Ass. Adv. Sci.* **30**, 105.
Rose, S. M. (1957). Cellular interaction during differentiation. *Biol. Rev.* **32**, 351.
Rosenbaum, R. M. and Rolan, C. (1960). Intracellular digestion and hydrolytic enzymes in the phagocytes of planarians. *Biol. Bull.* **118**, 315.
Rothschild, M. (1936). Gigantism and variation in *Peringia ulvae* caused by infection with larval trematodes. *Jour. Mar. Biol. Ass.* **20**, 537.
Rothschild, M. and Clay, T. (1952). *Fleas, Flukes and Cuckoos* (Collins).
Sandground, J. H. (1929). A consideration of the relation of host specificity of helminth parasites and other metazoan parasites to age resistance and acquired immunity. *Parasitol.* **21**, 227.
Sengel, C. (1959). Can the caudal region of a planarian regenerate a pharynx. *Journ. Embryol. Exp. Morph.* **7**, 73.
Smyth, J. D. (1947). The physiology of tape worms. *Biol. Rev.* **22**, 214.
Smyth, J. D. (1956). Histochemical study of egg shell formation in trematodes and cestodes. *Exp. Parasitol.* **5**, 519.
Smyth, J. D. and Clegg, J. A. (1959). Egg shell formation in trematodes and cestodes. *Exp. Parasitol.* **8**, 286.
Stephenson, W. (1947). The physiology of *Fasciola hepatica. Parasitol.* **38**, 116, etc.
Stunkard, H. W. (1946). Interrelationship and taxonomy of the digenetic trematodes. *Biol. Rev.* **21**, 148.
Stunkard, H. W. (1959). The morphology and life history of the digenetic trematode *Asymphylodora amnicolae*; possible significance of progenesis for the phylogeny of the Digenea. *Biol. Bull.* **117**, 562.
Taliaferro, W. H. (1940). Mechanisms of acquired immunity to infection by parasitic worms. *Phys. Rev.* **20**, 469.
Thomas, J. D. (1958). Studies on the trematode *Phyllodistomum simile. Proc. Zool. Soc.* **130**, 397.
Ullyott, P. (1936). The behaviour of *Dendrocoelum lacteum. Jour. Exp. Biol.* **13**, 253.
Van Cleave, H. J. (1952). Speciation and formation of genera in Acanthocephala. *System. Zool.* **1**, 72.
Wardle, R. A. and McLeod, J. A. (1952). *The Biology of Tape Worms* (University of Minnesota Press).
Yamaguti, S. (1958). *Systema Helminthum* (Interscience).

NEMATODES

The nematodes have had a great deal of interest focused on their parasitic habit (Lapage) principally because such a large percentage of the world's population is infected by nematodes. The nematodes also show interesting changes of sex, the sex ratio in some cases being dependent on the population density (Christie). The economic aspect of nematodes, such as the root eelworms, is important in those countries that are sufficiently civilized to have removed most of the nematodes parasitic in man. The reactions of the root eelworms are discussed by Miles. The suggestion has been made by Baylis that the nematodes might be descended from the arthropods, possibly from some animals such as the linguatulid arachnids. Keilin disagrees with this suggestion and gives his reasons for doing so.

Baldwin, E. and Moyle, V. (1947). An isolated nerve muscle preparation from *Ascaris*. *Jour. Exp. Biol.* **23**, 277.
Baylis, H. A. (1924). Systematic position of nematodes. *Ann. Mag. Nat. Hist.* **13**, 165.
Bird, A. F. (1957). Chemical composition of nematode cuticle. *Exp. Parasitol.* **6**, 383.
Chitwood, B. G. and Chitwood, M. B. (1950). *An Introduction to Nematology* (Chitwood, Baltimore).
Christie, J. R. (1929). Some observations on sex in the Mermithidae. *Jour. Exp. Zool.* **53**, 59.
Entner, N. and Gonzalez, C. (1959). Fate of glucose in *Ascaris lumbricoides*. *Exp. Parasitol.* **8**, 471.
Ferguson, M. S. (1943). Migration and localization of an animal parasite in the host. *Jour. Exp. Zool.* **93**, 375.
Goodey, T. (1951). *Soil and Freshwater Nematodes* (Methuen).
Harris, J. E. and Crofton, H. D. (1957). Structure and function in the nematodes. *Jour. Exp. Biol.* **34**, 116.
Inglis, W. G. (1957). The comparative anatomy and systematic significance of the head in the nematode family Heterakidae. *Proc. Zool. Soc.* **128**, 133.
Jarman, M. (1959). Electrical activity in the muscle cells of *Ascaris*. *Nature, Lond.*, **184**, 1244.
Johnson, G. E. (1913). On the nematodes of the common earthworm. *Quart. Jour. Micr. Sci.* **58**, 605.
Keilin, D. (1926). Origin of nematodes. *Parasitol.* **18**, 370.
Lapage, G. (1937). *Nematodes Parasitic in Animals* (Methuen).
Lapage, G. (1947). The menace of the roundworm. *New Biol.* **3**, 49.
Miles, H. and Miles, M. (1954). Root eelworms. *New Biol.* **16**, 101.
Nicholas, W. L., Dougherty, E. C., Hansen, E. L. and Moses, V. (1960). The incorporation of ^{14}C from sodium acetate-2-^{14}C into the amino acids of the soil-inhabiting nematode, *Caenorhabditis briggsae*. *Jour. Exp. Biol.* **37**, 435.
Rogers, W. P. (1960). The physiology of the infective process of nematode parasites. *Proc. Roy. Soc.* B, **152**, 367.
Rogers, W. P. and Somerville, R. I. (1960). Physiology of the second ecdysis of parasitic nematodes. *Parasitol.* **50**, 329.
Scott, J. A. (1930). The biology of hookworms in their hosts. *Quart. Rev. Biol.* **5**, 79.
Stauffer, H. (1924). Die Lokomotion der Nematoden. *Zool. Jahrb. System.* **49**, 1.
Weinstein, P. P. and Jones, M. F. (1959). Development *in vitro* of some parasitic nematodes of vertebrates. *Ann. N.Y. Acad. Sci.* **77**, 137.

ROTIFERA

Rotifers have very interesting life cycles. In many ways they resemble the life cycles of the Cladocera and the Aphids, in that the animals go through several different morphological forms and change their mode of reproduction from sexual to asexual. The rotifers are interesting too since some species have a fixed number of cells for any particular organ. These points are discussed below.

Berg, K. (1934). Cyclic reproduction, sex determination and depression in Cladocera (and Rotifera). *Biol. Rev.* **9**, 139.
Beauchamp, P. de (1945). Le développement de *Ploeosoma*. *Bull. Soc. Zool. France*, **81**.

Campbell, R. S. (1941). Vertical distribution of the planktonic rotifers in Douglas Lake, Michigan, with special reference to depression and individuality. *Ecol. Monog.* **11**, 1.
Edmonson, W. T. (1959). The Rotifera. In *Fresh Water Biology*, by H. B. Ward and G. P. Whipple, new ed. edited by W. T. Edmonson (Wiley).
Miller, H. M. (1931). Alternation of generations in the rotifer, *Lecane inermis Biol. Bull.* **60**, 345.
Shull, A. F. (1929). Determination of types of individuals in aphids, rotifers and cladocerans. *Biol. Rev.* **4**, 218.
Van Cleave, H. (1922). Degree of constancy of nuclei in certain organs of *Hydatina senta. Biol. Bull.* **42**, 85.
Van Cleave, H. (1932). Cell constancy and its relation to body size. *Quart. Rev. Biol.* **7**, 59.

ENDOPROCTA, GASTROTRICHA, KINORHYNCHIA

The best account will be found in the appropriate volume of Bronn's *Klassen*.

Cori, C. I. (1936). Kamptozoa. *Bronns Klassen und Ordnungen.* Band 4, Abt. 2, Buch 4, pp. 1–119.
Remane, A. (1936). Gastrotricha and Kinorhynchia. *Bronns Klassen und Ordnungen.* Band 4, Abt. 2, Buch 1, pp. 1–382.
Sachs, M. (1955). Observations on the embryology of an aquatic Gastrotrich, *Lepidodermella squammata. Jour. Morph.* **96**, 473.

ANNELIDA

Being segmented, the annelids have provided material for the study of locomotion in meristically repeating muscle units (Gray). The nervous system, particularly the giant fibres, have been well studied (Bullock, Nicol), though the physiology of peripheral innervation is not so well known (Hoyle). Goodrich, in the last major paper he published, wrote a very fine account of the nephridia and coelomoducts in invertebrates, paying special attention to the annelids. Recently, Chapman and Newell have been analysing the hydraulic skeleton in annelids, i.e. the way in which the body fluid coordinates the action of circular and longitudinal muscles.

Allen, M. J. (1959). Embryological development of the polychaetous annelid, *Diopatra cuprea. Biol. Bull.* **116**, 339.
Anderson, D. T. (1959). Embryology of the polychaete *Scolopos armiger. Quart. Jour. Micr. Sci.* **100**, 89.
Arbit, J. (1957). Diurnal cycles and learning in earthworms. *Science*, **126**, 654.
Baldwin, E. and Yudkin, W. H. (1950). The annelid phosphagen with a note on a phosphagen in echinoderms and protochordates. *Proc. Roy. Soc.* B, **136**, 614.
Barcroft, J. and Barcroft, H. (1924). The blood pigment of *Arenicola. Proc. Roy. Soc.* B, **96**, 28.
Berrill, N. J. (1952). Regeneration and budding in worms. *Biol. Rev.* **27**, 401.
Bookhaut, C. G. (1957). The development of *Dasybranchus caducus* from eggs to the preadult. *Jour. Morph.* **100**, 141.
Borden, M. A. (1931). A study of respiration and the function of haemoglobin in *Planorbis corneus* and *Arenicola marina. Jour. Mar. Biol. Ass.* **17**, 709.
Bradbury, S. (1959). The botryoidal and vaso-fibrous tissue of the leech *Hirudo medicinalis. Quart. Jour. Micr. Sci.* **100**, 483.
Bullock, T. H. (1945). Functional organization of the giant fibre system of *Lumbricus. Jour. Neurophys.* **8**, 55.

Chapman, G. (1958). The hydraulic skeleton in the invertebrates. *Biol. Rev.* **33**, 338.
Chapman, G. and Newell, G. (1947). The role of the body fluids in relation to movement in the soft-bodied invertebrates. *Proc. Roy. Soc.* B, **134**, 431.
Clark, R. B. (1956). *Capitella capitata* as a commensal; with a bibliography of parasitism and commensalism in polychaetes. *Ann. Mag. Nat. Hist.* **9**, 433.
Clark, R. B. (1956). The blood vascular system of *Nephthys*. *Quart. Jour. Micr. Sci.* **97**, 235.
Clark, R. B. (1961). The origin and formation of the heteronereis. *Biol. Rev.* **36**, 199.
Dales, R. P. (1957). Some quantitative aspects of feeding in sabellid and serpulid fanworms. *Jour. Mar. Biol. Ass.* **36**, 309; **34**, 55.
Dales, R. P. (1957). Role of coelomic cells in food storage and transport in certain polychaetes. *Jour. Mar. Biol. Ass.* **36**, 91.
Darwin, C. (1881). *The Formation of Vegetable Mould through the Action of Worms, with Observations on their Habits* (John Murray).
Ennor, A. H. and Morrison, J. F. (1958). Biochemistry of the phosphagens and related guanidines. *Physiol. Rev.* **38**, 631.
Gatenby, J. B. and Dalton, A. J. (1959). Spermiogenesis in *Lumbricus herculeus*. *Jour. Biophys. Biochem. Cytol.* **6**, 45.
Goodrich, E. S. (1946). The study of nephridia and genital ducts since 1895. *Quart. Jour. Micr. Sci.* **86**, 115.
Gray, J. (1939). The kinetics of locomotion of *Nereis diversicolor*. *Jour. Exp. Biol.* **16**, 9.
Gray, J. and Lissmann, H. W. (1938). Locomotory reflexes in the earthworm. *Jour. Exp. Biol.* **15**, 506.
Gray, J., Lissmann, H. W. and Pumphrey, R. J. (1938). The mechanism of locomotion in the leech *Hirudo medicinalis*. *Jour. Exp. Biol.* **15**, 408.
Hagedorn, J. (1958). Neurosecretion and the brain of the leech, *Theromyzon rude*. *Jour. Morph.* **102**, 55.
Hanson, J. (1949). The histology of the blood system in Oligochaetes and Polychaetes. *Biol. Rev.* **24**, 127.
Hedley, R. H. (1956). Studies on Serpulid tube formation. *Quart. Jour. Micr. Sci.* **97**, 411.
Horridge, G. A. (1959). Analysis of the rapid responses of *Nereis* and *Harmathoe*. *Proc. Roy. Soc.* B, **150**, 245.
Hoyle, G. (1957). *The Comparative Physiology of the Nervous Control of Muscular Contraction* (Cambridge University Press).
Jones, J. D. (1955). Observations on the respiratory physiology and haemoglobin of the polychaete genus *Nephthys*. *Jour. Exp. Biol.* **32**, 110.
Jones, J. D. (1962). The functions of the respiratory pigments of invertebrates. *Problems in Biology* (Pergamon Press).
Johnson, M. L. (1941). The respiratory function of the haemoglobin in the earthworm. *Jour. Exp. Biol.* **18**, 266.
Kennedy, G. Y. and Dale, R. P. (1958). The function of the heart body in polychaetes. *Jour. Mar. Biol. Ass.* **37**, 15.
Kerkut, G. A. (1960). Phosphagens. In *Implications of Evolution* (Pergamon Press).
Kiyoshi, H. (1959). The fine structure of the giant nerve fibres of the earthworm. *Jour. Biochem. Biophys. Cytol.* **6**, 61.
Laverack, M. S. (1961). Tactile and chemical perception in earthworms. *Comp. Biochem. Physiol.* **2**, 22.
Manwell, C. (1960). Histological specificity of respiratory pigments. *Comp. Biochem. Physiol.* **1**, 267.
Manwell, C. (1960). Comparative physiology of blood pigments. *Ann. Rev. Physiol.* **22**, 191.
Mann, K. H. (1953). The segmentation of leeches. *Biol. Rev.* **28**, 1.
Mann, K. H. (1961). *The Leeches* (Pergamon Press).

Needham, A. E. (1957). Components of nitrogenous excretion in the earthworms *Lumbricus terrestris* and *Eisenia foetida*. *Jour. Exp. Biol.* **34**, 425; **37**, 775.
Needham, A. E. (1958). Pattern of nitrogenous excretion during regeneration in Oligochaetes. *Jour. Exp. Zool.* **138**, 369.
Newell, G. (1950). Role of coelomic fluid in the movements of earthworms. *Jour. Exp. Biol.* **27**, 110.
Nicol, E. A. T. (1931). The feeding mechanism, formation of the tube and physiology of digestion of *Sabella pavonina*. *Trans. Roy. Soc. Edinb.* **56**, 537.
Nicol, J. A. C. (1948). The giant axons of annelids. *Quart. Rev. Biol.* **23**, 291.
Nicol, J. A. C. (1960). The regulation of light emission in animals. *Biol. Rev.* **35**, 1.
Prosser, C. L. (1934). The nervous system of the earthworm. *Quart. Rev. Biol.* **9**, 181.
Prosser, C. L. and Sperelakis, N. (1959). Electrical evidence for dual innervation of muscle fibres in the sipunculid *Golfingia (Phascolosoma)*. *Jour. Cell. Comp. Physiol.* **54**, 129.
Ralph, C. L. (1957). Persistent rhythms of activity and oxygen consumption in the earthworm. *Physiol. Zool.* **30**, 41.
Ratner, S. C. and Miller, K. R. (1959). Effect of spacing and ganglionic removal on conditioning in earthworms. *Jour. Comp. Physiol. Psychol.* **52**, 667.
Roots, B. I. (1956). The earthworm. *New Biol.* **21**, 102.
Roots, B. I. (1956). The water relations of earthworms. *Jour. Exp. Biol.* **33**, 29.
Roots, B. I. (1960). Chloragogenous tissue of earthworms. *Comp. Biochem. Physiol.* **1**, 218.
Reynoldson, T. B. (1948). An ecological study of the enchytraid worm population of sewage bacteria beds. *Jour. Anim. Ecol.* **17**, 27.
Robertson, J. D. (1936). The function of the calciferous glands of earthworms. *Jour. Exp. Biol.* **13**, 279.
Simpson, M. (1959). The saccular apparatus in the brain of *Glycera dibranchiata*. *Jour. Morph.* **104**, 561.
Smith, J. E. (1956). The nervous anatomy of the body segments of nereid polychaetes. *Phil. Trans.* B, **240**, 135.
Smith, R. I. (1958). Reproductive patterns in nereid polychaetes. *System. Zool.* **7**, 60.
Southward, A. J. and Southward, E. C. (1958). The breeding of *Arenicola ecaudata* and *A. branchialis* at Plymouth. *Jour. Mar. Biol. Ass.* **37**, 267.
Stephenson, J. (1930). *The Oligochaeta* (Oxford University Press).
Svendsen, J. A. (1957). The behaviour of Lumbricids under moorland conditions. *Jour. Anim. Ecol.* **26**, 423.
Wells, G. P. (1950). Spontaneous activity cycles in polychaete worms. *Symp. Soc. Exp. Biol.* **4**, 127.
Wells, G. P. (1950). The anatomy of the body wall and appendages of *Arenicola*. *Jour. Mar. Biol. Ass.* **29**, 1.
Wells, G. P. (1957). The life of the lugworm. *New Biol.* **22**, 39.
Wilczynski, J. Z. (1960). On egg dimorphism and sex determination in *Bonellia viridis*. *Jour. Exp. Zool.* **143**, 61.
Wilson, D. M. (1960). Nervous control of movement in annelids. *Jour. Exp. Biol.* **37**, 46.

ARTHROPODA

The arthropods are distinguished from the annelids by their hard exoskeleton. The formation and function of this integument is discussed by Richards. The major questions concerning the relationships of the various groups within the Arthropoda are discussed by Snodgrass, and also by Butt, and by Tiegs and Manton. The very general account of the structure and evolution of the locomotory mechanisms in the different groups will be found in Manton.

Butt, F. H. (1960). Head development in the arthropods. *Biol. Rev.* **35**, 43.
Edney, E. B. (1957). *The Water Relations of Terrestrial Arthropods* (Cambridge University Press).
Krijgsman, B. J. (1952). Contractile and pacemaker mechanisms in the heart of arthropods. *Biol. Rev.* **27**, 320.
Lees, A. D. (1956). *The Physiology of Diapause in Arthropods* (Cambridge University Press).
Manton, S. M. (1953). Locomotory habits and the evolution of the larger arthropodan groups. *Symp. Soc. Exp. Biol.* **7**, 339.
Manton, S. M. (1958). Habits of life and design of body in arthropods. *Jour. Linn. Soc. (Zool.)*, **44**, 58.
Manton, S. M. (1960). Concerning head development in the arthropods. *Biol. Rev.* **35**, 265.
Richards, G. (1951). *The Integument of Arthropods* (University of Minnesota Press).
Snodgrass, R. E. (1938). Evolution of the Annelida, Onychophora, and Arthropoda. *Smithsonian Misc. Coll.* **97**, 1–159.
Snodgrass, R. E. (1951). *Comparative Studies on the Head of the Mandibulate Arthropods* (Comstock Publ. Co.).
Tiegs, O. W. and Manton, S. M. (1958). The evolution of the Arthropoda. *Biol. Rev.* **33**, 255.

ONYCHOPHORA

These problematical animals have excited considerable zoological interest. They have characteristics between those of the annelids and the arthropods though as Manton and Ramsay and also Morrison point out, they lose water much more quickly than the annelids. Manton has made a series of studies on the embryology of *Peripatopsis* and from her results discusses the characteristics of the primitive arthropod. She also points out (1958) that the onychophorans, with their flexible body, are able to squeeze into various niches and so increase their survival chances.

There is a very interesting fossil, *Aysheaia*, which is found in marine deposits and which is believed to be a fossil onychophoran, or at any rate something very much like one. This is discussed by Hutchinson.

Andrews, A. (1933). *Peripatus* in Jamaica. *Quart. Rev. Biol.* **8**, 155.
Brues, C. T. (1923). The geographical distribution of the Onychophora. *Amer. Nat.* **57**, 210.
Hutchinson, G. E. (1930). Restudy of some Burgess Shale fossils. *Proc. U.S. Nat. Mus.* **78**, no. 11, 14–22.
Manton, S. M. and Ramsay, J. A. (1937). The control of water loss in *Peripatopsis*. *Jour. Exp. Biol.* **14**, 470.
Manton, S. M. (1937). The feeding, digestion, excretion and food storage of *Peripatopsis*. *Phil. Trans.* B, **227**, 411.
Manton, S. M. (1949). The early embryonic stages of *Peripatopsis* and some general considerations concerning the morphology and phylogeny of the Arthropoda. *Phil. Trans.* B, **233**, 483.
Manton, S. M. (1950). The locomotion of *Peripatus*. *Jour. Linn. Soc. (Zool.)*, **41**, 529.
Manton, S. M. (1958). Habits of life and design of body in arthropods. *Jour. Linn. Soc. (Zool.)*, **44**, 58.
Morrison, P. R. (1946). Water loss and oxygen consumption by *Peripatus*. *Biol. Bull.* **91**, 181.
Sedgwick, A. (1888). A monograph on the genus *Peripatus*. *Quart. Jour. Micr. Sci.* **28**, 431.

TRILOBITA

A good account of the trilobites will be found in Schrock and Twenhofel. Trilobites being extinct have come in for a great deal of phylogenetic speculation. Garstang and Gurney have put forward the suggestion that there were two types of trilobites, one with variable segmentation and the other with a fixed number of segments. The former, it is suggested, gave rise to the entomostracan Crustacea, whilst the second gave rise to the Malacostraca. Stormer, on the other hand, has made a study of the trilobite limb and has decided that the trilobites did not give rise to the Crustacea but instead are more closely related to the arachnids. His point of view is countered by Calman and also by Garstang. Perhaps the most useful account of the problems of placing the trilobites in relation to the other Arthropoda will be found in Tiegs and Manton's review.

Calman, W. T. (1939). The structure of the trilobites. *Nature, Lond.*, **141**, 1077.
Garstang, W. (1940). Stormer on the appendages of the trilobites. *Ann. Mag. Nat. Hist.* **6**, ser. 2, 59.
Garstang, W. and Gurney, R. (1938). The descent of the Crustacea from the trilobites and their larval relations. In *Evolution, Essays Presented to Goodrich*, pp. 271–86 (Oxford University Press).
Shrock, R. and Twenhofel, W. (1953). *Principles of Invertebrate Paleontology* (McGraw-Hill).
Stormer, L. (1933). Are the trilobites related to arachnids? *Amer. Jour. Sci.* **26** (ser. 5), 147.
Stormer, L. (1944). On the relationship and phylogeny of fossil and recent arachnomorphs. *Skrift. norske Vid.-Akad. Oslo* (Math.-Naturv. Klasse), **1**, 158.
Stubblefield, C. J. (1936). Cephalic sutures and their bearing on the current classification of the trilobites. *Biol. Rev.* **11**, 407.
Tiegs, O. W. and Manton, S. M. (1958). The evolution of the Arthropoda. *Biol. Rev.* **33**, 255 (see p. 279).
Whittington, H. B. (1957). The ontogeny of trilobites. *Biol. Rev.* **32**, 421.

CRUSTACEA

Interest in crustacean systematics has been revived by the discovery of two new problematical crustaceans, *Hutchinsoniella*, described by Sanders, and *Derocheilocaris*, described by Pennak and Zinn, and also by Chappuis and Debouteville. It has been suggested that these two animals represent two new subclasses of the Crustacea; *Hutchinsoniella* being placed in the subclass Cephalocarida, a group more primitive than the Branchiopoda; whilst *Derocheilocaris* is placed in the Mystacocarida, a group coming between the Branchiopoda and the Copepoda. There is some doubt as to their precise position; Sanders, Chappuis, and also Dahl suggest that these new animals are fairly distinct from the already known subclasses, whilst Tiegs and Manton think that *Hutchinsoniella* can be placed either within or very close to the Branchiopoda.

Hutchinsoniella (Fig. 1A) is a small crustacean (2·8 mm. long) that lives in soft mud. It has a small carapace that extends back to include the first thoracic segment. There is a biramous antenna. There are no palps on the mandible, a single pair of biramous maxillae, ten pairs of biramous limbs, and segments 11–18 lack limbs. There is a telson with a caudal fork. The animals are hermaphrodite, the testes and ovary being separate. They have three broods a year, each of two larvae. There are certain resemblances between *Hutchinsoniella* and *Lepidocaris*, the primitive fossil Branchiopod (Fig. 1B and p. 373), though Sanders thinks it is more primitive than *Lepidocaris*.

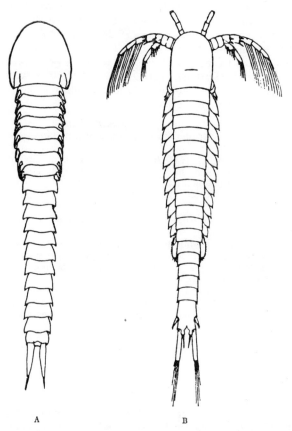

Fig. 1. A, *Hutchinsoniella macracantha*, a cephalocaridan; B, *Lepidocaris rhyniensis*. (From Sanders.)

Derocheilocaris (Fig. 2) is a small crustacean (1 mm. long) found living between sand grains. It swims rather like a nauplius larva by means of its biramous antenna and the mandibles. The anterior part of the head, bearing the antennules, is separated off dorsally by a groove from the rest of the head. The eyes are reduced. It uses its mandibular gnathobase, maxillule, maxilla, maxillipeds and labrum for feeding. The next four segments carry simple appendages; then there are six segments that are without appendages. There is a forked tail. Dahl suggests that *Derocheilocaris* represents an early divergence from the copepod line, though it is not a direct ancestor of the modern copepods.

There has recently been a valuable publication summarizing much of our knowledge of crustacean physiology. It is called *The Physiology of Crustacea*, edited by T. H. Waterman (Academic Press). The two volumes refer to most

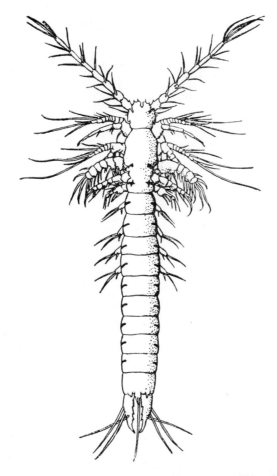

Fig. 2. *Derocheilocaris remanei*, a mystacocarid crustacean. (From Chappuis.)

of the relevant information on the various aspects of crustacean physiology. The first volume is concerned with metabolism, respiration, nutrition, sex, excretion, moulting and regeneration. The second volume includes sensory physiology, nerve muscle control, central nervous system, neurohumours, and behaviour.

Other topics are mentioned in the references given below; thus the physiology of woodlice, with particular reference to the problems of terrestrial life, is discussed by Edney. There is also the problem of choosing a suitable habitat by barnacle larvae (Crisp and Barnes), the changes in sex and shape of *Daphnia* (Banta *et al.*, Berg, Coker), the parasitic forms (Day), parasitic castration and sex determination (Ichikawa, Reinhard), the peculiar nerve muscle system (Katz, Hoyle), the complexity of hormonal control of colour change (Carlisle and Knowles) and the behaviour of planktonic forms (Hardy, Cushing, Harris, Marshall and Orr).

Allen, E. J. (1895). The reproduction of the lobster. *Jour. Mar. Biol. Ass.* **4**, 60.

Armstrong, J. C. (1949). The systematic position of the crustacean genus *Derocheilocaris* and the status of the subclass Mystacocarida. *Amer. Mus. Novit.* **1413**, 1.

Banta, A. M., Brown, L. A. and Ingle, L. (1939). *Studies on the Physiology, Genetics and Evolution of some Cladocera.* Paper 39 (Carneg. Inst. Wash. Dept. Genet.).

Berg, K. (1934). Cyclic reproduction, sex determination and depression in the Cladocera. *Biol. Rev.* **9**, 139.

Borradaile, L. A. (1926). On the primitive phyllopodium. *Ann. Mag. Nat. Hist.* **18**, 16.

Bovbjerg, R. V. (1953). Dominance order in the crayfish *Orconectes virilis*. *Physiol. Zool.* **26**, 173.

Brooks, J. L. (1946). Cyclomorphosis in *Daphnia*. *Ecol. Monog.* **16**, 409.

Brown, F. A. (1944). Hormones in the Crustacea. *Quart. Rev. Biol.* **19**, 32, 118.

Bursell, E. (1955). Cutaneous respiration in woodlice. *Jour. Exp. Biol.* **32**, 256.

Cannon, H. G. (1928). On the feeding mechanism of the copepods, *Calanus finmarchicus* and *Diaptomus gracilis*. *Jour. Exp. Biol.* **6**, 131.

Canon, H. G. and Manton, S. M. (1927). On the feeding mechanisms of a mysid crustacean *Hemimysis lamornae*. *Trans. Roy. Soc. Edinb.* **55**, 219.

Carlisle, D. B. and Knowles, F. (1959). *Endocrine Control in Crustaceans* (Cambridge University Press).

Chappuis, P.-A. and Debouteville, C. D. (1954). Morphologie des Mystacocarides. *Arch. Zool. exp. gen.* **91**, 7.

Cohen, M. J. (1960). The response patterns of single receptors in the crustacean statocyst. *Proc. Roy. Soc.* B, **152**, 30.

Coker, R. (1939). The problem of cyclomorphosis in *Daphnia*. *Quart. Rev. Biol.* **14**, 137.

Costlow, J. D. (1956). Shell development in *Balanus improvisus*. *Jour. Morph.* **99**, 359.

Crisp, D. J. and Barnes, H. (1954). The orientation and distribution of barnacles at settlement with particular reference to surface contours. *Jour. Anim. Ecol.* **23**, 142.

Cushing, D. J. (1951). The vertical migration of planktonic Crustacea. *Biol. Rev.* **26**, 158.

Dahl, E. (1956). Some crustacean relationships. In *Bertil Hanstrom Festschrift* (University of Lund).

Day, J. (1935). The life history of *Sacculina*. *Quart. Jour. Micr. Sci.* **77**, 549.

Dennell, R. (1947). The occurrence and significance of phenolic hardening in the newly formed cuticle of decapod crustaceans. *Proc. Roy. Soc.* B, **134**, 485.

Durand, J. B. (1960). Limb regeneration and endocrine activity in the crayfish. *Biol. Bull.* **118**, 250.

Edney, E. B. (1945). Woodlice and the land habitat. *Biol. Rev.* **29**, 185.

Fox, H. M. (1948). The haemoglobin of *Daphnia*. *Proc. Roy. Soc.* B, **135**, 195.

Fryer, G. (1957). The feeding mechanism of some freshwater cyclopoid copepods. *Proc. Zool. Soc.* **129**, 1.

Fryer, G. (1960). The spermatophore of *Dolops ranarum* (Crustacea, Branchiura), their structure, formation and transfer. *Quart. Jour. Micr. Sci.* **101**, 407.

Garstang, W. and Gurney, R. (1938). The descent of the Crustacea from trilobites and their larval relations. In *Evolution, Essays Presented to Goodrich*, pp. 271–286 (Oxford University Press).

LITERATURE

Gauld, D. T. (1959). Swimming and feeding in crustacean larvae; the nauplius larva. *Proc. Zool. Soc.* **132**, 31.
Glaessner, M. F. (1957). Evolutionary trends in the Crustacea. *Evolution*, **11**, 178.
Gilchrist, B. and Green, J. (1960). The pigments of *Artemia*. *Proc. Roy. Soc.* B, **152**, 118.
Green, J. (1955). Haemoglobin in the fat cells of *Daphnia*. *Quart. Jour. Micr. Sci.* **96**, 173.
Hardy, A. C. (1938). Change and choice, a study in pelagic ecology. In *Evolution, Essays Presented to Goodrich*, pp. 139–60 (Oxford University Press).
Harris, J. E. and Mason, P. (1956). Vertical migration in eyeless *Daphnia*. *Proc. Roy. Soc.* B, **145**, 280.
Hartog, M. (1888). The morphology of *Cyclops* and the relations of the copepods. *Trans. Linn. Soc.* **5**, 1.
Holmes, W. (1942). The giant myelinated nerve fibres of the prawn. *Phil. Trans. Roy. Soc.* B, **231**, 293.
Hoyle, G. (1957). *Comparative Physiology of the Nervous Control of Muscular Contraction* (Cambridge University Press).
Huxley, T. H. (1880). *The Crayfish* (Kegan Paul).
Huxley, J. S. and Richards, O. W. (1931). Relative growth rates of the abdomen and carapace of the shore crab *Carcinus maenas*. *Jour. Mar. Biol. Ass.* **17**, 1001.
Ichikawa, A. and Yanagimachi, R. (1960). Studies on the sexual organization of the Rhizocephala. *Annot. Zool. Jap.* **33**, 42.
Katz, B. (1949). Neuromuscular transmission in invertebrates. *Biol. Rev.* **24**, 1.
Knowles, F. G. and Carlisle, D. B. (1956). Endocrine control in the Crustacea. *Biol. Rev.* **31**, 396.
Linder, J. H. (1959). Studies on the fresh water fairy shrimp, *Chirocephalopsis bundyi*. *Jour. Morph.* **104**, 1.
Lockwood, A. P. M. (1960). Some effect of temperature and concentration of the medium on ionic regulation of the isopod *Asellus aquaticus*. *Jour. Exp. Biol.* **37**, 614.
MacArthur, J. W. and Baillie, W. T. (1929). Metabolic activity and the duration of life in *Daphnia*. *Jour. Exp. Zool.* **53**, 221.
Marshall, S. M. and Orr, A. P. (1955). *The Biology of 'Calanus finmarchicus'* (Oliver and Boyd).
McWhinnie, M. A. and Saller, S. P. (1960). Analysis of blood sugars in the crayfish *Orconectes virilis*. *Comp. Biochem. Physiol.* **1**, 110.
Moore, H. B. (1935). The biology of *Balanus balanoides*. *Jour. Mar. Biol. Ass.* **20**, 263.
Naier, S. G. (1956). On the embryology of the isopod *Irona*. *Jour. Emb. Exp. Morph.* **4**, 1.
Naylor, E. (1955). The diet and feeding mechanism of *Idotea*. *Jour. Mar. Biol. Ass.* **34**, 347.
Naylor, E. (1958). Tidal and diurnal rhythms of locomotory activity in *Carcinus maenas*. *Jour. Exp. Biol.* **37**, 481; **35**, 602.
Needham, A. E. (1949). Growth and regeneration in *Asellus aquaticus* in relation to age, sex and season. *Jour. Exp. Zool.* **112**, 49.
Nicholls, A. G. (1931). Studies on *Ligia oceanica*. *Jour. Mar. Biol. Ass.* **17**, 655.
Pannikar, L. (1941). Osmoregulation in some palaeomonid prawns. *Jour. Mar. Biol. Ass.* **25**, 317.
Pennak, R. W. and Zinn, D. J. (1943). Mystacocarida, a new order of Crustacea from intertidal beeches of Massachusetts and Connecticut. *Smithson. Misc. Coll.* **103**, 1.
Raymont, J. E. G. and Gross, F. (1942). Feeding and breeding of *Calanus finmarchicus* in laboratory conditions. *Trans. Roy. Soc. Edinb.* **61**, 267.
Reinhard, E. G. (1949). Experiments on the determination and differentiation of sex in the bopyrid *Stegophryxus hyptius*. *Biol. Bull.* **96**, 17.

Sanders, H. L. (1957). Cephalocarida and crustacean phylogeny. *System. Zool.* **6**, 112.
Scudamore, H. H. (1948). Factors influencing moulting and the sexual cycles in the crayfish. *Biol. Bull.* **95**, 229.
Shaw, J. (1960). The absorption of sodium ions by the crayfish *Astacus pallipes*. *Jour. Exp. Biol.* **37**, 534; **38**, 135.
Slobodkin, L. B. (1954). Population dynamics in *Daphnia obtusata*. *Ecol. Monog.* **24**, 69.
Snodgrass, R. E. (1956). Crustacean metamorphoses. *Smithson. Misc. Coll.* **131**, no. 10, 1.
Teal, J. M. (1959). Respiration of crabs in Georgia salt marshes and its relation to their ecology. *Physiol. Zool.* **32**, 1.
Tiegs, O. W. and Manton, S. M. (1958). The evolution of the Arthropoda. *Biol. Rev.* **33**, 255.
Waterman, T. H., ed. (1960). *The Physiology of Crustacea.* 2 vols. (Academic Press).
Wiersma, C. A. G. (1952). Neurons of arthropods. *Cold Spring Harbor Symp.* **17**, 155.
Wiersma, C. A. G. (1958). On the functional connections of single units in the central nervous system of the crayfish, *Procambarus clarkii*. *Jour. Comp. Neurol.* **110**, 421.
Yasuzumi, G. (1960). Spermatid differentiation in the crab *Eriocheir japonicus*. *Jour. Biophys. Biochem. Cytol.* **7**, 73.
Yonge, C. M. (1924). The feeding mechanism, digestion and assimilation in *Nephrops norvegicus*. *Jour. Exp. Biol.* **1**, 343.

MYRIAPODA

Within recent years a considerable amount of interest has been taken in the structure and range of forms of the Myriapoda. They are not a very closely linked group of animals and in addition to the diplopods and the chilopods already mentioned in the text there are two small but important groups of animals that may be classified as myriapods. These are the Pauropoda and the Symphyla.

Tiegs (1947) divided the basic arthropods that gave rise to the insect line into three groups. The hypothetical Monognatha, which had only one pair of jaws; *Peripatus* would come into this group. The Dignatha which had two pairs of jaws; this group would include the Diplopoda and the Pauropoda. The Trignatha which had three pairs of jaws; this would include the Chilopoda, Symphyla, and Insecta. The Insecta and Symphyla differ from the other Trignatha in that their second maxilla became fused to form a new lower lip to the preoral cavity, within which the first maxilla open. These animals are grouped to form the Labiata.

This classification based on head structure does not imply any close relationship between the Chilopoda and the Insecta, but Tiegs does think that there are some links between the basic insect stock such as the Thysanura or Diplura, and the Symphyla.

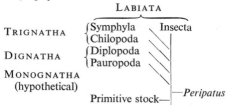

Pauropoda

Minute 'myriapods', the eggs of which hatch into a larva with three pairs of legs. There are four larval stages and the adult has twelve somites and ten pairs of appendages, nine of which are ambulatory (Fig. 3). A collum is present. The body segments are arranged in the 'diplopod' pattern so that

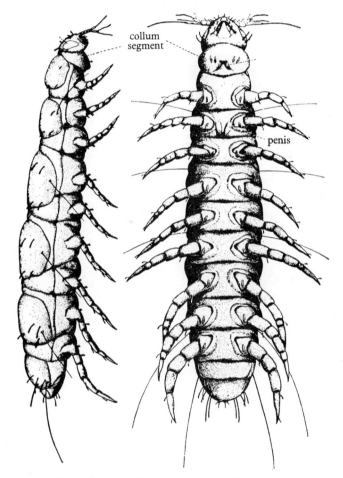

Fig. 3. *Pauropus silvaticus*, adult male. (From Tiegs.)

two somites and two pairs of legs form one segment (Fig. 3). The adults have their genital opening on the third segment (progoneate). They differ from the Diplopoda in that they only have twelve segments, they have their antenna branched, there are no trachea, there are no eyes, the mandible only has one segment, they have five pairs of long tactile setae attached to the side of the body. Examples are *Pauropus* and *Eurypauropus*.

Pauropods live in damp places, underneath decaying leaves and logs, where they are often found in considerable numbers. Starling calculated that there were more than two million per acre in Duke Forest, North Carolina.

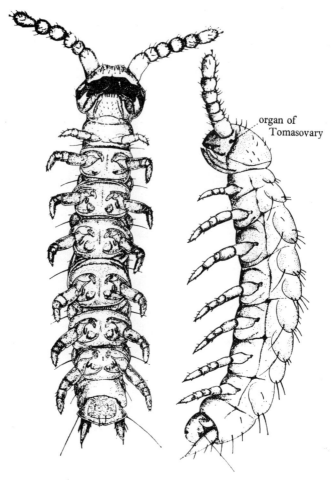

Fig. 4. *Hanseniella agilis*. First instar larva, ventral and lateral view. (From Tiegs.)

The pauropods are generally more primitive in the organization than the diplopods; the most important difference between the two groups being the structure of the mandibles; the pauropods only have one segment to the mandible, the diplopods have two. Otherwise there are many similarities. The segmental composition of the head, the legless collum, the primary ventral position of the gonads, the opening of the gonopore on the third body segment, the emergence of the larva with less than the adult number of segments (anamorphic development), are all similarities between the Pauropoda and the Diplopoda.

Symphyla

Small 'myriapods', the eggs of which hatch into a larva with six or seven pairs of legs (Fig. 4). There are six larval stages and the adult has sixteen pairs of appendages, twelve of which are ambulatory (Fig. 5). The body has true segments, not diplosomites, as in the Diplopoda and the Pauropoda. In fact the number of tergal plates (15) is greater than the number of pairs of legs instead of being less as in the diplopods. The adults have their genital ducts on the third body segment. The head has a long multi-segmented

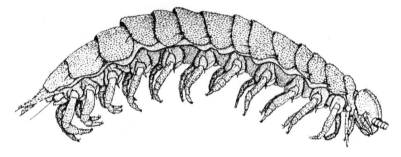

Fig. 5. *Hanseniella agilis.* Adult. Note that the antenna is much longer than shown in the figure. (From Tiegs.)

antenna and there are true spiracles opening on the head. The mandible has one segment. The basal segments of the third to twelfth legs have a movable stylet and a protrusible sac at their base. There is a pair of tail-like processes on the last somite. Spinning glands open on to the tail. Malpighian tubules are present. The animals live in damp places, under stones and amongst leaves. Examples are *Scutigerella, Scolopendrella,* and *Hanseniella* (Fig. 5).

The Symphyla are considered by Tiegs to be related quite closely to the Insecta and particularly the Thysanura. There is a similarity between the body form of the Symphyla and the Diplura, especially those that live under stones and rotting leaves. Both groups have a simple multi-articulated antennae, two pairs of maxillae, and a superlinguae. The Symphyla on segments 3–12 have an evertible sac and a small movable stylus. A similar structure is present in some segments of the Thysanura. Malpighian tubules are present in both groups; there being two in the Symphyla and 4–20 in the Thysanura. The locomotory gait of the Symphyla is basically similar to that of the Thysanura and other insects (Manton).

On the other hand there are differences between the two groups. Cleavage is holoblastic in the Symphyla and superficial in the Thysanura. This is possibly related to the amount of yolk in the egg. There is a post-antennal organ (Organ of Tomasovary) and a dorsal organ present in the Symphyla; both are absent in the Thysanura. The mandibles are segmented in the

Symphyla and non-segmented in the Thysanura. The gonads are ventral to the gut in the Symphyla and dorsal in the Thysanura. The Symphyla are progoneate, whilst the Thysanura are opisthogoneate.

Progoneate and Opisthogoneate

At one time the classification of the myriapods and the insects was based on the position of the gonopore. It was considered that the progoneate and the opisthogoneate conditions had evolved separately from a common ancestor in whom the gonads extended over many segments, and who shed the gonoproducts through segmental gonoducts.

In the Symphyla the embryos are opisthogoneate but the adult develops the progoneate condition. The opisthogoneate chilopods and Insecta have a mesodermal gonoduct, whilst Tiegs has shown that the progoneate Symphyla and Pauropoda have an ectodermal gonoduct. For these reasons it is probable that the progoneate condition is a secondary condition, possibly associated with anamorphosis: the hatching of the larval form without the full number of segments. In these animals the gonopore could thus develop away from the region of growth at the posterior end of the animal. The fact that some chilopods are anamorphic yet have a terminal genital opening does not completely invalidate this line of argument since there are at least five instars between completion of anamorphosis and the attainment of sexual maturity, and it is in these instars that the external genitalia arise. Sexual maturity thus takes place only in the adult chilopod. In the Symphyla, on the other hand, the genital ducts are well formed and even functioning before the most posterior segments are developed. In the pauropods, the genital ducts arise long before the posterior segments have completed their development, but as yet, there is no evidence that the gonads are functioning before the end of anamorphosis. Further details will be found in the papers of Tiegs.

Hansen, H. J. (1903). The genera and species of the Order Symphyla. *Quart. Jour. Micr. Sci.* **47**, 1.
Manton, S. M. (1952). The locomotion of the Chilopoda and Pauropoda. *Jour. Linn. Soc. (Zool.)*, **42**, 118.
Manton, S. M. (1954). The structure, habits and evolution of the Diplopoda. *Jour. Linn. Soc. (Zool.)*, **42**, 299.
Manton, S. M. (1958). Habits and evolution of the Lyssiopetaloidea (Diplopoda), some principles of leg design in Diplopoda and Chilopoda and limb structure of Diplopoda. *Jour. Linn. Soc. (Zool.)*, **43**, 487.
Starling, J. H. (1943). Pauropoda from the Duke Forest. *Proc. Ent. Soc. Wash.* **45**, 183.
Tiegs, O. W. (1945). The post embryonic development of *Hanseniella agilis* (Symphyla). *Quart. Jour. Micr. Sci.* **85**, 191.
Tiegs, O. W. (1947). The development and affinities of the Pauropoda based on a study of *Pauropus silvaticus*. *Quart. Jour. Micr. Sci.* **88**, 165.
Tiegs, O. W. and Manton, S. M. (1958). The evolution of the arthropoda. *Biol. Rev.* **33**, 255.

INSECTA

In such a large class it is only possible to refer to a few topics. Entomology is a subject in its own right and the specialist entomologist will know the works of Imms, Grassé, Roeder, and Wigglesworth as well as the major entomological journals.

The complex behaviour of bees is discussed in the books of von Frisch, Butler and Ribbands. Ford provides a full account of the biology of butterflies and another volume on moths. Williams has summarized the literature on insect migration; Lees that on insect diapause; Wigglesworth—ecdysis and metamorphosis; Pringle—insect flight; Richards—social insects. A useful introduction to entomological literature is provided by Chamberlain.

Bourgogne, J. (1951). *Traité de Zoologie* (Masson, Paris).
Brues, C.T. (1946). *Insect Dietary* (Harvard University Press).
Butler, C. G. (1954). *The World of the Honeybee* (Collins, London).
Buxton, P. (1955). *Natural History of the Tsetse Fly* (H. K. Lewis).
Campbell, F. L. (1956). *Conference on the Physiology of Insect Development* (Cambridge University Press).
Chamberlain, W. J. (1952). *Entomological Nomenclature and Literature* (W. C. Brown, Iowa).
Comstock, J. H. (1918). *The Wings of Insects* (Comstock, New York).
Corbet, P. S., Longfield, C. and Moore, N. W. (1960). *Dragonflies* (Collins).
Demerec, M., ed. (1950). *Biology of Drosophila* (Wiley, New York).
Dennell, R. (1958). The hardening of the insect cuticle. *Biol. Rev.* **33**, 178.
Eastham, L. E. S. and Eassa, Y. E. E. (1955). The feeding mechanism of the butterfly, *Pieris brassicae* L. *Phil. Trans. Roy. Soc. B*, **239**, 1.
Ford, E. B. (1946). *Butterflies* (Collins, London).
Ford, E. B. (1955). *Moths* (Collins, London).
Gilmour, D. (1960). *Biochemistry of Insects* (Academic Press, New York).
Graham-Smith, G. S. (1930). Further observations on the anatomy and function of the proboscis of the blowfly, *Calliphora erythrocephala* L. *Parasitol.* **22**, 47.
Grassé, Pierre-P. (1949). *Traité de Zoologie.* Volumes 9 and 10 (Masson, Paris).
Harker, J. E. (1958). Diurnal rhythms in the animal kingdom. *Biol. Rev.* **33**, 1.
Imms, A. D. (1957). *A General Textbook of Entomology.* Revised by O. W. Richards and R. G. Davies (Methuen).
Imms, A. D. (1959). *Outlines of Entomology.* Revised by O. W. Richards and R. G. Davies (Methuen).
Jeannel, R. (1960). *Introduction to Entomology.* Translated by J. H. Oldroyd (Hutchinson).
Lameere, A. (1938). *Précis de Zoologie* (Université de Bruxelles).
Lees, A. D. (1955). *The Physiology of Diapause in Arthropods* (Cambridge University Press).
Mansour, K. and Mansour-Bek, J. J. (1934). The digestion of wood by insects and the supposed role of micro-organisms. *Biol. Rev.* **9**, 363.
Martinov, A. B. (1938). Studies on the geological and phylogenetic history of the orders of winged insects. *Trav. Inst. Palaeont. Acad. Sci. U.S.S.R. (Leningrad)* 7, No. 4.
Mellanby, K. M. (1939). The functions of insect blood. *Biol. Rev.* **14**, 243.
Pringle, J. W. S. (1957). *Insect Flight* (Cambridge University Press).
Ribbands, R. (1953). *The Behaviour and Social Life of the Honey Bee* (Hale Publ. Co., Hapeville, Ga.).
Richards, O. W. (1953). *The Social Insects* (Macdonald).
Robinson, G. G. (1939). The mouthparts and their function in the female mosquito, *Anopheles maculipennis*. *Parasitol.* **31**, 212.
Roeder, K., ed. (1953). *Insect Physiology* (Wiley).
Seguy, E. (1950). *La Biologie des Diptères* (Lechevalier).
Snodgrass, R. E. (1935). *The Principles of Insect Morphology* (McGraw-Hill, New York).
Snodgrass, R. E. (1954). Insect metamorphosis. *Smithson. Misc. Coll.* **122**, no. 9.
Steinhaus, E. and Smith, R., eds. (1956–). *Annual Review of Entomology* (Annual Reviews, Inc.).

Thorpe, W. H. and Crisp, D. J. (1946–8). Studies in plastron respiration. *Jour. Exp. Biol.* **24**, 227, 270, 310; **26**, 219.
Thorpe, W. H. (1950). Plastron respiration in aquatic insects. *Biol. Rev.* **25**, 344.
Ulrich, W. (1927). Strepsiptera. *Biologie der Tiere Deutschlands*, pp. 23–41.
von Frisch, K. (1950). *Bees, their Vision, Chemical Senses and Language* (Cornell, New York).
Weber, H. (1930). *Biologie der Hemipteren* (Springer, Berlin).
Weber, H. (1933). *Lehrbuch der Entomologie* (Fischer, Jena).
Wheeler, W. M. (1928). *The Social Insects* (Kegan Paul, London).
Wigglesworth, V. B. (1953). *The Principles of Insect Physiology* (Methuen, London).
Wigglesworth, V. B. (1954). *The Physiology of Insect Metamorphosis* (Cambridge University Press).
Williams, C. B. (1958). *Insect Migration* (Collins, London).

ARACHNIDA

The early literature on *Limulus* is given by Kingsley and Lankester. The resemblance of *Limulus* to the fossil bony fish, *Bothriolepis*, was noticed both by Gaskell and Patten. They derived the vertebrates from the arachnids, though in somewhat different ways, and they each wrote a book on the subject. Though their views concerning vertebrate ancestry are not at present fashionable, the books make very stimulating reading. Bristowe's volumes, *The Comity of Spiders*, are full of interesting information about the numbers, distribution and habits of spiders. Similar information can be found in Savory.

Alexander, A. J. (1959). Courtship and mating in the Buthid scorpions. *Proc. Zool. Soc.* **133**, 145.
Arthur, D. (1961). *Ticks and Disease* (Pergamon Press).
Baerg, W. (1928). The life cycle and mating habits of the Tarantula. *Quart. Rev. Biol.* **3**, 109.
Baerg, W. (1958). *The Tarantula* (University of Kansas Press).
Baker, E. W. and Wharton, G. W. (1952). *An Introduction to Acarology* (Macmillan, New York).
Bristowe, W. S. (1949). The distribution of harvestmen in Great Britain and Ireland with notes on their names, enemies and food. *Jour. Anim. Ecol.* **18**, 100.
Bristowe, W. S. (1939, 1941). *The Comity of Spiders*. 2 vols. (Ray Society).
Bristowe, W. S. (1958). *The World of Spiders* (Collins).
Browning, H. C. (1942). The integument and moult cycle of *Tegenaria atrica*. *Proc. Roy. Soc.* B, **131**, 65.
Comstock, H. (1948). *The Spider Book*. Revised by W. Gertsch (Comstock Publ. Co.).
D'Amour, F., Becker, F. and Riper, W. (1936). The black widow spider. *Quart. Rev. Biol.* **11**, 123.
Davies, M. E. and Edney, E. B. (1952). The evaporation of water from spiders. *Jour. Exp. Biol.* **29**, 571.
Dillon, L. S. (1952). The mycology of the araneid leg. *Jour. Morph.* **90**, 467.
Ellis, C. H. (1944). The mechanism of extension in the legs of spiders. *Biol. Bull.* **86**, 41.
Freeman, J. A. G. (1946). The distribution of spiders and mites up to 300 feet in the air. *Jour. Anim. Ecol.* **15**, 69.
Gaskell, W. H. (1908). *The Origin of the Vertebrates* (Longmans).
Hedgepeth, J. W. (1947). On the evolutionary significance of the pycnogonids. *Smithson. Misc. Publ.* **106**, no. 18.
Hoyle, G. (1958). Studies on neuromuscular transmission in *Limulus*. *Biol. Bull.* **115**, 209.

Hughes, T. E. (1959). *Mites or The Acari* (University of London Press).
Jones, B. M. (1954). On the role of the integument in acarine development and its bearing on pupa formation. *Quart. Jour. Micr. Sci.* **95**, 169.
Jones, B. M. (1956). The British harvest mite. *New Biol.* **21**, 74.
Kingsley, J. S. (1892–3). The embryology of *Limulus. Jour. Morph.* **7**, 35; **8**, 195.
Lankester, E. R. (1881). *Limulus*, an arachnid. *Quart. Jour. Micr. Sci.* **21**, 504.
Lockett, G. H. and Millidge, A. F. (1951–3). *British Spiders.* 2 vols. (Ray Society).
Parry, D. (1960). The small leg nerves of spiders and a probable mechanoreceptor. *Quart. Jour. Micr. Sci.* **101**, 1.
Patten, W. (1912). *The Evolution of the Vertebrates and their Kin* (Blakiston).
Petrunkewitch, A. (1912). The circulatory system and segmentation in arachnids. *Jour. Morph.* **36**, 157.
Petrunkewitch, A. (1952). Principles of classification as illustrated by studies of Arachnida. *System. Zool.* **1**, 1.
Pringle, J. W. S. (1955). The function of the lyriform organs of arachnids. *Jour. Exp. Biol.* **32**, 270.
Savory, T. H. (1928). *The Biology of Spiders* (Sidgwick and Jackson).
Savory, T. H. (1949). Recent work on spiders. *New Biol.* **7**, 30.
Vachon, M. (1953). The biology of scorpions. *Endeavour*, **12**, 80.
Walcott, C. and van der Kloot, W. G. (1959). The physiology of the spider vibration receptor. *Jour. Exp. Zool.* **141**, 191.

MOLLUSCA

Perhaps the most interesting recent development in molluscan studies has been the description of *Neopilina's* morphology by Lemche and Wingstrand. Fig. 6 is taken from their paper and illustrates the general arrangement of the internal symmetry. Another interesting paper is that by Cox and Rees where they discuss the properties of *Tamanovalva* which is a gastropod with a typical bivalve shell. The topic of torsion is fully discussed by Garstang and Yonge, whilst Young gives an interesting account of the giant nerves in the squid and also the functioning of the octopus brain. The molluscs show interesting changes of sex (Coe), larval forms (Garstang), digestion and organization of the stomach (Graham), methods of swimming (Morton and Holme), and muscles which are capable of maintaining tension for long periods of time (Hoyle).

Allen, J. A. (1960). The ligament of the Lucinacea (Eulamellibranch). *Quart. Jour. Micr. Sci.* **101**, 25.
Arello, E. L. (1960). Factors affecting the ciliary activity on the gill of the mussel *Mytilus edulis. Physiol. Zool.* **33**, 120.
Arey, L. B. (1921). Encystment in *Glochidia. Jour. Exp. Zool.* **33**, 463.
Bailey, K. and Warboys, B. D. (1960). The lamellibranch crystalline style. *Biochem. Jour.* **76**, 487.
Barnes, H. F. (1949). The slugs in our gardens. *New Biol.* **6**, 29.
Barnes, G. (1955). The behaviour of *Anodonta cygnea* and its neurophysiological basis. *Jour. Exp. Biol.* **32**, 158.
Barnes, H. F. and Weil, J. W. (1944–5). Slugs in gardens, their numbers, activities and distribution. *Jour. Anim. Ecol.* **13**, 140; **14**, 71.
Baylor, E. R. (1959). The responses of snails to polarized light. *Jour. Exp. Biol.* **36**, 369.
Berry, Q. (1928). Cephalopod adaptations. *Quart. Rev. Biol.* **3**, 92.
Bidder, A. M. (1950). The digestive mechanisms of European squids. *Quart. Jour. Micr. Sci.* **91**, 1.

Block, D. and Hew, H. (1960). Spermatogenesis in the pulmonate snail *Helix aspersa* with special reference to histone transfer. *Jour. Biophys. Biochem. Cytol.* **7**, 515.

Boycott, B. B. and Young, J. Z. (1950). The comparative study of learning. *Symp. Soc. Exp. Biol.* **4**, 432.

Coe, W. R. (1936). Sequence of functional sexual phases in *Teredo*. *Biol. Bull.* **71**, 122.

Coe, W. R. (1953). Influences of association, isolation and nutrition on the sexuality of the snail *Crepidula*. *Jour. Exp. Zool.* **122**, 5.

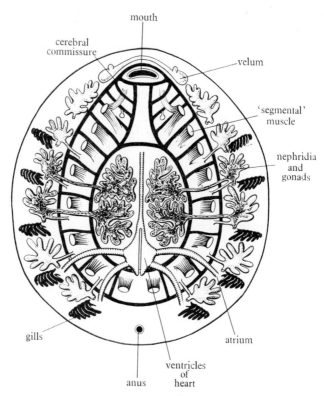

Fig. 6. *Neopilina galatheae*. Diagram of the 'segmented' organs. (From Lemche and Wingstrand.)

Cox, L. R. and Rees, W. J. (1960). A bivalve gastropod. *Nature, Lond.*, **185**, 749.

Crofts, D. (1937). The development of *Haliotis tuberculata*. *Phil. Trans. Roy. Soc.* B, **125**, 711.

Crofts, D. (1955). Muscle morphogenesis in primitive gastropods and its relation to torsion. *Proc. Zool. Soc.* **125**, 711.

Dean, B. (1901). Studies on the living *Nautilus*. *Amer. Nat.* **35**, 819.

Drew, G. A. (1907). Habits and locomotion of *Ensis*. *Biol. Bull.* **12**, 127.

Duncan, C. J. (1959). The life cycle and ecology of the freshwater snail *Physa fontinalis*. *Jour. Anim. Ecol.* **28**, 97.

Duncan, C. J. (1960). The evolution of the pulmonate genital system. *Proc. Zool. Soc.* **134**, 601; **131**, 55.

Eales, N. B. (1950). Torsion in gastropods. *Proc. Malacol. Soc.* **28**, 53.
Eckstein, B. and Abraham, B. (1959). Succinic dehydrogenase activity in aestivating and active snails *Helix hierosolyima*. *Phys. Zool.* **32**, 210.
Fischer, P.-H. (1950). *Vie et mœurs des mollusques* (Payot, Paris).
Freeman, J. A. (1960). Influence of carbonic anhydrase inhibitors on shell growth of fresh water snail *Physa heterostropha*. *Biol. Bull.* **118**, 412.
Galli, D. R. and Giese, A. C. (1959). Carbohydrate digestion in a herbivorous snail, *Tegula funebralis*. *Jour. Exp. Zool.* **140**, 415.
Garstang, W. (1928). Origin and evolution of larval forms. *Presidential Address of Brit. Ass.* Section D, p. 77.
George, W. C. and Ferguson, H. J. (1950). The blood of gastropod molluscs. *Jour. Morph.* **86**, 315.
Glaser, O. C. (1910). Nematocysts of aeolid molluscs. *Jour. Exp. Zool.* **9**, 117.
Graham, A. (1949). The molluscan stomach. *Trans. Roy. Soc. Edinb.* **61**, 737.
Graham, A. (1953). Form and function in the molluscs. A symposium. *Proc. Linn. Soc.* **164**, 213.
Holmes, W. (1940). The colour changes and colour patterns of *Sepia officinalis*. *Proc. Zool. Soc.* **110**, A, 17.
Howes, N. and Wells, G. P. (1934). The water relations of snails and slugs. *Jour. Exp. Biol.* **11**, 344.
Hoyle, G. (1957). *Comparative Physiology of the Nervous Control of Muscular Contraction* (Cambridge University Press).
Joysey, K. A. (1959). The evolution of the Liassic oysters, *Ostrea-Gryphaea*. *Biol. Rev.* **34**, 297.
Kawai, K. (1959). The cytochrome system in marine lamellibranch tissues. *Biol. Bull.* **117**, 125.
Kennedy, D. (1960). Neural photoreceptor in a lamellibranch mollusc. *Jour. Gen. Physiol.* **44**, 277.
Kerkut, G. A. and Laverack, M. S. (1957). The respiration of *Helix pomatia*, a balance sheet. *Jour. Exp. Biol.* **34**, 97.
Krijgsman, B. J. and Divaris, G. A. (1955). Contractile and pacemaker mechanisms of the heart of molluscs. *Biol. Rev.* **30**, 1.
Lavine, T. F. (1946). A study of the enzymatic and other properties of the crystalline style of clams. *Jour. Cell. Comp. Physiol.* **28**, 183.
Lemche, H. (1957). A new living deep-sea mollusc of the Cambro-Devonian class Monoplacophora. *Nature, Lond.*, **179**, 413.
Lemche, H. and Wingstrand, K. G. (1959). The anatomy of *Neopilina galatheae*. *Galathea Rep.* **3**, 7.
Lissmann, H. W. (1945–6). The mechanisms of locomotion in gastropod molluscs. *Jour. Exp. Biol.* **21**, 58; **22**, 37.
Martin, A. W., Harrison, F. M., Hustar, M. J. and Stewart, D. M. (1958). The blood volume of some representative molluscs. *Jour. Exp. Biol.* **35**, 260.
McMahon, P., Von Brand, T. and Nolan, M. O. (1957). Observations on polysaccharides of aquatic snails. *J. Cell. Comp. Physiol.* **50**, 219.
Miller, A. K. (1943). Cambro-Ordovician cephalopods. *Biol. Rev.* **18**, 98.
Milton, A. S. (1959). Choline acetylase in gill plates of *Mytilus*. *Proc. Roy. Soc.* B, **150**, 240.
Moore, H. B. (1936). The biology of *Purpura lapilis*. *Jour. Mar. Biol. Ass.* **21**, 61.
Morton, J. E. (1958). *Molluscs* (Hutchinson).
Morton, J. E. (1960). The functions of the gut in ciliary feeders. *Biol. Rev.* **35**, 92.
Morton, J. E. and Holme, N. A. (1955). Swimming of *Akera bullata*. *Jour. Mar. Biol. Ass.* **34**, 101.
Myers, F. L. and Northcote, D. H. (1958). A survey of the enzymes from the gastro-intestinal tract of *Helix pomatia*. *Jour. Exp. Biol.* **35**, 639.

Orton, J. H. (1937). *Oyster Biology and Oyster Culture* (Arnold).
Owen, G. (1958). Shell form, pallial attachment and the ligament in the Bivalvia. *Proc. Zool. Soc.* **131**, 637.
Petersen, R. P. (1959). The anatomy and histology of the reproductive system of *Octopus bimaculoides*. *Jour. Morph.* **104**, 61.
Raven, Ch. (1958). *Morphogenesis, the Analysis of Molluscan Development* (Pergamon Press).
Ridewood, W. G. (1903). On the structure of the gills of the lamellibranchs. *Phil. Trans. Roy. Soc.* B, **195**, 147.
Sato, M., Tamasigem, M. and Ozeki, M. (1960). Electrical activity of retractor muscle of snail. *Jap. Jour. Physiol.* **10**, 85.
Schindewolf, O. (1934). Concerning the evolution of the Cephalopoda. *Biol. Rev.* **9**, 458.
Snyder, L. and Crozier, W. J. (1922). Ctenidial variation in chitons. *Biol. Bull.* **43**, 246.
Sollas, I. B. (1907). The molluscan radula, its chemical composition, and some points in its development. *Quart. Jour. Micr. Sci.* **51**, 115.
Thompson, T. E. (1958). The natural history, embryology, larval biology and postlarval development of *Aldaria proxima* (Opisthob.). *Phil. Trans. Roy. Soc.* B, **242**, 1.
Thompson, T. E. (1960). Defensive acid secretion in marine gastropods. *Jour. Mar. Biol. Ass.* **39**, 115.
Welsh, J. H. (1956). Cardioregulators of *Cyprina* and *Buccinum*. *Jour. Mar. Biol. Ass.* **35**, 193.
Wiersma, C. A. G. (1952). Comparative physiology of invertebrate muscle. *Ann. Rev. Physiol.* **14**, 159.
Wilczynski, J. Z. (1959). Sex behaviour and sex determination in *Crepidula fornicata*. *Jour. Exp. Biol.* **36**, 34.
Wilson, D. P. and Wilson, M. A. (1956). A contribution to the biology of *Ianthina janthina*. *Jour. Mar. Biol. Ass.* **35**, 291.
Yoneda, M. (1960). Force exerted by a single cilium of *Mytilus edulis*. *Jour. Exp. Biol.* **37**, 461.
Yonge, C. M. (1947). The pallial organs in the aspidobranch gastropods and their evolution throughout the Mollusca. *Phil. Trans. Roy. Soc.* B, **232**, 443.
Yonge, C. M. (1950). *Oysters* (Collins).
Yonge, C. M. (1955). Adaptations to rockboring in *Botula* and *Lithophaga* with a discussion of the evolution of this habit. *Quart. Jour. Micr. Sci.* **96**, 383.
Young, J. Z. (1938). The evolution of the nervous system and the relationship of organisms and environment. In *Evolution, Essays Presented to Goodrich*, pp. 179–204 (Oxford University Press).
Young, J. Z. (1939). Fused neurones and synaptic contacts in the giant nerve fibres of cephalopods. *Phil. Trans. Roy. Soc.* B, **229**, 465.
Young, J. Z. (1961). Learning and discrimination in the octopus. *Biol. Rev.* **36**, 32.

CHAETOGNATHA

Gunther, R. T. (1907). The Chaetognatha, or primitive molluscs. *Quart. Jour. Micr. Sci.* **51**, 357.
Parry, D. A. (1944). Structure and function of the gut in *Spadella cephaloptera* and *Sagitta setosa*. *Jour. Mar. Biol. Ass.* **26**, 16.
Pierce, E. I. and Orton, J. H. (1939). *Sagitta* as an indicator of water movements in the Irish Sea. *Nature, Lond.*, **144**, 784.
Russell, F. S. (1932). On the biology of *Sagitta*. *Jour. Mar. Biol. Ass.* **18**, 131, 555.

BRACHIOPODA

These are dealt with in Shrock and Twenhofel's text-book of invertebrate palaeontology.

Atkins, D. (1960). The ciliary feeding mechanisms of the Megathyridae (Brachiopoda and the growth stages of the lophophores. *Jour. Mar. Biol. Ass.* **39**, 459.
Elliot, G. F. (1953). Brachial development and evolution in the terebratelloid brachiopods. *Biol. Rev.* **28**, 261.
Muir-Wood, H. M. (1955). *A History of the Classification of the Phylum Brachiopoda* (British Museum, Natural History, London).
Williams, A. (1956). The calcareous shell of Brachiopoda and its importance to their classification. *Biol. Rev.* **31**, 243.

ECTOPROCTA AND ENDOPROCTA

Atkins, D. (1932). The ciliary feeding mechanism of the endoproct polyzoans and a comparison with that of the ectoproct polyzoans. *Quart. Jour. Micr. Sci.* **75**, 393.

ECHINODERMATA

The radial symmetry of the echinoderms leads to certain problems concerning the control of their tube feet. These and other problems relating to their nervous system are discussed by Jennings, Smith and Kerkut. The properties and functions of the tube feet are given by Nichols. The echinoderms are considered as fairly close relatives to the vertebrates. This is supported by Spencer, Gislen and Gregory. On the other hand, Fell thinks that some of the resemblances are due to convergence.

Horstadius has carried out some very elegant experiments on the embryological development of the echinoid embryo.

Allee, W. C. (1927). Studies in animal aggregation. *Jour. Exp. Zool.* **48**, 475.
Anderson, J. M. (1960). Histological studies on the digestive system of a starfish, *Henricia*, with a note on Tiedemann's pouches. *Biol. Bull.* **119**, 371.
Boolotian, R. A. and Giese, A. C. (1959). Clotting of echinoderm coelomic fluid. *Jour. Exp. Zool.* **140**, 207.
Budington, R. A. (1942). The ciliary transport system of *Asterias forbesi*. *Biol. Bull.* **83**, 438.
Fell, H. B. (1948). Echinoderm embryology and the origin of the chordates. *Biol. Rev.* **23**, 81.
Gemmills, J. F. (1919). The rhythmic pulsation in the madreporite vesicle of young ophiuroids. *Quart. Jour. Micr. Sci.* **63**, 537.
Gislen, T. (1930). Affinities between echinoderms, enteropneusta, and chordates. *Zool. Bidrag.* **12**, 197.
Gregory, W. K. (1946). The role of motile larvae and fixed adults in the origin of the vertebrates. *Quart. Rev. Biol.* **21**, 248.
Hancock, D. A. (1955). The feeding behaviour of starfish on Essex oyster beds. *Jour. Mar. Biol. Ass.* **34**, 313.
Harvey, E. B. (1956). *The American Arbacia and other Sea Urchins* (Princeton University Press).
Hawkins, H. L. (1931). The first echinoid. *Biol. Rev.* **6**, 443.
Horstadius, S. (1939). The mechanics of sea-urchin development studied by operative methods. *Biol. Rev.* **14**, 132.
Horstadius, S. (1957). On the regulation of bilateral symmetry in plutei with exchanged meridional halves and in giant plutei. *Jour. Emb. Exp. Morph.* **5**, 60.

Jennings, H. S. (1907). Behaviour of the starfish *Asterias forreri*. *Univ. Calif. Publ. Zool.* **4**, 53.
Kerkut, G. A. (1954–5). The mechanisms of co-ordination of the starfish tube feet. *Behaviour*, **6**, 206; **8**, 112.
Kille, F. R. (1942). Regeneration of the reproductive system following binary fission in the sea cucumber *Holothuria parvula*. *Biol. Bull.* **83**, 55.
Millot, N. and Yoshida, M. (1960). The shadow reaction of *Diademia antillarum*. *Jour. Exp. Biol.* **37**, 363.
Mortenson, T. H. (1927). *Handbook of Echinoderms of the British Isles* (Oxford University Press).
Nichols, D. (1958). Changes in the chalk heart-urchin *Micraster* interpreted in relation to living forms. *Phil. Trans. Roy. Soc.* B, **242**, 347.
Nichols, D. (1960). Histology and activities of the tube feet of *Antedon bifida*. *Quart. Jour. Micr. Sci.* **101**, 105.
Pople, W. and Ewer, D. W. (1955). Circum-oral conduction in *Cucumaria*. *Jour. Exp. Biol.* **32**, 59; **31**, 114.
Smith, J. E. (1950). Some observations on the nervous mechanisms underlying the behaviour of starfish. *Symp. Soc. Exp. Biol.* **4**, 196.
Smith, J. E. (1950). The motor nervous system of the starfish *Astropecten irregularis* with special reference to the innervation of the tube feet and ampullae. *Phil. Trans. Roy. Soc.* B, **234**, 521.
Spencer, W. K. (1938). Some aspects of evolution in echinoderms. In *Evolution, Essays Presented to Goodrich*, pp. 287–384 (Oxford University Press).

PROTOCHORDATA
Pogonophora

The major recent development within the protochordates has been the elucidation of the structure of the small tubicolous forms related to *Siboglinum*; these are now placed in a new group, the Pogonophora. These animals, as can be seen from p. 714, were placed in the Hemichorda near to the Pterobranchiates. The Russian worker, A. V. Ivanov, thinks that they deserve separation from the Hemichorda and should be placed in a separate phylum, 'The Pogonophora'.

The characteristics of the Pogonophora are as follows: They are Chordate animals without a tail, atrium or bony tissue. The central nervous system is partly or wholly on the surface of the body (Fig. 7); there are three primary segments of the coelom and these are retained in the adult as three externally visible regions of the body.

They differ from the other three classes of the Hemichorda in that (1) the tentacles are borne on the first segment (see Figs. 8 and 9); (2) the main nervous system is on the first segment (see Figs. 7, 8 and 9); (3) they have no mouth, digestive tract or anus.

Other points of interest are that the first segment has a pair of coelomoducts which act as nephridia; the gonoducts are in the third segment together with the gonads; the larvae are ciliated and have chaetae; the existence of the notochord has not yet been proven.

The pogonophorans are filamentous animals and range in size from 5 mm. to 35 cm. in length and up to 3 mm. in diameter. They are tubicolous, the chitinous tube being secreted by, and usually much longer than, the body. They live upright in the mud on the ocean bed and may be at considerable depths (8000 m.), though *Oligobrachia* has been taken at 124 m. Over nine genera and twenty-two species are at present known, the main structural variation being the number of tentacles (Fig. 10). The first segment bears the tentacles which are hollow and well supplied with blood vessels and contain

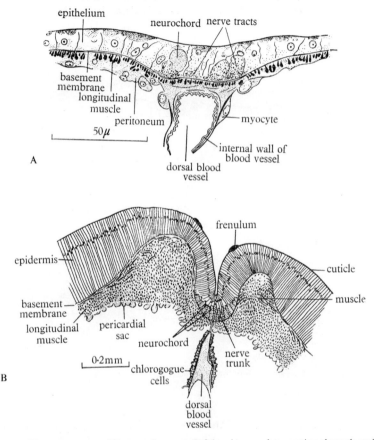

Fig. 7. Nervous system of Pogonophora. A, *Polybrachia annulata*, section through main nerve tract of metasome, showing 'giant fibres' and nerve cord; B, *Spirobrachia grandis* section through main nerve tract of mesosome. (From Ivanov.)

extensions of the coelom of the first segment. The coelom of the first segment connects to the outside by means of a pair of ciliated nephridia (Fig. 11).

The second segment has a pair of coelomic cavities but no ducts. The third segment has a pair of coelomic cavities and a pair of gonoducts. The sexes are separate and the gonads lie in the third segment. The testes are usually posterior whilst the ovaries are anterior. The eggs are large (5 mm.) and rich in yolk. The young develop in the adult's tube. Cleavage is holoblastic and the coelom forms enterocoelically. The larvae are ciliated and provided with chaetae. They have no gut but their interior is filled with vacuolated cells and the yolk becomes used up.

The adults and embryos have no digestive system and their method of feeding is not yet understood. Ivanov suggests that they are filter-feeders and that the food is trapped in a network of filaments and possibly slime and then digested externally. The digested food is then absorbed through the

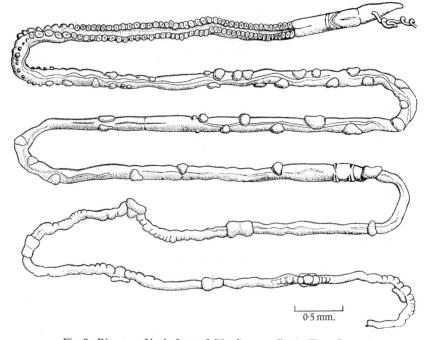

Fig. 8. Diagram of body form of *Siboglinum caulleryi*. (From Ivanov.)

tentacles and transported to the rest of the body through the well-developed blood system. Full details of the group are given by Ivanov and also by Hyman.

Pogonophora references

Hyman, L. H. (1959). Phylum Pogonophora. In *The Invertebrates*, vol. 5, pp. 208–227 (McGraw-Hill).

Ivanov, A. V. (1954). New Pogonophora from the Far Eastern Seas. *System. Zool.* **3**, 68.

Ivanov, A. V. (1955). The main features of the organization of the Pogonophora. *System. Zool.* **4**, 170.

Ivanov, A. V. (1956). On the systematic position of the Pogonophora. *System. Zool.* **5**, 165.

Ivanov, A. V. (1959). The nervous system of Pogonophora. *System. Zool.* **8**, 96.

Ivanov, A. V. (1959). The Pogonophora. In *Traité de Zoologie*, ed. by P.-P. Grassé, vol. 5, fascicle 2, pp. 1521–1622.

Ivanov, A. V. (1960). *Pogonophori* (in Russian). Fauna of Russia, no. 75. *Publ. Acad. Sci. U.S.S.R.* 269 pp.
Jagersten, G. (1956). Investigations on *Siboglinum ekmani* encountered in Skagerak with some general remarks on the group Pogonophora. *Zool. Bidrag.* **31**, 211.
Jagersten, G. (1957). Larva of *Siboglinum*. *Zool. Bidrag.* **32**, 67.
Southward, E. (1959). Two new species of Pogonophora from the north-east Atlantic. *Jour. Mar. Biol. Ass.* **38**, 439.

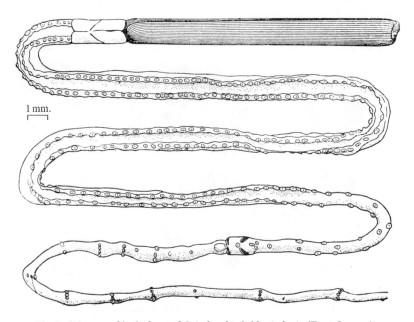

Fig. 9. Diagram of body form of *Spirobrachia beklemischevi*. (From Ivanov.)

TUNICATES

The tunicates have amazing power of regeneration and budding. They seem to ignore the germ layer theory and can rebuild a new body from almost any germ layer (Berrill). They also have a tadpole larva which can be made to metamorphose at different rates by chemical treatment (Glaser). The graptolites were till recently of uncertain affinities but, as the paper by Kozlowski indicates, they are most likely related to the pterobranchiates. Since the protochordates are looked upon as showing chordate characteristics, it is not surprising that they are regarded as ancestors of the vertebrates. This is discussed by Garstang, Gregory, Berrill and Bone.

Abbott, D. P. (1955). Larval structure and activity on the ascidian *Metandrocarpa taylori*. *Jour. Morph.* **97**, 569.
Berrill, N. J. (1945). Size and organization in the development of Ascidians. *Essays on Growth and Form presented to D'Arcy Thompson*, pp. 231–59 (Oxford University Press).
Berrill, N. J. (1950). *The Tunicata* (Ray Society).
Berrill, N. J. (1951). Regeneration and budding in the tunicates. *Biol. Rev.* **26**, 456.

Berrill, N. J. (1955). *The Origin of the Vertebrates* (Oxford University Press).
Bone, Q. (1960). The origin of the chordates. *Jour. Linn. Soc. (Zool.)*, **44**, 252.
Bullock, T. H. (1945). The anatomical organization of the nervous system of the Enteropneusta. *Quart. Jour. Micr. Sci.* **86**, 55.
Eaton, T. H. (1953). Pedomorphosis, an approach to the Chordate-Echinoderm problem. *System. Zool.* **2**, 1.
Endean, R. (1960). The blood cells of the ascidian, *Phallusia mammillata*. *Quart. Jour. Micr. Sci.* **101**, 177.
Garstang, W. (1929). The morphology of the Tunicata and its bearing on the phylogeny of the Chordata. *Quart. Jour. Micr. Sci.* **72**, 51.
Gilchrist, J. D. (1917). On the development of *Cephalodiscus*. *Quart. Jour. Micr. Sci.* **62**, 189.

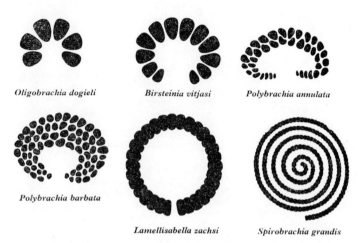

Fig. 10. Diagrammatic transverse section through tentacles of various pogonophorans indicating the variety of number. (From Ivanov.)

Glaser, O. and Anslow, G. A. (1949). Copper and ascidian metamorphosis. *Jour. Exp. Zool.* **111**, 117.
Goodbody, I. (1957). Nitrogen excretion in Ascidiacea. *Jour. Exp. Biol.* **34**, 297.
Gregory, W. K. (1946). The role of motile larvae and fixed adults in the origin of the vertebrates. *Quart. Rev. Biol.* **21**, 348.
Horst, van der C. J. (1935). *Planctosphaera* and *Tornaria*. *Quart. Jour. Micr. Sci.* **78**, 605.
Jorgensen, C. B. (1950). Quantitative aspects of filter feeding in invertebrates. *Biol. Rev.* **30**, 391.
Knight-Jones, E. W. (1952). The nervous system of *Saccoglossus cambrensis*. *Phil. Trans. Roy. Soc.* B, **236**, 315.
Knight-Jones, E. W. (1953). The feeding of *Saccoglossus*. *Proc. Zool. Soc.* **123**, 637.
Kozlowski, R. (1947). The affinities of the graptolites. *Biol. Rev.* **22**, 93.
Krijgsman, B. J. (1956). Contractile and pacemaker mechanisms in the heart of tunicates. *Biol. Rev.* **31**, 288.
Newell, G. E. (1951). The stomochord of Enteropneusta. *Proc. Zool. Soc.* **121**, 741.
Orton, J. H. (1913). The ciliary mechanism on the gills and mode of feeding in *Amphioxus*, ascidians and *Solenomya togata*. *Jour. Mar. Biol. Ass.* **10**, 19.

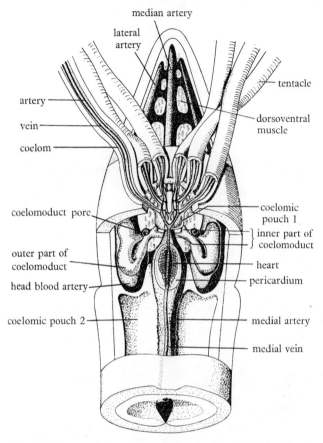

Fig. 11. Diagram of the structure of *Oligobrachia dogieli*, showing the blood system, and coelomic system. (From Ivanov.)

Trason, W. B. (1957). Larval structure and development of the oozoid of the ascidian *Euherdmania claviformis*. *Jour. Morph.* **100**, 509.

Watkins, B. J. (1958). Regeneration of buds in *Botryllus*. *Biol. Bull.* **115**, 147.

Webb, D. A. (1939). Observations on the blood of certain ascidians with special reference to the biochemistry of Vanadium. *Jour. Exp. Biol.* **16**, 499.

INDEX

The names of genera and species are printed in italics.
The figures in bold type refer to illustrations on the page indicated.
An asterisk (*) signifies that some of the series of figures which immediately follow it refer to particular genera only.
When the first of several figures after a word is followed by a semi-colon, it indicates the principal reference to the subject.

Abambulacral surface, 670
Abdomen, of Arachnida, see Opisthosoma; of Arthropoda, 320; of Crustacea*, 320, 344, 347, 349, 371, 373, 377, 381, 383, 384, 392, 396, 397, 403, 405, 408, 410, 413, 417, 419; of Insecta*, 320, 438, 467, 493, 526; of *Iulus*, 424; of Polychaeta, 279; of Tunicata, 724
Abdominal limbs, of Crustacea*, 344–5, 347, 355, 373, 397, 399, 403, 406, 414, 419; of Insecta, 438, 467, 475, 476; of *Iulus*, 424
Abdominal region, of *Balanoglossus*, 710
Aboral aspect, side, or surface, 140; of Echinodermata, 670
Aboral sinus system, 672, 674, 696
Acantharia, 84, 86
Acanthobdella, 311; 268, 305, 306, **309**
Acanthobdellidae, 306, 311
Acanthocephala, 253; 191, 228, 235
Acanthocystis, 47
Acanthograptus, 718
Acanthometra elastica, 86, **85**
Acarina, 567–73
Accessory nidamental gland, 640
Acephalous larva, 506
Acerentomon doderoi, 476, **477**
Aciculum, 274
Acineta, 115; 11, **12**, **114**
Acmaea, 596, 599
Acmaeidae, 612
Acoela, 207, 208
Acoelomate Triploblastica, 191, 228
Acontia, 184, **183**
Acrobeles, 238
Acrothoracica, 393
Actaeon, **614**, 615
Actinia, 182; *equina*, 183
Actinoceras, 650
Actinodactylella, **210**, 211
Actinometra, 699
Actinomma, 86, **85**
Actinomyxidea, 103
Actinophrys, 87; 38; *sol*, **86**
Actinosphaerium, 87; 13, 28, 29, 33, 47; *eichorni*, **49**
Actinotrocha larva, 668; **142**
Actinozoa, 176
Actinula larva, 157, **158**

Actipylaea (Acantharia), 84, 86
Adambulacral ossicles, 682
Adambulacral spines, 682
Adductor muscles, of Cirripedia, 389; of claws, 140; of Conchostraca, 368; of Crustacea, 349; of Lamellibranchiata*, 140, 622, 631, 633; of Leptostraca, 399
Adelea, 94
Adephaga, 490
Adoral wreath, 106
Aega, 405; *psora*, **405**
Aeolosoma, 305
Aeschna, 455, 472; *cyanea*, **482**
Aesthetascs (Olfactory hairs), 357; **358**
Aesthetes, 592
Afferent branchial vein, 628
Afferent canals, see Inhalant canals
Agametes, 42
Agamogony, 43, 91
Agamonts, 43
Aggregata, 93
Akera bullata, **617**
Alary muscles, **446**
Alcippe, 393
Alcyonaria, 176
Alcyonidium, **659**
Alcyonium, 176; *digitatum*, 176–9
Alepas, 393
Algae, Symbiotic, 50, **183**, 187
Alimentary canal, 128; of *Antedon*, 701; of Arachnida, 537; of Araneida, 557; of *Arenicola*, 283; of *Argas*, 569; of Arthropoda, 325; of Asteroidea, 682; of *Balanoglossus*, 711; of Brachiopoda, 663; of *Carcinus*, 415; of Chaetognatha, 666; of *Chirocephalus*, 371; of *Ciona*, 722; of Crustacea, 359, **359**; of *Cyclops*, 385; of *Daphnia*, 379; of Echinodermata, 672; of *Echinus*, 690; of *Gammarus*, 406; of Gastrotricha, 260; of *Helix*, 608; of Hirudinea, 306; of Holothuroidea, 696; of Insecta, 439; of *Iulus*, 425; of Lamellibranchiata, 627; of *Lepas*, 391; of *Ligia*, 403; of *Limulus*, 547; of *Lithobius*, 422; of Mollusca, 585; of *Nebalia*, 400; of Nematoda, 240; of Nematomorpha, 252; of Nemertea, 232; of Oligochaeta, 298; of Ophiuroidea, 686; of Pantopoda, 574; of *Peripatus*, 333; of *Phoronis*, 668; of

Alimentary canal (*cont.*)
  Polyzoa, 264; 653; of Pterobranchiata, 714, **713**; of Rotifera, 257; of Salpidae, 739; of Scorpionidea, 543; of *Sepia*, 642; of Stomatopoda, 400; of *Stylaria*, **304**; of Tardigrada, 576; **575**; of *Testacella*, 618; of the trochosphere, 295. *See also* Digestive system
Allantonema, 252
Allogromia, 83; 75; *oviformis*, **81**
Alloiocoela, 207
Allolobophora, 298, **301**, 302
Alternation of generations, *see* Life cycle
Alveolar layer, 106
Alveoli, of digestive gland, 608; of protoplasm, 13
Amaurobius, **557**
Amblyomma, **569**
Ambulacra, 670, 693
Ambulacral grooves, 670, 682, 700
Ambulacral ossicles, 682, 685, 700
Ambulacral surface, 670
Ametabola (Apterygota), 475; 465
Amino-acids, 26, 58, 62, 128, 443
Amiskwia, 668
Amitoses of Protozoa, 30
Ammonoidea, 650; 647
Amnion, Insect, **463**; Nemertean, **233**, 234
Amoeba, 75; *discoides*, **76**; *dubia*, **76**; *proteus*, **76**
Amoebina, 74
Ameoboid movement, 16
Amoebulae, 43
Amphiblastula larva, **121**
Amphilina, 221; **219**
Amphineura, 589
Amphinucleoli, 30
Amphioxus, 706, **707**, 723
Amphipneustic, 506
Amphipoda, 405
Amphitrite, 274
Amphiura, 687
Amphoridea (Hydrophoridea), 704
Ampulla of stone canal, 693
Ampullae, of Hydrocorallinae, 163, **163**; of tube feet, 672, **679**
Amusium, **584**
Anaerobic animals, 136
Anal papilla, 700
Anaphothrips striatus, 527
Anaspides, 400; *tasmaniae*, **398**
Ancestral group of Protozoa, 51
Ancylostoma, 245, **245, 246**, 248
Andrena, 523
Animal Kingdom, 1, 51
Animal pole, 291
Anisogamy, 36; **35**. *See also* Gametes
Anisoptera, 481
Anisospores, 84
Annelid cross, 294, **292**, 588
Annelida, 266
Annuli, of leeches, 305

Annulus of dinoflagellates, 60
Anodonta, 623, **624**, 628, **629**, 630, 633
Anomia, **634**
Anomopoda, 375
Anomura, 410
Anopheles, 508, **503**; *maculipennis*, **454, 455**, **501**
Anoplophrya, 108; *prolifera*, **111**
Anoplura, 526
Anostraca, 370
Antedon, 699; *bifida*, **699**; *rosaceu*, 699
Antennae, 318, 319, 320; First of Crustacea, *see* Antennules; of Crustacea (second pair), 344–5, 348, 354, 368, 373, 375, 376, 384, 397, 399, 402, 403, 413; of Insecta, 432, 475, 476; of Myriapoda, 420, 424; of Trilobita, 339; so called, of Rotifera, 256; of *Peripatus* (Preantennae), 330
Antennal glands, 361; **360**, 382, 415
Antennules, 344–5, 347, 354, 370, 373, 376, 384, 397, 413
Anterior aorta, of *Carcinus* (ophthalmic artery), **415**; of *Helix*, 607; of Insecta, 445; of Lamellibranchiata, 628; of Scorpionidea, 542; of *Sepia*, 641
Anterior cervical groove, **398**
Anterolateral edge, 411
Antheridia, 65, **64**
Anthobothrium, **225**, 224
Anthomedusae, 157; 153. *See also* Gymnoblastea
Anthonomus grandis, 490
Anthophora, 522
Anthozoa (Actinozoa), 176
Anthrenus museorum, 443
Antipathes, 396
Anurida maritima, 476
Anus, 128; of Amphineura, 589; of *Balanoglossus*, 710; of Brachiopoda, 660; of Echinodermata*, 682, 684, 687, 690, 693, 701; of Echiuroidea, 314; of Gasteropoda, 598, 616; of Gastrotricha, 260; of *Haliotis*, **601**, 612; of *Helix*, 606; of *Lepas*, 390; of Mollusca, 580; of Nemertea, 228; of Opisthobranchiata, 616; of *Peripatus*, 332; of Polychaeta, 294; of Pterobranchiata, 714; of *Sepia*, 640; of Sipunculoidea, 316. *See also* Alimentary canal
Aorta, of Anostraca, 364; of *Helix*, 607. *See also* Anterior aorta, Posterior aorta
Aperture, *see* Opening
Aphaniptera (Siphonaptera), 511
Aphasmidia, 236
Aphididae, 532
Aphids, 532, 446
Aphis,**438**,442; *rumicis*,532,**532**;*saliceti*,533
Aphodius, 491
Aphrodite, 273, **275, 277**
Apical nervous system, 676; of Crinoidea, 701; of Echinodermata, 676; of Holothuroidea, 696

Apical organ, 294, 658
Apical rosette of trochosphere, 294
Apis, 514, **515**, 516, **435**; *mellifica*, **440**, 522
Apistobuthus, 535
Aplacophora, 594
Aplysia, 616; **615**
Apocrita, 519
Apoda, 393
Apodemes, 355
Apodous larvae, *see* Larvae
Apoidea, 522
Apoplastid phytomastigina, 55
Apopyles, 118
Appendages, Paired, *see* Limbs
Appendicularia larva, 726
Apposition image, 324
Apseudes, 403, **409**
Apterygota, 475
Apus, cancriformis, 373, **374**
Aquatic oligochaeta, 302
Arachnida, 535
Aranea, 554, **555**, **558**
Araneida, 549
Arcella, discoideus, 78, **79**; 14, 75
Archenteron, 127; of polychaete embryo, 294
Archiannelida, 311
Archicerebrum, of Lankester, *see* Procerebrum, *sensu stricto*
Archigetes, 225, **225**
Architeuthis, 636
Arenaceous shells, 80
Arenicola, 282; **275**, **285**, **287**; *marina*, 282, **284**
Argas, 570; **569**; *persicus*, **571**, 572
Argonauta, 637, 645
Argulus, 389; *americanus*, **388**; *foliaceus*, **388**, 389
Argyroneta, 554; **539**
Arion, 618
Arista, 510
Aristocystites, 704
Aristotle's lantern, of *Echinus esculentus*, 690, **691**
Armadillidium, 404
Arms, of Brachiopoda, 662; of Crinoidea, 699; of Dibranchiata, 638; of Echinodermata, 670; of Echinoderm larvae, 678; of Pterobranchiata, 714
Artemia salina, 373; 370
Arterial system, of *Carcinus*, 415; of *Helix*, 607; of Lamellibranchiata, 628; of *Lithobius*, 423; of Malacostraca, 364; of Scorpionidea, 542. *See also* Aorta, Vascular system
Artery, Antennary, Dorsal abdominal (posterior aorta), Gastric, Hepatic, Ophthalmic (anterior aorta), Sternal, Ventral abdominal, Ventral thoracic, of Crustacea, Decapoda, **346**; **415**, **416**; Cephalic, Supraneural, of Chilopoda, **423**; Gastro-intestinal, Hepatic, Pallial,

Terminal, of Lamellibranchiata, 628, **629**; Lateral, Supraneural, of Arachnida*, 542, 548; Pedal, of Mollusca, 605. *See also* Aorta
Arthrobranchiae, **363**, 412
Arthropoda, 317
Articulamentum, 592
Ascaridata, 236
Ascaris, lumbricoides, 236, **237**, **241**, 243
Ascidia, 734
Ascidia (Insecta), 445
Ascidiacea, 733
Ascidiae compositae, **733**
Ascidian tadpole, 726; **725**
Ascon grade, 118
Ascopus, 257
Ascothoracica, 396
Asellus, 404; *aquaticus*, **406**
Asexual reproduction, 6; of Aquatic oligochaetes, 304, 305; of Metazoa, *see* Budding, Strobilation; of Porifera, 121; of Protozoa, *see* Agamogony, Schizogony, Sporogony; of Turbellaria, 206
Aspidiotus perniciosus, **532**
Aspidobranchiata (Diotocardia), 611
Aspidochirotae, 697
Aspidogaster, **212**
Aspirigera, (Holotricha), 108
Asplanchna, 258
Astacura, 410
Astacus, 345, **350**, 351, **353**, **356**, **358**, **359**, **361**, **363**, **364**, **410**; *fluviatilis*, 346
Astarte, **584**
Astasia, 35
Asterias, 681; **672**, **678**, **679**, **680**; *rubens*, 683; *vulgaris*, **673**
Asterina, **684**; 677
Asteroidea, 681
Asterope, 277; **274**, **286**
Astomata, 108
Astrea, **620**
Astroides, **184**
Astropecten, 684, **684**, **687**
Atractonema, 252
Atractylis, 149
Atrial opening, 720
Atrial siphon, 721
Atrium, Genital, *see* Genital atrium, of Cephalochorda, 724; of Tunicata, 724, 726
Atropus pulsatoria, 525
Atypus, 567, **566**, 535
Aulactinia, 182
Aulactinium, 87; *actinastrum*, 85
Aurelia aurita, 170–5; Strobilation of, **174**, 176
Auricles, of *Arenicola*, 283; of Gasteropoda, 596, 601; of Lamellibranchiata, 628; of Mollusca, 578; of *Sepia*, 641
Auriculae, 690
Auricular organs, 200
Auricularia larva, 677, **676**

INDEX

Autogamy, 37, **88**
Autolytus, 289, **290**
Autosomes, 533
Autotomy, 355
Autozooids, 180
Avenella, **659**
Avicularia, 656, **657**, **658**
Axelsonia, 475
Axial filament, 87
Axial organ, **679**, **683**; of Crinoidea, 702; of Holothuroidea, 696
Axial sinus, 672; of Echinoidea, 693
Axon, 134
Axopodia, 16, 87
Axostyles, 25, **71**

Babesia (*Piroplasma*), 100, 571
Bacillus pestis, 513
Baculites chicoensis, **650**, 651
Badhamia, 89
Balanoglossus, 709
Balantidium, 110; *entozoon*, **111**
Balanus, 393; **391**
Barnea, **584**
Basal disc, 184
Basal granule, 17, 19, **20**, **22**
Basal ossicles, 700
Basement membrane, 194
Baseodiscus, 228, **233**
Basilar plate, 424
Basipodite, **352**, **413**
Basommatophora, 618
Bathynella, 401
Bdellocephala, 204
Bdelloid rotifers, 260
Bdelloura, 192
Beaks of Cephalopoda, 642
Beania, **659**
Beds, Mussel, 631
Behaviour, of Protozoa, 44
Belemnites, **645**
Belemnoidea, 646
Bembex, 522
Beroe, 187, **189**
Bibio, 506, **508**
Bilateral cleavage, 143
Bilateral symmetry, 140, 680; of Actinozoa, 178; of Ciliata, 11; of Echinoderm larvae, 678–9; of Echinoidea, 670; of Holothuroidea, 670; of Metazoa, 140
Bilharzia (*Schistosoma*), 217
Bimeria, 149
Binary fission of Protozoa, 33
Bipalium kewense, 208
Bipinnaria larva, 677
Biramous limb (Stenopodium), 351
Birgus, 419
Bladder worm (*Cysticercus*), 224
Blastocoele, 127
Blastoidea, 704
Blastopore, 127
Blastostyles, 155

Blastozoite (Blastozooid), 731, 737
Blastozooid, 731, 737
Blastula, 127; of Coelenterata, 152; of Echinodermata, 677; of *Obelia*, 155
Blatta, 433, **441**, 483
Blepharoplast (basal granule: kinetosome), 19–24
Blood, 129; of *Arenicola*, 283; of *Ciona*, 725; of Crustacea, 364; of Insecta, 446; of Polychaeta, 272; of Scorpion, 542
Blood vessels, *see* Arterial system, Artery, Vascular system, of Echinodermata, *see* Lacunar system,
Bodo, 67; *saltans*, **67**; *sulcatus*, **45**; kinesis of, **45**
Body, of Amphipoda, 405; of *Arenicola*, 282; of *Balanoglossus*, **709**; of *Cephalodiscus* and *Rhabdopleura*, 714; of Crustacea, 347; of Ctenophora, 187; of Echinodermata, 670; of Isopoda, 403; of Medusa, 151, 157; of Metazoa, 126; of polyp, 150; of Porifera, 116; of Protozoa, 11; of Tubicolous Polychaeta, 278. *See also* Symmetry
Body cavity, 129; of Nematoda, 236; Perivisceral, *see* Perivisceral cavity; Primary, *see* Haemocoele; Secondary, *see* Coelom
Body wall, of Hirudinea, 307; of Holothuroidea, 696; of Metazoa, 126, 127; of polyps, 146, 150; of Rotifera, 255
Bombus, 522
Bombyx mori, 497
Bonellia, 314, **314**
Boophilus annulatus, 571
Bopyrus, 405; *fougerouxi*, **404**
Bothriotaenia, 225
Botryllus, 734, **734**; *violaceus*, **733**
Botryocrinus, 670
Botryoidal tissue, 307
Bougainvillea, 155; *fructuosa*, **156**
Brachial ossicles, 700
Brachial skeleton, 663
Brachiolaria larva, 679
Brachioles, 704
Brachiopoda, 660; Development of, 665, **665**
Brachyura, 410
Brachyurous type, 410
Brain, 133; of Acanthocephala, 253; of Acoelomate Triploblastica, 193; of Arthropods, 322; of Branchiopoda, **321**; of *Ciona*, 726; of Crustacea, 356; of Insecta, 458; of Nematomorpha, 252; of Polychaeta, 271; of *Sepia*, 642, 643. *See also* Ganglion, Cerebral
Branchia, of Phyllopodium, 369; of *Salpa*, *see* 'Gill'
Branchiae, *see* Gills
Branchial chamber, 722
Branchial hearts, veins, 641
Branchial opening, 720
Branchiobdellidae, 311; **306**
Branchiogenital region, 710

INDEX

Branchiopoda, 368
Branchiura, 388
Breathing, see Respiratory movements
Brisinga, 684, **687**
Brown body, 654, **658**
Bruchophagus funebris, **520**
Bryograptus, 719
Bryozoa (name for Polyzoa = Endoprocta, 263 + Ectoprocta, 652)
Buccal capsule, 247
Buccal cavity, of *Balanoglossus*, 710; of *Helix*, 608; **607**; of Insecta, 439; of Nematoda, 240; of Tunicata, 722. See also Alimentary canal
Buccal mass, of *Helix*, 608; of *Sepia*, 642
Buccal tube feet, 690
Buccinum, 612; **603**
Budding, of *Alcyonaria*, 179; of *Cysticercus*, 224; of *Hydratuba*, 174; of Hydrozoa, 166; of Madreporaria, 185; **184**; of *Microstomum*, 206; of Polychaeta, 289, **290**; of Protozoa, 113; 87; of Pterobranchiata, 714, 717; of *Stylaria*, 304; of Tunicata, 728. See also Colonies
Bugula, 656, **657**
Bulimus, 621
Bulla, 615; **614**
Bunodes, 549
Bursa copulatrix, of Insecta, 457, **457**; of Turbellaria, 203
Buthus, Internal anatomy of, **543**; *carpathicus*, embryo of, **538**
Byssus pit, 631

Caddis flies (Trichoptera), 494
Cadophore, 739; **738**
Caeca, Mesenteric, see Mesenteric caeca
Caecum, of Echinoidea, 690; of Ectoprocta, 653; of *Sepia*, 642
Caenis, 480
Caenogenetic features, 141
Calabar swellings, 249
Calandra, 490
Calanus, 385; **384**
Calathus, 252
Calcarea, 122
Calcareous ring in Holothuroidea, 696
Caligus, 387
Callidina, 260; **259**
Calliphora, 502, **504, 505**, 511
Calonympha, 19, 21, **20**
Calosoma, 490; *semilaeve*, **490**
Calotermes militaris, 485
Calymma, 84
Calyptoblastea (Leptomedusae), 154
Calyptomera, 380; 375
Calyx, 699
Camallanata, 236
Campodeiform larva, 467; **466**; of Coleoptera, 489
Canal system, of Medusae, 171, **171, 172**; of Sponges, 118, **119**

Cancer, 419; **417**
Capillaria, 236
Capillitium, 89
Capitellidae, **285**
Capitulum, 389, **571**
Caprella, 408; *grandimana*, **409**
Capsidae, 534
Captacula, 621, **622**
Carabus, 490; *violaceus*, 490
Carapace, 349; of Branchiopoda, 368; of Calyptomera, 375; of *Carcinus*, 411; of Cirripedia, 389; of Cladocera, 375; of Conchostraca, 375; of Gymnomera, 380; of *Lepas*, 389; of *Leptodora*, 381; of Malacostraca, 397; of *Nebalia*, 399; of Ostracoda, 382; of Peracarida, 401
Carchesium, 112; *epistylidis*, 112
Carcinus, **414**; *maenas*, 410, **411, 413, 415, 416**
Cardiomya, 634
Cardium, 623
Cardo, 433
Caridea, 410
Caridoid facies, 399
Carina, 389, **391**
Carinella, 228
Carinoma, 228
Carmarina, **159**
Carotin, 55
Carpoidea, 704
Carpopodite, **413**
Carteria, **50**, 62
Caryogamy, 37
Caryophyllaceus, 225
Caryophyllia, 184
Carysome, 30
Cases of Trichoptera, **495**
Cassiopeia, 175
Catch fibres, 140
Caudal furca (rami), 355
Cavolinia, 616, **615**
Cecidomyidae, 509
Cells, 10; assuming various functions, 126; Corneagen, 323; Flame, 200, **201, 255**; Interstitial, **146**, 167, 206; Iris, 323; Lasso, 187; Musculo-epithelial, 146; of Porifera, 116, **118**; Pole, 127, 300; Sensory, 133, see also Sense organs; Somatic of Volvocina, 65; **13**; Thread, 146; Yellow, 271. See also Choanocytes, Myoblast, Oenocytes, Pinacocytes, Porocytes, etc.
Cellular animals, 10, 126; structure, 126–7
Cellulases, 128, 443, 585, 608, 635
Cellulose, 14, 62, 720
Central capsule, 84; **48, 85**
Central nervous system, 133–5; of Annelida, 270; of Chordata, 707; of Tunicata, 726. See also Brain, Ganglion, Nervous system
Centroderes, 262
Centrodorsal ossicle, 699

INDEX

Cepedea, 73
Cephalization, 141; of Crustacea, 347; of Polychaeta, 275
Cephalochorda, 707. *See also Amphioxus*
Cephalodasys, 261
Cephalodiscus, 714
Cephalopoda, 636
Cephalothorax, of Arachnida, *see* Protosoma; of Brachyura, 410; of Copepoda, 384; of *Ligia*, 403
Cephalothrix, 228
Cerambycidae larvae, 443
Cerata, 616; **615**
Ceratium, 61; *macroceras*, **60**
Ceratophyllus fasciatus, 513
Cerci anales, 438
Cerebral ganglia, *see* Brain; Ganglion, Cerebral,
Cerebral organ, 232
Cerebral vesicle, 727
Cerebratulus, 228, 234
Cervical groove, 350
Cervical sclerites, 433
Cestoda, 219
Cestus veneris, 190, **189**
Ceuthorrhynchus, 491
Chaetae, 266, **268**; of *Acanthobdella*, 311; of Archiannelida, 313; of Echiuroidea, 314, **315**; of Oligochaeta, 305; of Polychaeta, 274; of brachiopod larva, 665, **665**
Chaetoderma, 596
Chaetognatha, 666
Chaetonotus, 261
Chaetopoda, 269
Chaetopterus, 280; 279; *pergamentaceus*, **278**
Chama, **634**
Chambered organ, 701; **700**
Chambers, of cephalopod shells, 647; of Foraminifera shells, 78, **82**
Chaos (*Amoeba*), 75, **76**
Cheeks of Trilobita, 338
Cheilostomata, 652
Cheimatobia, 500
Chela, 355
Cheleutoptera, 486
Chelifer, 535, 544
Chilaria, 537
Chilina, **614**
Chilo, 500
Chilomonas, 58
Chilopoda, 420
Chione, **584**
Chirocephalus, 370; **321**, 343, 347; *diaphanus*, 370
Chironomidae, 508
Chironomus, 446, 455
Chitin, 428
Chiton, 591, **592**
Chlamydomonas, 62; 33; *angulosa*, **34**; *brauni*, **35**; *euchlora*, 36; *longistigma*, **34**; *media*, **35**; *steini*, 36
Chlamydospores, 43

Chloeon, 480
Chlorocruorin, 130, 272
Chloromonadina, 60
Chlorophyll, 55
Chloroplasts, 55
Chlorops taeniopus, 511
Choanocytes, 116, **118**
Choanoflagellata (Choanoflagellidae), 69
Chondracanthus, 387; *gibbosus*, **386**
Chonotricha, 113
Chordata, 3, 706
Chorion, of egg of Insects, 456
Chromatophores, of Cephalopoda, 639; of Crustacea, 357; of Phytomastigina (*passim*), 55; 35
Chromidium (Chromatin), 29
Chromoplasts, *see* Chromatophores of Phytomastigina,
Chrysamoeba, 57; *radians*, 56
Chrysaora, **169**
Chrysidella, 58; *schaudinni*, **59**
Chrysis, 521
Chrysomelidae, 491
Chrysomonadina, 57
Chrysops, 502, 249; *caecutiens*, **509**; *dimidiata*, 510
Cicada septendecim, 531
Cicindela, 443, 490
Cilia, 17; of Ciliata, **19**, 106; of *Ciona* pharynx, 723; **723**; of Coelenterata, 171, 177, 187; of mollusc gills, 624-7; of Turbellaria, 193. *See also* Epidermis
Ciliary junctions, **624**, 625
Ciliata, 104
Ciliated band, of *Dipleurula*, 677; **676**; of *Tornaria*, 713; **711**
Ciliated funnel, *see* Dorsal tubercle
Ciliated organ, 287; **267**
Ciliated pits, 200
Ciliated ring, *see* Ciliated band, Metatroch, Prototroch, Velum
Ciliophora, 103
Ciliospores, 43
Cimex, 534
Comarocystites, 704
Cingulum, 256
Ciona intestinalis, 720
Circulation of blood, in Anostraca, 364; in Arachnida, 542; in *Arenicola*, 284; in Arthropoda, 326; in *Balanoglossus*, 711; in Chaetopoda, 272; in *Ciona*, 725; in Copepoda, 385; in Echinodermata, 675, **681**; in Entomostraca, 364; in Insecta, 445; in Malacostraca, 415; in Mollusca, 582; in Nemertea, 232; in Ostracoda, 382; in Scorpionidea, 542. *See also* Vascular system
Circulation of food, in *Alcyonium*, 178; in *Aurelia*, 172; in *Daphnia*, 379; in *Obelia*, 155
Circulatory system, *see* Vascular system
Circulus venosus (circular vein), 608; **605**

Circumoesophageal connectives, see Nervous system
Cirratulus, **283**
Cirrhal ossicles, 701
Cirri, of Crinoidea, 700; of Protozoa, 17, 106; of Thoracica, 390
Cirripedia, 389
Cirrodrilus, **306**
Cladocera, 375
Classification, of Acarina, 568; of Actinozoa, 176; of Annelida, 266; of Arachnida, 535; of Arthropoda, 317; of Ascidiacea, 734; of Brachiopoda, 660; of Cephalopoda, 636, 647; of Coelenterata, 144; of Ciliata, 104; of Crustacea, 341; of Ctenophora, 187; of Echinodermata, 669; of Echinoidea, 693; of Ectoprocta, 652; of Eucestoda, 224; of Gasteropoda, 596; of Graptolita, 718; of Hemichorda, 708; of Hirudinea, 306; of Holothuria, 697; of Insecta, 470; of Lamellibranchiata, 623; of Mastigophora, 53; of Mollusca, 578; of Mycetozoa, 89; of Nematoda, 236; of Nemertea, 228; of Oligochaeta, 297; of Onychophora, 329; of Parazoa, 122; of Pelmatozoa, 704; of Polychaeta, 273; of Phytomastigina, 53; of Platyhelminthes, 192; of Protochordata, 706; of Protozoa, 52; of Pulmonata, 618; of Radiolaria, 84; of Sarcodina, 74; of Scyphozoa, 169; of Sporozoa, 90; of Trematoda, 211; of Urochorda, 720; of Zoomastigina, 53
Clathrina, 122
Clathrulina, 88; **15**
Clavelina, 729, 733
Clavularia, 179
Cleaners, see Scavengers
Cleavage of ovum, affected by yolk, 143; Centrolecithal, 328, 366, 462; Determinate, 291, 726; of Archiannelida, 291; of Polychaeta, 291; of Polyclada, 291; of Mollusca, 291, 588; of Nemertea, 291; of Arthropoda, 328, 366, 462; of *Balanoglossus*, 713; of *Ciona*, 726; of Coelenterata, 152; of Crustacea, 366; of Ctenophora, 189; of Echinodermata, 677; of *Pyrosoma*, 737; Radial, 1, 677; Spiral, 1, 291
Climacograptus, 717
Cliona, 124
Clitellum, 297; 269
Cloaca, of Holothuroidea, 696
Coelomoducts, of Arthropoda, 327; of Crustacea, 362; of Mollusca, 582, **591**; of Polychaeta, 285
Coeloplana, 190
Coenosarc, of Calyptoblastea, 155; of Polyzoa (Ectoprocta), 655
Colacium, 60
Coleoptera, 489; three types of, **491**
Coleopterous larvae, **466**

Collar, of *Balanoglossus*, 709; of *Chaetopterus*, 280; of Choanoflagellata, 69; of Pterobranchiata, 714
Collar cavities, of *Balanoglossus*. 709
Collar cells (Choanocytes), 116, **118**
Collar pores, of *Balanoglossus*, 710
Collembola, 475
Colleterial glands, 458
Colletes, 516
Collinia, 108; 37
Colloblasts (Lasso cells), 187
Collozoum, 85; *inerme*, **48**
Collum, 424, **425**
Colon, see Large intestine
Colonies, of Alcyonaria, 179; of *Carchesium*, 112; of Hydroids, 154, 162, 163, 168; of Polyzoa (Ectoprocta), 652; of Protozoa, 11; of *Rhabdopleura*, 714, **715**, 717; of Siphonophora, 163; of *Syllis ramosa*, 289, **290**; of Tunicata, 728, 729, 733; of Volvocina, 62; of Zoantharia, 185
Colpidium, 108
Colpoda, 108; *steini*, **111**
Columella muscle, 605
Compasses, 690
Compensation sac, 656, **656**
Complemental males, 393
Compound ascidians, 733
Compound eyes, see Eyes
Conchostraca, 375
Conjugants, 36
Conjugation, 36, 7. See also Syngamy
Contarinia nasturtii, **509**; *purivora*, 509
Contractile vacuoles, 27–8; of *Euglena*, 28; of Heliozoa, 28; of Protozoa, 27
Conus, **611**, 620
Convoluta roscoffensis, 207, **50**; *henseni*, 207
Copepoda, 383
Copeus, 256
Copidosoma gelechiae, 456
Copromonas, 59; *subtilis*, **59**
Coprozoic Protozoa, 47
Copulatory bursa, 243; spicules, 243
Corallium, **179**; *rubrum*, 179
Corals, of Alcyonaria, 179; of Hydrocorallinae, 162; of Madreporaria, 185; of Polyzoa, 655
Corbula, 634
Cordylophora, 161
Corethra, 443
Corixa, 534
Cormidium, 163–4
Cornularia, 179
Corona, 688
Corpuscles, 130. See also Blood
Cortex, of Ciliata, 107; of Porifera, 119; of Protozoa, 14
Corticata, 51
Corymbites cupreus, 491
Coryne, 575
Cossidae, 500

Cossus, 500
Cotton spinner (Holothuria), 694
Cotyloplana, 192
Course of circulation, *see* Circulation of blood
Coxa, 433
Coxal glands, of Arachnida, 541; of Arthropoda, 326
Coxopodite, **352, 413**
Crangon, 417
Crania, 665; **664**
Craspedacusta, 162
Craspedochilus, 591
Craterolophus, 170
Cremaster, 497
Crenobia (*Planaria*), 208
Crepidula, 586 **590**, 620
Criconema, **238**
Crinoidea, 699
Crisia, **657**; 652
Cristatella, 652
Crithidia, 68; **67, 68**
Crop, of Gasteropoda, 608; of Insecta, 439; of Opisthobranchiata, 616
Crotchets, 303
Crural glands, 334
Crustacea, 340
Crustacean-insect-myriapod section, 319
Cryptocerata, 534
Cryptomitoses, 30, **31**
Cryptomonadina, 58
Cryptomonas, 58, **59**
Cryptoniscus, 405; *paguri*, **404**
Cryptus obscurus, **520**
Crystal cells (Vitrellae), 323
Crystalline style, 585, 627, **627**
Ctenidia, of Cephalopoda*, 636, **637, 638**, 640, 641, 647, 649; of *Chiton*, **593**; 592; of Gasteropoda*, 596, **602**, 603, 616; of Lamellibranchiata, 623, **623, 624**, 625, 626, **626**, 628; of Mollusca, 582, **590**
Ctenidial circulation of Lamellibranchiata, 628, **629**
Ctenocephalides canis, **512**
Ctenophora, 187; 144
Ctenoplana, 190
Ctenopoda, 375
Ctenostomata, 652
Cucullanus, **242**
Cucumaria, 697, **698**
Culex fatigans, 508
Culicidae, 508
Cumacea, 402
Cunina, 162
Cup-shaped organs, 310
Curculionidae, 491
Cursoria (Dictyoptera), 483
Cuspidaria, 623
Cuticle, 6, 11, 133, of Annelida, 266; of Arthropoda, 320; of Crustacea, 347; of Nematoda, 235, 237; of Protozoa, 14; of Rotifera, 255; of Trematoda, 211

Cuvierian organs, 696
Cyamus, 408, **409**
Cyanea arctica, 169
Cyathomonas, 58, 55, 14; *truncata*, **59**
Cyathozooid, 737; **736**
Cyclestheria hislopi, **352**
Cyclomyaria (Doliolids), 739
Cyclophyllidea, 225; 224
Cycloporus papillosus, **206**, 208
Cyclops, 383, **383**, **384**; 210, 223 250, 344, 366, **367**
'*Cyclops*' larvae, 385, 386
Cyclorrhapha, 510
Cyclosalpa, 737
Cyclospora, 37
Cyclostomata, 652
Cyphonautes larva, 658, **660**
Cypraea, **620**
Cypridina, 382
Cypris, 382, **382**, 344
'*Cypris*' larva, 392, **392**, 366
Cyrtoceras, 650
Cysticamera, 718
Cysticercus, 224; *pisiformis*, 224
Cystoflagellata (*Noctiluca*), 62; **61**
Cystoidea, 704
Cysts, 14, 28. *See also* Gamocysts, Oocysts, Sporocysts
Cytogamy, 39
Cytorhinus mundulus, 534
Cytostome, 106

Dactylopodite, **413**
Dactylozooids, 164, **164**, **165**
Dalyellia viridis, 207; **196**, 209, 211
Daphnia, 375; 344, **378**, **379**, 380
Dart, 252
Dart sac, 609, **609**
Davainea, **225**
Dead men's fingers (*Alcyonium digitatum*), 176
Deamination, 128
Death, 6
Decapoda, Cephalopoda, 636; Crustacea 408
Deep oral nervous system, 677
Deima, 697, **698**
Demodex folliculorum, **576**
Demospongiae, 124
Dendrites, 134
Dendritic tentacles, 696
Dendrochirotae, 697
Dendrocoelum lacteum, 208
Dendrocometes, 115; *paradoxus*, 114
Dendrocystis, 704
Dendroid graptolites, 718
Dendron, 134
Dense nuclei, 30
Dentalium, 621, **622**
Depastrum, 170
Depression, 40
Deraecoris fasciolus, **529**

Dermal layer, 116
Dermaptera, 488
Dermis, 133
Dermomuscular tube, 139
Desmoscolex, 238
Determinate cleavage, 143, 726
Detorsion, 602, 614
Deutocerebrum, 430
Deutomerite, 97
Development, 141. See also Embryology, Larvae, Life cycle
Diastopora, **659**
Diastylis, 402; *stygia*, **401**
Diazona, 728, **728**
Dibranchiata, 636
Dicranura, 497
Dictyonema, 718
Dictyoptera, 483
Dicyclical rotifers, 258
Didymograptus, **719**; *v-fractus*, **719**
Difflugia, 80; 14; *urceolata*, **79**
Digestion, 128; by Acarina, 570; by *Alcyonium*, 178; by *Arenicola*, 283; by Arthropoda, 325; by *Aurelia*, 172; by Coelenterata, 146; by Crustacea, 360; by *Daphnia*, 379; by *Helix*, 608; by Insecta, 440; by Lamellibranchiata, 627; by Oligochaeta, 298; by *Ostrea*, 633; by *Physalia*, 165; by Protozoa, 26; by Rotifera, 257; by *Teredo*, 635; by Turbellaria, 195; by Zoantharia, 183. See also Alimentary canal, Circulation of food, External digestion
Digestive caeca, see Digestive gland
Digestive gland, of Arachnida, 537; of Brachiopoda, 663; of Mollusca*, 597, 608, 627, 642. See also Liver, Mesenteric caeca
Digestive system, of *Alcyonium*, 178; of *Aurelia*, 172; of Ctenophora, 188; of Medusae, 151; of Platyhelminthes, 195; of Zoantharia, 183. See also Alimentary canal, Enteron
Dimorpha, 87; *mutans*, **86**
Dimorphic shells, **82, 83**
Dinamoebidium, 62
Dinobryon, 57; **56**
Dinoflagellata, 60
Dinophilus, 314; **312**
Dinophysinae, 61
Dinothrix, 62
Dioctophymata, 236
Diodora, 612, **601, 603**
Diotocardia, 611
Diphyllidea, 224
Diphyllobothrium, 222, 223; *latum*, 225
Diphyes, 164
Dipleurula, 673, **674**
Diploblastica, 126. See also Coelenterata
Diplograptus foliaceus, **719**
Diplomonadina, 70
Diplopoda, 423

Diploporida, 704
Diplostraca, 375
Diplozoon, 215; **212**
Diptera, 500
Dipylidium caninum, 226
Direct wing muscles, 437, **439**
Directives, 182
Discomedusae, 170
Dissosteira carolina, **448**
Distephanus, 58; **56**
Distomum macrostomum, 217
Distyla, **259**
Division of protozoan nuclei, 30, **31**
Docoglossate, **611**
Dolichoglossus kowalevskii, **709**
Doliolidae, 739; 735
Doliolum, 739; **738**
Donax, **627**
Doris, 616; **615**
Dorsal and ventral, 140; aspects of bilateral animals, 140; aspects of Ciliata, 106; aspects of Holothuroidea, 694; aspects of *Sepia*, 639; 'blood vessels' of Echinoidea and Holothuroidea, 675; mesenteries of Alcyonaria, 177; structures of radial animals, 140
Dorsal antenna, 256
Dorsal blood vessel, of Arthropoda, see Aorta; of *Balanoglossus*, 711; of Polychaeta, 272; of Rhynchobdellidae, 308
Dorsal 'blood vessels' of Echinodermata, 675
Dorsal cirrus, 274; lamina, 733; organ, 350; pores, 298; shield, 350; siphon, 625; tubercle, 722
Dorsolateral antennae, 256
Draconematid, 238
Dracunculus, 249
Drag line, 563
Dreissensia, **587**
Drepanophorus, 231, **232, 233**
Drift net of *Physalia*, 165
Drosophila, 445
Ductus communis, 202
Dugesia, 208; **198**
Dysdercus, 534
Dytiscus, 490, 433, 446, 454, **457**

Earthworms, 298
Ecardines, 660
Ecdyonurus, 479
Ecdysis, see Moulting
Echinaster sentus, **682**
Echinobothrium, 224; **223**
Echinocardium, 694; *cordatum*, **694**
Echinococcus (*Taenia echinococcus*), 226; **225**
Echinocyamus pusillus, 693
Echinoderes, 262
Echinodermata, 669
Echinoidea, 687
Echinopluteus, 678; **676**

Echinus, 688; 669; **688**, 690, **691, 692**, 693; *esculentus*, 688, **691**; *miliaris*, 692
Echiniscus, 536
Echiuroidea, 314
Echiurus, 314
Ectoderm, 126; 1, 133, 144, 146, 152, 155, 172, 178, 193, 194, 231, 235, 255, 286, 291
Ectoneural system of Crinoidea, 702
Ectoplasm, 14, 25, 51, 75, 84, 97
Ectoprocta, 652
Edrioaster, 704; *bigsbyi*, **702**
Edrioasteroidea, 704
Edwardsia, 182, **182**
Effectors, 133
Efferent canals (exhalant canals), 118; **117**
Egg sac, 385
Eggs and egg laying, of Arachnida*, 543, 548, **564**, 565, 566, 571; of Arthropoda, 328; of *Balanoglossus*, 713; of Chaetognatha, 666; of *Ciona*, 726; of Cnidaria, 152; of Crustacea*, 366, 373, 375, 379, 385, 389, 402, **407**; of Ctenophora, 189; of *Dinophilus*, 314; of Echinodermata, 677; of Hirudinea, 311; of Insecta*, 458, 462, 481, 483, 485, 494, **512**, 516, 518, 520, 521, 526, 527, 533; of *Iulus*, 426; of *Lithobius*, 423; of Mollusca*, 588, 610; of Nematoda, 243; of Oligochaeta, 300; of Pantopoda, 575; of *Peripatus*, 334; of Platyhelminthes*, 203, 204, 217, 218, 222; of Polychaeta, 288; of Rotifera, 258; of Thaliacea, 737
Eimeria, 92; *schubergi*, **93**
Ejaculatory duct, 457
Elasipoda, 697
Eledone, **643**
Eleutherozoa, 669, 681
Elphidium, 83; **82, 83**
Elytra of Coleoptera, 436, 489, **490**; of Polychaeta, 274, 275
Emarginula, **590**
Embia major, 488
Embioptera, 488
Embryology (*S. str.*) 141; of Arachnida, **538**, 541; of Arthropoda, **322**, 328; of *Asterias*, **673**, 674, 677, **678**; of Brachiopoda, **665**, 665; of Chaetognatha, 667, **667**; of Chordata, 707, **707**; of *Ciona*, 726; **725**; of Coelenterata, 152, **152**; of Crustacea, 328, **365**, 366; of Echinodermata, 677; of Insecta, 462, **463**; of *Lumbricus* (*Eisenia*), 300; of *Peripatus*, 334; 332, **335, 336**; of Polyzoa, 265; 660, **660**; of Tardigrada, 577; of Thaliacea, 737; **736**; of trochospheres, 294
Embryonic fission, *see* Polyembryony
Emplectonema, 228
End gut, *see* Hind gut
End sac, 361
Endocyclica, 693
Endoderm, 126; 1, 128; of Coelenterata, 144, 146, 152, 166. *See also* Enteron

Endoderm lamella, 151
Endomixis, 39; **38**
Endophragmal skeleton, 355
Endoplasm, 14
Endopodite of Crustacea, 351; of Trilobita, 339; of Xiphosura, 547. *See also* Limbs
Endoprocta, 263
Endopterygota, 489
Endopterygote wing formation, 467
Endosome, 29
Endosternite, 355
Ensis, **634, 635**
Entamoeba, 75; *coli*, 75, **77**; *dysenteriae*, 75; *histolytica*, 75, **77**
Enterobius vermicularis, 248; 236, 239
Enteron, 144, 151, **151**, 152, 155. *See also* Archenteron, Digestive system
Enteropneusta, 709
Entocoeles, 182
Entodiniomorpha, 110
Entodinium, 110; *caudatum*, **111**
Entomostraca, 347
Envelope cells, 101
Environment, 4
Enzymes, Digestive, 128, 442
Eolis, 616; **615**
Epeira, 563, **566**, 567
Ephelota, 115; *gemmipara*, 114
Ephemera, 480; *vulgata*, **479**
Ephemeroptera, 479
Ephestia, 500
Ephippium, 379
Ephyra larva, 174, **174**, 176
Epibolic gastrulation, 143, 293, **586**
Epibranchial spaces, of Decapoda, 412; of Lamellibranchiata, 625
Epicardial cavity, 724; diverticula, 724
Epicardium, 735; 724
Epicuticle, 320
Epidermis, 133; in several phyla, *see* names of Phyla. *See also* Ectoderm
Epimerite, 97
Epineural canal, 670; of Echinoidea, **692**; of Ophiuroidea, 685
Epipharynx, 430. *See also* Mouth parts of Insecta
Epiphragma, 610
Epipodites, 351, **352, 353**. *See also* Gills of Crustacea, Metepipodites, Oostegites, Proepipodites
Epistome of Decapoda, 412; of Ectoprocta, 652, 655
Epistylis, 112
Epizoanthus, 124
Equitant whorls, 83
Erichthus larva, **367**
Eriocrania, 496
Eristalis, 510
Eruciform larvae, **466**, 467; of Coleoptera, 489; of Symphyta, 518
Estheria, 375; *obliqua*, 376
Euanostraca, 373

Eucarida, 408
Eucephalous larva, 506, **508**
Eucoila, 519
Eudendrium, 159, **161**
Eudorina, 64, **13**
Euglena, 58, **59**; 28; *viridis*, 58
Euglenoid movement, 58; 17
Euglenoidina, 58
Euglypha, **80**; *alveolata*, **79**
Eugregarinaria, 97
Euherdmania, 730
Eulalia, 277; **274**
Eulamellibranchiata, 633
Eumitosis, 30
Eunice, 217
Eupagurus, 417; *bernhardus*, **418**
Euphausiacea, 408
Euplectella, 123
Eupomatus, Trochosphere of, 295
Eurypelma, 567
Eurypterida, 544
Eurypterus, 546
Eutyphoeus, 298
Evadne, 381
Evolution, 141
Exarate pupae, 467
Excreta, *see* Excretion
Excretion, 137, **137**; by Arthropoda, 326; by Crustacea, 361–3; by Echinodermata, 674; by Metazoa, 137; by Mollusca, 606; by Protozoa, 27; by Tunicata, 726. *See also* Excretory organs
Excretory organs, 137; of Arachnida, 326, 541; of Arthropoda, 326; of *Balanoglossus*, 711; of Crustacea, 361–3; **360**, **361**. *See also* Antennal glands, Maxillary glands of Insecta, 444; of Metazoa, 137; of Myriapoda, 326; of Nematoda, 242; of Nemertea, 232; of Onychophora, 326; of Platyhelminthes, 200; of Polychaeta, 272, 285–7; of Rotifera, 257. *See also* Coelomoducts, Coxal glands, Glomerulus, Kidneys, Malpighian tubules, Nephridia
Exerythrocyte stage (*Plasmodium*), 95
Exhalant canals of Porifera, 118
Exhalant passage of *Carcinus*, 412
Exites, 351
Exocoeles, 182
Exocyclica, 693
Exogamous syngamy, 37
Epipodite, of Crustacea, 351; of Trilobita, 339; of Xiphosura, 547. *See also* Limbs
Exopterygota, 479, 525
Exopterygote development of wings, 467
External digestion, by Araneida, 557; by Insecta, 443; by Oligochaeta, 298; by Rhizostomeae, 175; by Turbellaria, 195
External medium, *see* Medium
Extracellular digestion, 128
Extra-thecal zone, 184, **184**
Exumbrellar surface, 170, 175

Eyes, Compound, 323–5; Crustacean median, 357, 358; of Arachnida, 548, 550, 560, 563; of Arthropoda, 323; of Ascidian tadpole, 727; of Chaetognatha, 666; of Chaetopoda, 271, 274, 291; of Crustacea*, 357, **358**, 372, 374, 382, **383**, **388, 392, 325**; of Hirudinea, 310; of Hydrozoa, 159, **159**; of Insecta, 461; of Mollusca, 592, **644**; of Myriapoda, 423, 424; of Nemertea, **230**; of Onychophora, 337; of Polychaeta, 274, 291; of Trilobita, 338; of Turbellaria, 198, **199**. *See also* Eye spots
Eye spots, of Asteroidea, 677; of Protozoa, 25
Eye stalk, 413

Facial suture, 338
Falciform young, 98
Fasciola, 215; *hepatica*, **216**, 218
Fat body, 445
Favia, **186**
Fecampia, **209**
Feeding, of Actinozoa, 183, 187; of Araneida, 557; of Asteroidea, 682, **682**; of *Balanoglossus*, 711; of Brachiopoda, 662, **664**; of Branchiopoda, 369–71, **372**, **379**; of *Carcinus*, 414; of Cephalopoda, 642; of Chaetognatha, 666; of *Chaetopterus*, **278**, 280; of Chordata, 707; of Ciliata, 106; of *Ciona*, 724; of Copepoda, 386; 385; of Crinoidea (*Antedon*), 700; of Crustacea, 359; of *Cypris*, 382; of Diotocardia, 612; of Echinoidea, 690, 693; of Errant Polychaeta, 274; of Filter feeding Malacostraca, 400; of Gastrotricha, 261; of Holothuroidea, 696; of Holozoic Mastigophora, 65, 67, 69, 70; of *Hydatina*, 256; of Hydrocorallinae, 163; of Lamellibranchiata, 625; of *Lepas*, 390; of Mollusca, 585; of Nematoda, 245, 241; of Nemertea, 231; of *Oikopleura*, 732; of Ophiuroidea, 687; of Opisthobranchiata, 616; **615**; of *Physalia*, 165; of Polyzoa, 653; of Protozoa, 25, 26; of Pulmonata, 618; of *Sabella*, 280, **281**; of Sarcodina, 75; of Streptoneura, **611**; of Suctoria, 113; of *Temnocephala*, 210; of Trilobita, 339; of Tubicolous Polychaeta, 280; **278**, 281; of Turbellaria, 195; of Veliger, 589; of Zoantharia, 187

Feet, of *Histriobdella*, 277, **288**; of Onychophora, 330, **330**
Female gametes, 36; of Metazoa, *see* Eggs; of Porifera, 118; of Protozoa, 36
Femur, 433
Filaria, 249, **250**, 247
Filibranchiata, 631
Filograna, 274
Filopodia, 16
Finger and toe disease (*Plasmodiophora*), 89

INDEX 801

Fission, of Metazoa, *see* Budding, Strobilation; of Protozoa, *see* Fission of Protozoa
Fisson of Protozoa, 33; Binary*, 33, 43, 52, 59, 61, 75, 78, 80, 84, 87, 113; by budding, 33, 78, 87, 113; Longitudinal, 33, 52, 61; Multiple, 33, 43, 75, 78, 91; Oblique, 52, 61; Pseudotransverse, 33, 34; Radial, 33, 34; Repeated, 33, 52, 61, 63; Transverse, 33, 52. *See also* Plasmotomy
Fissurella (*Diodora*), 612; **601, 603**
Fixation disc, 680
Flabellum, 351
Flagella, 17, **18**
Flagellata (Mastigophora), 52
Flagellated chambers, 118
Flagellispores (Flagellulae), 43
Flagellulae, 43
Flagellum, of *Helix*, 609
Flame cells, 200, **201,** 255
Flatworms (Platyhelminthes), 191
Floscularia, 259
Flustra, 655, **659**
Foettingeria, 24; **22**
Follicle cells, 645
Follicles, Gonadial, 456
Food, 128. *See also* Feeding
Food groove, of Branchiopoda, 369; of *Chirocephalus,* 371; of Lamellibranchiata, 626
Foot, Molluscan, 580; of Amphineura, 589, 592, 594; of Cephalopoda, 639, 648; of Gasteropoda, 605, 614, **615,** 616, **617;** of Lamellibranchiata, 625, 628, 630, 631, 635; of Scaphopoda 621, 622
Foot of *Hydatina,* 256
Foragers (Bee workers), 516, 522
Foraminifera, 75
Forceps, 488
Forcipomya, 508
Forcipulate, 682
Fore gut, *see* Stomodaeum
Forficula auricularia, 488, **488**
Formica fusca, 521; *sanguinea,* 521
Formicoidea, 521
Fossil Invertebrates, 744; Arachnida, 543, 544, 548; Brachiopoda, 661; Cephalopoda, 650; Chaetognatha (*Amiskwia*), 668; Corals, 181; Echinodermata, 703-4; Foraminifera, **82**; Graptolita, 716; Insecta, 468; Lipostraca, 373; *Peripatus,* 337; Polyplacophora, 593; Radiolaria, **85;** Trilobita, 338
Frenulum, 435, 496, 500
Frilled organ, 221
Front of *Carcinus,* 411
Frontal appendage, 370; cilia, 626; **624;** horns, 392; organs, 357; surface, 655; suture, 510;
Functional nervous unit, 134
Fungia, 187; **186**
Fungus gardens, 521
Funiculus, 653

Funnel, of Cephalopoda, 639
Furca, caudal, *see* Caudal furca
Furcula, 475

Galathea, 419
Galea, 433
Galeodes, 567, **568**
Galleria, 500
Gametes, 36; of Ciliophora, 36; of Cnidosporidia, 101; of Foraminifera, 36, 78, 83; of Heliozoa, 38, 87, **88**; of Mastigophora, **35,** 36, 52; of Metazoa, *see* Eggs, Spermatozoa; of *Monocystis,* 37, 98, **98**; of Mycetozoa, 89; of *Opalina,* 73; of Protozoa, 36; of Radiolaria, 84, 85; of Telosporidia, 37, 92, 93, 94, 95, 97; of Volvocina, **35,** 36, 52, 62, 63, 64, 65
Gammarus, 405, 406, **407,** 408; 345, **434;** *neglectus,* **407**
Gamocysts, 28
Gamogony, 42, 43
Gamonts, 43
Ganglia, of Ophiuroidea, 686; of ventral nerve cord, 271, 322, 335, **356,** 406, 415, **416, 560**
Ganglia, System of, *see* Nervous system
Ganglion, Antennal, 322, 355, 430; Brachial, 642, 643; Cerebral (Supraoesophageal), 232, 271, 335, 423, 458, 543, 560, **619** 622, **630,** 642, **643,** 727, *see also* Brain; Gastric, 643; Inferior buccal, 643; Infundibular, 642, 643; of *Ciona,* 726; of Rhizocephala, 356, 394; Pedal, 578, **580, 619, 622, 630;** Pleural, 578, **580, 619, 622, 630;** Prostomial, *see* Cerebral; Suboesophageal, 271, **416,** 423, 458, 543, 560, **619,** 630, **643;** Superior buccal, 643; Supraoesophageal, *see* Cerebral; Trunk, of *Appendicularia* larva, 727; Visceral, 357, 458, 581
Gasteropoda, 596
Gasterostomum, 219; *fimbriatum,* **218**
Gasterozooids, of Doliolida, 739; **738;** of Siphonophora, 164, **164, 165**
Gastral layer, 116
Gastric cavity, 151; filaments, 171; glands, 257; mill, 359; shield, 627
Gastrodes, 190
Gastrophilus, 446; *equi,* 511
Gastropus, **259**
Gastrotricha, 260
Gastrovascular cavity, 172
Gastrula, 127; of Echinodermata, 677; of Mollusca, 586; of Polychaeta, 294
Gastrulation, 143; of Arthropoda, 328; of Crustacea, 328, 366; of Echinodermata, 677; of Insecta, 462; of Mollusca, 586, **586;** of Nematomorpha, 252; of Nemertea, 233; of *Obelia,* 155; of Polychaeta, 293
Gecarcinus, 419; 416
Gemmules, 121

Generative organs, General Morphology of, 129, 131, 140; of Amphineura, **593**; of Arachnida*, **538**, 541, 543, 548, 554; of Archiannelida, 314; of Arthropoda, 327; of *Balanoglossus*, 712; of Brachiopoda, 663; of Chaetognatha, 666; of Chilopoda, 423; of Cnidaria*, 152, 157, **161**, 164, 173, 178; of Crustacea, 364, 373, 379, 385, 391, 403, **407**; 416; of Ctenophora, 189; of Diplopoda, 426; of Echinodermata, 675; of Hirudinea, 310; of Insecta, 456; of Lamellibranchiata, 630; **591**; of Nematoda, 242; of Nemertea, 232; of Oligochaeta, 298, 303; of Onychophora, 333; of Opisthobranchiata, 616; of Platyhelminthes, 227; **226**, 202, 222, **196**, **203**, **216**, **218**; of Polychaeta, 299; 285, 271; of Pulmonata, 608, **609**; of *Rhabditis*, 242; of Rotifera, 257; of Scaphopoda, 621; of *Sepia*, 644; of Streptoneura*, 603, 612; of Tunicata, 725, **731**
Generative pore, *see* Generative organs
Genital aperture, opening, organs, systems, *see* Generative organs.
Genital atrium, of *Helix*, 609; of Platyhelminthes, 202; of *Stylaria*, 303
Genital bursae, 687; canal, 701; coelom, 641; **591**; cords, 702; ducts, *see* Gonoducts; operculum, 541; plate, 688; pleurae, 710; stolon, 675. *See also* Axial organ
Genital rachis, 675; of Crinoidea, 702; of Echinoidea, 693
Geodesmus, 208
Geometridae, 500
Geonemertes, 228, 235
Geotrupes, 443
Gephyrea, 269
Gerardia, 396
Germarium, **457**; 207, 257
Germs (gametes), *see* Gametes; of Cnidospores, 101
Geryonia, 162
Giardia, 70; *intestinalis*, 71
Gid, 224
'Gill' of *Salpa*, 739; **737**
Gill books, 537, **539**, 540, 542, 546
Gill chamber of Decapoda, 412
Gill clefts, of ascidian tadpole, 726; **725**; of *Balanoglossus*, 710; of *Cephalodiscus*, 714; of Chordata, 707; of Thalicea, 735; of Tunicata, 722
Gill filaments, lamellae, plates, 623–5
Gill pouches, slits, *see* Gill clefts
Gills, of Arthropoda, 325; of Asteroidea, 674; of Crustacea, **353**, 363, 412; *see also* Epipodites; of Echinodermata, 674; of Echinoidea, 674, 688, 690; of *Limulus*, 546; **539**; of Mollusca, *see* Ctenidia; of Polychaeta, 272, **284**; of Salpida, 739; **737**; Tracheal, *see* Tracheal gills
Girdle of Polyplacophora, 592

Gizzard, of Earthworms, 298; of Insecta, *see* Proventriculus
Glabella, 338
Glands, Aciniform, 563; Aggregate, 563; Ampuliform, 563; Antennal, *see* Antennal glands; Maxillary, *see* Maxillary glands; Mucous, 605; Oesophageal, 298; of Alimentary canal, *see* Alimentary canal, Digestive gland, Liver; Pedal, 605; Prostate, *see* Prostate glands; Pyriform, 563; shell of Platyhelminthes, 202, **216**, **226**; Spermiducal, *see* Prostate glands; Spinning, 554, 563; Tubuliform, 563
Globigerina, 84; *bulloides*, 15
Globigerinidae, 80
Glochidium, 630
Glomeris, 425
Glomerulus, 711
Glossae, 433
Glossina, 511; 505, **507**; 442, 457
Glossiphonia, 311; 307, **308**, **309**
Glycera, 285, **288**
Glycogen, 136
Glyptoscorpius, 545, 546
Gnathobase, in Arachnida, 546; in Crustacea, 351; in Trilobita, 339
Gnathobdellidae, 311
Gnathochilarium, 424, **425**
Gnathostoma, 236, **250**
Gonads, 140; of Arthropoda, 327; of *Balanoglossus*, 712; of Chaetopoda, **299**; 298; of Coelenterata, 155, 157, **158**, **161**, 173, 178; of Crustacea, 365; of Echinodermata, 675; of Nemertea, 228, 232. *See also* Generative organs
Gonapophyses, 457
Gonoducts, 131, **132**. *See also* Generative organs; Nephridia: Coelomoducts
Gonophore, 159; **158**
Gonopods, 421
Gonothecae, 155, **161**
Gonozoids, of Doliolida, 739; of Siphonophora, 164, **165**
Gordius robustus, 252, **253**
Gorgonacea, 179
Gorgonocephalus, **687**
Gorgonorhynchus, 231
Götte's larva, 206
Grantia, 122; 120
Graptolita, 716
Greeffiella, **238**
Gregarina longa, 99, **99**
Grub, 467
Gryllotalpa, 488
Gryllus, 488
Guard, 646, **645**
Gullet, of Metazoa (Oesophagus), *see* Alimentary canal; of Phytomastigina, 55; of Protozoa, 25
Gunda segmentata (*Procerodes*), 208
Gymnoblastea, 154. *See also* Anthomedusae

INDEX

Gymnocerata, 534
Gymnolaemata, 652
Gymnomera, 380
Gymnomyxa, 51
Gymnospores, 43
Gyrinus, 490
Gyroceras, 650
Gyrocotyle, 221; **220**
Gyrodactylus, 215, **212**

Haemadipsa, 311
Haematochrome, 55, 62
Haematococcus, 62
Haemocoel, 129; of Arthropoda, **322**, 326; of Echinodermata, 675; of *Helix*, 607; of Insecta, 445; of Mollusca, 578; of *Peripatus*, 332, 333; of Rotifera, 255. See also Vascular system
Haemocyanin, 130, 364, 446, 542, 578
Haemoglobin, 130, 232, 272, 364, 446
Haemogregarina, 94
Haemonchus, 248, 250
Haemopis, 311
Haemoproteus, 95
Haemosporidia, 94
Halcampa, 183
Halecium, 149
Halesus guttatipennis, **495**
Halichondria, 124
Haliclystus, 170
Halicryptus, 263; **262**
Haliotis, 611; 601, **601**, 603, **603**
Halistemma, 164
Hamitermes silvestris, **484**
Hamula, 475
Hamuli, 435
Haplopoda, 381
Haplosporidia, 103
Haplosporidium, 103; *limnodrili*, **31**
Harmolita, 519
Harrimania, 708
Hatschek's pit, 708
Head, 141; of Arthropoda, 320; of Chaetopoda, 270, 271, 274; of Crustacea, 348, **354**; of Insecta, 430; of Mollusca, 580; of Myriapoda, 420, 424, **425**; of Onychophora, 330; of Trilobita, 339
Head cavity of Chordates, 707
Head foot, 605, 639
Head kidney, **295**
Heart, 129; of Arachnida*, 542, **543**, 544, 559; of Arthropoda, 326, **322**; of Balanoglossus, 711; of Brachiopoda, 663; of *Ciona*, 725; of Crustacea, 364; of Insecta, 445; of Mollusca, 607, 596, 641; of Onychophora, 332, **336**
Hearts, Branchial, of *Arenicola*, 283; of Oligochaeta, 302, **304**; of *Sepia*, 641; of Polychaeta, 272
Heat, 5
Hectocotylus, 645
Heliolites, 181

Heliopora, 181, **181**
Heliosphaera, 85; *inermis*, **85**
Heliozoa, 87
Helix, 604–10; *aspersa*, 604; *pomatia*, **605**, **607**, **609**
Helkesimastix, 65
Hemiaspis, 537, **545**, 549
Hemichorda, 708
Hemimetabola, 479
Hemimyaria, 738
Hemimysis, **401**
Hemiptera (Rhynchota), 527
Hepatic diverticula, *see* Mesenteric caeca
Hepatopancreas, *see* Digestive gland
Hepialidae, 497
Hepialis humuli, 499
Heptagenia, Nymphal stages of, **480**
Hermaea, 616
Hermaphroditism, of Chaetognatha, 666; of Crustacea, 365; of Ctenophora, 189; of Gastrotricha, 260; of Hirudinea, 310; of *Icerya*, 455; of Mollusca, 608, 614, 618; of Nematoda, 243; of Oligochaeta, 298; of Platyhelmithes, 203, **196**, **216**, 219, **221**, 227; of Polyzoa, 265, **654**; of *Protodrilus*, 313; of Protozoa, 37; of Tunicata, 725, **731**. *See also* Mutual fertilization
Herpetomonas, 68, **68**
Herpyllobius, 387
Hesione, 285
Heterocotylea, 211
Heterodera, 251, **251**
Heterometabola, 465, 483. See also Exopterygota
Heteronemertini, 228, **229**
Heteronereis, 289, **291**
Heteroneura, 499
Heteropoda, 614; **613**
Heteroptera, 527, 534
Heterotricha (Spirotricha), 110
Hexactinellida, 123
Hexactinian, 183
Hexamita, 70, **12**
Hexapoda (Insecta), 427
Hind gut, *see* Proctodaeum
Hinge of Lamellibranchiata, 622, 631
Hippobosca, 511
Hippospongia, 125
Hippurites, **635**
Hirudinea, 305
Hirudo, 311; **307**, **309**, **310**
Histriobdella, **288**; 277
Hofstenia, 208; **192**
Holectypoidea, 693
Holgametes, 36, 37
Hologamy, 36
Holometabola, 465. *See also* Endopterygota
Holophytic nutrition (Photosynthesis), 25; of Phytomastigina, 54; of Protozoa, 25
Holophytic protozoa, *see* Holophytic nutrition

804 INDEX

Holothuria, 694, 697, 698
Holothuroidea, 694
Holothyrus, 568
Holotricha, 108
Holozoic nutrition, 25; of Phytomastigina, 54–5; of Zoomastigina, 55. *See also* Digestion, Feeding
Homarus, 417
Homoneura, 497
Homoptera, 527, 531
Hood of Tetrabranchiata, 648
Hoplocampa testudinea, **499**
Hoplocarida, 400
Hormiphora plumosa, 187, **188**
Hormones, 129, 134, 359, 395
Hormurus, 535
Houses, of Larvacea, 732; of Protozoa, 14; of Pterobranchiata (cases), **712, 715**
Hyalonema, 123
Hydatid cyst, 224
Hydatina, 256; *senta*, **257**
Hydra, 162; 160, 166, **167**, 168; *attenuata*, 147; *viridis*, 187
Hydrachna, **573**
Hydractinia, 162
Hydranths, 154, 155, 156, 160. *See also* Polyps of Hydrozoa
Hydratuba, 174–5
Hydrida, 145
Hydrocoele, *see* Water vascular system
Hydrocorallina, 162
Hydroids, 153
Hydrophyllium, 164
Hydropsyche, **495**
Hydroptila, **495**
Hydrorhiza, 154
Hydrospires, 704
Hydrotheca, 155
Hydrozoa, 153
Hydrurus, 57; **56**
Hylobius, 252
Hymenolepis nana, 226
Hymenoptera, 513
Hymenostomata, 108
Hyperparasitism, by *Cryptoniscus*, 405 **404**; by Hymenoptera, 520
Hyperpharyngeal band, 724
Hypobosca (*Hippobosca*), 511
Hypobranchial space, 412
Hypoderma lineatum, 511; **509**
Hypognathous, 472, 528
Hypopharynx, 433
Hypostoma, *see* Labrum
Hypostome, 569
Hypotricha, 110
Hypsibus, 536

Icerya purchasi, 455, 492
Ichthyophthirius, 108
Ideal malacostracan, 397
Ideal Mollusc, 580
Idiochromatin, 32

Idotea, 404
Ileum, *see* Small intestine
Imaginal disc, 467
Imperforate Foraminifera, 80
Incisor process, 397
Indirect wing muscles, 437, **439**
Infraciliature, 17
Infero-marginal ossicles, 682
Inhalant canals, 118
Inhibition, 134, 327
Ink sac, 640
Inostemma, 520
Insecta, 427
Instar, 465
Intercalary segment, 420, 424, 430
Interlamellar concrescences, 625; septum 632; spaces, 625
Internal environment, 4
Internal gills, 690
Internal longitudinal bars, 722
Internal madreporite, 673
Internal medium, 129. *See also* Blood
Internal sac of Ectoprocta, 658, **660**
Internal skeleton, of Alcyonaria, 178, 179, 180, 181; of Crustacea, 355; of Echinodermata*, 681, 685, 688, 699; of Metazoa, 138; of Porifera, 117; of Protozoa, 11, 14; of Triploblastica, 138
Interradial mesenteries, 170
Interradii of Echinoderms, 670
Intertentacular organ, 653
Intestine, *see* Alimentary canal
Intracellular body cavity, 237
Intracellular digestion, 128, 146, 195, 197, 257, 570, 608, 627
Invertebrata, 1, 3
Ips, 252
Irregular echinoids, 693
Ischiopodite, **413**
Isogamy, 36, 37. *See also* Gametes
Isopoda, 403
Isoptera, 483
Isospores, 84
Iulus, 424, 425, 426; *terrestris*, **424**
Ixodes, **570**, 571

Japyx, 476
Jaws, of Arthropoda, 317, *see also* Mouth parts; of Chaetognatha, 666; of Echinoidea, 690, **691**; of *Helix*, 608; **607**; of Onychophora, 330; 319; of *Sepia*, 642; **640**
Jelly, of Coelenterata, 146, 178; of Porifera, 117, 121
Jugal lobe, 435

Kakothrips robustus, 527
Keratosa, 125
Kerona, 110, 148
Kidney of molluscs, 582, **591, 593**, 594, **595**, 597, 603, 606, 612, 621, 628, 640. *See also* Renal openings

INDEX 805

Kineses, 45
Kinetodesma, 22
Kinetonucleus, 17. *See also* Parabasal body
Kinetosome, 22
Kinety, 22
Koenenia, 567, **568**

Labial hooks, 481
Labial palps, of Insecta, *see* Mouth parts of Insects; of Lamellibranchiata, 625; of Protobranchiata, 630
Labidoplax (Synapta), 697, **698**
Labidura, 489
Labium (Second maxillae of Insecta), *see* Maxillae
Labrum, of Crustacea, 355; of Insecta, **431**; 430; of Trilobita, 339. *See also* Mouthparts
Lacinia, 433; **432**
Lacinia mobilis, 402
Lacunar system, 675; of *Antedon*, 702; of Echinoidea, 657, 693; of Holothuroidea, 675, 696
Lacunar tissue, 675
Lambis, **620**
Lamblia (Giardia), 70, **71**
Lamellibranchiata, 622
Lampyris, 448
Languets, 724, **722**; **721**
Lanice, **287**
Lankesterella, 95
Lantern coelom, 690
Large intestine of Insecta, 442
Larvacea, 732
Larvae, 141, **142**, 143; *Actinotrocha*, 668, **668**; **142**; *Actinula*, 157, **158**; Amphiblastula, 121; *Appendicularia*, 726, **725**; **142**; Argulid, 389; *Auricularia*, 679; **676**; *Bipinnaria*, 677; **676**; *Brachiolaria*, 679; Brachiopoda, 665, **665**; **142**; Crinoid, 679; **676**; 'Cyclops', 386; *Cyphonautes*, 658, **660**; 'Cypris', 392, **392**; *Dipleurula*, 673, **674**, 677; Echinoderm, 677; **676**; *Echinopluteus*, 678; **676**; *Ephyra*, 174, **174**; *Erichthus*, 367; Euphausid, 408; *Glochidium*, 630; *Gordius*, 252, **253**; Götte's, 206; Insect, 467, *see also* names of Orders; *Megalopa*, 367, 416; *Metanauplius*, 366; Müller's, 206, **206**; **142**; *Mysis*, 399, *see also* Schizopod; *Nauplius*, 366, 367; **142**; *see also Nauplius* larva; Nematode, 245; *Ophiopluteus*, 678; **676**; Pentastomida, 577; **576**; *Phyllosoma*, 367, 417; *Pilidium*, **142**, 233, **233**; *Planula*, 152, **152**, 190, *see also Planula* larva; *Pluteus*, 678; **676**; Porifera, 121, **121**; *Protaspis*, 339, **339**; Rhabditoid, 247; Schizopod, 399; 367; Stomatopod, 367; *Tornaria*, 713, **711**; **142**; 'Trilobite' of *Limulus*, 548; Trochosphere, 294, **295**; **142**; *Veliger*, 589, **587**; **142**; *Zoaea*, 366, 367, *see also Zoaea* larva

Larval arms of Echinodermata, 677, 678
Larval nephridia, 295
Lasso cells, 187
Lasso of nematocyst, **147**
Lateral Cilia, 626; **624**
Lateral-frontal cilia, 626; **624**
Lateral lines of Nematoda, 235
Laura, 396; *gerardiae*, **395**
Laurer's canal, 227; 210
Leander, 416
Leda, **634**
Legs, of Arachnida*, 536, 544, 546, 551, 553; of Arthropoda, 317, 318, 320; of Insecta, 433, **435**; of Malacostraca, 350, 350, 397, **352**, **353**; of Myriapoda, 421, 424; of Onychophora, 330; *see also* Feet
Leishmania, 68, **68**
Lemnisci, 253, **254**
Leodice, 289, **290**; *fucata*, 289; *viridis*, 289
Lepadocrinus, 704
Lepas, 389–93, **392**; *anatifera*, **390**, **391**
Lephthyphantes, **550**
Lepidocaris, 373
Lepidonotus, 276
Lepidoptera, 495
Lepidurus, 373; *glacialis*, **376**
Lepisma, 476; *saccharina*, 476, **478**
Leptocoris trivittatus, **528**
Leptodora, 381; *kindti*, **381**
Leptomedusae, 144, 155, **154**; *see also* Calyptoblastea
Leptomonas (Herpetomonas), 68, **68**
Leptoplana, 208
Leptostraca, 399
Leptospironympha, 21, **21**
Lernaea, 387, **387**
Lernanthropus, 364
Leucandra, 122; 118, **120**
Leucifer, 416; **409**
Leucochrysis, 58
Leucon grade, 118, **119**
Leucosolenia, 122
Levuana iridescens, 511
Libellula, 455
Lieberkuhnia, 80; *wagneri*, **79**
Life cycle, of Actinomyxidea, 103, 101; of Aphididae, 533; of Cecidomyidae, 456; of Cestoda, 222; of Cladocera, 379, **380**; of Cnidosporidea, 101, **102**; of Coccidea, 92; of Coccidiomorpha, 92; of Coelenterata, 152, 175; of Doliolida, 739; **738**; of Foraminifera, 83; of Gregarinidea, 96, 97; of Haemosporidia, 95; **94**; of Hydrozoa, 152, 154–66; of Malacocotylea, 215; of Mycetozoa, 89; of Neosporidia, 101; of Piroplasmidea, 100; of Polythalamia, 80, 83; of Radiolaria, 84; of Rotifera, 260; of Scyphomedusae, 173, 175; of Sporozoa, **44**, 91–100; of Telosporidia, 93; **94**; of Trypanosomidae, 68–9; of Tunicata, 726

51

Life history, of Alcyonaria, 178; of Ascidians, 726; of Brachiopoda, 665; of Chaetognatha, 667; of Copepoda, 385, 386–7; of Crustacea, 366; of Echinodermata, 677–80; of Heterocotylea, 213; of Insecta, 465, *see also* names of Orders; of Leptostraca, 399; of Malacostraca, 399; of Mollusca, 589, 610; of Nematoda, 248–9, **250**; of Nemertea, 233; of *Pelagia*, 175; of Polychaeta, 289; of Polyzoa, Ectoprocta, 660, Endoprocta 265; of Porifera, 116, 121; of Protozoa, 42, **44**; of Siphonophora, 163; of Trilobita, 339. *See also* Embryology, Larvae, Life cycle
Ligament of Acanthocephala, 254
Light, 5
Ligia, 403; *oceanica*, **404**
Ligula, 433
Ligula, **225**
Limacina, **617**, 616
Limax amoebae, 75
Limbs, of Arachnida*, 535, 536, **537, 538**, 541, 544, 546, 551–4; of Arthropoda, 318, 319, 320; of Crustacea, 343–7, 350, **350**; of Onychophora, 330; of Trilobita, 339. *See also* Abdominal limbs, Antennae, Antennules, Legs, Mandibles, Maxillae, Maxillules, Mouth parts, Pleopods, Thoracic limbs, Trunk limbs, Uropods
Limnaea, 618, **620**; *abyssalis*, 618; *peregra*, **619**
Limnocnida, 162
Limnocodium (*Craspedacusta*), 162
Limnophilus, 494
Linguatula taenioides, 577; **576**
Lingula, 660–6, **664, 665**
Lipkea, 170
Lipostraca, 373
Lithobius, 420, 421, **421**, 422, 423; *forficatus*, **422**
Lithocampe tschernyschevi, **85**
Lithocircus, 87; *annularis*, **50**
Lithodes, 419; *maia*, **418**
Littorina, 596, 613; *rudis*, 613
Lituites, 650
Liver, 128; of Arachnida, 559; of Crustacea, 359; of *Helix*, 608; of *Sepia*, 642. *See also* Digestive gland, Mesenteric caeca
Living chamber of *Nautilus*, 647
Lizzia, 159; *koellikeri*, **159**
Loa loa, 249, 250, 510
Lobophyllium, 187
Lobopodia, 16
Locomotion of Protozoa, 16
Locusta, migratoria, 487
Loimia, **278**
Loligo, 647; **646**
Longitudinal band, *see* Ciliated band
Longitudinal fission in Protozoa, 33
Lophohelia, **184**, 185

Lophophore, of Brachiopoda, 662; of Ectoprocta, 652; of Endoprocta, 263; of *Phoronis*, 668
Loxodes, 108
Loxosoma, 265
Lucernaria, 170; **169**
Luciae (Pyrosomatidae), 737
Lucifer (*Leucifer*), 416; 409
Lucilia, 443
Lucina, **584**
Lumbricidae, 298
Lumbriculus, 305; **301**, 303); *variegatus*, **305**
Lumbricus, 297, 298, 300; **268, 270, 303, 304, 307**; *terrestris*, 299
Lung, 606, 618
Lung books, 538, **539**, 540, 542, **559**
Lycosa, 563

Macrobiotus, 576; **575**
Macrocheles, 568
Macrocorixa, 446
Macrodasys, 261
Macrogametes, 36. *See also* Female gametes
Macromeres, 291
Macrotrista angularis, **532**
Macrurous type, 410
Madreporic vesicle, 672, 674
Madreporite, 672, 673, **679**, 687, 693
Magellania flavescens, 661, **662, 663**
Maia, 419
Malacobdella, 234, **234**
Malacocotylea, 215
Malacostraca, 396
Malaria, 95, 96
Male eggs of Rotifera, 258, 260
Male gametes, 36; of Metazoa, *see* Spermatozoa; of Porifera, 118; of Protozoa, 36, **36**, 64, 65, 95
Malletia, **590**
Mallophaga, 525
Malpighian capsules, 138
Malpighian tubules, 444; of Arachnida, 543, 560; of Insecta, 444, **444**; 445; **440** of Myriapoda, 422, 425
Mandibles, 318; of Crustacea, 318, 334–5, 350, 354, **413**, 414; of Insecta, 318, **432, 433**; of Myriapoda, 318, 420, 422, 424. *See also* Mouth parts
Mandibular groove, 348
Mandibular palps, 354, 382, 385, 390, 397, 403, 406, 414, 415
Mantis, 483; **477**
Mantle, of Brachiopoda, 662, 663, 664; of Cirripedia, 389, **395**, 396; of Mollusca*, 578, 580, **581, 583**, 589, 592, 612, 616, 628, 639, 648; of Tunicata, 721
Mantle cavity, or groove, of Brachiopoda, 662; of Cirripedia, 389; of Mollusca*, 578, 596, 598–600, 606, 613, 614, **615**, 618, 622, 625, 636, 639
Mantle flap or fold, *see* Mantle
Manubrium, 151, **151**, 155, 157, **158**

Margellium, 159, **160**
Marginal anchors, 170
Maricola, 208
Mass provisioning, 516
Mastax, 256; **255, 257**
Mastigamoeba, 65; *aspersa*, 67
Mastigobranchiae, 412
Mastigophora, 52
Maxillae (Both pairs of), of Crustacea, 344–5, 347, **352, 353**, 354; of Insecta, 433; **432, 434**; of Myriapoda, 420, 421, **422**. *See also* Mouth parts
Maxillae, First of Crustacea, *see* Maxillules, Mouth parts
Maxillae, Second of Crustacea, 344–5, 347, 354. *See also* Mouth parts
Maxillary glands, 361; **360**
Maxillipeds, 318; of Crustacea*, 344–5, 347, 354, **384**, 385, **413**, 414; of *Lithobius*
Maxillules*, 344–5, 347, 354, 370, **384, 413**, 414. *See also* Mouth parts
Meandrina, **186**
Mecoptera, 494
Medium, 4. *See also* Internal medium
Medusa, The, 151
Medusae, 151; of Hydrozoa, 155–63; of Scyphozoa, 170–6
Megachile, 522
Megachromosomes, 32
Megalopa larva, **367**, 416
Megaloptera, 492
Megalospheric form, 80, **82, 83**
Meganephridium (Holonephridium), 301
Meganucleus, 32, 37, 103
Megascolecidae, 298
Melanin(s), 640; 94, 357
Melicerta, 259, **259**
Meloë, 489
Meloidae, 491
Melolontha, 491; **466**
Melophagus, 511
Membranellae, 17, 106
Membranipora, 655, **659**
Menopon pallidum, 526, **526**
Menthus, **544**
Mentum, 433
Mermis nigrescens, 251
Merodon, 510
Merogametes, 36
Meropodite, **413**
Merozoites (Schizozoites), **42**, 43
Mesenchyme, 127, 129; of Acoelomata, 191; of Ctenophora, 189; of Echinodermata, 677; **673**; of Hirudinea, 308; of Nemertea, 233; of Platyhelminthes, 195; of trochosphere, 295
Mesenteric caeca, of *Aphrodite*, 276; **275, 277**; of Arachnids, 537, 559; of Crustacea*, 359, 371, 379, 391, 400, 403, 406, 415; of Echinodermata*, 672, 682, 690, 701; of Hirudinea, **308, 309**; of Insecta, 440. *See also* Digestive gland, Liver
Mesenteric filaments, 183, **183**
Mesenteries, of Actinozoa, **182**; 176, **177**, 181; of Holothuroidea, 696; of Polychaeta, 296; of Scyphozoa, 170; **169**
Mesenteron (Mid gut), 128; of Crustacea, 359; of Hirudinea, 306; of Insecta, 440; of Nematoda, 240
Mesoblast, 294. *See also* Mesoderm
Mesoblastic somites, *see* Mesoderm segments
Mesocerebrum, *see* Deutocerebrum
Mesoderm, 127, 129; in the trochosphere, 295; 127; of Arachnida, 540; **538**; of Arthropoda, 328; of Chordata, **707**; of Insecta, **463**; 462. *See also* Mesenchyme, Mesoderm segments, Mesothelium
Mesoderm segments (Mesoblastic somites), of Annelids, 296; of Arachnida, 540; **538**; of Arthropoda, 328; of Chaetopoda, 270, 300, **300**; of Chordata, 708; **707**; of Onychophora, 332, **335**
Mesogloea, 144. *See also* Structureless lamella
Mesosoma, 320, 535, 537, 541, 544
Mesostigmata, 568
Mesostoma, 204; *ehrenbergi*, 207; *quadrangulare*, 207
Mesothelium, 127, 128. *See also* Mesoderm segments
Metabola (Pterygota), 465
Metaboly, 17, 58
Metacerebrum, *see* Tritocerebrum
Metacestode stage, 223, 224
Metachronal rhythm, 17, **19**
Metameric segmentation, 295; 140
Metamorphosis, *see* Life history
Metanauplius larva, 366
Metapneustic, 506
Metasicula, 718; **717**
Metasoma, 320, 535, 541, 544
Metasome, 384
Metasternite, 541
Metastoma, of Crustacea, 355; of Eurypterida, **545**; of Trilobita, 339
Metatroch, 295
Metazoa, 126
Metazoaea larva, 399
Metepipodites, *see* Branchia
Metridium, 186
Miastor, 456, 509
Microbisium, 535
Microchromosomes, 32
Microdina, **259**
Microfilaria diurna, 249; *nocturna*, 249
Microgametes, 36, *see* Male gametes
Microhydra, 162
Micromeres, 291
Micronephridia, 301
Micronuclei, 108; 32, **36**, 39, **42**
Micropterygidae, 496, 497

Micropteryx, 496, 497
Microspheric form, 80, **82, 83**
Microsporidia, 102
Microstomum lineare, 206, 207, 221
Mid gut, *see* Mesenteron
Mid-gut caeca, *see* Mesenteric caeca
Milk glands of tsetse fly, 456
Millepora, 163, **163**
Mimosella, **659**
Mitoses of Protozoa, 30, **31**
Mitra, **620**
Molar process, 397
Mollusca, 578; Types of, 580
Molpadida, 697
Monas, 65, **67**; 18; *vulgaris*, **67**
Monaxonida, 124
Moniezia, 193
Monocyclical rotifers, 258
Monocystis, 99; **98**; *lumbrici*, 99; *magna*, 99
Monodiscus, 211
Monograptus, 719
Monomorium minimum, **520**
Mononchus, 240
Monoplacophora, 593
Monopylaea, 84
Monosiga, 70; *brevipes*, **69**
Monothalamia, 78
Monotocardia, 612
Monstrilla, 386
Montacuta, **590**
Montipora, **186**
Morchellium, 728
Mosaic vision, 324
Motile organs of Protozoa, 16
Moulting, 245, 320, 467
Mouth, Position and shape of, in Arthropoda, 320; in Ascidian tadpole, 726; **721**; in *Balanoglossus*, 710, **710**; in Brachiopoda, 663; in Chilopoda, 421; in Coelenterata*, 150, 151, **154**, 157, 163, 175, 177, 187; in Echinodermata*, 682, 690, 693; in *Helix*, 608; in Hirudinea, 306; in *Hydatina*, 256; in Insecta, 433; in *Lepas*, 390; in *Peripatus*, 332; in Platyhelminthes, 195; in Protozoa, 25; in Trilobita, 339; in Triploblastica, 128
Mouth, *see also* Alimentary canal
Mouth parts (limb-jaws and lips), of Arthropoda, 317; of Crustacea, 354; of Insecta*, **432**, 433, **434**, 495, **498, 499**, 502–7, **503, 513, 515, 530**; of Myriapoda, 421, **422**, 424, **425**; of Onychophora, 330, **330**; 319
Mucous glands of *Helix*, 605
Muggiaea, 164, **164**
Müller's larva, 206, **206**; **142**
Multicilia, 65, 22
Multiple fission of Protozoa, 33
Murex, 612, **620**
Musca, 511; 506; *domestica*, **508**

Muscle(s), Alary, **446**; Adductor, *see* Adductor muscles; Columella, 605; Retractor, *see* Retractor muscles. *See also* Musculature
Muscle fibres, of Arthropoda, 326; of Chaetognatha, 666; of Coelenterata, 146, 182, 189; of Nematoda, 235, 239; of *Pecten*, 633; **632**; of *Peripatus*, 326; of Platyhelminthes, 194, 195
Muscular gland organ, 203, **203**
Musculature, of Actinozoa, 182, **183**, 186; of Ascidian tadpole, 726; of Brachiopoda, 662; of Cephalopoda, 639, 644, 649; of Chaetognatha, 666; of Chaetopoda, **270**, **275**, 280, 282; of *Ciona*, 721; of Crinoidea, 703; of Ctenophora, 189; of Echinoidea, 690; of Gasteropoda, 605; of gill books and lungs, **539**, 540; of Hirudinea, 307; of Holothuroidea, 694, 696; of Medusae, 172; of Metazoa, 140; of Nematoda, 235, 239; of Nemertea, 228, **229**, 230; of Onychophora, 329; of Ophiuroidea, 686; **139**; of Polyzoa, **656, 657**; of Platyhelminthes, 195; **194**; of Rotifera, 256; of Thaliacea, 735; of trochosphere, 296; of wings, 437, **438, 439**
Mutations, 40
Mutilla, 521
Mutual fertilization, by Ciliophora, **36**, 38; by *Helix*, 609; by Oligochaeta, 299; by Platyhelminthes, 203, 215
Mya, **584**
Mycetozoa, 89
Mygale, 557
Myoblast, 195
Myonemes, 17
Myophrisks, 86; **85**
Myopsida, 636
Myrianida, 289
Myriapoda, 420
Mysidacea, 402
Mysis, 402; **401**; *relicta*, **401**
'*Mysis*' larva, 367, *see also* Schizopod larva
Mytilus, 631; **624**, 625, **626**, 628, **629**, **635**
Myxicola, **283**
Myxobolus, 102, **102**
Myxospongiae, 125
Myxosporidia, 102
Myzostomum, **288**; 277

Nacreous layer, 582; **581**, **583**
Naegleria, 75; *bistadialis*, **74**
Narcomedusae, 162
Nassellaria, 84
Nauplius larvae, 366, **367**, 399, **394**
Nausithoe, 170
Nautiloidea, 647
Nautilus, 647; **637**, **644**, **648**, 649, 650; *macromphalus*, 649
Nebalia, 399; *bipes*, **398**

INDEX

Neck of Cestoda, 222
Neck gland, *see* Dorsal organ; organ (Nuchal organ)
Nectocalyces, 164
Needham's sac, 645
Nematocysts, **16**, 147–9, **147**
Nematoda, 235
Nematomorpha, 252
Nematus ribesii, 519
Nemertea, 228
Neodasys, 261
Neoechinorhynchus, 254, **254**
Neomenia, 594
Neopilina, 594, **595**
Neosporidia, 101
Neoteny, 141
Neotermes, 485
Nepa, 534
Nephridia, Nephridial system, 285, 286, 287, 301
Nephrocytes, 445
Nephromixia, 287; **285**
Nephrops, 417
Nephrostome, 272
Nereis, 274; **267**, **273**, **291**, 293
Nerilla, 313; **312**
Nerve-cord, *see* Nervous system
Nerve fibre, 134
Nerve net, 134; of Acoelomate Triploblastica, 191; of *Balanoglossus*, 711; of Coelenterata, 149; of Echinodermata, 672, 675; of *Helix*, 605; of Platyhelminthes, 197; Origin of centres and nerves in, 134
Nerve rings, of Echinodermata, 676; of Medusae, 152
Nerves, 134. *See also* Nervous system
Nervous system, of Acanthocephala, 253; of Annelida, 270, 271, **296**, **303**; of Arachnida*, 543, 544, 548, 560, **561**, **562**; of Arthropoda, 322; **321**; of Brachiopoda, 663; of Chaetognatha, 666; of Chordata, 707; of *Ciona*, 726; of Coelenterata, 149; of Crustacea, **321**, 326, **327**, 355; of Echinodermata, 675; of Gastrotricha, 260; of Hemichorda, 710; of Insecta, 458; of *Lithobius*, 423; of Mollusca, 578, **607**, **614**, **619**, **630**; of Nematoda, 241; of Nematomorpha, 252; of Nemertea, 231, **234**; of Onychophora, 335; of Platyhelminthes, 197; of Polyzoa, 263, 654; of Rotifera, 256; **255**
Neuronemertes, 235; **234**
Neurones, 134
Neuropodium, 274; **273**, **275**, **278**, **284**
Neuropteroidea, 492
Neuroterus, 519
Nidamental glands, 640
Noctiluca, 62; **61**
Noctuidae, 500
Nodosaria, 83; *hispida*, **15**
Nodus of Odonata, 481

Nomada, 522
Non-cellular animals, 10
Notochord, of Ascidian tadpole, 726; **725** of Hemichorda, 711; **710**, **713**
Notodelphys, 386
Notomastus, **285**
Notonecta, 534
Notopodium, 274; **273**, **275**, **279**, **284**
Notops, **259**
Notoptera, 486
Notostraca, 373
Novius cardinalis, 492
Nuchal sense organ (Neck organ), 350
Nucleariae, 80
Nuclei, in Metazoa and Protozoa, 10; of pansporoblasts, 101; of Protozoa, 29; Plurality of, in Protozoa, 13; Position of, in choanocytes, 122
Nucleic acid, 29
Nucleoli, 30
'Nucleus', of Thaliacea, 739
Nucula, 630; **580**, 623, **635**
Nuda, 187
Nudibranchiata, 615
Nummulites, 84; *laevigatus*, 82
Nuptial chamber, 485
Nurses, 588, 739
Nutrition, 25, 128; of Mastigophora, 52; of Phytomastigina, 54; of Protozoa, 25; of symbionts, 50. *See also* Holophytic nutrition, Holozoic nutrition, Saprophytic nutrition
Nycteribia, 511
Nyctiphanes, 408; *norwegica*, 398
Nyctotherus, 110; *cordiformis*, 111
Nymphon, 574, **575**
Nymphs, 465

Obelia, 154–7, **154**
Oblique fission, 33
Obtect pupae, 467
Ochromonas, 57, **56**
Octobothrium, **212**; 211
Octomitus (Hexamita), 70; **12**
Octopoda, 637
Octopus, 646, 647, **646**
Ocular plate, 688
Ocypus olens, 490, **491**
Odonata, 481
Odontoblasts, 608; **607**
Odontocerum, **495**
Odontosyllis, 289
Odynerus, 521
Oegopsida, 636
Oenocytes, 445
Oesophageal bulbs, 240
Oesophageal pouches, 298
Oesophagus, *see* Alimentary canal
Oesophagostomum, 242
Oikomonas, 67; *termo*, **67**
Oikopleura, 732; **731**; *albicans*, **732**
Olfactory hairs, 357, 358

Oligochaeta, 296
Oligolophus spinosus, 574, **574**
Oligotricha, 110
Oliva, **583**, 620
Olynthus, 116, **117**
Ommatidium, 323-4, **325**
Onchocerca, 236
Onchopalpida, 568
Onchosphere, 224
Onychophora, 329
Onychopoda, 381
Oocysts, 28; of Gregarinidae*, 97, 98
Oodinium, 62; *poucheti*, **50**
Oogamy, 36
Ookinete, 95; **94**
Oostegites, 401, **407**
Ootheca, 458
Ootype, 202
Oozooid, 737
Opalina, 73; 22; *ranarum*, **73**
Opening (Aperture), Atrial, see Atrial opening; Excretory, see Excretory organs, Renal openings; Genital, see Generative organs; of Mantle cavity, see Mantle. See also Anus, Mouth, Oscula, Ostia, Pneumostome, Pores
Operculum, 280
Ophiocomina, 687, **687**
Ophiopluteus, 678
Ophiothrix, 687
Ophiura, **685**, 687
Ophiuroidea, 685
Ophryocystis, 97, **97**
Opisthobranchiata, 614
Opisthosoma, 535. See also Mesosome, Metasoma,
Opisthoteuthis, 647
Optic lobes of Crustacea, 356
Oral aspect (side, or surface), 140; of Echinodermata, 670, **671**
Oral cone, 151; disc, 177; siphon, 721; valves, 700
Organisms, 5
Oria, **283**
Ornithodorus moubata, 571
Orthoceras, 650
Orthognathous, 472
Orthoptera, 486
Orthorrhapha, 507
Oscarella, 125
Oscinus, 511
Oscula, of Porifera, 116, **119**, **120**; of Radiolaria (pores), 84, **85**
Osphradia, 612
Ossicles, **139**; of Echinodermata, 672; Adambulacral, 682; Ambulacral, 682, **685**, **692**; Basal, 700; Brachial, 700; Centrodorsal, 699, 700; Cirrhal, 701; Infero-marginal, 682; of Holothuroidea, 697; Pinnulary, 700; Radial, 700; Rosette, 700; Supero-marginal, 682. See also Auriculae, Skeletal plates

Ossicles, System of, in Asteroidea, 681; in Crinoidea, 699; in Echinoidea (Plates), 688; in Holothuroidea (Calcareous ring), 696; in Ophiuroidea, 685
Ostia, of heart, 326, 364, **415**, 445, 542, **543**; of sponges, 118
Ostracoda, 382
Ostrea, 633, **634**, **635**; *edulis*, **632**
Otocelis, 192
Otocyst, **358**, see Statocysts,
Otoplana, 208
Ova, 7. See also Eggs
Ovarian lamella, 392
Ovarioles, 457, **457**
Ovary, see Generative organs
Ovicell, 657, **657**
Oviducts, see Generative organs
Ovigerous frenum, 392; legs, 574
Ovipositor, of Insecta*, 438, 486, 514, 520; of Phalangida, 573
Ovotestis of *Helix*, 608, **609**
Oxidation, 136
Oxypleurites, 569
Oxyuris (*Enterobius vermicularis*), 248; 236, 239

Pachytylus migratorius, 487
Paedogamy, 87
Paedogenesis, 456
Paired limbs, see Limbs
Palaeonemertini, 228
Palaeophonus, 543
Palinura, 410
Palinurus, 417
Pallial arteries and circulation of Lamellibranchiata, 628, **629**
Pallial budding, 731; gills, 612; sinus, 663
Palliovisceral cords, 589
Palmella, 55
Palmipes, **687**
Palolo worm (*Leodice viridis*), 289, **290**
Palpigrada, 535
Palps, Labial, see Labial palps; Mandibular, see Mandibular palps; of Acarina, see Pedipalps; of Polychaeta*, 276, 277
Paludicola, 298
Paludina, 613; 143; *vivipara*, **598**
Pamphilus, **520**
Pancreatic tissue, **640**
Pandorina, 62, **63**
Panorpa communis, 494
Pansporoblasts, 101
Panthalis, 275
Pantopoda, 574
Papilio, 500
Papilionoidea, 500
Parabasal body, 17, 22
Paracordodes, **253**; 252
Paractinopoda (Synaptida), 697
Paragaster, 116
Paraglossae, 433
Paragnatha, 355, see also Mouth parts

INDEX

Paramecium, 108; 28, **36**, 38, **38**, 39, 40, 105; *aurelia*, 39; *caudatum*, 27
Paramitoses, 30, **31**
Paranebalia, 399
Paraoesophageal (Circumoesophageal connectives), *see* Nervous system
Parapodia, 274; of *Aplysia*, 616; **615**; of Polychaeta, 274–7 (*passim*),
Parasites, 746. *See also* Parasitic habits
Parasitic castration, 395; by Isopoda, 405; by Rhizocephala, 395
Parasitic habits, of Acanthocephala, 254; of Acarina, 569; of Anoplura, 527; of Aphaniptera (Siphonaptera), 512; of Branchiobdellidae, 311; of Ciliata*, 108, **109**, 110, **111**, 112, 113; of Cirripedia, 393–6; of Copepoda, 386; of Crustacea, 355; of *Cyamus*, 408; of Dinoflagellata, 62; of Diptera, 511; of *Entamoeba*, 75; of Hemiptera, 534; of Hirudinea, 311; of *Histriobdella*, 277; of Hymenoptera, 518 (*passim*); of Isopoda, 405; of Mallophaga, 526; of Nematoda, 245–51 (*passim*); of Nematomorpha, 252; of Pentastomida, 577; of Platyhelminthes, 208, 210, 211, 219; of Polymastigina, 70; of Protozoa, 49; of Sporozoa, 91; of Trypanosomidae, 68–9
Parazoa (Porifera), 116
Parenchyma, Parenchymatous tissue, 195
Parthenogenesis, 7; of Crustacea*, 365, 375; 379, 382; of Gastrotricha, 260; of Insecta, 456, 521, 527, 533; of Protozoa, 39; of Rotifera, 258, 260
Parthenogonidia, 65
Patella, 612; **603**, **611**; *coerulea*, **587**
Peachia, 183; **182**
Pecten, 631, **634**, **635**, **644**; *maximus*, 631, **632**; *opercularis*, 631; *tenuicostatus*, 632
Pectines, **537**, **538**, 541, **542**
Pectinbranchiata (Monotocardia), 612
Pedal cords, 589; sinus, 628
Pedalion, **259**
Pedicellariae, 672, **683**, **692**; Crossed, 682; Gemmiform, 690, **692**; Ophiocephalous, 690, **692**; of Asteroidea, 682; of Echinoidea, 690; Tridactyle, 690, **692**; Trifoliate, 690, **692**; Uncrossed, of Asteroidea, 682
Pedicellina, 265; 264
Pediculus humanus, 527; **526**
Peduncle, *see* Stalk
Pelagia, 175
Pelagothuria, **697**, **698**
Pelagothurida, 697, **698**
Pellicle, 14
Pelmatozoa, 699
Pelomyxa, 75; *palustris*, **77**
Pen of *Loligo*, 647
Penaeidea, 410
Penaeus, **352**, **362**, 367, 416
Penilia, 375

Penis, *see* Generative organs
Pennaria, 162; **161**
Pennatula, 180
Pennatulacea, 180
Pentacrinus, 703
Pentastomida, 577
Pentatomidae, 534
Pentremites, 704
Peptonephridia, 302
Peracarida, 401
Peranema, 58; *trichophorum*, **59**
Perforate Foraminifera, 80
Peribranchial cavities, 724
Pericardial sinus, *see* Pericardium
Pericardium, of Arthropoda, 326; **322**; of Cephalopoda, 641; of *Ciona*, 725; of Crustacea, 364; of Enteropneusta, 711; of Insecta, 445; of Mollusca, 582, **591**, **593**; of Snail, 606
Perichaetine, 298
Perihaemal coelom of Enteropneusta, 710; system of Echinodermata, 672
Periostracum, of Brachiopoda, 662, **663**; of Mollusca, 582; **581**, **583**
Peripatus, 329–37; *capensis*, **330**, **331**, **332**, **334**, **335**, **336**
Peripharyngeal band, 722
Periplaneta, 483
Peripneustic larva, 506
Periproct, 688
Peripsocus phaepterus, 525
Peripylaea, 84
Perisarc, 155
Peristome, of Ciliata, 106; of Echinoidea, 688
Peristomial cirri, 274
Peristomium, 274
Peritoneum, 129, 133, 271
Peritricha, 112
Peritrophic membrane, of Crustacea, 361; of Insecta, 441; of Onychophora, 333
Perivisceral cavity, of *Acanthobdella*, 311; of Arthropoda, **322**; of Chaetopoda, 266; of *Ciona*, 724; of Echinodermata, 672; of Mollusca, 578; of Rotifera, 255. *See also* Coelom, Haemocoele
Perla maxima, 486
Pernicious malaria, 96
Perophora, 734; **729**, 730, **733**
Perradial, 170
Petrobius maritimus, 476, **478**
Phaenoserphus, 520; *viator*, **517**
Phaeococcus, 58
Phaeodaria (Tripylaea), 84
Phaeodium, 87
Phagocata, 192, **209**
Phalangida, 573
Phalangium, 536
Pharynx, *see* Alimentary canal
Phascolosoma, 316
Phasmidia, 236
Pheretima, 302

Philodinidae, 258
Philonexis, 645
Phlebotomus, 68
Pholas, 623, **635**
Pholcus, **551**, **566**, 567
Phoronidea, 668
Phoronis, 668
Phorozooids, 739
Phosphorescent Protozoa, 27
Photogenic organs of Insecta, 447
Photosynthesis, *see* Holophytic nutrition
Phoxichilidium femoratum, 575
Phragmocone, 646; **645**
Phronima, 408, **409**
Phryganea, 495
Phylactolaemata, 652
Phyllobius urticae, **466**
Phyllobranchiae, 412
Phylloceras, 647; *heterophyllum*, 649
Phyllodistomum, 212
Phyllodoce, **285**, **286**
Phyllopoda, 370
Phyllopodium, 351
Phyllosoma larva, **367**, 417
Phyllotreta, 491
Phylloxera vastatrix, 533
Physalia, 164, **165**
Phytomastigina, 54
Pieridae, 496
Pieris, 500
Pigments, Blood, 130. *See also* Chlorocruorin, Haemocyanin, Haemoglobin; of Crustacea, 357; of Haemosporidia, 94; of Lepidoptera, 496; of Phytomastigina, 55. *See also* Chromatophores, Melanins
Pilema, 175, **175**
Pilidium larva, 233, **233**
Pinacocytes, 116, **118**
Pinnate tentacles, 176
Pinnulary ossicles, 700
Pinnules, 699
Pirata, **558**
Piroplasma, 100
Piroplasmidea, 100
Placenta, of Onychophora, **335**; 334; of Salpidae, 737; of Scorpionidea, 543
Plagiostomum lemani, 208
Planaria, 208, **198**, **200**, **209**; *alpina*, 208; *lactea*, 199; *lugubris*, **199**, 199
Planipennia, 493
Planocera reticulata, 206
Planorbis, 618
Planula larva, 152, **152**, 155, 157, 173, 185, 190
Plasmodia, 13
Plasmodiophora, 89
Plasmodium, 95; **94**; *falciparum*, 95; *malariae*, 96; *vivax*, 95
Plasmodroma, 51
Plasmogamy, 37
Plasmotomy, 33
Plastin, 29

Platyctenea, 190
Platyhedra gossypiella, **497**
Platyhelminthes, 191
Plecoptera, 485
Pleodorina, 65; *californica*, **13**; *illinoiensis*, 13
Pleopods, 397. *See also* Abdominal limbs of Crustacea
Plesiocaris vagicollis, **464**
Plesiocoris, 534
Pleurobrachia, 187; *pileus*, 187
Pleurobranchiae, 363
Pleuron, 347
Pleurotomaria, **590**, **620**
Plicate canals, 628
Plodia, 500
Ploesoma, **259**
Plumatella, 652; *fungosa*, **654**
Plumularia, 160, **161**
Pluteus larva, 678; **676**
Pneumatophore, 163
Pneumostome, of Arachnida, 540; of Pulmonata, 606, 618, **621**
Podia (Tube feet), 671
Podical plates, 438
Podobranchiae, 408
Podocoryne, 162
Podocyrtis schomburgki, **85**
Podophrya, 113
Podura aquatica, 476
Pole capsules, 101, **102**
Polian vesicles, 684, **695**, **685**
Polistes aurifer, **520**
Polycelis nigra, 208, **209**
Polychaeta, 269
Polycirrus, 272
Polycladida, 208
Polyclinum, 733
Polydisc strobilation, 175
Polyembryony, of Hymenoptera, 456; of Polyzoa (Ectoprocta), 660
Polyenergid nuclei, 30
Polygordius, 313; **296** 312,
Polykrikos, 61; 60, **16**
Polymastigina, 70
Polyp, 150, **151**
Polyphaga, 490
Polyphemus, 381
Polypide, 652
Polyplacophora, 589
Polyps, of Actinozoa, 176; of Hydrozoa, 150; of Scyphozoa, 174
Polystomella (*Elphidium*), 83, **83**, **82**
Polystomum, 212, 213, 214, 215; *integerrimum*, **213**
Polythalamia, 80
Polytoma, 62; *uvella*, **34**
Polytricha, 110
Polyzoa (Endoprocta, 263, Ectoprocta, 652)
Pomatoceros, **279**; 272, **275**, **276**, **283**
Pontobdella, **308**, 311
Porcellana, **417**, 419; **367**

Pore plate of Radiolaria, 84
Pore-rhombs, **702**, 704
Pores, Collar, *see* Collar pores; Dorsal, *see* Dorsal pores; of Porifera, 116; Proboscis, *see* Proboscis pore; Water (Hydropore), 704. *See also* Madreporite
Porifera, 116
Porites, 185
Porocytes, 117
Poromya, 623
Porthetria dispar, 497
Portuguese man-of-war (*Physalia*), 164
Posterior aorta, of Arachnida, 542; *Astacus*, **346**; of *Carcinus*, **415**; of *Helix*, 607; of Lamellibranchiata, 628; of *Sepia*, **641**
Posterior interradius of Echinoidea, 693
Posterolateral arms of *Plutei*, 679
Posterolateral edge of *Carcinus*, 411
Postsegmental region of Crustacea, 349
Preantennae, 318, 330
Preantennal somite, 318, 430
Prebranchial zone, 722
Precheliceral somite (segment), 536
Pregenital somite (segment) of Arachnida, 536
Prementum, 433
Preoral lobe of Echinodermata, 673, 677
Preoral region, of Annelida, 266. *See also* Prostomium; of Arthropoda, 318, 319; of Crustacea, 344–5, 347; of Echinodermata, *see* Preoral lobe; of Enteropneusta, *see* Proboscis. *See also* Presegmental region, Preoral somites
Preoral somites, of Arachnida, 536; 318; of Arthropoda, 318, 319; of Chilopoda, 318, 420; of Crustacea, 347; 318; of Onychophora, 330; 318
Presegmental region of Crustacea, 347
Primary body cavity, *see* Haemocoele
Primary embryo (Ectoprocta), 660
Prismatic layer, of Brachiopod shell, **663**; 662; of Molluscan shell, 582; **581**
Proboscis, of Acarina, 567, **571**; of *Bonellia* 314, **314**; of Branchiura, 389; of *Buccinum*, 612; of Enteropneusta, 709; of Hemiptera, 528; of Lepidoptera, 495; of Nemertea, 230; of Pantopoda, 574, **575**; of Polychaeta, 274; of Rhynchobdellidae, 306; of Suctorial Copepoda, 386
Proboscis cavity, of *Balanoglossus*, 709; of Chordata, 707
Proboscis complex of *Balanoglossus*, 709–710; pore of *Balanoglossus*, 710
Proboscis sheath, of *Buccinum*, 612; of Nemertea, 230; of Rhynchobdellidae, 306
Probuds, 739
Procerebrum, 322
Procerodes lobata, 208
Proctodaeum, of Arthropods, 325; of Crustacea, 359; of trochosphere, 295

Proctotrypidae, 520
Prognathous, 472, 528
Progressive feeding, 516
Prolegs, 467
Proliferating region, of Syllidae, 289; of Tapeworms (neck), 222
Proostracum, **645**, 646
Propolis, 522
Prorodon, 108; *teres*, 109
Prorhynchus, 208
Prosicula, 718
Prosobranchiata (Streptoneura), 610
Prosoma ('Cephalothorax' of Arachnida), 535
Prosopyles, 118
Prostate glands of Oligochaeta, 298; of *Sepia*, 645
Prostomium, of Annelida, 266; of Echiuroidea, 314; of Hirudinea, 306; of Polychaeta, 274; of Sipunculoidea, 316
Protaspis larva, 339
Proteolepas, 393
Prothetelous, 523
Protobranchiata, 630
Protocerebrum, 322
Protochordata, 706
Protoclypeastroida, 693
Protococcaceae, 57
Protoconch, 647
Protodonata, 469
Protodrilus, 313; *chaetifer*, 313
Protoephemeroptera, 469
Protohymenoptera, 469
Protomerite, 97
Protomonadina, 65
Protoparce, 446
Protoplasm of Protozoa, 13
Protopodite, 339, 351
Prototroch, of *Cyphonautes*, 658; of *Pilidium*, 233; of trochosphere, 294
Protozoa, 10
Protozoa and Metazoa, Connexion between, 51
Protura, 476
Proventriculus of *Carcinus*, 415; of Crustacea, 359; of Earthworms, *see* Gizzard; of Insecta, 440; of *Gammarus*, 406; of *Ligia*, 403
Pseudococcus, 531
Pseudocolonies of Protozoa, 13
Pseudonavicellae, 98
Pseudophyllidea, 225
Pseudopodia, of Amoebina, 75; of Foraminifera, 78; of Heliozoa, 87; of *Hydra*, 146; of *Mastigamoeba*, 65, **67**; of Mycetozoa, 90; of Radiolaria, 84
Pseudopodiospores (Amoebulae), 43
Pseudoscorpionidea, 535, 544
Pseudotracheae, 502
Pseudotransverse fission, **34**
Pseudovelum, 172
Psocoptera, 525

Psychropotes, 669
Pterobranchiata, 714
Pterodrilus, **306**
Pteropoda, 616, **617**
Pterostichus, **466**, 517
Pterotrachea, 614; **613**
Pterygota, 479
Pterygotus, 546; *osiliensis*, **545**
Ptilinum, 506
Ptychodera, 708
Ptychomyia remota, 511
Pulex irritans, 513
Pulmonata, 618
Pulvillus, 433
Pupa, 465, 467, **455, 508**
'Pupa', of Holothuroidea, 679; of *Lernaea*, 387
Pupae, Coarctate, Exarate and Obtect, 467, 506
Puparium, 467, 506, **508**
Pure lines, 40
Pycnogonum littorale, 575
Pycnophyes, 262
Pygidium, 338
Pyloric caeca, of Asteroidea, 682; of Insecta, 441
Pyloric chamber, 359
Pyralidae, 500
Pyrenoids, 55
Pyriform organ, 658, 660
Pyrosoma, 737; **736**
Pyrosomatidae, 737
Pyrrhocoridae, 534

Quadrant, 291
Quartan ague, 96
Quartets of micromeres, 291

Rachiglossate tooth, **611**
Radial 'blood vessel', 675; nerve (aboral), 676; nerve (ectoneural), 676; ossicles, 700; perihaemal vessel, 675; water vessel, 672. *See also* Water vascular system
Radial cleavage, 1, **2**, 143, **292**
Radial fission, 33
Radial symmetry, 140, 670; of Actinozoa, 178; of Cnidaria, 151; of Echinodermata, 670, 680
Radii of Echinodermata, 670, **692**
Radiolaria, 84
Radula, 608; 585, 594, 596, **607, 611,** 612, 614, 616, 621
Raphidioptera, 493
Receptaculum seminis, *see* Spermatheca
Receptors, 133
Rectal caeca, 682
Rectum, *see* Alimentary canal
Reduction division (Meiosis), 32, **44**
Reduviidae, 534
Reflex arc, 134, **270**
'Reflexes', in Metazoa, 134; in Protozoa (Behaviour), 44

Regeneration, in Coelenterates, 166; in Crustacea, 355; in Turbellaria, 206
Regular sea urchins (Endocyclica), 693
Relation of Protozoa to their Environment, 46
Relicts, 402
Renal openings (apertures, papillae), of Mollusca, 582, **591, 593,** 596
Renopericardial openings (apertures, canals), 606, 621, 641
Repeated fission of Protozoa, 33
Reproduction, 6. *See also* Asexual reproduction, Sexual reproduction, of Protozoa, *see* Fission
Reproductive aperture, organs, *see* Generative organs
Reserve materials, 26
Respiration, 130, 135, **135,** 136; of Arthropoda, 325; of Crustacea, 363; of *Cyclops,* 385; of Lamellibranchiata, 628; of Protozoa, 28; of Tubiculous Polychaeta, 280. *See also* Respiratory movements, Respiratory organs
Respiratory movements, of *Aphrodite*, 276; of Arachnida, 540; of Branchiopoda, 369; of *Carcinus*, 412; of Crustacea, 363; of *Cyclops*, 385; of Gasteropoda, 612; of Insecta, 448, 453, 455; of *Mysis*, 402; of Pulmonata, 606, 618; of Tubiculous Polychaeta, 280; of *Tubifex*, 305
Respiratory organs, of Arachnida, 537; of Arthropoda, 325; of Branchiopoda, 369; of Chaetopoda, 272, 276, **278,** 284; of Crustacea, 363; of Echinodermata, 674; of Holothuroidea, 696; of Lamellibranchiata, 628; of *Ligia*, 403. *See also* Gills, Lung, Tracheae
Respiratory pigments, *see* Pigments, Blood
Respiratory trees, 696; **695**
Resting cysts, 28, 43; eggs, 258. *See also* Winter eggs; phase of Phytomastigina, 55
Rete mirabile, 297
Retinaculum, 435, 496
Retinulae, 323
Retractor muscle(s), of penis, 609; of proboscis, 230; of stomach, 682; of tentacles, 696
Retral processes, 83
Retropharyngeal band, 723
Rhabdammina, 83; *abyssorum*, **15**
Rhabdias, **242**
Rhabditata, 236
Rhabdites, 193
Rhabditis, 237; **239**
Rhabditoid larva, 247
Rhabdocoela, 207
Rhabdoliths, 58
Rhabdom, 323
Rhabdomeres, 323
Rhabdopleura, 714; **712, 713, 715, 716,** 717
Rhagodes, 567
Rhipidoglossate teeth, 611

INDEX 815

Rhizocephala, 393
Rhizochrysis, 57
Rhizocrinus, 703; **702**
Rhizomastigina, 65
Rhizoplasts, 17
Rhizopoda (Sarcodina), 74
Rhizopodia, 16
Rhizostoma, 175
Rhizostomeae, 170
Rhodites, 519
Rhodnius, **444**; prolixus, 534
Rhombifera, 704
Rhyacophila, 495
Rhynchobdellidae, 311
Rhynchocoel, 230
Rhynchodaeum, 231
Rhynchodemus terrestris, 208
Rhynchota, 527
Rhyssa, 519
Ring canal of Polyzoa (Ectoprocta), 654
Rings (Nervous, Water vascular, etc.), of Echinodermata, 672, 675. See also Nervous system, Water vascular system, etc.
Ripe proglottis, 222; **221**
Rods of eyes of Arthropoda, 323
Rosette ossicle, 700
Rostellum, 222
Rostellaria, **620**
Rostrum, of Cephalopoda, 647; of Crustacea, 350
Rotifer, 258
Rotifera, 254
Rotulae, 690
Royal pair, 484

Sabella, 280, **281**, **283**
Saccharicida, 534
Saccocirrus, 313, **313**
Sacculina, 393, **394**
Sagitta, 667, see also Chaetognatha; bipunctata, 667
Salicornaria, **659**
Salivary glands, of Arachnida, 557; of Hirudinea, 307; of Insecta, 440, 442; of Lithobius, 422; **421**; of Onychophora, 333; of Sepia, 642
Salpa, 739; **737**
Salpidae, 738
Saltatoria, 486–7
Salticus, 567; **566**
Sao hirsuta, **339**
Saprophytic nutrition, of Phytomastigina, 54; of Protozoa, 25
Saprophytic Protozoa, see Saprophytic nutrition
Saprozoic, see Saprophytic
Sarcocystis, **103**; lindemanni, **103**
Sarcodina, 74
Sarcophaga, 443
Sarcoptes, 572; **570**
Sarcoptiformes, 569
Sarcosporidia, 103

Sarsia, **164**, 166
Saxicava, 634
Scallops (Pecten), 631, **632**, **634**, **635**
Scalpellum, 392; vulgare, **391**
Scaphites, 651
Scaphopoda, 621
Scarabaeus thomsoni, **491**
Scarabeidae, 491
Scent scales, 496
Schellackia, 95
Schistocephalus gasterostei, 224
Schistosoma, 217, **217**
Schizocystis, 96
Schizogony, **42**, 43
Schizogregarinaria, 96
Schizopod larva, **367**, 399
Schizopoda, 399
Schizozoites, 97
Sclerotium, 89
Scolex, 222
Scolopendra, 421, 318
Scorpio, 543; swammerdami, **542**
Scorpionidea, 541
Scutariella, 211
Scutigera, 422, **423**
Scutum, 389
Scyphistoma, 174
Scyphomedusae, 169
Scyphozoa (Scyphomedusae), 169
Scytodes, **555**
Scytomonas (Copromonas), 59
Secondary body cavity, see Coelom
Secondary embryos, 660
Segmental organs, 285. See also Coelomoducts, Nephridia
Segmentation, 140; of Annelida, 266, 270, 295; of Arthropoda, 320, 326; of Cestoda, 222, 140; of Chordata, 708; of Vertebrata, 140, 708; suggested by certain organs in Mollusca, 582, 594, 649
Segmentation of the Ovum, see Cleavage
Segments of the body, see Somites
Seison, **259**
Self-fertilization, 243, 609
Seminal groove, of Oligochaeta, 299; of Opisthobranchiata, 616
Seminal receptacle of Sagitta, 666
Seminal vesicles (Vesiculae seminales), of Chaetognatha, 666; of Insecta, 457; of Oligochaeta, 298; of Platyhelminthes, 202
Sense organs, of Araneida, 560; **555**, 554; of Ascidian tadpole, 727; of Chaetopoda, 271; of Coelenterata, 155, **159**, 173; of Crustacea, 357; of Echinodermata, 677; of Hirudinea, 310; of Insecta, 458; of Mollusca*, 644, 592, 606; of Onychophora, 335; of Platyhelminthes, 198; of Protozoa, 25; of Rotifera, 256. See also Eyes
Sensillae, 460, **460**
Sepia officinalis, 637–45
Sepioidea, 636

Sepiola, 636
Septa, of Chaetopoda*, 282, 296, 303; of shell of Tetrabranchiata, 647; of Zoantharia, 184
Septibranchiata, 623
Sergestes, 367; *arcticus*, 353
Serpula intestinalis, 299
Sertularia, 161, **161**
Seuratia, **238**
Sexual congress, *see* Sexual differences and sexual behaviour, Mutual fertilization
Sexual differences and sexual behaviour, of Arachnida, 551, 562; of Archiannelida, 313; of *Balanoglossus*, 712; of *Bonellia*, 314; of Cephalopoda, 645; of Coelenterata, 152; of Crustacea, 364; of Echinodermata, 675; of Insecta, 456, 484, 521, 522; of Myriapoda, 423, 426; of Nematoda, 242; of Onychophora, 333; of Pantopoda, 575; of Polychaeta, 289; of Protozoa, 36; of Rotifera, 257. *See also* Generative organs
Sexual reproduction, of Metazoa, *see* Generative organs; of Protozoa, 36. *See also* Life history
Sexuparae, 533
Shape of body, *see* Body
Shell of Crustacea, *see* Carapace
Shell of Echinoidea, *see* Corona
Shell glands, of Crustacea, *see* Maxillary glands; of Platyhelminthes, 202
Shell ligament, 622
Shell types. Arenaceous, 80; imperforate, 80; Perforate, 80
Shells, of Brachiopoda, 662; of Crustacea, *see* Carapace; of Foraminifera, 78–80; of Mollusca, 583, **584**, **620**, **634**; of Protozoa, 14
Shield-shaped tentacles, 696
Sialis, 493, 254; *lutaria*, **493**
Siboglinum, 714, **715**
Sicula, 718
Sida, 375
Silicoflagellata, 58
Silicoflagellidae (Silicoflagellata), 58
Silver fish (*Lepisma saccharina*), 476, **478**
Simocephalus, 375; *sima*, 377
Simuliidae, 509
Simulium, 455
Sinus system of Hirudinea, 308
Sinuses, Haemal, of Arachnida, 542; of Arthropoda, 326; of Crustacea, 364; of *Helix*, 607; of Lamellibranchiata, 628; of *Lumbriculus*, 305; of *Pomatoceros*, 272, **276**
Siphon, of Echinoidea, 690; of Gasteropoda, 606; of suctorial Crustacea, *see* Proboscis
Siphonaptera, 511
Siphonoglyphes, 177, 181
Siphonophora, 163
Siphonozooids, 180

Siphons of Lamellibranchiata, 625
Siphuncle, 648
Sipunculoidea, 316
Sipunculus, 316
Sirex gigas, 518–19
Size of Protozoa, 11
Skeletal plates, of Echnoidea, 688, **689**; of Ophiuroidea, 685, **685**
Skeletogenous layer, 117
Skeleton, 138–9; External, Corals, Cuticle, Perisarc, Shell; Internal, *see* Internal skeleton
Skin, 133
Skull of Cephalopoda, 642
Slimonia, 545
Small intestine of Insecta, 442
Social life, of Hymenoptera, 521–2; of Isoptera, 484
Soldiers of Isoptera, **484**
Solenia, 178
Solenocyte, 286, **286**
Somatoblast, 294
Somite, First, of Arthropoda, 318
Somites (body segments), Series of, in Arthropoda, 318; in Crustacea, 344–5; in Polychaeta*, 274, 282, 295, 296. *See also* Tagmata
Somites, mesoblastic, *see* Mesoderm segments
Spatangoida, 693
Spatangus, 694
Species, estimated number of, 3
Sperm pouch of Chaetognatha, 666
Sperm sacs, *see* Seminal vesicles
Spermatheca (Receptaculum seminis), of *Helix*, 610; of Insecta, 458; of Nematoda, 243; of Platyhelminthes, 202
Spermatic atrium, 303
Spermatophores, of Crustacea, 366; of *Helix*, 609; of *Peripatus*, 333; of *Sepia*, 645
Spermatozoa, of Crustacea, 365; of Hirudinea, 310; of Nematoda, 243
Spermiducal glands of Oligochaeta, 298
Sphaeractinomyxon, 103
Sphaerella (*Haematococcus*), 62
Sphaerophrya, 114, **114**
Sphecoidea, 522
Spherularia, 252; **251**
Sphex, 522
Sphingidae, 522
Sphinx, 446
Spicules, of Alcyonaria, 178; of Porifera, 117; of Radiolaria, 84
Spines, of Echinodermata, 672; of Echnoidea, 690; of Ophiuroidea, 685; of Starfish, 682
Spinnerets, 554
Spiracles, of Arthropoda, *see* Stigmata; of Blastoidea, 704
Spiral cleavage, 291; 1, 143
Spirochona, 113; *gemmipara*, **109**

INDEX

Spirographis, 274
Spirostomum, 110; *ambiguum*, **109**
Spirotricha, 110
Spirula, 636, 646, **645**
Spirulirostra, 647; **645**
Spondylus, **584**
Sponges (Porifera), 116
Spongilla, 125; **121**
Spongillidae, 121
Spongin, 119
Sporangium, 89
Spore cases, 43, 97, 99, 101
Spores, 43, 62, 75, 84, 89, 93, 97, 98, 99, 101, 103
Sporoblasts, 91, 95
Sporocysts, 93, 98
Sporogony, 42
Sporont, 43
Sporozoa, 90
Sporozoites, 43, 92, 93, 95
Spumellaria (Peripylaea), 84
Squilla, 400; *mantis*, **398**
Stainers, 534
Stalk (Peduncle), of Brachiopoda, 664; of Cirripedia, 389; of Pelmatozoa, 703; 699; of Protozoa, 13; of Pterobranchiata, 714, **712, 713, 715**; Proboscis, 709
Stalked gland organ, 203
Staphylinidae, 491
Staphylocystis, 224
Statoblasts of Polyzoa (Ectoprocta), 657
Statocysts (Otocysts), of Calyptoblastea, **159**; of Crustacea, 357, **358**; of Turbellaria, 208
Statolith of Ascidian tadpole, 727, **725**
Stauromedusae, 170
Stegomyia, 508
Stenopodium, 351
Stentor, 110; *coeruleus*, 109
Stephanoceros, 259, **259**
Sterna, of Arachnida, 541, 545; of Crustacea, 347; of Myriapoda, 421, 424
Sternites, *see* Sterna
Stewart's organs, 690
Stigma of Odonata, 481
Stigmata (Spiracles), of Arachnida, 540; of Insecta, 449, **451**; of Myriapoda, 423, 424; of Onychophora (spiracle), 333; **332**
Stigmata of Tunicata, 722
Stimuli, Effect of, on Protozoa, 44
Stipe, 717
Stipes, 433
Stolon, of Alcyonaria, 179; of Hydratuba, 174; of Hydrozoa (Hydrorhiza), 154; of Tunicata, 730
Stomach, 128. *See also* Alimentary canal
Stomatopoda, 400
Stomodaeum (Fore gut), of Anthozoa, 176, 177, 181; of Arthropoda, 325; of *Ciona*, 722; of Crustacea, 359; of Ctenophora, 187; of Insecta, 439; of Nematoda, 240; of Onychophora, 333; of Rotifera, 257;
of Tricladida (pharynx), 195; of Trochosphere, 295
Stone canal, 673
Strepsiptera, 523
Streptoneura, 610
Strobilation, 174, 176; of *Aurelia*, 174; of Cestoda, 222
Strongylata, 236
Strongyloid larva, 247
Strongyloides stercoralis, 248
Structureless lamella (Mesogloea), 144, 176
Stylaria, 304; *proboscidea*, **304**
Stylets, of Hemiptera (Rhynchota), 529, **530, 531**; of Nemertea, 231, **232**
Stylochus, 206
Stylommatophora, 618
Stylonychia, 110; *mytilus*, **107**
Stylops, **523**
Subchela, 355
Subchelate limbs, 400, 406, 417, 536, 549
Subdermal cavities, 120
Subgenital pits, 171
Subimago, 479
Submentum, 433
Subneural gland of Tunicata, 722
Suboesophageal ganglion, *see* Ganglion, Suboesophageal
Substratum, 4
Subtentacular canals, 701
Subumbral pit, 170
Subumbrellar cavity, 155; ectoderm of Medusa, 155; ectoderm of trochosphere, 294; musculature, 155
Suctoria, 113
Sulcus, 60
Summer eggs of Cladocera, 379; of *Mesostomum*, 204
Superlinguae, 433
Supero-marginal ossicles, 682
Superposition image, 324
Supraoesophageal ganglia, *see* Ganglion, Cerebral
Surface, of Ciliata, 106; of Protozoa, 14. *See also* Ectoplasm
Suture line of ammonoid shell, 651; **649, 650**
Swarm spores, 43
Swarming of Polychaeta, 289
Sycon, 122; **121**; *raphanus*, **121**
Sycon grade, 118, **119**
Syllis, 289; **274**; *ramosa*, 289, **290**
Symbiosis, 50, 485
Symmetry, 140; of Actinozoa, 178; of Echinodermata, 669, 680; of Metazoa, 140; of Protozoa, 11. *See also* Bilateral symmetry, Radial symmetry
Sympathetic system, of Crustacea, 357; of Insecta, 458
Symphyta, 518
Symplasts, 13
Synagoga, 296, **395**
Synapse, 134
Synapta (*Labidoplax*), 697, **698**

Synaptida, 697
Syncarida, 400
Syncytia, 13. *See also* Plasmodia
Syngamy, 36; of Ciliophora, 36; of Dinoflagellata, 61; of Mastigophora, 36, 52; of Sarcodina, **83, 88**; of Sporozoa, 37, 92; of Volvocina, 36, **35, 63, 64**; of Zoomastigina, 65
Syracosphaera, 58; **56**
Syringopora, 181
Syrphus, 510
Syzygy, 92

Tabanus, 502
Tachardia lacca, 532
Tachinidae, 511
Tachypleus, 535
Tactile organs, *see* Sense organs
Taenia, coenurus, 224; *echinococcus* (*Echinococcus*), 225; *serrata*, 224; *solium*, **221**
Taeniothrips inconsequens, 527
Tagelus, 635
Tagmata, of Arachnida, 320; 318; of Crustacea, 320; 318, 347; of Insecta, 320; 318; of Myriapoda, 320; 318; of Onychophora, 320. *See also* Abdomen, Cephalothorax, Head, Mesosoma, Metasoma, Opisthosoma, Prosoma, Pygidium, Thorax, Trunk
Tail, of Chaetognatha, 666; of Chordata, 708; of Tunicata, 726
Tail fan, 397
Tanaidacea, 402
Tanais, 403
Tardigrada, 576
Tarsus, 433
Taxes, 46
Tealia, 187
Tectibranchiata, 615
Teeth, of Echinoidea, 690; of Ophiuroidea, 686. *See also* Radula
Tegenaria, **552**
Tegmen, 700
Tegmentum, 592
Tegmina, 486
Telosporidia, 91; Reduction division of, 44; 32
Telson, 318; of Arachnida, 537; of Crustacea, 344–5, 347; of *Lithobius*, 421; of Trilobita, 339
Temnocephala, 210, **210**
Temnocephalea, 210
Temperature, 5
Tentacle sheath, 655
Tentacles, of polyp and medusa, 151, 154, 155. *See also* Tentaculocysts
Tentacles, of Ctenophora, 187; of Gasteropoda, 606; of Holothuroidea, 696; of Hydrozoa, 154–7; of *Nautilus*, **648, 649**; of Polychaeta, 278, 279, 280, **281, 283**; of Polyzoa, 264; 654; of Scyphomedusae, 170; of Suctoria 113; of Turbellaria, 200

Tentaculata, 187
Tentaculocysts, 172, **173**
Terataspis, 338, **337**
Terebella, 274, 279, **283**
Terebratula, 661–3
Teredo, 633, **633, 635**
Terga (Tergites), 347
Terga of *Lepas*, 389
Tergo-sternal muscles, 449
Terminal arborization of axon, 134; organ, nephridial, **201**; tentacle of Echinodermata, 677
Terricolae, 297
Tertian ague, 95
Test of Tunicata, 721
Testacella, 618
Testes, *see* Generative organs
Testicardines, 660
Tetrablemma, 549
Tetrabranchiata, 647
Tetracotyle, **212**
Tetractinellida, 125
Tetragraptus, 719, **719**
Tetrahymena, **24**
Tetraphyllidea, 224
Tetrarhynchidae, 224
Textrix, 550
Thalassema, 316; *misakiensis*, 316; *neptuni*, 316; *taenioides*, 316
Thalassicolla, 85; *pelagica*, **48**
Thaliacea, 735
Theca, of corals, 184; of Pelmatozoa, 704
Theridium, **538**
Thompsonia, 395, **395**
Thoracic limbs, of Crustacea, 349, **352, 353**, 354; 344–5; of Eucarida and Pericarida, 402, 403, 405, 406, 408, 414. *See also* Legs, Maxillipeds
Thoracic membrane, 280
Thoracica, 389
Thorax, of Arthropoda, 320; of Crustacea, 349; of Insecta, 433; of *Iulus*, 423; of Polychaeta, 279
'Thorax' of Tunicata, 724
Thysanoptera, 527
Thysanozoon, 208
Thysanura, 476
Tibia, 433
Ticks, 567
Tiedemann's bodies, 684, 690
Tinea biselliella, 500; 443
Tintinnidium, 110; *inquilinum*, **111**
Tintinnina, 110
Tipula, **99**; *ocracea*, **509**
Tocophrya quadripartita, **114**
Todarodes sagittarius, 647
Tomoceros, 475
Tomopteris, **288**
Tornaria larva, 713; **711**
Torsion of Gasteropoda, 598
Toxiglossate teeth, **611**

INDEX

Trabea, 558
Trabeculae, of *Ciona*, 724; of coelom of Crinoidea, 701
Tracheae, 448; of Arachnida, 540; **539**, **556**; of Insecta, 448; of Myriapoda, 420, 423, 425; of Onychophora, 333; **332**; of *Velella*, 166; of Woodlice, 363, 404. *See also* Stigmata, Tracheal gills
Tracheal gills, 455, **480**
Trachelas, **552**
Tracheoles, 448
Trachomedusae, 162
Trachylina, 162
Transverse fission of Protozoa, 33
Trematoda, 211
Triaenophorus, 225
Triangle of odonate wing, 481
Triarthrus becki, 338
Triatoma, 69
Trichinella spiralis, 249
Trichurata, 236
Trichuris, 236
Trichobranchiae, 412
Trichodina, 112, 148; *pediculus*, 111
Trichomonas, 70, 71
Trichonympha, 40; *collaris*, **72**
Trichoptera, 494
Trichosphaerium, 80
Tricladida, 208
Tridacna, **584**
Trilobita, 338
Trilobite stage of Xiphosura, 548
Tripedalia, 170
Triploblastic animals (Triploblastica), 126
Tripylaea, 84
Tritocerebrum, 323
Triungulin, 489
Trochal disc, 256
Trochanter, 433
Trochodiscus longispinus, **85**
Trochosphere larva, of Annelida, 2, 294; of Mollusca, 3, 589; of Polychaeta, 294; of *Polygordius*, 296; of Polyzoa, 265, 658
Trochostoma, 697, **698**
Trochus, 256
Trochus, 585
Trombidiformes, 568
Trophallaxis, 516
Trophi, 257
Trophochromatin, 32
Trophozoite, 91
Trunk, of Arthropoda, 320; of Crustacea, 349; of Trilobita, 338. *See also* Abdomen, Thorax
'Trunk' of Ascidian tadpole, 726
'Trunk' ganglion, 727
Trunk limbs, of Crustacea, 340, 344–5; of Onychophora, 330. *See also* Abdominal limbs, Thoracic limbs
Trunk segments of Polychaeta, 274. *See also* Abdomen, Thorax
Tryblidium, 594

Trypanosoma, 68; *brucei*, 69; *cruzi*, 69; *equiperdum*, 68; *gambiense*, 69; *lewisi*, 69; *rhodesiense*, 69
Trypanosomidae, 68
Trypanosyllis, 289, **290**
Tryphaena pronuba, 500
Tube feet (Podia), 671
Tubifex, 305
Tubipora, 181
Tubularia, 157, **158**
Tunicata, 720
Turbellaria, 207
Tylenchus devastatrix, 251, *dispar*, 251; *tritici*, 250
Typhlosole, 298
Tyroglyphus, 572; **571**

Umbo of Brachiopoda, 662
Umbrella of trochosphere, 294
Umbrellar surfaces, etc., of Medusae, *see* Exumbrellar, Subumbrellar
Uncini, 279
Uncoiling of Cephalopoda, 651
Undulating membranes, of ciliates, 17, 106, **107**; of flagellates, 17, **68**
Uniramous limbs of Crustacea, 351
Urnatella, 265; **264**
Urochorda (Tunicata), 720
Uropods, 397
Urosome, 384
Uterus, of Cestoda, 222; of *Chirocephalus*, 373; of *Cyclops*, 385; of *Paludina*, 614; of Platyhelminthes, 203, **226**; of *Rhabditis*, 243; of Scorpionidea, 543; of Trematoda, **213, 216**

Vacuolaria, 60
Vacuoles, 13; Contractile, 27, 54, 58, 75, 87, 89, 104; Food, 26; Gas, 78; Hydrostatic, of ectoplasm, 13; of Dinoflagellata, 60
Vagina, *see* Generative organs
Vahlkampfia, 75
Valves of shell, of Brachiopoda, 662; of Conchostraca, 375, **376**; of Lamellibranchiata*, 631, **634**; of *Lepas*, 389; of Ostracoda, 382
Vanadis, **285, 286**
Vas deferens, *see* Generative organs
Vasa efferentia, of Insecta, 456; of Platyhelminthes, 202
Vascular system, of Anostraca, 364; of Arthropoda, 326; of *Balanoglossus*, 711; of *Carcinus*, 415; of Chaetopoda, 272; of Ciona, 725; of Crustacea, 364; of Insecta, 445; of Lamellibranchiata, 628; of *Lernanthropus*, 364; of *Limulus*, 548; of Malacostraca, 364; of Myriapoda, 423, 425; of Nemertea, 232; of Scorpionidea, 542
'Vascular' system of Echinodermata, *see* Lacunar system of Scyphomedusae, 171

820 INDEX

'Vascular' tissue of Echinodermata, *see* Lacunar tissue
Vegetative phase, 43
Vegetative pole, 291
Vein, Abdominal, 641; Afferent branchial, 628; Branchial, 641; Efferent branchial, 628; Genital, 641; Ink sac, 641; Longitudinal, of kidney, 628; Pulmonary, 606. *See also* Circulus venosus, Sinuses, Vena cava
Velella, 165, **165**
Veliger larva, 589; **587**
Velum, of Medusae, 155; of Rotifera, 256; of *Veliger* larva, 589
Vena cava, 641
Ventilators, 522
Ventral, *see* Dorsal and ventral
Ventral blood vessel, of *Balanoglossus*, 711; of Chaetopoda, 272; of Rhynchobdellidae, 308
Ventral 'blood vessels' of Echinodermata, 675, 697
Ventral cirrus, 274; mesenteries of Alcyonaria, 177, **182**; of Polychaeta, 274; midline of Nematoda, 235; plate of trochosphere, 295; siphon, 625; tube of Collembola, 475
Ventricle, *see* Heart
Venus' Girdle (*Cestus veneris*), 190; **189**
Vermes, 191
Vermicularia, **620**
Vertebrae of Ophiuroidea, 685
Vertebrata, 3, 4, 707
Vesiculae seminales, *see* Generative organs
Vesicular nuclei, 30
Vespa, 521; *crabro*, **519**; *germanica*, 522; *vulgaris*, 522
Vespoidea, 521
Vestibulata, 108
Vestibule, 25. *See also* Gullet
Vibracula, 656
Visceral clefts, 707
Visceral hump, 597
Visceral mass, 394
Visceral nerves, *see* Sympathetic system
Vitellarium, of Platyhelminthes, 202; of Rotifera, 257
Vitelline ducts, 202
Vitrellae, 323

Voluta, **620**
Volvocina, 62; reduction division of, 33
Volvox, 65; **64**
Vorticella, 112; **12, 112**

Waldheimia, 662
Water pore of Echinodermata, 674
Water vascular ring, *see* Water vascular system
Water vascular system, 672
Wilsonema, **238**
Wings, 434
Winter eggs, 204, 258
Wire worm (*Iulus terrestris*), 424
Workers, of Ants, 521; of Isoptera, 484; of Wasps, 521
Wuchereria, 249, 250

Xanthophyll, 55
Xanthoplasts, 55
Xenocoeloma, 388
Xenopsylla cheopis, 513
Xestobium, 443; *rufovillosum*, 490
Xiphosura, 546
Xylophaga, 634, **634**

Yellow cells of Chaetopoda, 271
'Yellow cells', Symbiotic, *see* Zooxanthellae
Yolk gland, *see* Vitellarium
Young, *see* Life history
Yungia, 208

Xammara tympanum, **529**

Zelleriella, 73
Zoaea larva, 366, **367**
Zoantharia, 181
Zones of fission, 304
Zoochlorellae, 57, 187, 207
Zooecium, 652
Zooids, of Coelenterata, *see* Hydranth, Polyp; of Polyzoa, *see* Polypide; of Protozoa, 11; of *Rhabdopleura*, 714; of Tunicata, 728; of Volvocina, 63–5
Zoomastigina, 65
Zooxanthellae, 57, **50**
Zoraptera, 485
Zygoptera, 481
Zygote*, 43, 54, 59, 63, 88